Grundlehren der mathematischen Wissenschaften 322

A Series of Comprehensive Studies in Mathematics

Springer
Berlin
Heidelberg
New York
Barcelona
Hong Kong
London
Milan
Paris
Singapore
Tokyo

Jürgen Neukirch

Algebraic Number Theory

Translated from the German
by Norbert Schappacher

With 16 Figures

Springer

Jürgen Neukirch†

Translator:

Norbert Schappacher

U.F.R. de Mathématique et d'Informatique
Université Louis Pasteur
7, rue René Descartes
F-67084 Strasbourg, France
e-mail: schappa@math.u-strasbg.fr

The original German edition was published in 1992 under the title
Algebraische Zahlentheorie
ISBN 3-540-54273-6 Springer-Verlag Berlin Heidelberg New York

Library of Congress Cataloging-in-Publication Data

Neukirch, Jürgen, 1937-. [Algebraische zahlentheorie. English] Algebraic number theory / Jürgen Neukirch; translated from the German by Norbert Schappacher.
p. cm. - (Grundlehren der mathematischen Wissenschaften; 322
Includes bibliographical references and index.
ISBN 3-540-65399-6 (hc.: alk. paper)
1. Algebraic number theory. I. Title II. Series.
QA247.N51713 1999 512'.74-dc21 99-21030 CIP

Mathematics Subject Classification (1991): 11-XX, 14-XX

ISSN 0072-7830
ISBN 3-540-65399-6 Springer-Verlag Berlin Heidelberg New York

Cover design: MetaDesign plus GmbH, Berlin
Photocomposed from the translator's LaTeX files after editing and reformatting by Raymond Seroul, Strasbourg
SPIN: 11561965 41/3111 - 5 4 3 Printed on acid-free paper

Foreword

It is a very sad moment for me to write this "Geleitwort" to the English translation of Jürgen Neukirch's book on Algebraic Number Theory. It would have been so much better, if he could have done this himself.

But it is also very difficult for me to write this "Geleitwort": The book contains Neukirch's Preface to the German edition. There he himself speaks about his intentions, the content of the book and his personal view of the subject. What else can be said?

It becomes clear from his Preface that Number Theory was Neukirch's favorite subject in mathematics. He was enthusiastic about it, and he was also able to implant this enthusiasm into the minds of his students.

He attracted them, they gathered around him in Regensburg. He told them that the subject and its beauty justified the highest effort and so they were always eager and motivated to discuss and to learn the newest developments in number theory and arithmetic algebraic geometry. I remember very well the many occasions when this equipe showed up in the meetings of the "Oberwolfach Arbeitsgemeinschaft" and demonstrated their strength (mathematically and on the soccer field).

During the meetings of the "Oberwolfach Arbeitsgemeinschaft" people come together to learn a subject which is not necessarily their own speciality. Always at the end, when the most difficult talks had to be delivered, the Regensburg crew took over. In the meantime many members of this team teach at German universities.

We find this charisma of Jürgen Neukirch in the book. It will be a motivating source for young students to study Algebraic Number Theory, and I am sure that it will attract many of them.

At Neukirch's funeral his daughter Christiane recited the poem which she often heard from her father: *Herr von Ribbeck auf Ribbeck im Havelland* by Theodor Fontane. It tells the story of a nobleman who always generously gives away the pears from his garden to the children. When he dies he asks for a pear to be put in his grave, so that later the children can pick the pears from the growing tree.

This is – I believe – a good way of thinking of Neukirch's book: There are seeds in it for a tree to grow from which the "children" can pick fruits in the time to come.

G. Harder

Translator's Note

When I first accepted Jürgen Neukirch's request to translate his *Algebraische Zahlentheorie*, back in 1991, no-one imagined that he would not live to see the English edition. He did see the raw version of the translation (I gave him the last chapters in the Fall of 1996), and he still had time to go carefully through the first four chapters of it.

The bulk of the text consists of detailed technical mathematical prose and was thus straightforward to translate, even though the author's desire to integrate involved arguments and displayed formulae into comprehensive sentences could not simply be copied into English. However, Jürgen Neukirch had peppered his book with more meditative paragraphs which make rather serious use of the German language. When I started to work on the translation, he warned me that in every one of these passages, he was not seeking poetic beauty, but only the precisely adequate expression of an idea. It is for the reader to judge whether I managed to render his ideas faithfully.

There is one neologism that I propose in this translation, with Jürgen Neukirch's blessing: I call *replete* divisor, ideal, etc., what is usually called Arakelov divisor, etc. (a terminology that Neukirch had avoided in the German edition). Time will deliver its verdict.

I am much indebted to Frazer Jarvis for going through my entire manuscript, thus saving the English language from various infractions. But needless to say, I alone am responsible for all deficiencies that remain.

After Jürgen Neukirch's untimely death early in 1997, it was Ms Eva-Maria Strobel who took it upon herself to finish as best she could what Jürgen Neukirch had to leave undone. She had already applied her infinite care and patience to the original German book, and she had assisted Jürgen Neukirch in proofreading the first four chapters of the translation. Without her knowledge, responsibility, and energy, this book would not be what it is. In particular, a fair number of small corrections and modifications of the German original that had been accumulated thanks to attentive readers, were taken into account for this English edition. Kay Wingberg graciously helped to check a few of them. We sincerely hope that the book published here would have made its author happy.

Hearty thanks go to Raymond Seroul, Strasbourg, for applying his wonderful expertise of TeX to the final preparation of the camera-ready manuscript.

Thanks go to the Springer staff for seeing this project through until it was finally completed. Among them I want to thank especially Joachim Heinze for interfering rarely, but effectively, over the years, with the realization of this translation.

Strasbourg, March 1999 *Norbert Schappacher*

Preface to the German Edition

Number Theory, among the mathematical disciplines, occupies an idealized position, similar to the one that mathematics holds among the sciences. Under no obligation to serve needs that do not originate within itself, it is essentially autonomous in setting its goals, and thus manages to protect its undisturbed harmony. The possibility of formulating its basic problems simply, the peculiar clarity of its statements, the arcane touch in its laws, be they discovered or undiscovered, merely divined; last but not least, the charm of its particularly satisfactory ways of reasoning – all these features have at all times attracted to number theory a community of dedicated followers.

But different number theorists may dedicate themselves differently to their science. Some will push the theoretical development only as far as is necessary for the concrete result they desire. Others will strive for a more universal, conceptual clarity, never tiring of searching for the deep-lying reasons behind the apparent variety of arithmetic phenomena. Both attitudes are justified, and they grow particularly effective through the mutual inspirational influence they exert on one another. Several beautiful textbooks illustrate the success of the first attitude, which is oriented towards specific problems. Among them, let us pick out in particular *Number Theory* by *S.I.* Borevicz and *I.R.* Šafarevič [14]: a book which is extremely rich in content, yet easy to read, and which we especially recommend to the reader.

The present book was conceived with a different objective in mind. It does provide the student with an essentially self-contained introduction to the theory of algebraic number fields, presupposing only basic algebra (it starts with the equation $2 = 1 + 1$). But unlike the textbooks alluded to above, it progressively emphasizes theoretical aspects that rely on modern concepts. Still, in doing so, a special effort is made to limit the amount of abstraction used, in order that the reader should not lose sight of the concrete goals of number theory proper. The desire to present number theory as much as possible from a unified theoretical point of view seems imperative today, as a result of the revolutionary development that number theory has undergone in the last decades in conjunction with 'arithmetic algebraic geometry'. The immense success that this new geometric perspective has brought about – for instance, in the context of the Weil conjectures, the Mordell conjecture, of problems related to the conjectures of Birch and Swinnerton-Dyer – is largely based on the unconditional and universal application of the conceptual approach.

It is true that those impressive results can hardly be touched upon in this book because they require higher dimensional theories, whereas the book deliberately confines itself to the theory of algebraic number fields, i.e., to the 1-dimensional case. But I thought it necessary to present the theory in a way which takes these developments into account, taking them as the distant focus, borrowing emphases and arguments from the higher point of view, thus integrating the theory of algebraic number fields into the higher dimensional theory – or at least avoiding any obstruction to such an integration. This is why I preferred, whenever it was feasible, the functorial point of view and the more far-reaching argument to the clever trick, and made a particular effort to place geometric interpretation to the fore, in the spirit of the theory of algebraic curves.

Let me forego the usual habit of describing the content of each individual chapter in this foreword; simply turning pages will yield the same information in a more entertaining manner. I would however like to emphasize a few basic principles that have guided me while writing the book. The first chapter lays down the foundations of the global theory and the second of the local theory of algebraic number fields. These foundations are finally summed up in the first three sections of chapter III, the aim of which is to present the perfect analogy of the classical notions and results with the theory of algebraic curves and the idea of the Riemann-Roch theorem. The presentation is dominated by "Arakelov's point of view", which has acquired much importance in recent years. It is probably the first time that this approach, with all its intricate normalizations, has received an extensive treatment in a textbook. But I finally decided not to employ the term "Arakelov divisor" although it is now widely used. This would have entailed attaching the name of *Arakelov* to many other concepts, introducing too heavy a terminology for this elementary material. My decision seemed all the more justified as *Arakelov* himself introduced his divisors only for arithmetic surfaces. The corresponding idea in the number field case goes back to *Hasse*, and is clearly highlighted for instance in *S. Lang*'s textbook [94].

It was not without hesitation that I decided to include *Class Field Theory* in chapters IV–VI. Since my book [107] on this subject had been published not long before, another treatment of this theory posed obvious questions. But in the end, after long consideration, there was simply no other choice. A sourcebook on algebraic number fields without the crowning conclusion of class field theory with its important consequences for the theory of L-series would have appeared like a torso, suffering from an unacceptable lack of completeness. This also gave me the opportunity to modify and emend my earlier treatment, to enrich that somewhat dry presentation with quite a few examples, to refer ahead with some remarks, and to add beneficial exercises.

A lot of work went into the last chapter on zeta functions and L-series. These functions have gained central importance in recent decades, but textbooks do not pay sufficient attention to them. I did not, however, include *Tate*'s approach to Hecke L-series, which is based on harmonic analysis, although it would have suited the more conceptual orientation of the book perfectly well. In fact, the clarity of *Tate*'s own presentation could hardly be improved upon, and it has also been sufficiently repeated in other places. Instead I have preferred to turn back to *Hecke*'s approach, which is not easy to understand in the original version, but for all its various advantages cried out for a modern treatment. This having been done, there was the obvious opportunity of giving a thorough presentation of *Artin*'s L-series with their functional equation – which surprisingly has not been undertaken in any existing textbook.

It was a difficult decision to exclude *Iwasawa Theory*, a relatively recent theory totally germane to algebraic number fields, the subject of this book. Since it mirrors important geometric properties of algebraic curves, it would have been a particularly beautiful vindication of our oft-repeated thesis that number theory is geometry. I do believe, however, that in this case the geometric aspect becomes truly convincing only if one uses *étale cohomology* – which can neither be assumed nor reasonably developed here. Perhaps the dissatisfaction with this exclusion will be strong enough to bring about a sequel to the present volume, devoted to the cohomology of algebraic number fields.

From the very start the book was not just intended as a modern sourcebook on algebraic number theory, but also as a convenient textbook for a course. This intention was increasingly jeopardized by the unexpected growth of the material which had to be covered in view of the intrinsic necessities of the theory. Yet I think that the book has not lost that character. In fact, it has passed a first test in this respect. With a bit of careful planning, the basic content of the first three chapters can easily be presented in one academic year (if possible including infinite Galois theory). The following term will then provide scarce, yet sufficient room for the class field theory of chapters IV–VI.

Sections 11–14 of chapter I may mostly be dropped from an introductory course. Although the results of section 12 on *orders* are irrelevant for the sequel, I consider its insertion in the book particularly important. For one thing, orders constitute the rings of multipliers, which play an eminent role in many diophantine problems. But most importantly, they represent the analogues of *singular* algebraic curves. As cohomology theory becomes increasingly important for algebraic number fields, and since this is even more true of *algebraic K-theory*, which cannot be constructed without singular schemes, the time has come to give orders an adequate treatment.

In chapter II, the special treatment of henselian fields in section 6 may be restricted to complete valued fields, and thus joined with section 4. If pressed for time, section 10 on higher ramification may be omitted completely.

The first three sections of chapter III should be presented in the lectures since they highlight a new approach to classical results of algebraic number theory. The subsequent theory concerning the theorem of Grothendieck-Riemann-Roch is a nice subject for a student seminar rather than for an introductory course.

Finally, in presenting class field theory, it saves considerable time if the students are already familiar with profinite groups and infinite Galois theory. Sections 4–7 of chapter V, on formal groups, Lubin-Tate theory and the theory of higher ramification may be omitted. Cutting out even more, chapter V, 3, on the Hilbert symbol, and VI, 7 and 8, still leaves a fully-fledged theory, which is however unsatisfactory because it remains in the abstract realm, and is never linked to classical problems.

A word on the exercises at the end of the sections. Some of them are not so much exercises, but additional remarks which did not fit well into the main text. The reader is encouraged to prove his versatility in looking up the literature. I should also point out that I have not actually done all the exercises myself, so that there might be occasional mistakes in the way they are posed. If such a case arises, it is for the reader to find the correct formulation. May the reader's reaction to such a possible slip of the author be mitigated by Goethe's distich:

> "Irrtum verläßt uns nie, doch ziehet ein höher Bedürfnis
> Immer den strebenden Geist leise zur Wahrheit hinan." *

During the writing of this book I have been helped in many ways. I thank the Springer Verlag for considering my wishes with generosity. My students *I. Kausz*, *B. Köck*, *P. Kölcze*, *Th. Moser*, *M. Spiess* have critically examined larger or smaller parts, which led to numerous improvements and made it possible to avoid mistakes and ambiguities. To my friends *W.-D. Geyer*, *G. Tamme*, and *K. Wingberg* I owe much valuable advice from which the book has profited, and it was *C. Deninger* and *U. Jannsen* who suggested that I give a new treatment of Hecke's theory of theta series and L-series. I owe a great debt of gratitude to Mrs. *Eva-Maria Strobel*. She drew the pictures and helped me with the proofreading and the formatting of the text, never tiring of going into the minutest detail. Let me heartily thank all those who assisted me, and also those who are not named here. Tremendous thanks are due to Mrs. *Martina Hertl* who did the typesetting of the manuscript in TeX. That the book can appear is

* Error is ever with us. Yet some angelic need
Gently coaxes our striving mind upwards, towards truth.
(Translation suggested by *Barry Mazur*.)

essentially due to her competence, to the unfailing and kind willingness with which she worked through the long handwritten manuscript, and through the many modifications, additions, and corrections, always prepared to give her best.

Regensburg, February 1992 *Jürgen Neukirch*

Table of Contents

Chapter I

Algebraic Integers

§ 1. The Gaussian Integers

The equations

$$2 = 1 + 1, \; 5 = 1 + 4, \; 13 = 4 + 9, \; 17 = 1 + 16, \; 29 = 4 + 25, \; 37 = 1 + 36$$

show the first prime numbers that can be represented as a sum of two squares. Except for 2, they are all $\equiv 1 \bmod 4$, and it is true in general that any odd prime number of the form $p = a^2 + b^2$ satisfies $p \equiv 1 \bmod 4$, because perfect squares are $\equiv 0$ or $\equiv 1 \bmod 4$. This is obvious. What is not obvious is the remarkable fact that the converse also holds:

(1.1) Theorem. *For all prime numbers $p \neq 2$, one has:*

$$p = a^2 + b^2 \quad (a, b \in \mathbb{Z}) \quad \Longleftrightarrow \quad p \equiv 1 \bmod 4 .$$

The natural explanation of this arithmetic law concerning the ring $\mathbb{Z}$ of rational integers is found in the larger domain of the **gaussian integers**

$$\mathbb{Z}[i] = \{ a + bi \mid a, b \in \mathbb{Z} \}, \quad i = \sqrt{-1} .$$

In this ring, the equation $p = x^2 + y^2$ turns into the product decomposition

$$p = (x + iy)(x - iy) ,$$

so that the problem is now when and how a prime number $p \in \mathbb{Z}$ factors in $\mathbb{Z}[i]$. The answer to this question is based on the following result about unique factorization in $\mathbb{Z}[i]$.

(1.2) Proposition. *The ring $\mathbb{Z}[i]$ is euclidean, therefore in particular factorial.*

Proof: We show that $\mathbb{Z}[i]$ is euclidean with respect to the function $\mathbb{Z}[i] \to \mathbb{N} \cup \{0\}$, $\alpha \mapsto |\alpha|^2$. So, for $\alpha, \beta \in \mathbb{Z}[i]$, $\beta \neq 0$, one has to verify the existence of gaussian integers γ, ρ such that

$$\alpha = \gamma\beta + \rho \quad \text{and} \quad |\rho|^2 < |\beta|^2 .$$

It clearly suffices to find $\gamma \in \mathbb{Z}[i]$ such that $\left| \frac{\alpha}{\beta} - \gamma \right| < 1$. Now, the gaussian integers form a **lattice** in the complex plane $\mathbb{C}$ (the points with integer coordinates with respect to the basis $1, i$). The complex number $\frac{\alpha}{\beta}$ lies in some mesh of the lattice and its distance from the nearest lattice point is not greater than half the length of the diagonal of the mesh, i.e. $\frac{1}{2}\sqrt{2}$. Therefore there exists an element $\gamma \in \mathbb{Z}[i]$ with $\left|\frac{\alpha}{\beta} - \gamma\right| \leq \frac{1}{2}\sqrt{2} < 1$. □

Based on this result about the ring $\mathbb{Z}[i]$, theorem (1.1) now follows like this: it is sufficient to show that a prime number $p \equiv 1 \bmod 4$ of $\mathbb{Z}$ does not remain a prime element in the ring $\mathbb{Z}[i]$. Indeed, if this is proved, then there exists a decomposition

$$p = \alpha \cdot \beta$$

into two non-units α, β of $\mathbb{Z}[i]$. The **norm** of $z = x + iy$ is defined by

$$N(x+iy) = (x+iy)(x-iy) = x^2 + y^2,$$

i.e., by $N(z) = |z|^2$. It is multiplicative, so that one has

$$p^2 = N(\alpha) \cdot N(\beta).$$

Since α and β are not units, it follows that $N(\alpha), N(\beta) \neq 1$ (see exercise 1), and therefore $p = N(\alpha) = a^2 + b^2$, where we put $\alpha = a + bi$.

Finally, in order to prove that a rational prime of the form $p = 1 + 4n$ cannot be a prime element in $\mathbb{Z}[i]$, we note that the congruence

$$-1 \equiv x^2 \bmod p$$

admits a solution, namely $x = (2n)!$. Indeed, since $-1 \equiv (p-1)! \bmod p$ by Wilson's theorem, one has

$$\begin{aligned} -1 \equiv (p-1)! &= \big[1 \cdot 2 \cdots (2n)\big]\big[(p-1)(p-2)\cdots(p-2n)\big] \\ &\equiv \big[(2n)!\big]\big[(-1)^{2n}(2n)!\big] = \big[(2n)!\big]^2 \bmod p. \end{aligned}$$

Thus we have $p \,|\, x^2 + 1 = (x+i)(x-i)$. But since $\frac{x}{p} \pm \frac{i}{p} \notin \mathbb{Z}[i]$, p does not divide any of the factors $x+i, x-i$, and is therefore not a prime element in the factorial ring $\mathbb{Z}[i]$.

The example of the equation $p = x^2 + y^2$ shows that even quite elementary questions about rational integers may lead to the consideration of higher domains of integers. But it was not so much for this equation that we have introduced the ring $\mathbb{Z}[i]$, but rather in order to preface the general theory of algebraic integers with a concrete example. For the same reason we will now look at this ring a bit more closely.

When developing the theory of divisibility for a ring, two basic problems are most prominent: on the one hand, to determine the **units** of the ring in question, on the other, its **prime elements**. The answer to the first question in the present case is particularly easy. A number $\alpha = a + bi \in \mathbb{Z}[i]$ is a unit if and only if its norm is 1:

$$N(\alpha) := (a + ib)(a - ib) = a^2 + b^2 = 1$$

(exercise 1), i.e., if either $a^2 = 1$, $b^2 = 0$, or $a^2 = 0$, $b^2 = 1$. We thus obtain the

(1.3) Proposition. *The group of units of the ring $\mathbb{Z}[i]$ consists of the fourth roots of unity,*

$$\mathbb{Z}[i]^* = \{1, \ -1, i, \ -i\}.$$

In order to answer the question for primes, i.e., irreducible elements of the ring $\mathbb{Z}[i]$, we first recall that two elements α, β in a ring are called **associated**, symbolically $\alpha \sim \beta$, if they differ only by a unit factor, and that every element associated to an irreducible element π is also irreducible. Using theorem (1.1) we obtain the following precise list of all prime numbers of $\mathbb{Z}[i]$.

(1.4) Theorem. *The prime elements π of $\mathbb{Z}[i]$, up to associated elements, are given as follows.*

(1) $\pi = 1 + i$,

(2) $\pi = a + bi$ *with* $a^2 + b^2 = p$, $p \equiv 1 \bmod 4$, $a > |b| > 0$,

(3) $\pi = p$, $p \equiv 3 \bmod 4$.

Here, p denotes a prime number of $\mathbb{Z}$.

Proof: Numbers as in (1) or (2) are prime because a decomposition $\pi = \alpha \cdot \beta$ in $\mathbb{Z}[i]$ implies an equation

$$p = N(\pi) = N(\alpha) \cdot N(\beta),$$

with some prime number p. Hence either $N(\alpha) = 1$ or $N(\beta) = 1$, so that either α or β is a unit.

Numbers $\pi = p$, where $p \equiv 3 \bmod 4$, are prime in $\mathbb{Z}[i]$, because a decomposition $p = \alpha \cdot \beta$ into non-units α, β would imply that $p^2 = N(\alpha) \cdot N(\beta)$, so that $p = N(\alpha) = N(a + bi) = a^2 + b^2$, which according to (1.1) would yield $p \equiv 1 \bmod 4$.

This being said, we have to check that an arbitrary prime element π of $\mathbb{Z}[i]$ is associated to one of those listed. First of all, the decomposition

$$N(\pi) = \pi \cdot \bar{\pi} = p_1 \cdots p_r,$$

with rational primes p_i, shows that $\pi \mid p$ for some $p = p_i$. This gives $N(\pi) \mid N(p) = p^2$, so that either $N(\pi) = p$ or $N(\pi) = p^2$. In the case $N(\pi) = p$ we get $\pi = a + bi$ with $a^2 + b^2 = p$, so π is of type (2) or, if $p = 2$, it is associated to $1 + i$. On the other hand, if $N(\pi) = p^2$, then π is associated to p since p/π is an integer with norm one and thus a unit. Moreover, $p \equiv 3 \bmod 4$ has to hold in this case because otherwise we would have $p = 2$ or $p \equiv 1 \bmod 4$ and because of (1.1) $p = a^2 + b^2 = (a + bi)(a - bi)$ could not be prime. This completes the proof. □

The proposition also settles completely the question of how prime numbers $p \in \mathbb{Z}$ decompose in $\mathbb{Z}[i]$. The prime $2 = (1+i)(1-i)$ is associated to the square of the prime element $1+i$. Indeed, the identity $1 - i = -i(1+i)$ shows that $2 \sim (1+i)^2$. The prime numbers $p \equiv 1 \bmod 4$ split into two conjugate prime factors

$$p = (a + bi)(a - bi),$$

and the prime numbers $p \equiv 3 \bmod 4$ remain prime in $\mathbb{Z}[i]$.

The gaussian integers play the same rôle in the field

$$\mathbb{Q}(i) = \{a + bi \mid a, b \in \mathbb{Q}\}$$

as the rational integers do in the field $\mathbb{Q}$. So they should be viewed as the "integers" in $\mathbb{Q}(i)$. This notion of integrality is relative to the coordinates of the basis $1, i$. However, we also have the following characterization of the gaussian integers, which is independent of a choice of basis.

(1.5) Proposition. *$\mathbb{Z}[i]$ consists precisely of those elements of the extension field $\mathbb{Q}(i)$ of $\mathbb{Q}$ which satisfy a monic polynomial equation*

$$x^2 + ax + b = 0$$

with coefficients $a, b \in \mathbb{Z}$.

Proof: An element $\alpha = c + id \in \mathbb{Q}(i)$ is a zero of the polynomial

$$x^2 + ax + b \in \mathbb{Q}[x] \quad \text{with} \quad a = -2c,\ b = c^2 + d^2.$$

If c and d are rational integers, then so are a and b. Conversely, if a and b are integers, then so are $2c$ and $2d$. From $(2c)^2 + (2d)^2 = 4b \equiv 0 \bmod 4$ it follows that $(2c)^2 \equiv (2d)^2 \equiv 0 \bmod 4$, since squares are always $\equiv 0$ or $\equiv 1$. Hence c and d are integers. □

The last proposition leads us to the general notion of an algebraic integer as being an element satisfying a monic polynomial equation with rational integer coefficients. For the domain of the gaussian integers we have obtained in this section a complete answer to the question of the units, the question of prime elements, and to the question of unique factorization.

These questions indicate already the fundamental problems in the general theory of algebraic integers. But the answers we found in the special case $\mathbb{Z}[i]$ are not typical. Novel features will present themselves instead.

Exercise 1. $\alpha \in \mathbb{Z}[i]$ is a unit if and only if $N(\alpha) = 1$.

Exercise 2. Show that, in the ring $\mathbb{Z}[i]$, the relation $\alpha\beta = \varepsilon\gamma^n$, for α, β relatively prime numbers and ε a unit, implies $\alpha = \varepsilon'\xi^n$ and $\beta = \varepsilon''\eta^n$, with ε', ε'' units.

Exercise 3. Show that the integer solutions of the equation

$$x^2 + y^2 = z^2$$

such that $x, y, z > 0$ and $(x, y, z) = 1$ ("pythagorean triples") are all given, up to possible permutation of x and y, by the formulæ

$$x = u^2 - v^2, \quad y = 2uv, \quad z = u^2 + v^2,$$

where $u, v \in \mathbb{Z}$, $u > v > 0$, $(u, v) = 1$, u, v not both odd.

Hint: Use exercise 2 to show that necessarily $x + iy = \varepsilon\alpha^2$ with a unit ε and with $\alpha = u + iv \in \mathbb{Z}[i]$.

Exercise 4. Show that the ring $\mathbb{Z}[i]$ cannot be ordered.

Exercise 5. Show that the only units of the ring $\mathbb{Z}[\sqrt{-d}] = \mathbb{Z} + \mathbb{Z}\sqrt{-d}$, for any rational integer $d > 1$, are ± 1.

Exercise 6. Show that the ring $\mathbb{Z}[\sqrt{d}] = \mathbb{Z} + \mathbb{Z}\sqrt{d}$, for any squarefree rational integer $d > 1$, has infinitely many units.

Exercise 7. Show that the ring $\mathbb{Z}[\sqrt{2}] = \mathbb{Z} + \mathbb{Z}\sqrt{2}$ is euclidean. Show furthermore that its units are given by $\pm(1 + \sqrt{2})^n$, $n \in \mathbb{Z}$, and determine its prime elements.

§ 2. Integrality

An **algebraic number field** is a finite field extension K of $\mathbb{Q}$. The elements of K are called **algebraic numbers**. An algebraic number is called **integral**, or an **algebraic integer**, if it is a zero of a monic polynomial $f(x) \in \mathbb{Z}[x]$. This notion of integrality applies not only to algebraic numbers, but occurs in many different contexts and therefore has to be treated in full generality.

In what follows, rings are always understood to be commutative rings with 1.

(2.1) Definition. *Let $A \subseteq B$ be an extension of rings. An element $b \in B$ is called* **integral** *over A, if it satisfies a monic equation*

$$x^n + a_1 x^{n-1} + \cdots + a_n = 0, \quad n \geq 1,$$

with coefficients $a_i \in A$. The ring B is called **integral** *over A if all elements $b \in B$ are integral over A.*

It is desirable, but strangely enough not immediately obvious, that the sum and the product of two elements which are integral over A are again integral. This will be a consequence of the following abstract reinterpretation of the notion of integrality.

(2.2) Proposition. *Finitely many elements $b_1, \dots, b_n \in B$ are all integral over A if and only if the ring $A[b_1, \dots, b_n]$ viewed as an A-module is finitely generated.*

To prove this we make use of the following result of linear algebra.

(2.3) Proposition (Row-Column Expansion). *Let $A = (a_{ij})$ be an $(r \times r)$-matrix with entries in an arbitrary ring, and let $A^* = (a_{ij}^*)$ be the adjoint matrix, i.e., $a_{ij}^* = (-1)^{i+j} \det(A_{ij})$, where the matrix A_{ij} is obtained from A by deleting the i-th column and the j-th row. Then one has*

$$AA^* = A^*A = \det(A)E,$$

where E denotes the unit matrix of rank r. For any vector $x = (x_1, \dots, x_r)$, this yields the implication

$$Ax = 0 \implies (\det A)x = 0.$$

Proof of proposition (2.2): Let $b \in B$ be integral over A and $f(x) \in A[x]$ a monic polynomial of degree $n \geq 1$ such that $f(b) = 0$. For an arbitrary polynomial $g(x) \in A[x]$ we may then write

$$g(x) = q(x)f(x) + r(x),$$

with $q(x), r(x) \in A[x]$ and $\deg(r(x)) < n$, so that one has

$$g(b) = r(b) = a_0 + a_1 b + \cdots + a_{n-1}b^{n-1}.$$

Thus $A[b]$ is generated as A-module by $1, b, \dots, b^{n-1}$.

More generally, if $b_1, \ldots, b_n \in B$ are integral over A, then the fact that $A[b_1, \ldots, b_n]$ is of finite type over A follows by induction on n. Indeed, since b_n is integral over $R = A[b_1, \ldots, b_{n-1}]$, what we have just shown implies that $R[b_n] = A[b_1, \ldots, b_n]$ is finitely generated over R, hence also over A, if we assume, by induction, that R is an A-module of finite type.

Conversely, assume that the A-module $A[b_1, \ldots, b_n]$ is finitely generated and that $\omega_1, \ldots, \omega_r$ is a system of generators. Then, for any element $b \in A[b_1, \ldots, b_n]$, one finds that

$$b\,\omega_i = \sum_{j=1}^{r} a_{ij}\omega_j\,, \quad i = 1, \ldots, r\,, \quad a_{ij} \in A\,.$$

From (2.3) we see that $\det(bE - (a_{ij}))\,\omega_i = 0$, $i = 1, \ldots, r$ (here E is the unit matrix of rank r), and since 1 can be written $1 = c_1\omega_1 + \cdots + c_r\omega_r$, the identity $\det(bE - (a_{ij})) = 0$ gives us a monic equation for b with coefficients in A. This shows that b is indeed integral over A. □

According to this proposition, if $b_1, \ldots, b_n \in B$ are integral over A, then so is *any* element b of $A[b_1, \ldots, b_n]$, because $A[b_1, \ldots, b_n, b] = A[b_1, \ldots, b_n]$ is a finitely generated A-module. In particular, given two integral elements $b_1, b_2 \in B$, then $b_1 + b_2$ and $b_1 b_2$ are also integral over A. At the same time we obtain the

(2.4) Proposition. *Let $A \subseteq B \subseteq C$ be two ring extensions. If C is integral over B and B is integral over A, then C is integral over A.*

Proof: Take $c \in C$, and let $c^n + b_1 c^{n-1} + \cdots + b_n = 0$ be an equation with coefficients in B. Write $R = A[b_1, \ldots, b_n]$. Then $R[c]$ is a finitely generated R-module. If B is integral over A, then $R[c]$ is even finitely generated over A, since R is finitely generated over A. Thus c is integral over A. □

From what we have proven, the set of all elements

$$\bar{A} = \{ b \in B \mid b \text{ integral over } A \}$$

in a ring extension $A \subseteq B$ forms a ring. It is called the **integral closure** of A in B. A is said to be **integrally closed** in B if $A = \bar{A}$. It is immediate from (2.4) that the integral closure $\bar{A}$ is itself integrally closed in B. If A is an integral domain with field of fractions K, then the integral closure $\bar{A}$ of A in K is called the **normalization** of A, and A is simply called integrally closed if $A = \bar{A}$. For instance, every **factorial** ring is integrally closed.

In fact, if $a/b \in K$ $(a, b \in A)$ is integral over A, i.e.,

$$(a/b)^n + a_1(a/b)^{n-1} + \cdots + a_n = 0,$$

with $a_i \in A$, then

$$a^n + a_1 b a^{n-1} + \cdots + a_n b^n = 0.$$

Therefore each prime element π which divides b also divides a. Assuming a/b to be reduced, this implies $a/b \in A$.

We now turn to a more specialized situation. Let A be an integral domain which is integrally closed, K its field of fractions, $L|K$ a finite field extension, and B the integral closure of A in L. According to (2.4), B is automatically integrally closed. Each element $\beta \in L$ is of the form

$$\beta = \frac{b}{a}, \quad b \in B, \ a \in A,$$

because if

$$a_n\beta^n + \cdots + a_1\beta + a_0 = 0, \quad a_i \in A, \ a_n \neq 0,$$

then $b = a_n\beta$ is integral over A, an integral equation

$$(a_n\beta)^n + \cdots + a_1'(a_n\beta) + a_0' = 0, \quad a_i' \in A,$$

being obtained from the equation for β by multiplication by a_n^{n-1}. Furthermore, the fact that A is integrally closed has the effect that an element $\beta \in L$ is integral over A if and only if its **minimal polynomial** $p(x)$ takes its coefficients in A. In fact, let β be a zero of the monic polynomial $g(x) \in A[x]$. Then $p(x)$ divides $g(x)$ in $K[x]$, so that all zeroes $\beta_1, \ldots, \beta_n$ of $p(x)$ are integral over A, hence the same holds for all the coefficients, in other words $p(x) \in A[x]$.

The trace and the norm in the field extension $L|K$ furnish important tools for the study of the integral elements in L. We recall the

(2.5) Definition. *The* **trace** *and* **norm** *of an element $x \in L$ are defined to be the trace and determinant, respectively, of the endomorphism*

$$T_x : L \to L, \quad T_x(\alpha) = x\alpha,$$

of the K-vector space L:

$$Tr_{L|K}(x) = Tr(T_x), \quad N_{L|K}(x) = \det(T_x).$$

In the characteristic polynomial

$$f_x(t) = \det(t\,\mathrm{id} - T_x) = t^n - a_1 t^{n-1} + \ldots + (-1)^n a_n \in K[t]$$

of T_x, $n = [L : K]$, we recognize the trace and the norm as

$$a_1 = Tr_{L|K}(x) \quad \text{and} \quad a_n = N_{L|K}(x).$$

Since $T_{x+y} = T_x + T_y$ and $T_{xy} = T_x \circ T_y$, we obtain homomorphisms

$$Tr_{L|K} : L \longrightarrow K \quad \text{and} \quad N_{L|K} : L^* \longrightarrow K^*.$$

In the case where the extension $L|K$ is separable, the trace and norm admit the following Galois-theoretic interpretation.

(2.6) Proposition. *If $L|K$ is a separable extension and $\sigma : L \to \overline{K}$ varies over the different K-embeddings of L into an algebraic closure $\overline{K}$ of K, then we have*

(i) $\quad f_x(t) = \prod_\sigma (t - \sigma x)$,

(ii) $Tr_{L|K}(x) = \sum_\sigma \sigma x$,

(iii) $N_{L|K}(x) = \prod_\sigma \sigma x$.

Proof: The characteristic polynomial $f_x(t)$ is a power

$$f_x(t) = p_x(t)^d, \quad d = \big[L : K(x)\big],$$

of the minimal polynomial

$$p_x(t) = t^m + c_1 t^{m-1} + \cdots + c_m, \quad m = \big[K(x) : K\big],$$

of x. In fact, $1, x, \ldots, x^{m-1}$ is a basis of $K(x)|K$, and if $\alpha_1, \ldots, \alpha_d$ is a basis of $L|K(x)$, then

$$\alpha_1, \alpha_1 x, \ldots, \alpha_1 x^{m-1}; \ \ldots; \ \alpha_d, \alpha_d x, \ldots, \alpha_d x^{m-1}$$

is a basis of $L|K$. The matrix of the linear transformation $T_x : y \mapsto xy$ with respect to this basis has obviously only blocks along the diagonal, each of them equal to

$$\begin{pmatrix} 0 & 1 & 0 & \cdots & 0 \\ 0 & 0 & 1 & \cdots & 0 \\ \cdots & \cdots & \cdots & \cdots & \cdots \\ 0 & 0 & 0 & \cdots & 1 \\ -c_m & -c_{m-1} & -c_{m-2} & \cdots & -c_1 \end{pmatrix}.$$

The corresponding characteristic polynomial is easily checked to be

$$t^m + c_1 t^{m-1} + \cdots + c_m = p_x(t),$$

so that finally $f_x(t) = p_x(t)^d$.

The set $\mathrm{Hom}_K(L, \overline{K})$ of all K-embeddings of L is partitioned by the equivalence relation

$$\sigma \sim \tau \iff \sigma x = \tau x$$

into m equivalence classes of d elements each. If $\sigma_1, \ldots, \sigma_m$ is a system of representatives, then we find

$$p_x(t) = \prod_{i=1}^{m} (t - \sigma_i x),$$

and $f_x(t) = \prod_{i=1}^m (t - \sigma_i x)^d = \prod_{i=1}^m \prod_{\sigma \sim \sigma_i} (t - \sigma x) = \prod_\sigma (t - \sigma x)$. This proves (i), and therefore also (ii) and (iii), after Vietà. □

(2.7) Corollary. *In a tower of finite field extensions $K \subseteq L \subseteq M$, one has*

$$Tr_{L|K} \circ Tr_{M|L} = Tr_{M|K}\,, \quad N_{L|K} \circ N_{M|L} = N_{M|K}\,.$$

Proof: We assume that $M|K$ is separable. The set $\mathrm{Hom}_K(M, \overline{K})$ of K-embeddings of M is partitioned by the relation

$$\sigma \sim \tau \iff \sigma|_L = \tau|_L$$

into $m = [L : K]$ equivalence classes. If $\sigma_1, \ldots, \sigma_m$ is a system of representatives, then $\mathrm{Hom}_K(L, \overline{K}) = \{\sigma_i|_L \mid i = 1, \ldots, m\}$, and we find

$$\begin{aligned} Tr_{M|K}(x) &= \sum_{i=1}^m \sum_{\sigma \sim \sigma_i} \sigma x = \sum_{i=1}^m Tr_{\sigma_i M|\sigma_i L}(\sigma_i x) = \sum_{i=1}^m \sigma_i\, Tr_{M|L}(x) \\ &= Tr_{L|K}\big(Tr_{M|L}(x)\big)\,. \end{aligned}$$

Likewise for the norm.

We will not need the inseparable case for the sequel. However it follows easily from what we have shown above, by passing to the maximal separable subextension $M^s|K$. Indeed, for the inseparable degree $[M : K]_i = [M : M^s]$ one has $[M : K]_i = [M : L]_i\,[L : K]_i$ and

$$Tr_{M|K}(x) = [M : K]_i\, Tr_{M^s|K}(x), \quad N_{M|K}(x) = N_{M^s|K}(x)^{[M:K]_i}$$

(see [143], vol. I, chap. II, § 10). □

The **discriminant** of a basis $\alpha_1, \ldots, \alpha_n$ of a separable extension $L|K$ is defined by

$$d(\alpha_1, \ldots, \alpha_n) = \det((\sigma_i \alpha_j))^2,$$

where σ_i, $i = 1, \ldots, n$, varies over the K-embeddings $L \to \overline{K}$. Because of the relation

$$Tr_{L|K}(\alpha_i \alpha_j) = \sum_k (\sigma_k \alpha_i)(\sigma_k \alpha_j),$$

the matrix $(Tr_{L|K}(\alpha_i \alpha_j))$ is the product of the matrices $(\sigma_k \alpha_i)^t$ and $(\sigma_k \alpha_j)$. Thus one may also write

$$d(\alpha_1, \ldots, \alpha_n) = \det\bigl(Tr_{L|K}(\alpha_i \alpha_j)\bigr).$$

In the special case of a basis of type $1, \theta, \ldots, \theta^{n-1}$ one gets

$$d(1, \theta, \ldots, \theta^{n-1}) = \prod_{i<j} (\theta_i - \theta_j)^2,$$

where $\theta_i = \sigma_i \theta$. This is seen by successively multiplying each of the first $(n-1)$ columns in the **Vandermonde matrix**

$$\begin{pmatrix} 1 & \theta_1 & \theta_1^2 & \cdots & \theta_1^{n-1} \\ 1 & \theta_2 & \theta_2^2 & \cdots & \theta_2^{n-1} \\ \cdots & \cdots & \cdots & \cdots & \cdots \\ 1 & \theta_n & \theta_n^2 & \cdots & \theta_n^{n-1} \end{pmatrix}$$

by θ_1 and subtracting it from the following.

(2.8) Proposition. *If $L|K$ is separable and $\alpha_1, \ldots, \alpha_n$ is a basis, then the discriminant*

$$d(\alpha_1, \ldots, \alpha_n) \neq 0,$$

and

$$(x, y) = Tr_{L|K}(xy)$$

is a nondegenerate bilinear form on the K-vector space L.

Proof: We first show that the bilinear form $(x, y) = Tr(xy)$ is nondegenerate. Let θ be a primitive element for $L|K$, i.e., $L = K(\theta)$. Then $1, \theta, \ldots, \theta^{n-1}$ is a basis with respect to which the form (x, y) is given by the matrix $M = (Tr_{L|K}(\theta^{i-1}\theta^{j-1}))_{i,\, j=1, \ldots, n}$. It is nondegenerate because, for $\theta_i = \sigma_i \theta$, we have

$$\det(M) = d(1, \theta, \ldots, \theta^{n-1}) = \prod_{i<j} (\theta_i - \theta_j)^2 \neq 0.$$

If $\alpha_1, \ldots, \alpha_n$ is an arbitrary basis of $L|K$, then the bilinear form (x, y) with respect to this basis is given by the matrix $M = (Tr_{L|K}(\alpha_i \alpha_j))$. From the above it follows that $d(\alpha_1, \ldots, \alpha_n) = \det(M) \neq 0$. □

After this review from the theory of fields, we return to the integrally closed integral domain A with field of fractions K, and to its integral closure B in the finite separable extension $L|K$. If $x \in B$ is an integral element of L, then all of its conjugates σx are also integral. Taking into account that A is integrally closed, i.e., $A = B \cap K$, (2.6) implies that

$$Tr_{L|K}(x), \quad N_{L|K}(x) \in A.$$

Furthermore, for the group of units of B over A, we obtain the relation

$$x \in B^* \iff N_{L|K}(x) \in A^*.$$

For if $aN_{L|K}(x) = 1$, $a \in A$, then $1 = a\prod_\sigma \sigma x = yx$ for some $y \in B$. The discriminant is often useful because of the following

(2.9) Lemma. *Let $\alpha_1, \ldots, \alpha_n$ be a basis of $L|K$ which is contained in B, of discriminant $d = d(\alpha_1, \ldots, \alpha_n)$. Then one has*

$$dB \subseteq A\alpha_1 + \cdots + A\alpha_n.$$

Proof: If $\alpha = a_1\alpha_1 + \cdots + a_n\alpha_n \in B$, $a_j \in K$, then the a_j are a solution of the system of linear equations

$$Tr_{L|K}(\alpha_i\alpha) = \sum_j Tr_{L|K}(\alpha_i\alpha_j)a_j,$$

and, as $Tr_{L|K}(\alpha_i\alpha) \in A$, they are given as the quotient of an element of A by the determinant $\det(Tr_{L|K}(\alpha_i\alpha_j)) = d$. Therefore $da_j \in A$, and thus

$$d\alpha \in A\alpha_1 + \cdots + A\alpha_n. \qquad \square$$

A system of elements $\omega_1, \ldots, \omega_n \in B$ such that each $b \in B$ can be written uniquely as a linear combination

$$b = a_1\omega_1 + \cdots + a_n\omega_n$$

with coefficients $a_i \in A$, is called an **integral basis** of B over A (or: an A-basis of B). Since such an integral basis is automatically a basis of $L|K$, its length n always equals the degree $[L:K]$ of the field extension. The existence of an integral basis signifies that B is a **free A-module** of rank $n = [L:K]$. In general, such an integral basis does not exist. If, however, A is a principal ideal domain, then one has the following more general

(2.10) Proposition. *If $L|K$ is separable and A is a principal ideal domain, then every finitely generated B-submodule $M \neq 0$ of L is a free A-module of rank $[L:K]$. In particular, B admits an integral basis over A.*

Proof: Let $M \neq 0$ be a finitely generated B-submodule of L and $\alpha_1, \dots, \alpha_n$ a basis of $L|K$. Multiplying by an element of A, we may arrange for the α_i to lie in B. By (2.9), we then have $dB \subseteq A\alpha_1 + \cdots + A\alpha_n$, in particular, $\text{rank}(B) \leq [L : K]$, and since a system of generators of the A-module B is also a system of generators of the K-module L, we have $\text{rank}(B) = [L : K]$. Let $\mu_1, \dots, \mu_r \in M$ be a system of generators of the B-module M. There exists an $a \in A$, $a \neq 0$, such that $a\mu_i \in B$, $i = 1, \dots, r$, so that $aM \subseteq B$. Then

$$adM \subseteq dB \subseteq A\alpha_1 + \cdots + A\alpha_n = M_0 .$$

According to the main theorem on finitely generated modules over principal ideal domains, since M_0 is a free A-module, so is adM, and hence also M. Finally,

$$[L : K] = \text{rank}(B) \leq \text{rank}(M) = \text{rank}(adM) \leq \text{rank}(M_0) = [L : K],$$

hence $\text{rank}(M) = [L : K]$. □

It is in general a difficult problem to produce integral bases. In concrete situations it can also be an important one. This is why the following proposition is interesting. Instead of integral bases of the integral closure B of A in L, we will now simply speak of integral bases of the extension $L|K$.

(2.11) Proposition. *Let $L|K$ and $L'|K$ be two Galois extensions of degree n, resp. n', such that $L \cap L' = K$. Let $\omega_1, \dots, \omega_n$, resp. $\omega'_1, \dots, \omega'_{n'}$, be an integral basis of $L \mid K$, resp. $L'|K$, with discriminant d, resp. d'. Suppose that d and d' are relatively prime in the sense that $xd + x'd' = 1$, for suitable $x, x' \in A$. Then $\omega_i\,\omega'_j$ is an integral basis of LL', of discriminant $d^{n'}d'^{n}$.*

Proof: As $L \cap L' = K$, we have $[LL' : K] = nn'$, so the nn' products $\omega_i\omega'_j$ do form a basis of $LL'|K$. Now let α be an integral element of LL', and write

$$\alpha = \sum_{i,j} a_{ij}\,\omega_i\,\omega'_j\,, \quad a_{ij} \in K\,.$$

We have to show that $a_{ij} \in A$. Put $\beta_j = \sum_i a_{ij}\,\omega_i$. Let $G(LL'|L') = \{\sigma_1, \dots, \sigma_n\}$ and $G(LL'|L) = \{\sigma'_1, \dots, \sigma'_{n'}\}$. Thus

$$G(LL'|K) = \left\{\sigma_k\sigma'_\ell \;\middle|\; k = 1, \dots, n,\ \ell = 1, \dots, n'\right\}.$$

Putting

$$T = (\sigma'_\ell\omega'_j), \quad a = (\sigma'_1\alpha, \dots, \sigma'_{n'}\alpha)^t, \quad b = (\beta_1, \dots, \beta_{n'})^t,$$

one finds $\det(T)^2 = d'$ and

$$a = T\,b\,.$$

Write T^* for the adjoint matrix of T. Then row-column expansion (2.3) gives

$$\det(T)\,b = T^* a\,.$$

Since T^* and a have integral entries in LL', the multiple $d'b$ has integral entries in L, namely $d'\beta_j = \sum_i d'a_{ij}\,\omega_i$. Thus $d'a_{ij} \in A$. Swapping the rôles of (ω_i) and (ω'_j), one checks in the same manner that $da_{ij} \in A$, so that

$$a_{ij} = xda_{ij} + x'd'a_{ij} \in A\,.$$

Therefore $\omega_i\,\omega'_j$ is indeed an integral basis of $LL'|K$. We compute the discriminant Δ of this integral basis. Since $G(LL'|K) = \{\sigma_k\sigma'_\ell \mid k = 1, \dots, n,\ \ell = 1, \dots, n'\}$, it is the square of the determinant of the $(nn' \times nn')$-matrix

$$M = (\sigma_k\sigma'_\ell\,\omega_i\omega'_j) = (\sigma_k\omega_i\,\sigma'_\ell\omega'_j)\,.$$

This matrix is itself an $(n' \times n')$-matrix with entries $(n \times n)$-matrices of which the (ℓ, j)-entry is the matrix $Q\sigma'_\ell\omega'_j$ where $Q = (\sigma_k\omega_i)$. In other words,

$$M = \begin{pmatrix} Q & & 0 \\ & \ddots & \\ 0 & & Q \end{pmatrix} \begin{pmatrix} E\sigma'_1\omega'_1 & \cdots & E\sigma'_{n'}\omega'_1 \\ \vdots & & \vdots \\ E\sigma'_1\omega'_{n'} & \cdots & E\sigma'_{n'}\omega'_{n'} \end{pmatrix}.$$

Here E denotes the $(n \times n)$-unit matrix. By changing indices the second matrix may be transformed to look like the first one. This yields

$$\Delta = \det(M)^2 = \det(Q)^{2n'} \det\big((\sigma'_\ell\omega'_j)\big)^{2n} = d^{n'} d'^{n}\,. \qquad \square$$

Remark: It follows from the proof that the proposition is valid for arbitrary separable extensions (not necessarily Galois), if one assumes instead of $L \cap L' = K$ that L and L' are linearly disjoint.

The chief application of our considerations on integrality will concern the integral closure $\mathcal{O}_K \subseteq K$ of $\mathbb{Z} \subseteq \mathbb{Q}$ in an algebraic number field K. By proposition (2.10), every finitely generated $\mathcal{O}_K$-submodule $\mathfrak{a}$ of K admits a $\mathbb{Z}$-basis $\alpha_1, \dots, \alpha_n$,

$$\mathfrak{a} = \mathbb{Z}\alpha_1 + \cdots + \mathbb{Z}\alpha_n\,.$$

The discriminant

$$d(\alpha_1, \dots, \alpha_n) = \det\big((\sigma_i\alpha_j)\big)^2$$

is independent of the choice of a $\mathbb{Z}$-basis; if $\alpha'_1, \dots, \alpha'_n$ is another basis, then the base change matrix $T = (a_{ij})$, $\alpha'_i = \sum_j a_{ij}\alpha_j$, as well as its inverse, has integral entries. It therefore has determinant ± 1, so that indeed

$$d(\alpha'_1, \dots, \alpha'_n) = \det(T)^2\, d(\alpha_1, \dots, \alpha_n) = d(\alpha_1, \dots, \alpha_n).$$

We may therefore write

$$d(\mathfrak{a}) = d(\alpha_1, \dots, \alpha_n).$$

In the special case of an integral basis $\omega_1, \dots, \omega_n$ of $\mathcal{O}_K$ we obtain the **discriminant of the algebraic number field** K,

$$d_K = d(\mathcal{O}_K) = d(\omega_1, \dots, \omega_n).$$

In general, one has the

(2.12) Proposition. *If $\mathfrak{a} \subseteq \mathfrak{a}'$ are two nonzero finitely generated $\mathcal{O}_K$-submodules of K, then the index $(\mathfrak{a}' : \mathfrak{a})$ is finite and satisfies*

$$d(\mathfrak{a}) = (\mathfrak{a}' : \mathfrak{a})^2\, d(\mathfrak{a}').$$

All we have to show is that the index $(\mathfrak{a}' : \mathfrak{a})$ equals the absolute value of the determinant of the base change matrix passing from a $\mathbb{Z}$-basis of $\mathfrak{a}$ to a $\mathbb{Z}$-basis of $\mathfrak{a}'$. This proof is part of the well-known theory of finitely generated $\mathbb{Z}$-modules.

Exercise 1. Is $\frac{3+2\sqrt{6}}{1-\sqrt{6}}$ an algebraic integer?

Exercise 2. Show that, if the integral domain A is integrally closed, then so is the polynomial ring $A[t]$.

Exercise 3. In the polynomial ring $A = \mathbb{Q}[X, Y]$, consider the principal ideal $\mathfrak{p} = (X^2 - Y^3)$. Show that $\mathfrak{p}$ is a prime ideal, but $A/\mathfrak{p}$ is not integrally closed.

Exercise 4. Let D be a squarefree rational integer $\neq 0, 1$ and d the discriminant of the quadratic number field $K = \mathbb{Q}(\sqrt{D})$. Show that

$$d = D, \qquad \text{if } D \equiv 1 \bmod 4,$$

$$d = 4D, \qquad \text{if } D \equiv 2 \text{ or } 3 \bmod 4,$$

and that an integral basis of K is given by $\{1, \sqrt{D}\}$ in the second case, by $\{1, \frac{1}{2}(1+\sqrt{D})\}$ in the first case, and by $\{1, \frac{1}{2}(d+\sqrt{d})\}$ in both cases.

Exercise 5. Show that $\{1, \sqrt[3]{2}, \sqrt[3]{2^2}\}$ is an integral basis of $\mathbb{Q}(\sqrt[3]{2})$.

Exercise 6. Show that $\{1, \theta, \frac{1}{2}(\theta + \theta^2)\}$ is an integral basis of $\mathbb{Q}(\theta)$, $\theta^3 - \theta - 4 = 0$.

Exercise 7. The discriminant d_K of an algebraic number field K is always $\equiv 0 \bmod 4$ or $\equiv 1 \bmod 4$ (*Stickelberger's discriminant relation*).

Hint: The determinant $\det(\sigma_i\omega_j)$ of an integral basis ω_j is a sum of terms, each prefixed by a positive or a negative sign. Writing P, resp. N, for the sum of the positive, resp. negative terms, one finds $d_K = (P - N)^2 = (P + N)^2 - 4PN$.

§ 3. Ideals

Being a generalization of the ring $\mathbb{Z} \subseteq \mathbb{Q}$, the ring $\mathcal{O}_K$ of integers of an algebraic number field K is at the center of all our considerations. As in $\mathbb{Z}$, every non-unit $\alpha \neq 0$ can be factored in $\mathcal{O}_K$ into a product of irreducible elements. For if α is not itself irreducible, then it can be written as a product of two non-units $\alpha = \beta\gamma$. Then by § 2, one has

$$1 < |N_{K|\mathbb{Q}}(\beta)| < |N_{K|\mathbb{Q}}(\alpha)|, \quad 1 < |N_{K|\mathbb{Q}}(\gamma)| < |N_{K|\mathbb{Q}}(\alpha)|,$$

and the prime decomposition of α follows by induction from those of β and γ. However, contrary to what happens in the rings $\mathbb{Z}$ and $\mathbb{Z}[i]$, the uniqueness of prime factorization does not hold in general.

Example: The ring of integers of the field $K = \mathbb{Q}(\sqrt{-5})$ is given by § 2, exercise 4, as $\mathcal{O}_K = \mathbb{Z} + \mathbb{Z}\sqrt{-5}$. In this ring, the rational integer 21 can be decomposed in two ways,

$$21 = 3 \cdot 7 = (1 + 2\sqrt{-5}) \cdot (1 - 2\sqrt{-5}).$$

All factors occurring here are irreducible in $\mathcal{O}_K$. For if one had, for instance, $3 = \alpha\beta$, with α, β non-units, then $9 = N_{K|\mathbb{Q}}(\alpha)\, N_{K|\mathbb{Q}}(\beta)$ would imply $N_{K|\mathbb{Q}}(\alpha) = \pm 3$. But the equation

$$N_{K|\mathbb{Q}}(x + y\sqrt{-5}) = x^2 + 5y^2 = \pm 3$$

has no solutions in $\mathbb{Z}$. In the same way it is seen that 7, $1 + 2\sqrt{-5}$, and $1 - 2\sqrt{-5}$ are irreducible. As the fractions

$$\frac{1 \pm 2\sqrt{-5}}{3}, \quad \frac{1 \pm 2\sqrt{-5}}{7}$$

do not belong to $\mathcal{O}_K$, the numbers 3 and 7 are not associated to $1 + 2\sqrt{-5}$ or $1 - 2\sqrt{-5}$. The two prime factorizations of 21 are therefore essentially different.

Realizing the failure of unique factorization in general has led to one of the grand events in the history of number theory, the discovery of ideal theory by *Eduard Kummer*. Inspired by the discovery of complex numbers, Kummer's idea was that the integers of K would have to admit an embedding into a bigger domain of "ideal numbers" where **unique** factorization into "ideal prime numbers" would hold. For instance, in the example of

$$21 = 3 \cdot 7 = (1 + 2\sqrt{-5}\,)(1 - 2\sqrt{-5}\,),$$

the factors on the right would be composed of ideal prime numbers $\mathfrak{p}_1, \mathfrak{p}_2, \mathfrak{p}_3, \mathfrak{p}_4$, subject to the rules

$$3 = \mathfrak{p}_1\mathfrak{p}_2, \quad 7 = \mathfrak{p}_3\mathfrak{p}_4, \quad 1 + 2\sqrt{-5} = \mathfrak{p}_1\mathfrak{p}_3, \quad 1 - 2\sqrt{-5} = \mathfrak{p}_2\mathfrak{p}_4.$$

This would resolve the above non-uniqueness into the wonderfully unique factorization

$$21 = (\mathfrak{p}_1\mathfrak{p}_2)(\mathfrak{p}_3\mathfrak{p}_4) = (\mathfrak{p}_1\mathfrak{p}_3)(\mathfrak{p}_2\mathfrak{p}_4).$$

Kummer's concept of "ideal numbers" was later replaced by that of **ideals** of the ring $\mathcal{O}_K$. The reason for this is easily seen: whatever an ideal number $\mathfrak{a}$ should be defined to be, it ought to be linked to certain numbers $a \in \mathcal{O}_K$ by a divisibility relation $\mathfrak{a} \,|\, a$ satisfying the following rules, for $a, b, \lambda \in \mathcal{O}_K$,

$$\mathfrak{a} \,|\, a \;\text{ and }\; \mathfrak{a} \,|\, b \;\Rightarrow\; \mathfrak{a} \,|\, a \pm b; \qquad \mathfrak{a} \,|\, a \;\Rightarrow\; \mathfrak{a} \,|\, \lambda a.$$

And an ideal number $\mathfrak{a}$ should be determined by the totality of its divisors in $\mathcal{O}_K$

$$\underline{\mathfrak{a}} = \big\{ a \in \mathcal{O}_K \;\big|\; \mathfrak{a} \,|\, a \big\}.$$

But in view of the rules for divisibility, this set is an ideal of $\mathcal{O}_K$.

This is the reason why *Richard Dedekind* re-introduced Kummer's "ideal numbers" as being the ideals of $\mathcal{O}_K$. Once this is done, the divisibility relation $\mathfrak{a} \,|\, a$ can simply be defined by the inclusion $a \in \mathfrak{a}$, and more generally the divisibility relation $\mathfrak{a} \,|\, \mathfrak{b}$ between two ideals by $\mathfrak{b} \subseteq \mathfrak{a}$. In what follows, we will study this notion of divisibility more closely. The basic theorem here is the following.

(3.1) Theorem. *The ring $\mathcal{O}_K$ is noetherian, integrally closed, and every prime ideal $\mathfrak{p} \neq 0$ is a maximal ideal.*

Proof: $\mathcal{O}_K$ is noetherian because every ideal $\mathfrak{a}$ is a finitely generated $\mathbb{Z}$-module by (2.10), and therefore *a fortiori* a finitely generated $\mathcal{O}_K$-module. By § 2, $\mathcal{O}_K$ is also integrally closed, being the integral closure of $\mathbb{Z}$ in K. It thus remains to show that each prime ideal $\mathfrak{p} \neq 0$ is maximal. Now, $\mathfrak{p} \cap \mathbb{Z}$ is a nonzero prime ideal (p) in $\mathbb{Z}$: the primality is clear, and if $y \in \mathfrak{p}$, $y \neq 0$, and

$$y^n + a_1 y^{n-1} + \cdots + a_n = 0$$

is an equation for y with $a_i \in \mathbb{Z}$, $a_n \neq 0$, then $a_n \in \mathfrak{p} \cap \mathbb{Z}$. The integral domain $\overline{\mathcal{O}} = \mathcal{O}_K/\mathfrak{p}$ arises from $\kappa = \mathbb{Z}/p\mathbb{Z}$ by adjoining algebraic elements and is therefore again a field (recall the fact that $\kappa[\alpha] = \kappa(\alpha)$, if α is algebraic). Therefore $\mathfrak{p}$ is a maximal ideal. □

The three properties of the ring $\mathcal{O}_K$ which we have just proven lay the foundation of the whole theory of divisibility of its ideals. This theory was developed by Dedekind, which suggested the following

(3.2) Definition. *A noetherian, integrally closed integral domain in which every nonzero prime ideal is maximal is called a* **Dedekind domain**.

Just as the rings of the form $\mathcal{O}_K$ may be viewed as generalizations of the ring $\mathbb{Z}$, the Dedekind domains may be viewed as generalized principal ideal domains. Indeed, if A is a principal ideal domain with field of fractions K, and $L|K$ is a finite field extension, then the integral closure B of A in L is, in general, not a principal ideal domain, but always a Dedekind domain, as we shall show further on.

Instead of the ring $\mathcal{O}_K$ we will now consider an arbitrary Dedekind domain $\mathcal{O}$, and we denote by K the field of fractions of $\mathcal{O}$. Given two ideals $\mathfrak{a}$ and $\mathfrak{b}$ of $\mathcal{O}$ (or more generally of an arbitrary ring), the divisibility relation $\mathfrak{a}|\mathfrak{b}$ is defined by $\mathfrak{b} \subseteq \mathfrak{a}$, and the sum of the ideals by

$$\mathfrak{a} + \mathfrak{b} = \{a + b \mid a \in \mathfrak{a},\ b \in \mathfrak{b}\}.$$

This is the smallest ideal containing $\mathfrak{a}$ as well as $\mathfrak{b}$, in other words, it is the greatest common divisor $\gcd(\mathfrak{a}, \mathfrak{b})$ of $\mathfrak{a}$ and $\mathfrak{b}$. By the same token the intersection $\mathfrak{a} \cap \mathfrak{b}$ is the lcm (least common multiple) of $\mathfrak{a}$ and $\mathfrak{b}$. We define the **product** of $\mathfrak{a}$ and $\mathfrak{b}$ by

$$\mathfrak{a}\mathfrak{b} = \Big\{ \sum_i a_i b_i \;\Big|\; a_i \in \mathfrak{a},\ b_i \in \mathfrak{b}\Big\}.$$

With respect to this multiplication the ideals of $\mathcal{O}$ will grant us what the elements alone may refuse to provide: the **unique prime factorization**.

(3.3) Theorem. *Every ideal $\mathfrak{a}$ of $\mathcal{O}$ different from (0) and (1) admits a factorization*

$$\mathfrak{a} = \mathfrak{p}_1 \cdots \mathfrak{p}_r$$

into nonzero prime ideals $\mathfrak{p}_i$ of $\mathcal{O}$ which is unique up to the order of the factors.

This theorem is of course perfectly in line with the invention of "ideal numbers". Still, the fact that it holds is remarkable because its proof is far from straightforward, and unveils a deeper principle governing the arithmetic in $\mathcal{O}$. We prepare the proof proper by two lemmas.

(3.4) Lemma. *For every ideal* $\mathfrak{a} \neq 0$ *of* $\mathcal{O}$ *there exist nonzero prime ideals* $\mathfrak{p}_1, \mathfrak{p}_2, \ldots, \mathfrak{p}_r$ *such that*

$$\mathfrak{a} \supseteq \mathfrak{p}_1\mathfrak{p}_2 \cdots \mathfrak{p}_r .$$

Proof: Suppose the set $\mathfrak{M}$ of those ideals which do not fulfill this condition is nonempty. As $\mathcal{O}$ is noetherian, every ascending chain of ideals becomes stationary. Therefore $\mathfrak{M}$ is inductively ordered with respect to inclusion and thus admits a maximal element $\mathfrak{a}$. This cannot be a prime ideal, so there exist elements $b_1, b_2 \in \mathcal{O}$ such that $b_1 b_2 \in \mathfrak{a}$, but $b_1, b_2 \notin \mathfrak{a}$. Put $\mathfrak{a}_1 = (b_1) + \mathfrak{a}$, $\mathfrak{a}_2 = (b_2) + \mathfrak{a}$. Then $\mathfrak{a} \subsetneqq \mathfrak{a}_1$, $\mathfrak{a} \subsetneqq \mathfrak{a}_2$ and $\mathfrak{a}_1\mathfrak{a}_2 \subseteq \mathfrak{a}$. By the maximality of $\mathfrak{a}$, both $\mathfrak{a}_1$ and $\mathfrak{a}_2$ contain a product of prime ideals, and the product of these products is contained in $\mathfrak{a}$, a contradiction. □

(3.5) Lemma. *Let* $\mathfrak{p}$ *be a prime ideal of* $\mathcal{O}$ *and define*

$$\mathfrak{p}^{-1} = \{x \in K \mid x\mathfrak{p} \subseteq \mathcal{O}\} .$$

Then one has $\mathfrak{a}\mathfrak{p}^{-1} := \{\sum_i a_i x_i \mid a_i \in \mathfrak{a},\ x_i \in \mathfrak{p}^{-1}\} \neq \mathfrak{a}$, *for every ideal* $\mathfrak{a} \neq 0$.

Proof: Let $a \in \mathfrak{p}$, $a \neq 0$, and $\mathfrak{p}_1\mathfrak{p}_2 \cdots \mathfrak{p}_r \subseteq (a) \subseteq \mathfrak{p}$, with r as small as possible. Then one of the $\mathfrak{p}_i$, say $\mathfrak{p}_1$, is contained in $\mathfrak{p}$, and so $\mathfrak{p}_1 = \mathfrak{p}$ because $\mathfrak{p}_1$ is a maximal ideal. (Indeed, if none of the $\mathfrak{p}_i$ were contained in $\mathfrak{p}$, then for every i there would exist $a_i \in \mathfrak{p}_i \smallsetminus \mathfrak{p}$ such that $a_1 \cdots a_r \in \mathfrak{p}$. But $\mathfrak{p}$ is prime.) Since $\mathfrak{p}_2 \cdots \mathfrak{p}_r \not\subseteq (a)$, there exists $b \in \mathfrak{p}_2 \cdots \mathfrak{p}_r$ such that $b \notin a\mathcal{O}$, i.e., $a^{-1}b \notin \mathcal{O}$. On the other hand we have $b\mathfrak{p} \subseteq (a)$, i.e., $a^{-1}b\mathfrak{p} \subseteq \mathcal{O}$, and thus $a^{-1}b \in \mathfrak{p}^{-1}$. It follows that $\mathfrak{p}^{-1} \neq \mathcal{O}$.

Now let $\mathfrak{a} \neq 0$ be an ideal of $\mathcal{O}$ and $\alpha_1, \ldots, \alpha_n$ a system of generators. Let us assume that $\mathfrak{a}\mathfrak{p}^{-1} = \mathfrak{a}$. Then for every $x \in \mathfrak{p}^{-1}$,

$$x\alpha_i = \sum_j a_{ij}\alpha_j , \quad a_{ij} \in \mathcal{O} .$$

Writing A for the matrix $(x\delta_{ij} - a_{ij})$ we obtain $A(\alpha_1, \ldots, \alpha_n)^t = 0$. By (2.3), the determinant $d = \det(A)$ satisfies $d\alpha_1 = \cdots = d\alpha_n = 0$ and thus $d = 0$. It follows that x is integral over $\mathcal{O}$, being a zero of the monic polynomial $f(X) = \det(X\delta_{ij} - a_{ij}) \in \mathcal{O}[X]$. Therefore $x \in \mathcal{O}$. This means that $\mathfrak{p}^{-1} = \mathcal{O}$, a contradiction. □

Proof of (3.3): I. Existence of the prime ideal factorization. Let $\mathfrak{M}$ be the set of all ideals different from (0) and (1) which do not admit a prime ideal decomposition. If $\mathfrak{M}$ is nonempty, then we argue as for (3.4) that there exists a maximal element $\mathfrak{a}$ in $\mathfrak{M}$. It is contained in a maximal ideal $\mathfrak{p}$, and the inclusion $\mathcal{O} \subseteq \mathfrak{p}^{-1}$ gives us

$$\mathfrak{a} \subseteq \mathfrak{a}\,\mathfrak{p}^{-1} \subseteq \mathfrak{p}\,\mathfrak{p}^{-1} \subseteq \mathcal{O}.$$

By (3.5), one has $\mathfrak{a} \subsetneqq \mathfrak{a}\,\mathfrak{p}^{-1}$ and $\mathfrak{p} \subsetneqq \mathfrak{p}\,\mathfrak{p}^{-1} \subseteq \mathcal{O}$. Since $\mathfrak{p}$ is a maximal ideal, it follows that $\mathfrak{p}\,\mathfrak{p}^{-1} = \mathcal{O}$. In view of the maximality of $\mathfrak{a}$ in $\mathfrak{M}$ and since $\mathfrak{a} \neq \mathfrak{p}$, i.e., $\mathfrak{a}\,\mathfrak{p}^{-1} \neq \mathcal{O}$, the ideal $\mathfrak{a}\,\mathfrak{p}^{-1}$ admits a prime ideal decomposition $\mathfrak{a}\,\mathfrak{p}^{-1} = \mathfrak{p}_1 \cdots \mathfrak{p}_r$, and so does $\mathfrak{a} = \mathfrak{a}\,\mathfrak{p}^{-1}\,\mathfrak{p} = \mathfrak{p}_1 \cdots \mathfrak{p}_r \mathfrak{p}$, a contradiction.

II. Uniqueness of the prime ideal factorization. For a prime ideal $\mathfrak{p}$ one has: $\mathfrak{a}\,\mathfrak{b} \subseteq \mathfrak{p} \Rightarrow \mathfrak{a} \subseteq \mathfrak{p}$ or $\mathfrak{b} \subseteq \mathfrak{p}$, i.e., $\mathfrak{p} \mid \mathfrak{a}\,\mathfrak{b} \Rightarrow \mathfrak{p} \mid \mathfrak{a}$ or $\mathfrak{p} \mid \mathfrak{b}$. Let

$$\mathfrak{a} = \mathfrak{p}_1\,\mathfrak{p}_2 \cdots \mathfrak{p}_r = \mathfrak{q}_1\,\mathfrak{q}_2 \cdots \mathfrak{q}_s$$

be two prime ideal factorizations of $\mathfrak{a}$. Then $\mathfrak{p}_1$ divides a factor $\mathfrak{q}_i$, say $\mathfrak{q}_1$, and being maximal equals $\mathfrak{q}_1$. We multiply by $\mathfrak{p}_1^{-1}$ and obtain, in view of $\mathfrak{p}_1 \neq \mathfrak{p}_1\,\mathfrak{p}_1^{-1} = \mathcal{O}$, that

$$\mathfrak{p}_2 \cdots \mathfrak{p}_r = \mathfrak{q}_2 \cdots \mathfrak{q}_s\,.$$

Continuing like this we see that $r = s$ and, possibly after renumbering, $\mathfrak{p}_i = \mathfrak{q}_i$, for all $i = 1, \ldots, r$. □

Grouping together the occurrences of the same prime ideals in the prime ideal factorization of an ideal $\mathfrak{a} \neq 0$ of $\mathcal{O}$, gives a product representation

$$\mathfrak{a} = \mathfrak{p}_1^{\nu_1} \cdots \mathfrak{p}_r^{\nu_r}\,, \qquad \nu_i > 0\,.$$

In the sequel such an identity will be automatically understood to signify that the $\mathfrak{p}_i$ are pairwise distinct. If in particular $\mathfrak{a}$ is a principal ideal (a), then – following the tradition which tends to attribute to the ideals the rôle of "ideal numbers" – we will write with a slight abuse of notation

$$a = \mathfrak{p}_1^{\nu_1} \cdots \mathfrak{p}_r^{\nu_r}\,.$$

Similarly, the notation $\mathfrak{a} \mid a$ is often used instead of $\mathfrak{a} \mid (a)$ and $(\mathfrak{a}, \mathfrak{b}) = 1$ is written for two relatively prime ideals, instead of the correct formula $(\mathfrak{a}, \mathfrak{b}) = \mathfrak{a} + \mathfrak{b} = \mathcal{O}$. For a product $\mathfrak{a} = \mathfrak{a}_1 \cdots \mathfrak{a}_n$ of relatively prime ideals $\mathfrak{a}_1, \ldots, \mathfrak{a}_n$, one has an analogue of the well-known "Chinese Remainder Theorem" from elementary number theory. We may formulate this result for an arbitrary ring taking into account that

$$\mathfrak{a} = \bigcap_{i=1}^{n} \mathfrak{a}_i\,.$$

Indeed, since $\mathfrak{a}_i \mid \mathfrak{a}$, $i = 1, \ldots, n$, we find on the one hand that $\mathfrak{a} \subseteq \bigcap_{i=1}^{n} \mathfrak{a}_i$, and for $a \in \bigcap_i \mathfrak{a}_i$ we find that $\mathfrak{a}_i \mid a$, and therefore, the factors being relatively prime, we get $\mathfrak{a} = \mathfrak{a}_1 \cdots \mathfrak{a}_n \mid a$, i.e., $a \in \mathfrak{a}$.

(3.6) Chinese Remainder Theorem. *Let $\mathfrak{a}_1, \dots, \mathfrak{a}_n$ be ideals in a ring $\mathcal{O}$ such that $\mathfrak{a}_i + \mathfrak{a}_j = \mathcal{O}$ for $i \neq j$. Then, if $\mathfrak{a} = \bigcap_{i=1}^n \mathfrak{a}_i$, one has*

$$\mathcal{O}/\mathfrak{a} \cong \bigoplus_{i=1}^n \mathcal{O}/\mathfrak{a}_i\,.$$

Proof: The canonical homomorphism

$$\mathcal{O} \longrightarrow \bigoplus_{i=1}^n \mathcal{O}/\mathfrak{a}_i\,, \quad a \longmapsto \bigoplus_{i=1}^n a \bmod \mathfrak{a}_i\,,$$

has kernel $\mathfrak{a} = \bigcap_i \mathfrak{a}_i$. It therefore suffices to show that it is surjective. For this, let $x_i \bmod \mathfrak{a}_i \in \mathcal{O}/\mathfrak{a}_i$, $i = 1, \dots, n$, be given. If $n = 2$, we may write $1 = a_1 + a_2$, $a_i \in \mathfrak{a}_i$, and putting $x = x_2 a_1 + x_1 a_2$ we get $x \equiv x_i \bmod \mathfrak{a}_i$, $i = 1, 2$.

If $n > 2$, we may find as before an element $y_1 \in \mathcal{O}$ such that

$$y_1 \equiv 1 \bmod \mathfrak{a}_1\,, \quad y_1 \equiv 0 \bmod \bigcap_{i=2}^n \mathfrak{a}_i\,,$$

and, by the same token, elements $y_2, \dots, y_n$ such that

$$y_i \equiv 1 \bmod \mathfrak{a}_i\,, \quad y_i \equiv 0 \bmod \mathfrak{a}_j \quad \text{for } i \neq j\,.$$

Putting $x = x_1 y_1 + \cdots + x_n y_n$ we find $x \equiv x_i \bmod \mathfrak{a}_i$, $i = 1, \dots, n$. This proves the surjectivity. □

Now let $\mathcal{O}$ be again a Dedekind domain. Just as for nonzero numbers, we may obtain **inverses** for the nonzero ideals of $\mathcal{O}$ by introducing the notion of fractional ideal in the field of fractions K.

(3.7) Definition. *A* **fractional ideal** *of K is a finitely generated $\mathcal{O}$-submodule $\mathfrak{a} \neq 0$ of K.*

For instance, an element $a \in K^*$ defines the fractional "principal ideal" $(a) = a\mathcal{O}$. Obviously, since $\mathcal{O}$ is noetherian, an $\mathcal{O}$-submodule $\mathfrak{a} \neq 0$ of K is a fractional ideal if and only if there exists $c \in \mathcal{O}$, $c \neq 0$, such that $c\mathfrak{a} \subseteq \mathcal{O}$ is an ideal of the ring $\mathcal{O}$. Fractional ideals are multiplied in the same way as ideals in $\mathcal{O}$. For distinction the latter may henceforth be called **integral ideals** of K.

(3.8) Proposition. *The fractional ideals form an abelian group, the* **ideal group** *J_K of K. The identity element is $(1) = \mathcal{O}$, and the inverse of $\mathfrak{a}$ is*

$$\mathfrak{a}^{-1} = \{x \in K \mid x\mathfrak{a} \subseteq \mathcal{O}\}\,.$$

Proof: One obviously has associativity, commutativity and $\mathfrak{a}(1) = \mathfrak{a}$. For a prime ideal $\mathfrak{p}$, (3.5) says that $\mathfrak{p} \subsetneqq \mathfrak{p}\mathfrak{p}^{-1}$ and therefore $\mathfrak{p}\mathfrak{p}^{-1} = \mathcal{O}$ because $\mathfrak{p}$ is maximal. Consequently, if $\mathfrak{a} = \mathfrak{p}_1 \cdots \mathfrak{p}_r$ is an integral ideal, then $\mathfrak{b} = \mathfrak{p}_1^{-1} \cdots \mathfrak{p}_r^{-1}$ is an inverse. $\mathfrak{b}\mathfrak{a} = \mathcal{O}$ implies that $\mathfrak{b} \subseteq \mathfrak{a}^{-1}$. Conversely, if $x\mathfrak{a} \subseteq \mathcal{O}$, then $x\mathfrak{a}\mathfrak{b} \subseteq \mathfrak{b}$, so $x \in \mathfrak{b}$ because $\mathfrak{a}\mathfrak{b} = \mathcal{O}$. Thus we have $\mathfrak{b} = \mathfrak{a}^{-1}$. Finally, if $\mathfrak{a}$ is an arbitrary fractional ideal and $c \in \mathcal{O}$, $c \neq 0$, is such that $c\mathfrak{a} \subseteq \mathcal{O}$, then $(c\mathfrak{a})^{-1} = c^{-1}\mathfrak{a}^{-1}$ is the inverse of $c\mathfrak{a}$, so $\mathfrak{a}\mathfrak{a}^{-1} = \mathcal{O}$. □

(3.9) Corollary. *Every fractional ideal $\mathfrak{a}$ admits a unique representation as a product*

$$\mathfrak{a} = \prod_{\mathfrak{p}} \mathfrak{p}^{\nu_{\mathfrak{p}}}$$

with $\nu_{\mathfrak{p}} \in \mathbb{Z}$ and $\nu_{\mathfrak{p}} = 0$ for almost all $\mathfrak{p}$. In other words, J_K is the free abelian group on the set of nonzero prime ideals $\mathfrak{p}$ of $\mathcal{O}$.

Proof: Every fractional ideal $\mathfrak{a}$ is a quotient $\mathfrak{a} = \mathfrak{b}/\mathfrak{c}$ of two integral ideals $\mathfrak{b}$ and $\mathfrak{c}$, which by (3.3) have a prime decomposition. Therefore $\mathfrak{a}$ has a prime decomposition of the type stated in the corollary. By (3.3), it is unique if $\mathfrak{a}$ is integral, and therefore clearly also in general. □

The fractional principal ideals $(a) = a\mathcal{O}$, $a \in K^*$, form a subgroup of the group of ideals J_K, which will be denoted P_K. The quotient group

$$Cl_K = J_K / P_K$$

is called the **ideal class group**, or **class group** for short, of K. Along with the group of units $\mathcal{O}^*$ of $\mathcal{O}$, it fits into the exact sequence

$$1 \longrightarrow \mathcal{O}^* \longrightarrow K^* \longrightarrow J_K \longrightarrow Cl_K \longrightarrow 1,$$

where the arrow in the middle is given by $a \mapsto (a)$. So the class group Cl_K measures the expansion that takes place when we pass from numbers to ideals, whereas the unit group $\mathcal{O}^*$ measures the contraction in the same process. This immediately raises the problem of understanding these groups $\mathcal{O}^*$ and Cl_K more thoroughly. For general Dedekind domains they may turn out to be completely arbitrary groups. For the ring $\mathcal{O}_K$ of integers in a number field K, however, one obtains important finiteness theorems, which are fundamental for the further development of number theory. But these results cannot be had for nothing. They will be obtained by viewing the numbers geometrically as lattice points in space. For this we will now prepare the necessary concepts, which all come from linear algebra.

Exercise 1. Decompose $33+11\sqrt{-7}$ into irreducible integral elements of $\mathbb{Q}(\sqrt{-7})$.

Exercise 2. Show that

$$54 = 2\cdot 3^3 = \frac{13+\sqrt{-47}}{2}\cdot\frac{13-\sqrt{-47}}{2}$$

are two essentially different decompositions into irreducible integral elements of $\mathbb{Q}(\sqrt{-47})$.

Exercise 3. Let d be squarefree and p a prime number not dividing $2d$. Let $\mathcal{O}$ be the ring of integers of $\mathbb{Q}(\sqrt{d})$. Show that $(p)=p\mathcal{O}$ is a prime ideal of $\mathcal{O}$ if and only if the congruence $x^2\equiv d \bmod p$ has no solution.

Exercise 4. A Dedekind domain with a finite number of prime ideals is a principal ideal domain.

Hint: If $\mathfrak{a}=\mathfrak{p}_1^{\nu_1}\cdots\mathfrak{p}_r^{\nu_r}\neq 0$ is an ideal, then choose elements $\pi_i\in\mathfrak{p}_i\smallsetminus\mathfrak{p}_i^2$ and apply the Chinese remainder theorem for the cosets $\pi_i^{\nu_i} \bmod \mathfrak{p}_i^{\nu_i+1}$.

Exercise 5. The quotient ring $\mathcal{O}/\mathfrak{a}$ of a Dedekind domain by an ideal $\mathfrak{a}\neq 0$ is a principal ideal domain.

Hint: For $\mathfrak{a}=\mathfrak{p}^n$ the only proper ideals of $\mathcal{O}/\mathfrak{a}$ are given by $\mathfrak{p}/\mathfrak{p}^n, \dots, \mathfrak{p}^{n-1}/\mathfrak{p}^n$. Choose $\pi\in\mathfrak{p}\smallsetminus\mathfrak{p}^2$ and show that $\mathfrak{p}^\nu=\mathcal{O}\pi^\nu+\mathfrak{p}^n$.

Exercise 6. Every ideal of a Dedekind domain can be generated by two elements.

Hint: Use exercise 5.

Exercise 7. In a noetherian ring R in which every prime ideal is maximal, each descending chain of ideals $\mathfrak{a}_1\supseteq\mathfrak{a}_2\supseteq\cdots$ becomes stationary.

Hint: Show as in (3.4) that (0) is a product $\mathfrak{p}_1\cdots\mathfrak{p}_r$ of prime ideals and that the chain $R\supseteq\mathfrak{p}_1\supseteq\mathfrak{p}_1\mathfrak{p}_2\supseteq\cdots\supseteq\mathfrak{p}_1\cdots\mathfrak{p}_r=(0)$ can be refined into a composition series.

Exercise 8. Let $\mathfrak{m}$ be a nonzero integral ideal of the Dedekind domain $\mathcal{O}$. Show that in every ideal class of Cl_K, there exists an integral ideal prime to $\mathfrak{m}$.

Exercise 9. Let $\mathcal{O}$ be an integral domain in which all nonzero ideals admit a unique factorization into prime ideals. Show that $\mathcal{O}$ is a Dedekind domain.

Exercise 10. The fractional ideals $\mathfrak{a}$ of a Dedekind domain $\mathcal{O}$ are projective $\mathcal{O}$-modules, i.e., given any surjective homomorphism $M\xrightarrow{f}N$ of $\mathcal{O}$-modules, each homomorphism $\mathfrak{a}\xrightarrow{g}N$ can be lifted to a homomorphism $h:\mathfrak{a}\to M$ such that $f\circ h=g$.

§ 4. Lattices

In § 1, when solving the basic problems concerning the gaussian integers, we used at a crucial place the inclusion

$$\mathbb{Z}[i]\subseteq\mathbb{C}$$

and considered the integers of $\mathbb{Q}(i)$ as lattice points in the complex plane. This point of view has been generalized to arbitrary number fields by *Hermann Minkowski* (1864–1909) and has led to results which make up an essential part of the foundations of algebraic number theory. In order to develop Minkowski's theory we first have to introduce the general notion of lattice and study some of its basic properties.

(4.1) Definition. *Let V be an n-dimensional $\mathbb{R}$-vector space. A* **lattice** *in V is a subgroup of the form*

$$\Gamma = \mathbb{Z}v_1 + \cdots + \mathbb{Z}v_m$$

with linearly independent vectors $v_1, \ldots, v_m$ of V. The m-tuple $(v_1, \ldots, v_m)$ is called a **basis** *and the set*

$$\Phi = \{x_1 v_1 + \cdots + x_m v_m \mid x_i \in \mathbb{R},\ 0 \le x_i < 1\}$$

a **fundamental mesh** *of the lattice. The lattice is called* **complete** *or a $\mathbb{Z}$-* **structure** *of V, if $m = n$.*

The completeness of the lattice is obviously tantamount to the fact that the set of all translates $\Phi + \gamma$, $\gamma \in \Gamma$, of the fundamental mesh covers the entire space V.

The above definition makes use of a choice of linearly independent vectors. But we will need a characterization of lattices which is independent of such a choice. Note that, first of all, a lattice is a finitely generated subgroup of V. But not every finitely generated subgroup is a lattice – for instance $\mathbb{Z} + \mathbb{Z}\sqrt{2} \subseteq \mathbb{R}$ is not. But each lattice $\Gamma = \mathbb{Z}v_1 + \cdots + \mathbb{Z}v_m$ has the special property of being a **discrete** subgroup of V. This is to say that every point $\gamma \in \Gamma$ is an isolated point in the sense that there exists a neighbourhood which contains no other points of Γ. In fact, if

$$\gamma = a_1 v_1 + \cdots + a_m v_m \in \Gamma,$$

then, extending $v_1, \ldots, v_m$ to a basis $v_1, \ldots, v_n$ of V, the set

$$\{x_1 v_1 + \cdots + x_n v_n \mid x_i \in \mathbb{R},\ |a_i - x_i| < 1 \text{ for } i = 1, \ldots, m\}$$

clearly is such a neighbourhood. This property is indeed characteristic.

(4.2) Proposition. *A subgroup $\Gamma \subseteq V$ is a lattice if and only if it is discrete.*

Proof: Let Γ be a discrete subgroup of V. Then Γ is closed. For let U be an arbitrary neighbourhood of 0. Then there exists a neighbourhood $U' \subseteq U$ of 0 such that every difference of elements of U' lies in U. If there were an $x \notin \Gamma$ belonging to the closure of Γ, then we could find in the neighbourhood $x+U'$ of x two distinct elements $\gamma_1, \gamma_2 \in \Gamma$, so that $0 \neq \gamma_1 - \gamma_2 \in U' - U' \subseteq U$. Thus 0 would not be an isolated point, a contradiction.

Let V_0 be the linear subspace of V which is spanned by the set Γ, and let m be its dimension. Then we may choose a basis $u_1, \dots, u_m$ of V_0 which is contained in Γ, and form the complete lattice

$$\Gamma_0 = \mathbb{Z}u_1 + \cdots + \mathbb{Z}u_m \subseteq \Gamma$$

of V_0. We claim that the index $(\Gamma : \Gamma_0)$ is finite. To see this, let $\gamma_i \in \Gamma$ vary over a system of representatives of the cosets in Γ/Γ_0. Since Γ_0 is complete in V_0, the translates $\Phi_0 + \gamma$, $\gamma \in \Gamma_0$, of the fundamental mesh

$$\Phi_0 = \big\{ x_1 u_1 + \cdots + x_m u_m \;\big|\; x_i \in \mathbb{R},\ 0 \le x_i < 1 \big\}$$

cover the entire space V_0. We may therefore write

$$\gamma_i = \mu_i + \gamma_{0i}\,, \qquad \mu_i \in \Phi_0\,, \qquad \gamma_{0i} \in \Gamma_0 \subseteq V_0\,.$$

As the $\mu_i = \gamma_i - \gamma_{0i} \in \Gamma$ lie discretely in the bounded set Φ_0, they have to be finite in number. In fact, the intersection of Γ with the closure of Φ_0 is compact and discrete, hence finite.

Putting now $q = (\Gamma : \Gamma_0)$, we have $q\Gamma \subseteq \Gamma_0$, whence

$$\Gamma \subseteq \frac{1}{q}\Gamma_0 = \mathbb{Z}\Big(\frac{1}{q}u_1\Big) + \cdots + \mathbb{Z}\Big(\frac{1}{q}u_m\Big).$$

By the main theorem on finitely generated abelian groups, Γ therefore admits a $\mathbb{Z}$-basis $v_1, \dots, v_r$, $r \le m$, i.e., $\Gamma = \mathbb{Z}v_1 + \cdots + \mathbb{Z}v_r$. The vectors $v_1, \dots, v_r$ are also $\mathbb{R}$-linearly independent because they span the m-dimensional space V_0. This shows that Γ is a lattice. □

Next we prove a criterion which will tell us when a lattice in the space V – given, say, as a discrete subgroup $\Gamma \subseteq V$ – is complete.

(4.3) Lemma. *A lattice Γ in V is complete if and only if there exists a bounded subset $M \subseteq V$ such that the collection of all translates $M + \gamma$, $\gamma \in \Gamma$, covers the entire space V.*

Proof: If $\Gamma = \mathbb{Z}v_1 + \cdots + \mathbb{Z}v_n$ is complete, then one may take M to be the fundamental mesh $\Phi = \{x_1v_1 + \cdots + x_nv_n \mid 0 \le x_i < 1\}$.

Conversely, let M be a bounded subset of V whose translates $M + \gamma$, for $\gamma \in \Gamma$, cover V. Let V_0 be the subspace spanned by Γ. We have to show that $V = V_0$. So let $v \in V$. Since $V = \bigcup_{\gamma\in\Gamma}(M + \gamma)$ we may write, for each $\nu \in \mathbb{N}$,

$$\nu v = a_\nu + \gamma_\nu\,, \quad a_\nu \in M\,, \quad \gamma_\nu \in \Gamma \subseteq V_0\,.$$

Since M is bounded, $\frac{1}{\nu}a_\nu$ converges to zero, and since V_0 is closed,

$$v = \lim_{\nu\to\infty} \frac{1}{\nu}a_\nu + \lim_{\nu\to\infty} \frac{1}{\nu}\gamma_\nu = \lim_{\nu\to\infty} \frac{1}{\nu}\gamma_\nu \in V_0\,. \qquad \square$$

Now let V be a *euclidean* vector space, i.e., an $\mathbb{R}$-vector space of finite dimension n equipped with a symmetric, positive definite bilinear form

$$\langle\ ,\ \rangle : V \times V \longrightarrow \mathbb{R}\,.$$

Then we have on V a notion of volume – more precisely a Haar measure. The cube spanned by an orthonormal basis $e_1, \ldots, e_n$ has volume 1, and more generally, the parallelepiped spanned by n linearly independent vectors $v_1, \ldots, v_n$,

$$\Phi = \left\{x_1v_1 + \cdots + x_nv_n \,\middle|\, x_i \in \mathbb{R},\ 0 \le x_i < 1\right\}$$

has volume

$$\operatorname{vol}(\Phi) = |\det A|\,,$$

where $A = (a_{ik})$ is the matrix of the base change from $e_1, \ldots, e_n$ to $v_1, \ldots, v_n$, so that $v_i = \sum_k a_{ik}e_k$. Since

$$\big(\langle v_i, v_j\rangle\big) = \Big(\sum_{k,\ell} a_{ik}\, a_{j\ell} \langle e_k, e_\ell\rangle\Big) = \Big(\sum_k a_{ik}\, a_{jk}\Big) = AA^t,$$

we also have the invariant notation

$$\operatorname{vol}(\Phi) = \big|\det(\langle v_i, v_j\rangle)\big|^{1/2}.$$

Let Γ be the lattice spanned by $v_1, \ldots, v_n$. Then Φ is a fundamental mesh of Γ, and we write for short

$$\operatorname{vol}(\Gamma) = \operatorname{vol}(\Phi)\,.$$

This does not depend on the choice of a basis $v_1, \ldots, v_n$ of the lattice because the transition matrix passing to a different basis, as well as its inverse, has integer coefficients, and therefore has determinant ± 1 so that the set Φ is transformed into a set of the same volume.

We now come to the most important theorem about lattices. A subset X of V is called *centrally symmetric*, if, given any point $x \in X$, the point $-x$ also belongs to X. It is called *convex* if, given any two points $x, y \in X$, the whole line segment $\{ty + (1-t)x \mid 0 \le t \le 1\}$ joining x with y is contained in X. With these definitions we have

(4.4) Minkowski's Lattice Point Theorem. *Let Γ be a complete lattice in the euclidean vector space V and X a centrally symmetric, convex subset of V. Suppose that*

$$\mathrm{vol}(X) > 2^n \,\mathrm{vol}(\Gamma).$$

Then X contains at least one nonzero lattice point $\gamma \in \Gamma$.

Proof: It is enough to show that there exist two distinct lattice points $\gamma_1, \gamma_2 \in \Gamma$ such that

$$\left(\tfrac{1}{2}X + \gamma_1\right) \cap \left(\tfrac{1}{2}X + \gamma_2\right) \neq \emptyset.$$

In fact, choosing a point in this intersection,

$$\frac{1}{2}x_1 + \gamma_1 = \frac{1}{2}x_2 + \gamma_2, \quad x_1, x_2 \in X,$$

we obtain an element

$$\gamma = \gamma_1 - \gamma_2 = \frac{1}{2}x_2 - \frac{1}{2}x_1,$$

which is the center of the line segment joining x_2 and $-x_1$, and therefore belongs to $X \cap \Gamma$.

Now, if the sets $\frac{1}{2}X + \gamma$, $\gamma \in \Gamma$, were pairwise disjoint, then the same would be true of their intersections $\Phi \cap (\frac{1}{2}X + \gamma)$ with a fundamental mesh Φ of Γ, i.e., we would have

$$\mathrm{vol}(\Phi) \geq \sum_{\gamma \in \Gamma} \mathrm{vol}\left(\Phi \cap \left(\tfrac{1}{2}X + \gamma\right)\right).$$

But translation of $\Phi \cap (\frac{1}{2}X + \gamma)$ by $-\gamma$ creates the set $(\Phi - \gamma) \cap \frac{1}{2}X$ of equal volume, and the $\Phi - \gamma$, $\gamma \in \Gamma$, cover the entire space V, therefore also the set $\frac{1}{2}X$. Consequently we would obtain

$$\mathrm{vol}(\Phi) \geq \sum_{\gamma \in \Gamma} \mathrm{vol}\left((\Phi - \gamma) \cap \tfrac{1}{2}X\right) = \mathrm{vol}\left(\tfrac{1}{2}X\right) = \frac{1}{2^n}\,\mathrm{vol}(X),$$

which contradicts the hypothesis. □

Exercise 1. Show that a lattice Γ in $\mathbb{R}^n$ is complete if and only if the quotient $\mathbb{R}^n/\Gamma$ is compact.

Exercise 2. Show that Minkowski's lattice point theorem cannot be improved, by giving an example of a centrally symmetric convex set $X \subseteq V$ such that $\mathrm{vol}(X) = 2^n \,\mathrm{vol}(\Gamma)$ which does not contain any nonzero point of the lattice Γ. If X is compact, however, then the statement (4.4) does remain true in the case of equality.

Exercise 3 (Minkowski's Theorem on Linear Forms). Let

$$L_i(x_1, \dots, x_n) = \sum_{j=1}^{n} a_{ij} x_j, \quad i = 1, \dots, n,$$

be real linear forms such that $\det(a_{ij}) \neq 0$, and let $c_1, \dots, c_n$ be positive real numbers such that $c_1 \cdots c_n > |\det(a_{ij})|$. Show that there exist integers $m_1, \dots, m_n \in \mathbb{Z}$ such that

$$|L_i(m_1, \dots, m_n)| < c_i, \quad i = 1, \dots, n.$$

Hint: Use Minkowski's lattice point theorem.

§ 5. Minkowski Theory

The basic idea in Minkowski's treatment of an algebraic number field $K|\mathbb{Q}$ of degree n is to interpret its numbers as points in n-dimensional space. This explains why his theory has been called "Geometry of Numbers." It seems appropriate, however, to follow the current trend and call it "Minkowski Theory" instead, because in the meantime a geometric approach to number theory has been developed which is quite different in nature and much more comprehensive. We will explain this in § 13. In the present section, we consider the canonical mapping

$$j : K \longrightarrow K_{\mathbb{C}} := \prod_{\tau} \mathbb{C}, \quad a \longmapsto ja = (\tau a),$$

which results from the n complex embeddings $\tau : K \to \mathbb{C}$. The $\mathbb{C}$-vector space $K_{\mathbb{C}}$ is equipped with the *hermitian scalar product*

$$(*) \qquad \langle x, y \rangle = \sum_{\tau} x_\tau \bar{y}_\tau .$$

Let us recall that a hermitian scalar product is given by a form $H(x, y)$ which is linear in the first variable and satisfies $\overline{H(x,y)} = H(y,x)$ as well as $H(x,x) > 0$ for $x \neq 0$. In the sequel we always view $K_{\mathbb{C}}$ as a hermitian space, with respect to the "standard metric" $(*)$.

The Galois group $G(\mathbb{C}|\mathbb{R})$ is generated by complex conjugation

$$F : z \longmapsto \bar{z} .$$

The notation F will be justified only later (see chap. III, § 4). F acts on the one hand on the factors of the product $\prod_\tau \mathbb{C}$, but on the other hand it also acts on the indexing set of τ's; to each embedding $\tau : K \to \mathbb{C}$ corresponds its complex conjugate $\bar{\tau} : K \to \mathbb{C}$. Altogether, this defines an involution

$$F : K_{\mathbb{C}} \longrightarrow K_{\mathbb{C}}$$

which, on the points $z = (z_\tau) \in K_{\mathbb{C}}$, is given by

$$(Fz)_\tau = \bar{z}_{\bar{\tau}}.$$

The scalar product $\langle\ ,\ \rangle$ is equivariant under F, that is

$$\langle Fx, Fy\rangle = F\langle x, y\rangle .$$

Finally, we have on the $\mathbb{C}$-vector space $K_{\mathbb{C}} = \prod_\tau \mathbb{C}$ the linear map

$$Tr : K_{\mathbb{C}} \longrightarrow \mathbb{C},$$

given as the sum of the coordinates. It is also F-invariant. The composite

$$K \xrightarrow{\ j\ } K_{\mathbb{C}} \xrightarrow{\ Tr\ } \mathbb{C}$$

gives the usual trace of $K\,|\,\mathbb{Q}$ (see (2.6), (ii)),

$$Tr_{K|\mathbb{Q}}(a) = Tr(ja).$$

We now concentrate on the $\mathbb{R}$-vector space

$$K_{\mathbb{R}} = K_{\mathbb{C}}^{+} = \Big[\prod_\tau \mathbb{C}\Big]^{+}$$

consisting of the $G(\mathbb{C}\,|\,\mathbb{R})$-invariant, i.e., F-invariant, points of $K_{\mathbb{C}}$. These are the points (z_τ) such that $z_{\bar{\tau}} = \bar{z}_\tau$. An explicit description of $K_{\mathbb{R}}$ will be given anon. Since $\bar{\tau}a = \overline{\tau a}$ for $a \in K$, one has $F(ja) = ja$. This yields a mapping

$$j : K \longrightarrow K_{\mathbb{R}}.$$

The restriction of the hermitian scalar product $\langle\ ,\ \rangle$ from $K_{\mathbb{C}}$ to $K_{\mathbb{R}}$ gives a scalar product

$$\langle\ ,\ \rangle : K_{\mathbb{R}} \times K_{\mathbb{R}} \longrightarrow \mathbb{R}$$

on the $\mathbb{R}$-vector space $K_{\mathbb{R}}$. Indeed, for $x, y \in K_{\mathbb{R}}$, one has $\langle x, y\rangle \in \mathbb{R}$ in view of the relations $F\langle x, y\rangle = \langle Fx, Fy\rangle = \langle x, y\rangle$, $\langle x, y\rangle = \overline{\langle x, y\rangle} = \langle y, x\rangle$, and, in any case, $\langle x, x\rangle > 0$ for $x \neq 0$.

We call the *euclidean* vector space

$$K_{\mathbb{R}} = \Big[\prod_\tau \mathbb{C}\Big]^{+}$$

the **Minkowski space**, its scalar product $\langle\ ,\ \rangle$ the **canonical metric**, and the associated Haar measure (see §4, p. 26) the **canonical measure**. Since $Tr \circ F = F \circ Tr$ we have on $K_{\mathbb{R}}$ the $\mathbb{R}$-linear map

$$Tr : K_{\mathbb{R}} \longrightarrow \mathbb{R},$$

and its composite with $j : K \to K_{\mathbb{R}}$ is again the usual trace of $K|\mathbb{Q}$,

$$Tr_{K|\mathbb{Q}}(a) = Tr(ja).$$

Remark: We mention in passing – it will not be used in the sequel – that the mapping $j : K \to K_{\mathbb{R}}$ identifies the vector space $K_{\mathbb{R}}$ with the tensor product $K \otimes_{\mathbb{Q}} \mathbb{R}$,

$$K \otimes_{\mathbb{Q}} \mathbb{R} \xrightarrow{\sim} K_{\mathbb{R}}, \quad a \otimes x \longmapsto (ja)x.$$

Likewise, $K \otimes_{\mathbb{Q}} \mathbb{C} \xrightarrow{\sim} K_{\mathbb{C}}$. In this approach, the inclusion $K_{\mathbb{R}} \subseteq K_{\mathbb{C}}$ corresponds to the canonical mapping $K \otimes_{\mathbb{Q}} \mathbb{R} \to K \otimes_{\mathbb{Q}} \mathbb{C}$ which is induced by the inclusion $\mathbb{R} \hookrightarrow \mathbb{C}$. F corresponds to $F(a \otimes z) = a \otimes \bar{z}$.

An explicit description of the Minkowski space $K_{\mathbb{R}}$ can be given in the following manner. Some of the embeddings $\tau : K \to \mathbb{C}$ are real in that they land already in $\mathbb{R}$, and others are complex, i.e., not real. Let

$$\rho_1, \dots, \rho_r : K \longrightarrow \mathbb{R}$$

be the real embeddings. The complex ones come in pairs

$$\sigma_1, \bar{\sigma}_1, \dots, \sigma_s, \bar{\sigma}_s : K \longrightarrow \mathbb{C}$$

of complex conjugate embeddings. Thus $n = r + 2s$. We choose from each pair some fixed complex embedding, and let ρ vary over the family of real embeddings and σ over the family of chosen complex embeddings. Since F leaves the ρ invariant, but exchanges the $\sigma, \bar{\sigma}$, we have

$$K_{\mathbb{R}} = \Big\{ (z_\tau) \in \prod_\tau \mathbb{C} \;\Big|\; z_\rho \in \mathbb{R},\ z_{\bar{\sigma}} = \bar{z}_\sigma \Big\}.$$

This gives the

(5.1) Proposition. *There is an isomorphism*

$$f : K_{\mathbb{R}} \longrightarrow \prod_\tau \mathbb{R} = \mathbb{R}^{r+2s}$$

given by the rule $(z_\tau) \mapsto (x_\tau)$ *where*

$$x_\rho = z_\rho, \quad x_\sigma = \mathrm{Re}(z_\sigma), \quad x_{\bar{\sigma}} = \mathrm{Im}(z_\sigma).$$

This isomorphism transforms the canonical metric $\langle\ ,\ \rangle$ *into the scalar product*

$$(x, y) = \sum_\tau \alpha_\tau x_\tau y_\tau,$$

where $\alpha_\tau = 1$, *resp.* $\alpha_\tau = 2$, *if* τ *is real, resp. complex.*

Proof: The map is clearly an isomorphism. If $z = (z_\tau) = (x_\tau + iy_\tau)$, $z' = (z'_\tau) = (x'_\tau + iy'_\tau) \in K_{\mathbb{R}}$, then $z_\rho \bar{z}'_\rho = x_\rho x'_\rho$, and in view of $y_\sigma = x_{\bar{\sigma}}$ and $y'_\sigma = x'_{\bar{\sigma}}$, one gets

$$z_\sigma \bar{z}'_\sigma + z_{\bar{\sigma}} \bar{z}'_{\bar{\sigma}} = z_\sigma \bar{z}'_\sigma + \bar{z}_\sigma z'_\sigma = 2\,\mathrm{Re}(z_\sigma \bar{z}'_\sigma) = 2(x_\sigma x'_\sigma + x_{\bar{\sigma}} x'_{\bar{\sigma}}).$$

This proves the claim concerning the scalar products. □

The scalar product $(x, y) = \sum_\tau \alpha_\tau x_\tau y_\tau$ transfers the canonical measure from $K_{\mathbb{R}}$ to $\mathbb{R}^{r+2s}$. It obviously differs from the standard Lebesgue measure by

$$\mathrm{vol}_{\text{canonical}}(X) = 2^s \,\mathrm{vol}_{\text{Lebesgue}}\big(f(X)\big).$$

Minkowski himself worked with the Lebesgue measure on $\mathbb{R}^{r+2s}$, and most textbooks follow suit. The corresponding measure on $K_{\mathbb{R}}$ is the one determined by the scalar product

$$(x, y) = \sum_\tau \frac{1}{\alpha_\tau} x_\tau \bar{y}_\tau.$$

This scalar product may therefore be called the **Minkowski metric** on $K_{\mathbb{R}}$. But we will systematically work with the canonical metric, and denote by vol the corresponding canonical measure.

The mapping $j : K \to K_{\mathbb{R}}$ gives us the following lattices in Minkowski space $K_{\mathbb{R}}$.

(5.2) Proposition. *If $\mathfrak{a} \neq 0$ is an ideal of $\mathcal{O}_K$, then $\Gamma = j\mathfrak{a}$ is a complete lattice in $K_{\mathbb{R}}$. Its fundamental mesh has volume*

$$\mathrm{vol}(\Gamma) = \sqrt{|d_K|}\,(\mathcal{O}_K : \mathfrak{a}).$$

Proof: Let $\alpha_1, \dots, \alpha_n$ be a $\mathbb{Z}$-basis of $\mathfrak{a}$, so that $\Gamma = \mathbb{Z} j\alpha_1 + \dots + \mathbb{Z} j\alpha_n$. We choose a numbering of the embeddings $\tau : K \to \mathbb{C}$, $\tau_1, \dots, \tau_n$, and form the matrix $A = (\tau_\ell \alpha_i)$. Then, according to (2.12), we have

$$d(\mathfrak{a}) = d(\alpha_1, \dots, \alpha_n) = (\det A)^2 = (\mathcal{O}_K : \mathfrak{a})^2 d(\mathcal{O}_K) = (\mathcal{O}_K : \mathfrak{a})^2 d_K,$$

and on the other hand

$$\big(\langle j\alpha_i, j\alpha_k\rangle\big) = \Big(\sum_{\ell=1}^{n} \tau_\ell \alpha_i \bar{\tau}_\ell \alpha_k\Big) = A\bar{A}^t.$$

This indeed yields

$$\mathrm{vol}(\Gamma) = \big|\det(\langle j\alpha_i, j\alpha_k\rangle)\big|^{1/2} = |\det A| = \sqrt{|d_K|}\,(\mathcal{O}_K : \mathfrak{a}). \qquad \square$$

Using this proposition, Minkowski's lattice point theorem now gives the following result, which is what we chiefly intend to use in our applications to number theory.

(5.3) Theorem. *Let $\mathfrak{a} \neq 0$ be an integral ideal of K, and let $c_\tau > 0$, for $\tau \in \mathrm{Hom}(K,\mathbb{C})$, be real numbers such that $c_\tau = c_{\overline{\tau}}$ and*

$$\prod_\tau c_\tau > A(\mathcal{O}_K : \mathfrak{a}),$$

where $A = \left(\frac{2}{\pi}\right)^s \sqrt{|d_K|}$. Then there exists $a \in \mathfrak{a}$, $a \neq 0$, such that

$$|\tau a| < c_\tau \quad \textit{for all} \quad \tau \in \mathrm{Hom}(K,\mathbb{C}).$$

Proof: The set $X = \{(z_\tau) \in K_\mathbb{R} \mid |z_\tau| < c_\tau\}$ is centrally symmetric and convex. Its volume $\mathrm{vol}(X)$ can be computed via the map (5.1)

$$f : K_\mathbb{R} \xrightarrow{\sim} \prod_\tau \mathbb{R}, \quad (z_\tau) \longmapsto (x_\tau),$$

given by $x_\rho = z_\rho$, $x_\sigma = \mathrm{Re}(z_\sigma)$, $x_{\overline{\sigma}} = \mathrm{Im}(z_\sigma)$. It comes out to be 2^s times the Lebesgue-volume of the image

$$f(X) = \Big\{ (x_\tau) \in \prod_\tau \mathbb{R} \;\Big|\; |x_\rho| < c_\rho,\ x_\sigma^2 + x_{\overline{\sigma}}^2 < c_\sigma^2 \Big\}.$$

This gives

$$\mathrm{vol}(X) = 2^s\, \mathrm{vol}_{\text{Lebesgue}}\big(f(X)\big) = 2^s \prod_\rho (2c_\rho) \prod_\sigma (\pi c_\sigma^2) = 2^{r+s}\pi^s \prod_\tau c_\tau .$$

Now using (5.2), we obtain

$$\mathrm{vol}(X) > 2^{r+s}\pi^s \left(\frac{2}{\pi}\right)^s \sqrt{|d_K|}\,(\mathcal{O}_K : \mathfrak{a}) = 2^n\, \mathrm{vol}(\Gamma).$$

Thus the hypothesis of Minkowski's lattice point theorem is satisfied. So there does indeed exist a lattice point $ja \in X$, $a \neq 0$, $a \in \mathfrak{a}$; in other words $|\tau a| < c_\tau$. □

There is also a **multiplicative version** of Minkowski theory. It is based on the homomorphism

$$j : K^* \longrightarrow K_\mathbb{C}^* = \prod_\tau \mathbb{C}^*.$$

The multiplicative group $K_\mathbb{C}^*$ admits the homomorphism

$$N : K_\mathbb{C}^* \longrightarrow \mathbb{C}^*$$

given by the product of the coordinates. The composite

$$K^* \xrightarrow{j} K_\mathbb{C}^* \xrightarrow{N} \mathbb{C}^*$$

is the usual norm of $K|\mathbb{Q}$,

$$N_{K|\mathbb{Q}}(a) = N(ja).$$

In order to produce a lattice from the multiplicative theory, we use the logarithm to pass from multiplicative to additive groups

$$\ell : \mathbb{C}^* \longrightarrow \mathbb{R}, \quad z \longmapsto \log|z|.$$

It induces a surjective homomorphism

$$\ell : K_{\mathbb{C}}^* \longrightarrow \prod_{\tau} \mathbb{R},$$

and we obtain the commutative diagram

$$\begin{array}{ccccc} K^* & \xrightarrow{\;j\;} & K_{\mathbb{C}}^* & \xrightarrow{\;\ell\;} & \prod_\tau \mathbb{R} \\ {\scriptstyle N_{K|\mathbb{Q}}}\downarrow & & \downarrow{\scriptstyle N} & & \downarrow{\scriptstyle Tr} \\ \mathbb{Q}^* & \longrightarrow & \mathbb{C}^* & \xrightarrow{\;\ell\;} & \mathbb{R}. \end{array}$$

The involution $F \in G(\mathbb{C}|\mathbb{R})$ acts on all groups in this diagram, trivially on K^*, on $K_{\mathbb{C}}^*$ as before, and on the points $x = (x_\tau) \in \prod_\tau \mathbb{R}$ by $(Fx)_\tau = x_{\bar\tau}$. One clearly has

$$F \circ j = j, \quad F \circ \ell = \ell \circ F, \quad N \circ F = F \circ N, \quad Tr \circ F = Tr,$$

i.e., the homomorphisms of the diagram are $G(\mathbb{C}|\mathbb{R})$-homomorphisms. We now pass everywhere to the fixed modules under $G(\mathbb{C}|\mathbb{R})$ and obtain the diagram

$$\begin{array}{ccccc} K^* & \xrightarrow{\;j\;} & K_{\mathbb{R}}^* & \xrightarrow{\;\ell\;} & \left[\prod_\tau \mathbb{R}\right]^+ \\ {\scriptstyle N_{K|\mathbb{Q}}}\downarrow & & \downarrow{\scriptstyle N} & & \downarrow{\scriptstyle Tr} \\ \mathbb{Q}^* & \longrightarrow & \mathbb{R}^* & \xrightarrow{\;\ell\;} & \mathbb{R}. \end{array}$$

The $\mathbb{R}$-vector space $\left[\prod_\tau \mathbb{R}\right]^+$ is explicitly given as follows. Separate as before the embeddings $\tau : K \to \mathbb{C}$ into real ones, $\rho_1, \dots, \rho_r$, and pairs of complex conjugate ones, $\sigma_1, \bar\sigma_1, \dots, \sigma_s, \bar\sigma_s$. We obtain a decomposition which is analogous to the one we saw above for $\left[\prod_\tau \mathbb{C}\right]^+$,

$$\Big[\prod_\tau \mathbb{R}\Big]^+ = \prod_\rho \mathbb{R} \times \prod_\sigma [\,\mathbb{R} \times \mathbb{R}\,]^+.$$

The factor $[\,\mathbb{R} \times \mathbb{R}\,]^+$ now consists of the points (x,x), and we identify it with $\mathbb{R}$ by the map $(x,x) \mapsto 2x$. In this way we obtain an isomorphism

$$\Big[\prod_\tau \mathbb{R}\Big]^+ \cong \mathbb{R}^{r+s},$$

which again transforms the map $Tr : \big[\prod_\tau \mathbb{R}\big]^+ \to \mathbb{R}$ into the usual map

$$Tr : \mathbb{R}^{r+s} \longrightarrow \mathbb{R}$$

given by the sum of the coordinates. Identifying $\big[\prod_\tau \mathbb{R}\big]^+$ with $\mathbb{R}^{r+s}$, the homomorphism

$$\ell : K_\mathbb{R}^* \longrightarrow \mathbb{R}^{r+s}$$

is given by

$$\ell(x) = \big(\log|x_{\rho_1}|, \ldots, \log|x_{\rho_r}|, \log|x_{\sigma_1}|^2, \ldots, \log|x_{\sigma_s}|^2\big),$$

where we write $x \in K_\mathbb{R}^* \subseteq \prod_\tau \mathbb{C}^*$ as $x = (x_\tau)$.

Exercise 1. Write down a constant A which depends only on K such that every integral ideal $\mathfrak{a} \neq 0$ of K contains an element $a \neq 0$ satisfying

$$|\tau a| < A(\mathcal{O}_K : \mathfrak{a})^{1/n} \quad \text{for all } \tau \in \operatorname{Hom}(K, \mathbb{C}),\ n = [K : \mathbb{Q}].$$

Exercise 2. Show that the convex, centrally symmetric set

$$X = \Big\{ (z_\tau) \in K_\mathbb{R} \;\Big|\; \sum_\tau |z_\tau| < t \Big\}$$

has volume $\operatorname{vol}(X) = 2^r \pi^s \frac{t^n}{n!}$ (see chap. III, (2.15)).

Exercise 3. Show that in every ideal $\mathfrak{a} \neq 0$ of $\mathcal{O}_K$ there exists an $a \neq 0$ such that

$$|N_{K|\mathbb{Q}}(a)| \le M(\mathcal{O}_K : \mathfrak{a}),$$

where $M = \dfrac{n!}{n^n}\Big(\dfrac{4}{\pi}\Big)^s\sqrt{|d_K|}$ (the so-called **Minkowski bound**).

Hint: Use exercise 2 to proceed as in (5.3), and make use of the inequality between arithmetic and geometric means,

$$\frac{1}{n}\sum_\tau |z_\tau| \ge \Big(\prod_\tau |z_\tau|\Big)^{1/n}.$$

§ 6. The Class Number

As a first application of Minkowski theory, we are going to show that the ideal class group $Cl_K = J_K/P_K$ of an algebraic number field K is finite. In order to count the ideals $\mathfrak{a} \neq 0$ of the ring $\mathcal{O}_K$ we consider their **absolute norm**

$$\mathfrak{N}(\mathfrak{a}) = (\mathcal{O}_K : \mathfrak{a}).$$

(Throughout this book the case of the zero ideal $\mathfrak{a} = 0$ is often tacitly excluded, when its consideration would visibly make no sense.) This index is finite by (2.12), and the name is justified by the special case of a principal ideal (α) of $\mathcal{O}_K$, where we have the identity

$$\mathfrak{N}\big((\alpha)\big) = \big|N_{K|\mathbb{Q}}(\alpha)\big| .$$

Indeed, if $\omega_1, \dots, \omega_n$ is a $\mathbb{Z}$-basis of $\mathcal{O}_K$, then $\alpha\,\omega_1, \dots, \alpha\,\omega_n$ is a $\mathbb{Z}$-basis of $(\alpha) = \alpha\mathcal{O}_K$, and if $A = (a_{ij})$ denotes the transition matrix, $\alpha\,\omega_i = \sum a_{ij}\,\omega_j$, then, as was pointed out already in §2, one has $|\det(A)| = (\mathcal{O}_K : (\alpha))$ as well as $\det(A) = N_{K|\mathbb{Q}}(\alpha)$ by definition.

(6.1) Proposition. *If* $\mathfrak{a} = \mathfrak{p}_1^{\nu_1} \cdots \mathfrak{p}_r^{\nu_r}$ *is the prime factorization of an ideal* $\mathfrak{a} \neq 0$, *then one has*

$$\mathfrak{N}(\mathfrak{a}) = \mathfrak{N}(\mathfrak{p}_1)^{\nu_1} \cdots \mathfrak{N}(\mathfrak{p}_r)^{\nu_r} .$$

Proof: By the Chinese remainder theorem (3.6), one has

$$\mathcal{O}_K/\mathfrak{a} = \mathcal{O}_K/\mathfrak{p}_1^{\nu_1} \oplus \cdots \oplus \mathcal{O}_K/\mathfrak{p}_r^{\nu_r} .$$

We are thus reduced to considering the case where $\mathfrak{a}$ is a prime power $\mathfrak{p}^\nu$. In the chain

$$\mathfrak{p} \supseteq \mathfrak{p}^2 \supseteq \cdots \supseteq \mathfrak{p}^\nu$$

one has $\mathfrak{p}^i \neq \mathfrak{p}^{i+1}$ because of the unique prime factorization, and each quotient $\mathfrak{p}^i/\mathfrak{p}^{i+1}$ is an $\mathcal{O}_K/\mathfrak{p}$-vector space of dimension 1. In fact, if $a \in \mathfrak{p}^i \smallsetminus \mathfrak{p}^{i+1}$ and $\mathfrak{b} = (a) + \mathfrak{p}^{i+1}$, then $\mathfrak{p}^i \supseteq \mathfrak{b} \supsetneqq \mathfrak{p}^{i+1}$ and consequently $\mathfrak{p}^i = \mathfrak{b}$, because otherwise $\mathfrak{b}' = \mathfrak{b}\mathfrak{p}^{-i}$ would be a proper divisor of $\mathfrak{p} = \mathfrak{p}^{i+1}\mathfrak{p}^{-i}$. Thus $\bar{a} = a \bmod \mathfrak{p}^{i+1}$ is a basis of the $\mathcal{O}_K/\mathfrak{p}$-vector space $\mathfrak{p}^i/\mathfrak{p}^{i+1}$. So we have $\mathfrak{p}^i/\mathfrak{p}^{i+1} \cong \mathcal{O}_K/\mathfrak{p}$ and therefore

$$\mathfrak{N}(\mathfrak{p}^\nu) = (\mathcal{O}_K : \mathfrak{p}^\nu) = (\mathcal{O}_K : \mathfrak{p})(\mathfrak{p} : \mathfrak{p}^2) \cdots (\mathfrak{p}^{\nu-1} : \mathfrak{p}^\nu) = \mathfrak{N}(\mathfrak{p})^\nu . \qquad \square$$

The proposition immediately implies the multiplicativity

$$\mathfrak{N}(\mathfrak{a}\mathfrak{b}) = \mathfrak{N}(\mathfrak{a})\mathfrak{N}(\mathfrak{b})$$

of the absolute norm. It may therefore be extended to a homomorphism

$$\mathfrak{N} : J_K \longrightarrow \mathbb{R}_+^*$$

defined on all fractional ideals $\mathfrak{a} = \prod_\mathfrak{p} \mathfrak{p}^{\nu_\mathfrak{p}}$, $\nu_\mathfrak{p} \in \mathbb{Z}$. The following lemma, a consequence of (5.3), is crucial for the finiteness of the ideal class group.

(6.2) Lemma. *In every ideal* $\mathfrak{a} \neq 0$ *of* $\mathcal{O}_K$ *there exists an* $a \in \mathfrak{a}$, $a \neq 0$, *such that*

$$\big|N_{K|\mathbb{Q}}(a)\big| \le \Big(\frac{2}{\pi}\Big)^s \sqrt{|d_K|}\,\mathfrak{N}(\mathfrak{a}) .$$

Proof: Given $\varepsilon > 0$, we choose positive real numbers c_τ, for $\tau \in \mathrm{Hom}(K, \mathbb{C})$, such that $c_\tau = c_{\bar\tau}$ and

$$\prod_\tau c_\tau = \left(\frac{2}{\pi}\right)^s \sqrt{|d_K|}\,\mathfrak{N}(\mathfrak{a}) + \varepsilon.$$

Then by (5.3) we find an element $a \in \mathfrak{a}$, $a \neq 0$, satisfying $|\tau a| < c_\tau$. Thus

$$\big|N_{K|\mathbb{Q}}(a)\big| = \prod_\tau |\tau a| < \left(\frac{2}{\pi}\right)^s \sqrt{|d_K|}\,\mathfrak{N}(\mathfrak{a}) + \varepsilon\,.$$

This being true for all $\varepsilon > 0$ and since $|N_{K|\mathbb{Q}}(a)|$ is always a positive integer, there has to exist an $a \in \mathfrak{a}$, $a \neq 0$, such that

$$\big|N_{K|\mathbb{Q}}(a)\big| \le \left(\frac{2}{\pi}\right)^s \sqrt{|d_K|}\,\mathfrak{N}(\mathfrak{a})\,. \qquad \square$$

(6.3) Theorem. *The ideal class group $Cl_K = J_K/P_K$ is finite. Its order*

$$h_K = (J_K : P_K)$$

is called the **class number** *of K.*

Proof: If $\mathfrak{p} \neq 0$ is a prime ideal of $\mathcal{O}_K$ and $\mathfrak{p} \cap \mathbb{Z} = p\mathbb{Z}$, then $\mathcal{O}_K/\mathfrak{p}$ is a finite field extension of $\mathbb{Z}/p\mathbb{Z}$ of degree, say, $f \geq 1$, and we have

$$\mathfrak{N}(\mathfrak{p}) = p^f\,.$$

Given p, there are only finitely many prime ideals $\mathfrak{p}$ such that $\mathfrak{p} \cap \mathbb{Z} = p\mathbb{Z}$, because this means that $\mathfrak{p} \,|\, (p)$. It follows that there are only finitely many prime ideals $\mathfrak{p}$ of bounded absolute norm. Since every integral ideal admits a representation $\mathfrak{a} = \mathfrak{p}_1^{\nu_1} \cdots \mathfrak{p}_r^{\nu_r}$ where $\nu_i > 0$ and

$$\mathfrak{N}(\mathfrak{a}) = \mathfrak{N}(\mathfrak{p}_1)^{\nu_1} \cdots \mathfrak{N}(\mathfrak{p}_r)^{\nu_r},$$

there are altogether only a finite number of ideals $\mathfrak{a}$ of $\mathcal{O}_K$ with bounded absolute norm $\mathfrak{N}(\mathfrak{a}) \le M$.

It therefore suffices to show that each class $[\mathfrak{a}] \in Cl_K$ contains an integral ideal $\mathfrak{a}_1$ satisfying

$$\mathfrak{N}(\mathfrak{a}_1) \le M = \left(\frac{2}{\pi}\right)^s \sqrt{|d_K|}\,.$$

For this, choose an arbitrary representative $\mathfrak{a}$ of the class, and a $\gamma \in \mathcal{O}_K$, $\gamma \neq 0$, such that $\mathfrak{b} = \gamma\,\mathfrak{a}^{-1} \subseteq \mathcal{O}_K$. By (6.2), there exists $\alpha \in \mathfrak{b}$, $\alpha \neq 0$, such that

$$\big|N_{K|\mathbb{Q}}(\alpha)\big| \cdot \mathfrak{N}(\mathfrak{b})^{-1} = \mathfrak{N}\big((\alpha)\mathfrak{b}^{-1}\big) = \mathfrak{N}(\alpha\mathfrak{b}^{-1}) \le M\,.$$

The ideal $\mathfrak{a}_1 = \alpha\mathfrak{b}^{-1} = \alpha\gamma^{-1}\mathfrak{a} \in [\mathfrak{a}]$ therefore has the required property. $\square$

The theorem of the finiteness of the class number h_K means that passing from numbers to ideals has not thrust us into unlimited new territory. The most favourable case occurs of course when $h_K = 1$. This means that $\mathcal{O}_K$ is a principal ideal domain, i.e., that prime factorization of elements in the classical sense holds. In general, however, one has $h_K > 1$. For instance, we know now that the only imaginary quadratic fields $\mathbb{Q}(\sqrt{d})$, d squarefree and < 0, which have class number 1 are those with

$$d = -1,\ -2,\ -3,\ -7,\ -11,\ -19,\ -43,\ -67,\ -163.$$

Among real quadratic fields, class number 1 is more common. In the range $2 \le d < 100$ for instance, it occurs for

$$\begin{aligned} d = {} & 2, 3, 5, 6, 7, 11, 13, 14, 17, 19, 21, 22, 23, 29, \\ & 31, 33, 37, 38, 41, 43, 46, 47, 53, 57, 59, 61, \\ & 62, 67, 69, 71, 73, 77, 83, 86, 89, 93, 94, 97. \end{aligned}$$

It is conjectured that there are infinitely many real quadratic fields of class number 1. But we do not even yet know whether there are infinitely many algebraic number fields (of arbitrary degree) with class number 1. It was found time and again in innumerable investigations that the ideal class groups Cl_K behave completely unpredictably, both in their size and their structure. An exception to this lack of rule is *Kenkichi Iwasawa*'s discovery that the p-part of the class number of the field of p^n-th roots of unity obeys a very strict law when n varies (see [136], th. 13.13).

In the case of the field of p-th roots of unity, the question whether the class number is divisible by p has played a very important special rôle because it is intimately linked to the celebrated **Fermat's Last Theorem** according to which the equation

$$x^p + y^p = z^p$$

for $p \ge 3$ has no solutions in integers $\neq 0$. In a similar way as the sums of two squares $x^2+y^2 = (x+iy)(x-iy)$ lead to studying the gaussian integers, the decomposition of x^p+y^p by means of a p-th root of unity $\zeta \neq 1$ leads to a problem in the ring $\mathbb{Z}[\zeta]$ of integers of $\mathbb{Q}(\zeta)$. The equation $y^p = z^p - x^p$ there turns into the identity

$$y \cdot y \cdots y = (z-x)(z-\zeta x)\cdots(z-\zeta^{p-1}x).$$

Thus, assuming the existence of a solution, one obtains two multiplicative decompositions of the same number in $\mathbb{Z}[\zeta]$. One can show that this contradicts the unique factorization – provided that this holds in the ring $\mathbb{Z}[\zeta]$. Supposing erroneously that this was the case in general – in other words that the class number h_p of the field $\mathbb{Q}(\zeta)$ were always equal to 1 – some actually thought they had proved "Fermat's Last Theorem" in this way. *KUMMER*, however, did not fall into this trap. Instead, he proved that the arguments we have indicated can be salvaged if one only assumes $p \nmid h_p$ instead of $h_p = 1$. In this case he called a prime number p **regular**, otherwise **irregular**. He even showed that p is regular if and only if the numerators of the **Bernoulli numbers** $B_2, B_4, \ldots, B_{p-3}$ are not divisible by p. Among the first 25 prime numbers < 100 only three are irregular: 37, 59, and 67. We still do not know today whether there are infinitely many regular prime numbers.

The connection with Fermat's last theorem has at last become obsolete. Following a surprising discovery by the mathematician *GERHARD FREY*, who established a link with the theory of *elliptic curves*, it was *KENNETH RIBET*, who managed to reduce Fermat's statement to another, much more important conjecture, the **Taniyama-Shimura-Weil Conjecture**. This was proved in sufficient generality in 1994 by *ANDREW WILES*, after many years of work, and with a helping hand from *RICHARD TAYLOR*. See [144].

The regular and irregular prime numbers do however continue to be important.

Exercise 1. How many integral ideals $\mathfrak{a}$ are there with the given norm $\mathfrak{N}(\mathfrak{a}) = n$?

Exercise 2. Show that the quadratic fields with discriminant 5, 8, 11, -3, -4, -7, -8, -11 have class number 1.

Exercise 3. Show that in every ideal class of an algebraic number field K of degree n, there exists an integral ideal $\mathfrak{a}$ such that

$$\mathfrak{N}(\mathfrak{a}) \leq \frac{n!}{n^n}\Big(\frac{4}{\pi}\Big)^s\sqrt{|d_K|}\,.$$

Hint: Using exercise 3, §5, proceed as in the proof of (6.3).

Exercise 4. Show that the absolute value of the discriminant $|d_K|$ is > 1 for every algebraic number field $K \neq \mathbb{Q}$ (Minkowski's theorem on the discriminant, see chap. III, (2.17)).

Exercise 5. Show that the absolute value of the discriminant $|d_K|$ tends to ∞ with the degree n of the field.

Exercise 6. Let $\mathfrak{a}$ be an integral ideal of K and $\mathfrak{a}^m = (a)$. Show that $\mathfrak{a}$ becomes a principal ideal in the field $L = K(\sqrt[m]{a}\,)$, in the sense that $\mathfrak{a}\mathcal{O}_L = (\alpha)$.

Exercise 7. Show that, for every number field K, there exists a finite extension L such that every ideal of K becomes a principal ideal.

§ 7. Dirichlet's Unit Theorem

After considering the ideal class group Cl_K, we now turn to the second main problem posed by the ring $\mathcal{O}_K$ of integers of an algebraic number field K, the group of units $\mathcal{O}_K^*$. It contains the finite group $\mu(K)$ of the roots of unity that lie in K, but in general is not itself finite. Its size is in fact determined by the number r of real embeddings $\rho : K \to \mathbb{R}$ and the number s of pairs $\sigma, \overline{\sigma} : K \to \mathbb{C}$ of complex conjugate embeddings. In order to describe the group, we use the diagram which was set up in § 5:

$$\begin{array}{ccccc} K^* & \xrightarrow{\ j\ } & K_{\mathbb{R}}^* & \xrightarrow{\ \ell\ } & \big[\prod_\tau \mathbb{R}\big]^+ \\ {\scriptstyle N_{K|\mathbb{Q}}}\downarrow & & \downarrow{\scriptstyle N} & & \downarrow{\scriptstyle Tr} \\ \mathbb{Q}^* & \longrightarrow & \mathbb{R}^* & \xrightarrow{\ \log|\ |\ } & \mathbb{R}\,. \end{array}$$

In the upper part of the diagram we consider the subgroups

$$\mathcal{O}_K^* = \big\{\varepsilon \in \mathcal{O}_K \mid N_{K|\mathbb{Q}}(\varepsilon) = \pm 1\big\}, \quad \text{the group of units,}$$

$$S = \big\{y \in K_{\mathbb{R}}^* \mid N(y) = \pm 1\big\}, \quad \text{the "norm-one surface",}$$

$$H = \big\{x \in \big[\prod_\tau \mathbb{R}\big]^+ \mid Tr(x) = 0\big\}, \quad \text{the "trace-zero hyperplane".}$$

We obtain the homomorphisms

$$\mathcal{O}_K^* \xrightarrow{\ j\ } S \xrightarrow{\ \ell\ } H$$

and the composite $\lambda := \ell \circ j : \mathcal{O}_K^* \to H$. The image will be denoted by

$$\Gamma = \lambda(\mathcal{O}_K^*) \subseteq H,$$

and we obtain the

(7.1) Proposition. *The sequence*

$$1 \longrightarrow \mu(K) \longrightarrow \mathcal{O}_K^* \xrightarrow{\ \lambda\ } \Gamma \longrightarrow 0$$

is exact.

Proof: We have to show that $\mu(K)$ is the kernel of λ. For $\zeta \in \mu(K)$ and $\tau : K \to \mathbb{C}$ any embedding, we find $\log|\tau\zeta| = \log 1 = 0$, so that certainly $\mu(K) \subseteq \ker(\lambda)$. Conversely, let $\varepsilon \in \mathcal{O}_K^*$ be an element in the kernel, so that $\lambda(\varepsilon) = \ell(j\varepsilon) = 0$. This means that $|\tau\varepsilon| = 1$ for each embedding

$\tau : K \to \mathbb{C}$, so that $j\varepsilon = (\tau\varepsilon)$ lies in a bounded domain of the $\mathbb{R}$-vector space $K_{\mathbb{R}}$. On the other hand, $j\varepsilon$ is a point of the lattice $j\mathcal{O}_K$ of $K_{\mathbb{R}}$ (see (5.2)). Therefore the kernel of λ can contain only a finite number of elements, and thus, being a finite group, contains only roots of unity in K^*. □

Given this proposition, it remains to determine the group Γ. For this, we need the following

(7.2) Lemma. *Up to multiplication by units there are only finitely many elements $\alpha \in \mathcal{O}_K$ of given norm $N_{K|\mathbb{Q}}(\alpha) = a$.*

Proof: Let $a \in \mathbb{Z}$, $a > 1$. In every one of the finitely many cosets of $\mathcal{O}_K/a\mathcal{O}_K$ there exists, up to multiplication by units, at most one element α such that $|N(\alpha)| = |N_{K|\mathbb{Q}}(\alpha)| = a$. For if $\beta = \alpha + a\gamma$, $\gamma \in \mathcal{O}_K$, is another one, then

$$\frac{\alpha}{\beta} = 1 \pm \frac{N(\beta)}{\beta}\gamma \in \mathcal{O}_K$$

because $N(\beta)/\beta \in \mathcal{O}_K$, and by the same token $\frac{\beta}{\alpha} = 1 \pm \frac{N(\alpha)}{\alpha}\gamma \in \mathcal{O}_K$, i.e., β is associated to α. Therefore, up to multiplication by units, there are at most $(\mathcal{O}_K : a\mathcal{O}_K)$ elements of norm $\pm a$. □

(7.3) Theorem. *The group Γ is a complete lattice in the $(r+s-1)$-dimensional vector space H, and is therefore isomorphic to $\mathbb{Z}^{r+s-1}$.*

Proof: We first show that $\Gamma = \lambda(\mathcal{O}_K^*)$ is a lattice in H, i.e., a discrete subgroup. The mapping $\lambda : \mathcal{O}_K^* \to H$ arises by restricting the mapping

$$K^* \xrightarrow{j} \prod_\tau \mathbb{C}^* \xrightarrow{\ell} \prod_\tau \mathbb{R},$$

and it suffices to show that, for any $c > 0$, the bounded domain $\{(x_\tau) \in \prod_\tau \mathbb{R} \mid |x_\tau| \le c\}$ contains only finitely many points of $\Gamma = \ell(j\mathcal{O}_K^*)$. Since $\ell((z_\tau)) = (\log|z_\tau|)$, the preimage of this domain with respect to ℓ is the bounded domain

$$\Big\{(z_\tau) \in \prod_\tau \mathbb{C}^* \;\Big|\; e^{-c} \le |z_\tau| \le e^c\Big\}.$$

It contains only finitely many elements of the set $j\mathcal{O}_K^*$ because this is a subset of the lattice $j\mathcal{O}_K$ in $\big[\prod_\tau \mathbb{C}\big]^+$ (see (5.2)). Therefore Γ is a lattice.

We now show that Γ is a complete lattice in H. This is the principal claim of the theorem. We apply the criterion (4.3). So we have to find a bounded set $M \subseteq H$ such that

$$H = \bigcup_{\gamma \in \Gamma} (M + \gamma).$$

We construct this set through its preimage with respect to the surjective homomorphism

$$\ell : S \longrightarrow H.$$

More precisely, we will construct a bounded set T in the norm-one surface S, the *multiplicative* translations $Tj\varepsilon$, $\varepsilon \in \mathcal{O}_K^*$, of which cover all of S:

$$S = \bigcup_{\varepsilon \in \mathcal{O}_K^*} Tj\varepsilon.$$

For $x = (x_\tau) \in T$, it will follow that the absolute values $|x_\tau|$ are bounded from above and also away from zero, because $\prod_\tau |x_\tau| = 1$. Thus $M = \ell(T)$ will also be bounded. We choose real numbers $c_\tau > 0$, for $\tau \in \operatorname{Hom}(K, \mathbb{C})$, satisfying $c_\tau = c_{\bar\tau}$ and

$$C = \prod_\tau c_\tau > \left(\frac{2}{\pi}\right)^s \sqrt{|d_K|},$$

and we consider the set

$$X = \left\{ (z_\tau) \in K_{\mathbb{R}} \;\middle|\; |z_\tau| < c_\tau \right\}.$$

For an arbitrary point $y = (y_\tau) \in S$, it follows that

$$Xy = \left\{ (z_\tau) \in K_{\mathbb{R}} \;\middle|\; |z_\tau| < c'_\tau \right\}$$

where $c'_\tau = c_\tau |y_\tau|$, and one has $c'_\tau = c'_{\bar\tau}$ and $\prod_\tau c'_\tau = \prod_\tau c_\tau = C$ because $\prod_\tau |y_\tau| = |N(y)| = 1$. Then, by (5.3), there is a point

$$ja = (\tau a) \in Xy, \quad a \in \mathcal{O}_K, \quad a \neq 0.$$

Now, according to lemma (7.2), we may pick a system $\alpha_1, \ldots, \alpha_N \in \mathcal{O}_K$, $\alpha_i \neq 0$, in such a way that every $a \in \mathcal{O}_K$ with $0 < |N_{K|\mathbb{Q}}(a)| \le C$ is associated to one of these numbers. The set

$$T = S \cap \bigcup_{i=1}^{N} X(j\alpha_i)^{-1}$$

then has the required property: since X is bounded, so is $X(j\alpha_i)^{-1}$ and therefore also T, and we have

$$S = \bigcup_{\varepsilon \in \mathcal{O}_K^*} Tj\varepsilon.$$

In fact, if $y \in S$, we find by the above an $a \in \mathcal{O}_K$, $a \neq 0$, such that $ja \in Xy^{-1}$, so $ja = xy^{-1}$ for some $x \in X$. Since

$$|N_{K|\mathbb{Q}}(a)| = |N(xy^{-1})| = |N(x)| < \prod_\tau c_\tau = C,$$

a is associated to some α_i, $\alpha_i = \varepsilon a$, $\varepsilon \in \mathcal{O}_K^*$. Consequently

$$y = xja^{-1} = xj(\alpha_i^{-1}\varepsilon).$$

Since $y, j\varepsilon \in S$, one finds $xj\alpha_i^{-1} \in S \cap Xj\alpha_i^{-1} \subseteq T$, and thus $y \in Tj\varepsilon$. □

From proposition (7.1) and theorem (7.3) we immediately deduce **Dirichlet's unit theorem** in its classical form.

(7.4) Theorem. *The group of units $\mathcal{O}_K^*$ of $\mathcal{O}_K$ is the direct product of the finite cyclic group $\mu(K)$ and a free abelian group of rank $r+s-1$.*

In other words: there exist units $\varepsilon_1, \dots, \varepsilon_t$, $t = r+s-1$, called **fundamental units**, such that any other unit ε can be written uniquely as a product

$$\varepsilon = \zeta\, \varepsilon_1^{\nu_1} \cdots \varepsilon_t^{\nu_t}$$

with a root of unity ζ and integers ν_i.

Proof: In the exact sequence

$$1 \longrightarrow \mu(K) \longrightarrow \mathcal{O}_K^* \xrightarrow{\lambda} \Gamma \longrightarrow 0$$

Γ is a free abelian group of rank $t = r+s-1$ by (7.3). Let $v_1, \dots, v_t$ be a $\mathbb{Z}$-basis of Γ, let $\varepsilon_1, \dots, \varepsilon_t \in \mathcal{O}_K^*$ be preimages of the v_i, and let $A \subseteq \mathcal{O}_K^*$ be the subgroup generated by the ε_i. Then A is mapped isomorphically onto Γ by λ, i.e., one has $\mu(K) \cap A = \{1\}$ and therefore $\mathcal{O}_K^* = \mu(K) \times A$. □

Identifying $\left[\prod_\tau \mathbb{R}\right]^+ = \mathbb{R}^{r+s}$ (see §5, p. 33), H becomes a subspace of the euclidean space $\mathbb{R}^{r+s}$ and thus itself a euclidean space. We may therefore speak of the volume of the fundamental mesh $\mathrm{vol}(\lambda(\mathcal{O}_K^*))$ of the unit lattice $\Gamma = \lambda(\mathcal{O}_K^*) \subseteq H$, and will now compute it. Let $\varepsilon_1, \dots, \varepsilon_t$, $t = r+s-1$, be a system of fundamental units and Φ the fundamental mesh of the unit lattice $\lambda(\mathcal{O}_K^*)$, spanned by the vectors $\lambda(\varepsilon_1), \dots, \lambda(\varepsilon_t) \in H$. The vector

$$\lambda_0 = \frac{1}{\sqrt{r+s}}(1, \dots, 1) \in \mathbb{R}^{r+s}$$

is obviously orthogonal to H and has length 1. The t-dimensional volume of Φ therefore equals the $(t+1)$-dimensional volume of the parallelepiped spanned by $\lambda_0, \lambda(\varepsilon_1), \dots, \lambda(\varepsilon_t)$ in $\mathbb{R}^{t+1}$. But this has volume

$$\pm\det\begin{pmatrix} \lambda_{01} & \lambda_1(\varepsilon_1) & \cdots & \lambda_1(\varepsilon_t) \\ \vdots & \vdots & & \vdots \\ \lambda_{0t+1} & \lambda_{t+1}(\varepsilon_1) & \cdots & \lambda_{t+1}(\varepsilon_t) \end{pmatrix}.$$

Adding all rows to a fixed one, say the i-th row, this row has only zeroes, except for the first entry, which equals $\sqrt{r+s}$. We therefore get the

(7.5) Proposition. *The volume of the fundamental mesh of the unit lattice $\lambda(\mathcal{O}_K^*)$ in H is*

$$\operatorname{vol}\big(\lambda(\mathcal{O}_K^*)\big) = \sqrt{r+s}\,R\,,$$

where R is the absolute value of the determinant of an arbitrary minor of rank $t = r+s-1$ of the following matrix:

$$\begin{pmatrix} \lambda_1(\varepsilon_1) & \cdots & \lambda_1(\varepsilon_t) \\ \vdots & & \vdots \\ \lambda_{t+1}(\varepsilon_1) & \cdots & \lambda_{t+1}(\varepsilon_t) \end{pmatrix}.$$

This absolute value R is called the **regulator** *of the field K.*

The importance of the regulator will only be demonstrated later (see chap. VII, § 5).

Exercise 1. Let $D > 1$ be a squarefree integer and d the discriminant of the real quadratic number field $K = \mathbb{Q}(\sqrt{D})$ (see § 2, exercise 4). Let x_1, y_1 be the uniquely determined rational integer solution of the equation

$$x^2 - dy^2 = -4,$$

or – in case this equation has no rational integer solutions – of the equation

$$x^2 - dy^2 = 4,$$

for which $x_1, y_1 > 0$ are as small as possible. Then

$$\varepsilon_1 = \frac{x_1 + y_1\sqrt{d}}{2}$$

is a fundamental unit of K. (The pair of equations $x^2 - dy^2 = \pm 4$ is called **Pell's equation**.)

Exercise 2. Check the following table of fundamental units ε_1 for $\mathbb{Q}(\sqrt{D})$:

D	2	3	5	6	7	10
ε_1	$1+\sqrt{2}$	$2+\sqrt{3}$	$(1+\sqrt{5})/2$	$5+2\sqrt{6}$	$8+3\sqrt{7}$	$3+\sqrt{10}$

Hint: Check one by one for $y = 1, 2, 3, \ldots$, whether one of the numbers $dy^2 \mp 4$ is a square x^2. By the unit theorem this is bound to happen, with the plus sign. However, for fixed y, let preference be given to the minus sign. Then the first case, in this order, where $dy_1^2 \mp 4 = x_1^2$, gives the fundamental unit $\varepsilon_1 = (x_1 + y_1\sqrt{d})/2$.

Exercise 3. The Battle of Hastings (October 14, 1066).
"The men of Harold stood well together, as their wont was, and formed thirteen squares, with a like number of men in every square thereof, and woe to the hardy Norman who ventured to enter their redoubts; for a single blow of a Saxon war-hatched would break his lance and cut through his coat of mail... When Harold threw himself into the fray the Saxons were one mighty square of men, shouting the battle-cries, 'Ut!', 'Olicrosse!', 'Godemite!'." [Fictitious historical text, following essentially problem no. 129 in: H.E. Dundeney, *Amusements in Mathematics*, 1917 (Dover reprints 1958 and 1970).]
Question. How many troops does this suggest Harold II had at the battle of Hastings?

Exercise 4. Let ζ be a primitive p-th root of unity, p an odd prime number. Show that $\mathbb{Z}[\zeta]^* = (\zeta)\mathbb{Z}[\zeta + \zeta^{-1}]^*$. Show that $\mathbb{Z}[\zeta]^* = \{\pm\zeta^k(1+\zeta)^n \mid 0 \le k < 5,\ n \in \mathbb{Z}\}$, if $p = 5$.

Exercise 5. Let ζ be a primitive m-th root of unity, $m \ge 3$. Show that the numbers $\frac{1-\zeta^k}{1-\zeta}$ for $(k, m) = 1$ are units in the ring of integers of the field $\mathbb{Q}(\zeta)$. The subgroup of the group of units they generate is called the group of **cyclotomic units**.

Exercise 6. Let K be a totally real number field, i.e., $X = \mathrm{Hom}(K, \mathbb{C}) = \mathrm{Hom}(K, \mathbb{R})$, and let T be a proper nonempty subset of X. Then there exists a unit ε satisfying $0 < \tau\varepsilon < 1$ for $\tau \in T$, and $\tau\varepsilon > 1$ for $\tau \notin T$.

Hint: Apply Minkowski's lattice point theorem to the unit lattice in trace-zero space.

§ 8. Extensions of Dedekind Domains

Having studied the ideal class group and the group of units of the ring $\mathcal{O}_K$ of integers of a number field K, we now propose to make a first survey of the set of prime ideals of $\mathcal{O}_K$. They are often referred to as the prime ideals of K – an imprecise manner of speaking which is, however, not likely to cause any misunderstanding.

Every prime ideal $\mathfrak{p} \neq 0$ of $\mathcal{O}_K$ contains a rational prime number p (see § 3, p. 17) and is therefore a divisor of the ideal $p\mathcal{O}_K$. Hence the question arises as to how a prime number p factors into prime ideals of the ring $\mathcal{O}_K$. We treat this problem in a more general context, starting from an arbitrary Dedekind domain $\mathcal{O}$ at the base instead of $\mathbb{Z}$, and taking instead of $\mathcal{O}_K$ the integral closure $\mathcal{O}$ of $\mathcal{O}$ in a finite extension of its field of fractions.

(8.1) Proposition. *Let $\mathcal{o}$ be a Dedekind domain with field of fractions K, let $L|K$ be a finite extension of K and $\mathcal{O}$ the integral closure of $\mathcal{o}$ in L. Then $\mathcal{O}$ is again a Dedekind domain.*

Proof: Being the integral closure of $\mathcal{o}$, $\mathcal{O}$ is integrally closed. The fact that the nonzero prime ideals $\mathfrak{P}$ of $\mathcal{O}$ are maximal is proved similarly as in the case $\mathcal{o} = \mathbb{Z}$ (see (3.1)): $\mathfrak{p} = \mathfrak{P} \cap \mathcal{o}$ is a nonzero prime ideal of $\mathcal{o}$. Thus the integral domain $\mathcal{O}/\mathfrak{P}$ is an extension of the field $\mathcal{o}/\mathfrak{p}$, and therefore has itself to be a field, because if it were not, then it would admit a nonzero prime ideal whose intersection with $\mathcal{o}/\mathfrak{p}$ would again be a nonzero prime ideal in $\mathcal{o}/\mathfrak{p}$. It remains to show that $\mathcal{O}$ is noetherian. In the case that is of chief interest to us, namely, if $L|K$ is a separable extension, the proof is very easy. Let $\alpha_1, \dots, \alpha_n$ be a basis of $L|K$ contained in $\mathcal{O}$, of discriminant $d = d(\alpha_1, \dots, \alpha_n)$. Then $d \neq 0$ by (2.8), and (2.9) tells us that $\mathcal{O}$ is contained in the finitely generated $\mathcal{o}$-module $\mathcal{o}\alpha_1/d + \cdots + \mathcal{o}\alpha_n/d$. Every ideal of $\mathcal{O}$ is also contained in this finitely generated $\mathcal{o}$-module, and therefore is itself an $\mathcal{o}$-module of finite type, hence *a fortiori* a finitely generated $\mathcal{O}$-module. This shows that $\mathcal{O}$ is noetherian, provided $L|K$ is separable. We ask the reader's permission to content ourselves for the time being with this case. We shall come back to the general case on a more convenient occasion. In fact, we shall give the proof in a more general framework in § 12 (see (12.8)). □

For a prime ideal $\mathfrak{p}$ of $\mathcal{o}$ one always has

$$\mathfrak{p}\mathcal{O} \neq \mathcal{O}.$$

In fact, let $\pi \in \mathfrak{p} \setminus \mathfrak{p}^2$ ($\mathfrak{p} \neq 0$), so that $\pi\mathcal{o} = \mathfrak{p}\mathfrak{a}$ with $\mathfrak{p} \nmid \mathfrak{a}$, hence $\mathfrak{p} + \mathfrak{a} = \mathcal{o}$. Writing $1 = b + s$, with $b \in \mathfrak{p}$ and $s \in \mathfrak{a}$, we find $s \notin \mathfrak{p}$ and $s\mathfrak{p} \subseteq \mathfrak{p}\mathfrak{a} = \pi\mathcal{o}$. If one had $\mathfrak{p}\mathcal{O} = \mathcal{O}$, then it would follow that $s\mathcal{O} = s\mathfrak{p}\mathcal{O} \subseteq \pi\mathcal{O}$, so that $s = \pi x$ for some $x \in \mathcal{O} \cap K = \mathcal{o}$, i.e., $s \in \mathfrak{p}$, a contradiction.

A prime ideal $\mathfrak{p} \neq 0$ of the ring $\mathcal{o}$ decomposes in $\mathcal{O}$ in a unique way into a product of prime ideals,

$$\mathfrak{p}\mathcal{O} = \mathfrak{P}_1^{e_1} \cdots \mathfrak{P}_r^{e_r}.$$

Instead of $\mathfrak{p}\mathcal{O}$ we will often write simply $\mathfrak{p}$. The prime ideals $\mathfrak{P}_i$ occurring in the decomposition are precisely those prime ideals $\mathfrak{P}$ of $\mathcal{O}$ which lie over $\mathfrak{p}$ in the sense that one has the relation

$$\mathfrak{p} = \mathfrak{P} \cap \mathcal{o}.$$

This we also denote for short by $\mathfrak{P}\,|\,\mathfrak{p}$, and we call $\mathfrak{P}$ a prime divisor of $\mathfrak{p}$. The exponent e_i is called the **ramification index**, and the degree of the field extension

$$f_i = [\mathcal{O}/\mathfrak{P}_i : \mathcal{o}/\mathfrak{p}]$$

is called the **inertia degree** of $\mathfrak{P}_i$ over $\mathfrak{p}$. If the extension $L|K$ is separable, the numbers e_i, f_i and the degree $n = [L : K]$ are connected by the following law.

(8.2) Proposition. *Let $L|K$ be separable. Then we have the* **fundamental identity**

$$\sum_{i=1}^{r} e_i f_i = n\,.$$

Proof: The proof is based on the Chinese remainder theorem

$$\mathcal{O}/\mathfrak{p}\mathcal{O} \cong \bigoplus_{i=1}^{r} \mathcal{O}/\mathfrak{P}_i^{e_i}\,.$$

$\mathcal{O}/\mathfrak{p}\mathcal{O}$ and $\mathcal{O}/\mathfrak{P}_i^{e_i}$ are vector spaces over the field $\kappa = o/\mathfrak{p}$, and it suffices to show that

$$\dim_\kappa(\mathcal{O}/\mathfrak{p}\mathcal{O}) = n \quad \text{and} \quad \dim_\kappa(\mathcal{O}/\mathfrak{P}_i^{e_i}) = e_i f_i\,.$$

In order to prove the first identity, let $\omega_1, \dots, \omega_m \in \mathcal{O}$ be representatives of a basis $\overline{\omega}_1, \dots, \overline{\omega}_m$ of $\mathcal{O}/\mathfrak{p}\mathcal{O}$ over κ (we have seen in the proof of (8.1) that $\mathcal{O}$ is a finitely generated o-module, so certainly $\dim_\kappa(\mathcal{O}/\mathfrak{p}\mathcal{O}) < \infty$). It is sufficient to show that $\omega_1, \dots, \omega_m$ is a basis of $L|K$. Assume the $\omega_1, \dots, \omega_m$ are linearly dependent over K, and hence also over o. Then there are elements $a_1, \dots, a_m \in o$ not all zero such that

$$a_1\omega_1 + \cdots + a_m\omega_m = 0\,.$$

Consider the ideal $\mathfrak{a} = (a_1, \dots, a_m)$ of o and find $a \in \mathfrak{a}^{-1}$ such that $a \notin \mathfrak{a}^{-1}\mathfrak{p}$, hence $a\mathfrak{a} \not\subseteq \mathfrak{p}$. Then the elements $aa_1, \dots, aa_m$ lie in o, but not all belong to $\mathfrak{p}$. The congruence

$$aa_1\omega_1 + \cdots + aa_m\omega_m \equiv 0 \bmod \mathfrak{p}$$

thus gives us a linear dependence among the $\overline{\omega}_1, \dots, \overline{\omega}_m$ over κ, a contradiction. The $\omega_1, \dots, \omega_m$ are therefore linearly independent over K.

In order to show that the ω_i are a basis of $L|K$, we consider the o-modules $M = o\omega_1 + \cdots + o\omega_m$ and $N = \mathcal{O}/M$. Since $\mathcal{O} = M + \mathfrak{p}\mathcal{O}$, we have $\mathfrak{p}N = N$. As $L|K$ is separable, $\mathcal{O}$, and hence also N, are finitely generated o-modules (see p. 45). If $\alpha_1, \dots, \alpha_s$ is a system of generators of N, then

$$\alpha_i = \sum_j a_{ij}\alpha_j \quad \text{for } a_{ij} \in \mathfrak{p}\,.$$

Let A be the matrix $(a_{ij}) - I$, where I is the unit matrix of rank s, and let B be the **adjoint** matrix of A, whose entries are the minors of rank $(s-1)$

of A. Then one has $A(\alpha_1, \dots, \alpha_s)^t = 0$ and $BA = dI$, with $d = \det(A)$, (see (2.3)). Hence

$$0 = BA(\alpha_1, \dots, \alpha_s)^t = (d\alpha_1, \dots, d\alpha_s)^t,$$

and therefore $dN = 0$, i.e., $d\mathcal{O} \subseteq M = \mathcal{o}\omega_1 + \cdots + \mathcal{o}\omega_m$. We have $d \neq 0$, because expanding the determinant $d = \det((a_{ij}) - I)$ we find $d \equiv (-1)^s \bmod \mathfrak{p}$ because $a_{ij} \in \mathfrak{p}$. It follows that $L = dL = K\omega_1 + \cdots + K\omega_m$. $\omega_1, \dots, \omega_m$ is therefore indeed a basis of $L|K$.

In order to prove the second identity, let us consider the descending chain

$$\mathcal{O}/\mathfrak{P}_i^{e_i} \supseteq \mathfrak{P}_i/\mathfrak{P}_i^{e_i} \supseteq \cdots \supseteq \mathfrak{P}_i^{e_i-1}/\mathfrak{P}_i^{e_i} \supseteq (0)$$

of κ-vector spaces. The successive quotients $\mathfrak{P}_i^\nu/\mathfrak{P}_i^{\nu+1}$ in this chain are isomorphic to $\mathcal{O}/\mathfrak{P}_i$, for if $\alpha \in \mathfrak{P}_i^\nu \setminus \mathfrak{P}_i^{\nu+1}$, then the homomorphism

$$\mathcal{O} \longrightarrow \mathfrak{P}_i^\nu/\mathfrak{P}_i^{\nu+1}, \quad a \longmapsto a\alpha,$$

has kernel $\mathfrak{P}_i$ and is surjective because $\mathfrak{P}_i^\nu$ is the gcd of $\mathfrak{P}_i^{\nu+1}$ and $(\alpha) = \alpha\mathcal{O}$ so that $\mathfrak{P}_i^\nu = \alpha\mathcal{O} + \mathfrak{P}_i^{\nu+1}$. Since $f_i = [\mathcal{O}/\mathfrak{P}_i : \kappa]$, we obtain $\dim_\kappa(\mathfrak{P}_i^\nu/\mathfrak{P}_i^{\nu+1}) = f_i$ and therefore

$$\dim_\kappa(\mathcal{O}/\mathfrak{P}_i^{e_i}) = \sum_{\nu=0}^{e_i-1} \dim_\kappa(\mathfrak{P}_i^\nu/\mathfrak{P}_i^{\nu+1}) = e_i f_i . \qquad \square$$

Suppose now that the separable extension $L|K$ is given by a primitive element $\theta \in \mathcal{O}$ with minimal polynomial

$$p(X) \in \mathcal{o}[X],$$

so that $L = K(\theta)$. We may then deduce a result about the nature of the decomposition of $\mathfrak{p}$ in $\mathcal{O}$ which, albeit not complete, does show characteristic phenomena and a striking simplicity. It is incomplete in that a finite number of prime ideals are excluded; only those relatively prime to the **conductor** of the ring $\mathcal{o}[\theta]$ can be considered. This conductor is defined to be the biggest ideal $\mathfrak{F}$ of $\mathcal{O}$ which is contained in $\mathcal{o}[\theta]$. In other words

$$\mathfrak{F} = \{\alpha \in \mathcal{O} \mid \alpha\mathcal{O} \subseteq \mathcal{o}[\theta]\}.$$

Since $\mathcal{O}$ is a finitely generated $\mathcal{o}$-module (see proof of (8.1)), one has $\mathfrak{F} \neq 0$.

(8.3) Proposition. *Let $\mathfrak{p}$ be a prime ideal of $\mathcal{o}$ which is relatively prime to the conductor $\mathfrak{F}$ of $\mathcal{o}[\theta]$, and let*

$$\bar{p}(X) = \bar{p}_1(X)^{e_1} \cdots \bar{p}_r(X)^{e_r}$$

be the factorization of the polynomial $\overline{p}(X) = p(X) \bmod \mathfrak{p}$ *into irreducibles* $\overline{p}_i(X) = p_i(X) \bmod \mathfrak{p}$ *over the residue class field* $\mathcal{o}/\mathfrak{p}$, *with all* $p_i(X) \in \mathcal{o}[X]$ *monic. Then*

$$\mathfrak{P}_i = \mathfrak{p}\mathcal{O} + p_i(\theta)\mathcal{O}, \quad i = 1, \dots, r,$$

are the different prime ideals of $\mathcal{O}$ *above* $\mathfrak{p}$. *The inertia degree* f_i *of* $\mathfrak{P}_i$ *is the degree of* $\overline{p}_i(X)$, *and one has*

$$\mathfrak{p} = \mathfrak{P}_1^{e_1} \cdots \mathfrak{P}_r^{e_r}.$$

Proof: Writing $\mathcal{O}' = \mathcal{o}[\theta]$ and $\overline{\mathcal{o}} = \mathcal{o}/\mathfrak{p}$, we have a canonical isomorphism

$$\mathcal{O}/\mathfrak{p}\mathcal{O} \cong \mathcal{O}'/\mathfrak{p}\mathcal{O}' \cong \overline{\mathcal{o}}[X]/\big(\overline{p}(X)\big).$$

The first isomorphism follows from the relative primality $\mathfrak{p}\mathcal{O} + \mathfrak{F} = \mathcal{O}$. As $\mathfrak{F} \subseteq \mathcal{O}'$, it follows that $\mathcal{O} = \mathfrak{p}\mathcal{O} + \mathcal{O}'$, i.e., the homomorphism $\mathcal{O}' \to \mathcal{O}/\mathfrak{p}\mathcal{O}$ is surjective. It has kernel $\mathfrak{p}\mathcal{O} \cap \mathcal{O}'$, which equals $\mathfrak{p}\mathcal{O}'$. Since $(\mathfrak{p}, \mathfrak{F} \cap \mathcal{o}) = 1$, it follows that $\mathfrak{p}\mathcal{O} \cap \mathcal{O}' = (\mathfrak{p} + \mathfrak{F})(\mathfrak{p}\mathcal{O} \cap \mathcal{O}') \subseteq \mathfrak{p}\mathcal{O}'$.

The second isomorphism is deduced from the surjective homomorphism

$$\mathcal{o}[X] \longrightarrow \overline{\mathcal{o}}[X]/\big(\overline{p}(X)\big).$$

Its kernel is the ideal generated by $\mathfrak{p}$ and $p(X)$, and in view of $\mathcal{O}' = \mathcal{o}[\theta] = \mathcal{o}[X]/(p(X))$, we have $\mathcal{O}'/\mathfrak{p}\mathcal{O}' \cong \overline{\mathcal{o}}[X]/(\overline{p}(X))$.

Since $\overline{p}(X) = \prod_{i=1}^r \overline{p}_i(X)^{e_i}$, the Chinese remainder theorem finally gives the isomorphism

$$\overline{\mathcal{o}}[X]/\big(\overline{p}(X)\big) \cong \bigoplus_{i=1}^{r} \overline{\mathcal{o}}[X]\big/(\overline{p}_i(X))^{e_i}.$$

This shows that the prime ideals of the ring $R = \overline{\mathcal{o}}[X]/(\overline{p}(X))$ are the principal ideals $(\overline{p}_i)$ generated by the $\overline{p}_i(X) \bmod \overline{p}(X)$, for $i = 1, \dots, r$, that the degree $[R/(\overline{p}_i) : \overline{\mathcal{o}}]$ equals the degree of the polynomial $\overline{p}_i(X)$, and that

$$(0) = (\overline{p}) = \bigcap_{i=1}^{r} (\overline{p}_i)^{e_i}.$$

In view of the isomorphism $\overline{\mathcal{o}}[X]/(\overline{p}(X)) \cong \mathcal{O}/\mathfrak{p}\mathcal{O}$, $f(X) \mapsto f(\theta)$, the same situation holds in the ring $\overline{\mathcal{O}} = \mathcal{O}/\mathfrak{p}\mathcal{O}$. Thus the prime ideals $\overline{\mathfrak{P}}_i$ of $\overline{\mathcal{O}}$ correspond to the prime ideals $(\overline{p}_i)$, and they are the principal ideals generated by the $p_i(\theta) \bmod \mathfrak{p}\mathcal{O}$. The degree $[\,\overline{\mathcal{O}}/\overline{\mathfrak{P}}_i : \overline{\mathcal{o}}]$ is the degree of the polynomial $\overline{p}_i(X)$, and we have $(0) = \bigcap_{i=1}^r \overline{\mathfrak{P}}_i^{e_i}$. Now let $\mathfrak{P}_i = \mathfrak{p}\mathcal{O} + p_i(\theta)\mathcal{O}$ be the preimage of $\overline{\mathfrak{P}}_i$ with respect to the canonical homomorphism

$$\mathcal{O} \longrightarrow \mathcal{O}/\mathfrak{p}\mathcal{O}.$$

Then $\mathfrak{P}_i$, for $i = 1, \dots, r$, varies over the prime ideals of $\mathcal{O}$ above $\mathfrak{p}$. $f_i = [\,\mathcal{O}/\mathfrak{P}_i : \mathcal{o}/\mathfrak{p}]$ is the degree of the polynomial $\overline{p}_i(X)$. Furthermore $\mathfrak{P}_i^{e_i}$ is the preimage of $\overline{\mathfrak{P}}_i^{e_i}$ (because $e_i = \#\{\overline{\mathfrak{P}}^\nu \mid \nu \in \mathbb{N}\}$), and $\mathfrak{p}\mathcal{O} \supseteq \bigcap_{i=1}^r \mathfrak{P}_i^{e_i}$, so that $\mathfrak{p}\,\mathcal{O} \,|\, \prod_{i=1}^r \mathfrak{P}_i^{e_i}$ and therefore $\mathfrak{p}\mathcal{O} = \prod_{i=1}^r \mathfrak{P}_i^{e_i}$ because $\sum e_i f_i = n$. □

The prime ideal $\mathfrak{p}$ is said to **split completely** (or to be **totally split**) in L, if in the decomposition

$$\mathfrak{p} = \mathfrak{P}_1^{e_1} \cdots \mathfrak{P}_r^{e_r},$$

one has $r = n = [L : K]$, so that $e_i = f_i = 1$ for all $i = 1, \ldots, r$. $\mathfrak{p}$ is called **nonsplit**, or **indecomposed**, if $r = 1$, i.e., if there is only a single prime ideal of L over $\mathfrak{p}$. From the fundamental identity

$$\sum_{i=1}^{r} e_i f_i = n$$

we now understand the name of inertia degree: the smaller this degree is, the more the ideal $\mathfrak{p}$ will be tend to factor into different prime ideals.

The prime ideal $\mathfrak{P}_i$ in the decomposition $\mathfrak{p} = \prod_{i=1}^{r} \mathfrak{P}_i^{e_i}$ is called **unramified** over $\mathcal{o}$ (or over K) if $e_i = 1$ and if the residue class field extension $\mathcal{O}/\mathfrak{P}_i | \mathcal{o}/\mathfrak{p}$ is separable. If not, it is called **ramified**, and **totally ramified** if furthermore $f_i = 1$. The prime ideal $\mathfrak{p}$ is called unramified if all $\mathfrak{P}_i$ are unramified, otherwise it is called ramified. The extension $L|K$ itself is called unramified if all prime ideals $\mathfrak{p}$ of K are unramified in L.

The case where a prime ideal $\mathfrak{p}$ of K is ramified in L is an exceptional phenomenon. In fact, we have the

(8.4) Proposition. *If $L|K$ is separable, then there are only finitely many prime ideals of K which are ramified in L.*

Proof: Let $\theta \in \mathcal{O}$ be a primitive element for L, i.e., $L = K(\theta)$, and let $p(X) \in \mathcal{o}[X]$ be its minimal polynomial. Let

$$d = d(1, \theta, \ldots, \theta^{n-1}) = \prod_{i<j} (\theta_i - \theta_j)^2 \in \mathcal{o}$$

be the discriminant of $p(X)$ (see § 2, p. 11). Then every prime ideal $\mathfrak{p}$ of K which is relatively prime to d and to the conductor $\mathfrak{F}$ of $\mathcal{o}[\theta]$ is unramified. In fact, by (8.3), the ramification indices e_i equal 1 as soon as they are equal to 1 in the factorization of $\overline{p}(X) = p(X) \bmod \mathfrak{p}$ in $\mathcal{o}/\mathfrak{p}$, so certainly if $\overline{p}(X)$ has no multiple roots. But this is the case since the discriminant $\overline{d} = d \bmod \mathfrak{p}$ of $\overline{p}(X)$ is nonzero. The residue class field extensions $\mathcal{O}/\mathfrak{P}_i | \mathcal{o}/\mathfrak{p}$ are generated by $\overline{\theta} = \theta \bmod \mathfrak{P}_i$ and therefore separable. Hence $\mathfrak{p}$ is unramified. □

The precise description of the ramified prime ideals is given by the **discriminant** of $\mathcal{O}|\mathcal{o}$. It is defined to be the ideal $\mathfrak{d}$ of $\mathcal{o}$ which is generated by the discriminants $d(\omega_1, \ldots, \omega_n)$ of all bases $\omega_1, \ldots, \omega_n$ of $L|K$ contained in $\mathcal{O}$. We will show in chapter III, §2 that the prime divisors of $\mathfrak{d}$ are exactly the prime ideals which ramify in L.

Example: The law of decomposition of prime numbers p in a **quadratic number field** $\mathbb{Q}(\sqrt{a})$ is intimately related to Gauss's famous **quadratic reciprocity law**. The latter concerns the problem of integer solutions of the equation

$$x^2 + by = a, \quad (a, b \in \mathbb{Z}),$$

the simplest among the nontrivial diophantine equations. The theory of this equation reduces immediately to the case where b is an odd prime number p and $(a, p) = 1$ (exercise 6). Let us assume this for the sequel. We are then facing the question as to whether a is a **quadratic residue** mod p, i.e., whether the congruence

$$x^2 \equiv a \bmod p$$

does or does not have a solution. In other words, we want to know if the equation $\bar{x}^2 = \bar{a}$, for a given element $\bar{a} = a \bmod p \in \mathbb{F}_p^*$, admits a solution in the field $\mathbb{F}_p$ or not. For this one introduces the **Legendre symbol** $\left(\frac{a}{p}\right)$, which, for every rational number a relatively prime to p, is defined to be $\left(\frac{a}{p}\right) = 1$ or -1, according as $x^2 \equiv a \bmod p$ has or does not have a solution. This symbol is multiplicative,

$$\left(\frac{ab}{p}\right) = \left(\frac{a}{p}\right)\left(\frac{b}{p}\right).$$

This is because the group $\mathbb{F}_p^*$ is cyclic of order $p-1$ and the subgroup $\mathbb{F}_p^{*2}$ of squares has index 2, i.e., $\mathbb{F}_p^*/\mathbb{F}_p^{*2} \cong \mathbb{Z}/2\mathbb{Z}$. Since $\left(\frac{a}{p}\right) = 1 \iff \bar{a} \in \mathbb{F}_p^{*2}$, one also has

$$\left(\frac{a}{p}\right) \equiv a^{\frac{p-1}{2}} \bmod p.$$

In the case of squarefree a, the Legendre symbol $\left(\frac{a}{p}\right)$ bears the following relation with prime factorization. $\left(\frac{a}{p}\right) = 1$ signifies that

$$x^2 - a \equiv (x - \alpha)(x + \alpha) \bmod p$$

for some $\alpha \in \mathbb{Z}$. The conductor of $\mathbb{Z}[\sqrt{a}\,]$ in the ring of integers of $\mathbb{Q}(\sqrt{a})$ is a divisor of 2 (see §2, exercise 4). We may therefore apply proposition (8.3) and obtain the

(8.5) Proposition. *For squarefree a and $(p, 2a) = 1$, we have the equivalence*

$$\left(\frac{a}{p}\right) = 1 \iff p \text{ is totally split in } \mathbb{Q}(\sqrt{a}\,).$$

For the Legendre symbol, one has the following remarkable law, which like none other has left its mark on the development of algebraic number theory.

(8.6) Theorem (Gauss's Reciprocity Law). *For two distinct odd prime numbers ℓ and p, the following identity holds:*

$$\Big(\frac{\ell}{p}\Big)\Big(\frac{p}{\ell}\Big) = (-1)^{\frac{\ell-1}{2}\frac{p-1}{2}}.$$

One also has the two "supplementary theorems"

$$\Big(\frac{-1}{p}\Big) = (-1)^{\frac{p-1}{2}}, \quad \Big(\frac{2}{p}\Big) = (-1)^{\frac{p^2-1}{8}}.$$

Proof: $\big(\frac{-1}{p}\big) \equiv (-1)^{\frac{p-1}{2}} \bmod p$ implies $\big(\frac{-1}{p}\big) = (-1)^{\frac{p-1}{2}}$ since $p \neq 2$.

In order to determine $\big(\frac{2}{p}\big)$, we work in the ring $\mathbb{Z}[i]$ of gaussian integers. Since $(1+i)^2 = 2i$, we find

$$(1+i)^p = (1+i)\big((1+i)^2\big)^{\frac{p-1}{2}} = (1+i)i^{\frac{p-1}{2}}2^{\frac{p-1}{2}},$$

and since $(1+i)^p \equiv 1+i^p \bmod p$ and $\big(\frac{2}{p}\big) \equiv 2^{\frac{p-1}{2}} \bmod p$, it follows that

$$\Big(\frac{2}{p}\Big)(1+i)\,i^{\frac{p-1}{2}} \equiv 1+i(-1)^{\frac{p-1}{2}} \bmod p.$$

From this, an easy computation yields

$$\Big(\frac{2}{p}\Big) \equiv (-1)^{\frac{p-1}{4}} \bmod p, \quad \text{resp.} \quad \Big(\frac{2}{p}\Big) \equiv (-1)^{\frac{p+1}{4}} \bmod p,$$

if $\frac{p-1}{2}$ is even, resp. odd. Since $\frac{p^2-1}{8} = \frac{p-1}{4}\frac{p+1}{2} = \frac{p+1}{4}\frac{p-1}{2}$, we deduce $\big(\frac{2}{p}\big) = (-1)^{\frac{p^2-1}{8}}$.

In order to prove the first formula, we work in the ring $\mathbb{Z}[\zeta]$, where ζ is a primitive ℓ-th root of unity. We consider the **Gauss sum**

$$\tau = \sum_{a\in(\mathbb{Z}/\ell\mathbb{Z})^*} \Big(\frac{a}{\ell}\Big)\zeta^a$$

and show that

$$\tau^2 = \Big(\frac{-1}{\ell}\Big)\ell.$$

For this, let a and b vary over the group $(\mathbb{Z}/\ell\mathbb{Z})^*$, put $c = ab^{-1}$ and deduce from the identity $\left(\frac{b}{\ell}\right) = \left(\frac{b^{-1}}{\ell}\right)$ that

$$\left(\frac{-1}{\ell}\right)\tau^2 = \sum_{a,b}\left(\frac{-ab}{\ell}\right)\zeta^{a+b} = \sum_{a,b}\left(\frac{ab^{-1}}{\ell}\right)\zeta^{a-b} = \sum_{b,c}\left(\frac{c}{\ell}\right)\zeta^{bc-b}$$
$$= \sum_{c\neq 1}\left(\frac{c}{\ell}\right)\sum_b \zeta^{b(c-1)} + \sum_b\left(\frac{1}{\ell}\right).$$

Now $\sum_c\left(\frac{c}{\ell}\right) = 0$, as one sees by multiplying the sum with a symbol $\left(\frac{x}{\ell}\right) = -1$, and putting $\xi = \zeta^{c-1}$ gives $\sum_b \zeta^{b(c-1)} = \xi + \xi^2 + \cdots + \xi^{\ell-1} = -1$, from which we indeed find that

$$\left(\frac{-1}{\ell}\right)\tau^2 = (-1)(-1) + \ell - 1 = \ell.$$

This, together with the congruence $\left(\frac{\ell}{p}\right) \equiv \ell^{\frac{p-1}{2}} \bmod p$ and the identity $\left(\frac{-1}{\ell}\right) = (-1)^{\frac{\ell-1}{2}}$, implies

$$\tau^p = \tau(\tau^2)^{\frac{p-1}{2}} \equiv \tau(-1)^{\frac{\ell-1}{2}\frac{p-1}{2}}\left(\frac{\ell}{p}\right) \bmod p.$$

On the other hand one has

$$\tau^p \equiv \sum_a\left(\frac{a}{\ell}\right)\zeta^{ap} \equiv \left(\frac{p}{\ell}\right)\sum_a\left(\frac{ap}{\ell}\right)\zeta^{ap} \equiv \left(\frac{p}{\ell}\right)\tau \bmod p,$$

so that

$$\tau\left(\frac{p}{\ell}\right) \equiv \tau(-1)^{\frac{\ell-1}{2}\frac{p-1}{2}}\left(\frac{\ell}{p}\right) \bmod p.$$

Multiplying by τ and dividing by $\pm\ell$ yields the claim. □

We have proved Gauss's reciprocity law by a rather contrived calculation. In § 10, however, we will recognize the true reason why it holds in the law of decomposition of primes in the field $\mathbb{Q}(\zeta)$ of ℓ-th roots of unity. The Gauss sums do have a higher theoretical significance, though, as will become apparent later (see VII, § 2 and § 6).

Exercise 1. If $\mathfrak{a}$ and $\mathfrak{b}$ are ideals of $\mathcal{O}$, then one has $\mathfrak{a} = \mathfrak{a}\,\mathcal{O} \cap \mathcal{O}$ and $\mathfrak{a} \mid \mathfrak{b} \Longleftrightarrow \mathfrak{a}\mathcal{O} \mid \mathfrak{b}\,\mathcal{O}$.

Exercise 2. For every integral ideal $\mathfrak{A}$ of $\mathcal{O}$, there exists a $\theta \in \mathcal{O}$ such that the conductor $\mathfrak{F} = \{\alpha \in \mathcal{O} \mid \alpha\mathcal{O} \subseteq o[\theta]\}$ is prime to $\mathfrak{A}$ and such that $L = K(\theta)$.

Exercise 3. If a prime ideal $\mathfrak{p}$ of K is totally split in two separable extensions $L|K$ and $L'|K$, then it is also totally split in the composite extension.

Exercise 4. A prime ideal $\mathfrak{p}$ of K is totally split in the separable extension $L|K$ if and only if it is totally split in the Galois closure $N|K$ of $L|K$.

Exercise 5. For a number field K the statement of proposition (8.3) concerning the prime decomposition in the extension $K(\theta)$ holds for all prime ideals $\mathfrak{p} \nmid (\mathcal{O} : \mathcal{O}[\theta])$.

Exercise 6. Given a positive integer $b > 1$, an integer a relatively prime to b is a quadratic residue mod b if and only if it is a quadratic residue modulo each prime divisor p of b, and if $a \equiv 1 \bmod 4$ when $4|b$, $8 \nmid b$, resp. $a \equiv 1 \bmod 8$ when $8|b$.

Exercise 7. Let $(a, p) = 1$ and $a\nu \equiv r_\nu \bmod p$, $\nu = 1, \dots, p-1$, $0 < r_\nu < p$. Then the r_ν give a permutation π of the numbers $1, \dots, p-1$. Show that $\operatorname{sgn}\pi = \left(\frac{a}{p}\right)$.

Exercise 8. Let $a_n = \dfrac{\varepsilon^n - \varepsilon'^n}{\sqrt{5}}$, where $\varepsilon = \dfrac{1+\sqrt{5}}{2}$, $\varepsilon' = \dfrac{1-\sqrt{5}}{2}$ (a_n is the n-th Fibonacci number). If p is a prime number $\neq 2, 5$, then one has

$$a_p \equiv \left(\frac{p}{5}\right) \bmod p .$$

Exercise 9. Study the Legendre symbol $\left(\frac{3}{p}\right)$ as a function of $p > 3$. Show that the property of 3 being a quadratic residue or nonresidue mod p depends only on the class of p mod 12.

Exercise 10. Show that the number of solutions of $x^2 \equiv a \bmod p$ equals $1 + \left(\frac{a}{p}\right)$.

Exercise 11. Show that the number of solutions of the congruence $ax^2 + bx + c \equiv 0 \bmod p$, where $(a, p) = 1$, equals $1 + \left(\frac{b^2 - 4ac}{p}\right)$.

§ 9. Hilbert's Ramification Theory

The question of prime decomposition in a finite extension $L|K$ takes a particularly interesting and important turn once we assume $L|K$ to be a Galois extension. The prime ideals are then subject to the action of the Galois group

$$G = G(L|K).$$

The "ramification theory" that arises from this assumption has been introduced into number theory by *David Hilbert* (1862–1943). Given a in the ring $\mathcal{O}$ of integral elements of L, the conjugate σa, for every $\sigma \in G$, also belongs to $\mathcal{O}$, i.e., G acts on $\mathcal{O}$. If $\mathfrak{P}$ is a prime ideal of $\mathcal{O}$ above $\mathfrak{p}$, then so is $\sigma\mathfrak{P}$, for each $\sigma \in G$, because

$$\sigma\mathfrak{P} \cap o = \sigma(\mathfrak{P} \cap o) = \sigma\mathfrak{p} = \mathfrak{p} .$$

The ideals $\sigma\mathfrak{P}$, for $\sigma \in G$, are called the prime ideals **conjugate** to $\mathfrak{P}$.

(9.1) Proposition. *The Galois group G acts transitively on the set of all prime ideals $\mathfrak{P}$ of $\mathcal{O}$ lying above $\mathfrak{p}$, i.e., these prime ideals are all conjugates of each other.*

Proof: Let $\mathfrak{P}$ and $\mathfrak{P}'$ be two prime ideals above $\mathfrak{p}$. Assume $\mathfrak{P}' \neq \sigma\mathfrak{P}$ for any $\sigma \in G$. By the Chinese remainder theorem there exists $x \in \mathcal{O}$ such that

$$x \equiv 0 \bmod \mathfrak{P}' \quad \text{and} \quad x \equiv 1 \bmod \sigma\mathfrak{P} \quad \text{for all} \quad \sigma \in G .$$

Then the norm $N_{L|K}(x) = \prod_{\sigma \in G} \sigma x$ belongs to $\mathfrak{P}' \cap o = \mathfrak{p}$. On the other hand, $x \notin \sigma\mathfrak{P}$ for any $\sigma \in G$, hence $\sigma x \notin \mathfrak{P}$ for any $\sigma \in G$. Consequently $\prod_{\sigma \in G} \sigma x \notin \mathfrak{P} \cap o = \mathfrak{p}$, a contradiction. □

(9.2) Definition. *If $\mathfrak{P}$ is a prime ideal of $\mathcal{O}$, then the subgroup*

$$G_{\mathfrak{P}} = \{\sigma \in G \mid \sigma\mathfrak{P} = \mathfrak{P}\}$$

is called the **decomposition group** *of $\mathfrak{P}$ over K. The fixed field*

$$Z_{\mathfrak{P}} = \{x \in L \mid \sigma x = x \quad \text{for all } \sigma \in G_{\mathfrak{P}}\}$$

is called the **decomposition field** *of $\mathfrak{P}$ over K.*

The decomposition group encodes in group-theoretic language the number of different prime ideals into which a prime ideal $\mathfrak{p}$ of o decomposes in $\mathcal{O}$. For if $\mathfrak{P}$ is one of them and σ varies over a system of representatives of the cosets in $G/G_{\mathfrak{P}}$, then $\sigma\mathfrak{P}$ varies over the different prime ideals above $\mathfrak{p}$, each one occurring precisely once, i.e., their number equals the index $(G : G_{\mathfrak{P}})$. In particular, one has

$$G_{\mathfrak{P}} = 1 \iff Z_{\mathfrak{P}} = L \iff \mathfrak{p} \text{ is totally split,}$$

$$G_{\mathfrak{P}} = G \iff Z_{\mathfrak{P}} = K \iff \mathfrak{p} \text{ is nonsplit.}$$

The decomposition group of a prime ideal $\sigma\mathfrak{P}$ conjugate to $\mathfrak{P}$ is the conjugate subgroup

$$G_{\sigma\mathfrak{P}} = \sigma G_{\mathfrak{P}} \sigma^{-1} .$$

In fact, for $\tau \in G$, one has the equivalences

$$\tau \in G_{\sigma\mathfrak{P}} \iff \tau\sigma\mathfrak{P} = \sigma\mathfrak{P} \iff \sigma^{-1}\tau\sigma\mathfrak{P} = \mathfrak{P}$$
$$\iff \sigma^{-1}\tau\sigma \in G_{\mathfrak{P}} \iff \tau \in \sigma G_{\mathfrak{P}} \sigma^{-1} .$$

Remark: The decomposition group regulates the prime decomposition also in the case of a non-Galois extension. For subgroups U and V of a group G, consider the equivalence relation in G defined by

$$\sigma \sim \sigma' \iff \sigma' = u\sigma v \quad \text{for } u \in U,\, v \in V .$$

The corresponding equivalence classes

$$U\sigma V = \{u\sigma v \mid u \in U,\ v \in V\}$$

are called the **double cosets** of G modd U, V. The set of these double cosets, which form a partition of G, is denoted $U\backslash G/V$.

Now let $L|K$ be an arbitrary separable extension, and embed it into a Galois extension $N|K$ with Galois group G. In G, consider the subgroup $H = G(N|L)$. Let $\mathfrak{p}$ be a prime ideal of K and $P_{\mathfrak{p}}$ the set of prime ideals of L above $\mathfrak{p}$. If $\mathfrak{P}$ is a prime ideal of N above $\mathfrak{p}$, then the rule

$$H\backslash G/G_{\mathfrak{P}} \longrightarrow P_{\mathfrak{p}}, \quad H\sigma G_{\mathfrak{P}} \longmapsto \sigma\mathfrak{P} \cap L,$$

gives a well-defined bijection. The proof is left to the reader.

In the Galois case, the inertia degrees $f_1, \dots, f_r$ and the ramification indices $e_1, \dots, e_r$ in the prime decomposition

$$\mathfrak{p} = \mathfrak{P}_1^{e_1} \cdots \mathfrak{P}_r^{e_r}$$

of a prime ideal $\mathfrak{p}$ of K are both independent of i,

$$f_1 = \cdots = f_r = f, \quad e_1 = \cdots = e_r = e.$$

In fact, writing $\mathfrak{P} = \mathfrak{P}_1$, we find $\mathfrak{P}_i = \sigma_i\mathfrak{P}$ for suitable $\sigma_i \in G$, and the isomorphism $\sigma_i : \mathcal{O} \to \mathcal{O}$ induces an isomorphism

$$\mathcal{O}/\mathfrak{P} \xrightarrow{\sim} \mathcal{O}/\sigma_i\mathfrak{P}, \quad a \bmod \mathfrak{P} \longmapsto \sigma_i a \bmod \sigma_i\mathfrak{P},$$

so that

$$f_i = [\,\mathcal{O}/\sigma_i\mathfrak{P} : o/\mathfrak{p}] = [\,\mathcal{O}/\mathfrak{P} : o/\mathfrak{p}], \quad i = 1, \dots, r.$$

Furthermore, since $\sigma_i(\mathfrak{p}\mathcal{O}) = \mathfrak{p}\mathcal{O}$, we deduce from

$$\mathfrak{P}^\nu \mid \mathfrak{p}\mathcal{O} \iff \sigma_i(\mathfrak{P}^\nu) \mid \sigma_i(\mathfrak{p}\mathcal{O}) \iff (\sigma_i\mathfrak{P})^\nu \mid \mathfrak{p}\,\mathcal{O}$$

the equality of the e_i, $i = 1, \dots, r$. Thus the prime decomposition of $\mathfrak{p}$ in $\mathcal{O}$ takes on the following simple form in the Galois case:

$$\mathfrak{p} = \Big(\prod_\sigma \sigma\mathfrak{P}\Big)^e,$$

where σ varies over a system of representatives of $G/G_{\mathfrak{P}}$. The decomposition field $Z_{\mathfrak{P}}$ of $\mathfrak{P}$ over K has the following significance for the decomposition of $\mathfrak{p}$ and the invariants e and f.

(9.3) Proposition. *Let $\mathfrak{P}_Z = \mathfrak{P} \cap Z_{\mathfrak{P}}$ be the prime ideal of $Z_{\mathfrak{P}}$ below $\mathfrak{P}$. Then we have:*

(i) *$\mathfrak{P}_Z$ is nonsplit in L, i.e., $\mathfrak{P}$ is the only prime ideal of L above $\mathfrak{P}_Z$.*

(ii) *$\mathfrak{P}$ over $Z_{\mathfrak{P}}$ has ramification index e and inertia degree f.*

(iii) *The ramification index and the inertia degree of $\mathfrak{P}_Z$ over K both equal 1.*

Proof: (i) Since $G(L|Z_\mathfrak{P}) = G_\mathfrak{P}$, the prime ideals above $\mathfrak{P}_Z$ are the $\sigma\mathfrak{P}$, for $\sigma \in G(L|Z_\mathfrak{P})$, and they are all equal to $\mathfrak{P}$.
(ii) Since in the Galois case, ramification indices and inertia degrees are independent of the prime divisor, the fundamental identity in this case reads

$$n = efr\,,$$

where $n := \#G, r = (G : G_\mathfrak{P})$. We see therefore that $\#G_\mathfrak{P} = [L : Z_\mathfrak{P}] = ef$. Let e', resp. e'', be the ramification index of $\mathfrak{P}$ over $Z_\mathfrak{P}$, resp. of $\mathfrak{P}_Z$ over K. Then $\mathfrak{p} = \mathfrak{P}_Z^{e''} \cdot \ldots$ in $Z_\mathfrak{P}$ and $\mathfrak{P}_Z = \mathfrak{P}^{e'}$ in L, so that $\mathfrak{p} = \mathfrak{P}^{e''e'} \cdot \ldots$, i.e., $e = e'e''$. One also obviously gets the analogous identity for the inertia degrees $f = f'f''$. The fundamental identity for the decomposition of $\mathfrak{P}_Z$ in L then reads $[L : Z_\mathfrak{P}] = e'f'$, i.e., we have $e'f' = ef$, and therefore $e' = e$, $f' = f$, $e'' = f'' = 1$. □

The ramification index e and the inertia degree f admit a further interesting group-theoretic interpretation. Since $\sigma\mathcal{O} = \mathcal{O}$ and $\sigma\mathfrak{P} = \mathfrak{P}$, every $\sigma \in G_\mathfrak{P}$ induces an automorphism

$$\bar{\sigma} : \mathcal{O}/\mathfrak{P} \longrightarrow \mathcal{O}/\mathfrak{P}, \quad a \bmod \mathfrak{P} \longmapsto \sigma a \bmod \mathfrak{P}\,,$$

of the residue class field $\mathcal{O}/\mathfrak{P}$. Putting $\kappa(\mathfrak{P}) = \mathcal{O}/\mathfrak{P}$ and $\kappa(\mathfrak{p}) = o/\mathfrak{p}$, we obtain the

(9.4) Proposition. *The extension $\kappa(\mathfrak{P})|\kappa(\mathfrak{p})$ is normal and admits a surjective homomorphism*

$$G_\mathfrak{P} \longrightarrow G\big(\kappa(\mathfrak{P})|\kappa(\mathfrak{p})\big)\,.$$

Proof: The inertia degree of $\mathfrak{P}_Z$ over K equals 1, i.e., $Z_\mathfrak{P}$ has the same residue class field $\kappa(\mathfrak{p})$ as K with respect to $\mathfrak{p}$. Therefore we may, and do, assume that $Z_\mathfrak{P} = K$, i.e., $G_\mathfrak{P} = G$. Let $\theta \in \mathcal{O}$ be a representative of an element $\bar{\theta} \in \kappa(\mathfrak{P})$ and $f(X)$, resp. $\bar{g}(X)$, the minimal polynomial of θ over K, resp. of $\bar{\theta}$ over $\kappa(\mathfrak{p})$. Then $\bar{\theta} = \theta \bmod \mathfrak{P}$ is a zero of the polynomial $\bar{f}(X) = f(X) \bmod \mathfrak{p}$, i.e., $\bar{g}(X)$ divides $\bar{f}(X)$. Since $L|K$ is normal, $f(X)$ splits over $\mathcal{O}$ into linear factors. Hence $\bar{f}(X)$ splits into linear factors over $\kappa(\mathfrak{P})$, and the same is true of $\bar{g}(X)$. In other words, $\kappa(\mathfrak{P})|\kappa(\mathfrak{p})$ is a normal extension.

Now let $\bar{\theta}$ be a primitive element for the maximal separable subextension of $\kappa(\mathfrak{P})|\kappa(\mathfrak{p})$ and

$$\bar{\sigma} \in G\big(\kappa(\mathfrak{P})|\kappa(\mathfrak{p})\big) = G\big(\kappa(\mathfrak{p})(\bar{\theta})|\kappa(\mathfrak{p})\big)\,.$$

Then $\bar{\sigma}\,\bar{\theta}$ is a root of $\bar{g}(X)$, and hence of $\bar{f}(X)$, i.e., there exists a zero θ' of $f(X)$ such that $\theta' \equiv \bar{\sigma}\,\bar{\theta} \mod \mathfrak{P}$. θ' is a conjugate of θ, i.e., $\theta' = \sigma\,\theta$ for some $\sigma \in G(L|K)$. Since $\sigma\,\theta \equiv \bar{\sigma}\,\bar{\theta} \mod \mathfrak{P}$, the automorphism σ is mapped by the homomorphism in question to $\bar{\sigma}$. This proves the surjectivity. □

(9.5) Definition. *The kernel $I_{\mathfrak{P}} \subseteq G_{\mathfrak{P}}$ of the homomorphism*

$$G_{\mathfrak{P}} \longrightarrow G\big(\kappa(\mathfrak{P})|\kappa(\mathfrak{p})\big)$$

is called the **inertia group** *of $\mathfrak{P}$ over K. The fixed field*

$$T_{\mathfrak{P}} = \big\{x \in L \;\big|\; \sigma x = x \quad \text{for all } \sigma \in I_{\mathfrak{P}}\big\}$$

is called the **inertia field** *of $\mathfrak{P}$ over K.*

This inertia field $T_{\mathfrak{P}}$ appears in the tower of fields

$$K \subseteq Z_{\mathfrak{P}} \subseteq T_{\mathfrak{P}} \subseteq L,$$

and we have the exact sequence

$$1 \longrightarrow I_{\mathfrak{P}} \longrightarrow G_{\mathfrak{P}} \longrightarrow G\big(\kappa(\mathfrak{P})|\kappa(\mathfrak{p})\big) \longrightarrow 1.$$

Its properties are expressed in the

(9.6) Proposition. *The extension $T_{\mathfrak{P}}|Z_{\mathfrak{P}}$ is normal, and one has*

$$G(T_{\mathfrak{P}}|Z_{\mathfrak{P}}) \cong G\big(\kappa(\mathfrak{P})|\kappa(\mathfrak{p})\big), \quad G(L|T_{\mathfrak{P}}) = I_{\mathfrak{P}}.$$

If the residue field extension $\kappa(\mathfrak{P})|\kappa(\mathfrak{p})$ is separable, then one has

$$\#I_{\mathfrak{P}} = [L : T_{\mathfrak{P}}] = e, \quad (G_{\mathfrak{P}} : I_{\mathfrak{P}}) = [T_{\mathfrak{P}} : Z_{\mathfrak{P}}] = f.$$

In this case one finds for the prime ideal $\mathfrak{P}_T$ of $T_{\mathfrak{P}}$ below $\mathfrak{P}$:

(i) *The ramification index of $\mathfrak{P}$ over $\mathfrak{P}_T$ is e and the inertia degree is 1.*

(ii) *The ramification index of $\mathfrak{P}_T$ over $\mathfrak{P}_Z$ is 1, and the inertia degree is f.*

Proof: The first two claims follow from the identity $\#G_{\mathfrak{P}} = ef$. So we only have to show statements (i) and (ii). Using the fundamental identity, they all follow from $\kappa(\mathfrak{P}_T) = \kappa(\mathfrak{P})$. As the inertia group $I_{\mathfrak{P}}$ of $\mathfrak{P}$ over K is also the inertia group of $\mathfrak{P}$ over $T_{\mathfrak{P}}$, it follows from an application of proposition (9.4) to the extension $L|T_{\mathfrak{P}}$ that $G(\kappa(\mathfrak{P})|\kappa(\mathfrak{P}_T)) = 1$, hence $\kappa(\mathfrak{P}_T) = \kappa(\mathfrak{P})$. □

In the diagram

$$K \xrightarrow[1]{1} Z_{\mathfrak{P}} \xrightarrow[f]{1} T_{\mathfrak{P}} \xrightarrow[1]{e} L$$

we have indicated the ramification indices of the individual field extensions on top, and the inertia degrees on the bottom. In the special case where the residue field extension $\kappa(\mathfrak{P})|\kappa(\mathfrak{p})$ is separable we find

$$I_{\mathfrak{P}} = 1 \iff T_{\mathfrak{P}} = L \iff \mathfrak{p} \text{ is unramified in } L\,.$$

In this case the Galois group $G(\kappa(\mathfrak{P})|\kappa(\mathfrak{p})) \cong G_{\mathfrak{P}}$ of the residue class field extension may be viewed as a subgroup of $G = G(L|K)$.

Hilbert's ramification theory, with its various refinements and generalizations, belongs naturally to the theory of valuations, which we will develop in the next chapter (see chap. II, § 9).

Exercise 1. If $L|K$ is a Galois extension of algebraic number fields with noncyclic Galois group, then there are at most finitely many nonsplit prime ideals of K.

Exercise 2. If $L|K$ is a Galois extension of algebraic number fields, and $\mathfrak{P}$ a prime ideal which is unramified over K (i.e., $\mathfrak{p} = \mathfrak{P} \cap K$ is unramified in L), then there is one and only one automorphism $\varphi_{\mathfrak{P}} \in G(L|K)$ such that

$$\varphi_{\mathfrak{P}} a \equiv a^q \bmod \mathfrak{P} \quad \text{for all } a \in \mathcal{O}\,,$$

where $q = [\kappa(\mathfrak{P}) : \kappa(\mathfrak{p})]$. It is called the **Frobenius automorphism**. The decomposition group $G_{\mathfrak{P}}$ is cyclic and $\varphi_{\mathfrak{P}}$ is a generator of $G_{\mathfrak{P}}$.

Exercise 3. Let $L|K$ be a solvable extension of prime degree p (not necessarily Galois). If the unramified prime ideal $\mathfrak{p}$ in L has two prime factors $\mathfrak{P}$ and $\mathfrak{P}'$ of degree 1, then it is already totally split (theorem of *F.K. Schmidt*).

Hint: Use the following result of *Galois* (see [75], chap. II, § 3): if G is a transitive solvable permutation group of prime degree p, then there is no nontrivial permutation $\sigma \in G$ which fixes two distinct letters.

Exercise 4. Let $L|K$ be a finite (not necessarily Galois) extension of algebraic number fields and $N|K$ the normal closure of $L|K$. Show that a prime ideal $\mathfrak{p}$ of K is totally split in L if and only if it is totally split in N.

Hint: Use the double coset decomposition $H\backslash G/G_{\mathfrak{P}}$, where $G = G(N|K)$, $H = G(N|L)$ and $G_{\mathfrak{P}}$ is the decomposition group of a prime ideal $\mathfrak{P}$ over $\mathfrak{p}$.

§ 10. Cyclotomic Fields

The concepts and results of the theory as far as it has now been developed have reached a degree of abstraction which we will now balance by something more concrete. We will put the insights of the general theory to the task and make them more explicit in the example of the **n-th cyclotomic field** $\mathbb{Q}(\zeta)$, where ζ is a **primitive n-th root of unity**. Among all number fields, this field occupies a special, central place. So studying it does not only furnish a worthwhile example but in fact an essential building block for the further theory.

It will be our first goal to determine explicitly the ring of integers of the field $\mathbb{Q}(\zeta)$. For this we need the

(10.1) Lemma. *Let n be a prime power ℓ^ν and put $\lambda = 1 - \zeta$. Then the principal ideal (λ) in the ring $\mathcal{O}$ of integers of $\mathbb{Q}(\zeta)$ is a prime ideal of degree 1, and we have*

$$\ell\,\mathcal{O} = (\lambda)^d, \quad \textit{where} \quad d = \varphi(\ell^\nu) = [\mathbb{Q}(\zeta) : \mathbb{Q}].$$

Furthermore, the basis $1, \zeta, \dots, \zeta^{d-1}$ of $\mathbb{Q}(\zeta)|\mathbb{Q}$ has the discriminant

$$d(1, \zeta, \dots, \zeta^{d-1}) = \pm \ell^s, \quad s = \ell^{\nu-1}(\nu\ell - \nu - 1).$$

Proof: The minimal polynomial of ζ over $\mathbb{Q}$ is the n-th cyclotomic polynomial

$$\phi_n(X) = (X^{\ell^\nu} - 1)/(X^{\ell^{\nu-1}} - 1) = X^{\ell^{\nu-1}(\ell-1)} + \dots + X^{\ell^{\nu-1}} + 1.$$

Putting $X = 1$, we obtain the identity

$$\ell = \prod_{g \in (\mathbb{Z}/n\mathbb{Z})^*} (1 - \zeta^g).$$

But $1 - \zeta^g = \varepsilon_g(1 - \zeta)$, for the algebraic integer $\varepsilon_g = \frac{1 - \zeta^g}{1 - \zeta} = 1 + \zeta + \dots + \zeta^{g-1}$. If g' is an integer such that $gg' \equiv 1 \bmod \ell^\nu$, then

$$\frac{1 - \zeta}{1 - \zeta^g} = \frac{1 - (\zeta^g)^{g'}}{1 - \zeta^g} = 1 + \zeta^g + \dots + (\zeta^g)^{g'-1}$$

is integral as well, i.e., ε_g is a unit. Consequently $\ell = \varepsilon(1 - \zeta)^{\varphi(\ell^\nu)}$, with the unit $\varepsilon = \prod_g \varepsilon_g$, hence $\ell\,\mathcal{O} = (\lambda)^{\varphi(\ell^\nu)}$. Since $[\mathbb{Q}(\zeta) : \mathbb{Q}] = \varphi(\ell^\nu)$, the fundamental identity (8.2) shows that (λ) is a prime ideal of degree 1.

Let $\zeta = \zeta_1, \dots, \zeta_d$ be the conjugates of ζ. Then the cyclotomic polynomial is $\phi_n(X) = \prod_{i=1}^d (X - \zeta_i)$ and (see §2, p. 11)

$$\pm d(1, \zeta, \dots, \zeta^{d-1}) = \prod_{i \neq j} (\zeta_i - \zeta_j) = \prod_{i=1}^d \phi_n'(\zeta_i) = N_{\mathbb{Q}(\zeta)|\mathbb{Q}}\big(\phi_n'(\zeta)\big).$$

Differentiating the equation

$$(X^{\ell^{\nu-1}}-1)\phi_n(X)=X^{\ell^\nu}-1$$

and substituting ζ for X yields

$$(\xi-1)\phi_n'(\zeta)=\ell^\nu\zeta^{-1},$$

with the primitive ℓ-th root of unity $\xi=\zeta^{\ell^{\nu-1}}$. But $N_{\mathbb{Q}(\xi)|\mathbb{Q}}(\xi-1)=\pm\ell$, so that

$$N_{\mathbb{Q}(\zeta)|\mathbb{Q}}(\xi-1)=N_{\mathbb{Q}(\xi)|\mathbb{Q}}(\xi-1)^{\ell^{\nu-1}}=\pm\ell^{\ell^{\nu-1}}.$$

Observing that ζ^{-1} has norm ±1 we obtain

$$d(1,\zeta,\ldots,\zeta^{d-1})=\pm N_{\mathbb{Q}(\zeta)|\mathbb{Q}}\big(\phi_n'(\zeta)\big)=\pm\ell^{\nu\ell^{\nu-1}(\ell-1)-\ell^{\nu-1}}=\pm\ell^s$$

with $s=\ell^{\nu-1}(\nu\ell-\nu-1)$. □

The ring of integers of $\mathbb{Q}(\zeta)$ is now determined, for arbitrary n, as follows.

(10.2) Proposition. *A $\mathbb{Z}$-basis of the ring $\mathcal{O}$ of integers of $\mathbb{Q}(\zeta)$ is given by $1,\zeta,\ldots,\zeta^{d-1}$, with $d=\varphi(n)$, in other words,*

$$\mathcal{O}=\mathbb{Z}+\mathbb{Z}\zeta+\cdots+\mathbb{Z}\zeta^{d-1}=\mathbb{Z}[\zeta].$$

Proof: We first prove the proposition in the case where n is a prime power ℓ^ν. Since $d(1,\zeta,\ldots,\zeta^{d-1})=\pm\ell^s$, (2.9) gives us

$$\ell^s\mathcal{O}\subseteq\mathbb{Z}[\zeta]\subseteq\mathcal{O}.$$

Putting $\lambda=1-\zeta$, lemma (10.1) tells us that $\mathcal{O}/\lambda\mathcal{O}\cong\mathbb{Z}/\ell\mathbb{Z}$, so that $\mathcal{O}=\mathbb{Z}+\lambda\mathcal{O}$, and *a fortiori*

$$\lambda\mathcal{O}+\mathbb{Z}[\zeta]=\mathcal{O}.$$

Multiplying this by λ and substituting the result $\lambda\mathcal{O}=\lambda^2\mathcal{O}+\lambda\mathbb{Z}[\zeta]$, we obtain

$$\lambda^2\mathcal{O}+\mathbb{Z}[\zeta]=\mathcal{O}.$$

Iterating this procedure, we find

$$\lambda^t\mathcal{O}+\mathbb{Z}[\zeta]=\mathcal{O}\quad\text{for all}\quad t\geq1.$$

For $t = s\,\varphi(\ell^\nu)$ this implies, in view of $\ell\,\mathcal{O} = \lambda^{\varphi(\ell^\nu)}\mathcal{O}$ (see (10.1)), that

$$\mathcal{O} = \lambda^t\mathcal{O} + \mathbb{Z}[\zeta] = \ell^s\mathcal{O} + \mathbb{Z}[\zeta] = \mathbb{Z}[\zeta].$$

In the general case, let $n = \ell_1^{\nu_1}\cdots\ell_r^{\nu_r}$. Then $\zeta_i = \zeta^{n/\ell_i^{\nu_i}}$ is a primitive $\ell_i^{\nu_i}$-th root of unity, and one has

$$\mathbb{Q}(\zeta) = \mathbb{Q}(\zeta_1)\cdots\mathbb{Q}(\zeta_r)$$

and $\mathbb{Q}(\zeta_1)\cdots\mathbb{Q}(\zeta_{i-1})\cap\mathbb{Q}(\zeta_i) = \mathbb{Q}$. By what we have just seen, for each $i = 1, \dots, r$, the elements $1, \zeta_i, \dots, \zeta_i^{d_i-1}$, where $d_i = \varphi(\ell_i^{\nu_i})$, form an integral basis of $\mathbb{Q}(\zeta_i)|\mathbb{Q}$. Since the discriminants $d(1, \zeta_i, \dots, \zeta_i^{d_i-1}) = \pm\ell_i^{s_i}$ are pairwise relatively prime, we conclude successively from (2.11) that the elements $\zeta_1^{j_1}\cdots\zeta_r^{j_r}$, with $j_i = 0, \dots, d_i - 1$, form an integral basis of $\mathbb{Q}(\zeta)|\mathbb{Q}$. But each one of these elements is a power of ζ. Therefore every $\alpha \in \mathcal{O}$ may be written as a polynomial $\alpha = f(\zeta)$ with coefficients in $\mathbb{Z}$. Since ζ has degree $\varphi(n)$ over $\mathbb{Q}$, the degree of the polynomial $f(\zeta)$ may be reduced to $\varphi(n) - 1$. In this way one obtains a representation

$$\alpha = a_0 + a_1\zeta + \cdots + a_{\varphi(n)-1}\zeta^{\varphi(n)-1}.$$

Thus $1, \zeta, \dots, \zeta^{\varphi(n)-1}$ is indeed an integral basis. □

Knowing that $\mathbb{Z}[\zeta]$ is the ring of integers of the field $\mathbb{Q}(\zeta)$ we are now in a position to state explicitly the law of decomposition of prime numbers p into prime ideals of $\mathbb{Q}(\zeta)$. It is of the most beautiful simplicity.

(10.3) Proposition. *Let $n = \prod_p p^{\nu_p}$ be the prime factorization of n and, for every prime number p, let f_p be the smallest positive integer such that*

$$p^{f_p} \equiv 1 \bmod n/p^{\nu_p}.$$

Then one has in $\mathbb{Q}(\zeta)$ the factorization

$$p = (\mathfrak{p}_1\cdots\mathfrak{p}_r)^{\varphi(p^{\nu_p})},$$

where $\mathfrak{p}_1, \dots, \mathfrak{p}_r$ are distinct prime ideals, all of degree f_p.

Proof: Since $\mathcal{O} = \mathbb{Z}[\zeta]$, the conductor of $\mathbb{Z}[\zeta]$ equals 1, and we may apply proposition (8.3) to any prime number p. As a consequence, every p

decomposes into prime ideals in exactly the same way as the minimal polynomial $\phi_n(X)$ of ζ factors into irreducible polynomials mod p. All we have to show is therefore that

$$\phi_n(X) \equiv \big(p_1(X)\cdots p_r(X)\big)^{\varphi(p^{\nu_p})} \bmod p\,,$$

where $p_1(X), \ldots, p_r(X)$ are distinct irreducible polynomials over $\mathbb{Z}/p\mathbb{Z}$ of degree f_p. In order to see this, put $n = p^{\nu_p}m$. As ξ_i, resp. η_j, varies over primitive roots of unity of order m, resp. p^{ν_p}, the products $\xi_i\eta_j$ vary precisely over the primitive n-th roots of unity, i.e., one has the decomposition over $\mathcal{O}$:

$$\phi_n(X) = \prod_{i,j}(X - \xi_i\eta_j)\,.$$

Since $X^{p^{\nu_p}} - 1 \equiv (X-1)^{p^{\nu_p}} \bmod p$, one has $\eta_j \equiv 1 \bmod \mathfrak{p}$, for any prime ideal $\mathfrak{p}\,|\,p$. In other words,

$$\phi_n(X) \equiv \prod_i (X - \xi_i)^{\varphi(p^{\nu_p})} = \phi_m(X)^{\varphi(p^{\nu_p})} \bmod \mathfrak{p}\,.$$

This implies the congruence

$$\phi_n(X) \equiv \phi_m(X)^{\varphi(p^{\nu_p})} \bmod p\,.$$

Observing that f_p is the smallest positive integer such that $p^{f_p} \equiv 1 \bmod m$, it is obvious that this congruence reduces us to the case where $p \nmid n$, and hence $\varphi(p^{\nu_p}) = \varphi(1) = 1$.

As the characteristic p of $\mathcal{O}/\mathfrak{p}$ does not divide n, the polynomials $X^n - 1$ and nX^{n-1} have no common root in $\mathcal{O}/\mathfrak{p}$. So $X^n - 1 \bmod \mathfrak{p}$ has no multiple roots. We therefore see that passing to the quotient $\mathcal{O} \to \mathcal{O}/\mathfrak{p}$ maps the group μ_n of n-th roots of unity bijectively onto the group of n-th roots of unity of $\mathcal{O}/\mathfrak{p}$. In particular, the primitive n-th root of unity ζ modulo $\mathfrak{p}$ remains a primitive n-th root of unity. The smallest extension field of $\mathbb{F}_p = \mathbb{Z}/p\mathbb{Z}$ containing it is the field $\mathbb{F}_{p^{f_p}}$, because its multiplicative group $\mathbb{F}^*_{p^{f_p}}$ is cyclic of order $p^{f_p}-1$. $\mathbb{F}_{p^{f_p}}$ is therefore the field of decomposition of the reduced cyclotomic polynomial

$$\overline{\phi}_n(X) = \phi_n(X) \bmod p\,.$$

Being a divisor of $X^n - 1 \bmod p$, this polynomial has no multiple roots, and if

$$\overline{\phi}_n(X) = \overline{p}_1(X)\cdots\overline{p}_r(X)$$

is its factorization into irreducibles over $\mathbb{F}_p$, then every $\overline{p}_i(X)$ is the minimal polynomial of a primitive n-th root of unity $\overline{\xi} \in \mathbb{F}^*_{p^{f_p}}$. Its degree is therefore f_p. This proves the proposition. □

Let us emphasize two special cases of the above law of decomposition:

(10.4) Corollary. *A prime number p is ramified in $\mathbb{Q}(\zeta)$ if and only if*

$$n \equiv 0 \bmod p,$$

except in the case where $p = 2 = (4, n)$. A prime number $p \neq 2$ is totally split in $\mathbb{Q}(\zeta)$ if and only if

$$p \equiv 1 \bmod n.$$

The completeness of these results concerning the integral basis and the decomposition of primes in the field $\mathbb{Q}(\zeta)$ will not be matched by our study of the group of units and the ideal class group. The problems arising in this context are in fact among the most difficult problems posed by algebraic number theory. At the same time one encounters here plenty of astonishing laws which are the subject of a theory which has been developed only recently, **Iwasawa theory**.

The law of decomposition (10.3) in the cyclotomic field provides the proper explanation of Gauss's reciprocity law (8.6). This is based on the following

(10.5) Proposition. *Let ℓ and p be odd prime numbers, $\ell^* = (-1)^{\frac{\ell-1}{2}}\ell$, and ζ a primitive ℓ-th root of unity. Then one has:*

$$p \text{ is totally split in } \mathbb{Q}(\sqrt{\ell^*}) \iff p \text{ splits in } \mathbb{Q}(\zeta) \text{ into an even number of prime ideals.}$$

Proof: The little computation in § 8, p. 51 has shown us that $\ell^* = \tau^2$ with $\tau = \sum_{a \in (\mathbb{Z}/\ell\mathbb{Z})^*} \left(\frac{a}{\ell}\right)\zeta^a$, so that $\mathbb{Q}(\sqrt{\ell^*}) \subseteq \mathbb{Q}(\zeta)$. If p is totally split in $\mathbb{Q}(\sqrt{\ell^*})$, say $p = \mathfrak{p}_1\mathfrak{p}_2$, then some automorphism σ of $\mathbb{Q}(\zeta)$ such that $\sigma\mathfrak{p}_1 = \mathfrak{p}_2$ transforms the set of all prime ideals lying above $\mathfrak{p}_1$ bijectively into the set of prime ideals above $\mathfrak{p}_2$. Therefore the number of prime ideals of $\mathbb{Q}(\zeta)$ above p is even. Now assume conversely that this is the case. Then the index of the decomposition group $G_{\mathfrak{p}}$, or in other words, the degree $[Z_{\mathfrak{p}} : \mathbb{Q}]$ of the decomposition field of a prime ideal $\mathfrak{p}$ of $\mathbb{Q}(\zeta)$ over p, is even. Since $G(\mathbb{Q}(\zeta)|\mathbb{Q})$ is cyclic, it follows that $\mathbb{Q}(\sqrt{\ell^*}) \subseteq Z_{\mathfrak{p}}$. The inertia degree of $\mathfrak{p} \cap Z_{\mathfrak{p}}$ over $\mathbb{Q}$ is 1 by (9.3), hence also the inertia degree of $\mathfrak{p} \cap \mathbb{Q}(\sqrt{\ell^*})$. This implies that p is totally split in $\mathbb{Q}(\sqrt{\ell^*})$. □

From this proposition we obtain the reciprocity law for two odd prime numbers ℓ and p,

$$\left(\frac{\ell}{p}\right)\left(\frac{p}{\ell}\right) = (-1)^{\frac{\ell-1}{2}\frac{p-1}{2}}$$

as follows. It suffices to show that

$$\left(\frac{\ell^*}{p}\right) = \left(\frac{p}{\ell}\right).$$

In fact, the completely elementary result $\left(\frac{-1}{p}\right) = (-1)^{\frac{p-1}{2}}$ (see §8, p. 51) then gives

$$\left(\frac{p}{\ell}\right) = \left(\frac{\ell^*}{p}\right) = \left(\frac{-1}{p}\right)^{\frac{\ell-1}{2}} \left(\frac{\ell}{p}\right) = \left(\frac{\ell}{p}\right)(-1)^{\frac{p-1}{2}\frac{\ell-1}{2}}.$$

By (8.5) and (10.5), we know that $\left(\frac{\ell^*}{p}\right) = 1$ if and only if p decomposes in the field $\mathbb{Q}(\zeta)$ of ℓ-th roots of unity into an even number of prime ideals. By (10.3), this number is $r = \frac{\ell - 1}{f}$, where f is the smallest positive integer such that $p^f \equiv 1 \bmod \ell$, i.e., r is even if and only if f is a divisor of $\frac{\ell-1}{2}$. But this is tantamount to the condition $p^{(\ell-1)/2} \equiv 1 \bmod \ell$. Since an element in the cyclic group $\mathbb{F}_\ell^*$ has an order dividing $\frac{\ell-1}{2}$ if and only if it belongs to $\mathbb{F}_\ell^{*2}$, the last congruence is equivalent to $\left(\frac{p}{\ell}\right) = 1$. So we do have $\left(\frac{\ell^*}{p}\right) = \left(\frac{p}{\ell}\right)$ as claimed.

Historically, Gauss's reciprocity law marked the beginning of algebraic number theory. It was discovered by *Euler*, but first proven by *Gauss*. The quest for similar laws concerning higher power residues, i.e., the congruences $x^n \equiv a \bmod p$, with $n > 2$, dominated number theory for a long time. Since this problem required working with the n-th cyclotomic field, *Kummer*'s attempts to solve it led to his seminal discovery of ideal theory. We have developed the basics of this theory in the preceding sections and tested it successfully in the example of cyclotomic fields. The further development of this theory has led to a totally comprehensive generalization of Gauss's reciprocity law, **Artin's reciprocity law**, one of the high points in the history of number theory, and of compelling charm. This law is the main theorem of **class field theory**, which we will develop in chapters IV–VI.

Exercise 1. (Dirichlet's Prime Number Theorem). For every natural number n there are infinitely many prime numbers $p \equiv 1 \bmod n$.

Hint: Assume there are only finitely many. Let P be their product and consider the n-th cyclotomic polynomial ϕ_n. Not all numbers $\phi_n(xnP)$, for $x \in \mathbb{Z}$, can equal 1. Let $p \mid \phi_n(xnP)$ for suitable x. Deduce a contradiction from this. (Dirichlet's prime number theorem is valid more generally for prime numbers $p \equiv a \bmod n$, provided $(a,n) = 1$ (see VII, (5.14) and VII, § 13)).

Exercise 2. For every finite abelian group A there exists a Galois extension $L|\mathbb{Q}$ with Galois group $G(L|\mathbb{Q}) \cong A$.

Hint: Use exercise 1.

Exercise 3. Every quadratic number field $\mathbb{Q}(\sqrt{d})$ is contained in some cyclotomic field $\mathbb{Q}(\zeta_n)$, ζ_n a primitive n-th root of unity.

Exercise 4. Describe the quadratic subfields of $\mathbb{Q}(\zeta_n)|\mathbb{Q}$, in the case where n is odd.

Exercise 5. Show that $\mathbb{Q}(\sqrt{-1})$, $\mathbb{Q}(\sqrt{2})$, $\mathbb{Q}(\sqrt{-2})$ are the quadratic subfields of $\mathbb{Q}(\zeta_n)|\mathbb{Q}$ for $n = 2^q$, $q \geq 3$.

§ 11. Localization

To "localize" means to form quotients, the most familiar case being the passage from an integral domain A to its field of fractions

$$K = \left\{ \frac{a}{b} \;\middle|\; a \in A,\ b \in A \smallsetminus \{0\} \right\}.$$

More generally, choosing instead of $A \smallsetminus \{0\}$ any nonempty $S \subseteq A \smallsetminus \{0\}$ which is closed under multiplication, one again obtains a ring structure on the set

$$AS^{-1} = \left\{ \frac{a}{s} \in K \;\middle|\; a \in A,\ s \in S \right\}.$$

The most important special case of such a multiplicative subset is the complement $S = A \smallsetminus \mathfrak{p}$ of a prime ideal $\mathfrak{p}$ of A. In this case one writes $A_{\mathfrak{p}}$ instead of AS^{-1}, and one calls the ring $A_{\mathfrak{p}}$ the **localization** of A at $\mathfrak{p}$. When dealing with problems that involve a single prime ideal $\mathfrak{p}$ of A at a time it is often expedient to replace A by the localization $A_{\mathfrak{p}}$. This procedure forgets everything that has nothing to do with $\mathfrak{p}$, and brings out more clearly all the properties concerning $\mathfrak{p}$. For instance, the mapping

$$\mathfrak{q} \longmapsto \mathfrak{q}A_{\mathfrak{p}}$$

gives a 1–1-correspondence between the prime ideals $\mathfrak{q} \subseteq \mathfrak{p}$ of A and the prime ideals of $A_{\mathfrak{p}}$. More generally for any multiplicative set S, one has the

(11.1) Proposition. *The mappings*

$$\mathfrak{q} \longmapsto \mathfrak{q}S^{-1} \quad \textit{and} \quad \mathfrak{Q} \longmapsto \mathfrak{Q} \cap A$$

are mutually inverse 1–1-correspondences between the prime ideals $\mathfrak{q} \subseteq A \smallsetminus S$ of A and the prime ideals $\mathfrak{Q}$ of AS^{-1}.

Proof: If $\mathfrak{q} \subseteq A \smallsetminus S$ is a prime ideal of A, then

$$\mathfrak{Q} = \mathfrak{q}S^{-1} = \left\{ \frac{q}{s} \mid q \in \mathfrak{q},\ s \in S \right\}$$

is a prime ideal of AS^{-1}. Indeed, in obvious notation, the relation $\frac{a}{s}\frac{a'}{s'} \in \mathfrak{Q}$, i.e., $\frac{aa'}{ss'} = \frac{q}{s''}$, implies that $s''aa' = qss' \in \mathfrak{q}$. Therefore $aa' \in \mathfrak{q}$, because $s'' \notin \mathfrak{q}$, and hence a or a' belong to $\mathfrak{q}$, which shows that $\frac{a}{s}$ or $\frac{a'}{s'}$ belong to $\mathfrak{Q}$. Furthermore one has

$$\mathfrak{q} = \mathfrak{Q} \cap A,$$

since $\frac{q}{s} = a \in \mathfrak{Q} \cap A$ implies $q = as \in \mathfrak{q}$, whence $a \in \mathfrak{q}$ because $s \notin \mathfrak{q}$.

Conversely, let $\mathfrak{Q}$ be an arbitrary prime ideal of AS^{-1}. Then $\mathfrak{q} = \mathfrak{Q} \cap A$ is obviously a prime ideal of A, and one has $\mathfrak{q} \subseteq A \smallsetminus S$. In fact, if $\mathfrak{q}$ were to contain an $s \in S$, then we would have $1 = s \cdot \frac{1}{s} \in \mathfrak{Q}$ because $\frac{1}{s} \in AS^{-1}$. Furthermore one has

$$\mathfrak{Q} = \mathfrak{q}S^{-1}.$$

For if $\frac{a}{s} \in \mathfrak{Q}$, then $a = \frac{a}{s} \cdot s \in \mathfrak{Q} \cap A = \mathfrak{q}$, hence $\frac{a}{s} = a\frac{1}{s} \in \mathfrak{q}S^{-1}$. The mappings $\mathfrak{q} \mapsto \mathfrak{q}S^{-1}$ and $\mathfrak{Q} \mapsto \mathfrak{Q} \cap A$ are therefore inverses of each other, which proves the proposition. □

Usually S will be the complement of a union $\bigcup_{\mathfrak{p}\in X} \mathfrak{p}$ over a set X of prime ideals of A. In this case one writes

$$A(X) = \left\{ \frac{f}{g} \mid f, g \in A,\ g \not\equiv 0 \bmod \mathfrak{p} \text{ for } \mathfrak{p} \in X \right\}$$

instead of AS^{-1}. The prime ideals of $A(X)$ correspond by (11.1) 1–1 to the prime ideals of A which are contained in $\bigcup_{\mathfrak{p}\in X} \mathfrak{p}$, all the others are being eliminated when passing from A to $A(X)$. For instance, if X is finite or omits only finitely many prime ideals of A, then only the prime ideals from X survive in $A(X)$.

In the case that X consists of only one prime ideal $\mathfrak{p}$, the ring $A(X)$ is the localization

$$A_{\mathfrak{p}} = \left\{ \frac{f}{g} \mid f, g \in A,\ g \not\equiv 0 \bmod \mathfrak{p} \right\}$$

of A at $\mathfrak{p}$. Here we have the

(11.2) Corollary. *If $\mathfrak{p}$ is a prime ideal of A, then $A_{\mathfrak{p}}$ is a* **local ring**, *i.e., $A_{\mathfrak{p}}$ has a unique maximal ideal, namely $\mathfrak{m}_{\mathfrak{p}} = \mathfrak{p}A_{\mathfrak{p}}$. There is a canonical embedding*

$$A/\mathfrak{p} \hookrightarrow A_{\mathfrak{p}}/\mathfrak{m}_{\mathfrak{p}},$$

identifying $A_{\mathfrak{p}}/\mathfrak{m}_{\mathfrak{p}}$ with the field of fractions of $A/\mathfrak{p}$. In particular, if $\mathfrak{p}$ is a maximal ideal of A, then one has

$$A/\mathfrak{p}^n \cong A_{\mathfrak{p}}/\mathfrak{m}_{\mathfrak{p}}^n \quad \text{for } n \geq 1.$$

Proof: Since the ideals of $A_\mathfrak{p}$ correspond 1–1 to the ideals of A contained in $\mathfrak{p}$, the ideal $\mathfrak{m}_\mathfrak{p} = \mathfrak{p}A_\mathfrak{p}$ is the unique maximal ideal. Let us consider the homomorphism

$$f : A/\mathfrak{p}^n \longrightarrow A_\mathfrak{p}/\mathfrak{m}_\mathfrak{p}^n, \quad a \bmod \mathfrak{p}^n \longmapsto a \bmod \mathfrak{m}_\mathfrak{p}^n.$$

For $n = 1$, f is injective because $\mathfrak{p} = \mathfrak{m}_\mathfrak{p} \cap A$. Hence $A_\mathfrak{p}/\mathfrak{m}_\mathfrak{p}A_\mathfrak{p}$ becomes the field of fractions of $A/\mathfrak{p}$. Let $\mathfrak{p}$ be maximal and $n \geq 1$. For every $s \in A \smallsetminus \mathfrak{p}$ one has $\mathfrak{p}^n + sA = A$, i.e., $\bar{s} = s \bmod \mathfrak{p}^n$ is a unit in $A/\mathfrak{p}^n$. For $n = 1$ this is clear from the maximality of $\mathfrak{p}$, and for $n \geq 1$ it follows by induction: $A = \mathfrak{p}^{n-1} + sA \Rightarrow \mathfrak{p} = \mathfrak{p}A = \mathfrak{p}(\mathfrak{p}^{n-1} + sA) \subsetneqq \mathfrak{p}^n + sA \Rightarrow \mathfrak{p}^n + sA = A$.

Injectivity of f: let $a \in A$ be such that $a \in \mathfrak{m}_\mathfrak{p}^n$, i.e., $a = b/s$ with $b \in \mathfrak{p}^n$, $s \notin \mathfrak{p}$. Then $as = b \in \mathfrak{p}^n$, so that $\bar{a}\,\bar{s} = 0$ in $A/\mathfrak{p}^n$, and hence $\bar{a} = 0$ in $A/\mathfrak{p}^n$.

Surjectivity of f: let $a/s \in A_\mathfrak{p}$, $a \in A$, $s \notin \mathfrak{p}$. Then by the above, there exists an $a' \in A$ such that $a \equiv a's \bmod \mathfrak{p}^n$. Therefore $a/s \equiv a' \bmod \mathfrak{p}^n A_\mathfrak{p}$, i.e., $a/s \bmod \mathfrak{m}_\mathfrak{p}^n$ lies in the image of f. □

In a local ring with maximal ideal $\mathfrak{m}$, every element $a \notin \mathfrak{m}$ is a unit. Indeed, since the principal ideal (a) is not contained in any other maximal ideal, it has to be the whole ring. So we have

$$A^* = A \smallsetminus \mathfrak{m}.$$

The simplest local rings, except for fields, are **discrete valuation rings**.

(11.3) Definition. *A discrete valuation ring is a principal ideal domain $\mathcal{O}$ with a unique maximal ideal $\mathfrak{p} \neq 0$.*

The maximal ideal is of the form $\mathfrak{p} = (\pi) = \pi\mathcal{O}$, for some prime element π. Since every element not contained in $\mathfrak{p}$ is a unit, it follows that, up to associated elements, π is the only prime element of $\mathcal{O}$. Every nonzero element of $\mathcal{O}$ may therefore be written as $\varepsilon\pi^n$, for some $\varepsilon \in \mathcal{O}^*$, and $n \geq 0$. More generally, every element $a \neq 0$ of the field of fractions K may be uniquely written as

$$a = \varepsilon\pi^n, \quad \varepsilon \in \mathcal{O}^*, \quad n \in \mathbb{Z}.$$

The exponent n is called the **valuation** of a. It is denoted $v(a)$, and it is obviously characterized by the equation

$$(a) = \mathfrak{p}^{v(a)}.$$

The valuation is a function

$$v : K^* \longrightarrow \mathbb{Z}.$$

Extending it to K by the convention $v(0) = \infty$, a simple calculation shows that it satisfies the conditions

$$v(ab) = v(a) + v(b), \quad v(a+b) \geq \min\{v(a), v(b)\}.$$

This innocuous looking function gives rise to a theory which will occupy all of the next chapter.

The discrete valuation rings arise as localizations of Dedekind domains. This is a consequence of the

(11.4) Proposition. *If $\mathcal{O}$ is a Dedekind domain, and $S \subseteq \mathcal{O} \smallsetminus \{0\}$ is a multiplicative subset, then $\mathcal{O}S^{-1}$ is also a Dedekind domain.*

Proof: Let $\mathfrak{A}$ be an ideal of $\mathcal{O}S^{-1}$ and $\mathfrak{a} = \mathfrak{A} \cap \mathcal{O}$. Then $\mathfrak{A} = \mathfrak{a}S^{-1}$, because if $\frac{a}{s} \in \mathfrak{A}$, $a \in \mathcal{O}$ and $s \in S$, then one has $a = s \cdot \frac{a}{s} \in \mathfrak{A} \cap \mathcal{O} = \mathfrak{a}$, so that $\frac{a}{s} = a \cdot \frac{1}{s} \in \mathfrak{a}S^{-1}$. As $\mathfrak{a}$ is finitely generated, so is $\mathfrak{A}$, i.e., $\mathcal{O}S^{-1}$ is noetherian. It follows from (11.1) that every prime ideal of $\mathcal{O}S^{-1}$ is maximal, because this holds in $\mathcal{O}$. Finally, $\mathcal{O}S^{-1}$ is integrally closed, for if $x \in K$ satisfies the equation

$$x^n + \frac{a_1}{s_1}x^{n-1} + \cdots + \frac{a_n}{s_n} = 0$$

with coefficients $\frac{a_i}{s_i} \in \mathcal{O}S^{-1}$, then multiplying it with the n-th power of $s = s_1 \ldots s_n$ shows that sx is integral over $\mathcal{O}$, whence $sx \in \mathcal{O}$ and therefore $x \in \mathcal{O}S^{-1}$. This shows that $\mathcal{O}S^{-1}$ is a Dedekind domain. □

(11.5) Proposition. *Let $\mathcal{O}$ be a noetherian integral domain. $\mathcal{O}$ is a Dedekind domain if and only if, for all prime ideals $\mathfrak{p} \neq 0$, the localizations $\mathcal{O}_\mathfrak{p}$ are discrete valuation rings.*

Proof: If $\mathcal{O}$ is a Dedekind domain, then so are the localizations $\mathcal{O}_\mathfrak{p}$. The maximal ideal $\mathfrak{m} = \mathfrak{p}\mathcal{O}_\mathfrak{p}$ is the only nonzero prime ideal of $\mathcal{O}_\mathfrak{p}$. Therefore, choosing any $\pi \in \mathfrak{m} \smallsetminus \mathfrak{m}^2$, one necessarily finds $(\pi) = \mathfrak{m}$, and furthermore $\mathfrak{m}^n = (\pi^n)$. Thus $\mathcal{O}_\mathfrak{p}$ is a principal ideal domain, and hence a discrete valuation ring.

Letting $\mathfrak{p}$ vary over all prime ideals $\neq 0$ of $\mathcal{O}$, we find in any case that

$$\mathcal{O} = \bigcap_{\mathfrak{p}} \mathcal{O}_\mathfrak{p}.$$

For if $\frac{a}{b} \in \bigcap_\mathfrak{p} \mathcal{O}_\mathfrak{p}$, with $a, b \in \mathcal{O}$, then

$$\mathfrak{a} = \{x \in \mathcal{O} \mid xa \in b\mathcal{O}\}$$

is an ideal which cannot be contained in any prime ideal of $\mathcal{O}$. In fact, for any $\mathfrak{p}$, we may write $\frac{a}{b} = \frac{c}{s}$ with $c \in \mathcal{O}$, $s \notin \mathfrak{p}$, so that $sa = bc$, hence $s \in \mathfrak{a} \smallsetminus \mathfrak{p}$. As $\mathfrak{a}$ is not contained in any maximal ideal, it follows that $\mathfrak{a} = \mathcal{O}$, hence $a = 1 \cdot a \in b\,\mathcal{O}$, i.e., $\frac{a}{b} \in \mathcal{O}$.

Suppose now that the $\mathcal{O}_\mathfrak{p}$ are discrete valuation rings. Being principal ideal domains, they are integrally closed (see § 2), so $\mathcal{O} = \bigcap_\mathfrak{p} \mathcal{O}_\mathfrak{p}$ is also integrally closed. Finally, from (11.1) it follows that every prime ideal $\mathfrak{p} \neq 0$ of $\mathcal{O}$ is maximal because this is so in $\mathcal{O}_\mathfrak{p}$. Therefore $\mathcal{O}$ is a Dedekind domain. □

For a Dedekind domain $\mathcal{O}$, we have for each prime ideal $\mathfrak{p} \neq 0$ the discrete valuation ring $\mathcal{O}_\mathfrak{p}$ and the corresponding valuation

$$v_\mathfrak{p} : K^* \longrightarrow \mathbb{Z}$$

of the field of fractions. The significance of these valuations lies in their relation to the prime ideal factorization. If $x \in K^*$ and

$$(x) = \prod_\mathfrak{p} \mathfrak{p}^{\nu_\mathfrak{p}}$$

is the prime factorization of the principal ideal (x), then, for each $\mathfrak{p}$, one has

$$\nu_\mathfrak{p} = v_\mathfrak{p}(x).$$

In fact, for a fixed prime ideal $\mathfrak{q} \neq 0$ of $\mathcal{O}$, the first equation above implies (because $\mathfrak{p}\,\mathcal{O}_\mathfrak{q} = \mathcal{O}_\mathfrak{q}$ for $\mathfrak{p} \neq \mathfrak{q}$) that

$$x\mathcal{O}_\mathfrak{q} = \Big(\prod_\mathfrak{p} \mathfrak{p}^{\nu_\mathfrak{p}}\Big)\,\mathcal{O}_\mathfrak{q} = \mathfrak{q}^{\nu_\mathfrak{q}}\mathcal{O}_\mathfrak{q} = \mathfrak{m}_\mathfrak{q}^{\nu_\mathfrak{q}}.$$

Hence indeed $v_\mathfrak{q}(x) = \nu_\mathfrak{q}$. In view of this relation, the valuations $v_\mathfrak{p}$ are also called **exponential valuations**.

The reader should check that the localization of the ring $\mathbb{Z}$ at the prime ideal $(p) = p\mathbb{Z}$ is given by

$$\mathbb{Z}_{(p)} = \Big\{\frac{a}{b} \mid a, b \in \mathbb{Z},\ p \nmid b\Big\}.$$

The maximal ideal $p\mathbb{Z}_{(p)}$ consists of all fractions a/b satisfying $p \mid a$, $p \nmid b$, and the group of units consists of all fractions a/b satisfying $p \nmid ab$. The valuation associated to $\mathbb{Z}_{(p)}$,

$$v_p : \mathbb{Q} \longrightarrow \mathbb{Z} \cup \{\infty\},$$

is called the ***p*-adic valuation** of $\mathbb{Q}$. The valuation $v_p(x)$ of an element $x \in \mathbb{Q}^*$ is given by

$$v_p(x) = \nu,$$

where $x = p^\nu a/b$ with integers a, b relatively prime to p.

To end this section, we now want to compare a Dedekind domain $\mathcal{O}$ to the ring

$$\mathcal{O}(X) = \left\{ \frac{f}{g} \mid f, g \in \mathcal{O},\ g \not\equiv 0 \bmod \mathfrak{p} \text{ for } \mathfrak{p} \in X \right\},$$

where X is a set of prime ideals $\neq 0$ of $\mathcal{O}$ which contains almost all prime ideals of $\mathcal{O}$. By (11.1), the prime ideals $\neq 0$ of $\mathcal{O}(X)$ are given as $\mathfrak{p}_X = \mathfrak{p}\mathcal{O}(X)$, for $\mathfrak{p} \in X$, and it is easily checked that $\mathcal{O}$ and $\mathcal{O}(X)$ have the same localizations

$$\mathcal{O}_\mathfrak{p} = \mathcal{O}(X)_{\mathfrak{p}_X}.$$

We denote by $Cl(\mathcal{O})$, resp. $Cl(\mathcal{O}(X))$, the ideal class groups of $\mathcal{O}$, resp. $\mathcal{O}(X)$. They, as well as the groups of units $\mathcal{O}^*$ and $\mathcal{O}(X)^*$, are related by the following

(11.6) Proposition. *There is a canonical exact sequence*

$$1 \longrightarrow \mathcal{O}^* \longrightarrow \mathcal{O}(X)^* \longrightarrow \bigoplus_{\mathfrak{p} \notin X} K^*/\mathcal{O}_\mathfrak{p}^* \longrightarrow Cl(\mathcal{O}) \longrightarrow Cl(\mathcal{O}(X)) \longrightarrow 1,$$

and one has $K^*/\mathcal{O}_\mathfrak{p}^* \cong \mathbb{Z}$.

Proof: The first arrow is inclusion and the second one is induced by the inclusion $\mathcal{O}(X)^* \to K^*$, followed by the projections $K^* \to K^*/\mathcal{O}_\mathfrak{p}^*$. If $a \in \mathcal{O}(X)^*$ belongs to the kernel, then $a \in \mathcal{O}_\mathfrak{p}$ for $\mathfrak{p} \notin X$, and also for $\mathfrak{p} \in X$ because $\mathcal{O}_\mathfrak{p} = \mathcal{O}(X)_{\mathfrak{p}_X}$, hence $a \in \bigcap_\mathfrak{p} \mathcal{O}_\mathfrak{p}^* = \mathcal{O}^*$ (see the argument in the proof of (11.5)). This shows the exactness at $\mathcal{O}(X)^*$. The arrow

$$\bigoplus_{\mathfrak{p} \notin X} K^*/\mathcal{O}_\mathfrak{p}^* \longrightarrow Cl(\mathcal{O})$$

is induced by mapping

$$\bigoplus_{\mathfrak{p} \notin X} \alpha_\mathfrak{p} \bmod \mathcal{O}_\mathfrak{p}^* \longmapsto \prod_{\mathfrak{p} \notin X} \mathfrak{p}^{v_\mathfrak{p}(\alpha_\mathfrak{p})},$$

where $v_\mathfrak{p} : K^* \to \mathbb{Z}$ is the exponential valuation of K associated to $\mathcal{O}_\mathfrak{p}$. Let $\bigoplus_{\mathfrak{p} \notin X} \alpha_\mathfrak{p} \bmod \mathcal{O}_\mathfrak{p}^*$ be an element in the kernel, i.e.,

$$\prod_{\mathfrak{p} \notin X} \mathfrak{p}^{v_\mathfrak{p}(\alpha_\mathfrak{p})} = (\alpha) = \prod_\mathfrak{p} \mathfrak{p}^{v_\mathfrak{p}(\alpha)},$$

for some $\alpha \in K^*$. Because of unique prime factorization, this means that $v_\mathfrak{p}(\alpha) = 0$ for $\mathfrak{p} \in X$, and $v_\mathfrak{p}(\alpha_\mathfrak{p}) = v_\mathfrak{p}(\alpha)$ for $\mathfrak{p} \notin X$. It follows that $\alpha \in \bigcap_{\mathfrak{p} \in X} \mathcal{O}_\mathfrak{p}^* = \mathcal{O}(X)^*$ and $\alpha \equiv \alpha_\mathfrak{p} \bmod \mathcal{O}_\mathfrak{p}^*$. This shows exactness in the middle. The arrow

$$Cl(\mathcal{O}) \longrightarrow Cl(\mathcal{O}(X))$$

comes from mapping $\mathfrak{a} \mapsto \mathfrak{a}\mathcal{O}(X)$. The classes of prime ideals $\mathfrak{p} \in X$ are mapped onto the classes of prime ideals of $\mathcal{O}(X)$. Since $Cl(\mathcal{O}(X))$ is generated by these classes, the arrow is surjective. For $\mathfrak{p} \notin X$ we have $\mathfrak{p}\mathcal{O}(X) = (1)$, and this means that the kernel consists of the classes of the ideals $\prod_{\mathfrak{p} \notin X} \mathfrak{p}^{\nu_\mathfrak{p}}$. This, however, is visibly the image of the preceding arrow. Therefore the whole sequence is exact. Finally, the valuation $v_\mathfrak{p} : K^* \to \mathbb{Z}$ produces the isomorphism $K^*/\mathcal{O}_\mathfrak{p}^* \cong \mathbb{Z}$. □

For the ring of integers $\mathcal{O}_K$ of an algebraic number field K, the proposition yields the following results. Let S denote a finite set of prime ideals of $\mathcal{O}_K$ (not any more a multiplicative subset), and let X be the set of all prime ideals that do not belong to S. We put

$$\mathcal{O}_K^S = \mathcal{O}_K(X).$$

The units of this ring are called the **S-units**, and the group $Cl_K^S = Cl(\mathcal{O}_K^S)$ the **S-class group** of K.

(11.7) Corollary. *For the group $K^S = (\mathcal{O}_K^S)^*$ of S-units of K there is an isomorphism*

$$K^S \cong \mu(K) \times \mathbb{Z}^{\#S+r+s-1},$$

where r and s are defined as in § 5, p. 30.

Proof: The torsion subgroup of K^S is the group $\mu(K)$ of roots of unity in K. Since $Cl(\mathcal{O})$ is finite, we obtain the following identities from the exact sequence (11.6) and from (7.4):

$$\operatorname{rank}(K^S) = \operatorname{rank}(\mathcal{O}_K^*) + \operatorname{rank}\Big(\bigoplus_{\mathfrak{p}\in S} \mathbb{Z}\Big) = \#S + r + s - 1.$$

This proves the corollary. □

(11.8) Corollary. *The S-class group $Cl_K^S = Cl(\mathcal{O}_K^S)$ is finite.*

Exercise 1. Let A be an arbitrary ring, not necessarily an integral domain, let M be an A-module and S a multiplicatively closed subset of A such that $0 \notin S$. In $M \times S$ consider the equivalence relation

$$(m, s) \sim (m', s') \iff \exists s'' \in S \text{ such that } s''(s'm - sm') = 0.$$

Show that the set M_S of equivalence classes $\overline{(m, s)}$ forms an A-module, and that $M \to M_S$, $a \mapsto \overline{(a, 1)}$, is a homomorphism. In particular, A_S is a ring. It is called the **localization** of A with respect to S.

Exercise 2. Show that, in the above situation, the prime ideals of A_S correspond 1–1 to the prime ideals of A which are disjoint from S. If $\mathfrak{p} \subseteq A$ and $\mathfrak{p}_S \subseteq A_S$ correspond in this way, then $A_S/\mathfrak{p}_S$ is the localization of $A/\mathfrak{p}$ with respect to the image of S.

Exercise 3. Let $f : M \to N$ be a homomorphism of A-modules. Then the following conditions are equivalent:

(i) f is injective (surjective).

(ii) $f_\mathfrak{p} : M_\mathfrak{p} \to N_\mathfrak{p}$ is injective (surjective) for every prime ideal $\mathfrak{p}$.

(iii) $f_\mathfrak{m} : M_\mathfrak{m} \to N_\mathfrak{m}$ is injective (surjective) for every maximal ideal $\mathfrak{m}$.

Exercise 4. Let S and T be two multiplicative subsets of A, and T^* the image of T in A_S. Then one has $A_{ST} \cong (A_S)_{T^*}$.

Exercise 5. Let $f : A \to B$ be a homomorphism of rings and S a multiplicatively closed subset such that $f(S) \subseteq B^*$. Then f induces a homomorphism $A_S \to B$.

Exercise 6. Let A be an integral domain. If the localization A_S is integral over A, then $A_S = A$.

Exercise 7 (Nakayama's Lemma). Let A be a local ring with maximal ideal $\mathfrak{m}$, let M be an A-module and $N \subseteq M$ a submodule such that M/N is finitely generated. Then one has the implication:

$$M = N + \mathfrak{m}M \implies M = N .$$

§ 12. Orders

The ring $\mathcal{O}_K$ of integers of an algebraic number field K is our chief interest because of its excellent property of being a Dedekind domain. Due to important theoretical as well as practical circumstances, however, one is pushed to devise a theory of greater generality which comprises also the theory of rings of algebraic integers which, like the ring

$$\mathcal{O} = \mathbb{Z} + \mathbb{Z}\sqrt{5} \subseteq \mathbb{Q}(\sqrt{5}),$$

are not necessarily integrally closed. These rings are the so-called **orders**.

(12.1) Definition. *Let $K|\mathbb{Q}$ be an algebraic number field of degree n. An order of K is a subring $\mathcal{O}$ of $\mathcal{O}_K$ which contains an integral basis of length n. The ring $\mathcal{O}_K$ is called the* **maximal order** *of K.*

In concrete terms, orders are obtained as rings of the form

$$\mathcal{O} = \mathbb{Z}[\alpha_1, \ldots, \alpha_r],$$

where $\alpha_1, \ldots, \alpha_r$ are integers such that $K = \mathbb{Q}(\alpha_1, \ldots, \alpha_r)$. Being a submodule of the free $\mathbb{Z}$-module $\mathcal{O}_K$, $\mathcal{O}$ does of course admit a $\mathbb{Z}$-basis which, as $\mathbb{Q}\mathcal{O} = K$, has to be at the same time a basis of $K \mid \mathbb{Q}$, and therefore has length n. Orders arise often as rings of multipliers, and as such have their practical applications. For instance, if $\alpha_1, \ldots, \alpha_n$ is any basis of $K \mid \mathbb{Q}$ and $M = \mathbb{Z}\alpha_1 + \cdots + \mathbb{Z}\alpha_n$, then

$$\mathcal{O}_M = \{\alpha \in K \mid \alpha M \subseteq M\}$$

is an order. The theoretical significance of orders, however, lies in the fact that they admit "singularities", which are excluded as long as only Dedekind domains with their "regular" localizations $\mathcal{O}_{\mathfrak{p}}$ are considered. We will explain what this means in the next section.

In the preceding section we studied the localizations of a Dedekind domain $\mathcal{O}_K$. They are extension rings of $\mathcal{O}_K$ which are integrally closed, yet no longer integral over $\mathbb{Z}$. Now we study orders. They are subrings of $\mathcal{O}_K$ which are integral over $\mathbb{Z}$, yet no longer integrally closed. As a common generalization of both types of rings let us consider for now all **one-dimensional noetherian integral domains**. These are the noetherian integral domains in which every prime ideal $\mathfrak{p} \neq 0$ is a maximal ideal. The term "one-dimensional" refers to the general definition of the **Krull dimension** of a ring as being the maximal length d of a chain of prime ideals $\mathfrak{p}_0 \subsetneqq \mathfrak{p}_1 \subsetneqq \cdots \subsetneqq \mathfrak{p}_d$.

(12.2) Proposition. *An order $\mathcal{O}$ of K is a one-dimensional noetherian integral domain.*

Proof: Since $\mathcal{O}$ is a finitely generated $\mathbb{Z}$-module of rank $n = [K : \mathbb{Q}]$, every ideal $\mathfrak{a}$ is also a finitely generated $\mathbb{Z}$-module, and *a fortiori* a finitely generated $\mathcal{O}$-module. This shows that $\mathcal{O}$ is noetherian. If $\mathfrak{p} \neq 0$ is a prime ideal and $a \in \mathfrak{p} \cap \mathbb{Z}$, $a \neq 0$, then $a\mathcal{O} \subseteq \mathfrak{p} \subseteq \mathcal{O}$, i.e., $\mathfrak{p}$ and $\mathcal{O}$ have the same rank n. Therefore $\mathcal{O}/\mathfrak{p}$ is a finite integral domain, hence a field, and thus $\mathfrak{p}$ is a maximal ideal. □

In what follows, we always let $\mathcal{O}$ be a one-dimensional noetherian integral domain and K its field of fractions. We set out by proving the following stronger version of the Chinese remainder theorem.

(12.3) Proposition. *If $\mathfrak{a} \neq 0$ is an ideal of $\mathcal{O}$, then*

$$\mathcal{O}/\mathfrak{a} \cong \bigoplus_{\mathfrak{p}} \mathcal{O}_{\mathfrak{p}}/\mathfrak{a}\mathcal{O}_{\mathfrak{p}} = \bigoplus_{\mathfrak{p}\supseteq\mathfrak{a}} \mathcal{O}_{\mathfrak{p}}/\mathfrak{a}\mathcal{O}_{\mathfrak{p}} .$$

Proof: Let $\tilde{\mathfrak{a}}_{\mathfrak{p}} = \mathcal{O} \cap \mathfrak{a}\mathcal{O}_{\mathfrak{p}}$. For almost all $\mathfrak{p}$ one has $\mathfrak{p} \not\supseteq \mathfrak{a}$ and therefore $\mathfrak{a}\mathcal{O}_{\mathfrak{p}} = \mathcal{O}_{\mathfrak{p}}$, hence $\tilde{\mathfrak{a}}_{\mathfrak{p}} = \mathcal{O}$. Furthermore, one has $\mathfrak{a} = \bigcap_{\mathfrak{p}} \tilde{\mathfrak{a}}_{\mathfrak{p}} = \bigcap_{\mathfrak{p}\supseteq\mathfrak{a}} \tilde{\mathfrak{a}}_{\mathfrak{p}}$. Indeed, for any $a \in \bigcap_{\mathfrak{p}} \tilde{\mathfrak{a}}_{\mathfrak{p}}$, the ideal $\mathfrak{b} = \{x \in \mathcal{O} \mid xa \in \mathfrak{a}\}$ does not belong to any of the maximal ideals $\mathfrak{p}$ (in fact, one has $s_{\mathfrak{p}}a \in \mathfrak{a}$ for any $s_{\mathfrak{p}} \notin \mathfrak{p}$), consequently, $\mathfrak{b} = \mathcal{O}$, i.e., $a = 1 \cdot a \in \mathfrak{a}$, as claimed. (11.1) implies that, if $\mathfrak{p} \supseteq \mathfrak{a}$, then $\mathfrak{p}$ is the only prime ideal containing $\tilde{\mathfrak{a}}_{\mathfrak{p}}$. Therefore, given two distinct prime ideals $\mathfrak{p}$ and $\mathfrak{q}$ of $\mathcal{O}$, the ideal $\tilde{\mathfrak{a}}_{\mathfrak{p}}+\tilde{\mathfrak{a}}_{\mathfrak{q}}$ cannot be contained in any maximal ideal, whence $\tilde{\mathfrak{a}}_{\mathfrak{p}} + \tilde{\mathfrak{a}}_{\mathfrak{q}} = \mathcal{O}$. The Chinese remainder theorem (3.6) now gives the isomorphism

$$\mathcal{O}/\mathfrak{a} \cong \bigoplus_{\mathfrak{p}\supseteq\mathfrak{a}} \mathcal{O}/\tilde{\mathfrak{a}}_{\mathfrak{p}} ,$$

and we have $\mathcal{O}/\tilde{\mathfrak{a}}_{\mathfrak{p}} = \mathcal{O}_{\mathfrak{p}}/\mathfrak{a}\mathcal{O}_{\mathfrak{p}}$, because $\bar{\mathfrak{p}} = \mathfrak{p} \bmod \tilde{\mathfrak{a}}_{\mathfrak{p}}$ is the only maximal ideal of $\mathcal{O}/\tilde{\mathfrak{a}}_{\mathfrak{p}}$. □

For the ring $\mathcal{O}$, the fractional ideals of $\mathcal{O}$, in other words, the finitely generated nonzero $\mathcal{O}$-submodules of the field of fractions K, no longer form a group – unless $\mathcal{O}$ happens to be Dedekind. The way out is to restrict attention to the **invertible ideals**, i.e., to those fractional ideals $\mathfrak{a}$ of $\mathcal{O}$ for which there exists a fractional ideal $\mathfrak{b}$ such that

$$\mathfrak{a}\mathfrak{b} = \mathcal{O} .$$

These form an abelian group, for trivial reasons. The inverse of $\mathfrak{a}$ is still the fractional ideal

$$\mathfrak{a}^{-1} = \{x \in K \mid x\mathfrak{a} \subseteq \mathcal{O}\} ,$$

because it is the biggest ideal such that $\mathfrak{a}\mathfrak{a}^{-1} \subseteq \mathcal{O}$. The invertible ideals of $\mathcal{O}$ may be characterized as those fractional ideals which are "locally" principal:

(12.4) Proposition. *A fractional ideal $\mathfrak{a}$ of $\mathcal{O}$ is invertible if and only if, for every prime ideal $\mathfrak{p} \neq 0$,*

$$\mathfrak{a}_{\mathfrak{p}} = \mathfrak{a}\mathcal{O}_{\mathfrak{p}}$$

is a fractional principal ideal of $\mathcal{O}_{\mathfrak{p}}$.

Proof: Let $\mathfrak{a}$ be an invertible ideal and $\mathfrak{a}\mathfrak{b} = \mathcal{O}$. Then $1 = \sum_{i=1}^{r} a_i b_i$ with $a_i \in \mathfrak{a}$, $b_i \in \mathfrak{b}$, and not all $a_i b_i \in \mathcal{O}_\mathfrak{p}$ can lie in the maximal ideal $\mathfrak{p}\mathcal{O}_\mathfrak{p}$. Suppose $a_1 b_1$ is a unit in $\mathcal{O}_\mathfrak{p}$. Then $\mathfrak{a}_\mathfrak{p} = a_1 \mathcal{O}_\mathfrak{p}$ because, for $x \in \mathfrak{a}_\mathfrak{p}$, $x b_1 \in \mathfrak{a}_\mathfrak{p}\mathfrak{b} = \mathcal{O}_\mathfrak{p}$, hence $x = x b_1 (b_1 a_1)^{-1} a_1 \in a_1 \mathcal{O}_\mathfrak{p}$.

Conversely, assume $\mathfrak{a}_\mathfrak{p} = \mathfrak{a}\mathcal{O}_\mathfrak{p}$ is a principal ideal $a_\mathfrak{p}\mathcal{O}_\mathfrak{p}$, $a_\mathfrak{p} \in K^*$, for every $\mathfrak{p}$. Then we may and do assume that $a_\mathfrak{p} \in \mathfrak{a}$. We claim that the fractional ideal $\mathfrak{a}^{-1} = \{x \in K \mid x\mathfrak{a} \subseteq \mathcal{O}\}$ is an inverse for $\mathfrak{a}$. If this were not the case, then we would have a maximal ideal $\mathfrak{p}$ such that $\mathfrak{a}\mathfrak{a}^{-1} \subseteq \mathfrak{p} \subset \mathcal{O}$. Let $a_1, \dots, a_n$ be generators of $\mathfrak{a}$. As $a_i \in a_\mathfrak{p}\mathcal{O}_\mathfrak{p}$, we may write $a_i = a_\mathfrak{p}\frac{b_i}{s_i}$, with $b_i \in \mathcal{O}$, $s_i \in \mathcal{O} \smallsetminus \mathfrak{p}$. Then $s_i a_i \in a_\mathfrak{p}\mathcal{O}$. Putting $s = s_1 \cdots s_n$, we have $s a_i \in a_\mathfrak{p}\mathcal{O}$ for $i = 1, \dots, n$, hence $s a_\mathfrak{p}^{-1}\mathfrak{a} \subseteq \mathcal{O}$ and therefore $s a_\mathfrak{p}^{-1} \in \mathfrak{a}^{-1}$. Consequently, $s = s a_\mathfrak{p}^{-1} a_\mathfrak{p} \in \mathfrak{a}^{-1}\mathfrak{a} \subseteq \mathfrak{p}$, a contradiction. □

We denote the group of invertible ideals of $\mathcal{O}$ by $J(\mathcal{O})$. It contains the group $P(\mathcal{O})$ of fractional principal ideals $a\mathcal{O}$, $a \in K^*$.

(12.5) Definition. *The quotient group*

$$Pic(\mathcal{O}) = J(\mathcal{O})/P(\mathcal{O})$$

is called the **Picard group** *of the ring* $\mathcal{O}$.

In the case where $\mathcal{O}$ is a Dedekind domain, the Picard group is of course nothing but the ideal class group Cl_K. In general, we have the following description for $J(\mathcal{O})$ and $Pic(\mathcal{O})$.

(12.6) Proposition. *The correspondence* $\mathfrak{a} \mapsto (\mathfrak{a}_\mathfrak{p}) = (\mathfrak{a}\mathcal{O}_\mathfrak{p})$ *yields an isomorphism*

$$J(\mathcal{O}) \cong \bigoplus_\mathfrak{p} P(\mathcal{O}_\mathfrak{p}).$$

Identifying the subgroup $P(\mathcal{O})$ *with its image in the direct sum one gets*

$$Pic(\mathcal{O}) \cong \Big(\bigoplus_\mathfrak{p} P(\mathcal{O}_\mathfrak{p})\Big)/P(\mathcal{O}).$$

Proof: For every $\mathfrak{a} \in J(\mathcal{O})$, $\mathfrak{a}_\mathfrak{p} = \mathfrak{a}\mathcal{O}_\mathfrak{p}$ is a principal ideal by (12.4), and we have $\mathfrak{a}_\mathfrak{p} = \mathcal{O}_\mathfrak{p}$ for almost all $\mathfrak{p}$ because $\mathfrak{a}$ lies in only finitely many maximal ideals $\mathfrak{p}$. We therefore obtain a homomorphism

$$J(\mathcal{O}) \longrightarrow \bigoplus_\mathfrak{p} P(\mathcal{O}_\mathfrak{p}), \quad \mathfrak{a} \longmapsto (\mathfrak{a}_\mathfrak{p}).$$

It is injective, for if $\mathfrak{a}_\mathfrak{p} = \mathcal{O}_\mathfrak{p}$ for all $\mathfrak{p}$, then $\mathfrak{a} \subseteq \bigcap_\mathfrak{p} \mathcal{O}_\mathfrak{p} = \mathcal{O}$ (see the proof of (11.5)), and one has to have $\mathfrak{a} = \mathcal{O}$ because otherwise there would exist a maximal ideal $\mathfrak{p}$ such that $\mathfrak{a} \subseteq \mathfrak{p} \subset \mathcal{O}$, i.e., $\mathfrak{a}_\mathfrak{p} \subseteq \mathfrak{p}\mathcal{O}_\mathfrak{p} \neq \mathcal{O}_\mathfrak{p}$. In order to prove surjectivity, let $(a_\mathfrak{p}\mathcal{O}_\mathfrak{p}) \in \bigoplus_\mathfrak{p} P(\mathcal{O}_\mathfrak{p})$ be given. Then the $\mathcal{O}$-submodule

$$\mathfrak{a} = \bigcap_\mathfrak{p} a_\mathfrak{p}\mathcal{O}_\mathfrak{p}$$

of K is a fractional ideal. Indeed, since $a_\mathfrak{p}\mathcal{O}_\mathfrak{p} = \mathcal{O}_\mathfrak{p}$ for almost all $\mathfrak{p}$, there is some $c \in \mathcal{O}$ such that $ca_\mathfrak{p} \in \mathcal{O}_\mathfrak{p}$ for all $\mathfrak{p}$, i.e., $c\mathfrak{a} \subseteq \bigcap_\mathfrak{p} \mathcal{O}_\mathfrak{p} = \mathcal{O}$. We have to show that one has

$$\mathfrak{a}\mathcal{O}_\mathfrak{p} = a_\mathfrak{p}\mathcal{O}_\mathfrak{p}$$

for every $\mathfrak{p}$. The inclusion $\subseteq$ is trivial. In order to show that $a_\mathfrak{p}\mathcal{O}_\mathfrak{p} \subseteq \mathfrak{a}\mathcal{O}_\mathfrak{p}$, let us choose $c \in \mathcal{O}$, $c \neq 0$, such that $ca_\mathfrak{p}^{-1}a_\mathfrak{q} \in \mathcal{O}$ for the finitely many $\mathfrak{q}$ which satisfy $a_\mathfrak{p}^{-1}a_\mathfrak{q} \notin \mathcal{O}_\mathfrak{q}$. By the Chinese remainder theorem (12.3), we may find $a \in \mathcal{O}$ such that

$$a \equiv c \bmod \mathfrak{p} \quad \text{and} \quad a \in ca_\mathfrak{p}^{-1}a_\mathfrak{q}\mathcal{O}_\mathfrak{q} \quad \text{for} \quad \mathfrak{q} \neq \mathfrak{p}.$$

Then $\varepsilon = ac^{-1}$ is a unit in $\mathcal{O}_\mathfrak{p}$ and $a_\mathfrak{p}\varepsilon \in \bigcap_\mathfrak{q} a_\mathfrak{q}\mathcal{O}_\mathfrak{q} = \mathfrak{a}$, hence

$$a_\mathfrak{p}\mathcal{O}_\mathfrak{p} = (a_\mathfrak{p}\varepsilon)\mathcal{O}_\mathfrak{p} \subseteq \mathfrak{a}\mathcal{O}_\mathfrak{p}. \qquad \square$$

Passing from the ring $\mathcal{O}$ to its **normalization** $\tilde{\mathcal{O}}$, i.e., to the integral closure of $\mathcal{O}$ in K, one obtains a Dedekind domain. This is not all that easy to prove, however, because $\tilde{\mathcal{O}}$ is in general not a finitely generated $\mathcal{O}$-module. But at any rate we have the

(12.7) Lemma. *Let $\mathcal{O}$ be a one-dimensional noetherian integral domain and $\tilde{\mathcal{O}}$ its normalization. Then, for each ideal $\mathfrak{a} \neq 0$ of $\mathcal{O}$, the quotient $\tilde{\mathcal{O}}/\mathfrak{a}\tilde{\mathcal{O}}$ is a finitely generated $\mathcal{O}$-module.*

Proof: Let $a \in \mathfrak{a}$, $a \neq 0$. Then $\tilde{\mathcal{O}}/\mathfrak{a}\tilde{\mathcal{O}}$ is a quotient of $\tilde{\mathcal{O}}/a\tilde{\mathcal{O}}$. It thus suffices to show that $\tilde{\mathcal{O}}/a\tilde{\mathcal{O}}$ is a finitely generated $\mathcal{O}$-module. With this end, consider in $\mathcal{O}$ the descending chain of ideals containing $a\mathcal{O}$

$$\mathfrak{a}_m = (a^m\tilde{\mathcal{O}} \cap \mathcal{O}, a\mathcal{O}).$$

This chain becomes stationary. In fact, the prime ideals of the ring $\mathcal{O}/a\mathcal{O}$ are not only maximal but also minimal in the sense that $\mathcal{O}/a\mathcal{O}$ is a zero-dimensional noetherian ring. In such a ring every descending chain of ideals becomes stationary (see §3, exercise 7). If the chain $\bar{\mathfrak{a}}_m = \mathfrak{a}_m \bmod a\mathcal{O}$ is stationary at n, then so is the chain $\mathfrak{a}_m$. We show that, for this n, we have

$$\tilde{\mathcal{O}} \subseteq a^{-n}\mathcal{O} + a\tilde{\mathcal{O}}.$$

Let $\beta = \frac{b}{c} \in \tilde{\mathcal{O}}$, $b, c \in \mathcal{O}$. Apply the descending chain condition to the ring $\mathcal{O}/c\mathcal{O}$ and the chain of ideals $(\bar{a}^m)$, where $\bar{a} = a \bmod c\mathcal{O}$. Then $(\bar{a}^h) = (\bar{a}^{h+1})$, i.e., we find some $x \in \mathcal{O}$ such that $a^h \equiv xa^{h+1} \bmod c\mathcal{O}$, hence $(1 - xa)a^h \in c\mathcal{O}$, and therefore

$$\beta = \frac{b}{c}(1 - xa) + \beta xa = \frac{b}{a^h}\frac{(1 - xa)a^h}{c} + \beta xa \in a^{-h}\mathcal{O} + a\tilde{\mathcal{O}}.$$

Let h be the smallest positive integer such that $\beta \in a^{-h}\mathcal{O} + a\tilde{\mathcal{O}}$. It then suffices to show that $h \le n$. Assume $h > n$. Writing

$$(*) \qquad \beta = \frac{u}{a^h} + a\tilde{u} \quad \text{with } u \in \mathcal{O},\ \tilde{u} \in \tilde{\mathcal{O}},$$

we have $u = a^h(\beta - a\tilde{u}) \in a^h\tilde{\mathcal{O}} \cap \mathcal{O} \subseteq \mathfrak{a}_h = \mathfrak{a}_{h+1}$ because $h > n$, hence $u = a^{h+1}\tilde{u}' + au'$, $u' \in \mathcal{O}$, $\tilde{u}' \in \tilde{\mathcal{O}}$. Substituting this into $(*)$ gives

$$\beta = \frac{u'}{a^{h-1}} + a(\tilde{u} + \tilde{u}') \in a^{1-h}\mathcal{O} + a\tilde{\mathcal{O}}.$$

This contradicts the minimality of h. So we do have $\tilde{\mathcal{O}} \subseteq a^{-n}\mathcal{O} + a\tilde{\mathcal{O}}$.

$\tilde{\mathcal{O}}/a\tilde{\mathcal{O}}$ thus becomes a submodule of the $\mathcal{O}$-module $(a^{-n}\mathcal{O} + a\tilde{\mathcal{O}})/a\tilde{\mathcal{O}}$ generated by $a^{-n} \bmod a\tilde{\mathcal{O}}$. It is therefore itself a finitely generated $\mathcal{O}$-module, q.e.d. □

(12.8) Proposition (KRULL-AKIZUKI). *Let $\mathcal{O}$ be a one-dimensional noetherian integral domain with field of fractions K. Let $L|K$ be a finite extension and $\mathcal{O}$ the integral closure of $\mathcal{O}$ in L. Then $\mathcal{O}$ is a Dedekind domain.*

Proof: The facts that $\mathcal{O}$ is integrally closed and that every nonzero prime ideal is maximal, are deduced as in (3.1). It remains to show that $\mathcal{O}$ is noetherian. Let $\omega_1, \dots, \omega_n$ be a basis of $L|K$ which is contained in $\mathcal{O}$. Then the ring $\mathcal{O}_0 = \mathcal{O}[\omega_1, \dots, \omega_n]$ is a finitely generated $\mathcal{O}$-module and in particular is noetherian since $\mathcal{O}$ is noetherian. We argue as before that $\mathcal{O}_0$ is one-dimensional and are thus reduced to the case $L = K$. So let $\mathfrak{A}$ be an ideal of $\mathcal{O}$ and $a \in \mathfrak{A} \cap \mathcal{O}$, $a \neq 0$; then by the above lemma $\mathcal{O}/a\mathcal{O}$ is a finitely generated $\mathcal{O}$-module. Since $\mathcal{O}$ is noetherian, so is the $\mathcal{O}$-submodule $\mathfrak{A}/a\mathcal{O}$, and also the $\mathcal{O}$-module $\mathfrak{A}$. □

Remark: The above proof is taken from KAPLANSKY's book [82] (see also [101]). It shows at the same time that proposition (8.1), which we had proved only in the case of a separable extension $L|K$, is valid for general finite extensions of the field of fractions of a Dedekind domain.

Next we want to compare the one-dimensional noetherian integral domain $\mathcal{O}$ with its normalization $\tilde{\mathcal{O}}$. The fact that $\tilde{\mathcal{O}}$ is a Dedekind domain is evident and does not require the lengthy proof of (12.8) provided we make the following hypothesis:

(∗) $\mathcal{O}$ is an integral domain whose normalization $\tilde{\mathcal{O}}$ is a finitely generated $\mathcal{O}$-module.

This condition will be assumed for all that follows. It avoids pathological situations and is satisfied in all interesting cases, in particular for the orders in an algebraic number field.

The groups of units and the Picard groups of $\mathcal{O}$ and $\tilde{\mathcal{O}}$ are compared with each other by the following

(12.9) Proposition. *One has the canonical exact sequence*

$$1 \longrightarrow \mathcal{O}^* \longrightarrow \tilde{\mathcal{O}}^* \longrightarrow \bigoplus_{\mathfrak{p}} \tilde{\mathcal{O}}_{\mathfrak{p}}^*/\mathcal{O}_{\mathfrak{p}}^* \longrightarrow Pic(\mathcal{O}) \longrightarrow Pic(\tilde{\mathcal{O}}) \longrightarrow 1.$$

In the sum, $\mathfrak{p}$ varies over the prime ideals $\neq 0$ of $\mathcal{O}$ and $\tilde{\mathcal{O}}_{\mathfrak{p}}$ denotes the integral closure of $\mathcal{O}_{\mathfrak{p}}$ in K.

Proof: If $\tilde{\mathfrak{p}}$ varies over the prime ideals of $\tilde{\mathcal{O}}$, we know from (12.6) that

$$J(\tilde{\mathcal{O}}) \cong \bigoplus_{\tilde{\mathfrak{p}}} P(\tilde{\mathcal{O}}_{\tilde{\mathfrak{p}}}).$$

If $\mathfrak{p}$ is a prime ideal of $\mathcal{O}$, then $\mathfrak{p}\tilde{\mathcal{O}}$ splits in the Dedekind domain $\tilde{\mathcal{O}}$ into a product

$$\mathfrak{p}\tilde{\mathcal{O}} = \tilde{\mathfrak{p}}_1^{e_1} \cdots \tilde{\mathfrak{p}}_r^{e_r},$$

i.e., there are only finitely many prime ideals of $\tilde{\mathcal{O}}$ above $\mathfrak{p}$. The same holds for the integral closure $\tilde{\mathcal{O}}_{\mathfrak{p}}$ of $\mathcal{O}_{\mathfrak{p}}$. Since every nonzero prime ideal of $\tilde{\mathcal{O}}_{\mathfrak{p}}$ has to lie above $\mathfrak{p}\mathcal{O}_{\mathfrak{p}}$, the localization $\tilde{\mathcal{O}}_{\mathfrak{p}}$ has only a finite number of prime ideals and is therefore a principal ideal domain (see § 3, exercise 4). In view of (12.6), it follows that

$$P(\tilde{\mathcal{O}}_{\mathfrak{p}}) = J(\tilde{\mathcal{O}}_{\mathfrak{p}}) \cong \bigoplus_{\tilde{\mathfrak{p}} \supseteq \mathfrak{p}} P(\tilde{\mathcal{O}}_{\tilde{\mathfrak{p}}})$$

and therefore

$$J(\tilde{\mathcal{O}}) \cong \bigoplus_{\mathfrak{p}} \bigoplus_{\tilde{\mathfrak{p}} \supseteq \mathfrak{p}} P(\tilde{\mathcal{O}}_{\tilde{\mathfrak{p}}}) \cong \bigoplus_{\mathfrak{p}} P(\tilde{\mathcal{O}}_{\mathfrak{p}}).$$

Observing that $P(R) \cong K^*/R^*$ for any integral domain R with field of fractions K, we obtain the commutative exact diagram

$$\begin{array}{ccccccccc}
1 & \longrightarrow & K^*/\mathcal{O}^* & \longrightarrow & \bigoplus_{\mathfrak{p}} K^*/\mathcal{O}_{\mathfrak{p}}^* & \longrightarrow & Pic(\mathcal{O}) & \longrightarrow & 1 \\
 & & \downarrow{\scriptstyle\alpha} & & \downarrow{\scriptstyle\beta} & & \downarrow{\scriptstyle\gamma} & & \\
1 & \longrightarrow & K^*/\tilde{\mathcal{O}}^* & \longrightarrow & \bigoplus_{\mathfrak{p}} K^*/\tilde{\mathcal{O}}_{\mathfrak{p}}^* & \longrightarrow & Pic(\tilde{\mathcal{O}}) & \longrightarrow & 1.
\end{array}$$

For such a diagram one has in complete generality the well-known **snake lemma**: the diagram gives in a canonical way an exact sequence

$$\begin{aligned}
1 \longrightarrow \ker(\alpha) \longrightarrow \ker(\beta) \longrightarrow \ker(\gamma)& \\
\xrightarrow{\delta} \operatorname{coker}(\alpha) \longrightarrow \operatorname{coker}(\beta) \longrightarrow \operatorname{coker}(\gamma) \longrightarrow 1&
\end{aligned}$$

relating the kernels and cokernels of α, β, γ (see [23], chap. III, §3, lemma 3.3). In our particular case, α, β, and therefore also γ, are surjective, whereas

$$\ker(\alpha) = \tilde{\mathcal{O}}^*/\mathcal{O}^* \quad \text{and} \quad \ker(\beta) = \bigoplus_{\mathfrak{p}} \tilde{\mathcal{O}}_{\mathfrak{p}}^*/\mathcal{O}_{\mathfrak{p}}^*.$$

This then yields the exact sequence

$$1 \longrightarrow \mathcal{O}^* \longrightarrow \tilde{\mathcal{O}}^* \longrightarrow \bigoplus_{\mathfrak{p}} \tilde{\mathcal{O}}_{\mathfrak{p}}^*/\mathcal{O}_{\mathfrak{p}}^* \longrightarrow Pic(\mathcal{O}) \longrightarrow Pic(\tilde{\mathcal{O}}) \longrightarrow 1. \qquad \square$$

A prime ideal $\mathfrak{p} \neq 0$ of $\mathcal{O}$ is called **regular** if $\mathcal{O}_{\mathfrak{p}}$ is integrally closed, and thus a discrete valuation ring. For the regular prime ideals, the summands $\tilde{\mathcal{O}}_{\mathfrak{p}}^*/\mathcal{O}_{\mathfrak{p}}^*$ in (12.9) are trivial. There are only finitely many non-regular prime ideals of $\mathcal{O}$, namely the divisors of the **conductor** of $\mathcal{O}$. This is by definition the biggest ideal of $\tilde{\mathcal{O}}$ which is contained in $\mathcal{O}$, in other words,

$$\mathfrak{f} = \{ a \in \tilde{\mathcal{O}} \mid a\tilde{\mathcal{O}} \subseteq \mathcal{O} \}.$$

Since $\tilde{\mathcal{O}}$ is a finitely generated $\mathcal{O}$-module, we have $\mathfrak{f} \neq 0$.

(12.10) Proposition. *For any prime ideal $\mathfrak{p} \neq 0$ of $\mathcal{O}$ one has*

$$\mathfrak{p} \nmid \mathfrak{f} \iff \mathfrak{p} \textit{ is regular.}$$

If this is the case, then $\tilde{\mathfrak{p}} = \mathfrak{p}\tilde{\mathcal{O}}$ is a prime ideal of $\tilde{\mathcal{O}}$ and $\mathcal{O}_{\mathfrak{p}} = \tilde{\mathcal{O}}_{\tilde{\mathfrak{p}}}$.

Proof: Assume $\mathfrak{p} \nmid \mathfrak{f}$, i.e., $\mathfrak{p} \not\supseteq \mathfrak{f}$, and let $t \in \mathfrak{f} \smallsetminus \mathfrak{p}$. Then $t\tilde{\mathcal{O}} \subseteq \mathcal{O}$, hence $\tilde{\mathcal{O}} \subseteq \frac{1}{t}\mathcal{O} \subseteq \mathcal{O}_\mathfrak{p}$. If $\mathfrak{m} = \mathfrak{p}\mathcal{O}_\mathfrak{p}$ is the maximal ideal of $\mathcal{O}_\mathfrak{p}$ then, putting $\tilde{\mathfrak{p}} = \mathfrak{m} \cap \tilde{\mathcal{O}}$, $\tilde{\mathfrak{p}}$ is a prime ideal of $\tilde{\mathcal{O}}$ such that $\mathfrak{p} \subseteq \tilde{\mathfrak{p}} \cap \mathcal{O}$, hence $\mathfrak{p} = \tilde{\mathfrak{p}} \cap \mathcal{O}$ because $\mathfrak{p}$ is maximal. Trivially, $\mathcal{O}_\mathfrak{p} \subseteq \tilde{\mathcal{O}}_{\tilde{\mathfrak{p}}}$, and if conversely $\frac{a}{s} \in \tilde{\mathcal{O}}_{\tilde{\mathfrak{p}}}$, for $a \in \tilde{\mathcal{O}}$, $s \in \tilde{\mathcal{O}} \smallsetminus \tilde{\mathfrak{p}}$, then $ta \in \mathcal{O}$ and $ts \in \mathcal{O} \smallsetminus \mathfrak{p}$, hence $\frac{a}{s} = \frac{ta}{ts} \in \mathcal{O}_\mathfrak{p}$. Therefore $\mathcal{O}_\mathfrak{p} = \tilde{\mathcal{O}}_{\tilde{\mathfrak{p}}}$. Thus, by (11.5), $\mathcal{O}_\mathfrak{p}$ is a valuation ring, i.e., $\mathfrak{p}$ is regular.

One has furthermore that $\tilde{\mathfrak{p}} = \mathfrak{p}\tilde{\mathcal{O}}$. In fact, $\tilde{\mathfrak{p}}$ is the only prime ideal of $\tilde{\mathcal{O}}$ above $\mathfrak{p}$. For if $\tilde{\mathfrak{q}}$ is another one, then $\tilde{\mathcal{O}}_{\tilde{\mathfrak{p}}} = \mathcal{O}_\mathfrak{p} \subseteq \tilde{\mathcal{O}}_{\tilde{\mathfrak{q}}}$, and therefore

$$\tilde{\mathfrak{p}} = \tilde{\mathcal{O}} \cap \tilde{\mathfrak{p}}\mathcal{O}_{\tilde{\mathfrak{p}}} \subseteq \tilde{\mathcal{O}} \cap \tilde{\mathfrak{q}}\mathcal{O}_{\tilde{\mathfrak{q}}} = \tilde{\mathfrak{q}},$$

hence $\tilde{\mathfrak{p}} = \tilde{\mathfrak{q}}$. Consequently, $\mathfrak{p}\tilde{\mathcal{O}} = \tilde{\mathfrak{p}}^e$, with $e \geq 1$, and furthermore $\mathfrak{m} = \mathfrak{p}\mathcal{O}_\mathfrak{p} = (\mathfrak{p}\tilde{\mathcal{O}})\mathcal{O}_\mathfrak{p} = \tilde{\mathfrak{p}}^e\mathcal{O}_\mathfrak{p} = \mathfrak{m}^e$, i.e., $e = 1$ and thus $\tilde{\mathfrak{p}} = \mathfrak{p}\tilde{\mathcal{O}}$.

Conversely, assume $\mathcal{O}_\mathfrak{p}$ is a discrete valuation ring. Being a principal ideal domain, it is integrally closed, and since $\tilde{\mathcal{O}}$ is integral over $\mathcal{O}$, hence *a fortiori* over $\mathcal{O}_\mathfrak{p}$, we have $\tilde{\mathcal{O}} \subseteq \mathcal{O}_\mathfrak{p}$. Let $x_1, \dots, x_n$ be a system of generators of the $\mathcal{O}$-module $\tilde{\mathcal{O}}$. We may write $x_i = \frac{a_i}{s_i}$, with $a_i \in \mathcal{O}$, $s_i \in \mathcal{O} \smallsetminus \mathfrak{p}$. Setting $s = s_1 \cdots s_n \in \mathcal{O} \smallsetminus \mathfrak{p}$, we find $sx_1, \dots, sx_n \in \mathcal{O}$ and therefore $s\tilde{\mathcal{O}} \subseteq \mathcal{O}$, i.e., $s \in \mathfrak{f} \smallsetminus \mathfrak{p}$. It follows that $\mathfrak{p} \nmid \mathfrak{f}$. □

We now obtain the following simple description for the sum $\bigoplus_\mathfrak{p} \tilde{\mathcal{O}}_\mathfrak{p}^*/\mathcal{O}_\mathfrak{p}^*$ in (12.9).

(12.11) Proposition. $\bigoplus_\mathfrak{p} \tilde{\mathcal{O}}_\mathfrak{p}^*/\mathcal{O}_\mathfrak{p}^* \cong (\tilde{\mathcal{O}}/\mathfrak{f})^*/(\mathcal{O}/\mathfrak{f})^*$.

Proof: We apply the Chinese remainder theorem (12.3) repeatedly. We have

$$\text{(1)} \qquad \mathcal{O}/\mathfrak{f} \cong \bigoplus_\mathfrak{p} \mathcal{O}_\mathfrak{p}/\mathfrak{f}\mathcal{O}_\mathfrak{p}.$$

The integral closure $\tilde{\mathcal{O}}_\mathfrak{p}$ of $\mathcal{O}_\mathfrak{p}$ possesses only the finitely many prime ideals that lie above $\mathfrak{p}\mathcal{O}_\mathfrak{p}$. They give the localizations $\tilde{\mathcal{O}}_{\tilde{\mathfrak{p}}}$, where $\tilde{\mathfrak{p}}$ varies over the prime ideals above $\mathfrak{p}$ of the ring $\tilde{\mathcal{O}}$. At the same time, $\tilde{\mathcal{O}}_\mathfrak{p}$ is the localization of $\tilde{\mathcal{O}}$ with respect to the multiplicative subset $\tilde{\mathcal{O}} \smallsetminus \tilde{\mathfrak{p}}$. Since $\mathfrak{f}$ is an ideal of $\tilde{\mathcal{O}}$, it follows that $\mathfrak{f}\tilde{\mathcal{O}}_\mathfrak{p} = \mathfrak{f}\mathcal{O}_\mathfrak{p}$. The Chinese remainder theorem yields

$$\tilde{\mathcal{O}}_\mathfrak{p}/\mathfrak{f}\tilde{\mathcal{O}}_\mathfrak{p} \cong \bigoplus_{\tilde{\mathfrak{p}} \supseteq \mathfrak{p}} \tilde{\mathcal{O}}_{\tilde{\mathfrak{p}}}/\mathfrak{f}\tilde{\mathcal{O}}_{\tilde{\mathfrak{p}}}$$

and

$$\text{(2)} \qquad \tilde{\mathcal{O}}/\mathfrak{f} \cong \bigoplus_\mathfrak{p} \bigoplus_{\tilde{\mathfrak{p}} \supseteq \mathfrak{p}} \tilde{\mathcal{O}}_{\tilde{\mathfrak{p}}}/\mathfrak{f}\tilde{\mathcal{O}}_{\tilde{\mathfrak{p}}} \cong \bigoplus_\mathfrak{p} \tilde{\mathcal{O}}_\mathfrak{p}/\mathfrak{f}\tilde{\mathcal{O}}_\mathfrak{p}.$$

Passing to unit groups, we get from (1) and (2) that

$$(3) \qquad (\tilde{\mathcal{O}}/\mathfrak{f})^*/(\mathcal{O}/\mathfrak{f})^* \cong \bigoplus_{\mathfrak{p}} (\tilde{\mathcal{O}}_{\mathfrak{p}}/\mathfrak{f}\tilde{\mathcal{O}}_{\mathfrak{p}})^*/(\mathcal{O}_{\mathfrak{p}}/\mathfrak{f}\mathcal{O}_{\mathfrak{p}})^* .$$

For $\mathfrak{f} \subseteq \mathfrak{p}$ we now consider the homomorphism

$$\varphi : \tilde{\mathcal{O}}_{\mathfrak{p}}^* \to (\tilde{\mathcal{O}}_{\mathfrak{p}}/\mathfrak{f}\tilde{\mathcal{O}}_{\mathfrak{p}})^*/(\mathcal{O}_{\mathfrak{p}}/\mathfrak{f}\mathcal{O}_{\mathfrak{p}})^* .$$

It is surjective. In fact, if ε mod $\mathfrak{f}\tilde{\mathcal{O}}_{\mathfrak{p}}$ is a unit in $\tilde{\mathcal{O}}_{\mathfrak{p}}/\mathfrak{f}\tilde{\mathcal{O}}_{\mathfrak{p}}$, then ε is a unit in $\tilde{\mathcal{O}}_{\mathfrak{p}}$. This is so because the units in any ring are precisely those elements that are not contained in any maximal ideal, and the preimages of the maximal ideals of $\tilde{\mathcal{O}}_{\mathfrak{p}}/\mathfrak{f}\tilde{\mathcal{O}}_{\mathfrak{p}}$ give precisely all the maximal ideals of $\tilde{\mathcal{O}}_{\mathfrak{p}}$, since $\mathfrak{f}\tilde{\mathcal{O}}_{\mathfrak{p}} \subseteq \mathfrak{p}\tilde{\mathcal{O}}_{\mathfrak{p}}$. The kernel of φ is a subgroup of $\tilde{\mathcal{O}}_{\mathfrak{p}}^*$ which is contained in $\mathcal{O}_{\mathfrak{p}}$, and which contains $\mathcal{O}_{\mathfrak{p}}^*$. It is therefore equal to $\mathcal{O}_{\mathfrak{p}}^*$. We now conclude that

$$\tilde{\mathcal{O}}_{\mathfrak{p}}^*/\mathcal{O}_{\mathfrak{p}}^* \cong (\tilde{\mathcal{O}}_{\mathfrak{p}}/\mathfrak{f}\tilde{\mathcal{O}}_{\mathfrak{p}})^*/(\mathcal{O}_{\mathfrak{p}}/\mathfrak{f}\mathcal{O}_{\mathfrak{p}})^* .$$

This remains true also for $\mathfrak{p} \not\supseteq \mathfrak{f}$ because then both sides are equal to 1 according to (12.10). The claim of the proposition now follows from (3). □

Our study of one-dimensional noetherian integral domains was motivated by the consideration of *orders*. For them, (12.9) and (12.11) imply the following generalization of Dirichlet's unit theorem and of the theorem on the finiteness of the class group.

(12.12) Theorem. *Let $\mathcal{O}$ be an order in an algebraic number field K, $\mathcal{O}_K$ the maximal order, and $\mathfrak{f}$ the conductor of $\mathcal{O}$.*

Then the groups $\mathcal{O}_K^*/\mathcal{O}^*$ and $Pic(\mathcal{O})$ are finite and one has

$$\# Pic(\mathcal{O}) = \frac{h_K}{(\mathcal{O}_K^* : \mathcal{O}^*)} \frac{\#(\mathcal{O}_K/\mathfrak{f})^*}{\#(\mathcal{O}/\mathfrak{f})^*} ,$$

where h_K is the class number of K. In particular, one has that

$$\operatorname{rank}(\mathcal{O}^*) = \operatorname{rank}(\mathcal{O}_K^*) = r + s - 1 .$$

Proof: By (12.9) and (12.11), and since $Pic(\mathcal{O}_K) = Cl_K$, we have the exact sequence

$$1 \longrightarrow \mathcal{O}_K^*/\mathcal{O}^* \longrightarrow (\mathcal{O}_K/\mathfrak{f})^*/(\mathcal{O}/\mathfrak{f})^* \longrightarrow Pic(\mathcal{O}) \longrightarrow Cl_K \longrightarrow 1 .$$

This gives the claim. □

The definition of the Picard group of a one-dimensional noetherian integral domain $\mathcal{O}$ avoids the problem of the uniqueness of prime ideal decomposition by restricting attention to the invertible ideals, and thus leaving aside the information carried by noninvertibles. But there is another important generalization of the ideal class group which does take into account *all* prime ideals of $\mathcal{O}$. It is based on an artificial re-introduction of the uniqueness of prime decomposition. This group is called the **divisor class group**, or **Chow group** of $\mathcal{O}$. Its definition starts from the free abelian group

$$Div(\mathcal{O}) = \bigoplus_{\mathfrak{p}} \mathbb{Z}\mathfrak{p}$$

on the set of all maximal ideals $\mathfrak{p}$ of $\mathcal{O}$ (i.e., the set of all prime ideals $\neq 0$). This group is called the **divisor group** of $\mathcal{O}$. Its elements are formal sums

$$D = \sum_{\mathfrak{p}} n_{\mathfrak{p}}\mathfrak{p}$$

with $n_{\mathfrak{p}} \in \mathbb{Z}$ and $n_{\mathfrak{p}} = 0$ for almost all $\mathfrak{p}$, called **divisors** (or **0-cycles**). Corollary (3.9) simply says that, in the case of a Dedekind domain, the divisor group $Div(\mathcal{O})$ and the group of ideals are canonically isomorphic. The additive notation and the name of the group stem from function theory where divisors for analytic functions play the same rôle as ideals do for algebraic numbers (see chap. III, § 3).

In order to define the divisor class group we have to associate to every $f \in K^*$ a "principal divisor" $\operatorname{div}(f)$. We use the case of a Dedekind domain to guide us. There the principal ideal (f) was given by

$$(f) = \prod_{\mathfrak{p}} \mathfrak{p}^{v_{\mathfrak{p}}(f)},$$

where $v_{\mathfrak{p}} : K^* \to \mathbb{Z}$ is the $\mathfrak{p}$-adic exponential valuation associated to the valuation ring $\mathcal{O}_{\mathfrak{p}}$. In general, $\mathcal{O}_{\mathfrak{p}}$ is not anymore a discrete valuation ring. Nevertheless, $\mathcal{O}_{\mathfrak{p}}$ defines a homomorphism

$$\operatorname{ord}_{\mathfrak{p}} : K^* \to \mathbb{Z}$$

which generalizes the valuation function. If $f = a/b \in K^*$, with $a, b \in \mathcal{O}$, then we put

$$\operatorname{ord}_{\mathfrak{p}}(f) = \ell_{\mathcal{O}_{\mathfrak{p}}}(\mathcal{O}_{\mathfrak{p}}/a\mathcal{O}_{\mathfrak{p}}) - \ell_{\mathcal{O}_{\mathfrak{p}}}(\mathcal{O}_{\mathfrak{p}}/b\mathcal{O}_{\mathfrak{p}}),$$

where $\ell_{\mathcal{O}_{\mathfrak{p}}}(M)$ denotes the **length** of an $\mathcal{O}_{\mathfrak{p}}$-module M, i.e., the maximal length of a strictly decreasing chain

$$M = M_0 \supsetneqq M_1 \supsetneqq \cdots \supsetneqq M_\ell = 0$$

of $\mathcal{O}_{\mathfrak{p}}$-submodules. In the special case where $\mathcal{O}_{\mathfrak{p}}$ is a discrete valuation ring with maximal ideal $\mathfrak{m}$, the value $\nu = v_{\mathfrak{p}}(a)$ of $a \in \mathcal{O}_{\mathfrak{p}}$, for $a \neq 0$, is given by the equation

$$a\mathcal{O}_{\mathfrak{p}} = \mathfrak{m}^{\nu}.$$

It is equal to the length of the $\mathcal{O}_\mathfrak{p}$-module $\mathcal{O}_\mathfrak{p}/\mathfrak{m}^\nu$, because the longest chain of submodules is

$$\mathcal{O}_\mathfrak{p}/\mathfrak{m}^\nu \supset \mathfrak{m}/\mathfrak{m}^\nu \supset \cdots \supset \mathfrak{m}^\nu/\mathfrak{m}^\nu = (0).$$

Thus the function $\operatorname{ord}_\mathfrak{p}$ agrees with the exponential valuation $v_\mathfrak{p}$ in this case.

The property of the function $\operatorname{ord}_\mathfrak{p}$ to be a homomorphism follows from the fact (which is easily proved) that the length function $\ell_{\mathcal{O}_\mathfrak{p}}$ is multiplicative on short exact sequences of $\mathcal{O}_\mathfrak{p}$-modules.

Using the functions $\operatorname{ord}_\mathfrak{p} : K^* \to \mathbb{Z}$, we can now associate to every element $f \in K^*$ the divisor

$$\operatorname{div}(f) = \sum_\mathfrak{p} \operatorname{ord}_\mathfrak{p}(f)\mathfrak{p},$$

and thus obtain a canonical homomorphism

$$\operatorname{div} : K^* \longrightarrow Div(\mathcal{O}).$$

The elements $\operatorname{div}(f)$ are called **principal divisors**. They form a subgroup $\mathcal{P}(\mathcal{O})$ of $Div(\mathcal{O})$. Two divisors D and D' which differ only by a principal divisor are called **rationally equivalent**.

(12.13) Definition. *The quotient group*

$$CH^1(\mathcal{O}) = Div(\mathcal{O})/\mathcal{P}(\mathcal{O})$$

is called the **divisor class group** *or* **Chow group** *of* $\mathcal{O}$.

The Chow group is related to the Picard group by a canonical homomorphism

$$\operatorname{div} : Pic(\mathcal{O}) \longrightarrow CH^1(\mathcal{O})$$

which is defined as follows. If $\mathfrak{a}$ is an invertible ideal, then, by (12.4), $\mathfrak{a}\mathcal{O}_\mathfrak{p}$, for any prime ideal $\mathfrak{p} \neq 0$, is a principal ideal $a_\mathfrak{p}\mathcal{O}_\mathfrak{p}$, $a_\mathfrak{p} \in K^*$, and we put

$$\operatorname{div}(\mathfrak{a}) = \sum_\mathfrak{p} -\operatorname{ord}_\mathfrak{p}(a_\mathfrak{p})\mathfrak{p}.$$

This gives us a homomorphism

$$\operatorname{div} : J(\mathcal{O}) \longrightarrow Div(\mathcal{O})$$

of the ideal group $J(\mathcal{O})$ which takes principal ideals into principal divisors, and therefore induces a homomorphism

$$\operatorname{div} : Pic(\mathcal{O}) \longrightarrow CH^1(\mathcal{O}).$$

In the special case of a Dedekind domain we obtain:

(12.14) Proposition. *If $\mathcal{O}$ is a Dedekind domain, then*

$$\mathrm{div} : Pic(\mathcal{O}) \longrightarrow CH^1(\mathcal{O})$$

is an isomorphism.

Exercise 1. Show that

$$\mathbb{C}[X,Y]/(XY - X), \quad \mathbb{C}[X,Y]/(XY - 1),$$

$$\mathbb{C}[X,Y]/(X^2 - Y^3), \quad \mathbb{C}[X,Y]/(Y^2 - X^2 - X^3)$$

are one-dimensional noetherian rings. Which ones are integral domains? Determine their normalizations.

Hint: For instance in the last example, put $t = X/Y$ and show that the homomorphism $\mathbb{C}[X,Y] \to \mathbb{C}[t]$, $X \mapsto t^2 - 1$, $Y \mapsto t(t^2 - 1)$, has kernel $(Y^2 - X^2 - X^3)$.

Exercise 2. Let a and b be positive integers that are not perfect squares. Show that the fundamental unit of the order $\mathbb{Z} + \mathbb{Z}\sqrt{a}$ of the field $\mathbb{Q}(\sqrt{a})$ is also the fundamental unit of the order $\mathbb{Z} + \mathbb{Z}\sqrt{a} + \mathbb{Z}\sqrt{-b} + \mathbb{Z}\sqrt{a}\sqrt{-b}$ in the field $\mathbb{Q}(\sqrt{a}, \sqrt{-b})$.

Exercise 3. Let K be a number field of degree $n = [K : \mathbb{Q}]$. A complete module in K is a subgroup of the form

$$M = \mathbb{Z}\alpha_1 + \cdots + \mathbb{Z}\alpha_n ,$$

where $\alpha_1, \ldots, \alpha_n$ are linearly independent elements of K. Show that the ring of multipliers

$$\mathcal{O} = \{\alpha \in K \mid \alpha M \subseteq M\}$$

is an order in K, but in general not the maximal order.

Exercise 4. Determine the ring of multipliers $\mathcal{O}$ of the complete module $M = \mathbb{Z} + \mathbb{Z}\sqrt{2}$ in $\mathbb{Q}(\sqrt{2})$. Show that $\varepsilon = 1 + \sqrt{2}$ is a fundamental unit of $\mathcal{O}$. Determine all integer solutions of "Pell's equation"

$$x^2 - 2y^2 = 7 .$$

Hint: $N(x + y\sqrt{2}) = x^2 - 2y^2$, $N(3 + \sqrt{2}) = N(5 + 3\sqrt{2}) = 7$.

Exercise 5. In a one-dimensional noetherian integral domain the regular prime ideals $\neq 0$ are precisely the invertible prime ideals.

§ 13. One-dimensional Schemes

The first approach to the theory of algebraic number fields is dominated by the methods of arithmetic and algebra. But the theory may also be treated fundamentally from a geometric point of view, which will bring out novel aspects in a variety of ways. This geometric interpretation hinges on the possibility of viewing numbers as functions on a topological space.

In order to explain this, let us start from polynomials

$$f(x) = a_n x^n + \cdots + a_0$$

with complex coefficients $a_i \in \mathbb{C}$, which may be immediately interpreted as functions on the complex plane. This property may be formulated in a purely algebraic way as follows. Let $a \in \mathbb{C}$ be a point in the complex plane. The set of all functions $f(x)$ in the polynomial ring $\mathbb{C}[x]$ which vanish at the point a forms the maximal ideal $\mathfrak{p} = (x - a)$ of $\mathbb{C}[x]$. In this way the points of the complex plane correspond 1-1 to the maximal ideals of $\mathbb{C}[x]$. We denote the set of all these maximal ideals by

$$M = \mathrm{Max}\big(\mathbb{C}[x]\big).$$

We may view M as a new kind of space and may interpret the elements $f(x)$ of the ring $\mathbb{C}[x]$ as functions on M as follows. For every point $\mathfrak{p} = (x - a)$ of M we have the canonical isomorphism

$$\mathbb{C}[x]/\mathfrak{p} \xrightarrow{\sim} \mathbb{C},$$

which sends the residue class $f(x) \bmod \mathfrak{p}$ to $f(a)$. We may thus view this residue class

$$f(\mathfrak{p}) := f(x) \bmod \mathfrak{p} \in \kappa(\mathfrak{p})$$

in the residue class field $\kappa(\mathfrak{p}) = \mathbb{C}[x]/\mathfrak{p}$ as the "value" of f at the point $\mathfrak{p} \in M$. The topology on $\mathbb{C}$ cannot be transferred to M by algebraic means. All that can be salvaged algebraically are the point sets defined by equations of the form

$$f(x) = 0$$

(i.e., only the finite sets and M itself). These sets are defined to be the closed subsets. In the new formulation they are the sets

$$V(f) = \big\{\mathfrak{p} \in M \mid f(\mathfrak{p}) = 0\big\} = \big\{\mathfrak{p} \in M \mid \mathfrak{p} \supseteq (f(x))\big\}.$$

The algebraic interpretation of functions given above leads to the following geometric perception of completely general rings. For an arbitrary ring $\mathcal{O}$, one introduces the **spectrum**

$$X = \mathrm{Spec}(\mathcal{O})$$

as being the set of all prime ideals $\mathfrak{p}$ of $\mathcal{O}$. The **Zariski topology** on X is defined by stipulating that the sets

$$V(\mathfrak{a}) = \big\{\mathfrak{p} \mid \mathfrak{p} \supseteq \mathfrak{a}\big\}$$

be the closed sets, where $\mathfrak{a}$ varies over the ideals of $\mathcal{O}$. This does make X into a topological space (observe that $V(\mathfrak{a}) \cup V(\mathfrak{b}) = V(\mathfrak{a}\mathfrak{b})$) which, however, is usually not Hausdorff. The closed points correspond to the maximal ideals of $\mathcal{O}$.

The elements $f \in \mathcal{O}$ now play the rôle of functions on the topological space X: the "value" of f at the point $\mathfrak{p}$ is defined to be

$$f(\mathfrak{p}) := f \bmod \mathfrak{p}$$

and is an element of the residue class field $\kappa(\mathfrak{p})$, i.e., in the field of fractions of $\mathcal{O}/\mathfrak{p}$. So the values of f do not in general lie in a single field.

Admitting also the non-maximal prime ideals as non-closed points, turns out to be extremely useful – and has an intuitive reason as well. For instance in the case of the ring $\mathcal{O} = \mathbb{C}[x]$, the point $\mathfrak{p} = (0)$ has residue class field $\kappa(\mathfrak{p}) = \mathbb{C}(x)$. The "value" of a polynomial $f \in \mathbb{C}[x]$ at this point is $f(x)$ itself, viewed as an element of $\mathbb{C}(x)$. This element should be thought of as the value of f at the **unknown** place x – which one may imagine to be everywhere or nowhere at all. This intuition complies with the fact that the closure of the point $\mathfrak{p} = (0)$ in the Zariski topology of X is the total space X. This is why $\mathfrak{p}$ is also called the **generic point** of X.

Example: The space $X = \operatorname{Spec}(\mathbb{Z})$ may be represented by a line.

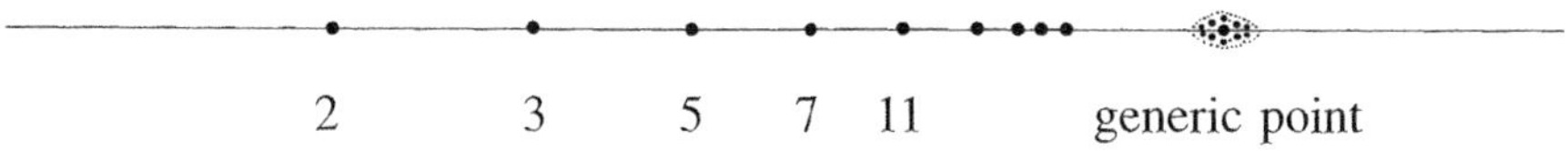

For every prime number one has a closed point, and there is also the generic point (0), the closure of which is the total space X. The nonempty open sets in X are obtained by throwing out finitely many prime numbers $p_1, \dots, p_n$. The integers $a \in \mathbb{Z}$ are viewed as functions on X by defining the value of a at the point (p) to be the residue class

$$a(p) = a \bmod p \in \mathbb{Z}/p\mathbb{Z}.$$

The fields of values are then

$$\mathbb{Z}/2\mathbb{Z},\ \mathbb{Z}/3\mathbb{Z},\ \mathbb{Z}/5\mathbb{Z},\ \mathbb{Z}/7\mathbb{Z},\ \mathbb{Z}/11\mathbb{Z}, \dots, \mathbb{Q}.$$

Thus every prime field occurs exactly once.

An important refinement of the geometric interpretation of elements of the ring $\mathcal{O}$ as functions on the space $X = \operatorname{Spec}(\mathcal{O})$ is obtained by forming the **structure sheaf** $\mathcal{O}_X$. This means the following. Let $U \neq \emptyset$ be an open subset of X. If $\mathcal{O}$ is a one-dimensional integral domain, then the ring of "regular functions" on U is given by

$$\mathcal{O}(U) = \left\{ \frac{f}{g} \mid g(\mathfrak{p}) \neq 0 \quad \text{for all } \mathfrak{p} \in U \right\},$$

in other words, it is the localization of $\mathcal{O}$ with respect to the multiplicative set $S = \mathcal{O} \setminus \bigcup_{\mathfrak{p}\in U} \mathfrak{p}$ (see § 11). In the general case, $\mathcal{O}(U)$ is defined to consist of all elements

$$s = (s_\mathfrak{p}) \in \prod_{\mathfrak{p}\in U} \mathcal{O}_\mathfrak{p}$$

which locally are quotients of two elements of $\mathcal{O}$. More precisely, this means that for every $\mathfrak{p} \in U$, there exists a neighbourhood $V \subseteq U$ of $\mathfrak{p}$, and elements $f, g \in \mathcal{O}$ such that, for each $\mathfrak{q} \in V$, one has $g(\mathfrak{q}) \neq 0$ and $s_\mathfrak{q} = f/g$ in $\mathcal{O}_\mathfrak{q}$. These quotients have to be understood in the more general sense of commutative algebra (see § 11, exercise 1). We leave it to the reader to check that one gets back the above definition in the case of a one-dimensional integral domain $\mathcal{O}$.

If $V \subseteq U$ are two open sets of X, then the projection

$$\prod_{\mathfrak{p}\in U} \mathcal{O}_\mathfrak{p} \longrightarrow \prod_{\mathfrak{p}\in V} \mathcal{O}_\mathfrak{p}$$

induces a homomorphism

$$\rho_{UV} : \mathcal{O}(U) \longrightarrow \mathcal{O}(V),$$

called the **restriction** from U to V. The system of rings $\mathcal{O}(U)$ and mappings ρ_{UV} is a **sheaf** on X. This notion means the following.

(13.1) Definition. *Let X be a topological space. A* **presheaf** *$\mathcal{F}$ of abelian groups (rings, etc.) consists of the following data.*

(1) *For every open set U, an abelian group (a ring, etc.) $\mathcal{F}(U)$ is given.*

(2) *For every inclusion $U \subseteq V$, a homomorphism $\rho_{UV} : \mathcal{F}(U) \to \mathcal{F}(V)$ is given, which is called restriction.*

These data are subject to the following conditions:

(a) $\mathcal{F}(\emptyset) = 0$,

(b) *ρ_{UU} is the identity* $\mathrm{id} : \mathcal{F}(U) \to \mathcal{F}(U)$,

(c) $\rho_{UW} = \rho_{VW} \circ \rho_{UV}$, *for open sets $W \subseteq V \subseteq U$.*

The elements $s \in \mathcal{F}(U)$ are called the **sections** of the presheaf $\mathcal{F}$ over U. If $V \subseteq U$, then one usually writes $\rho_{UV}(s) = s|_V$. The definition of a presheaf can be reformulated most concisely in the language of categories. The open sets of the topological space X form a category X_{top} in which only inclusions are admitted as morphisms. A presheaf of abelian groups (rings) is then simply a contravariant functor

$$\mathcal{F} : X_{\text{top}} \longrightarrow (ab),\ (rings)$$

into the category of abelian groups (resp. rings) such that $\mathcal{F}(\emptyset) = 0$.

(13.2) Definition. *A presheaf $\mathcal{F}$ on the topological space X is called a* **sheaf** *if, for all open coverings $\{U_i\}$ of the open sets U, one has:*

(i) *If $s, s' \in \mathcal{F}(U)$ are two sections such that $s|_{U_i} = s'|_{U_i}$ for all i, then $s = s'$.*

(ii) *If $s_i \in \mathcal{F}(U_i)$ is a family of sections such that*

$$s_i|_{U_i \cap U_j} = s_j|_{U_i \cap U_j}$$

for all i, j, then there exists a section $s \in \mathcal{F}(U)$ such that $s|_{U_i} = s_i$ for all i.

The **stalk** of the sheaf $\mathcal{F}$ at the point $x \in X$ is defined to be the direct limit (see chap. IV, § 2)

$$\mathcal{F}_x = \varinjlim_{U \ni x} \mathcal{F}(U),$$

where U varies over all open neighbourhoods of x. In other words, two sections $s_U \in \mathcal{F}(U)$ and $s_V \in \mathcal{F}(V)$ are called equivalent in the disjoint union $\bigcup_{U \ni x} \mathcal{F}(U)$ if there exists a neighbourhood $W \subseteq U \cap V$ of x such that $s_U|_W = s_V|_W$. The equivalence classes are called **germs** of sections at x. They are the elements of $\mathcal{F}_x$.

We now return to the spectrum $X = \operatorname{Spec}(\mathcal{O})$ of a ring $\mathcal{O}$ and obtain the

(13.3) Proposition. *The rings $\mathcal{O}(U)$, together with the restriction mappings ρ_{UV}, form a* **sheaf** *on X. It is denoted by $\mathcal{O}_X$ and called the* **structure sheaf** *on X. The stalk of $\mathcal{O}_X$ at the point $\mathfrak{p} \in X$ is the localization $\mathcal{O}_\mathfrak{p}$, i.e., $\mathcal{O}_{X,\mathfrak{p}} \cong \mathcal{O}_\mathfrak{p}$.*

The proof of this proposition follows immediately from the definitions. The couple $(X, \mathcal{O}_X)$ is called an **affine scheme**. Usually, however, the structure sheaf $\mathcal{O}_X$ is dropped from the notation. Now let

$$\varphi : \mathcal{O} \longrightarrow \mathcal{O}'$$

be a homomorphism of rings and $X = \operatorname{Spec}(\mathcal{O})$, $X' = \operatorname{Spec}(\mathcal{O}')$. Then φ induces a continuous map

$$f : X' \longrightarrow X, \quad f(\mathfrak{p}') := \varphi^{-1}(\mathfrak{p}'),$$

and, for every open subset U of X, a homomorphism

$$f_U^* : \mathcal{O}(U) \longrightarrow \mathcal{O}(U'), \quad s \longmapsto s \circ f|_{U'},$$

where $U' = f^{-1}(U)$. The maps f_U^* have the following two properties.

a) If $V \subseteq U$ are open sets, then the diagram

$$\begin{array}{ccc} \mathcal{O}(U) & \xrightarrow{f_U^*} & \mathcal{O}(U') \\ {\scriptstyle \rho_{UV}}\downarrow & & \downarrow{\scriptstyle \rho_{U'V'}} \\ \mathcal{O}(V) & \xrightarrow{f_V^*} & \mathcal{O}(V') \end{array}$$

is commutative.

b) For $\mathfrak{p}' \in U' \subseteq X'$ and $a \in \mathcal{O}(U)$ one has

$$a\big(f(\mathfrak{p}')\big) = 0 \implies f_U^*(a)(\mathfrak{p}') = 0 .$$

A continuous map $f : X' \to X$ together with a family of homomorphisms $f_U^* : \mathcal{O}(U) \to \mathcal{O}(U')$ which satisfy conditions a) and b) is called a **morphism** from the scheme X' to the scheme X. When referring to such a morphism, the maps f_U^* are usually not written explicitly. One can show that every morphism between two affine schemes $X' = \operatorname{Spec}(\mathcal{O}')$ and $X = \operatorname{Spec}(\mathcal{O})$ is induced in the way described above by a ring homomorphism $\varphi : \mathcal{O} \to \mathcal{O}'$.

The proofs of the above claims are easy, although some of them are a bit lengthy. The notion of scheme is the basis of a very extensive theory which occupies a central place in mathematics. As introductions into this important discipline let us recommend the books [51] and [104].

We will now confine ourselves to considering noetherian integral domains $\mathcal{O}$ of dimension ≤ 1, and propose to illustrate geometrically, via the scheme-theoretic interpretation, some of the facts treated in previous sections.

1. Fields. If K is a field, then the scheme $\operatorname{Spec}(K)$ consists of a single point (0) on top of which the field itself sits as the structure sheaf. One must not think that these one-point schemes are all the same because they differ essentially in their structure sheaves.

2. Valuation rings. If $\mathcal{O}$ is a discrete valuation ring with maximal ideal $\mathfrak{p}$, then the scheme $X = \operatorname{Spec}(\mathcal{O})$ consists of two points, the closed point $x = \mathfrak{p}$ with residue class field $\kappa(\mathfrak{p}) = \mathcal{O}/\mathfrak{p}$, and the generic point $\eta = (0)$ with residue class field $\kappa(\eta) = K$, the field of fractions of $\mathcal{O}$. One should think of X as a point x with an infinitesimal neighbourhood described by the generic point η:

X:

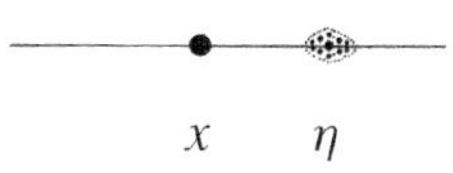

This intuition is justified by the following observation.

The discrete valuation rings arise as localizations

$$\mathcal{O}_{\mathfrak{p}} = \left\{ \frac{f}{g} \mid f, g \in \mathcal{O},\ g(\mathfrak{p}) \neq 0 \right\}$$

of Dedekind domains $\mathcal{O}$. There is no neighbourhood of $\mathfrak{p}$ in $X = \operatorname{Spec}(\mathcal{O})$ on which all functions $\frac{f}{g} \in \mathcal{O}_{\mathfrak{p}}$ are defined because, if $\mathcal{O}$ is not a local ring, we find by the Chinese remainder theorem for every point $\mathfrak{q} \neq \mathfrak{p}$, $\mathfrak{q} \neq 0$, an element $g \in \mathcal{O}$ satisfying $g \equiv 0 \bmod \mathfrak{q}$ and $g \equiv 1 \bmod \mathfrak{p}$. Then $\frac{1}{g} \in \mathcal{O}_{\mathfrak{p}}$ as a function is not defined at $\mathfrak{q}$. But every element $\frac{f}{g} \in \mathcal{O}_{\mathfrak{p}}$ is defined on a sufficiently small neighbourhood; hence one may say that all elements $\frac{f}{g}$ of the discrete valuation ring $\mathcal{O}_{\mathfrak{p}}$ are like functions defined on a "germ" of neighbourhoods of $\mathfrak{p}$. Thus $\operatorname{Spec}(\mathcal{O}_{\mathfrak{p}})$ may be thought of as such a "germ of neighbourhoods" of $\mathfrak{p}$.

We want to point out a small discrepancy of intuitions. Considering the spectrum of the one-dimensional ring $\mathbb{C}[x]$, the points of which constitute the complex plane, we will not want to visualize the infinitesimal neighbourhood $X_{\mathfrak{p}} = \operatorname{Spec}(\mathbb{C}[x]_{\mathfrak{p}})$ of a point $\mathfrak{p} = (x - a)$ as a small line segment, but rather as a little disc:

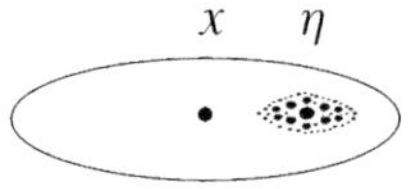

This two-dimensional nature is actually inherent in all discrete valuation rings with algebraically closed residue field. But the algebraic justification of this intuition is provided only by the introduction of a new topology, the **étale topology**, which is much finer than the Zariski topology (see [103], [132]).

3. Dedekind rings. The spectrum $X = \operatorname{Spec}(\mathcal{O})$ of a Dedekind domain $\mathcal{O}$ is visualized as a smooth curve. At each point $\mathfrak{p}$ one may consider the **localization** $\mathcal{O}_{\mathfrak{p}}$. The inclusion $\mathcal{O} \hookrightarrow \mathcal{O}_{\mathfrak{p}}$ induces a morphism

$$f : X_{\mathfrak{p}} = \operatorname{Spec}(\mathcal{O}_{\mathfrak{p}}) \longrightarrow X\,,$$

which extracts the scheme $X_{\mathfrak{p}}$ from X as an "infinitesimal neighbourhood" of $\mathfrak{p}$:

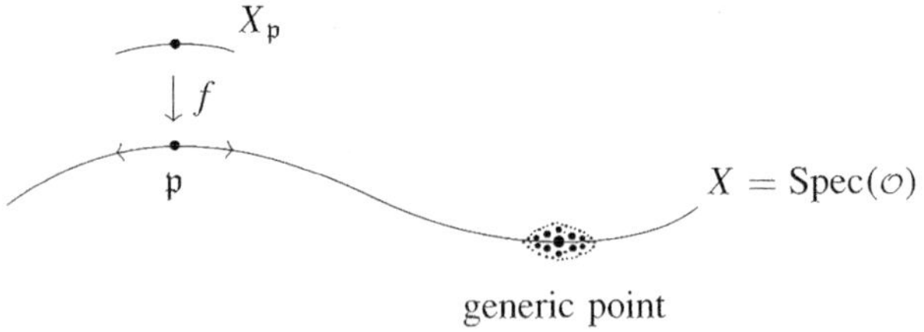

4. Singularities. We now consider a one-dimensional noetherian integral domain $\mathcal{O}$ which is not a Dedekind domain, *e.g.*, an order in an algebraic number field which is different from the maximal order. Again we view the scheme $X = \mathrm{Spec}(\mathcal{O})$ as a curve. But now the curve will not be everywhere smooth, but will have singularities at certain points.

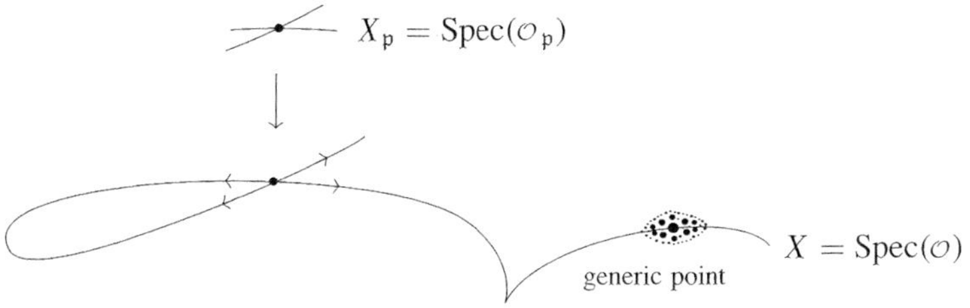

These are precisely the nongeneric points $\mathfrak{p}$ for which the localization $\mathcal{O}_{\mathfrak{p}}$ is no longer a discrete valuation ring, that is to say, the maximal ideal $\mathfrak{p}\mathcal{O}_{\mathfrak{p}}$ is not generated by a single element. For example, in the one-dimensional ring $\mathcal{O} = \mathbb{C}[x, y]/(y^2 - x^3)$, the closed points of the scheme X are given by the prime ideals

$$\mathfrak{p} = (x - a, y - b) \bmod (y^2 - x^3)$$

where (a, b) varies over the points of $\mathbb{C}^2$ which satisfy the equation

$$b^2 - a^3 = 0.$$

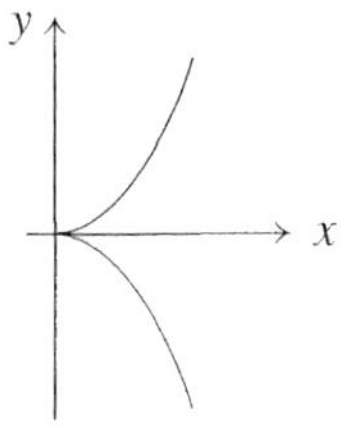

The only singular point is the origin. It corresponds to the maximal ideal $\mathfrak{p}_0 = (\bar{x}, \bar{y})$, where $\bar{x} = x \bmod (y^2 - x^3)$, $\bar{y} = y \bmod (y^2 - x^3) \in \mathcal{O}$. The maximal ideal $\mathfrak{p}_0\mathcal{O}_{\mathfrak{p}_0}$ of the local ring is generated by the elements $\bar{x}, \bar{y}$, and cannot be generated by a single element.

5. Normalization. Passing to the normalization $\tilde{\mathcal{O}}$ of a one-dimensional noetherian integral domain $\mathcal{O}$ means, in geometric terms, taking the *resolution* of the singularities that were just discussed. Indeed, if $X = \mathrm{Spec}(\mathcal{O})$ and

$\widetilde{X} = \operatorname{Spec}(\tilde{\mathcal{O}})$, then the inclusion $\mathcal{O} \hookrightarrow \tilde{\mathcal{O}}$ induces a morphism $f : \widetilde{X} \to X$.

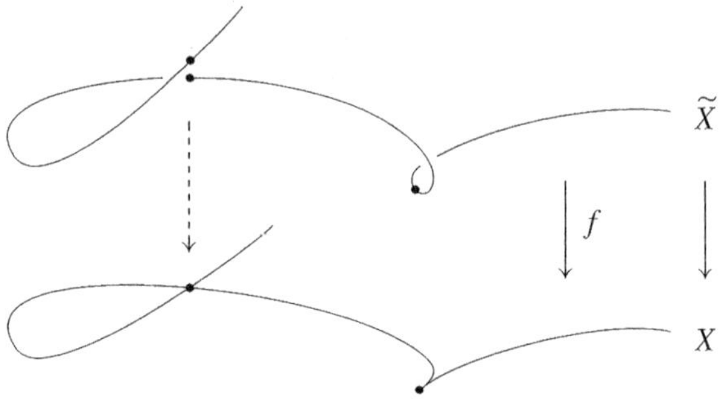

Since $\tilde{\mathcal{O}}$ is a Dedekind domain, the scheme $\widetilde{X}$ is to be considered as smooth. If $\mathfrak{p}\tilde{\mathcal{O}} = \tilde{\mathfrak{p}}_1^{e_r} \cdots \tilde{\mathfrak{p}}_r^{e_r}$ is the prime factorization of $\mathfrak{p}$ in $\tilde{\mathcal{O}}$, then $\tilde{\mathfrak{p}}_1, \dots, \tilde{\mathfrak{p}}_r$ are the different points of $\widetilde{X}$ that are mapped to $\mathfrak{p}$ by f. One can show that $\mathfrak{p}$ is a regular point of X – in the sense that $\mathcal{O}_\mathfrak{p}$ is a discrete valuation ring – if and only if $r = 1$, $e_1 = 1$ and $f_1 = (\tilde{\mathcal{O}}/\tilde{\mathfrak{p}}_1 : \mathcal{O}/\mathfrak{p}) = 1$.

6. Extensions. Let $\mathcal{O}$ be a Dedekind domain with field of fractions K. Let $L|K$ be a finite separable extension, and $\mathcal{O}$ the integral closure of $\mathcal{O}$ in L. Let $Y = \operatorname{Spec}(\mathcal{O})$, $X = \operatorname{Spec}(\mathcal{O})$, and

$$f : X \longrightarrow Y$$

the morphism induced by the inclusion $\mathcal{O} \hookrightarrow \mathcal{O}$. If $\mathfrak{p}$ is a maximal ideal of $\mathcal{O}$ and

$$\mathfrak{p}\mathcal{O} = \mathfrak{P}_1^{e_1} \cdots \mathfrak{P}_r^{e_r}$$

the prime decomposition of $\mathfrak{p}$ in $\mathcal{O}$, then $\mathfrak{P}_1, \dots, \mathfrak{P}_r$ are the different points of X which are mapped to $\mathfrak{p}$ by f. The morphism f is a "ramified covering." It is graphically represented by the following picture:

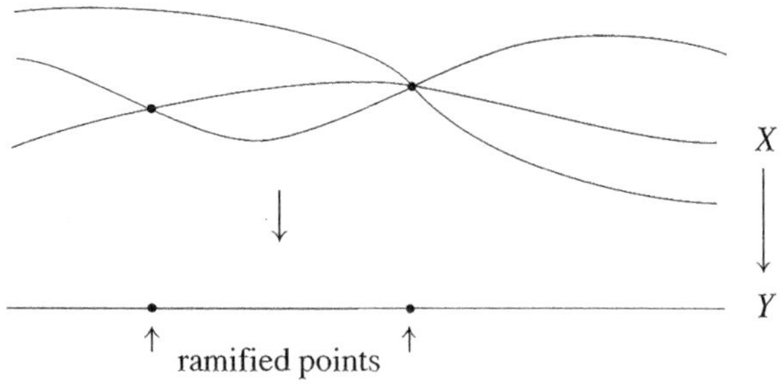

This picture, however, is a fair rendering of the algebraic situation only in the case where the residue class fields of $\mathcal{O}$ are algebraically closed (like for the ring $\mathbb{C}[x]$). Then, from the fundamental identity $\sum_i e_i f_i = n$, there are exactly $n = [L:K]$ points $\mathfrak{P}_1, \dots, \mathfrak{P}_n$ of X lying above each point $\mathfrak{p}$ of Y, except when $\mathfrak{p}$ is ramified in $\mathcal{O}$. At a point $\mathfrak{p}$ of ramification, several of the points $\mathfrak{P}_1, \dots, \mathfrak{P}_n$ coalesce. This also explains the terminology of ideals that "ramify."

If $L|K$ is Galois with Galois group $G = G(L|K)$, then every automorphism $\sigma \in G$ induces via $\sigma : \mathcal{O} \to \mathcal{O}$ an automorphism of schemes $\sigma : X \to X$. Since the ring $\mathcal{O}$ is fixed, the diagram

$$\begin{array}{ccc} X & \xrightarrow{\ \sigma\ } & X \\ & {\scriptstyle f}\searrow \quad \swarrow{\scriptstyle f} & \\ & Y & \end{array}$$

is commutative. Such an automorphism is called a **covering transformation** of the ramified covering X/Y. The group of covering transformations is denoted by $\mathrm{Aut}_Y(X)$. We thus have a canonical isomorphism

$$G(L|K) \cong \mathrm{Aut}_Y(X).$$

In chap. II, § 7, we will see that the composite of two unramified extensions of K is again unramified. The composite $\widetilde{K}$, taken inside some algebraic closure $\overline{K}$ of K, of *all* unramified extensions $L|K$ is called the **maximal unramified extension** of K. The integral closure $\tilde{\mathcal{O}}$ of $\mathcal{O}$ in $\widetilde{K}$ is still a one-dimensional integral domain, but in general no longer noetherian, and, as a rule, there will be infinitely many prime ideals lying above a given prime ideal $\mathfrak{p} \neq 0$ of $\mathcal{O}$. The scheme $\widetilde{Y} = \mathrm{Spec}(\tilde{\mathcal{O}})$ with the morphism

$$f : \widetilde{Y} \longrightarrow Y$$

is called the **universal covering** of Y. It plays the same rôle for schemes that the universal covering space $\widetilde{X} \to X$ of a topological space plays in topology. There the group of covering transformations $\mathrm{Aut}_X(\widetilde{X})$ is canonically isomorphic to the fundamental group $\pi_1(X)$. Therefore we define in our present context the **fundamental group** of the scheme Y by

$$\pi_1(Y) = \mathrm{Aut}_Y(\widetilde{Y}) = G(\widetilde{K}|K).$$

This establishes a first link of Galois theory with classical topology. This link is pursued much further in **étale topology**.

The geometric point of view of algebraic number fields explained in this section is corroborated very convincingly by the theory of function fields of algebraic curves over a finite field $\mathbb{F}_p$. In fact, a very close analogy exists between both theories.

§ 14. Function Fields

We conclude this chapter with a brief sketch of the theory of **function fields**. They represent a striking analogy with algebraic number fields, and since they are immediately related to geometry, they actually serve as an important model for the theory of algebraic number fields.

The ring $\mathbb{Z}$ of integers with its field of fractions $\mathbb{Q}$ exhibits obvious analogies with the polynomial ring $\mathbb{F}_p[t]$ over the field $\mathbb{F}_p$ with p elements and its field of fractions $\mathbb{F}_p(t)$. Like $\mathbb{Z}$, $\mathbb{F}_p[t]$ is also a principal ideal domain. The prime numbers correspond to the monic irreducible polynomials $p(t) \in \mathbb{F}_p[t]$. Like the prime numbers they have finite fields $\mathbb{F}_{p^d}$, $d = \deg(p(t))$, as their residue class rings. The difference is, however, that now all these fields have the same characteristic. The geometric character of the ring $\mathbb{F}_p[t]$ becomes much more apparent in that, for an element $f = f(t) \in \mathbb{F}_p[t]$, the value of f at a point $\mathfrak{p} = (p(t))$ of the affine scheme $X = \operatorname{Spec}(\mathbb{F}_p[t])$ is actually given by the value $f(a) \in \mathbb{F}_p$, if $p(t) = t - a$, or more generally by $f(\alpha) \in \mathbb{F}_{p^d}$, if $\alpha \in \mathbb{F}_{p^d}$ is a zero of $p(t)$. This is due to the isomorphism

$$\mathbb{F}_p[t]/\mathfrak{p} \xrightarrow{\sim} \mathbb{F}_{p^d},$$

which takes the residue class $f(\mathfrak{p}) = f \bmod \mathfrak{p}$ to $f(\alpha)$. In the analogy between, on the one hand, the progression of the prime numbers $2, 3, 5, 7, \dots$, and the growing of the cardinalities $p, p^2, p^3, p^4, \dots$ of the residue fields $\mathbb{F}_{p^d}$ on the other, resides one of the most profound mysteries of arithmetic.

One obtains the same arithmetic theory for the finite extensions K of $\mathbb{F}_p(t)$ as for algebraic number fields. This is clear from what we have developed for arbitrary one-dimensional noetherian integral domains. But the crucial difference with the number field case is seen in that the function field K hides away a finite number of further prime ideals, besides the prime ideals of $\mathcal{O}$, which must be taken into account in a fully-fledged development of the theory.

This phenomenon appears already for the rational function field $\mathbb{F}_p(t)$, where it is due to the fact that the choice of the unknown t which determines the ring of integrality $\mathbb{F}_p[t]$ is totally arbitrary. A different choice, say $t' = 1/t$, determines a completely different ring $\mathbb{F}_p[1/t]$, and thus completely different prime ideals. It is therefore crucial to build a theory which is independent of such choices. This may be done either via the theory of valuations, or scheme theoretically, i.e., in a geometric way.

Let us first sketch the more naïve method, via the theory of valuations. Let K be a finite extension of $\mathbb{F}_p(t)$ and $\mathcal{O}$ the integral closure of $\mathbb{F}_p[t]$ in K.

By § 11, for every prime ideal $\mathfrak{p} \neq 0$ of $\mathcal{O}$ there is an associated normalized discrete valuation, i.e., a surjective function

$$v_{\mathfrak{p}} : K \longrightarrow \mathbb{Z} \cup \{\infty\}$$

satisfying the properties

(i) $v_{\mathfrak{p}}(0) = \infty$,

(ii) $v_{\mathfrak{p}}(ab) = v_{\mathfrak{p}}(a) + v_{\mathfrak{p}}(b)$,

(iii) $v_{\mathfrak{p}}(a+b) \geq \min\{v_{\mathfrak{p}}(a), v_{\mathfrak{p}}(b)\}$.

The relation between the valuations and the prime decomposition in the Dedekind domain $\mathcal{O}$ is given by

$$(a) = \prod_{\mathfrak{p}} \mathfrak{p}^{v_{\mathfrak{p}}(a)} .$$

The definition of a discrete valuation of K does not require the subring $\mathcal{O}$ to be given in advance, and in fact, aside from those arising from $\mathcal{O}$, there are finitely many other discrete valuations of K. In the case of the field $\mathbb{F}_p(t)$ there is one more valuation, besides the ones associated to the prime ideals $\mathfrak{p} = (p(t))$ of $\mathbb{F}_p[t]$, namely, the **degree valuation** v_∞. For $\frac{f}{g} \in \mathbb{F}_p(t)$, $f, g \in \mathbb{F}_p[t]$, it is defined by

$$v_\infty\Big(\frac{f}{g}\Big) = \deg(g) - \deg(f).$$

It is associated to the prime ideal $\mathfrak{p} = y\mathbb{F}_p[y]$ of the ring $\mathbb{F}_p[y]$, where $y = 1/t$. One can show that this exhausts all normalized valuations of the field $\mathbb{F}_p(t)$.

For an arbitrary finite extension K of $\mathbb{F}_p(t)$, instead of restricting attention to prime ideals, one now considers all normalized discrete valuations $v_{\mathfrak{p}}$ of K in the above sense, where the index $\mathfrak{p}$ has kept only a symbolic value. As an analogue of the ideal group we form the "divisor group", i.e., the free abelian group generated by these symbols,

$$Div(K) = \Big\{ \sum_{\mathfrak{p}} n_{\mathfrak{p}}\mathfrak{p} \;\Big|\; n_{\mathfrak{p}} \in \mathbb{Z},\ n_{\mathfrak{p}} = 0 \text{ for almost all } \mathfrak{p} \Big\} .$$

We consider the mapping

$$\operatorname{div} : K^* \longrightarrow Div(K), \quad \operatorname{div}(f) = \sum_{\mathfrak{p}} v_{\mathfrak{p}}(f)\mathfrak{p},$$

the image of which is written $\mathcal{P}(K)$, and we define the divisor class group of K by

$$Cl(K) = Div(K)/\mathcal{P}(K).$$

Unlike the ideal class group of an algebraic number field, this group is not finite. Rather, one has the canonical homomorphism

$$\deg : Cl(K) \longrightarrow \mathbb{Z},$$

which associates to the class of $\mathfrak{p}$ the degree $\deg(\mathfrak{p}) = [\kappa(\mathfrak{p}) : \mathbb{F}_p]$ of the residue class field of the valuation ring of $\mathfrak{p}$, and which associates to the class of an arbitrary divisor $\mathfrak{a} = \sum_{\mathfrak{p}} n_{\mathfrak{p}}\mathfrak{p}$ the sum

$$\deg(\mathfrak{a}) = \sum_{\mathfrak{p}} n_{\mathfrak{p}} \deg(\mathfrak{p}).$$

For a principal divisor $\operatorname{div}(f)$, $f \in K^*$, we find by an easy calculation that $\deg(\operatorname{div}(f)) = 0$, so that the mapping deg is indeed well-defined. As an analogue of the finiteness of the class number of an algebraic number field, one obtains here the fact that, if not $Cl(K)$ itself, the kernel $Cl^0(K)$ of deg is finite. The infinitude of the class group of function fields must not be considered as strange. On the contrary, it is rather the finiteness in the number field case that should be regarded as a deficiency which calls for correction. The adequate appreciation of this situation and its amendment will be explained in chap. III, § 1.

The ideal, completely satisfactory framework for the theory of function fields is provided by the notion of **scheme**. In the last section we introduced affine schemes as pairs $(X, \mathcal{O}_X)$ consisting of a topological space $X = \operatorname{Spec}(\mathcal{O})$ and a sheaf of rings $\mathcal{O}_X$ on X. More generally, a scheme is a topological space X with a sheaf of rings $\mathcal{O}_X$ such that, for every point of X, there exists a neighbourhood U which, together with the restriction $\mathcal{O}_U$ of the sheaf $\mathcal{O}_X$ to U, is isomorphic to an affine scheme in the sense of § 13. This generalization of affine schemes is the correct notion for a function field K. It shows all prime ideals at once, and misses none.

In the case $K = \mathbb{F}_p(t)$ for instance, the corresponding scheme $(X, \mathcal{O}_X)$ is obtained by gluing the two rings $A = \mathbb{F}_p[u]$ and $B = \mathbb{F}_p[v]$, or more precisely the two affine schemes $U = \operatorname{Spec}(A)$ and $V = \operatorname{Spec}(B)$. Removing from U the point $\mathfrak{p}_0 = (u)$, and the point $\mathfrak{p}_\infty = (v)$ from V, one has $U - \{\mathfrak{p}_0\} = \operatorname{Spec}(\mathbb{F}_p[u, u^{-1}])$, $V - \{\mathfrak{p}_\infty\} = \operatorname{Spec}(\mathbb{F}_p[v, v^{-1}])$, and the isomorphism $f : \mathbb{F}_p[u, u^{-1}] \to \mathbb{F}_p[v, v^{-1}]$, $u \mapsto v^{-1}$, yields a bijection

$$\varphi : V - \{\mathfrak{p}_\infty\} \longrightarrow U - \{\mathfrak{p}_0\}, \quad \mathfrak{p} \longmapsto f^{-1}(\mathfrak{p}).$$

We now identify in the union $U \cup V$ the points of $V - \{\mathfrak{p}_\infty\}$ with those of $U - \{\mathfrak{p}_0\}$ by means of φ, and obtain a topological space X. It is immediately obvious how to obtain a sheaf of rings $\mathcal{O}_X$ on X from the two sheaves $\mathcal{O}_U$ and $\mathcal{O}_V$. Removing from X the point $\mathfrak{p}_\infty$, resp. $\mathfrak{p}_0$, one gets canonical isomorphisms

$$\big(X - \{\mathfrak{p}_\infty\}, \mathcal{O}_{X - \{\mathfrak{p}_\infty\}}\big) \cong (U, \mathcal{O}_U), \quad \big(X - \{\mathfrak{p}_0\}, \mathcal{O}_{X - \{\mathfrak{p}_0\}}\big) \cong (V, \mathcal{O}_V).$$

The pair $(X, \mathcal{O}_X)$ is the scheme corresponding to the field $\mathbb{F}_p(t)$. It is called the **projective line** over $\mathbb{F}_p$ and denoted $\mathbb{P}^1_{\mathbb{F}_p}$.

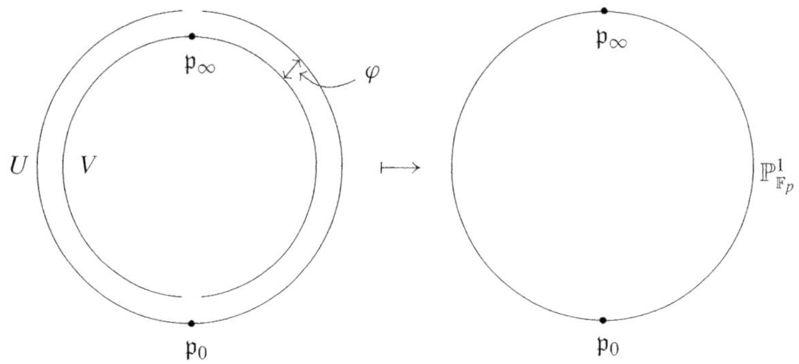

More generally, one may similarly associate a scheme $(X, \mathcal{O}_X)$ to an arbitrary extension $K|\mathbb{F}_p(t)$. For the precise description of this procedure we refer the reader to [51].

Chapter II

The Theory of Valuations

§ 1. The p-adic Numbers

The p-adic numbers were invented at the beginning of the twentieth century by the mathematician *Kurt Hensel* (1861–1941) with a view to introduce into number theory the powerful method of power series expansion which plays such a predominant rôle in function theory. The idea originated from the observation made in the last chapter that the numbers $f \in \mathbb{Z}$ may be viewed in analogy with the polynomials $f(z) \in \mathbb{C}[z]$ as functions on the space X of prime numbers in $\mathbb{Z}$, associating to them their "value" at the point $p \in X$, i.e., the element

$$f(p) := f \bmod p$$

in the residue class field $\kappa(p) = \mathbb{Z}/p\mathbb{Z}$.

This point of view suggests the further question: whether not only the "value" of the integer $f \in \mathbb{Z}$ at p, but also the higher derivatives of f can be reasonably defined. In the case of the polynomials $f(z) \in \mathbb{C}[z]$, the higher derivatives at the point $z = a$ are given by the coefficients of the expansion

$$f(z) = a_0 + a_1(z-a) + \cdots + a_n(z-a)^n,$$

and more generally, for rational functions $f(z) = \frac{g(z)}{h(z)} \in \mathbb{C}(z)$, with $g, h \in \mathbb{C}[z]$, they are defined by the Taylor expansion

$$f(z) = \sum_{\nu=0}^{\infty} a_\nu (z-a)^\nu,$$

provided there is no pole at $z = a$, i.e., as long as $(z-a) \nmid h(z)$. The fact that such an expansion can also be written down, relative to a prime number p in $\mathbb{Z}$, for any rational number $f \in \mathbb{Q}$ as long as it lies in the local ring

$$\mathbb{Z}_{(p)} = \left\{ \tfrac{g}{h} \mid g, h \in \mathbb{Z},\ p \nmid h \right\},$$

leads us to the notion of p-adic number. First, every positive integer $f \in \mathbb{N}$ admits a ***p*-adic expansion**

$$f = a_0 + a_1 p + \cdots + a_n p^n,$$

with coefficients a_i in $\{0, 1, \ldots, p-1\}$, i.e., in a fixed system of representatives of the "field of values" $\kappa(p) = \mathbb{F}_p$. This representation is clearly unique. It is computed explicitly by successively dividing by p, forming the following system of equations:

$$\begin{aligned} f &= a_0 + pf_1 , \\ f_1 &= a_1 + pf_2 , \\ &\vdots \\ f_{n-1} &= a_{n-1} + pf_n , \\ f_n &= a_n . \end{aligned}$$

Here $a_i \in \{0, 1, \ldots, p-1\}$ denotes the representative of $f_i \bmod p \in \mathbb{Z}/p\mathbb{Z}$. In concrete cases, one sometimes writes the number f simply as the sequence of digits $a_0, a_1 a_2 \ldots a_n$, for instance

$$\begin{aligned} 216 &= 0,0011011 \quad &(2\text{-adic}), \\ 216 &= 0,0022 \quad &(3\text{-adic}), \\ 216 &= 1,331 \quad &(5\text{-adic}). \end{aligned}$$

As soon as one tries to write down such p-adic expansions also for negative integers, let alone for fractions, one is forced to allow infinite series

$$\sum_{\nu=0}^{\infty} a_\nu p^\nu = a_0 + a_1 p + a_2 p^2 + \cdots .$$

This notation should at first be understood in a purely formal sense, i.e., $\sum_{\nu=0}^{\infty} a_\nu p^\nu$ simply stands for the sequence of partial sums

$$s_n = \sum_{\nu=0}^{n-1} a_\nu p^\nu , \quad n = 1, 2, \ldots$$

(1.1) Definition. *Fix a prime number p. A* ***p-adic integer*** *is a formal infinite series*

$$a_0 + a_1 p + a_2 p^2 + \cdots ,$$

where $0 \le a_i < p$, for all $i = 0, 1, 2, \ldots$ The set of all p-adic integers is denoted by $\mathbb{Z}_p$.

The p-adic expansion of an arbitrary number $f \in \mathbb{Z}_{(p)}$ results from the following proposition about the residue classes in $\mathbb{Z}/p^n\mathbb{Z}$.

(1.2) Proposition. *The residue classes $a \bmod p^n \in \mathbb{Z}/p^n\mathbb{Z}$ can be uniquely represented in the form*

$$a \equiv a_0 + a_1 p + a_2 p^2 + \cdots + a_{n-1} p^{n-1} \bmod p^n$$

where $0 \le a_i < p$ for $i = 0, \ldots, n-1$.

Proof (induction on n): This is clear for $n = 1$. Assume the statement is proved for $n-1$. Then we have a unique representation

$$a = a_0 + a_1 p + a_2 p^2 + \cdots + a_{n-2} p^{n-2} + g p^{n-1},$$

for some integer g. If $g \equiv a_{n-1} \bmod p$ such that $0 \le a_{n-1} < p$, then a_{n-1} is uniquely determined by a, and the congruence of the proposition holds. □

Every integer f and, more generally, every rational number $f \in \mathbb{Z}_{(p)}$ the denominator of which is not divisible by p, defines a sequence of residue classes

$$\bar{s}_n = f \bmod p^n \in \mathbb{Z}/p^n\mathbb{Z}, \quad n = 1, 2, \ldots,$$

for which we find, by the preceding proposition,

$$\begin{aligned} \bar{s}_1 &= a_0 \bmod p, \\ \bar{s}_2 &= a_0 + a_1 p \bmod p^2, \\ \bar{s}_3 &= a_0 + a_1 p + a_2 p^2 \bmod p^3, \quad \text{etc.}, \end{aligned}$$

with uniquely determined coefficients $a_0, a_1, a_2, \ldots \in \{0, 1, \ldots, p-1\}$ which keep their meaning from one line to the next. The sequence of numbers

$$s_n = a_0 + a_1 p + a_2 p^2 + \cdots + a_{n-1} p^{n-1}, \quad n = 1, 2, \ldots,$$

defines a p-adic integer

$$\sum_{\nu=0}^{\infty} a_\nu p^\nu \in \mathbb{Z}_p.$$

We call it the ***p*-adic expansion** of f.

In analogy with the Laurent series $f(z) = \sum_{\nu=-m}^{\infty} a_\nu (z-a)^\nu$, we now extend the domain of p-adic integers into that of the formal series

$$\sum_{\nu=-m}^{\infty} a_\nu p^\nu = a_{-m} p^{-m} + \cdots + a_{-1} p^{-1} + a_0 + a_1 p + \cdots,$$

where $m \in \mathbb{Z}$ and $0 \le a_\nu < p$. Such series we call simply ***p*-adic numbers** and we write $\mathbb{Q}_p$ for the set of all these p-adic numbers. If $f \in \mathbb{Q}$ is any rational number, then we write

$$f = \frac{g}{h} p^{-m} \quad \text{where } g, h \in \mathbb{Z}, \quad (gh, p) = 1,$$

and if

$$a_0 + a_1 p + a_2 p^2 + \cdots$$

is the p-adic expansion of $\frac{g}{h}$, then we attach to f the p-adic number

$$a_0 p^{-m} + a_1 p^{-m+1} + \cdots + a_m + a_{m+1} p + \cdots \in \mathbb{Q}_p$$

as its p-adic expansion.

In this way we obtain a canonical mapping

$$\mathbb{Q} \longrightarrow \mathbb{Q}_p,$$

which takes $\mathbb{Z}$ into $\mathbb{Z}_p$ and is injective. For if $a, b \in \mathbb{Z}$ have the same p-adic expansion, then $a - b$ is divisible by p^n for every n, and hence $a = b$. We now identify $\mathbb{Q}$ with its image in $\mathbb{Q}_p$, so that we may write $\mathbb{Q} \subseteq \mathbb{Q}_p$ and $\mathbb{Z} \subseteq \mathbb{Z}_p$. Thus, for every rational number $f \in \mathbb{Q}$, we obtain an identity

$$f = \sum_{\nu=-m}^{\infty} a_\nu p^\nu .$$

This establishes the arithmetic analogue of the function-theoretic power series expansion for which we were looking.

Examples: a) $-1 = (p-1) + (p-1)p + (p-1)p^2 + \cdots$.

In fact, we have

$$-1 = (p-1) + (p-1)p + \cdots + (p-1)p^{n-1} - p^n,$$

hence $-1 \equiv (p-1) + (p-1)p + \cdots + (p-1)p^{n-1} \bmod p^n$.

b) $\dfrac{1}{1-p} = 1 + p + p^2 + \cdots$.

In fact,

$$1 = (1 + p + \cdots + p^{n-1})(1-p) + p^n,$$

hence $$\frac{1}{1-p} \equiv 1 + p + \cdots + p^{n-1} \bmod p^n .$$

One can define addition and multiplication of p-adic numbers which turn $\mathbb{Z}_p$ into a ring, and $\mathbb{Q}_p$ into its field of fractions. However, the direct approach, defining sum and product via the usual carry-over rules for digits, as one does it when dealing with real numbers as decimal fractions, leads into complications. They disappear once we use another representation of the p-adic numbers $f = \sum_{\nu=0}^{\infty} a_\nu p^\nu$, viewing them not as sequences of sums of integers

$$s_n = \sum_{\nu=0}^{n-1} a_\nu p^\nu \in \mathbb{Z},$$

but rather as sequences of **residue classes**

$$\bar{s}_n = s_n \bmod p^n \in \mathbb{Z}/p^n\mathbb{Z}.$$

The terms of such a sequence lie in different rings $\mathbb{Z}/p^n\mathbb{Z}$, but these are related by the canonical projections

$$\mathbb{Z}/p\mathbb{Z} \xleftarrow{\lambda_1} \mathbb{Z}/p^2\mathbb{Z} \xleftarrow{\lambda_2} \mathbb{Z}/p^3\mathbb{Z} \xleftarrow{\lambda_3} \cdots$$

and we find

$$\lambda_n(\bar{s}_{n+1}) = \bar{s}_n .$$

In the direct product

$$\prod_{n=1}^{\infty} \mathbb{Z}/p^n\mathbb{Z} = \big\{ (x_n)_{n\in\mathbb{N}} \bigm| x_n \in \mathbb{Z}/p^n\mathbb{Z} \big\},$$

we now consider all elements $(x_n)_{n\in\mathbb{N}}$ with the property that

$$\lambda_n(x_{n+1}) = x_n \quad \text{for all} \quad n = 1, 2, \ldots$$

This set is called the **projective limit** of the rings $\mathbb{Z}/p^n\mathbb{Z}$ and is denoted by $\varprojlim_n \mathbb{Z}/p^n\mathbb{Z}$. In other words, we have

$$\varprojlim_n \mathbb{Z}/p^n\mathbb{Z} = \Big\{ (x_n)_{n\in\mathbb{N}} \in \prod_{n=1}^{\infty} \mathbb{Z}/p^n\mathbb{Z} \Bigm| \lambda_n(x_{n+1}) = x_n,\ n = 1, 2, \ldots \Big\} .$$

The modified representation of the p-adic numbers alluded to above now follows from the

(1.3) Proposition. *Associating to every p-adic integer*

$$f = \sum_{\nu=0}^{\infty} a_\nu p^\nu$$

the sequence $(\bar{s}_n)_{n\in\mathbb{N}}$ of residue classes

$$\bar{s}_n = \sum_{\nu=0}^{n-1} a_\nu p^\nu \bmod p^n \in \mathbb{Z}/p^n\mathbb{Z},$$

yields a bijection

$$\mathbb{Z}_p \xrightarrow{\sim} \varprojlim_n \mathbb{Z}/p^n\mathbb{Z} .$$

The proof is an immediate consequence of proposition (1.2). The projective limit $\varprojlim \mathbb{Z}/p^n\mathbb{Z}$ offers the advantage of being clearly a ring. In fact, it is a subring of the direct product $\prod_{n=1}^{\infty} \mathbb{Z}/p^n\mathbb{Z}$ where addition and multiplication are defined componentwise. We identify $\mathbb{Z}_p$ with $\varprojlim \mathbb{Z}/p^n\mathbb{Z}$ and obtain the **ring of p-adic integers** $\mathbb{Z}_p$.

Since every element $f \in \mathbb{Q}_p$ admits a representation

$$f = p^{-m} g$$

with $g \in \mathbb{Z}_p$, addition and multiplication extend from $\mathbb{Z}_p$ to $\mathbb{Q}_p$ and $\mathbb{Q}_p$ becomes the field of fractions of $\mathbb{Z}_p$.

In $\mathbb{Z}_p$, we found the rational integers $a \in \mathbb{Z}$ which were determined by the congruences

$$a \equiv a_0 + a_1 p + \cdots + a_{n-1} p^{n-1} \bmod p^n ,$$

$0 \le a_i < p$. Making the identification

$$\mathbb{Z}_p = \varprojlim_n \mathbb{Z}/p^n\mathbb{Z}$$

the subset $\mathbb{Z}$ is taken to the set of tuples

$$(a \bmod p,\ a \bmod p^2,\ a \bmod p^3,\ \ldots) \in \prod_{n=1}^{\infty} \mathbb{Z}/p^n\mathbb{Z}$$

and thereby is realized as a subring of $\mathbb{Z}_p$. We obtain $\mathbb{Q}$ as a subfield of the field $\mathbb{Q}_p$ of p-adic numbers in the same way.

Despite their origin in function-theoretic ideas, the p-adic numbers live up to their destiny entirely within arithmetic, more precisely at its classical heart, the **Diophantine equations**. Such an equation

$$F(x_1, \ldots, x_n) = 0$$

is given by a polynomial $F \in \mathbb{Z}[x_1, \ldots, x_n]$, and the question is whether it admits solutions in integers. This difficult problem can be weakened by considering, instead of the equation, all the congruences

$$F(x_1, \ldots, x_n) \equiv 0 \bmod m .$$

By the Chinese remainder theorem, this amounts to considering the congruences

$$F(x_1, \ldots, x_n) \equiv 0 \bmod p^{\nu}$$

modulo all prime powers. The hope is to obtain in this way information about the original equation. This plethora of congruences is now synthesized again into a single equation by means of the p-adic numbers. In fact, one has the

(1.4) Proposition. *Let $F(x_1, \dots, x_n)$ be a polynomial with integer coefficients, and fix a prime number p. The congruence*

$$F(x_1, \dots, x_n) \equiv 0 \bmod p^\nu$$

is solvable for arbitrary $\nu \geq 1$ if and only if the equation

$$F(x_1, \dots, x_n) = 0$$

is solvable in p-adic integers.

Proof: As established above, we view the ring $\mathbb{Z}_p$ as the projective limit

$$\mathbb{Z}_p = \varprojlim_\nu \mathbb{Z}/p^\nu\mathbb{Z} \subseteq \prod_{\nu=1}^{\infty} \mathbb{Z}/p^\nu\mathbb{Z}.$$

Viewed over the ring on the right, the equation $F = 0$ splits up into components over the individual rings $\mathbb{Z}/p^\nu\mathbb{Z}$, namely, the congruences

$$F(x_1, \dots, x_n) \equiv 0 \bmod p^\nu .$$

If now

$$(x_1, \dots, x_n) = \left(x_1^{(\nu)}, \dots, x_n^{(\nu)}\right)_{\nu\in\mathbb{N}} \in \mathbb{Z}_p^n,$$

with $(x_i^{(\nu)})_{\nu\in\mathbb{N}} \in \mathbb{Z}_p = \varprojlim_\nu \mathbb{Z}/p^\nu\mathbb{Z}$, is a p-adic solution of the equation $F(x_1, \dots, x_n) = 0$, then the congruences are solved by

$$F\left(x_1^{(\nu)}, \dots, x_n^{(\nu)}\right) \equiv 0 \bmod p^\nu , \quad \nu = 1, 2, \dots$$

Conversely, let a solution $(x_1^{(\nu)}, \dots, x_n^{(\nu)})$ of the congruence

$$F(x_1, \dots, x_n) \equiv 0 \bmod p^\nu$$

be given for every $\nu \geq 1$. If the elements $(x_i^{(\nu)})_{\nu\in\mathbb{N}} \in \prod_{\nu=1}^{\infty} \mathbb{Z}/p^\nu\mathbb{Z}$ are already in $\varprojlim \mathbb{Z}/p^\nu\mathbb{Z}$, for all $i = 1, \dots, n$, then we have a p-adic solution of the equation $F = 0$. But this is not automatically the case. We will therefore extract a subsequence from the sequence $(x_1^{(\nu)}, \dots, x_n^{(\nu)})$ which fits our needs. For simplicity of notation we only carry this out in the case $n = 1$, writing $x_\nu = x_1^{(\nu)}$. The general case follows exactly the same pattern.

In what follows, we view (x_ν) as a sequence in $\mathbb{Z}$. Since $\mathbb{Z}/p\mathbb{Z}$ is finite, there are infinitely many terms x_ν which mod p are congruent to the same element $y_1 \in \mathbb{Z}/p\mathbb{Z}$. Hence we may choose a subsequence $\{x_\nu^{(1)}\}$ of $\{x_\nu\}$ such that

$$x_\nu^{(1)} \equiv y_1 \bmod p \quad \text{and} \quad F\left(x_\nu^{(1)}\right) \equiv 0 \bmod p .$$

Likewise, we may extract from $\{x_\nu^{(1)}\}$ a subsequence $\{x_\nu^{(2)}\}$ such that

$$x_\nu^{(2)} \equiv y_2 \bmod p^2 \quad \text{and} \quad F\big(x_\nu^{(2)}\big) \equiv 0 \bmod p^2,$$

where $y_2 \in \mathbb{Z}/p^2\mathbb{Z}$ evidently satisfies $y_2 \equiv y_1 \bmod p$. Continuing in this way, we obtain for each $k \geq 1$ a subsequence $\{x_\nu^{(k)}\}$ of $\{x_\nu^{(k-1)}\}$ the terms of which satisfy the congruences

$$x_\nu^{(k)} \equiv y_k \bmod p^k \quad \text{and} \quad F(x_\nu^{(k)}) \equiv 0 \bmod p^k$$

for some $y_k \in \mathbb{Z}/p^k\mathbb{Z}$ such that

$$y_k \equiv y_{k-1} \bmod p^{k-1}.$$

The y_k define a p-adic integer $y = (y_k)_{k\in\mathbb{N}} \in \varprojlim_k \mathbb{Z}/p^k\mathbb{Z} = \mathbb{Z}_p$ satisfying

$$F(y_k) \equiv 0 \bmod p^k$$

for all $k \geq 1$. In other words, $F(y) = 0$. □

Exercise 1. A p-adic number $a = \sum_{\nu=-m}^{\infty} a_\nu p^\nu \in \mathbb{Q}_p$ is a rational number if and only if the sequence of digits is periodic (possibly with a finite string before the first period).

Hint: Write $p^m a = b + c\,\dfrac{p^\ell}{1-p^n}$, $0 \leq b < p^\ell$, $0 \leq c < p^n$.

Exercise 2. A p-adic integer $a = a_0 + a_1 p + a_2 p^2 + \cdots$ is a unit in the ring $\mathbb{Z}_p$ if and only if $a_0 \neq 0$.

Exercise 3. Show that the equation $x^2 = 2$ has a solution in $\mathbb{Z}_7$.

Exercise 4. Write the numbers $\frac{2}{3}$ and $-\frac{2}{3}$ as 5-adic numbers.

Exercise 5. The field $\mathbb{Q}_p$ of p-adic numbers has no automorphisms except the identity.

Exercise 6. How is the addition, subtraction, multiplication and division of rational numbers reflected in the representation by p-adic digits?

§ 2. The p-adic Absolute Value

The representation of a p-adic integer

$$(1) \qquad a_0 + a_1 p + a_2 p^2 + \cdots, \quad 0 \leq a_i < p,$$

resembles very much the decimal fraction representation

$$a_0 + a_1\Big(\frac{1}{10}\Big) + a_2\Big(\frac{1}{10}\Big)^2 + \cdots, \qquad 0 \le a_i < 10,$$

of a real number between 0 and 10. But it does not converge as the decimal fraction does. Nonetheless, the field $\mathbb{Q}_p$ of p-adic numbers can be constructed from the field $\mathbb{Q}$ in the same fashion as the field of real numbers $\mathbb{R}$. The key to this is to replace the ordinary absolute value by a new "p-adic" absolute value $|\ |_p$ with respect to which the series (1) converge so that the p-adic numbers appear in the usual manner as limits of Cauchy sequences of rational numbers. This approach was proposed by the Hungarian mathematician *J. Kürschak*. The p-adic absolute value $|\ |_p$ is defined as follows.

Let $a = \frac{b}{c}$, $b, c \in \mathbb{Z}$ be a nonzero rational number. We extract from b and from c as high a power of the prime number p as possible,

$$a = p^m \frac{b'}{c'}, \qquad (b'c', p) = 1, \tag{2}$$

and we put

$$|a|_p = \frac{1}{p^m}.$$

Thus the p-adic value no longer measures the size of a number $a \in \mathbb{N}$. Instead it becomes small if the number is divisible by a high power of p. This elaborates on the idea suggested in (1.4) that an integer has to be 0 if it is infinitely divisible by p. In particular, the summands of a p-adic series $a_0 + a_1 p + a_2 p^2 + \cdots$ form a sequence converging to 0 with respect to $|\ |_p$.

The exponent m in the representation (2) of the number a is denoted by $v_p(a)$, and one puts formally $v_p(0) = \infty$. This gives the function

$$v_p : \mathbb{Q} \longrightarrow \mathbb{Z} \cup \{\infty\},$$

which is easily checked to satisfy the properties

1) $v_p(a) = \infty \Longleftrightarrow a = 0$,

2) $v_p(ab) = v_p(a) + v_p(b)$,

3) $v_p(a+b) \ge \min\{v_p(a), v_p(b)\}$,

where $x + \infty = \infty$, $\infty + \infty = \infty$ and $\infty > x$, for all $x \in \mathbb{Z}$. The function v_p is called the ***p*-adic exponential valuation** of $\mathbb{Q}$. The ***p*-adic absolute value** is given by

$$|\ |_p : \mathbb{Q} \longrightarrow \mathbb{R}, \quad a \longmapsto |a|_p = p^{-v_p(a)}.$$

In view of 1), 2), 3), it satisfies the conditions of a *norm* on $\mathbb{Q}$:

1) $|a|_p = 0 \Longleftrightarrow a = 0$,

2) $|ab|_p = |a|_p |b|_p$,

3) $|a+b|_p \le \max\{|a|_p, |b|_p\} \le |a|_p + |b|_p$.

One can show that the absolute values $|\ |_p$ and $|\ |$ essentially exhaust all norms on $\mathbb{Q}$: any further norm is a power $|\ |_p^s$ or $|\ |^s$, for some real number $s > 0$ (see (3.7)). The usual absolute value $|\ |$ is denoted in this context by $|\ |_\infty$. The good reason for this will be explained in due course. In conjunction with the absolute values $|\ |_p$, it satisfies the following important **product formula**:

(2.1) Proposition. *For every rational number $a \neq 0$, one has*

$$\prod_p |a|_p = 1,$$

where p varies over all prime numbers as well as the symbol ∞.

Proof: In the prime factorization

$$a = \pm \prod_{p \neq \infty} p^{\nu_p}$$

of a, the exponent ν_p of p is precisely the exponential valuation $v_p(a)$ and the sign equals $\frac{a}{|a|_\infty}$. The equation therefore reads

$$a = \frac{a}{|a|_\infty} \prod_{p \neq \infty} \frac{1}{|a|_p},$$

so that one has indeed $\prod_p |a|_p = 1$. □

The notation $|\ |_\infty$ for the ordinary absolute value is motivated by the analogy of the field of rational numbers $\mathbb{Q}$ with the rational function field $k(t)$ over a finite field k, with which we started our considerations. Instead of $\mathbb{Z}$, we have inside $k(t)$ the polynomial ring $k[t]$, the prime ideals $\mathfrak{p} \neq 0$ of which are given by the monic irreducible polynomials $p(t) \in k[t]$. For every such $\mathfrak{p}$, one defines an absolute value

$$|\ |_{\mathfrak{p}} : k(t) \longrightarrow \mathbb{R}$$

as follows. Let $f(t) = \frac{g(t)}{h(t)}$, $g(t), h(t) \in k[t]$ be a nonzero rational function. We extract from $g(t)$ and $h(t)$ the highest possible power of the irreducible polynomial $p(t)$,

$$f(t) = p(t)^m \frac{\tilde{g}(t)}{\tilde{h}(t)}, \quad (\tilde{g}\tilde{h}, p) = 1,$$

and put

$$v_\mathfrak{p}(f) = m\,, \quad |f|_\mathfrak{p} = q_\mathfrak{p}^{-v_\mathfrak{p}(f)}\,,$$

where $q_\mathfrak{p} = q^{d_\mathfrak{p}}$, $d_\mathfrak{p}$ being the degree of the residue class field of $\mathfrak{p}$ over k and q a fixed real number > 1. Furthermore we put $v_\mathfrak{p}(0) = \infty$ and $|0|_\mathfrak{p} = 0$, and obtain for $v_\mathfrak{p}$ and $|\;|_\mathfrak{p}$ the same conditions 1), 2), 3) as for v_p and $|\;|_p$ above. In the case $\mathfrak{p} = (t - a)$ for $a \in k$, the valuation $v_\mathfrak{p}(f)$ is clearly the order of the zero, resp. pole, of the function $f = f(t)$ at $t = a$.

But for the function field $k(t)$, there is one more exponential valuation

$$v_\infty : k(t) \longrightarrow \mathbb{Z} \cup \{\infty\}\,,$$

namely

$$v_\infty(f) = \deg(h) - \deg(g)\,,$$

where $f = \frac{g}{h} \neq 0$, $g, h \in k[t]$. It describes the order of zero, resp. pole, of $f(t)$ at the point at infinity ∞, i.e., the order of zero, resp. pole, of the function $f(1/t)$ at the point $t = 0$. It is associated to the prime ideal $\mathfrak{p} = (1/t)$ of the ring $k[1/t] \subseteq k(t)$ in the same way as the exponential valuations $v_\mathfrak{p}$ are associated to the prime ideals $\mathfrak{p}$ of $k[t]$. Putting

$$|f|_\infty = q^{-v_\infty(f)}\,,$$

the unique factorization in $k(t)$ yields, as in (2.1) above, the formula

$$\prod_\mathfrak{p} |f|_\mathfrak{p} = 1\,,$$

where $\mathfrak{p}$ varies over the prime ideals of $k[t]$ as well as the symbol ∞, which now denotes the point at infinity (see chap. I, §14, p. 95).

In view of the product formula (2.1), the above consideration shows that the ordinary absolute value $|\;|$ of $\mathbb{Q}$ should be thought of as being associated to a virtual point at infinity. This point of view justifies the notation $|\;|_\infty$, obeys our constant *leitmotiv* to study numbers as functions from a geometric perspective, and it will fulfill the expectations thus raised in an ever growing and amazing manner. The decisive difference between the absolute value $|\;|_\infty$ of $\mathbb{Q}$ and the absolute value $|\;|_\infty$ of $k(t)$ is, however, that the former is not derived from any exponential valuation $v_\mathfrak{p}$ attached to a prime ideal.

Having introduced the p-adic absolute value $|\;|_p$ on the field $\mathbb{Q}$, let us now give a new definition of the field $\mathbb{Q}_p$ of p-adic numbers, imitating the construction of the field of real numbers. We will verify afterwards that this new, analytic construction does agree with Hensel's definition, which was motivated by function theory.

A **Cauchy sequence** with respect to $|\ |_p$ is by definition a sequence $\{x_n\}$ of rational numbers such that for every $\varepsilon > 0$, there exists a positive integer n_0 satisfying

$$|x_n - x_m|_p < \varepsilon \quad \text{for all} \quad n, m \geq n_0 .$$

Example: Every formal series

$$\sum_{\nu=0}^{\infty} a_\nu p^\nu, \quad 0 \leq a_\nu < p,$$

provides a Cauchy sequence via its partial sums

$$x_n = \sum_{\nu=0}^{n-1} a_\nu p^\nu,$$

because for $n > m$ one has

$$|x_n - x_m|_p = \Big| \sum_{\nu=m}^{n-1} a_\nu p^\nu \Big|_p \leq \max_{m \leq \nu < n} \big\{ |a_\nu p^\nu|_p \big\} \leq \frac{1}{p^m} .$$

A sequence $\{x_n\}$ in $\mathbb{Q}$ is called a **nullsequence** with respect to $|\ |_p$ if $|x_n|_p$ is a sequence converging to 0 in the usual sense.

Example: $1, p, p^2, p^3, \ldots$

The Cauchy sequences form a ring R, the nullsequences form a maximal ideal $\mathfrak{m}$, and we define afresh the field of p-adic numbers to be the residue class field

$$\mathbb{Q}_p := R/\mathfrak{m} .$$

We embed $\mathbb{Q}$ in $\mathbb{Q}_p$ by associating to every element $a \in \mathbb{Q}$ the residue class of the constant sequence $(a, a, a, \ldots)$. The p-adic absolute value $|\ |_p$ on $\mathbb{Q}$ is extended to $\mathbb{Q}_p$ by giving the element $x = \{x_n\} \bmod \mathfrak{m} \in R/\mathfrak{m}$ the absolute value

$$|x|_p := \lim_{n \to \infty} |x_n|_p \in \mathbb{R} .$$

This limit exists because $\{|x_n|_p\}$ is a Cauchy sequence in $\mathbb{R}$, and it is independent of the choice of the sequence $\{x_n\}$ within its class $\bmod \mathfrak{m}$ because any p-adic nullsequence $\{y_n\} \in \mathfrak{m}$ satisfies of course $\lim_{n \to \infty} |y_n|_p = 0$.

The p-adic exponential valuation v_p on $\mathbb{Q}$ extends to an exponential valuation

$$v_p : \mathbb{Q}_p \longrightarrow \mathbb{Z} \cup \{\infty\} .$$

In fact, if $x \in \mathbb{Q}_p$ is the class of the Cauchy sequence $\{x_n\}$ where $x_n \neq 0$, then

$$v_p(x_n) = -\log_p |x_n|_p$$

either diverges to ∞ or is a Cauchy sequence in $\mathbb{Z}$ which eventually must become constant for large n because $\mathbb{Z}$ is discrete. We put

$$v_p(x) = \lim_{n\to\infty} v_p(x_n) = v_p(x_n) \quad \text{for } n \geq n_0 .$$

Again we find for all $x \in \mathbb{Q}_p$ that

$$|x|_p = p^{-v_p(x)} .$$

As for the field of real numbers one proves the

(2.2) Proposition. *The field $\mathbb{Q}_p$ of p-adic numbers is* **complete** *with respect to the absolute value $|\ |_p$, i.e., every Cauchy sequence in $\mathbb{Q}_p$ converges with respect to $|\ |_p$.*

As well as the field $\mathbb{R}$, we thus obtain for each prime number p a new field $\mathbb{Q}_p$ with equal rights and standing, so that $\mathbb{Q}$ has given rise to the infinite family of fields

$$\mathbb{Q}_2,\ \mathbb{Q}_3,\ \mathbb{Q}_5,\ \mathbb{Q}_7,\ \mathbb{Q}_{11},\ \ldots,\ \mathbb{Q}_\infty = \mathbb{R}.$$

An important special property of the p-adic absolute values $|\ |_p$ lies in the fact that they do not only satisfy the usual triangle inequality, but also the stronger version

$$|x+y|_p \leq \max\{\,|x|_p, |y|_p\} .$$

This yields the following remarkable proposition, which gives us a new definition of the p-adic *integers*.

(2.3) Proposition. *The set*

$$\mathbb{Z}_p := \{x \in \mathbb{Q}_p \mid |x|_p \leq 1\}$$

is a subring of $\mathbb{Q}_p$. It is the closure with respect to $|\ |_p$ of the ring $\mathbb{Z}$ in the field $\mathbb{Q}_p$.

Proof: That $\mathbb{Z}_p$ is closed under addition and multiplication follows from

$$|x+y|_p \le \max\{|x|_p, |y|_p\} \quad \text{and} \quad |xy|_p = |x|_p|y|_p .$$

If $\{x_n\}$ is a Cauchy sequence in $\mathbb{Z}$ and $x = \lim_{n\to\infty} x_n$, then $|x_n|_p \le 1$ implies also $|x|_p \le 1$, hence $x \in \mathbb{Z}_p$. Conversely, let $x = \lim_{n\to\infty} x_n \in \mathbb{Z}_p$, for a Cauchy sequence $\{x_n\}$ in $\mathbb{Q}$. We saw above that one has $|x|_p = |x_n|_p \le 1$ for $n \ge n_0$, i.e., $x_n = \frac{a_n}{b_n}$, with $a_n, b_n \in \mathbb{Z}$, $(b_n, p) = 1$. Choosing for each $n \ge n_0$ a solution $y_n \in \mathbb{Z}$ of the congruence $b_n y_n \equiv a_n \bmod p^n$ yields $|x_n - y_n|_p \le \frac{1}{p^n}$ and hence $x = \lim_{n\to\infty} y_n$, so that x belongs to the closure of $\mathbb{Z}$. □

The group of units of $\mathbb{Z}_p$ is obviously

$$\mathbb{Z}_p^* = \{x \in \mathbb{Z}_p \mid |x|_p = 1\} .$$

Every element $x \in \mathbb{Q}_p^*$ admits a unique representation

$$x = p^m u \quad \text{with } m \in \mathbb{Z} \text{ and } u \in \mathbb{Z}_p^* .$$

For if $v_p(x) = m \in \mathbb{Z}$, then $v_p(xp^{-m}) = 0$, hence $|xp^{-m}|_p = 1$, i.e., $u = xp^{-m} \in \mathbb{Z}_p^*$. Furthermore we have the

(2.4) Proposition. *The nonzero ideals of the ring $\mathbb{Z}_p$ are the principal ideals*

$$p^n\mathbb{Z}_p = \{x \in \mathbb{Q}_p \mid v_p(x) \ge n\} ,$$

with $n \ge 0$, and one has

$$\mathbb{Z}_p/p^n\mathbb{Z}_p \cong \mathbb{Z}/p^n\mathbb{Z} .$$

Proof: Let $\mathfrak{a} \ne (0)$ be an ideal of $\mathbb{Z}_p$ and $x = p^m u$, $u \in \mathbb{Z}_p^*$, an element of $\mathfrak{a}$ with smallest possible m (since $|x|_p \le 1$, one has $m \ge 0$). Then $\mathfrak{a} = p^m\mathbb{Z}_p$ because $y = p^n u' \in \mathfrak{a}$, $u' \in \mathbb{Z}_p^*$, implies $n \ge m$, hence $y = (p^{n-m}u')p^m \in p^m\mathbb{Z}_p$. The homomorphism

$$\mathbb{Z} \longrightarrow \mathbb{Z}_p/p^n\mathbb{Z}_p, \quad a \longmapsto a \bmod p^n\mathbb{Z}_p ,$$

has kernel $p^n\mathbb{Z}$ and is surjective. Indeed, for every $x \in \mathbb{Z}_p$, there exists by (2.3) an $a \in \mathbb{Z}$ such that

$$|x-a|_p \le \frac{1}{p^n} ,$$

i.e., $v_p(x-a) \geq n$, therefore $x - a \in p^n\mathbb{Z}_p$ and hence $x \equiv a \bmod p^n\mathbb{Z}_p$. So we obtain an isomorphism

$$\mathbb{Z}_p/p^n\mathbb{Z}_p \cong \mathbb{Z}/p^n\mathbb{Z}. \qquad \square$$

We now want to establish the link with Hensel's definition of the ring $\mathbb{Z}_p$ and the field $\mathbb{Q}_p$ which was given in § 1. There we defined the p-adic integers as formal series

$$\sum_{\nu=0}^{\infty} a_\nu p^\nu, \quad 0 \leq a_\nu < p,$$

which we identified with sequences

$$\bar{s}_n = s_n \bmod p^n \in \mathbb{Z}/p^n\mathbb{Z}, \quad n = 1, 2, \ldots,$$

where s_n was the partial sum

$$s_n = \sum_{\nu=0}^{n-1} a_\nu p^\nu.$$

These sequences constituted the projective limit

$$\varprojlim_n \mathbb{Z}/p^n\mathbb{Z} = \Big\{ (x_n)_{n\in\mathbb{N}} \in \prod_{n=1}^{\infty} \mathbb{Z}/p^n\mathbb{Z} \;\Big|\; x_{n+1} \mapsto x_n \Big\}.$$

We viewed the p-adic integers as elements of this ring. Since

$$\mathbb{Z}_p/p^n\mathbb{Z}_p \cong \mathbb{Z}/p^n\mathbb{Z},$$

we obtain, for every $n \geq 1$, a surjective homomorphism

$$\mathbb{Z}_p \longrightarrow \mathbb{Z}/p^n\mathbb{Z}.$$

It is clear that the family of these homomorphisms yields a homomorphism

$$\mathbb{Z}_p \longrightarrow \varprojlim_n \mathbb{Z}/p^n\mathbb{Z}.$$

It is now possible to identify both definitions given for $\mathbb{Z}_p$ (and therefore also for $\mathbb{Q}_p$) via the

(2.5) Proposition. *The homomorphism*

$$\mathbb{Z}_p \longrightarrow \varprojlim_n \mathbb{Z}/p^n\mathbb{Z}$$

is an isomorphism.

Proof: If $x \in \mathbb{Z}_p$ is mapped to zero, this means that $x \in p^n\mathbb{Z}_p$ for all $n \geq 1$, i.e., $|x|_p \leq \frac{1}{p^n}$ for all $n \geq 1$, so that $|x|_p = 0$ and thus $x = 0$. This shows injectivity.

An element of $\varprojlim_n \mathbb{Z}/p^n\mathbb{Z}$ is given by a sequence of partial sums

$$s_n = \sum_{\nu=0}^{n-1} a_\nu p^\nu, \quad 0 \leq a_\nu < p.$$

We saw above that this sequence is a Cauchy sequence in $\mathbb{Z}_p$, and thus converges to an element

$$x = \sum_{\nu=0}^{\infty} a_\nu p^\nu \in \mathbb{Z}_p.$$

Since

$$x - s_n = \sum_{\nu=n}^{\infty} a_\nu p^\nu \in p^n\mathbb{Z}_p,$$

one has $x \equiv s_n \bmod p^n$ for all n, i.e., x is mapped to the element of $\varprojlim_n \mathbb{Z}/p^n\mathbb{Z}$ which is defined by the given sequence $(s_n)_{n\in\mathbb{N}}$. This shows surjectivity. □

We emphasize that the elements on the right hand side of the isomorphism

$$\mathbb{Z}_p \xrightarrow{\sim} \varprojlim_n \mathbb{Z}/p^n\mathbb{Z}$$

are given formally by sequences of partial sums

$$s_n = \sum_{\nu=0}^{n-1} a_\nu p^\nu, \quad n = 1, 2, \ldots$$

On the left, however, these sequences converge with respect to the absolute value and yield the elements of $\mathbb{Z}_p$ in the familiar way, as convergent infinite series

$$x = \sum_{\nu=0}^{\infty} a_\nu p^\nu.$$

Yet another, very elegant method to introduce the p-adic numbers comes about as follows. Let $\mathbb{Z}[[X]]$ denote the ring of all formal power series $\sum_{i=0}^{\infty} a_i X^i$ with integer coefficients. Then one has the

(2.6) Proposition. *There is a canonical isomorphism*

$$\mathbb{Z}_p \cong \mathbb{Z}[[X]]/(X - p).$$

Proof: Consider the visibly surjective homomorphism $\mathbb{Z}[[X]] \to \mathbb{Z}_p$ which to every formal power series $\sum_{\nu=0}^{\infty} a_\nu X^\nu$ associates the convergent series $\sum_{\nu=0}^{\infty} a_\nu p^\nu$. The principal ideal $(X - p)$ clearly belongs to the kernel of this mapping. In order to show that it is the whole kernel, let $f(X) = \sum_{\nu=0}^{\infty} a_\nu X^\nu$ be a power series such that $f(p) = \sum_{\nu=0}^{\infty} a_\nu p^\nu = 0$. Since $\mathbb{Z}_p/p^n\mathbb{Z}_p \cong \mathbb{Z}/p^n\mathbb{Z}$, this means that

$$a_0 + a_1 p + \cdots + a_{n-1} p^{n-1} \equiv 0 \bmod p^n$$

for all n. We put, for $n \geq 1$,

$$b_{n-1} = -\frac{1}{p^n}(a_0 + a_1 p + \cdots + a_{n-1} p^{n-1}).$$

Then we obtain successively

$$\begin{aligned} a_0 &= \quad\;\; - pb_0, \\ a_1 &= b_0 - pb_1, \\ a_2 &= b_1 - pb_2, \qquad \text{etc.} \end{aligned}$$

But this amounts to the equality

$$(a_0 + a_1 X + a_2 X^2 + \cdots) = (X - p)(b_0 + b_1 X + b_2 X^2 + \cdots),$$

i.e., $f(X)$ belongs to the principal ideal $(X - p)$. □

Exercise 1. $|x - y|_p \geq \big|\, |x|_p - |y|_p \,\big|$.

Exercise 2. Let n be a natural number, $n = a_0 + a_1 p + \cdots + a_{r-1} p^{r-1}$ its p-adic expansion, with $0 \leq a_i < p$, and $s = a_0 + a_1 + \cdots + a_{r-1}$. Show that $v_p(n!) = \dfrac{n - s}{p - 1}$.

Exercise 3. The sequence $1, \frac{1}{10}, \frac{1}{10^2}, \frac{1}{10^3}, \ldots$ does not converge in $\mathbb{Q}_p$, for any p.

Exercise 4. Let $\varepsilon \in 1 + p\mathbb{Z}_p$, and let $\alpha = a_0 + a_1 p + a_2 p^2 + \cdots$ be a p-adic integer, and write $s_n = a_0 + a_1 p + \cdots + a_{n-1} p^{n-1}$. Show that the sequence ε^{s_n} converges to a number ε^α in $1 + p\mathbb{Z}_p$. Show furthermore that $1 + p\mathbb{Z}_p$ is thus turned into a multiplicative $\mathbb{Z}_p$-module.

Exercise 5. For every $a \in \mathbb{Z}$, $(a, p) = 1$, the sequence $\{a^{p^n}\}_{n \in \mathbb{N}}$ converges in $\mathbb{Q}_p$.

Exercise 6. The fields $\mathbb{Q}_p$ and $\mathbb{Q}_q$ are not isomorphic, unless $p = q$.

Exercise 7. The algebraic closure of $\mathbb{Q}_p$ has infinite degree.

Exercise 8. In the ring $\mathbb{Z}_p[[X]]$ of formal power series $\sum_{\nu=0}^{\infty} a_\nu X^\nu$ over $\mathbb{Z}_p$, one has the following **division with remainder**. Let $f, g \in \mathbb{Z}_p[[X]]$ and let $f(X) = a_0 + a_1 X + \cdots$ such that $p | a_\nu$ for $\nu = 0, \ldots, n - 1$, but $p \nmid a_n$. Then one may write in a unique way

$$g = qf + r,$$

where $q \in \mathbb{Z}_p[[X]]$, and $r \in \mathbb{Z}_p[X]$ is a polynomial of degree $\leq n - 1$.

Hint: Let τ be the operator $\tau(\sum_{\nu=0}^{\infty} b_\nu X^\nu) = \sum_{\nu=n}^{\infty} b_\nu X^{\nu-n}$. Show that $U(X) = a_n + a_{n+1}X + \cdots = \tau(f(X))$ is a unit in $\mathbb{Z}_p[[X]]$ and write $f(X) = pP(X) + X^n U(X)$ with a polynomial $P(X)$ of degree $\leq n-1$. Show that

$$q(X) = \frac{1}{U(X)} \sum_{i=0}^{\infty} (-1)^i p^i \Big(\tau \circ \frac{P}{U}\Big)^i \circ \tau(g)$$

is a well-defined power series in $\mathbb{Z}_p[[X]]$ such that $\tau(qf) = \tau(g)$.

Exercise 9 (p-adic Weierstrass Preparation Theorem). Every nonzero power series

$$f(X) = \sum_{\nu=0}^{\infty} a_\nu X^\nu \in \mathbb{Z}_p[[X]]$$

admits a unique representation

$$f(X) = p^\mu P(X) U(X),$$

where $U(X)$ is a unit in $\mathbb{Z}_p[[X]]$ and $P(X) \in \mathbb{Z}_p[X]$ is a monic polynomial satisfying $P(X) \equiv X^n \bmod p$.

§ 3. Valuations

The procedure we performed in the previous section with the field $\mathbb{Q}$ in order to obtain the p-adic numbers can be generalized to arbitrary fields using the concept of (multiplicative) valuation.

(3.1) Definition. *A* **valuation** *of a field K is a function*

$$|\ |: K \to \mathbb{R}$$

enjoying the properties

(i) $|x| \geq 0$, *and* $|x| = 0 \iff x = 0$,

(ii) $|xy| = |x||y|$,

(iii) $|x+y| \leq |x| + |y|$ *"triangle inequality".*

We tacitly exclude in the sequel the case where $|\ |$ is the trivial valuation of K which satisfies $|x| = 1$ for all $x \neq 0$. Defining the distance between two points $x, y \in K$ by

$$d(x,y) = |x - y|$$

makes K into a metric space, and hence in particular a topological space.

(3.2) Definition. *Two valuations of K are called* **equivalent** *if they define the same topology on K.*

(3.3) Proposition. *Two valuations* $|\ |_1$ *and* $|\ |_2$ *on* K *are equivalent if and only if there exists a real number* $s > 0$ *such that one has*

$$|x|_1 = |x|_2^s$$

for all $x \in K$.

Proof: If $|\ |_1 = |\ |_2^s$, with $s > 0$, then $|\ |_1$ and $|\ |_2$ are obviously equivalent. For an arbitrary valuation $|\ |$ on K, the inequality $|x| < 1$ is tantamount to the condition that $\{x^n\}_{n\in\mathbb{N}}$ converges to zero in the topology defined by $|\ |$. Therefore if $|\ |_1$ and $|\ |_2$ are equivalent, one has the implication

$$(*) \qquad |x|_1 < 1 \implies |x|_2 < 1 .$$

Now let $y \in K$ be a fixed element satisfying $|y|_1 > 1$. Let $x \in K$, $x \neq 0$. Then $|x|_1 = |y|_1^\alpha$ for some $\alpha \in \mathbb{R}$. Let m_i/n_i be a sequence of rational numbers (with $n_i > 0$) which converges to α from above. Then we have $|x|_1 = |y|_1^\alpha < |y|_1^{m_i/n_i}$, hence

$$\left|\frac{x^{n_i}}{y^{m_i}}\right|_1 < 1 \implies \left|\frac{x^{n_i}}{y^{m_i}}\right|_2 < 1 ,$$

so that $|x|_2 \le |y|_2^{m_i/n_i}$, and thus $|x|_2 \le |y|_2^\alpha$. Using a sequence m_i/n_i which converges to α from below $(*)$ tells us that $|x|_2 \ge |y|_2^\alpha$. So we have $|x|_2 = |y|_2^\alpha$. For all $x \in K$, $x \neq 0$, we therefore get

$$\frac{\log|x|_1}{\log|x|_2} = \frac{\log|y|_1}{\log|y|_2} =: s ,$$

hence $|x|_1 = |x|_2^s$. But $|y|_1 > 1$ implies $|y|_2 > 1$, hence $s > 0$. □

The proof shows that the equivalence of $|\ |_1$ and $|\ |_2$ is also equivalent to the condition

$$|x|_1 < 1 \implies |x|_2 < 1 .$$

We use this for the proof of the following approximation theorem, which may be considered a variant of the Chinese remainder theorem.

(3.4) Approximation Theorem. *Let* $|\ |_1, \ldots, |\ |_n$ *be pairwise inequivalent valuations of the field* K *and let* $a_1, \ldots, a_n \in K$ *be given elements. Then for every* $\varepsilon > 0$ *there exists an* $x \in K$ *such that*

$$|x - a_i|_i < \varepsilon \quad \text{for all } i = 1, \ldots, n .$$

Proof: By the above remark, since $|\ |_1$ and $|\ |_n$ are inequivalent, there exists $\alpha \in K$ such that $|\alpha|_1 < 1$ and $|\alpha|_n \geq 1$. By the same token, there exists $\beta \in K$ such that $|\beta|_n < 1$ and $|\beta|_1 \geq 1$. Putting $y = \beta/\alpha$, one finds $|y|_1 > 1$ and $|y|_n < 1$.

We now prove by induction on n that there exists $z \in K$ such that

$$|z|_1 > 1 \quad \text{and} \quad |z|_j < 1 \quad \text{for } j = 2, \dots, n.$$

We have just done this for $n = 2$. Assume we have found $z \in K$ satisfying

$$|z|_1 > 1 \quad \text{and} \quad |z|_j < 1 \quad \text{for } j = 2, \dots, n-1.$$

If $|z|_n \leq 1$, then $z^m y$ will do, for m large. If however $|z|_n > 1$, the sequence $t_m = z^m/(1+z^m)$ will converge to 1 with respect to $|\ |_1$ and $|\ |_n$, and to 0 with respect to $|\ |_2, \dots, |\ |_{n-1}$. Hence, for m large, $t_m y$ will suffice.

The sequence $z^m/(1+z^m)$ converges to 1 with respect to $|\ |_1$ and to 0 with respect to $|\ |_2, \dots, |\ |_n$. For every i we may construct in this way a z_i which is very close to 1 with respect to $|\ |_i$, and very close to 0 with respect to $|\ |_j$ for $j \neq i$. The element

$$x = a_1 z_1 + \cdots + a_n z_n$$

then satisfies the statement of the approximation theorem. □

(3.5) Definition. *The valuation* $|\ |$ *is called* **nonarchimedean** *if* $|n|$ *stays bounded, for all* $n \in \mathbb{N}$. *Otherwise it is called* **archimedean**.

(3.6) Proposition. *The valuation* $|\ |$ *is nonarchimedean if and only if it satisfies the* **strong triangle inequality**

$$|x + y| \leq \max\{|x|, |y|\}.$$

Proof: If the strong triangle inequality holds, then one has

$$|n| = |1 + \cdots + 1| \leq 1.$$

Conversely, let $|n| \leq N$ for all $n \in \mathbb{N}$. Let $x, y \in K$ and suppose $|x| \geq |y|$. Then $|x|^\nu |y|^{n-\nu} \leq |x|^n$ for $\nu \geq 0$ and one gets

$$|x+y|^n \leq \sum_{\nu=0}^{n} \left|\binom{n}{\nu}\right| |x|^\nu |y|^{n-\nu} \leq N(n+1)|x|^n,$$

hence

$$|x+y| \leq N^{1/n}(1+n)^{1/n}|x| = N^{1/n}(1+n)^{1/n} \max\{|x|, |y|\},$$

and thus $|x+y| \leq \max\{|x|, |y|\}$ by letting $n \to \infty$. □

Remark: The strong triangle inequality immediately implies that
$$|x| \neq |y| \implies |x+y| = \max\{\,|x|,|y|\,\}\,.$$
One may extend the nonarchimedean valuation $|\ |$ of K to a valuation of the function field $K(t)$ in a canonical way by setting, for a polynomial $f(t) = a_0 + a_1 t + \cdots + a_n t^n$,
$$|f| = \max\{\,|a_0|,\ \ldots,\ |a_n|\,\}\,.$$
The triangle inequality $|f+g| \le \max\{|f|,|g|\}$ is immediate. The proof that $|fg| = |f||g|$ is the same as the proof of Gauss's lemma for polynomials over factorial rings once we replace the **content** of f in this lemma by the absolute value $|f|$.

For the field $\mathbb{Q}$, we have the usual absolute value $|\ |_\infty = |\ |$, this being the archimedean valuation, and for each prime number p the nonarchimedean valuation $|\ |_p$. As a matter of fact:

(3.7) Proposition. *Every valuation of $\mathbb{Q}$ is equivalent to one of the valuations $|\ |_p$ or $|\ |_\infty$.*

Proof: Let $\|\ \|$ be a nonarchimedean valuation of $\mathbb{Q}$. Then $\|n\| = \|1+\cdots+1\| \le 1$, and there is a prime number p such that $\|p\| < 1$ because, if not, unique prime factorization would imply $\|x\| = 1$ for all $x \in \mathbb{Q}^*$. The set
$$\mathfrak{a} = \{a \in \mathbb{Z} \mid \|a\| < 1\}$$
is an ideal of $\mathbb{Z}$ satisfying $p\mathbb{Z} \subseteq \mathfrak{a} \neq \mathbb{Z}$, and since $p\mathbb{Z}$ is a maximal ideal, we have $\mathfrak{a} = p\mathbb{Z}$. If now $a \in \mathbb{Z}$ and $a = bp^m$ with $p \nmid b$, so that $b \notin \mathfrak{a}$, then $\|b\| = 1$ and hence
$$\|a\| = \|p\|^m = |a|_p^s$$
where $s = -\log\|p\|/\log p$. Consequently $\|\ \|$ is equivalent to $|\ |_p$.

Now let $\|\ \|$ be archimedean. Then one has, for every two natural numbers $n, m > 1$,
$$(*) \qquad \|m\|^{1/\log m} = \|n\|^{1/\log n}.$$
In fact, we may write
$$m = a_0 + a_1 n + \cdots + a_r n^r$$
where $a_i \in \{0,1,\ldots,n-1\}$ and $n^r \le m$. Hence, observing that $r \le \log m/\log n$ and $\|a_i\| = \|1+\cdots+1\| \le a_i\,\|1\| \le n$, one gets the inequality
$$\|m\| \le \sum \|a_i\|\cdot\|n\|^i \le \sum \|a_i\|\cdot\|n\|^r \le \Big(1 + \frac{\log m}{\log n}\Big)n\cdot\|n\|^{\log m/\log n}.$$

Substituting here m^k for m, taking k-th roots on both sides, and letting k tend to ∞, one finally obtains

$$\|m\| \leq \|n\|^{\log m/\log n}, \quad \text{or} \quad \|m\|^{1/\log m} \leq \|n\|^{1/\log n}.$$

Swapping m with n gives the identity $(*)$. Putting $c = \|n\|^{1/\log n}$ we have $\|n\| = c^{\log n}$, and putting $c = e^s$ yields, for every positive rational number $x = a/b$,

$$\|x\| = e^{s\log x} = |x|^s .$$

Therefore $\| \ \|$ is equivalent to the usual absolute value $| \ |$ on $\mathbb{Q}$. □

Let $| \ |$ be a nonarchimedean valuation of the field K. Putting

$$v(x) = -\log|x| \quad \text{for } x \neq 0, \quad \text{and } v(0) = \infty,$$

we obtain a function

$$v : K \longrightarrow \mathbb{R} \cup \{\infty\}$$

verifying the properties

(i) $v(x) = \infty \Longleftrightarrow x = 0$,

(ii) $v(xy) = v(x) + v(y)$,

(iii) $v(x+y) \geq \min\{v(x), v(y)\}$,

where we fix the following conventions regarding elements $a \in \mathbb{R}$ and the symbol ∞: $a < \infty$, $a + \infty = \infty$, $\infty + \infty = \infty$.

A function v on K with these properties is called an **exponential valuation** of K. We exclude the case of the trivial function $v(x) = 0$ for $x \neq 0$, $v(0) = \infty$. Two exponential valuations v_1 and v_2 of K are called **equivalent** if $v_1 = sv_2$, for some real number $s > 0$. For every exponential valuation v we obtain a valuation in the sense of (3.1) by putting

$$|x| = q^{-v(x)},$$

for some fixed real number $q > 1$. To distinguish it from v, we call $| \ |$ an associated **multiplicative valuation**, or **absolute value**. Replacing v by an equivalent valuation sv (i.e., replacing q by $q' = q^s$) changes $| \ |$ into the equivalent multiplicative valuation $| \ |^s$. The conditions (i), (ii), (iii) immediately imply the

(3.8) Proposition. *The subset*

$$\mathcal{O} = \{x \in K \mid v(x) \geq 0\} = \{x \in K \mid |x| \leq 1\}$$

is a ring with group of units

$$\mathcal{O}^* = \{x \in K \mid v(x) = 0\} = \{x \in K \mid |x| = 1\}$$

and the unique maximal ideal

$$\mathfrak{p} = \{x \in K \mid v(x) > 0\} = \{x \in K \mid |x| < 1\} .$$

$\mathcal{O}$ is an integral domain with field of fractions K and has the property that, for every $x \in K$, either $x \in \mathcal{O}$ or $x^{-1} \in \mathcal{O}$. Such a ring is called a **valuation ring**. Its only maximal ideal is $\mathfrak{p} = \{x \in \mathcal{O} \mid x^{-1} \notin \mathcal{O}\}$. The field $\mathcal{O}/\mathfrak{p}$ is called the **residue class field** of $\mathcal{O}$. A valuation ring is always integrally closed. For if $x \in K$ is integral over $\mathcal{O}$, then there is an equation

$$x^n + a_1 x^{n-1} + \cdots + a_n = 0$$

with $a_i \in \mathcal{O}$ and the hypothesis $x \notin \mathcal{O}$, so that $x^{-1} \in \mathcal{O}$, would imply the contradiction $x = -a_1 - a_2 x^{-1} - \cdots - a_n (x^{-1})^{n-1} \in \mathcal{O}$.

An exponential valuation v is called **discrete** if it admits a smallest positive value s. In this case, one finds

$$v(K^*) = s\mathbb{Z}\,.$$

It is called **normalized** if $s = 1$. Dividing by s we may always pass to a normalized valuation without changing the invariants $\mathcal{O}, \mathcal{O}^*, \mathfrak{p}$. Having done so, an element

$$\pi \in \mathcal{O} \quad \text{such that} \quad v(\pi) = 1$$

is a **prime element**, and every element $x \in K^*$ admits a unique representation

$$x = u\,\pi^m$$

with $m \in \mathbb{Z}$ and $u \in \mathcal{O}^*$. For if $v(x) = m$, then $v(x\,\pi^{-m}) = 0$, hence $u = x\,\pi^{-m} \in \mathcal{O}^*$.

(3.9) Proposition. *If v is a discrete exponential valuation of K, then*

$$\mathcal{O} = \big\{x \in K \mid v(x) \geq 0\big\}$$

is a principal ideal domain, hence a discrete valuation ring (see I, (11.3)).

Suppose v is normalized. Then the nonzero ideals of $\mathcal{O}$ are given by

$$\mathfrak{p}^n = \pi^n \mathcal{O} = \big\{x \in K \mid v(x) \geq n\big\}, \quad n \geq 0,$$

where π is a prime element, i.e., $v(\pi) = 1$. One has

$$\mathfrak{p}^n/\mathfrak{p}^{n+1} \cong \mathcal{O}/\mathfrak{p}\,.$$

Proof: Let $\mathfrak{a} \neq 0$ be an ideal of $\mathcal{O}$ and $x \neq 0$ an element in $\mathfrak{a}$ with smallest possible value $v(x) = n$. Then $x = u\,\pi^n$, $u \in \mathcal{O}^*$, so that $\pi^n \mathcal{O} \subseteq \mathfrak{a}$. If $y = \varepsilon\,\pi^m \in \mathfrak{a}$ is arbitrary with $\varepsilon \in \mathcal{O}^*$, then $m = v(y) \geq n$, hence $y = (\varepsilon\,\pi^{m-n})\pi^n \in \pi^n \mathcal{O}$, so that $\mathfrak{a} = \pi^n \mathcal{O}$. The isomorphism

$$\mathfrak{p}^n/\mathfrak{p}^{n+1} \cong \mathcal{O}/\mathfrak{p}$$

results from the correspondence $a\pi^n \mapsto a \bmod \mathfrak{p}$. □

In a discretely valued field K the chain

$$\mathcal{O} \supseteq \mathfrak{p} \supseteq \mathfrak{p}^2 \supseteq \mathfrak{p}^3 \supseteq \cdots$$

consisting of the ideals of the valuation ring $\mathcal{O}$ forms a basis of neighbourhoods of the zero element. Indeed, if v is a normalized exponential valuation and $|\ | = q^{-v}$ $(q > 1)$ an associated multiplicative valuation, then

$$\mathfrak{p}^n = \left\{x \in K \;\middle|\; |x| < \frac{1}{q^{n-1}}\right\}.$$

As a basis of neighbourhoods of the element 1 of K^*, we obtain in the same way the descending chain

$$\mathcal{O}^* = U^{(0)} \supseteq U^{(1)} \supseteq U^{(2)} \supseteq \cdots$$

of subgroups

$$U^{(n)} = 1 + \mathfrak{p}^n = \left\{x \in K^* \;\middle|\; |1 - x| < \frac{1}{q^{n-1}}\right\}, \qquad n > 0,$$

of $\mathcal{O}^*$. (Observe that $1+\mathfrak{p}^n$ is closed under multiplication and that, if $x \in U^{(n)}$, then so is x^{-1} because $|1 - x^{-1}| = |x|^{-1}|x - 1| = |1 - x| < \frac{1}{q^{n-1}}$.) $U^{(n)}$ is called the n-th **higher unit group** and $U^{(1)}$ the group of **principal units**. Regarding the successive quotients of the chain of higher unit groups, we have the

(3.10) Proposition. *$\mathcal{O}^*/U^{(n)} \cong (\mathcal{O}/\mathfrak{p}^n)^*$ and $U^{(n)}/U^{(n+1)} \cong \mathcal{O}/\mathfrak{p}$, for $n \geq 1$.*

Proof: The first isomorphism is induced by the canonical and obviously surjective homomorphism

$$\mathcal{O}^* \longrightarrow (\mathcal{O}/\mathfrak{p}^n)^*, \quad u \longmapsto u \bmod \mathfrak{p}^n,$$

the kernel of which is $U^{(n)}$. The second isomorphism is given, once we choose a prime element π, by the surjective homomorphism

$$U^{(n)} = 1 + \pi^n\mathcal{O} \longrightarrow \mathcal{O}/\mathfrak{p}, \quad 1 + \pi^n a \longmapsto a \bmod \mathfrak{p},$$

which has kernel $U^{(n+1)}$. □

Exercise 1. Show that $|z| = (z\bar{z})^{1/2} = \sqrt{|N_{\mathbb{C}|\mathbb{R}}(z)|}$ is the only valuation of $\mathbb{C}$ which extends the absolute value $|\ |$ of $\mathbb{R}$.

Exercise 2. What is the relation between the Chinese remainder theorem and the approximation theorem (3.4)?

Exercise 3. Let k be a field and $K = k(t)$ the function field in one variable. Show that the valuations $v_{\mathfrak{p}}$ associated to the prime ideals $\mathfrak{p} = (p(t))$ of $k[t]$, together with the degree valuation v_∞, are the only valuations of K, up to equivalence. What are the residue class fields?

Exercise 4. Let $\mathcal{O}$ be an arbitrary valuation ring with field of fractions K, and let $\Gamma = K^*/\mathcal{O}^*$. Then Γ becomes a totally ordered group if we define $x \bmod \mathcal{O}^* \geq y \bmod \mathcal{O}^*$ to mean $x/y \in \mathcal{O}$.

Write Γ additively and show that the function

$$v : K \longrightarrow \Gamma \cup \{\infty\},$$

$v(0) = \infty$, $v(x) = x \bmod \mathcal{O}^*$ for $x \in K^*$, satisfies the conditions

1) $v(x) = \infty \Longrightarrow x = 0$,

2) $v(xy) = v(x) + v(y)$,

3) $v(x + y) \geq \min\{v(x), v(y)\}$.

v is called a **Krull valuation**.

§ 4. Completions

(4.1) Definition. *A valued field $(K, |\ |)$ is called* **complete** *if every Cauchy sequence $\{a_n\}_{n\in\mathbb{N}}$ in K converges to an element $a \in K$, i.e.,*

$$\lim_{n\to\infty} |a_n - a| = 0.$$

Here, as usual, we call $\{a_n\}_{n\in\mathbb{N}}$ a **Cauchy sequence** if for every $\varepsilon > 0$ there exists $N \in \mathbb{N}$ such that

$$|a_n - a_m| < \varepsilon \quad \text{for all} \quad n, m \geq N.$$

From any valued field $(K, |\ |)$ we get a complete valued field $(\widehat{K}, |\ |)$ by the process of **completion**. This completion is obtained in the same way as the field of real numbers is constructed from the field of rational numbers.

Take the ring R of all Cauchy sequences of $(K, |\ |)$, consider therein the maximal ideal $\mathfrak{m}$ of all nullsequences with respect to $|\ |$, and define

$$\widehat{K} = R/\mathfrak{m}.$$

One embeds the field K into $\widehat{K}$ by sending every $a \in K$ to the class of the constant Cauchy sequence $(a, a, a, \dots)$. The valuation $|\ |$ is extended from K to $\widehat{K}$ by giving the element $a \in \widehat{K}$ which is represented by the Cauchy sequence $\{a_n\}_{n\in\mathbb{N}}$ the absolute value

$$|a| = \lim_{n\to\infty} |a_n|.$$

This limit exists because $\big|\, |a_n| - |a_m| \,\big| \le |a_n - a_m|$ implies that $|a_n|$ is a Cauchy sequence of real numbers. As in the case of the field of real numbers, one proves that $\widehat{K}$ is complete with respect to the extended $|\ |$, and that each $a \in \widehat{K}$ is a limit of a sequence $\{a_n\}$ in K. Finally one proves the uniqueness of the completion $(\widehat{K}, |\ |)$: if $(\widehat{K}', |\ |')$ is another complete valued field that contains $(K, |\ |)$ as a dense subfield, then mapping

$$|\ |\text{-}\lim_{n\to\infty} a_n \quad\longmapsto\quad |\ |'\text{-}\lim_{n\to\infty} a_n$$

gives a K-isomorphism $\sigma : \widehat{K} \to \widehat{K}'$ such that $|a| = |\sigma a|'$.

The fields $\mathbb{R}$ and $\mathbb{C}$ are the most familiar examples of complete fields. They are complete with respect to an archimedean valuation. Amazingly enough, there are no others of this type. More precisely we have the

(4.2) Theorem (Ostrowski). *Let K be a field which is complete with respect to an archimedean valuation $|\ |$. Then there is an isomorphism σ from K onto $\mathbb{R}$ or $\mathbb{C}$ satisfying*

$$|a| = |\sigma a|^s \quad \textit{for all} \quad a \in K,$$

for some fixed $s \in (0, 1]$.

Proof: We may assume without loss of generality that $\mathbb{R} \subseteq K$ and that the valuation $|\ |$ of K is an extension of the usual absolute value of $\mathbb{R}$. In fact, replacing $|\ |$ by $|\ |^{s^{-1}}$ for a suitable $s > 0$, we may assume by (3.7) that the restriction of $|\ |$ to $\mathbb{Q}$ is equal to the usual absolute value. Then taking the closure $\widehat{\mathbb{Q}}$ in K we find that $\widehat{\mathbb{Q}}$ is complete with respect to the restriction of $|\ |$ to $\widehat{\mathbb{Q}}$, in other words, it is a completion of $(\mathbb{Q}, |\ |)$. In view of the uniqueness of completions, there is an isomorphism $\sigma : \mathbb{R} \to \widehat{\mathbb{Q}}$ such that $|a| = |\sigma a|$ as required.

In order to prove that $K = \mathbb{R}$ or $= \mathbb{C}$ we show that each $\xi \in K$ satisfies a quadratic equation over $\mathbb{R}$. For this, consider the continuous function $f : \mathbb{C} \to \mathbb{R}$ defined by

$$f(z) = \big|\xi^2 - (z + \bar{z})\xi + z\bar{z}\big|.$$

Note here that $z+\bar{z},\ z\bar{z} \in \mathbb{R} \subseteq K$. Since $\lim_{z\to\infty} f(z) = \infty$, $f(z)$ has a minimum m. The set

$$S = \{z \in \mathbb{C} \mid f(z) = m\}$$

is therefore nonempty, bounded, and closed, and there is a $z_0 \in S$ such that $|z_0| \geq |z|$ for all $z \in S$. It suffices to show that $m = 0$, because then one has the equation $\xi^2 - (z_0+\bar{z}_0)\xi + z_0\bar{z}_0 = 0$.

Assume $m > 0$. Consider the real polynomial

$$g(x) = x^2 - (z_0+\bar{z}_0)x + z_0\bar{z}_0 + \varepsilon,$$

where $0 < \varepsilon < m$, with the roots $z_1, \bar{z}_1 \in \mathbb{C}$. We have $z_1\bar{z}_1 = z_0\bar{z}_0 + \varepsilon$, hence $|z_1| > |z_0|$ and thus

$$f(z_1) > m.$$

For fixed $n \in \mathbb{N}$, consider on the other hand the real polynomial

$$G(x) = \big[g(x) - \varepsilon\big]^n - (-\varepsilon)^n = \prod_{i=1}^{2n}(x-\alpha_i) = \prod_{i=1}^{2n}(x-\bar{\alpha}_i)$$

with roots $\alpha_1, \dots, \alpha_{2n} \in \mathbb{C}$. It follows that $G(z_1) = 0$; say, $z_1 = \alpha_1$. We may substitute $\xi \in K$ into the polynomial

$$G(x)^2 = \prod_{i=1}^{2n}\big(x^2 - (\alpha_i + \bar{\alpha}_i)x + \alpha_i\bar{\alpha}_i\big)$$

and get

$$\big|G(\xi)\big|^2 = \prod_{i=1}^{2n} f(\alpha_i) \geq f(\alpha_1)m^{2n-1}.$$

From this and the inequality

$$\big|G(\xi)\big| \leq |\xi^2 - (z_0+\bar{z}_0)\xi + z_0\bar{z}_0|^n + |-\varepsilon|^n = f(z_0)^n + \varepsilon^n = m^n + \varepsilon^n,$$

it follows that $f(\alpha_1)m^{2n-1} \leq (m^n + \varepsilon^n)^2$ and hence

$$\frac{f(\alpha_1)}{m} \leq \Big(1 + \Big(\frac{\varepsilon}{m}\Big)^n\Big)^2.$$

For $n \to \infty$ we have $f(\alpha_1) \leq m$, which contradicts the inequality $f(\alpha_1) > m$ proved before. □

In view of *Ostrowski*'s theorem, we will henceforth restrict attention to the case of nonarchimedean valuations. In this case it is usually expedient – both with regard to the substance and to practical technique – to work with the exponential valuations v rather than the multiplicative valuations. So let v

be an exponential valuation of the field K. It is canonically continued to an exponential valuation $\hat{v}$ of the completion $\widehat{K}$ by setting

$$\hat{v}(a) = \lim_{n\to\infty} v(a_n),$$

where $a = \lim_{n\to\infty} a_n \in \widehat{K}$, $a_n \in K$. Observe here that the sequence $v(a_n)$ has to become stationary (provided $a \neq 0$) because, for $n \geq n_0$, one has $\hat{v}(a - a_n) > \hat{v}(a)$, so that it follows from the remark on p. 119

$$v(a_n) = \hat{v}(a_n - a + a) = \min\{\hat{v}(a_n - a), \hat{v}(a)\} = \hat{v}(a).$$

Therefore it follows that

$$v(K^*) = \hat{v}(\widehat{K}^*),$$

and if v is discrete and normalized, then so is the extension $\hat{v}$. In the nonarchimedean case, for a sequence $\{a_n\}_{n\in\mathbb{N}}$ to be a Cauchy sequence, it suffices that $a_{n+1} - a_n$ be a nullsequence. In fact, $v(a_n - a_m) \geq \min_{m\leq i<n}\{v(a_{i+1} - a_i)\}$. By the same token an infinite series $\sum_{\nu=0}^{\infty} a_\nu$ converges in $\widehat{K}$ if and only if the sequence of its terms a_ν is a nullsequence. The following proposition is proved exactly as its analogue, proposition (2.4), in the special case $(\mathbb{Q}, v_p)$.

(4.3) Proposition. *If $\mathcal{O} \subseteq K$, resp. $\widehat{\mathcal{O}} \subseteq \widehat{K}$, is the valuation ring of v, resp. of $\hat{v}$, and $\mathfrak{p}$, resp. $\widehat{\mathfrak{p}}$, is the maximal ideal, then one has*

$$\widehat{\mathcal{O}}/\widehat{\mathfrak{p}} \cong \mathcal{O}/\mathfrak{p}$$

and, if v is discrete, one has furthermore

$$\widehat{\mathcal{O}}/\widehat{\mathfrak{p}}^n \cong \mathcal{O}/\mathfrak{p}^n \quad \text{for} \quad n \geq 1.$$

Generalizing the p-adic expansion to the case of an arbitrary discrete valuation v of the field K, we have the

(4.4) Proposition. *Let $R \subseteq \mathcal{O}$ be a system of representatives for $\kappa = \mathcal{O}/\mathfrak{p}$ such that $0 \in R$, and let $\pi \in \mathcal{O}$ be a prime element. Then every $x \neq 0$ in $\widehat{K}$ admits a unique representation as a convergent series*

$$x = \pi^m(a_0 + a_1\pi + a_2\pi^2 + \cdots)$$

where $a_i \in R$, $a_0 \neq 0$, $m \in \mathbb{Z}$.

Proof: Let $x = \pi^m u$ with $u \in \hat{\mathcal{O}}^*$. Since $\hat{\mathcal{O}}/\hat{\mathfrak{p}} \cong \mathcal{O}/\mathfrak{p}$, the class $u \bmod \hat{\mathfrak{p}}$ has a unique representative $a_0 \in R$, $a_0 \neq 0$. We thus have $u = a_0 + \pi b_1$, for some $b_1 \in \hat{\mathcal{O}}$. Assume now that $a_0, \ldots, a_{n-1} \in R$ have been found, satisfying

$$u = a_0 + a_1\pi + \cdots + a_{n-1}\pi^{n-1} + \pi^n b_n$$

for some $b_n \in \hat{\mathcal{O}}$, and that the a_i are uniquely determined by this equation. Then the representative $a_n \in R$ of $b_n \bmod \pi\hat{\mathcal{O}} \in \hat{\mathcal{O}}/\hat{\mathfrak{p}} \cong \mathcal{O}/\mathfrak{p}$ is also uniquely determined by u and we have $b_n = a_n + \pi b_{n+1}$, for some $b_{n+1} \in \hat{\mathcal{O}}$. Hence

$$u = a_0 + a_1\pi + \cdots + a_{n-1}\pi^{n-1} + a_n\pi^n + \pi^{n+1}b_{n+1}.$$

In this way we find an infinite series $\sum_{\nu=0}^{\infty} a_\nu \pi^\nu$ which is uniquely determined by u. It converges to u because the remainder term $\pi^{n+1}b_{n+1}$ tends to zero. □

In the case of the field of rational numbers $\mathbb{Q}$ and the p-adic valuation v_p with its completion $\mathbb{Q}_p$, the numbers $0, 1, \ldots, p-1$ form a system of representatives R for the residue class field $\mathbb{Z}/p\mathbb{Z}$ of the valuation, and we get back the representation of p-adic numbers which has already been discussed in § 2:

$$x = p^m(a_0 + a_1 p + a_2 p^2 + \cdots),$$

where $0 \le a_i < p$ and $m \in \mathbb{Z}$.

In the case of the rational function field $k(t)$ and the valuation $v_\mathfrak{p}$ attached to a prime ideal $\mathfrak{p} = (t-a)$ of $k[t]$ (see § 2), we may take as a system of representatives R the field of coefficients k itself. The completion then turns out to be the **field of formal power series** $k((x))$, $x = t - a$, consisting of all formal Laurent series

$$f(t) = (t-a)^m(a_0 + a_1(t-a) + a_2(t-a)^2 + \cdots),$$

with $a_i \in k$ and $m \in \mathbb{Z}$. The motivating analogy of the beginning of this chapter, between power series and p-adic numbers, thus appears as two special instances of the same concrete mathematical situation.

In § 1 we identified the ring $\mathbb{Z}_p$ of p-adic integers as being the projective limit $\varprojlim_n \mathbb{Z}/p^n\mathbb{Z}$. We obtain a similar result in the general setting of valuation theory. To explain this, let K be complete with respect to a discrete valuation. Let $\mathcal{O}$ be the valuation ring with the maximal ideal $\mathfrak{p}$. We then have for every $n \ge 1$ the canonical homomorphisms

$$\mathcal{O} \longrightarrow \mathcal{O}/\mathfrak{p}^n$$

and

$$\mathcal{O}/\mathfrak{p} \xleftarrow{\lambda_1} \mathcal{O}/\mathfrak{p}^2 \xleftarrow{\lambda_2} \mathcal{O}/\mathfrak{p}^3 \xleftarrow{\lambda_3} \cdots.$$

This gives us a homomorphism

$$\mathcal{O} \longrightarrow \varprojlim_n \mathcal{O}/\mathfrak{p}^n$$

into the projective limit

$$\varprojlim_n \mathcal{O}/\mathfrak{p}^n = \Big\{ (x_n) \in \prod_{n=1}^{\infty} \mathcal{O}/\mathfrak{p}^n \mid \lambda_n(x_{n+1}) = x_n \Big\}.$$

Considering the rings $\mathcal{O}/\mathfrak{p}^n$ as topological rings, for the discrete topology, gives us the product topology on $\prod_{n=1}^{\infty} \mathcal{O}/\mathfrak{p}^n$, and the projective limit $\varprojlim_n \mathcal{O}/\mathfrak{p}^n$ becomes a topological ring in a canonical way, being a closed subset of the product (see chap. IV, § 2).

(4.5) Proposition. *The canonical mapping*

$$\mathcal{O} \longrightarrow \varprojlim_n \mathcal{O}/\mathfrak{p}^n$$

is an isomorphism and a homeomorphism. The same is true for the mapping

$$\mathcal{O}^* \longrightarrow \varprojlim_n \mathcal{O}^*/U^{(n)}.$$

Proof: The map is injective since its kernel is $\bigcap_{n=1}^{\infty} \mathfrak{p}^n = (0)$. To prove surjectivity, let $\mathfrak{p} = \pi\mathcal{O}$ and let $R \subseteq \mathcal{O}$, $R \ni 0$, be a system of representatives of $\mathcal{O}/\mathfrak{p}$. We saw in the proof of (4.4) (and in fact already in (1.2)) that the elements $a \bmod \mathfrak{p}^n \in \mathcal{O}/\mathfrak{p}^n$ can be given uniquely in the form

$$a \equiv a_0 + a_1\pi + \cdots + a_{n-1}\pi^{n-1} \bmod \mathfrak{p}^n,$$

where $a_i \in R$. Each element $s \in \varprojlim_n \mathcal{O}/\mathfrak{p}^n$ is therefore given by a sequence of sums

$$s_n = a_0 + a_1\pi + \cdots + a_{n-1}\pi^{n-1}, \quad n = 1, 2, \ldots,$$

with fixed coefficients $a_i \in R$, and it is thus the image of the element $x = \lim_{n\to\infty} s_n = \sum_{\nu=0}^{\infty} a_\nu \pi^\nu \in \mathcal{O}$.

The sets $P_n = \prod_{\nu>n} \mathcal{O}/\mathfrak{p}^\nu$ form a basis of neighbourhoods of the zero element of $\prod_{\nu=1}^{\infty} \mathcal{O}/\mathfrak{p}^\nu$. Under the bijection

$$\mathcal{O} \longrightarrow \varprojlim_\nu \mathcal{O}/\mathfrak{p}^\nu$$

the basis of neighbourhoods $\mathfrak{p}^n$ of zero in $\mathcal{O}$ is mapped onto the basis of neighbourhoods $P_n \cap \varprojlim_\nu \mathcal{O}/\mathfrak{p}^\nu$ of zero in $\varprojlim_\nu \mathcal{O}/\mathfrak{p}^\nu$. Thus the bijection is a homeomorphism. It induces an isomorphism and homeomorphism on the group of units

$$\mathcal{O}^* \cong (\varprojlim \mathcal{O}/\mathfrak{p}^n)^* \cong \varprojlim (\mathcal{O}/\mathfrak{p}^n)^* \cong \varprojlim \mathcal{O}^*/U^{(n)}. \qquad \square$$

One of our chief concerns will be to study the finite extensions $L|K$ of a complete valued field K. This means that we have to turn to the question of factoring algebraic equations

$$f(x) = a_n x^n + a_{n-1}x^{n-1} + \cdots + a_0 = 0$$

over complete valued fields. For this, Hensel's seminal "lemma" is of fundamental importance. Let K again be a field which is complete with respect to a nonarchimedean valuation $|\ |$. Let $\mathcal{O}$ be the corresponding valuation ring with maximal ideal $\mathfrak{p}$ and residue class field $\kappa = \mathcal{O}/\mathfrak{p}$. We call a polynomial $f(x) = a_0 + a_1 x + \cdots + a_n x^n \in \mathcal{O}[x]$ **primitive** if $f(x) \not\equiv 0 \bmod \mathfrak{p}$, i.e., if

$$|f| = \max\{|a_0|, \ldots, |a_n|\} = 1.$$

(4.6) Hensel's Lemma. *If a primitive polynomial $f(x) \in \mathcal{O}[x]$ admits modulo $\mathfrak{p}$ a factorization*

$$f(x) \equiv \bar{g}(x)\bar{h}(x) \bmod \mathfrak{p}$$

into relatively prime polynomials $\bar{g}, \bar{h} \in \kappa[x]$, then $f(x)$ admits a factorization

$$f(x) = g(x)h(x)$$

into polynomials $g, h \in \mathcal{O}[x]$ such that $\deg(g) = \deg(\bar{g})$ *and*

$$g(x) \equiv \bar{g}(x) \bmod \mathfrak{p} \quad \textit{and} \quad h(x) \equiv \bar{h}(x) \bmod \mathfrak{p}.$$

Proof: Let $d = \deg(f)$, $m = \deg(\bar{g})$, hence $d - m \geq \deg(\bar{h})$. Let g_0, $h_0 \in \mathcal{O}[x]$ be polynomials such that $g_0 \equiv \bar{g} \bmod \mathfrak{p}$, $h_0 \equiv \bar{h} \bmod \mathfrak{p}$ and $\deg(g_0) = m$, $\deg(h_0) \leq d - m$. Since $(\bar{g}, \bar{h}) = 1$, there exist polynomials $a(x), b(x) \in \mathcal{O}[x]$ satisfying $ag_0 + bh_0 \equiv 1 \bmod \mathfrak{p}$. Among the coefficients of the two polynomials $f - g_0h_0$ and $ag_0 + bh_0 - 1 \in \mathfrak{p}[x]$ we pick one with minimum value and call it π.

Let us look for the polynomials g and h in the following form:

$$g = g_0 + p_1\pi + p_2\pi^2 + \cdots,$$
$$h = h_0 + q_1\pi + q_2\pi^2 + \cdots,$$

where $p_i, q_i \in \mathcal{O}[x]$ are polynomials of degree $< m$, resp. $\leq d-m$. We then determine successively the polynomials

$$g_{n-1} = g_0 + p_1\pi + \cdots + p_{n-1}\pi^{n-1},$$
$$h_{n-1} = h_0 + q_1\pi + \cdots + q_{n-1}\pi^{n-1},$$

in such a way that one has

$$f \equiv g_{n-1}h_{n-1} \bmod \pi^n.$$

Passing to the limit as $n \to \infty$, we will finally obtain the identity $f = gh$. For $n = 1$ the congruence is satisfied in view of our choice of π. Let us assume that it is already established for some $n \geq 1$. Then, in view of the relation

$$g_n = g_{n-1} + p_n\pi^n, \quad h_n = h_{n-1} + q_n\pi^n,$$

the condition on g_n, h_n reduces to

$$f - g_{n-1}h_{n-1} \equiv (g_{n-1}q_n + h_{n-1}p_n)\pi^n \bmod \pi^{n+1}.$$

Dividing by π^n, this means

$$g_{n-1}q_n + h_{n-1}p_n \equiv g_0q_n + h_0p_n \equiv f_n \bmod \pi,$$

where $f_n = \pi^{-n}(f - g_{n-1}h_{n-1}) \in \mathcal{O}[x]$. Since $g_0a + h_0b \equiv 1 \bmod \pi$, one has

$$g_0af_n + h_0bf_n \equiv f_n \bmod \pi.$$

At this point we would like to put $q_n = af_n$ and $p_n = bf_n$, but the degrees might be too big. For this reason, we write

$$b(x)f_n(x) = q(x)g_0(x) + p_n(x),$$

where $\deg(p_n) < \deg(g_0) = m$. Since $g_0 \equiv \bar{g} \bmod \mathfrak{p}$ and $\deg(g_0) = \deg(\bar{g})$, the highest coefficient of g_0 is a unit; hence $q(x) \in \mathcal{O}[x]$ and we obtain the congruence

$$g_0(af_n + h_0q) + h_0p_n \equiv f_n \bmod \pi.$$

Omitting now from the polynomial $af_n + h_0q$ all coefficients divisible by π, we get a polynomial q_n such that $g_0q_n + h_0p_n \equiv f_n \bmod \pi$ and which, in view of $\deg(f_n) \leq d$, $\deg(g_0) = m$ and $\deg(h_0p_n) < (d-m) + m = d$, has degree $\leq d - m$ as required. □

Example: The polynomial $x^{p-1}-1 \in \mathbb{Z}_p[x]$ splits over the residue class field $\mathbb{Z}_p/p\mathbb{Z}_p = \mathbb{F}_p$ into distinct linear factors. Applying (repeatedly) Hensel's lemma, we see that it also splits into linear factors over $\mathbb{Z}_p$. We thus obtain the astonishing result that the field $\mathbb{Q}_p$ of p-adic numbers contains the $(p-1)$-th roots of unity. These, together with 0, even form a system of representatives for the residue class field, which is closed under multiplication.

(4.7) Corollary. *Let the field K be complete with respect to the nonarchimedean valuation $|\ |$. Then, for every irreducible polynomial $f(x) = a_0 + a_1x + \cdots + a_nx^n \in K[x]$ such that $a_0a_n \neq 0$, one has*

$$|f| = \max\{\,|a_0|, |a_n|\}\,.$$

In particular, $a_n = 1$ and $a_0 \in \mathcal{O}$ imply that $f \in \mathcal{O}[x]$.

Proof: After multiplying by a suitable element of K we may assume that $f \in \mathcal{O}[x]$ and $|f| = 1$. Let a_r be the first one among the coefficients $a_0, \ldots, a_n$ such that $|a_r| = 1$. In other words, we have

$$f(x) \equiv x^r(a_r + a_{r+1}x + \cdots + a_nx^{n-r}) \bmod \mathfrak{p}\,.$$

If one had $\max\{|a_0|, |a_n|\} < 1$, then $0 < r < n$ and the congruence would contradict Hensel's lemma. □

From this corollary we can now deduce the following theorem on extensions of valuations.

(4.8) Theorem. *Let K be complete with respect to the valuation $|\ |$. Then $|\ |$ may be extended in a unique way to a valuation of any given algebraic extension $L|K$. This extension is given by the formula*

$$|\alpha| = \sqrt[n]{|N_{L|K}(\alpha)|}\,,$$

when $L|K$ has finite degree n. In this case L is again complete.

Proof: If the valuation $|\ |$ is archimedean, then by Ostrowski's theorem, $K = \mathbb{R}$ or $\mathbb{C}$. We have $N_{\mathbb{C}|\mathbb{R}}(z) = z\bar{z} = |z|^2$ and the theorem is part of classical analysis. So let $|\ |$ be nonarchimedean. Since every algebraic extension $L|K$ is the union of its finite subextensions, we may assume that the degree $n = [L:K]$ is finite.

Existence of the extended valuation: let $\mathcal{o}$ be the valuation ring of K and $\mathcal{O}$ its integral closure in L. Then one has

$$(*)\qquad \mathcal{O} = \{\alpha \in L \mid N_{L|K}(\alpha) \in \mathcal{o}\}.$$

The implication $\alpha \in \mathcal{O} \Rightarrow N_{L|K}(\alpha) \in \mathcal{o}$ is evident (see chap. I, § 2, p. 12). Conversely, let $\alpha \in L^*$ and $N_{L|K}(\alpha) \in \mathcal{o}$. Let

$$f(x) = x^d + a_{d-1}x^{d-1} + \cdots + a_0 \in K[x]$$

be the minimal polynomial of α over K. Then $N_{L|K}(\alpha) = \pm a_0^m \in \mathcal{o}$, so that $|a_0| \le 1$, i.e., $a_0 \in \mathcal{o}$. By (4.7) this gives $f(x) \in \mathcal{o}[x]$, i.e., $\alpha \in \mathcal{O}$.

For the function $|\alpha| = \sqrt[n]{|N_{L|K}(\alpha)|}$, the conditions $|\alpha| = 0 \iff \alpha = 0$ and $|\alpha\beta| = |\alpha||\beta|$ are obvious. The strong triangle inequality

$$|\alpha + \beta| \le \max\{|\alpha|, |\beta|\}$$

reduces, after dividing by α or β, to the implication

$$|\alpha| \le 1 \implies |\alpha + 1| \le 1,$$

and then, by $(*)$, to $\alpha \in \mathcal{O} \Rightarrow \alpha + 1 \in \mathcal{O}$, which is trivially true. Thus the formula $|\alpha| = \sqrt[n]{|N_{L|K}(\alpha)|}$ does define a valuation of L and, restricted to K, it clearly gives back the given valuation. Equally obviously it has $\mathcal{O}$ as its valuation ring.

Uniqueness of the extended valuation: let $|\ |'$ be another extension with valuation ring $\mathcal{O}'$. Let $\mathfrak{P}$, resp. $\mathfrak{P}'$, be the maximal ideal of $\mathcal{O}$, resp. $\mathcal{O}'$. We show that $\mathcal{O} \subseteq \mathcal{O}'$. Let $\alpha \in \mathcal{O} \setminus \mathcal{O}'$ and let

$$f(x) = x^d + a_1x^{d-1} + \cdots + a_d$$

be the minimal polynomial of α over K. Then one has $a_1, \dots, a_d \in \mathcal{o}$ and $\alpha^{-1} \in \mathfrak{P}'$, hence $1 = -a_1\alpha^{-1} - \cdots - a_d(\alpha^{-1})^d \in \mathfrak{P}'$, a contradiction. This shows the inclusion $\mathcal{O} \subseteq \mathcal{O}'$. In other words, we have that $|\alpha| \le 1 \Rightarrow |\alpha|' \le 1$ and this implies that the valuations $|\ |$ and $|\ |'$ are equivalent. For if they were not, then the approximation theorem (3.4) would allow us to find an $\alpha \in L$ such that $|\alpha| \le 1 \Rightarrow |\alpha|' > 1$. Thus $|\ |$ and $|\ |'$ are equal because they agree on K.

The fact that L is again complete with respect to the extended valuation is deduced from the following general result. □

(4.9) Proposition. *Let K be complete with respect to the valuation $|\ |$ and let V be an n-dimensional normed vector space over K. Then, for any basis $v_1, \dots, v_n$ of V the maximum norm*

$$\|x_1v_1 + \cdots + x_nv_n\| = \max\{|x_1|, \dots, |x_n|\}$$

is equivalent to the given norm on V. In particular, V is complete and the isomorphism

$$K^n \longrightarrow V\,, \quad (x_1, \ldots, x_n) \longmapsto x_1 v_1 + \cdots + x_n v_n\,,$$

is a homeomorphism.

Proof: Let $v_1, \ldots, v_n$ be a basis and $\| \ \|$ be the corresponding maximum norm on V. It suffices to show that, for every norm $| \ |$ on V, there exist constants $\rho, \rho' > 0$ such that

$$\rho \|x\| \le |x| \le \rho' \|x\| \quad \text{for all} \quad x \in V\,.$$

Then the norm $| \ |$ defines the same topology on V as the norm $\| \ \|$, and we obtain the topological isomorphism $K^n \to V$, $(x_1, \ldots, x_n) \mapsto x_1 v_1 + \cdots + x_n v_n$. In fact, $\| \ \|$ is transformed into the maximum norm on K^n.

For ρ' we may obviously take $|v_1| + \cdots + |v_n|$. The existence of ρ is proved by induction on n. For $n = 1$ we may take $\rho = |v_1|$. Suppose that everything is proved for $(n-1)$-dimensional vector spaces. Let

$$V_i = K v_1 + \cdots + K v_{i-1} + K v_{i+1} + \cdots + K v_n\,,$$

so that $V = V_i + K v_i$. Then V_i is complete with respect to the restriction of $| \ |$ by induction, hence it is closed in V. Thus $V_i + v_i$ is also closed. Since $0 \notin \bigcup_{i=1}^n (V_i + v_i)$, there exists a neighbourhood of 0 which is disjoint from $\bigcup_{i=1}^n (V_i + v_i)$, i.e., there exists $\rho > 0$ such that

$$|w_i + v_i| \ge \rho \quad \text{for all} \quad w_i \in V_i \quad \text{and all} \quad i = 1, \ldots, n\,.$$

For $x = x_1 v_1 + \cdots + x_n v_n \neq 0$ and $|x_r| = \max\{|x_i|\}$, one finds

$$|x_r^{-1} x| = \left| \frac{x_1}{x_r} v_1 + \cdots + v_r + \cdots + \frac{x_n}{x_r} v_n \right| \ge \rho\,,$$

so that one has $|x| \ge \rho |x_r| = \rho \|x\|$. □

The fact that an exponential valuation v on K associated with $| \ |$ extends uniquely to L is a trivial consequence of theorem (4.8). The extension w is given by the formula

$$w(\alpha) = \frac{1}{n} v\big(N_{L|K}(\alpha)\big)$$

if $n = [L : K] < \infty$.

Exercise 1. An infinite algebraic extension of a complete field K is never complete.

Exercise 2. Let $X_0, X_1, \ldots$ be an infinite sequence of unknowns, p a fixed prime number and $W_n = X_0^{p^n} + pX_1^{p^{n-1}} + \cdots + p^n X_n$, $n \geq 0$. Show that there exist polynomials $S_0, S_1, \ldots;\ P_0, P_1, \ldots \in \mathbb{Z}[X_0, X_1, \ldots;\ Y_0, Y_1, \ldots]$ such that

$$W_n(S_0, S_1, \ldots) = W_n(X_0, X_1, \ldots) + W_n(Y_0, Y_1, \ldots),$$

$$W_n(P_0, P_1, \ldots) = W_n(X_0, X_1, \ldots) \cdot W_n(Y_0, Y_1, \ldots).$$

Exercise 3. Let A be a commutative ring. For $a = (a_0, a_1, \ldots)$, $b = (b_0, b_1, \ldots)$, $a_i, b_i \in A$, put

$$a + b = (S_0(a,b), S_1(a,b), \ldots), \quad a \cdot b = (P_0(a,b), P_1(a,b), \ldots).$$

Show that with these operations the vectors $a = (a_0, a_1, \ldots)$ form a commutative ring $W(A)$ with 1. It is called the **ring of Witt vectors** over A.

Exercise 4. Assume $pA = 0$. For every Witt vector $a = (a_0, a_1, \ldots) \in W(A)$ consider the "ghost components"

$$a^{(n)} = W_n(a) = a_0^{p^n} + pa_1^{p^{n-1}} + \cdots + p^n a_n$$

as well as the mappings $V, F : W(A) \to W(A)$ defined by

$$Va = (0, a_0, a_1, \ldots) \quad \text{and} \quad Fa = (a_0^p, a_1^p, \ldots),$$

called respectively "transfer" ("Verschiebung" in German) and "Frobenius". Show that

$$(Va)^{(n)} = pa^{(n-1)} \quad \text{and} \quad a^{(n)} = (Fa)^{(n)} + p^n a_n .$$

Exercise 5. Let k be a field of characteristic p. Then V is a homomorphism of the additive group of $W(k)$ and F is a ring homomorphism, and one has

$$VFa = FVa = pa .$$

Exercise 6. If k is a perfect field of characteristic p, then $W(k)$ is a complete discrete valuation ring with residue class field k.

§ 5. Local Fields

Among all complete (nonarchimedean) valued fields, those arising as completions of a **global field**, i.e., of a finite extension of either $\mathbb{Q}$ or $\mathbb{F}_p(t)$, have the most eminent relevance for number theory. The valuation on such a completion is discrete and has a finite residue class field, as we shall see shortly. In contrast to the global fields, all fields which are complete with respect to a discrete valuation and have a finite residue class field are called **local fields**. For such a local field, the normalized exponential valuation is denoted by $v_\mathfrak{p}$, and $|\ |_\mathfrak{p}$ denotes the absolute value normalized by

$$|x|_\mathfrak{p} = q^{-v_\mathfrak{p}(x)},$$

where q is the cardinality of the residue class field.

(5.1) Proposition. *A local field K is locally compact. Its valuation ring $\mathcal{O}$ is compact.*

Proof: By (4.5) we have $\mathcal{O} \cong \varprojlim \mathcal{O}/\mathfrak{p}^n$, both algebraically and topologically. Since $\mathfrak{p}^\nu/\mathfrak{p}^{\nu+1} \cong \mathcal{O}/\mathfrak{p}$ (see (3.9)), the rings $\mathcal{O}/\mathfrak{p}^n$ are finite, hence compact. Being a closed subset of the compact product $\prod_{n=1}^{\infty} \mathcal{O}/\mathfrak{p}^n$, it follows that the projective limit $\varprojlim \mathcal{O}/\mathfrak{p}^n$, and thus $\mathcal{O}$, is also compact. For every $a \in K$, the set $a + \mathcal{O}$ is an open, and at the same time compact neighbourhood, so that K is locally compact. □

In happy concord with the definition of global fields as the finite extensions of $\mathbb{Q}$ and $\mathbb{F}_p(t)$, we now obtain the following characterization of local fields.

(5.2) Proposition. *The local fields are precisely the finite extensions of the fields $\mathbb{Q}_p$ and $\mathbb{F}_p((t))$.*

Proof: A finite extension K of $k = \mathbb{Q}_p$ or $k = \mathbb{F}_p((t))$ is again complete, by (4.8), with respect to the extended valuation $|\alpha| = \sqrt[n]{|N_{K|k}(\alpha)|}$, which itself is obviously again discrete. Since $K|k$ is of finite degree, so is the residue class field extension $\kappa|\mathbb{F}_p$, for if $\bar{x}_1, \dots, \bar{x}_n \in \kappa$ are linearly independent, then any choice of preimages $x_1, \dots, x_n \in K$ is linearly independent over k. Indeed, dividing any nontrivial k-linear relation $\lambda_1 x_1 + \cdots + \lambda_n x_n = 0$, $\lambda_i \in k$, by the coefficient λ_i with biggest absolute value, yields a linear combination with coefficients in the valuation ring of k with 1 as i-th coefficient, from which we obtain a nontrivial relation $\bar{\lambda}_1 \bar{x}_1 + \cdots + \bar{\lambda}_n \bar{x}_n = 0$ by reducing to κ. Therefore K is a local field.

Conversely, let K be a local field, v its discrete exponential valuation, and p the characteristic of its residue class field κ. If K has characteristic 0, then the restriction of v to $\mathbb{Q}$ is equivalent to the p-adic valuation v_p of $\mathbb{Q}$ because $v(p) > 0$. In view of the completeness of K, the closure of $\mathbb{Q}$ in K is the completion of $\mathbb{Q}$ with respect to v_p, in other words $\mathbb{Q}_p \subseteq K$. The fact that $K|\mathbb{Q}_p$ is of finite degree results from the local compactness of the vector space K, by a general theorem of topological linear algebra (see [18], chap. I, §2, n° 4, th. 3), but it also follows from (6.8) below. If on the other hand the characteristic of K is not equal to zero, then it has to equal p. In this case we find $K = \kappa((t))$, for a prime element t of K (see p. 127), hence $\mathbb{F}_p((t)) \subseteq K$. In fact, if $\kappa = \mathbb{F}_p(\alpha)$ and $p(X) \in \mathbb{F}_p[X] \subseteq K[X]$ is the minimal polynomial of α over $\mathbb{F}_p$, then, by Hensel's lemma, $p(X)$ splits over K into linear factors. We may therefore view κ as a subfield of K, and then the elements of K turn out to be, by (4.4), the Laurent series in t with coefficients in κ. □

Remark: One can show that a field K which is locally compact with respect to a nondiscrete topology is isomorphic either to $\mathbb{R}$ or $\mathbb{C}$, or to a finite extension of $\mathbb{Q}_p$ or $\mathbb{F}_p((t))$, i.e., to a local field (see [137], chap. I, § 3).

We have just seen that the local fields of characteristic p are the **power series fields** $\mathbb{F}_q((t))$, with $q = p^f$. The local fields of characteristic 0, i.e., the finite extensions $K|\mathbb{Q}_p$ of the fields of p-adic numbers $\mathbb{Q}_p$, are called **$\mathfrak{p}$-adic number fields**. For them one has an *exponential function* and a *logarithm function*. In contrast to the real and complex case, however, the former is not defined on all of K, whereas the latter is given on the whole multiplicative group K^*. For the definition of the logarithm we make use of the following fact.

(5.3) Proposition. *The multiplicative group of a local field K admits the decomposition*

$$K^* = (\pi) \times \mu_{q-1} \times U^{(1)} .$$

Here π is a prime element, $(\pi) = \{\pi^k \mid k \in \mathbb{Z}\}$, $q = \#\kappa$ is the number of elements in the residue class field $\kappa = \mathcal{O}/\mathfrak{p}$, and $U^{(1)} = 1 + \mathfrak{p}$ is the group of principal units.

Proof: For every $\alpha \in K^*$, one has a unique representation $\alpha = \pi^n u$ with $n \in \mathbb{Z}$, $u \in \mathcal{O}^*$ so that $K^* = (\pi) \times \mathcal{O}^*$. Since the polynomial $X^{q-1} - 1$ splits into linear factors over K by Hensel's lemma, $\mathcal{O}^*$ contains the group μ_{q-1} of $(q-1)$-th roots of unity. The homomorphism $\mathcal{O}^* \to \kappa^*$, $u \mapsto u \bmod \mathfrak{p}$, has kernel $U^{(1)}$ and maps μ_{q-1} bijectively onto κ^*. Hence $\mathcal{O}^* = \mu_{q-1} \times U^{(1)}$. □

(5.4) Proposition. *For a $\mathfrak{p}$-adic number field K there is a uniquely determined continuous homomorphism*

$$\log : K^* \to K$$

such that $\log p = 0$ *which on principal units $(1 + x) \in U^{(1)}$ is given by the series*

$$\log(1 + x) = x - \frac{x^2}{2} + \frac{x^3}{3} - \cdots .$$

Proof: By § 4, we can think of the p-adic valuation v_p of $\mathbb{Q}_p$ as extended to K. Observing that $v_p(x) > 0$, so that $c = p^{v_p(x)} > 1$, and $p^{v_p(\nu)} \le \nu$, giving $v_p(\nu) \le \dfrac{\ln \nu}{\ln p}$ (with the usual logarithm), we compute the valuation of the terms x^ν/ν of the series,

$$v_p\Big(\frac{x^\nu}{\nu}\Big) = \nu v_p(x) - v_p(\nu) \ge \nu \frac{\ln c}{\ln p} - \frac{\ln \nu}{\ln p} = \frac{\ln(c^\nu/\nu)}{\ln p}.$$

This shows that x^ν/ν is a nullsequence, i.e., the logarithm series converges. It defines a homomorphism because

$$\log\big((1+x)(1+y)\big) = \log(1+x) + \log(1+y)$$

is an identity of formal power series and all series in it converge provided $1+x, 1+y \in U^{(1)}$.

For every $\alpha \in K^*$, choosing a prime element π, we have a unique representation

$$\alpha = \pi^{v_\mathfrak{p}(\alpha)}\omega(\alpha)\langle\alpha\rangle,$$

where $v_\mathfrak{p} = e v_p$ is the normalized valuation of K, $\omega(\alpha) \in \mu_{q-1}$, $\langle\alpha\rangle \in U^{(1)}$. As suggested by the equation $p = \pi^e\omega(p)\langle p\rangle$, we define $\log \pi = -\frac{1}{e}\log\langle p\rangle$ and thus obtain the homomorphism $\log : K^* \to K$ by

$$\log \alpha = v_\mathfrak{p}(\alpha)\log \pi + \log\langle\alpha\rangle.$$

It is obviously continuous and has the property that $\log p = 0$. If $\lambda : K^* \to K$ is any continuation of $\log : U^{(1)} \to K$ such that $\lambda(p) = 0$, then we find that $\lambda(\xi) = \frac{1}{q-1}\lambda(\xi^{q-1}) = 0$ for each $\xi \in \mu_{q-1}$. It follows that $0 = e\lambda(\pi) + \lambda(\langle p\rangle) = e\lambda(\pi) + \log\langle p\rangle$, so that $\lambda(\pi) = \log \pi$, and thus $\lambda(\alpha) = v_\mathfrak{p}(\alpha)\lambda(\pi) + \lambda(\langle\alpha\rangle) = v_\mathfrak{p}(\alpha)\log\pi + \log\langle\alpha\rangle = \log\alpha$, for all $\alpha \in K^*$. log is therefore uniquely determined and independent of the choice of π. □

(5.5) Proposition. *Let $K|\mathbb{Q}_p$ be a $\mathfrak{p}$-adic number field with valuation ring $\mathcal{O}$ and maximal ideal $\mathfrak{p}$, and let $p\mathcal{O} = \mathfrak{p}^e$. Then the power series*

$$\exp(x) = 1 + x + \frac{x^2}{2!} + \frac{x^3}{3!} + \cdots \quad \textit{and} \quad \log(1+z) = z - \frac{z^2}{2} + \frac{z^3}{3} - \cdots,$$

yield, for $n > \frac{e}{p-1}$, two mutually inverse isomorphisms (and homeomorphisms)

$$\mathfrak{p}^n \underset{\log}{\overset{\exp}{\rightleftarrows}} U^{(n)}.$$

We prepare the proof by the following elementary lemma.

(5.6) Lemma. *Let $\nu = \sum_{i=0}^{r} a_i p^i$, $0 \le a_i < p$, be the p-adic expansion of the natural number $\nu \in \mathbb{N}$. Then*

$$v_p(\nu!) = \frac{1}{p-1} \sum_{i=0}^{r} a_i (p^i - 1).$$

Proof: Let $[c]$ signify the biggest integer $\le c$. Then we have

$$\begin{aligned} [\nu/p] &= a_1 + a_2 p + \cdots + a_r p^{r-1}, \\ [\nu/p^2] &= \qquad a_2 + \cdots + a_r p^{r-2}, \\ &\vdots \\ [\nu/p^r] &= \qquad\qquad a_r . \end{aligned}$$

Now we count how many numbers $1, 2, \ldots, \nu$ are divisible by p, and then by p^2, etc. We find

$$v_p(\nu!) = [\nu/p] + \cdots + [\nu/p^r] = a_1 + (p+1)a_2 + \cdots + (p^{r-1} + \cdots + 1)a_r$$

and hence

$$(p-1)v_p(\nu!) = (p-1)a_1 + (p^2-1)a_2 + \cdots + (p^r - 1)a_r = \sum_{i=0}^{r} a_i (p^i - 1).$$

□

Proof of (5.5): We again think of the p-adic valuation v_p of $\mathbb{Q}_p$ as being extended to K. Then $v_{\mathfrak{p}} = e v_p$ is the normalized valuation of K. For every natural number $\nu > 1$, one has the estimate

$$\frac{v_p(\nu)}{\nu - 1} \le \frac{1}{p-1},$$

for if $\nu = p^a \nu_0$, with $(\nu_0, p) = 1$ and $a > 0$, then

$$\frac{v_p(\nu)}{\nu - 1} = \frac{a}{p^a \nu_0 - 1} \le \frac{a}{p^a - 1} = \frac{1}{p-1} \, \frac{a}{p^{a-1} + \cdots + p + 1} \le \frac{1}{p-1}.$$

For $v_p(z) > \frac{1}{p-1}$, $z \neq 0$, i.e., $v_{\mathfrak{p}}(z) > \frac{e}{p-1}$, this yields

$$v_p\Big(\frac{z^\nu}{\nu}\Big) - v_p(z) = (\nu - 1)v_p(z) - v_p(\nu) > (\nu - 1)\Big(\frac{1}{p-1} - \frac{v_p(\nu)}{\nu - 1}\Big) \ge 0,$$

and thus $v_{\mathfrak{p}}(\log(1+z)) = v_{\mathfrak{p}}(z)$. For $n > \frac{e}{p-1}$, log therefore maps $U^{(n)}$ into $\mathfrak{p}^n$.

For the exponential series $\sum_{\nu=0}^{\infty} x^\nu/\nu!$, we compute the valuations $v_p(x^\nu/\nu!)$ as follows. Writing, for $\nu > 0$,

$$\nu = a_0 + a_1 p + \cdots + a_r p^r, \quad 0 \le a_i < p,$$

we get from (5.6) that

$$v_p(\nu!) = \frac{1}{p-1}\sum_{i=0}^{r} a_i(p^i - 1) = \frac{1}{p-1}\big(\nu - (a_0 + a_1 + \cdots + a_r)\big).$$

Putting $s_\nu = a_0 + \cdots + a_r$ this becomes

$$v_p\Big(\frac{x^\nu}{\nu!}\Big) = \nu v_p(x) - \frac{\nu - s_\nu}{p-1} = \nu\Big(v_p(x) - \frac{1}{p-1}\Big) + \frac{s_\nu}{p-1}.$$

For $v_\mathfrak{p}(x) > \frac{e}{p-1}$, i.e., $v_p(x) > \frac{1}{p-1}$, this implies the convergence of the exponential series. If furthermore $x \neq 0$ and $\nu > 1$, then one has

$$v_p\Big(\frac{x^\nu}{\nu!}\Big) - v_p(x) = (\nu - 1)v_p(x) - \frac{\nu - 1}{p-1} + \frac{s_\nu - 1}{p-1} > \frac{s_\nu - 1}{p-1} \geq 0.$$

Therefore $v_\mathfrak{p}(\exp(x) - 1) = v_\mathfrak{p}(x)$, i.e., for $n > \frac{e}{p-1}$, exp maps the group $\mathfrak{p}^n$ into $U^{(n)}$. Furthermore, one has for $v_\mathfrak{p}(x)$, $v_\mathfrak{p}(z) > \frac{e}{p-1}$ that

$$\exp\log(1+z) = 1+z \quad \text{and} \quad \log\exp x = x,$$

for these are identities of formal power series and all of the series converge. This proves the proposition. □

For an arbitrary local field K, the group of principal units $U^{(1)}$ is a $\mathbb{Z}_p$-module (where $p = \operatorname{char}(\kappa)$) in a canonical way, i.e., for every $1 + x \in U^{(1)}$ and every $z \in \mathbb{Z}_p$, one has the power $(1+x)^z \in U^{(1)}$. This is a consequence of the fact that $U^{(1)}/U^{(n+1)}$ has order q^n for all n (where $q = \#\mathcal{O}/\mathfrak{p}$ – the reason for this is that $U^{(i)}/U^{(i+1)} \cong \mathcal{O}/\mathfrak{p}$, by (3.10), so that $U^{(1)}/U^{(n+1)}$ is a $\mathbb{Z}/q^n\mathbb{Z}$-module) and of the formulas

$$U^{(1)} = \varprojlim_n U^{(1)}/U^{(n+1)} \quad \text{and} \quad \mathbb{Z}_p = \varprojlim_n \mathbb{Z}/q^n\mathbb{Z}.$$

This obviously extends the $\mathbb{Z}$-module structure of $U^{(1)}$. The function

$$f(z) = (1+x)^z$$

is continuous because the congruence $z \equiv z' \bmod q^n\mathbb{Z}_p$ implies $(1+x)^z \equiv (1+x)^{z'} \bmod U^{(n+1)}$, so that the neighbourhood $z + q^n\mathbb{Z}_p$ of z is mapped to the neighbourhood $(1+x)^z U^{(n+1)}$ of $f(z)$. In particular, $(1+x)^z$ may be expressed as the limit

$$(1+x)^z = \lim_{i\to\infty}(1+x)^{z_i}$$

of ordinary powers $(1+x)^{z_i}$, $z_i \in \mathbb{Z}$, if $z = \lim_{i\to\infty} z_i$.

After this discussion we can now determine explicitly the structure of the locally compact multiplicative group K^* of a local field K.

(5.7) Proposition. *Let K be a local field and $q = p^f$ the number of elements in the residue class field. Then the following hold.*

(i) *If K has characteristic 0, then one has (both algebraically and topologically)*

$$K^* \cong \mathbb{Z} \oplus \mathbb{Z}/(q-1)\mathbb{Z} \oplus \mathbb{Z}/p^a\mathbb{Z} \oplus \mathbb{Z}_p^d,$$

where $a \geq 0$ and $d = [K : \mathbb{Q}_p]$.

(ii) *If K has characteristic p, then one has (both algebraically and topologically)*

$$K^* \cong \mathbb{Z} \oplus \mathbb{Z}/(q-1)\mathbb{Z} \oplus \mathbb{Z}_p^{\mathbb{N}}.$$

Proof: By (5.3) we have (both algebraically and topologically)

$$K^* = (\pi) \times \mu_{q-1} \times U^{(1)} \cong \mathbb{Z} \oplus \mathbb{Z}/(q-1)\mathbb{Z} \oplus U^{(1)}.$$

This reduces us to the computation of the $\mathbb{Z}_p$-module $U^{(1)}$.

(i) Assume $\operatorname{char}(K) = 0$. For n sufficiently big, (5.5) gives us the isomorphism

$$\log : U^{(n)} \longrightarrow \mathfrak{p}^n = \pi^n \mathcal{O} \cong \mathcal{O}.$$

Since log, exp, and $f(z) = (1+x)^z$ are continuous, this is a topological isomorphism of $\mathbb{Z}_p$-modules. By chap. I, (2.9), $\mathcal{O}$ admits an integral basis $\alpha_1, \ldots, \alpha_d$ over $\mathbb{Z}_p$, i.e., $\mathcal{O} = \mathbb{Z}_p\alpha_1 \oplus \cdots \oplus \mathbb{Z}_p\alpha_d \cong \mathbb{Z}_p^d$. Therefore $U^{(n)} \cong \mathbb{Z}_p^d$. Since the index $(U^{(1)} : U^{(n)})$ is finite and $U^{(n)}$ is a finitely generated $\mathbb{Z}_p$-module of rank d, so is $U^{(1)}$. The torsion subgroup of $U^{(1)}$ is the group μ_{p^a} of roots of unity in K of p-power order. By the main theorem on modules over principal ideal domains, there exists in $U^{(1)}$ a free, finitely generated, and therefore closed, $\mathbb{Z}_p$-submodule V of rank d such that

$$U^{(1)} = \mu_{p^a} \times V \cong \mathbb{Z}/p^a\mathbb{Z} \oplus \mathbb{Z}_p^d,$$

both algebraically and topologically.

(ii) If $\operatorname{char}(K) = p$, we have $K \cong \mathbb{F}_q((t))$ (see p. 127) and

$$U^{(1)} = 1 + \mathfrak{p} = 1 + t\,\mathbb{F}_q[[t]].$$

The following argument is taken from the book [79] of K. Iwasawa.

Let $\omega_1, \ldots, \omega_f$ be a basis of $\mathbb{F}_q|\mathbb{F}_p$. For every natural number n relatively prime to p we consider the continuous homomorphism

$$g_n : \mathbb{Z}_p^f \to U^{(n)}, \quad g_n(a_1, \ldots, a_f) = \prod_{i=1}^{f} (1 + \omega_i t^n)^{a_i}.$$

This function has the following properties. If $m = np^s$, $s \geq 0$, then

$$U^{(m)} = g_n(p^s\mathbb{Z}_p^f)U^{(m+1)} \tag{1}$$

and, for $\alpha = (a_1, \ldots, a_f) \in \mathbb{Z}_p^f$,

$$\alpha \notin p\mathbb{Z}_p^f \iff g_n(p^s\alpha) \notin U^{(m+1)}. \tag{2}$$

Indeed, for $\omega = \sum_{i=1}^f b_i\omega_i \in \mathbb{F}_q$, $b_i \in \mathbb{Z}$, $b_i \equiv a_i \bmod p$, we have

$$g_n(\alpha) \equiv \prod_{i=1}^f (1+\omega_i t^n)^{b_i} \equiv 1 + \omega t^n \bmod \mathfrak{p}^{n+1}$$

and hence, since we are in characteristic p,

$$g_n(p^s\alpha) = g_n(\alpha)^{p^s} \equiv 1 + \omega^{p^s} t^m \bmod \mathfrak{p}^{m+1}.$$

As α varies over the elements of $\mathbb{Z}_p^f$, ω, and thus also ω^{p^s}, varies over the elements of $\mathbb{F}_q$, and we get (1). Furthermore one has $g_n(p^s\alpha) \equiv 1 \bmod \mathfrak{p}^{m+1} \iff \omega = 0 \iff b_i \equiv 0 \bmod p$, for $i = 1, \ldots, f \iff a_i \equiv 0 \bmod p$, for $i = 1, \ldots, f \iff \alpha \in p\mathbb{Z}_p^f$, and this amounts to (2).

We now consider the continuous homomorphism of $\mathbb{Z}_p$-modules

$$g = \prod_{(n,p)=1} g_n : A = \prod_{(n,p)=1} \mathbb{Z}_p^f \longrightarrow U^{(1)},$$

where the product $\prod_{(n,p)=1} \mathbb{Z}_p^f$ is taken over all $n \geq 1$ such that $(n, p) = 1$, each factor being a copy of $\mathbb{Z}_p^f$. Observe that the product $g(\xi) = \prod g_n(\alpha_n)$ converges because $g_n(\alpha_n) \in U^{(n)}$. Let $m = np^s$, with $(n, p) = 1$, be any natural number. As $g_n(\mathbb{Z}_p^f) \subseteq g(A)$, it follows from (1) that each coset of $U^{(m)}/U^{(m+1)}$ is represented by an element of $g(A)$. This means that $g(A)$ is dense in $U^{(1)}$. Since A is compact and g is continuous, g is actually surjective.

On the other hand, let $\xi = (\ldots, \alpha_n, \ldots) \in A$, $\xi \neq 0$, i.e., $\alpha_n \neq 0$ for some n. Such an α_n is of the form $\alpha_n = p^s\beta_n$ with $s = s(\alpha_n) \geq 0$, and $\beta_n \in \mathbb{Z}_p^f \smallsetminus p\mathbb{Z}_p^f$. It now follows from (2) that

$$g_n(\alpha_n) \in U^{(m)}, \quad g_n(\alpha_n) \notin U^{(m+1)} \quad \text{for} \quad m = m(\alpha_n) = np^s.$$

Since the n are prime to p, all the $m(\alpha_n)$ have to be distinct, for all $\alpha_n \neq 0$. Let n be the natural number, prime to p and such that $\alpha_n \neq 0$, which satisfies $m(\alpha_n) < m(\alpha_{n'})$, for all $n' \neq n$ such that $\alpha_{n'} \neq 0$. Then one has, for all $n' \neq n$, that

$$g_{n'}(\alpha_{n'}) \in U^{(m+1)} \quad \text{where} \quad m = m(\alpha_n) < m(\alpha_{n'}).$$

Consequently

$$g(\xi) \equiv g_n(\alpha_n) \not\equiv 1 \bmod U^{(m+1)},$$

and so $g(\xi) \neq 1$. This shows the injectivity of g. Since $A = \mathbb{Z}_p^{\mathbb{N}}$, this proves the claim (ii). □

(5.8) Corollary. *If the natural number n is not divisible by the characteristic of K, then one finds the following indices for the subgroups of n-th powers K^{*n} and U^n in the multiplicative group K^* and in the unit group U:*

$$(K^* : K^{*n}) = n(U : U^n) = \frac{n}{|n|_{\mathfrak{p}}}\#\mu_n(K).$$

Proof: The first equality is a consequence of $K^* = (\pi) \times U$. By (5.7), we have

$$U \cong \mu(K) \times \mathbb{Z}_p^d, \quad \text{resp.} \quad U \cong \mu(K) \times \mathbb{Z}_p^{\mathbb{N}},$$

when char $(K) = 0$, resp. $p > 0$. From the exact sequence

$$1 \longrightarrow \mu_n(K) \longrightarrow \mu(K) \xrightarrow{n} \mu(K) \longrightarrow \mu(K)/\mu(K)^n \longrightarrow 1,$$

one has $\#\mu_n(K) = \#\mu(K)/\mu(K)^n$. When char$(K) = 0$, this gives:

$$(U : U^n) = \#\mu_n(K)\#(\mathbb{Z}_p/n\mathbb{Z}_p)^d = \#\mu_n(K)p^{dv_p(n)} = \#\mu_n(K)/|n|_{\mathfrak{p}},$$

and when char$(K) = p$ one gets simply $(U : U^n) = \#\mu_n(K) = \#\mu_n(K)/|n|_{\mathfrak{p}}$ because $(n, p) = 1$, i.e., $n\mathbb{Z}_p = \mathbb{Z}_p$. □

Exercise 1. The logarithm function can be continued to a continuous homomorphism $\log : \overline{\mathbb{Q}}_p^* \to \mathbb{Q}$, and the exponential function to a continuous homomorphism $\exp : \bar{\mathfrak{p}}^{\frac{1}{1-p}} \to \overline{\mathbb{Q}}_p^*$, where $\bar{\mathfrak{p}}^{\frac{1}{1-p}} = \{x \in \overline{\mathbb{Q}}_p \mid v_p(x) > \frac{1}{1-p}\}$ and v_p is the unique extension of the normalized valuation on $\mathbb{Q}_p$.

Exercise 2. Let $K|\mathbb{Q}_p$ be a $\mathfrak{p}$-adic number field. For $1 + x \in U^{(1)}$ and $z \in \mathbb{Z}_p$ one has

$$(1+x)^z = \sum_{\nu=0}^{\infty} \binom{z}{\nu} x^\nu .$$

The series converges even for $x \in K$ such that $v_{\mathfrak{p}}(x) > \dfrac{e}{p-1}$.

Exercise 3. Under the above hypotheses one has

$$(1+x)^z = \exp(z \log(1+x)) \quad \text{and} \quad \log(1+x)^z = z \log(1+x).$$

Exercise 4. For a $\mathfrak{p}$-adic number field K, every subgroup of finite index in K^* is both open and closed.

Exercise 5. If K is a $\mathfrak{p}$-adic number field, then the groups K^{*n}, for $n \in \mathbb{N}$, form a basis of neighbourhoods of 1 in K^*.

Exercise 6. Let K be a $\mathfrak{p}$-adic number field, $v_{\mathfrak{p}}$ the normalized exponential valuation of K, and dx the Haar measure on the locally compact additive group K, scaled so that $\int_{\mathcal{O}} dx = 1$. Then one has $v_{\mathfrak{p}}(a) = \int_{a\mathcal{O}} dx$. Furthermore,

$$I(f) = \int_{K \setminus \{0\}} f(x) \frac{dx}{|x|_{\mathfrak{p}}}$$

is a Haar measure on the locally compact group K^*.

§ 6. Henselian Fields

Most results on complete valued fields can be derived from Hensel's lemma alone, without the full strength of completeness. This lemma is valid in a much bigger class of nonarchimedean valued fields than the complete ones. For example, let (K, v) be a nonarchimedean valued field and $(\widehat{K}, \hat{v})$ its completion. Let $\mathcal{O}$, resp. $\widehat{\mathcal{O}}$, be the valuation rings of K, resp. $\widehat{K}$. We then consider the separable closure K_v of K in $\widehat{K}$, and the valuation ring $\mathcal{O}_v \subseteq K_v$ with maximal ideal $\mathfrak{p}_v$, which is associated to the restriction of $\hat{v}$ to K_v,

$$K \subseteq K_v \subseteq \widehat{K}, \quad \mathcal{O} \subseteq \mathcal{O}_v \subseteq \widehat{\mathcal{O}}.$$

Then Hensel's lemma holds in the ring $\mathcal{O}_v$ as well as in the ring $\widehat{\mathcal{O}}$ even though K_v will not, as a rule, be complete. When K_v is algebraically closed in $\widehat{K}$ – hence in particular $\operatorname{char}(K) = 0$ – this is immediately obvious (otherwise it follows from (6.6) and § 6, exercise 3 below). Indeed, by (4.3) we have

$$\mathcal{O}/\mathfrak{p} = \mathcal{O}_v/\mathfrak{p}_v = \widehat{\mathcal{O}}/\widehat{\mathfrak{p}},$$

and if a primitive polynomial $f(x) \in \mathcal{O}_v[x]$ splits over $\mathcal{O}_v/\mathfrak{p}_v$ into relatively prime factors $\bar{g}(x), \bar{h}(x)$, then we have by Hensel's lemma (4.6) a factorization in $\widehat{\mathcal{O}}$

$$f(x) = g(x)h(x)$$

such that $g \equiv \bar{g} \bmod \widehat{\mathfrak{p}}$, $h \equiv \bar{h} \bmod \widehat{\mathfrak{p}}$, $\deg(g) = \deg(\bar{g})$. But this factorization already takes place over $\mathcal{O}_v$ once the highest coefficient of g is chosen to be in $\mathcal{O}_v^*$, because the coefficients of f, and therefore also those of g and h are algebraic over K.

The valued field K_v is called the **henselization** of the field K with respect to v. It enjoys all the relevant algebraic properties of the completion $\widehat{K}$, but offers the advantage of being itself an algebraic extension of K which can also be obtained in a purely algebraic manner, without the analytic recourse to the completion (see § 9, exercise 4). The consequence is that taking the henselization of an infinite algebraic extension $L|K$ is possible within the category of algebraic extensions. Let us define in general:

(6.1) Definition. *A* **henselian field** *is a field with a nonarchimedean valuation v whose valuation ring $\mathcal{O}$ satisfies Hensel's lemma in the sense of (4.6). One also calls the valuation v or the valuation ring $\mathcal{O}$ henselian.*

(6.2) Theorem. *Let K be a henselian field with respect to the valuation $|\ |$. Then $|\ |$ admits one and only one extension to any given algebraic extension $L|K$. It is given by*

$$|\alpha| = \sqrt[n]{|N_{L|K}(\alpha)|},$$

if $L|K$ has finite degree n. In any case, the valuation ring of the extended valuation is the integral closure of the valuation ring of K in L.

The proof of this theorem is *verbatim* the same as in the case of a complete field (see (4.8)). What is remarkable about our current setting is that, conversely, the unique extendability also characterizes henselian fields. In order to prove this, we appeal to a method which allows us to express the valuations of the roots of a polynomial in terms of the valuations of the coefficients. It relies on the notion of **Newton polygon**, which arises as follows.

Let v be an arbitrary exponential valuation of the field K and let

$$f(x) = a_0 + a_1x + \cdots + a_nx^n \in K[x]$$

be a polynomial satisfying $a_0a_n \neq 0$. To each term a_ix^i we associate a point $(i, v(a_i)) \in \mathbb{R}^2$, ignoring however the point (i, ∞) if $a_i = 0$. We now take the lower convex envelope of the set of points

$$\{(0, v(a_0)), (1, v(a_1)), \ldots, (n, v(a_n))\}.$$

This produces a polygonal chain which is called the **Newton polygon** of $f(x)$.

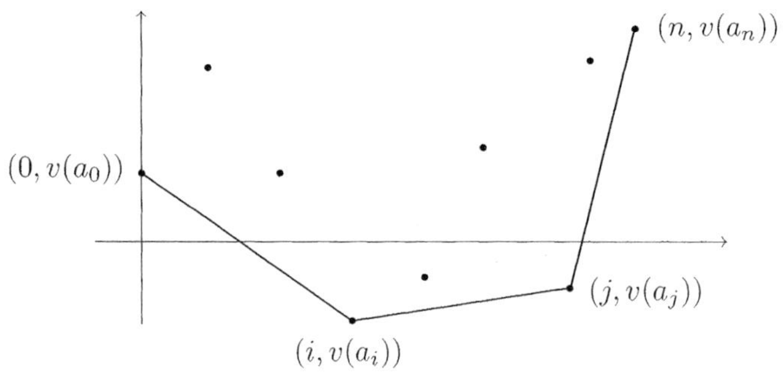

The polygon consists of a sequence of line segments $S_1, S_2, \ldots$ whose slopes are strictly increasing, and which are subject to the following

(6.3) Proposition. *Let $f(x) = a_0 + a_1x + \cdots + a_nx^n$, $a_0a_n \neq 0$, be a polynomial over the field K, v an exponential valuation of K, and w an extension to the splitting field L of f.*

If $(r, v(a_r)) \leftrightarrow (s, v(a_s))$ is a line segment of slope $-m$ occurring in the Newton polygon of f, then $f(x)$ has precisely $s - r$ roots $\alpha_1, \ldots, \alpha_{s-r}$ of value

$$w(\alpha_1) = \cdots = w(\alpha_{s-r}) = m\,.$$

Proof: Dividing by a_n only shifts the polygon up or down. Thus we may assume that $a_n = 1$. We number the roots $\alpha_1, \ldots, \alpha_n \in L$ of f in such a way that

$$\begin{aligned} w(\alpha_1) &= \cdots = w(\alpha_{s_1}) = m_1\,, \\ w(\alpha_{s_1+1}) &= \cdots = w(\alpha_{s_2}) = m_2\,, \\ \cdots \quad &\quad \cdots \quad \cdots \quad \cdots \\ w(\alpha_{s_t+1}) &= \cdots = w(\alpha_n) = m_{t+1}\,, \end{aligned}$$

where $m_1 < m_2 < \cdots < m_{t+1}$. Viewing the coefficients a_i as elementary symmetric functions of the roots α_j, we immediately find

$$\begin{aligned} v(a_n) &= v(1) = 0\,, \\ v(a_{n-1}) &\geq \min_i\{w(\alpha_i)\} = m_1\,, \\ v(a_{n-2}) &\geq \min_{i,j}\{w(\alpha_i\alpha_j)\} = 2m_1\,, \\ \cdots \quad & \quad \cdots \quad \cdots \quad \cdots \\ v(a_{n-s_1}) &= \min_{i_1,\ldots,i_{s_1}}\{w(\alpha_{i_1}\ldots\alpha_{i_{s_1}})\} = s_1m_1\,, \end{aligned}$$

the latter because the value of the term $\alpha_1 \ldots \alpha_{s_1}$ is smaller than that of all the others,

$$\begin{aligned} v(a_{n-s_1-1}) &\geq \min_{i_1,\ldots,i_{s_1+1}}\{w(\alpha_{i_1}\ldots\alpha_{i_{s_1+1}})\} = s_1m_1 + m_2\,, \\ v(a_{n-s_1-2}) &\geq \min_{i_1,\ldots,i_{s_1+2}}\{w(\alpha_{i_1}\ldots\alpha_{i_{s_1+2}})\} = s_1m_1 + 2m_2\,, \\ \cdots \quad & \quad \cdots \quad \cdots \quad \cdots \quad \cdots \\ v(a_{n-s_2}) &= \min_{i_1,\ldots,i_{s_2}}\{w(\alpha_{i_1}\ldots\alpha_{i_{s_2}})\} = s_1m_1 + (s_2 - s_1)m_2\,, \end{aligned}$$

and so on. From this result one concludes that the vertices of the Newton polygon, from right to left, are given by

$$(n, 0), \quad (n - s_1, s_1m_1), \quad (n - s_2, s_1m_1 + (s_2 - s_1)m_2), \ \ldots$$

The slope of the extreme right-hand line segment is

$$\frac{0 - s_1 m_1}{n - (n - s_1)} = -m_1,$$

and, proceeding further to the left,

$$\frac{(s_1 m_1 + \cdots + (s_j - s_{j-1})m_j) - (s_1 m_1 + \cdots + (s_{j+1} - s_j)m_{j+1})}{(n - s_j) - (n - s_{j+1})} = -m_{j+1}\,.$$

□

We emphasize that, according to the preceding proposition, the Newton polygon consists of precisely one segment if and only if the roots $\alpha_1, \ldots, \alpha_n$ of f all have the same value. In general, $f(x)$ factors into a product according to the slopes $-m_r < \cdots < -m_1$,

$$f(x) = a_n \prod_{j=1}^{r} f_j(x),$$

where

$$f_j(x) = \prod_{w(\alpha_i) = m_j} (x - \alpha_i).$$

Here the factor f_j corresponds to the $(r - j + 1)$-th segment of the Newton polygon, whose slope equals minus the value of the roots of f_j.

(6.4) Proposition. *If the valuation v admits a unique extension w to the splitting field L of f, then the factorization*

$$f(x) = a_n \prod_{j=1}^{r} f_j(x)$$

is defined already over K, i.e., $f_j(x) = \prod_{w(\alpha_i)=m_j}(x - \alpha_i) \in K[x]$.

Proof: We may clearly assume that $a_n = 1$. The statement is obvious when $f(x)$ is irreducible because then one has $\alpha_i = \sigma_i \alpha_1$ for some $\sigma_i \in G(L|K)$, and since, for any extension w of v, $w \circ \sigma_i$ is another one, the uniqueness implies that $w(\alpha_i) = w(\sigma_i \alpha_1) = m_1$, hence $f_1(x) = f(x)$.

The general case follows by induction on n. For $n = 1$ there is nothing to show. Let $p(x)$ be the minimal polynomial of α_1 and $g(x) = f(x)/p(x) \in K[x]$. Since all roots of $p(x)$ have the same value m_1, $p(x)$ is a divisor of $f_1(x)$. Let $g_1(x) = f_1(x)/p(x)$. The factorization of $g(x)$ according to the slopes is

$$g(x) = g_1(x) \prod_{j=2}^{r} f_j(x).$$

Since $\deg(g) < \deg(f)$, it follows that $f_j(x) \in K[x]$ for all $j = 1, \ldots, r$. □

If the polynomial f is irreducible, then, by the above factorization result, there is only *one* slope, i.e., the Newton polygon consists of a single segment. The values of all coefficients lie on or above this line segment and we get the

(6.5) Corollary. *Let $f(x) = a_0 + a_1 x + \cdots + a_n x^n \in K[x]$ be an irreducible polynomial with $a_n \neq 0$. Then, if $|\ |$ is a nonarchimedean valuation of K with a unique extension to the splitting field, one has*

$$|f| = \max\{\,|a_0|, |a_n|\}\,.$$

In (4.7) we deduced this result for complete fields from Hensel's lemma and thus obtained the uniqueness of the extended valuation. Here we obtain it, by contrast, as a consequence of the uniqueness of the extended valuation. We now proceed to deduce Hensel's lemma from the unique extendability.

(6.6) Theorem. *A nonarchimedean valued field $(K, |\ |)$ is henselian if and only if the valuation $|\ |$ can be uniquely extended to any algebraic extension.*

Proof: The fact that a henselian valuation $|\ |$ extends uniquely was dealt with in (6.2). Let us assume conversely that $|\ |$ admits one and only one extension to any given algebraic extension. We first show:

Let $f(x) = a_0 + a_1 x + \cdots + a_n x^n \in \mathcal{o}[x]$ be a primitive, irreducible polynomial such that $a_0 a_n \neq 0$, and let $\bar{f}(x) = f(x) \bmod \mathfrak{p} \in \kappa[x]$. Then we have $\deg(\bar{f}) = 0$ or $\deg(\bar{f}) = \deg(f)$, and we find

$$\bar{f}(x) = \bar{a}\,\bar{\varphi}(x)^m,$$

for some irreducible polynomial $\bar{\varphi}(x) \in \kappa[x]$ and a constant $\bar{a}$.

As f is irreducible, the Newton polygon is a single line segment and thus $|f| = \max\{|a_0|, |a_n|\}$. We may assume that a_n is a unit, because otherwise the Newton polygon is a segment which does not lie on the x-axis and this means that $\bar{f}(x) = \bar{a}_0$.

Let $L|K$ be the splitting field of $f(x)$ over K and $\mathcal{O}$ the valuation ring of the unique extension $|\ |$ to L, with maximal ideal $\mathfrak{P}$. For an arbitrary K-automorphism $\sigma \in G = G(L|K)$, we have $|\sigma\alpha| = |\alpha|$ for all $\alpha \in L$, because $|\ |$ and the composite $|\ | \circ \sigma$ extend the same valuation. This shows that $\sigma\mathcal{O} = \mathcal{O}$, $\sigma\mathfrak{P} = \mathfrak{P}$. If α is a zero of $f(x)$ and μ its multiplicity, then $\sigma\alpha \in \mathcal{O}$ for all $\sigma \in G$. Indeed, if $\alpha \notin \mathcal{O}$, then $\prod_\sigma |\sigma\alpha|^\mu = |\prod_\sigma \sigma\alpha|^\mu > 1$ would imply that the constant coefficient a_0 could not belong to $\mathcal{o}$. Thus every $\sigma \in G$ induces a κ-automorphism $\bar{\sigma}$ of $\mathcal{O}/\mathfrak{P}$, and the zeroes $\overline{\sigma\alpha} = \bar{\sigma}\bar{\alpha}$

of $\bar{f}(x)$ are all conjugate over κ. It follows that $\bar{f}(x) = \bar{a}\bar{\varphi}(x)^m$, if $\bar{\varphi}(x)$ is the minimal polynomial of $\bar{\alpha}$ over κ. Since $a_n \in \mathcal{O}^*$, we have furthermore that $\deg(\bar{f}) = \deg(f)$.

Let now $f(x) \in \mathcal{O}[x]$ be an arbitrary primitive polynomial, and let

$$f(x) = f_1(x) \cdots f_r(x)$$

be its factorization into irreducibles over K. Since $1 = |f| = \prod |f_i|$, multiplying the f_i by suitable constants yields $|f_i| = 1$. The $f_i(x)$ are therefore primitive, irreducible polynomials in $\mathcal{O}[x]$. It follows that

$$\bar{f}(x) = \bar{f}_1(x) \cdots \bar{f}_r(x),$$

where $\deg(\bar{f}_i) = 0$ or $\deg(\bar{f}_i) = \deg(f_i)$, and $\bar{f}_i$ is, up to a constant factor, the power of an irreducible polynomial. If $\bar{f} = \bar{g}\,\bar{h}$ is a factorization into relatively prime polynomials $\bar{g}, \bar{h} \in \kappa[x]$, then we must have

$$\bar{g} = \bar{a} \prod_{i \in I} \bar{f}_i, \quad \bar{h} = \bar{b} \prod_{j \in J} \bar{f}_j$$

where $\bar{a}, \bar{b} \in \kappa$ and $\{1, \dots, r\} = I \mathbin{\dot\cup} J$ and $\deg(\bar{f}_i) = \deg(f_i)$ for $i \in I$. We now put

$$g = a \prod_{i \in I} f_i, \quad h = b \prod_{j \in J} f_j,$$

for $a, b \in \mathcal{O}^*$ such that $a \equiv \bar{a}$, $b \equiv \bar{b} \bmod \mathfrak{p}$ and $f = gh$. □

We have introduced henselian fields by a condition of which the reader will find weaker versions in the literature, restricted to *monic polynomials* only. Both are equivalent as is shown by the following

(6.7) Proposition. *A nonarchimedean field (K, v) is henselian if any monic polynomial $f(x) \in \mathcal{O}[x]$ which splits over the residue class field $\kappa = \mathcal{O}/\mathfrak{p}$ as*

$$f(x) \equiv \bar{g}(x)\bar{h}(x) \bmod \mathfrak{p}$$

with relatively prime monic factors $\bar{g}(x), \bar{h}(x) \in \kappa[x]$, admits itself a splitting

$$f(x) = g(x)h(x)$$

into monic factors $g(x), h(x) \in \mathcal{O}[x]$ such that

$$g(x) \equiv \bar{g}(x) \bmod \mathfrak{p} \quad \textit{and} \quad h(x) \equiv \bar{h}(x) \bmod \mathfrak{p}.$$

Proof (*E. Nart*): We have just seen that the property of K to be henselian follows from the condition that the Newton polygon of every irreducible polynomial $f(x) = a_0 + a_1x + \cdots + a_nx^n \in K[x]$ is a single line segment. It is therefore sufficient to show this. We may assume that $a_n = 1$. Let $L|K$ be the splitting field of f. Then there is always an extension w of v to L. It is obtained for example by taking the completion $\widehat{K}$ of K, extending the valuation of $\widehat{K}$ in a unique way to a valuation $\bar{v}$ of the algebraic closure $\overline{\widehat{K}}$ of $\widehat{K}$, embedding L into $\overline{\widehat{K}}$, and restricting $\bar{v}$ to L. It is also possible to get the extension w directly, without passing through the completion. For this we refer to [93], chap. XII, §4, th. 1.

Assume now that the Newton polygon of f consists of more than one segment:

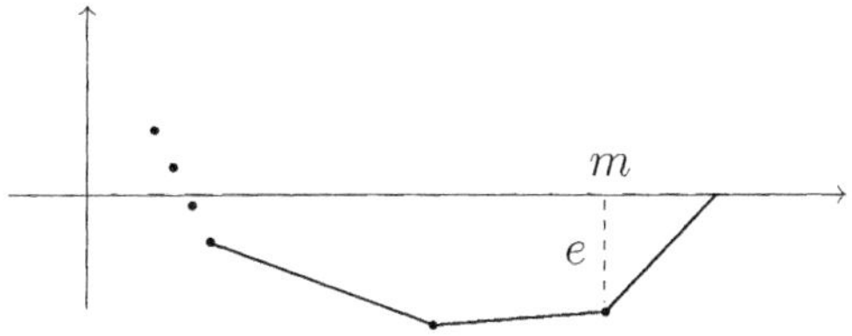

Let the last segment be given by the points (m,e) and $(n,0)$. If $e = 0$, we immediately have a contradiction. Because then we have $v(a_i) \geq 0$, so that $f(x) \in \mathcal{O}[x]$, and $a_0 \equiv \cdots \equiv a_{m-1} \equiv 0 \bmod \mathfrak{p}$, $a_m \not\equiv 0 \bmod \mathfrak{p}$. Therefore $f(x) \equiv (X^{n-m} + \cdots + a_m)X^m \bmod \mathfrak{p}$, with $m > 0$ because there is more than one segment. In view of the condition of the proposition this contradicts the irreducibility of f.

We will now reduce to $e = 0$ by a transformation. Let $\alpha \in L$ be a root of $f(x)$ of minimum value $w(\alpha)$ and let $a \in K$ such that $v(a) = e$. We consider the characteristic polynomial $g(x)$ of $a^{-1}\alpha^r \in K(\alpha)$, $r = n - m$. If $f(x) = \prod_{i=1}^n (x - \alpha_i)$, then $g(x) = \prod_{i=1}^n (x - \alpha_i^m a^{-1})$. Proposition (6.3) shows that the Newton polygon of $g(x)$ also has more than one segment, the last one of slope

$$-w(a^{-1}\alpha^r) = v(a) - rw(\alpha) = e - r\frac{e}{r} = 0.$$

Since $g(x)$ is a power of the minimal polynomial of $a^{-1}\alpha^r$, hence of an irreducible polynomial, this produces the same contradiction as before. □

Let K be a field which is henselian with respect to the exponential valuation v. If $L|K$ is a finite extension of degree n, then v extends uniquely to an exponential valuation w of L, namely

$$w(\alpha) = \frac{1}{n} v\big(N_{L|K}(\alpha)\big).$$

This follows from (6.2) by taking the logarithm. For the value groups and residue class fields of v and w, one gets the inclusions

$$v(K^*) \subseteq w(L^*) \quad \text{and} \quad \kappa \subseteq \lambda .$$

The index

$$e = e(w \mid v) = \big(w(L^*) : v(K^*)\big)$$

is called the **ramification index** of the extension $L|K$ and the degree

$$f = f(w \mid v) = [\lambda : \kappa]$$

is called the **inertia degree**. If v, and hence $w = \frac{1}{n} v \circ N_{L|K}$, is discrete and if $\mathcal{o}, \mathfrak{p}, \pi$, resp. $\mathcal{O}, \mathfrak{P}, \Pi$, are the valuation ring, the maximal ideal and a prime element of K, resp. L, then one has

$$e = \big(w(\Pi)\mathbb{Z} : v(\pi)\mathbb{Z}\big) ,$$

so that $v(\pi) = ew(\Pi)$, and we find

$$\pi = \varepsilon \Pi^e ,$$

for some unit $\varepsilon \in \mathcal{O}^*$. From this one deduces the familiar (see chap. I) interpretation of the ramification index: $\mathfrak{p}\,\mathcal{O} = \pi\,\mathcal{O} = \Pi^e\mathcal{O} = \mathfrak{P}^e$, or

$$\mathfrak{p} = \mathfrak{P}^e .$$

(6.8) Proposition. *One has* $[L : K] \geq ef$ *and the* **fundamental identity**

$$[L : K] = ef ,$$

if v is discrete and $L|K$ is separable.

Proof: Let $\omega_1, \dots, \omega_f$ be representatives of a basis of $\lambda|\kappa$ and let $\pi_0, \dots, \pi_{e-1} \in L^*$ be elements the values of which represent the various cosets in $w(L^*)/v(K^*)$ (the finiteness of e will be a consequence of what follows). If v is discrete, we may choose for instance $\pi_i = \Pi^i$. We show that the elements

$$\omega_j \pi_i , \quad j = 1, \dots, f , \quad i = 0, \dots, e-1 ,$$

are linearly independent over K, and in the discrete case form even a basis of $L|K$. Let

$$\sum_{i=0}^{e-1} \sum_{j=1}^{f} a_{ij} \omega_j \pi_i = 0$$

with $a_{ij} \in K$. Assume that not all $a_{ij} = 0$. Then there exist nonzero sums $s_i = \sum_{j=1}^{f} a_{ij}\omega_j$, and each time that $s_i \neq 0$ we find $w(s_i) \in v(K^*)$. In fact, dividing s_i by the coefficient $a_{i\nu}$ of minimum value, we get a linear combination of the $\omega_1, \dots, \omega_f$ with coefficients in the valuation ring $\mathcal{o} \subseteq K$ one of which equals 1. This linear combination is $\not\equiv 0 \bmod \mathfrak{P}$, hence a unit, so that $w(s_i) = w(a_{i\nu}) \in v(K^*)$.

In the sum $\sum_{i=0}^{e-1} s_i\pi_i$, two nonzero summands must have the same value, say $w(s_i\pi_i) = w(s_j\pi_j)$, $i \neq j$, because otherwise it could not be zero (observe that $w(x) \neq w(y) \Rightarrow w(x+y) = \min\{w(x), w(y)\}$). It follows that

$$w(\pi_i) = w(\pi_j) + w(s_j) - w(s_i) \equiv w(\pi_j) \bmod v(K^*),$$

a contradiction. This shows the linear independence of the $\omega_j\pi_i$. In particular, we have $ef \leq [L:K]$.

Assume now that v, and thus also w, is discrete and let Π be a prime element in the valuation ring $\mathcal{O}$ of w. We consider the $\mathcal{o}$-module

$$M = \sum_{i=0}^{e-1}\sum_{j=1}^{f} \mathcal{o}\omega_j\pi_i$$

where $\pi_i = \Pi^i$ and show that $M = \mathcal{O}$, i.e., $\{\omega_j\pi_i\}$ is even an integral basis of $\mathcal{O}$ over $\mathcal{o}$. We put

$$N = \sum_{j=1}^{f} \mathcal{o}\omega_j\,,$$

so that $M = N + \Pi N + \cdots + \Pi^{e-1}N$. We find that

$$\mathcal{O} = N + \Pi\mathcal{O}\,,$$

because, for $\alpha \in \mathcal{O}$, we have $\alpha \equiv a_1\omega_1 + \cdots + a_f\omega_f \bmod \Pi\mathcal{O}$, $a_i \in \mathcal{o}$. This implies

$$\mathcal{O} = N + \Pi(N + \Pi\mathcal{O}) = \cdots = N + \Pi N + \cdots + \Pi^{e-1}N + \Pi^e\mathcal{O}\,,$$

so that $\mathcal{O} = M + \mathfrak{P}^e = M + \mathfrak{p}\mathcal{O}$. Since $L|K$ is separable, $\mathcal{O}$ is a finitely generated $\mathcal{o}$-module (see chap. I, (2.11)), and we conclude $\mathcal{O} = M$ from Nakayama's lemma (chap. I, § 11, exercise 7). □

Remark: We had already proved the identity $[L:K] = ef$ in a somewhat different way in chap. I, (8.2), also in the case where v was discrete and $L|K$ separable. Both hypotheses are actually needed. But, strangely enough, the separability condition can be dropped once K is complete with respect to the discrete valuation. In this case, one deduces the equality $\mathcal{O} = M$ in the above proof from $\mathcal{O} = M + \mathfrak{p}\mathcal{O}$, not by means of Nakayama's lemma, but rather like this: as $\mathfrak{p}^i M \subseteq M$, we get successively

$$\mathcal{O} = M + \mathfrak{p}(M + \mathfrak{p}\mathcal{O}) = M + \mathfrak{p}^2\mathcal{O} = \cdots = M + \mathfrak{p}^\nu\mathcal{O}$$

for all $\nu \geq 1$, and since $\{\mathfrak{p}^\nu\mathcal{O}\}_{\nu\in\mathbb{N}}$ is a basis of neighbourhoods of zero in $\mathcal{O}$, M is dense in $\mathcal{O}$. Since $\mathcal{o}$ is closed in K, (4.9) implies that M is closed in $\mathcal{O}$, so that $M = \mathcal{O}$.

Exercise 1. In a henselian field the zeroes of a polynomial are continuous functions of its coefficients. More precisely, one has: let $f(x) \in K[x]$ be a monic polynomial of degree n and

$$f(x) = \prod_{i=1}^{r} (x - \alpha_i)^{m_i}$$

its decomposition into linear factors, with $m_i \geq 1$, $\alpha_i \neq \alpha_j$ for $i \neq j$. If the monic polynomial $g(x)$ of degree n has all coefficients sufficiently close to those of $f(x)$, then it has r roots $\beta_1, \dots, \beta_r$ which approximate the $\alpha_1, \dots, \alpha_r$ to any previously given precision.

Exercise 2 (Krasner's Lemma). Let $\alpha \in \overline{K}$ be separable over K and let $\alpha = \alpha_1, \dots, \alpha_n$ be its conjugates over K. If $\beta \in \overline{K}$ is such that

$$|\alpha - \beta| < |\alpha - \alpha_i| \quad \text{for} \quad i = 2, \dots, n,$$

then one has $K(\alpha) \subseteq K(\beta)$.

Exercise 3. A field which is henselian with respect to two inequivalent valuations is separably closed (Theorem of *F.K. Schmidt*).

Exercise 4. A separably closed field K is henselian with respect to any nonarchimedean valuation.

More generally, every valuation of K admits a unique extension to any purely inseparable extension $L|K$.

Hint: If $\alpha^p = a \in K$, one is forced to put $w(\alpha) = \frac{1}{p} v(a)$.

Exercise 5. Let K be a nonarchimedean valued field, $\mathcal{O}$ the valuation ring, and $\mathfrak{p}$ the maximal ideal. K is henselian if and only if every polynomial $f(x) = x^n + a_{n-1}x^{n-1} + \cdots + a_0 \in \mathcal{O}[x]$ such that $a_0 \in \mathfrak{p}$ and $a_1 \notin \mathfrak{p}$ has a zero $a \in \mathfrak{p}$.

Hint: The Newton polygon.

Remark: A local ring $\mathcal{O}$ with maximal ideal $\mathfrak{p}$ is called *henselian* if Hensel's lemma in the sense of (6.7) holds for it. A characterization of these rings which is important in algebraic geometry is the following:

A local ring $\mathcal{O}$ is henselian if and only if every finite commutative $\mathcal{O}$-algebra A splits into a direct product $A = \prod_{i=1}^{r} A_i$ of local rings A_i.

The proof is not straightforward, we refer to [103], chap. I, §4, th. 4.2.

§ 7. Unramified and Tamely Ramified Extensions

In this section we fix a base field K which is henselian with respect to a nonarchimedean valuation v or $|\ |$. As before, we denote the valuation ring, the maximal ideal and the residue class field by $\mathcal{O}, \mathfrak{p}, \kappa$, respectively. If $L|K$ is an algebraic extension, then the corresponding invariants are labelled $w, \mathcal{O}, \mathfrak{P}, \lambda$, respectively. An especially important rôle among these extensions is played by the unramified extensions, which are defined as follows.

(7.1) Definition. *A finite extension $L|K$ is called* **unramified** *if the extension $\lambda|\kappa$ of the residue class field is separable and one has*

$$[L:K] = [\lambda:\kappa].$$

An arbitrary algebraic extension $L|K$ is called unramified if it is a union of finite unramified subextensions.

Remark: This definition does not require K to be henselian; it applies in all cases where v extends uniquely to L.

(7.2) Proposition. *Let $L|K$ and $K'|K$ be two extensions inside an algebraic closure $\overline{K}|K$ and let $L' = LK'$. Then one has*

$$L|K \text{ unramified} \implies L'|K' \text{ unramified.}$$

Each subextension of an unramified extension is unramified.

Proof: The notations $\mathcal{o}, \mathfrak{p}, \kappa$; $\mathcal{o}', \mathfrak{p}', \kappa'$; $\mathcal{O}, \mathfrak{P}, \lambda$; $\mathcal{O}', \mathfrak{P}', \lambda'$ are self-explanatory. We may assume that $L|K$ is finite. Then $\lambda|\kappa$ is also finite and, being separable, is therefore generated by a primitive element $\overline{\alpha}$, $\lambda = \kappa(\overline{\alpha})$. Let $\alpha \in \mathcal{O}$ be a lifting, $f(x) \in \mathcal{o}[x]$ the minimal polynomial of α and $\overline{f}(x) = f(x) \bmod \mathfrak{p} \in \kappa[x]$. Since

$$[\lambda:\kappa] \le \deg(\overline{f}) = \deg(f) = [K(\alpha):K] \le [L:K] = [\lambda:\kappa],$$

one has $L = K(\alpha)$ and $\overline{f}(x)$ is the minimal polynomial of $\overline{\alpha}$ over κ.

We thus have $L' = K'(\alpha)$. In order to prove that $L'|K'$ is unramified, let $g(x) \in \mathcal{o}'[x]$ be the minimal polynomial of α over K' and $\overline{g}(x) = g(x) \bmod \mathfrak{p}' \in \kappa'[x]$. Being a factor of $\overline{f}(x)$, $\overline{g}(x)$ is separable and hence irreducible over κ', because otherwise $g(x)$ is reducible by Hensel's lemma. We obtain

$$[\lambda':\kappa'] \le [L':K'] = \deg(g) = \deg(\overline{g}) = [\kappa'(\overline{\alpha}):\kappa'] \le [\lambda':\kappa'].$$

This implies $[L':K'] = [\lambda':\kappa']$, i.e., $L'|K'$ is unramified.

If $L|K$ is a subextension of the unramified extension $L'|K$, then it follows from what we have just proved that $L'|L$ is unramified. Hence so is $L|K$, by the formula for the degree. □

(7.3) Corollary. *The composite of two unramified extensions of K is again unramified.*

Proof: It suffices to show this for two finite extensions $L|K$ and $L'|K$. $L|K$ is unramified, hence so is $LL'|L'$, by (7.2). This implies that $LL'|K$ is unramified as well because separability is transitive and the degrees of field (and residue field) extensions are multiplicative. □

(7.4) Definition. *Let $L|K$ be an algebraic extension. Then the composite $T|K$ of all unramified subextensions is called the* **maximal unramified subextension** *of $L|K$.*

(7.5) Proposition. *The residue class field of T is the separable closure λ_s of κ in the residue class field extension $\lambda|\kappa$ of $L|K$, whereas the value group of T equals that of K.*

Proof: Let λ_0 be the residue class field of T and assume $\bar{\alpha} \in \lambda$ is separable over κ. We have to show that $\bar{\alpha} \in \lambda_0$. Let $\bar{f}(x) \in \kappa[x]$ be the minimal polynomial of $\bar{\alpha}$ and $f(x) \in \mathcal{O}[x]$ a monic polynomial such that $\bar{f} = f \bmod \mathfrak{p}$. Then $f(x)$ is irreducible and by Hensel's lemma has a root α in L such that $\bar{\alpha} = \alpha \bmod \mathfrak{P}$, i.e., $[K(\alpha) : K] = [\kappa(\bar{\alpha}) : \kappa]$. This implies that $K(\alpha)|K$ is unramified, so that $K(\alpha) \subseteq T$, and thus $\bar{\alpha} \in \lambda_0$.

In order to prove $w(T^*) = v(K^*)$ we may suppose $L|K$ to be finite. The claim then follows from

$$[T : K] \geq \big(w(T^*) : v(K^*)\big)[\lambda_0 : \kappa] = \big(w(T^*) : v(K^*)\big)[T : K]. \qquad \square$$

The composite of all unramified extensions inside the algebraic closure $\overline{K}$ of K is simply called the maximal unramified extension $K_{nr}|K$ of K (nr = 'non ramifiée'). Its residue class field is the separable closure $\bar{\kappa}_s|\kappa$. K_{nr} contains all roots of unity of order m not divisible by the characteristic of κ because the separable polynomial $x^m - 1$ splits over $\bar{\kappa}_s$ and hence also over K_{nr}, by Hensel's lemma. If κ is a finite field, then the extension $K_{nr}|K$ is even generated by these roots of unity because they generate $\bar{\kappa}_s|\kappa$.

If the characteristic $p = \mathrm{char}(\kappa)$ of the residue class field is positive, then one has the following weaker notion accompanying that of an unramified extension.

(7.6) Definition. *An algebraic extension $L|K$ is called* **tamely ramified** *if the extension $\lambda|\kappa$ of the residue class fields is separable and one has $([L : T], p) = 1$. In the infinite case this latter condition is taken to mean that the degree of each finite subextension of $L|T$ is prime to p.*

As before, in this definition K need not be henselian. We apply it whenever the valuation v of K has a unique extension to L. When the fundamental identity $ef = [L:K]$ holds and $\lambda|\kappa$ is separable, to say that the extension is unramified, resp. tamely ramified, simply amounts to saying that $e = 1$, resp. $(e,p) = 1$.

(7.7) Proposition. *A finite extension $L|K$ is tamely ramified if and only if the extension $L|T$ is generated by radicals*

$$L = T\big(\sqrt[m_1]{a_1}, \ldots, \sqrt[m_r]{a_r}\big)$$

such that $(m_i, p) = 1$. In this case the fundamental identity always holds:

$$[L:K] = ef\,.$$

Proof: We may assume that $K = T$ because $L|K$ is obviously tamely ramified if and only if $L|T$ is tamely ramified, and if this is the case, then $[T:K] = [\lambda:\kappa] = f$. Let $L|K$ be tamely ramified, so that $\kappa = \lambda$ and $([L:K],p) = 1$. We first show that $e = 1$ implies $L = K$. Let $\alpha \in L \smallsetminus K$. Writing $\alpha = \alpha_1, \ldots, \alpha_m$ for the conjugates and $a = Tr(\alpha) = \sum_{i=1}^m \alpha_i$, the element $\beta = \alpha - \frac{1}{m}a \in L\smallsetminus K$ has trace $Tr(\beta) = \sum_{i=1}^m \beta_i = 0$. Since $v(K^*) = w(L^*)$, we may choose a $b \in K^*$ such that $v(b) = w(\beta)$ and obtain a unit $\varepsilon = \beta/b \in L\smallsetminus K$ with trace $\sum_{i=1}^m \varepsilon_i = 0$. But the conjugates ε_i have the same residue classes $\bar\varepsilon_i$ in λ, because $\lambda = \kappa$. Hence $0 = \sum_{i=1}^m \bar\varepsilon_i = m\bar\varepsilon$, and thus $m \equiv 0 \bmod p$, which contradicts $p \nmid [L:K]$ and $m \,|\, [L:K]$.

Now let $\omega_1, \ldots, \omega_r \in w(L^*)$ be a system of representatives for the quotient $w(L^*)/v(K^*)$ and m_i the order of $\omega_i \bmod v(K^*)$. Since $w(L^*) = \frac{1}{n}v(N_{L|K}(L^*)) \subseteq \frac{1}{n}v(K^*)$, where $n = [L:K]$, we have $m_i|n$, so that $(m_i,p) = 1$. Let $\gamma_i \in L^*$ be an element such that $w(\gamma_i) = \omega_i$. Then $w(\gamma_i^{m_i}) = v(c_i)$, with $c_i \in K$, so that $\gamma_i^{m_i} = c_i\varepsilon_i$ for some unit ε_i in L. As $\lambda = \kappa$ we may write $\varepsilon_i = b_i u_i$, where $b_i \in K$ and u_i is a unit in L which tends to 1 in λ. By Hensel's lemma the equation $x^{m_i} - u_i = 0$ has a solution $\beta_i \in L$. Putting $\alpha_i = \gamma_i\beta_i^{-1} \in L$, we find $w(\alpha_i) = \omega_i$ and

$$\alpha_i^{m_i} = a_i\,, \quad i = 1, \ldots, r\,,$$

where $a_i = c_i b_i \in K$, i.e., we have $K(\sqrt[m_1]{a_1}, \ldots, \sqrt[m_r]{a_r}) \subseteq L$. By construction, both fields have the same value group and the same residue class field. So, by what we proved first, we have

$$L = K\big(\sqrt[m_1]{a_1}, \ldots, \sqrt[m_r]{a_r}\big)\,.$$

The inequality $[L:K] \le e$ and thus, in view of (6.8), the equality $[L:K] = e$, now follows by induction on r. If $L_1 = K(\sqrt[m_1]{a_1})$, then

$\omega_1 \in w(L_1^*)$ yields

$$e(L_1|K) = \big(w(L_1^*) : v(K^*)\big) \geq m_1 \geq [L_1 : K].$$

Also $e(L|L_1) \geq [L : L_1]$, because $w(L^*)/w(L_1^*)$ is generated by the residue classes of $\omega_2, \ldots, \omega_r$. Thus

$$e = e(L|L_1)\, e(L_1|K) \geq [L : L_1][L_1 : K] = [L : K].$$

In order to prove that an extension $L = K(\sqrt[m_1]{a_1}, \ldots, \sqrt[m_r]{a_r})$ is tamely ramified, it suffices to look at the case $r = 1$, i.e., $L = K(\sqrt[m]{a})$, where $(m, p) = 1$. The general case then follows by induction. We may assume without loss of generality that κ is separably closed. This is seen by passing to the maximal unramified extension $K_1 = K_{nr}$, which has the separable closure $\kappa_1 = \bar{\kappa}_s$ of κ as its residue class field. We obtain the following diagram

$$\begin{array}{ccc} L & \text{———} & L_1 \\ | & & | \\ K & \text{———} & K_1, \end{array}$$

where $L \cap K_1 = T = K$ and $L_1 = K_1(\sqrt[m]{a})$. If now $L_1|K_1$ is tamely ramified, then $\lambda_1|\kappa_1$ is separable; hence $\lambda_1 = \kappa_1$ and $p \nmid [L_1 : K_1] = [L : K] = [L : T]$, i.e., $L|K$ is also tamely ramified.

Let $\alpha = \sqrt[m]{a}$. We may assume that $[L : K] = [K(\sqrt[m]{a}) : K] = m$. In fact, if d is the greatest divisor of m such that $a = a'^d$ for some $a' \in K^*$, and if $m' = m/d$, then $\alpha = \sqrt[m']{a'}$ and $[K(\sqrt[m']{a'}) : K] = m'$. Now let $n = \operatorname{ord}(w(\alpha) \bmod v(K^*))$. Since $mw(\alpha) = v(a) \in v(K^*)$, we have $m = dn$. Consequently $w(\alpha^n) = v(b)$, $b \in K^*$, and $v(b^d) = w(\alpha^m) = v(a)$; thus $\alpha^m = a = \varepsilon b^d$ for some unit ε in K. As $(d, p) = 1$, the equation $x^d - \varepsilon = 0$ splits over the separably closed residue field κ into distinct linear factors, hence also over K by Hensel's lemma. Therefore $\alpha^m = b^d = a$ for some new $b \in K^*$. Since $x^m - a$ is irreducible, we have $d = 1$, and hence $m = n$. Thus

$$e \geq n = [L : K] \geq ef \geq e,$$

in other words $f = 1$, and so $\lambda = \kappa$ and $p \nmid n = e$. This shows that $L|K$ is tamely ramified. □

(7.8) Corollary. *Let $L|K$ and $K'|K$ be two extensions inside the algebraic closure $\overline{K}|K$, and $L' = LK'$. Then we have:*

$$L|K \text{ tamely ramified} \implies L'|K' \text{ tamely ramified.}$$

Every subextension of a tamely ramified extension is tamely ramified.

Proof: We may assume without loss of generality that $L|K$ is finite and consider the diagram

$$\begin{array}{ccc} L & \text{———} & L' \\ | & & | \\ T & \text{———} & T' \\ | & & | \\ K & \text{———} & K'. \end{array}$$

The inclusion $T \subseteq T'$ follows from (7.2). If $L|K$ is tamely ramified, then $L = T(\sqrt[m_1]{a_1}, \dots, \sqrt[m_r]{a_r})$, $(m_i, p) = 1$; hence $L' = LK' = LT' = T'(\sqrt[m_1]{a_1}, \dots, \sqrt[m_r]{a_r})$, so that $L'|K'$ is also tamely ramified, by (7.7).

The claim concerning the subextensions follows exactly as in the unramified case. □

(7.9) Corollary. *The composite of tamely ramified extensions is tamely ramified.*

Proof: This follows from (7.8), exactly as (7.3) followed from (7.2) in the unramified case. □

(7.10) Definition. *Let $L|K$ be an algebraic extension. Then the composite $V|K$ of all tamely ramified subextensions is called the* **maximal tamely ramified** *subextension of $L|K$.*

Let $w(L^*)^{(p)}$ denote the subgroup of all elements $\omega \in w(L^*)$ such that $m\omega \in v(K^*)$ for some m satisfying $(m, p) = 1$. The quotient group $w(L^*)^{(p)}/v(K^*)$ then consists of all elements of $w(L^*)/v(K^*)$ whose order is prime to p.

(7.11) Proposition. *The maximal tamely ramified subextension $V|K$ of $L|K$ has value group $w(V^*) = w(L^*)^{(p)}$ and residue class field equal to the separable closure λ_s of κ in $\lambda|\kappa$.*

Proof: We may restrict to the case of a finite extension $L|K$. By passing from K to the maximal unramified subextension, we may assume by (7.5) that $\lambda_s = \kappa$. As $p \nmid e(V|K) = \#w(V^*)/v(K^*)$, we certainly have $w(V^*) \subseteq w(L^*)^{(p)}$. Conversely we find, as in the proof of (7.7), for every $\omega \in w(L^*)^{(p)}$ a radical $\alpha = \sqrt[m]{a} \in L$ such that $a \in K$, $(m, p) = 1$ and $w(\alpha) = \omega$, so that one has $\alpha \in V$, and $\omega \in w(V^*)$. □

The results obtained in this section may be summarized in the following picture:

$$\begin{array}{ccccccc} K & \subseteq & T & \subseteq & V & \subseteq & L \\ \kappa & \subseteq & \lambda_s & = & \lambda_s & \subseteq & \lambda \\ v(K^*) & = & w(T^*) & \subseteq & w(L^*)^{(p)} & \subseteq & w(L^*)\,. \end{array}$$

If $L|K$ is finite and $e = e'p^a$ where $(e', p) = 1$, then $[V : T] = e'$. The extension $L|K$ is called **totally** (or **purely**) **ramified** if $T = K$, and **wildly ramified** if it is not tamely ramified, i.e., if $V \neq L$.

Important Example: Consider the extension $\mathbb{Q}_p(\zeta)|\mathbb{Q}_p$ for a primitive n-th root of unity ζ. In the two cases $(n, p) = 1$ and $n = p^s$, this extension behaves completely differently. Let us first look at the case $(n, p) = 1$ and choose as our base field, instead of $\mathbb{Q}_p$, any discretely valued complete field K with finite residue class field $\kappa = \mathbb{F}_q$, with $q = p^r$.

(7.12) Proposition. *Let $L = K(\zeta)$, and let $\mathcal{O}|\mathcal{o}$, resp. $\lambda|\kappa$, be the extension of valuation rings, resp. residue class fields, of $L|K$. Suppose that $(n,\ p) = 1$. Then one has:*

(i) *The extension $L|K$ is unramified of degree f, where f is the smallest natural number such that $q^f \equiv 1 \bmod n$.*

(ii) *The Galois group $G(L|K)$ is canonically isomorphic to $G(\lambda|\kappa)$ and is generated by the automorphism $\varphi : \zeta \mapsto \zeta^q$.*

(iii) $\mathcal{O} = \mathcal{o}[\zeta]$.

Proof: (i) If $\phi(X)$ is the minimal polynomial of ζ over K, then the reduction $\bar{\phi}(X)$ is the minimal polynomial of $\bar{\zeta} = \zeta \bmod \mathfrak{P}$ over κ. Indeed, being a divisor of $X^n - \bar{1}$, $\bar{\phi}(X)$ is separable and by Hensel's lemma cannot split into factors. ϕ and $\bar{\phi}$ have the same degree, so that $[L : K] = [\kappa(\bar{\zeta}) : \kappa] = [\lambda : \kappa] =: f$. $L|K$ is therefore unramified. The polynomial $X^n - 1$ splits over $\mathcal{O}$ and thus (because $(n, p) = 1$) over λ into distinct linear factors, so that $\lambda = \mathbb{F}_{q^f}$ contains the group μ_n of n-th roots of unity and is generated by it. Consequently f is the smallest number such that $\mu_n \subseteq \mathbb{F}_{q^f}^*$, i.e., such that $n \,|\, q^f - 1$. This shows (i). (ii) results trivially from this.

(iii) Since $L|K$ is unramified, we have $\mathfrak{p}\mathcal{O} = \mathfrak{P}$, and since $1, \zeta, \dots, \zeta^{f-1}$ represents a basis of $\lambda|\kappa$, we have $\mathcal{O} = \mathcal{o}[\zeta] + \mathfrak{p}\mathcal{O}$, and $\mathcal{O} = \mathcal{o}[\zeta]$ by Nakayama's lemma. □

(7.13) Proposition. *Let ζ be a primitive p^m-th root of unity. Then one has:*

(i) *$\mathbb{Q}_p(\zeta)|\mathbb{Q}_p$ is totally ramified of degree $\varphi(p^m) = (p-1)p^{m-1}$.*

(ii) *$G(\mathbb{Q}_p(\zeta)|\mathbb{Q}_p) \cong (\mathbb{Z}/p^m\mathbb{Z})^*$.*

(iii) *$\mathbb{Z}_p[\zeta]$ is the valuation ring of $\mathbb{Q}_p(\zeta)$.*

(iv) *$1-\zeta$ is a prime element of $\mathbb{Z}_p[\zeta]$ with norm p.*

Proof: $\xi = \zeta^{p^{m-1}}$ is a primitive p-th root of unity, i.e.,

$$\xi^{p-1} + \xi^{p-2} + \cdots + 1 = 0, \quad \text{hence}$$
$$\zeta^{(p-1)p^{m-1}} + \zeta^{(p-2)p^{m-1}} + \cdots + 1 = 0.$$

Denoting by ϕ the polynomial on the left, $\zeta - 1$ is a root of the equation $\phi(X+1) = 0$. But this is irreducible because it satisfies Eisenstein's criterion: $\phi(1) = p$ and $\phi(X) \equiv (X^{p^m} - 1)/(X^{p^{m-1}} - 1) = (X-1)^{p^{m-1}(p-1)} \bmod p$. It follows that $[\mathbb{Q}_p(\zeta) : \mathbb{Q}_p] = \varphi(p^m)$. The canonical injection $G(\mathbb{Q}_p(\zeta)|\mathbb{Q}_p) \to (\mathbb{Z}/p^m\mathbb{Z})^*$, $\sigma \mapsto n(\sigma)$, where $\sigma\zeta = \zeta^{n(\sigma)}$, is therefore bijective, since both groups have order $\varphi(p^m)$. Thus

$$N_{\mathbb{Q}_p(\zeta)|\mathbb{Q}_p}(1-\zeta) = \prod_\sigma (1-\sigma\zeta) = \phi(1) = p.$$

Writing w for the extension of the normalized valuation v_p of $\mathbb{Q}_p$, we find furthermore that $\varphi(p^m)w(\zeta - 1) = v_p(p) = 1$, i.e., $\mathbb{Q}_p(\zeta)|\mathbb{Q}_p$ is totally ramified and $\zeta - 1$ is a prime element of $\mathbb{Q}_p(\zeta)$. As in the proof of (6.8), it follows that $\mathbb{Z}_p[\zeta - 1] = \mathbb{Z}_p[\zeta]$ is the valuation ring of $\mathbb{Q}_p(\zeta)$. This concludes the proof. □

If ζ_n is a primitive n-th root of unity and $n = n'p^m$, with $(n', p) = 1$, then propositions (7.12) and (7.13) yield the following result for the maximal unramified and the maximal tamely ramified extension:

$$\mathbb{Q}_p \subseteq T = \mathbb{Q}_p(\zeta_{n'}) \subseteq V = T(\zeta_p) \subseteq \mathbb{Q}_p(\zeta_n).$$

Exercise 1. The maximal unramified extension of $\mathbb{Q}_p$ is obtained by adjoining all roots of unity of order prime to p.

Exercise 2. Let K be henselian and $K_{nr}|K$ the maximal unramified extension. Show that the subextensions of $K_{nr}|K$ correspond 1–1 to the subextensions of the separable closure $\bar\kappa_s|\kappa$.

Exercise 3. Let $L|K$ be totally and tamely ramified, and let Δ, resp. Γ, be the value group of L, resp. K. Show that the intermediate fields of $L|K$ correspond 1–1 to the subgroups of Δ/Γ.

§ 8. Extensions of Valuations

Having seen that the henselian valuations extend uniquely to algebraic extensions we will now study the question of how a valuation v of a field K extends to an algebraic extension in general. So let v be an arbitrary archimedean or nonarchimedean valuation. There is a little discrepancy in notation here, because archimedean valuations manifest themselves only as absolute values while the letter v has hitherto been used for nonarchimedean exponential valuations. In spite of this, it will prove advantageous, and agrees with current usage, to employ the letter v simultaneously for both types of valuations, to denote the corresponding multiplicative valuation in both cases by $|\ |_v$ and the completion by K_v. Where confusion lurks, we will supply clarifying remarks.

For every valuation v of K we consider the completion K_v and an algebraic closure $\overline{K}_v$ of K_v. The canonical extension of v to K_v is again denoted by v and the unique extension of this latter valuation to $\overline{K}_v$ by $\bar{v}$.

Let $L|K$ be an algebraic extension. Choosing a K-embedding

$$\tau : L \longrightarrow \overline{K}_v ,$$

we obtain by restriction of $\bar{v}$ to τL an extension

$$w = \bar{v} \circ \tau$$

of the valuation v to L. In other words, if v, resp. $\bar{v}$, are given by the absolute values $|\ |_v$, resp. $|\ |_{\bar{v}}$, on K, K_v, resp. $\overline{K}_v$, where $|\ |_{\bar{v}}$ extends precisely the absolute value $|\ |_v$ of K_v, then we obtain on L the multiplicative valuation

$$|x|_w = |\tau x|_{\bar{v}} .$$

The mapping $\tau : L \to \overline{K}_v$ is obviously continuous with respect to this valuation. It extends in a unique way to a continuous K-embedding

$$\tau : L_w \longrightarrow \overline{K}_v ,$$

where, in the case of an infinite extension $L|K$, L_w does not mean the completion of L with respect to w, but the union $L_w = \bigcup_i L_{iw}$ of the completions L_{iw} of all finite subextensions $L_i|K$ of $L|K$. This union will be henceforth called the **localization** of L with respect to w. When $[L:K] < \infty$, τ is given by the rule

$$x = w\text{-}\lim_{n\to\infty} x_n \quad \longmapsto \quad \tau x := \bar{v}\text{-}\lim_{n\to\infty} \tau x_n ,$$

where $\{x_n\}_{n\in\mathbb{N}}$ is a w-Cauchy sequence in L, and hence $\{\tau x_n\}_{n\in\mathbb{N}}$ a $\bar{v}$-Cauchy sequence in $\overline{K}_v$. Note here that the sequence τx_n converges in the finite complete extension $\tau L \cdot K_v$ of K_v. We consider the diagram of fields

$$(*)\qquad \begin{array}{ccc} L & \text{---} & L_w \\ | & & | \\ | & & K_v \\ K & \diagup & \end{array}$$

The canonical extension of the valuation w from L to L_w is precisely the unique extension of the valuation v from K_v to the extension $L_w|K_v$. We have

$$L_w = LK_v,$$

because if $L|K$ is finite, then the field $LK_v \subseteq L_w$ is complete by (4.8), contains the field L and therefore has to be its completion. If $L_w|K_v$ has degree $n < \infty$, then, by (4.8), the absolute values corresponding to v and w satisfy the relation

$$|x|_w = \sqrt[n]{\left|N_{L_w|K_v}(x)\right|_v}.$$

The field diagram $(*)$ is of central importance for algebraic number theory. It shows the passage from the "global extension" $L|K$ to the "local extension" $L_w|K_v$ and thus represents one of the most important methods of algebraic number theory, the so-called **local-to-global principle**. This terminology arises from the case of a function field K, for example $K = \mathbb{C}(t)$, where the elements of the extension L are algebraic functions on a Riemann surface, hence on a *global* object, whereas passing to K_v and L_w signifies looking at power series expansions, i.e., the *local* study of functions. The diagram $(*)$ thus expresses in an abstract manner our original goal, to provide methods of function theory for use in the theory of numbers by means of valuations.

We saw that every K-embedding $\tau : L \to \overline{K}_v$ gave us an extension $w = \bar{v} \circ \tau$ of v. For every automorphism $\sigma \in G(\overline{K}_v|K_v)$ of $\overline{K}_v$ over K_v, we obtain with the composite

$$L \xrightarrow{\tau} \overline{K}_v \xrightarrow{\sigma} \overline{K}_v$$

a new K-embedding $\tau' = \sigma \circ \tau$ of L. It will be said to be *conjugate to τ over K_v*. The following result gives us a complete description of the possible extensions of v to L.

(8.1) Extension Theorem. *Let $L|K$ be an algebraic field extension and v a valuation of K. Then one has:*

(i) *Every extension* w *of the valuation* v *arises as the composite* $w = \bar{v} \circ \tau$ *for some* K*-embedding* $\tau : L \to \overline{K}_v$.

(ii) *Two extensions* $\bar{v} \circ \tau$ *and* $\bar{v} \circ \tau'$ *are equal if and only if* τ *and* τ' *are conjugate over* K_v.

Proof: (i) Let w be an extension of v to L and L_w the localization of the canonical valuation, which is again denoted by w. This is the unique extension of the valuation v from K_v to L_w. Choosing any K_v-embedding $\tau : L_w \to \overline{K}_v$, the valuation $\bar{v} \circ \tau$ has to coincide with w. The restriction of τ to L is therefore a K-embedding $\tau : L \to \overline{K}_v$ such that $w = \bar{v} \circ \tau$.

(ii) Let τ and $\sigma \circ \tau$, with $\sigma \in G(\overline{K}_v|K_v)$, be two embeddings of L conjugate over K_v. Since $\bar{v}$ is the only extension of the valuation v from K_v to $\overline{K}_v$, one has $\bar{v} = \bar{v} \circ \sigma$, and thus $\bar{v} \circ \tau = \bar{v} \circ (\sigma \circ \tau)$. The extensions induced to L by τ and by $\sigma \circ \tau$ are therefore the same.

Conversely, let $\tau, \tau' : L \to \overline{K}_v$ be two K-embeddings such that $\bar{v} \circ \tau = \bar{v} \circ \tau'$. Let $\sigma : \tau L \to \tau' L$ be the K-isomorphism $\sigma = \tau' \circ \tau^{-1}$. We can extend σ to a K_v-isomorphism

$$\sigma : \tau L \cdot K_v \longrightarrow \tau' L \cdot K_v .$$

Indeed, τL is dense in $\tau L \cdot K_v$, so every element $x \in \tau L \cdot K_v$ can be written as a limit

$$x = \lim_{n \to \infty} \tau x_n$$

for some sequence x_n which belongs to a finite subextension of L. As $\bar{v} \circ \tau = \bar{v} \circ \tau'$, the sequence $\tau' x_n = \sigma \tau x_n$ converges to an element

$$\sigma x := \lim_{n \to \infty} \sigma \tau x_n$$

in $\tau' L \cdot K_v$. Clearly the correspondence $x \mapsto \sigma x$ does not depend on the choice of a sequence $\{x_n\}$, and yields an isomorphism $\tau L \cdot K_v \xrightarrow{\sigma} \tau' L \cdot K_v$ which leaves K_v fixed. Extending σ to a K_v-automorphism $\tilde{\sigma} \in G(\overline{K}_v|K_v)$ gives $\tau' = \tilde{\sigma} \circ \tau$, so that τ and τ' are indeed conjugate over K_v. □

Those who prefer to be given an extension $L|K$ by an algebraic equation $f(X) = 0$ will appreciate the following concrete variant of the above extension theorem.

Let $L = K(\alpha)$ be generated by the zero α of an irreducible polynomial $f(X) \in K[X]$ and let

$$f(X) = f_1(X)^{m_1} \cdots f_r(X)^{m_r}$$

be the decomposition of $f(X)$ into irreducible factors $f_1(X), \dots, f_r(X)$ over the completion K_v. Of course, the m_i are one if f is separable. The K-embeddings $\tau : L \to \overline{K}_v$ are then given by the zeroes β of $f(X)$ which lie in $\overline{K}_v$:

$$\tau : L \longrightarrow \overline{K}_v, \quad \tau(\alpha) = \beta .$$

Two embeddings τ and τ' are conjugate over K_v if and only if the zeroes $\tau(\alpha)$ and $\tau'(\alpha)$ are conjugate over K_v, i.e., if they are zeroes of the same irreducible factor f_i. With (8.1), this gives the

(8.2) Proposition. *Suppose the extension $L|K$ is generated by the zero α of the irreducible polynomial $f(X) \in K[X]$.*

Then the valuations $w_1, \dots, w_r$ extending v to L correspond 1–1 to the irreducible factors $f_1, \dots, f_r$ in the decomposition

$$f(X) = f_1(X)^{m_1} \cdots f_r(X)^{m_r}$$

of f over the completion K_v.

The extended valuation w_i is explicitly obtained from the factor f_i as follows: let $\alpha_i \in \overline{K}_v$ be a zero of f_i and let

$$\tau_i : L \to \overline{K}_v, \quad \alpha \mapsto \alpha_i ,$$

be the corresponding K-embedding of L into $\overline{K}_v$. Then one has

$$w_i = \bar{v} \circ \tau_i .$$

τ_i extends to an isomorphism

$$\tau_i : L_{w_i} \xrightarrow{\sim} K_v(\alpha_i)$$

on the completion L_{w_i} of L with respect to w_i.

Let $L|K$ be again an arbitrary finite extension. We will write $w|v$ to indicate that w is an extension of the valuation v of K to L. The inclusions $L \hookrightarrow L_w$ induce homomorphisms $L \otimes_K K_v \to L_w$ via $a \otimes b \mapsto ab$, and hence a canonical homomorphism

$$\varphi : L \otimes_K K_v \longrightarrow \prod_{w|v} L_w .$$

To begin with, the tensor product is taken in the sense of vector spaces, i.e., the K-vector space L is lifted to a K_v-vector space $L \otimes_K K_v$. This latter, however, is in fact a K_v-algebra, with the multiplication $(a \otimes b)(a' \otimes b') = aa' \otimes bb'$, and φ is a homomorphism of K_v-algebras. This homomorphism is the subject of the

(8.3) Proposition. *If $L|K$ is separable, then $L \otimes_K K_v \cong \prod_{w|v} L_w$.*

Proof: Let α be a primitive element for $L|K$, so that $L = K(\alpha)$, and let $f(X) \in K[X]$ be its minimal polynomial. To every $w|v$, there corresponds an irreducible factor $f_w(X) \in K_v[X]$ of $f(X)$, and in view of the separability, we have $f(X) = \prod_{w|v} f_w(X)$. Consider all the L_w as embedded into an algebraic closure $\overline{K}_v$ of K_v and denote by α_w the image of α under $L \to L_w$. Then we find $L_w = K_v(\alpha_w)$ and $f_w(X)$ is the minimal polynomial of α_w over K_v. We now get a commutative diagram

$$\begin{array}{ccc} K_v[X]/(f) & \longrightarrow & \prod\limits_{w|v} K_v[X]/(f_w) \\ \downarrow & & \downarrow \\ L \otimes_K K_v & \longrightarrow & \prod\limits_{w|v} L_w, \end{array}$$

where the top arrow is an isomorphism by the Chinese remainder theorem. The arrow on the left is induced by $X \mapsto \alpha \otimes 1$ and is an isomorphism because $K[X]/(f) \cong K(\alpha) = L$. The arrow on the right is induced by $X \mapsto \alpha_w$ and is an isomorphism because $K_v[X]/(f_w) \cong K_v(\alpha_w) = L_w$. Hence the bottom arrow is an isomorphism as well. □

(8.4) Corollary. *If $L|K$ is separable, then one has*

$$[L:K] = \sum_{w|v} [L_w : K_v]$$

and

$$N_{L|K}(\alpha) = \prod_{w|v} N_{L_w|K_v}(\alpha), \qquad Tr_{L|K}(\alpha) = \sum_{w|v} Tr_{L_w|K_v}(\alpha).$$

Proof: The first equation results from (8.3) since $[L:K] = \dim_K(L) = \dim_{K_v}(L \otimes_K K_v)$. On both sides of the isomorphism

$$L \otimes_K K_v \cong \prod_{w|v} L_w$$

let us consider the endomorphism: multiplication by α. The characteristic polynomial of α on the K_v-vector space $L \otimes_K K_v$ is the same as that on the K-vector space L. Therefore

$$\text{char. polynomial}_{L|K}(\alpha) = \prod_{w|v} \text{char. polynomial}_{L_w|K_v}(\alpha).$$

This implies immediately the identities for the norm and the trace. □

If v is a nonarchimedean valuation, then we define, as in the henselian case, the **ramification index** of an extension $w|v$ by

$$e_w = \big(w(L^*) : v(K^*)\big)$$

and the **inertia degree** by

$$f_w = [\lambda_w : \kappa],$$

where λ_w, resp. κ, is the residue class field of w, resp. v. From (8.4) and (6.8), we obtain the **fundamental identity of valuation theory:**

(8.5) Proposition. *If v is discrete and $L|K$ separable, then*

$$\sum_{w|v} e_w f_w = [L : K].$$

This proposition repeats what we have already seen in chap. I, (8.2), working with the prime decomposition. If K is the field of fractions of a Dedekind domain $\mathcal{O}$, then to every nonzero prime ideal $\mathfrak{p}$ of $\mathcal{O}$ is associated the $\mathfrak{p}$**-adic valuation** $v_\mathfrak{p}$ of K, defined by $v_\mathfrak{p}(a) = \nu_\mathfrak{p}$, where $(a) = \prod_\mathfrak{p} \mathfrak{p}^{\nu_\mathfrak{p}}$ (see chap. I, § 11, p. 67). The valuation ring of $v_\mathfrak{p}$ is the localization $\mathcal{O}_\mathfrak{p}$. If $\mathcal{O}$ is the integral closure of $\mathcal{O}$ in L and if

$$\mathfrak{p}\mathcal{O} = \mathfrak{P}_1^{e_1} \cdots \mathfrak{P}_r^{e_r}$$

is the prime decomposition of $\mathfrak{p}$ in L, then the valuations $w_i = \frac{1}{e_i} v_{\mathfrak{P}_i}$, $i = 1, \dots, r$, are precisely the extensions of $v = v_\mathfrak{p}$ to L, e_i are the corresponding ramification indices and $f_i = [\mathcal{O}/\mathfrak{P}_i : \mathcal{O}/\mathfrak{p}]$ the inertia degrees. The fundamental identity

$$\sum_{i=1}^{r} e_i f_i = [L : K]$$

has thus been established in two different ways. The *raison d'être* of valuation theory, however, is not to reformulate ideal-theoretic knowledge, but rather, as has been stressed earlier, to provide the possibility of passing from the extension $L|K$ to the various completions $L_w|K_v$ where much simpler arithmetic laws apply. Let us also emphasize once more that completions may always be replaced with henselizations.

Exercise 1. Up to equivalence, the valuations of the field $\mathbb{Q}(\sqrt{5})$ are given as follows.

1) $|a + b\sqrt{5}|_1 = |a + b\sqrt{5}|$ and $|a + b\sqrt{5}|_2 = |a - b\sqrt{5}|$ are the archimedean valuations.

2) If $p = 2$ or 5 or a prime number $\neq 2, 5$ such that $\left(\frac{p}{5}\right) = -1$, then there is exactly one extension of $|\ \ |_p$ to $\mathbb{Q}(\sqrt{5})$, namely

$$|a + b\sqrt{5}|_{\mathfrak{p}} = |a^2 - 5b^2|_p^{1/2}.$$

3) If p is a prime number $\neq 2, 5$ such that $\left(\frac{p}{5}\right) = 1$, then there are two extensions of $|\ \ |_p$ to $\mathbb{Q}(\sqrt{5})$, namely

$$|a + b\sqrt{5}|_{\mathfrak{p}_1} = |a + b\gamma|_p, \quad \text{resp.} \quad |a + b\sqrt{5}|_{\mathfrak{p}_2} = |a - b\gamma|_p,$$

where γ is a solution of $x^2 - 5 = 0$ in $\mathbb{Q}_p$.

Exercise 2. Determine the valuations of the field $\mathbb{Q}(i)$ of the Gaussian numbers.

Exercise 3. How many extensions to $\mathbb{Q}(\sqrt[n]{2})$ does the archimedean absolute value $|\ \ |$ of $\mathbb{Q}$ admit?

Exercise 4. Let $L|K$ be a finite separable extension, $\mathcal{o}$ the valuation ring of a discrete valuation v and $\mathcal{O}$ its integral closure in L. If $w|v$ varies over the extensions of v to L and $\widehat{\mathcal{o}}_v$, resp. $\widehat{\mathcal{O}}_w$, are the valuation rings of the completions K_v, resp. L_w, then one has

$$\mathcal{O} \otimes_{\mathcal{o}} \widehat{\mathcal{o}}_v \cong \prod_{w|v} \widehat{\mathcal{O}}_w .$$

Exercise 5. How does proposition (8.2) relate to Dedekind's proposition, chap. I, (8.3)?

Exercise 6. Let $L|K$ be a finite field extension, v a nonarchimedean exponential valuation, and w an extension to L. If $\mathcal{O}$ is the integral closure of the valuation ring $\mathcal{o}$ of v in L, then the localization $\mathcal{O}_{\mathfrak{P}}$ of $\mathcal{O}$ at the prime ideal $\mathfrak{P} = \{\alpha \in \mathcal{O} \mid w(\alpha) > 0\}$ is the valuation ring of w.

§ 9. Galois Theory of Valuations

We now consider Galois extensions $L|K$ and study the effect of the Galois action on the extended valuations $w|v$. This leads to a direct generalization of "Hilbert's ramification theory" – see chap. I, § 9, where we studied, instead of valuations v, the prime ideals $\mathfrak{p}$ and their decomposition $\mathfrak{p} = \mathfrak{P}_1^{e_1} \cdots \mathfrak{P}_r^{e_r}$ in Galois extensions of algebraic number fields. The arguments stay the same, so we may be rather brief here. However, we formulate and prove all results for extensions that are not necessarily finite, using infinite Galois theory. The reader who happens not to know this theory should feel free to assume all extensions in this section to be finite. On the other hand, we treat infinite Galois theory also in chap. IV, § 1 below. Its main result can be put in a nutshell like this:

In the case of a Galois extension $L|K$ of infinite degree, the main theorem of ordinary Galois theory, concerning the 1–1 correspondence between the intermediate fields of $L|K$ and the subgroups of the Galois group $G(L|K)$ ceases to hold; there are more subgroups than intermediate fields. The correspondence can be salvaged, however, by considering a canonical topology on the group $G(L|K)$, the **Krull topology**. It is given by defining, for every $\sigma \in G(L|K)$, as a basis of neighbourhoods the cosets $\sigma G(L|M)$, where $M|K$ varies over the *finite* Galois subextensions of $L|K$. $G(L|K)$ is thus turned into a compact, Hausdorff topological group. The main theorem of Galois theory then has to be modified in the infinite case by the condition that the intermediate fields of $L|K$ correspond 1–1 to the *closed* subgroups of $G(L|K)$. Otherwise, everything goes through as in the finite case. So one tacitly restricts attention to *closed* subgroups, and accordingly to *continuous* homomorphisms of $G(L|K)$.

So let $L|K$ be an arbitrary, finite or infinite, Galois extension with Galois group $G = G(L|K)$. If v is an (archimedean or nonarchimedean) valuation of K and w an extension to L, then, for every $\sigma \in G$, $w \circ \sigma$ also extends v, so that the group G acts on the set of extensions $w|v$.

(9.1) Proposition. *The group G acts transitively on the set of extensions $w|v$, i.e., every two extensions are conjugate.*

Proof: Let w and w' be two extensions of v to L. Suppose $L|K$ is finite. If w and w' are not conjugate, then the sets

$$\{w \circ \sigma \mid \sigma \in G\} \quad \text{and} \quad \{w' \circ \sigma \mid \sigma \in G\}$$

would be disjoint. By the approximation theorem (3.4), we would be able to find an $x \in L$ such that

$$|\sigma x|_w < 1 \quad \text{and} \quad |\sigma x|_{w'} > 1$$

for all $\sigma \in G$. Then one would have for the norm $\alpha = N_{L|K}(x) = \prod_{\sigma\in G} \sigma x$ that $|\alpha|_v = \prod_\sigma |\sigma x|_w < 1$ and likewise $|\alpha|_v > 1$, a contradiction.

If $L|K$ is infinite, then we let $M|K$ vary over all finite Galois subextensions and consider the sets $X_M = \{\sigma \in G \mid w \circ \sigma|_M = w'|_M\}$. They are nonempty, as we have just seen, and also closed because, for $\sigma \in G \smallsetminus X_M$, the whole open neighbourhood $\sigma G(L|M)$ lies in the complement of X_M. We have $\bigcap_M X_M \neq \emptyset$, because otherwise the compactness of G would yield a relation $\bigcap_{i=1}^r X_{M_i} = \emptyset$ with finitely many M_i, and this is a contradiction because if $M = M_1 \cdots M_r$, then $X_M = \bigcap_{i=1}^r X_{M_i}$. □

(9.2) Definition. *The* **decomposition group** *of an extension w of v to L is defined by*

$$G_w = G_w(L|K) = \{\sigma \in G(L|K) \mid w \circ \sigma = w\}.$$

If v is a nonarchimedean valuation, then the decomposition group contains two further canonical subgroups

$$G_w \supseteq I_w \supseteq R_w\,,$$

which are defined as follows. Let $\mathcal{o}$, resp. $\mathcal{O}$, be the valuation ring, $\mathfrak{p}$, resp. $\mathfrak{P}$, the maximal ideal, and let $\kappa = \mathcal{o}/\mathfrak{p}$, resp. $\lambda = \mathcal{O}/\mathfrak{P}$, be the residue class field of v, resp. w.

(9.3) Definition. *The* **inertia group** *of $w|v$ is defined by*

$$I_w = I_w(L|K) = \big\{\sigma \in G_w \mid \sigma x \equiv x \bmod \mathfrak{P} \quad \textit{for all} \quad x \in \mathcal{O}\big\}$$

and the **ramification group** *by*

$$R_w = R_w(L|K) = \big\{\sigma \in G_w \mid \tfrac{\sigma x}{x} \equiv 1 \bmod \mathfrak{P} \quad \textit{for all} \quad x \in L^*\big\}\,.$$

Observe in this definition that, for $\sigma \in G_w$, the identity $w \circ \sigma = w$ implies that one always has $\sigma\mathcal{O} = \mathcal{O}$ and $\sigma x/x \in \mathcal{O}$, for all $x \in L^*$.

The subgroups G_w, I_w, R_w of $G = G(L|K)$, and in fact all canonical subgroups we will encounter in the sequel, are all *closed* in the Krull topology. The proof of this is routine in all cases. Let us just illustrate the model of the argument for the example of the decomposition group.

Let $\sigma \in G = G(L|K)$ be an element which belongs to the closure of G_w. This means that, in every neighbourhood $\sigma G(L|M)$, there is some element σ_M of G_w. Here $M|K$ varies over all finite Galois subextensions of $L|K$. Since $\sigma_M \in \sigma G(L|M)$, we have $\sigma_M|_M = \sigma|_M$, and $w \circ \sigma_M = w$ implies that $w \circ \sigma|_M = w \circ \sigma_M|_M = w|_M$. As L is the union of all the M, we get $w \circ \sigma = w$, so that $\sigma \in G_w$. This shows that the subgroup G_w is closed in G.

The groups G_w, I_w, R_w carry very significant information about the behaviour of the valuation v of K as it is extended to L. But before going into this, we will treat the functorial properties of the groups G_w, I_w, R_w.

Consider two Galois extensions $L|K$ and $L'|K'$ and a commutative diagram

$$\begin{array}{ccc} L & \xrightarrow{\ \tau\ } & L' \\ \uparrow & & \uparrow \\ K & \xrightarrow{\ \tau\ } & K' \end{array}$$

with homomorphisms τ which will typically be inclusions. They induce a homomorphism

$$\tau^* : G(L'|K') \longrightarrow G(L|K), \quad \tau^*(\sigma') = \tau^{-1}\sigma'\tau.$$

Observe here that, $L|K$ being normal, the same is true of $\tau L|\tau K$, and thus one has $\sigma'\tau L \subseteq \tau L$, so that composing with τ^{-1} makes sense.

Now let w' be a valuation of L', $v' = w'|_{K'}$ and $w = w' \circ \tau$, $v = w|_K$. Then we have the

(9.4) Proposition. $\tau^* : G(L'|K') \to G(L|K)$ *induces homomorphisms*

$$G_{w'}(L'|K') \longrightarrow G_w(L|K),$$

$$I_{w'}(L'|K') \longrightarrow I_w(L|K),$$

$$R_{w'}(L'|K') \longrightarrow R_w(L|K).$$

In the latter two cases, v is assumed to be nonarchimedean.

Proof: Let $\sigma' \in G_{w'}(L'|K')$ and $\sigma = \tau^*(\sigma')$. If $x \in L$, then one has

$$|x|_{w\circ\sigma} = |\sigma x|_w = |\tau^{-1}\sigma'\tau x|_w = |\sigma'\tau x|_{w'} = |\tau x|_{w'} = |x|_w,$$

so that $\sigma \in G_w(L|K)$. If $\sigma' \in I_{w'}(L'|K')$ and $x \in \mathcal{O}$, then

$$w(\sigma x - x) = w\big(\tau^{-1}(\sigma'\tau x - \tau x)\big) = w'\big(\sigma'(\tau x) - (\tau x)\big) > 0,$$

and $\sigma \in I_w(L|K)$. If $\sigma' \in R_{w'}(L'|K')$ and $x \in L^*$, then

$$w\Big(\frac{\sigma x}{x} - 1\Big) = w\Big(\tau^{-1}\Big(\frac{\sigma'\tau x}{\tau x} - 1\Big)\Big) = w'\Big(\frac{\sigma'\tau x}{\tau x} - 1\Big) > 0,$$

so that $\sigma \in R_w(L|K)$. □

If the two homomorphisms $\tau : L \to L'$ and $\tau : K \to K'$ are isomorphisms, then the homomorphisms (9.4) are of course isomorphisms. In particular, in the case $K = K'$, $L = L'$, we find for each $\tau \in G(L|K)$:

$$G_{w\circ\tau} = \tau^{-1}G_w\tau, \quad I_{w\circ\tau} = \tau^{-1}I_w\tau, \quad R_{w\circ\tau} = \tau^{-1}R_w\tau,$$

i.e., the decomposition, inertia, and ramification groups of conjugate valuations are conjugate.

Another special case arises from an intermediate field M of $L|K$ by the diagram

$$\begin{array}{ccc} L & = & L \\ | & & | \\ K & \hookrightarrow & M. \end{array}$$

τ^* then becomes the inclusion $G(L|M) \hookrightarrow G(L|K)$, and we trivially get the

(9.5) Proposition. *For the extensions $K \subseteq M \subseteq L$, one has*

$$G_w(L|M) = G_w(L|K) \cap G(L|M),$$
$$I_w(L|M) = I_w(L|K) \cap G(L|M),$$
$$R_w(L|M) = R_w(L|K) \cap G(L|M).$$

A particularly important special case of (9.4) occurs with the diagram

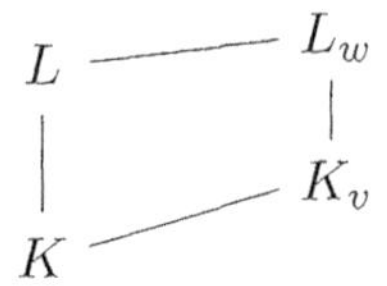

which can be associated to any extension of valuations $w|v$ of $L|K$. If $L|K$ is infinite, then L_w has to be read as the localization in the sense of §8, p. 160. (This distinction is rendered superfluous if we consider, as we may perfectly well do, the henselization of $L|K$.) Since in the local extension $L_w|K_v$ the extension of the valuation is unique, we denote the decomposition, inertia, and ramification groups simply by $G(L_w|K_v)$, $I(L_w|K_v)$, $R(L_w|K_v)$. In this case, the homomorphism τ^* is the restriction map

$$G(L_w|K_v) \longrightarrow G(L|K), \quad \sigma \longmapsto \sigma|_L,$$

and we have the

(9.6) Proposition.

$$G_w(L|K) \cong G(L_w|K_v),$$
$$I_w(L|K) \cong I(L_w|K_v),$$
$$R_w(L|K) \cong R(L_w|K_v).$$

Proof: The proposition derives from the fact that the decomposition group $G_w(L|K)$ consists precisely of those automorphisms $\sigma \in G(L|K)$ which are continuous with respect to the valuation w. Indeed, the continuity of the $\sigma \in G_w(L|K)$ is clear. For an arbitrary continuous automorphism σ, one has

$$|x|_w < 1 \quad \Longrightarrow \quad |\sigma x|_w = |x|_{w\circ\sigma} < 1,$$

because $|x|_w < 1$ means that x^n and hence also σx^n is a w-nullsequence, i.e., $|\sigma x|_w < 1$. By §3, p. 117, this implies that w and $w \circ \sigma$ are equivalent, and hence in fact equal because $w|_K = w \circ \sigma|_K$, so that $\sigma \in G_w(L|K)$.

Since L is dense in L_w, every $\sigma \in G_w(L|K)$ extends uniquely to a continuous K_v-automorphism $\hat{\sigma}$ of L_w and it is clear that $\hat{\sigma} \in I(L_w|K_v)$, resp. $\hat{\sigma} \in R(L_w|K_v)$, if $\sigma \in I_w(L|K)$, resp. $\sigma \in R_w(L|K)$. This proves the bijectivity of the mappings in question in all three cases. □

The above proposition reduces the problems concerning a single valuation of K to the local situation. We identify the decomposition group $G_w(L|K)$ with the Galois group of $L_w|K_v$ and write

$$G_w(L|K) = G(L_w|K_v),$$

and similarly $I_w(L|K) = I(L_w|K_v)$ and $R_w(L|K) = R(L_w|K_v)$.

We now explain the concrete meaning of the subgroups G_w, I_w, R_w of $G = G(L|K)$ for the field extension $L|K$.

The **decomposition group** G_w consists – as was shown in the proof of (9.6) – of all automorphisms $\sigma \in G$ that are **continuous** with respect to the valuation w. It controls the extension of v to L in a group-theoretic manner. Denoting by $G_w \backslash G$ the set of all right cosets $G_w\sigma$, by W_v the set of extensions of v to L and choosing a fixed extension w, we obtain a bijection

$$G_w \backslash G \xrightarrow{\sim} W_v, \quad G_w\sigma \longmapsto w\sigma.$$

In particular, the number $\# W_v$ of extensions equals the index $(G : G_w)$. As mentioned already in chap. I, §9 – and left for the reader to verify – the decomposition group also describes the way a valuation v extends to an arbitrary separable extension $L|K$. For this, we embed $L|K$ into a Galois extension $N|K$, choose an extension w of v to N, and put $G = G(N|K)$, $H = G(N|L)$, $G_w = G_w(N|K)$, to get a bijection

$$G_w \backslash G/H \xrightarrow{\sim} W_v, \quad G_w\sigma H \longmapsto w \circ \sigma|_L.$$

(9.7) Definition. *The fixed field of* G_w,

$$Z_w = Z_w(L|K) = \left\{ x \in L \;\middle|\; \sigma x = x \text{ for all } \sigma \in G_w \right\},$$

is called the **decomposition field** *of w over K.*

The rôle of the decomposition field in the extension $L|K$ is described by the following proposition.

(9.8) Proposition.

(i) *The restriction w_Z of w to Z_w extends uniquely to L.*

(ii) *If v is nonarchimedean, w_Z has the same residue class field and the same value group as v.*

(iii) $Z_w = L \cap K_v$ *(the intersection is taken inside L_w).*

Proof: (i) An arbitrary extension w' of w_Z to L is conjugate to w over Z_w; thus $w' = w \circ \sigma$, for some $\sigma \in G(L|Z_w) = G_w$, i.e., $w' = w$.

(iii) The identity $Z_w = L \cap K_v$ follows immediately from $G_w(L|K) \cong G(L_w|K_v)$.

(ii) Since K_v has the same residue class field and the same value group as K, the same holds true for $Z_w = L \cap K_v$. □

The **inertia group** I_w is defined only if w is a nonarchimedean valuation of L. It is the kernel of a canonical homomorphism of G_w. For if $\mathcal{O}$ is the valuation ring of w and $\mathfrak{P}$ the maximal ideal, then, since $\sigma\mathcal{O} = \mathcal{O}$ and $\sigma\mathfrak{P} = \mathfrak{P}$, every $\sigma \in G_w$ induces a κ-automorphism

$$\bar{\sigma} : \mathcal{O}/\mathfrak{P} \longrightarrow \mathcal{O}/\mathfrak{P}, \quad x \bmod \mathfrak{P} \longmapsto \sigma x \bmod \mathfrak{P},$$

of the residue class field λ, and we obtain a homomorphism

$$G_w \longrightarrow \mathrm{Aut}_\kappa(\lambda)$$

with kernel I_w.

(9.9) Proposition. *The residue class field extension $\lambda|\kappa$ is normal, and we have an exact sequence*

$$1 \longrightarrow I_w \longrightarrow G_w \longrightarrow G(\lambda|\kappa) \longrightarrow 1.$$

Proof: In the case of a finite Galois extension, we have proved this already in chap. I, (9.4). In the infinite case $\lambda|\kappa$ is normal since $L|K$, and hence also $\lambda|\kappa$, is the union of the finite normal subextensions. In order to prove the surjectivity of $f : G_w \to G(\lambda|\kappa)$ all one has to show is that $f(G_w)$ is dense in $G(\lambda|\kappa)$ because $f(G_w)$, being the continuous image of a compact set, is compact and hence closed. Let $\bar{\sigma} \in G(\lambda|\kappa)$ and $\bar{\sigma}G(\lambda|\mu)$ be a neighbourhood of $\bar{\sigma}$, where $\mu|\kappa$ is a finite Galois subextension of $\lambda|\kappa$. We have to show that this neighbourhood contains an element of the image $f(\tau)$, $\tau \in G_w$. Since Z_w has the residue class field κ, there exists a finite Galois subextension $M|Z_w$ of $L|Z_w$ whose residue class field $\overline{M}$ contains the field μ. As $G(M|Z_w) \to G(\overline{M}|\kappa)$ is surjective, the composite

$$G_w = G(L|Z_w) \longrightarrow G(M|Z_w) \longrightarrow G(\overline{M}|\kappa) \longrightarrow G(\mu|\kappa)$$

is also surjective, and if $\tau \in G_w$ is mapped to $\bar{\sigma}|_\mu$, then $f(\tau) \in \bar{\sigma}G(\lambda|\mu)$, as required. □

(9.10) Definition. *The fixed field of* I_w,

$$T_w = T_w(L|K) = \{x \in L \mid \sigma x = x \text{ for all } \sigma \in I_w\},$$

is called the **inertia field** *of* w *over* K.

For the inertia field, (9.9) gives us the isomorphism

$$G(T_w|Z_w) \cong G(\lambda|\kappa).$$

It has the following significance for the extension $L|K$.

(9.11) Proposition. $T_w|Z_w$ *is the maximal unramified subextension of* $L|Z_w$.

Proof: By (9.6), we may assume that $K = Z_w$ is henselian. Let $T|K$ be the maximal unramified subextension of $L|K$. It is Galois, since the conjugate extensions are also unramified. By (7.5), T has the residue class field λ_s, and we have an isomorphism

$$G(T|K) \xrightarrow{\sim} G(\lambda_s|\kappa).$$

Surjectivity follows from (9.9) and the injectivity from the fact that $T|K$ is unramified: every finite Galois subextension has the same degree as its residue class field extension. An element $\sigma \in G(L|K)$ therefore induces the identity on λ_s, i.e., on λ, if and only if it belongs to $G(L|T)$. Consequently, $G(L|T) = I_w$, hence $T = T_w$. □

If in particular K is a henselian field and $\overline{K}_s|K$ its separable closure, then the inertia field of this extension is *the* maximal unramified extension $T|K$ and has the separable closure $\overline{\kappa}_s|\kappa$ as its residue class field. The isomorphism

$$G(T|K) \cong G(\overline{\kappa}_s|\kappa)$$

shows that the unramified extensions of K correspond 1–1 to the separable extensions of κ.

Like the inertia group, the **ramification group** R_w is the kernel of a canonical homomorphism

$$I_w \to \chi(L|K),$$

where

$$\chi(L|K) = \operatorname{Hom}(\Delta/\Gamma, \lambda^*),$$

where $\Delta = w(L^*)$, and $\Gamma = v(K^*)$. If $\sigma \in I_w$, then the associated homomorphism

$$\chi_\sigma : \Delta/\Gamma \to \lambda^*$$

is given as follows: for $\bar{\delta} = \delta \bmod \Gamma \in \Delta/\Gamma$, choose an $x \in L^*$ such that $w(x) = \delta$ and put

$$\chi_\sigma(\bar{\delta}) = \frac{\sigma x}{x} \bmod \mathfrak{P}.$$

This definition is independent of the choice of the representative $\delta \in \bar{\delta}$ and of $x \in L^*$. For if $x' \in L^*$ is an element such that $w(x') \equiv w(x) \bmod \Gamma$, then $w(x') = w(xa)$, $a \in K^*$. Then $x' = xau$, $u \in \mathcal{O}^*$, and since $\sigma u/u \equiv 1 \bmod \mathfrak{P}$ (because $\sigma \in I_{\mathfrak{P}}$), one gets $\sigma x'/x' \equiv \sigma x/x \bmod \mathfrak{P}$.

One sees immediately that mapping $\sigma \mapsto \chi_\sigma$ is a homomorphism $I_w \to \chi(L|K)$ with kernel R_w.

(9.12) Proposition. *R_w is the unique p-Sylow subgroup of I_w.*

Remark: If $L|K$ is a finite extension, then it is clear what this means. In the infinite case it has to be understood in the sense of **profinite groups**, i.e., all finite quotient groups of R_w, resp. I_w/R_w, by closed normal subgroups have p-power order, resp. an order prime to p. In order to understand this better, we refer the reader to chap. IV, §2, exercise 3–5.

Proof of (9.12): By (9.6), we may assume that K is henselian. We restrict to the case where $L|K$ is a finite extension. The infinite case of the proposition follows formally from this.

If R_w were not a p-group, then we would find an element $\sigma \in R_w$ of prime order $\ell \neq p$. Let K' be the fixed field of σ and κ' its residue class field. We show that $\kappa' = \lambda$. Since $R_w \subseteq I_w$, we have that $T \subseteq K'$. Thus $\lambda_s \subseteq \kappa'$, so that $\lambda|\kappa'$ is purely inseparable and of p-power degree. On the other hand, the degree has to be a power of ℓ, for if $\bar{\alpha} \in \lambda$ and if $\alpha \in L$ is a lifting of $\bar{\alpha}$, and $f(x) \in K'[x]$ is the minimal polynomial of α over K', then $\bar{f}(x) = \bar{g}(x)^m$, where $\bar{g}(x) \in \kappa'[x]$ is the minimal polynomial of $\bar{\alpha}$ over κ', which has degree either 1 or ℓ, as this is so for $f(x)$. Thus we have indeed $\kappa' = \lambda$, so that $L|K'$ is tamely ramified, and by (7.7) is of the form $L = K'(\alpha)$ with $\alpha = \sqrt[\ell]{a}$, $a \in K'$. It follows that $\sigma\alpha = \zeta\alpha$, with a primitive ℓ-th root of unity $\zeta \in K'$. Since $\sigma \in R_w$, we have on the other hand $\sigma\alpha/\alpha = \zeta \equiv 1 \bmod \mathfrak{P}$, a contradiction. This proves that R_w is a p-group.

Since $p = \operatorname{char}(\lambda)$, the elements in λ^* have order prime to p, provided they are of finite order. The group $\chi(L|K) = \operatorname{Hom}(\Delta/\Gamma, \lambda^*)$ therefore has order prime to p. This also applies to the group $I_w/R_w \subseteq \chi(L|K)$, so that R_w is indeed the unique p-Sylow subgroup. $\square$

(9.13) Definition. *The fixed field of* R_w,

$$V_w = V_w(L|K) = \{x \in L \mid \sigma x = x \quad \text{for all} \quad \sigma \in R_w\},$$

is called the **ramification field** *of* w *over* K.

(9.14) Proposition. $V_w|Z_w$ *is the maximal tamely ramified subextension of* $L|Z_w$.

Proof: By (9.6) and the fact that the value group and residue class field do not change, we may assume that $K = Z_w$ is henselian. Let V_w be the fixed field of R_w. Since R_w is the p-Sylow subgroup of I_w, V_w is the union of all finite Galois subextensions of $L|T$ of degree prime to p. Therefore V_w contains the maximal tamely ramified extension V of T (and thus of Z_w). Since the degree of each finite subextension $M|V$ of $V_w|V$ is not divisible by p, the residue field extension of $M|V$ is separable (see the argument in the proof of (9.12)). Therefore $V_w|V$ is tamely ramified, and $V_w = V$. □

(9.15) Corollary. *We have the exact sequence*

$$1 \longrightarrow R_w \longrightarrow I_w \longrightarrow \chi(L|K) \longrightarrow 1.$$

Proof: By (9.6) we may assume, as we have already done several times before, that K is henselian. We restrict to considering the case of a finite extension $L|K$. In the infinite case the proof follows as in (9.9). We have already seen that R_w is the kernel of the arrow on the right. It therefore suffices to show that

$$(I_w : R_w) = [V_w : T_w] = \#\chi(L|K).$$

As $T_w|K$ is the maximal unramified subextension of $V_w|K$, $V_w|T_w$ has inertia degree 1. Thus, by (7.7),

$$[V_w : T_w] = \#\big(w(V_w^*)/w(T_w^*)\big).$$

Furthermore, by (7.5), we have $w(T_w^*) = v(K^*) =: \Gamma$, and putting $\Delta = w(L^*)$, we see that $w(V_w^*)/v(K^*)$ is the subgroup $\Delta^{(p)}/\Gamma$ of Δ/Γ consisting of all elements of order prime to p, where $p = \operatorname{char}(\kappa)$. Thus

$$[V_w : T_w] = \#(\Delta^{(p)}/\Gamma).$$

Since λ^* has no elements of order divisible by p, we have on the other hand that

$$\chi(L|K) = \operatorname{Hom}(\Delta/\Gamma, \lambda^*) = \operatorname{Hom}(\Delta^{(p)}/\Gamma, \lambda^*).$$

But (7.7) implies that λ^* contains the m-th roots of unity whenever $\Delta^{(p)}/\Gamma$ contains an element of order m, because then there is a Galois extension generated by radicals $T_w(\sqrt[m]{a})|T_w$ of degree m. This shows that $\chi(L|K)$ is the Pontryagin dual of the group $\Delta^{(p)}/\Gamma$ so that indeed

$$[V_w : T_w] = \#(\Delta^{(p)}/\Gamma) = \#\chi(L|K). \qquad \square$$

Exercise 1. Let K be a henselian field, $L|K$ a tamely ramified Galois extension, $G = G(L|K)$, $I = I(L|K)$ and $\Gamma = G/I = G(\lambda|\kappa)$. Then I is abelian and becomes a Γ-module by letting $\bar{\sigma} = \sigma I \in \Gamma$ operate on I via $\tau \mapsto \sigma\tau\sigma^{-1}$.

Show that there is a canonical isomorphism $I \cong \chi(L|K)$ of Γ-modules. Show furthermore that every tamely ramified extension can be embedded into a tamely ramified extension $L|K$, such that G is the semi-direct product of $\chi(L|K)$ with $G(\lambda|\kappa)$: $G \cong \chi(L|K) \rtimes G(\lambda|\kappa)$.

Hint: Use (7.7).

Exercise 2. The maximal tamely ramified abelian extension V of $\mathbb{Q}_p$ is finite over the maximal unramified abelian extension T of $\mathbb{Q}_p$.

Exercise 3. Show that the maximal unramified extension of the power series field $K = \mathbb{F}_p((t))$ is given by $T = \overline{\mathbb{F}}_p((t))$, where $\overline{\mathbb{F}}_p$ is the algebraic closure of $\mathbb{F}_p$, and the maximal tamely ramified extension by $T(\{\sqrt[m]{t} \mid m \in \mathbb{N}, (m,p) = 1\})$.

Exercise 4. Let v be a nonarchimedean valuation of the field K and let $\bar{v}$ be an extension to the separable closure $\bar{K}$ of K. Then the decomposition field $Z_{\bar{v}}$ of $\bar{v}$ over K is isomorphic to the henselization of K with respect to v, in the sense of §6, p. 143.

§ 10. Higher Ramification Groups

The inertia group and the ramification group inside the Galois group of valued fields are only the first terms in a whole series of subgroups that we are now going to study. We assume that $L|K$ is a finite Galois extension and that v_K is a discrete normalized valuation of K, with positive residue field characteristic p, which admits a *unique* extension w to L. We denote by $v_L = ew$ the associated normalized valuation of L.

(10.1) Definition. *For every real number $s \geq -1$ we define the* **s-th ramification group** *of $L|K$ by*

$$G_s = G_s(L|K) = \{\sigma \in G \mid v_L(\sigma a - a) \geq s+1 \text{ for all } a \in \mathcal{O}\}.$$

Clearly, $G_{-1} = G$, G_0 is the inertia group $I = I(L|K)$, and G_1 the ramification group $R = R(L|K)$ which have already been defined in (9.3). As

$$v_L(\tau^{-1}\sigma\tau a - a) = v_L\big(\tau^{-1}(\sigma\tau a - \tau a)\big) = v_L\big(\sigma(\tau a) - \tau a\big)$$

and $\tau\mathcal{O} = \mathcal{O}$, the ramification groups form a chain

$$G = G_{-1} \supseteq G_0 \supseteq G_1 \supseteq G_2 \supseteq \cdots$$

of normal subgroups of G. The quotients of this chain satisfy the

(10.2) Proposition. *Let $\pi_L \in \mathcal{O}$ be a prime element of L. For every integer $s \geq 0$, the mapping*

$$G_s/G_{s+1} \longrightarrow U_L^{(s)}/U_L^{(s+1)}, \quad \sigma \longmapsto \frac{\sigma\pi_L}{\pi_L},$$

is an injective homomorphism which is independent of the prime element π_L. Here $U_L^{(s)}$ denotes the s-th group of principal units of L, i.e., $U_L^{(0)} = \mathcal{O}^*$ and $U_L^{(s)} = 1 + \pi_L^s\mathcal{O}$, for $s \geq 1$.

We leave the elementary proof to the reader. Observe that one has $U_L^{(0)}/U_L^{(1)} \cong \lambda^*$ and $U_L^{(s)}/U_L^{(s+1)} \cong \lambda$, for $s \geq 1$. The factors G_s/G_{s+1} are therefore abelian groups of exponent p, for $s \geq 1$, and of order prime to p, for $s = 0$. In particular, we find again that the ramification group $R = G_1$ is the unique p-Sylow subgroup in the inertia group $I = G_0$.

We now study the behaviour of the higher ramification groups under change of fields. If only the base field K is changed, then we get directly from the definition of the ramification groups the following generalization of (9.5).

(10.3) Proposition. *If K' is an intermediate field of $L|K$, then one has, for all $s \geq -1$, that*

$$G_s(L|K') = G_s(L|K) \cap G(L|K').$$

Matters become much more complicated when we pass from $L|K$ to a Galois subextension $L'|K$. It is true that the ramification groups of $L|K$ are mapped under $G(L|K) \to G(L'|K)$ into the ramification groups of $L'|K$, but the indexing changes. For the precise description of the situation we need some preparation. We will assume for the sequel that the residue field extension $\lambda|\kappa$ of $L|K$ is **separable**.

(10.4) Lemma. *The ring extension $\mathcal{O}$ of $\mathcal{o}$ is monogenous, i.e., there exists an $x \in \mathcal{O}$ such that $\mathcal{O} = \mathcal{o}[x]$.*

Proof: As the residue field extension $\lambda|\kappa$ is separable by assumption, it admits a primitive element $\bar{x}$. Let $f(X) \in \mathcal{o}[X]$ be a lifting of the minimal polynomial $\bar{f}(X)$ of $\bar{x}$. Then there is a representative $x \in \mathcal{O}$ of $\bar{x}$ such that $\pi = f(x)$ is a prime element of $\mathcal{O}$. Indeed, if x is an arbitrary representative, then we certainly have $v_L(f(x)) \geq 1$ because $\bar{f}(\bar{x}) = 0$. If x itself is not as required, i.e., if $v_L(f(x)) \geq 2$, the representative $x + \pi_L$ will do. In fact, from Taylor's formula

$$f(x+\pi_L) = f(x) + f'(x)\pi_L + b\pi_L^2, \quad b \in \mathcal{O},$$

we obtain $v_L(f(x+\pi_L)) = 1$ since $f'(x) \in \mathcal{O}^*$, because $\bar{f}'(\bar{x}) \neq 0$. In the proof of (6.8), we saw that the

$$x^j\pi^i = x^j f(x)^i, \quad j = 0, \dots, f-1, \quad i = 0, \dots, e-1,$$

form an integral basis of $\mathcal{O}$ over $\mathcal{o}$. Hence indeed $\mathcal{O} = \mathcal{o}[x]$. □

For every $\sigma \in G$ we now put

$$i_{L|K}(\sigma) = v_L(\sigma x - x),$$

where $\mathcal{O} = \mathcal{o}[x]$. This definition does not depend on the choice of the generator x and we may write

$$G_s(L|K) = \big\{\sigma \in G \;\big|\; i_{L|K}(\sigma) \geq s+1\big\}.$$

Passing to a Galois subextension $L'|K$ of $L|K$, the numbers $i_{L|K}(\sigma)$ obey the following rule.

(10.5) Proposition. *If $e' = e_{L|L'}$ is the ramification index of $L|L'$, then*

$$i_{L'|K}(\sigma') = \frac{1}{e'} \sum_{\sigma|_{L'}=\sigma'} i_{L|K}(\sigma).$$

Proof: For $\sigma' = 1$ both sides are infinite. Let $\sigma' \neq 1$, and let $\mathcal{O} = \mathcal{o}[x]$ and $\mathcal{O}' = \mathcal{o}[y]$, with $\mathcal{O}'$ the valuation ring of L'. By definition, we have

$$e' i_{L'|K}(\sigma') = v_L(\sigma' y - y), \quad i_{L|K}(\sigma) = v_L(\sigma x - x).$$

We choose a fixed $\sigma \in G = G(L|K)$ such that $\sigma|_{L'} = \sigma'$. The other elements of G with image σ' in $G' = G(L'|K)$ are then given by $\sigma\tau$, $\tau \in H = G(L|L')$. It therefore suffices to show that the elements

$$a = \sigma y - y \quad \text{and} \quad b = \prod_{\tau \in H} (\sigma\tau x - x)$$

generate the same ideal in $\mathcal{O}$.

Let $f(X) \in \mathcal{O}'[X]$ be the minimal polynomial of x over L'. Then $f(X) = \prod_{\tau \in H}(X - \tau x)$. Letting σ act on the coefficients of f, we get the polynomial $(\sigma f)(X) = \prod_{\tau \in H}(X - \sigma\tau x)$. The coefficients of $\sigma f - f$ are all divisible by $a = \sigma y - y$. Hence a divides $(\sigma f)(x) - f(x) = \pm b$.

To show that conversely b is a divisor of a, we write y as a polynomial in x with coefficients in $\mathcal{o}$, $y = g(x)$. As x is a zero of the polynomial $g(X) - y \in \mathcal{O}'[X]$, we have

$$g(X) - y = f(X)h(X), \quad h(X) \in \mathcal{O}'[X].$$

Letting σ operate on the coefficients of both sides and then substituting $X = x$ yields $y - \sigma y = (\sigma f)(x)(\sigma h)(x) = \pm b(\sigma h)(x)$, i.e., b divides a. □

We now want to show that the ramification group $G_s(L|K)$ is mapped onto the ramification group $G_t(L'|K)$ by the projection

$$G(L|K) \longrightarrow G(L'|K),$$

where t is given by the function $\eta_{L|K} : [-1, \infty) \to [-1, \infty)$,

$$t = \eta_{L|K}(s) = \int_0^s \frac{dx}{(G_0 : G_x)}.$$

Here $(G_0 : G_x)$ is meant to denote the inverse $(G_x : G_0)^{-1}$ when $-1 \le x \le 0$, i.e., simply 1, if $-1 < x \le 0$. For $0 < m \le s \le m+1$, $m \in \mathbb{N}$, we have explicitly

$$\eta_{L|K}(s) = \frac{1}{g_0}\big(g_1 + g_2 + \cdots + g_m + (s-m)g_{m+1}\big), \quad g_i = \#G_i.$$

The function $\eta_{L|K}$ can be expressed in terms of the numbers $i_{L|K}(\sigma)$ as follows:

(10.6) Proposition. $\eta_{L|K}(s) = \frac{1}{g_0}\sum_{\sigma \in G} \min\big\{i_{L|K}(\sigma), s+1\big\} - 1.$

Proof: Let $\theta(s)$ be the function on the right-hand side. It is continuous and piecewise linear. One has $\theta(0) = \eta_{L|K}(0) = 0$, and if $m \ge -1$ is an integer and $m < s < m+1$, then

$$\theta'(s) = \frac{1}{g_0}\#\big\{\sigma \in G \mid i_{L|K}(\sigma) \ge m+2\big\} = \frac{1}{(G_0 : G_{m+1})} = \eta'_{L|K}(s).$$

Hence $\theta = \eta_{L|K}$. □

(10.7) Theorem (HERBRAND). *Let $L'|K$ be a Galois subextension of $L|K$ and $H = G(L|L')$. Then one has*

$$G_s(L|K)H/H = G_t(L'|K) \quad \text{where} \quad t = \eta_{L|L'}(s).$$

Proof: Let $G = G(L|K)$, $G' = G(L'|K)$. For every $\sigma' \in G'$, we choose an preimage $\sigma \in G$ of maximal value $i_{L|K}(\sigma)$ and show that

$$(*) \qquad i_{L'|K}(\sigma') - 1 = \eta_{L|L'}\big(i_{L|K}(\sigma) - 1\big).$$

Let $m = i_{L|K}(\sigma)$. If $\tau \in H$ belongs to $H_{m-1} = G_{m-1}(L|L')$, then $i_{L|K}(\tau) \geq m$, and $i_{L|K}(\sigma\tau) \geq m$, so that $i_{L|K}(\sigma\tau) = m$. If $\tau \notin H_{m-1}$, then $i_{L|K}(\tau) < m$ and $i_{L|K}(\sigma\tau) = i_{L|K}(\tau)$. In both cases we therefore find that $i_{L|K}(\sigma\tau) = \min\{i_{L|K}(\tau), m\}$. Applying (10.5), this gives

$$i_{L'|K}(\sigma') = \frac{1}{e'} \sum_{\tau \in H} \min\big\{i_{L|K}(\tau), m\big\}.$$

Since $i_{L|K}(\tau) = i_{L|L'}(\tau)$ and $e' = e_{L|L'} = \#H_0$, (10.6) gives the formula $(*)$, which in turn yields

$$\begin{aligned} \sigma' \in G_sH/H &\iff i_{L|K}(\sigma) - 1 \geq s \iff \eta_{L|L'}(i_{L|K}(\sigma) - 1) \geq \eta_{L|L'}(s) \\ &\iff i_{L'|K}(\sigma') - 1 \geq \eta_{L|L'}(s) \\ &\iff \sigma' \in G_t(L'|K), \quad t = \eta_{L|L'}(s). \end{aligned}$$

□

The function $\eta_{L|K}$ is by definition strictly increasing. Let the inverse function be $\psi_{L|K} : [-1,\infty) \to [-1,\infty)$. One defines the **upper numbering** of the ramification groups by

$$G^t(L|K) := G_s(L|K) \quad \text{where} \quad s = \psi_{L|K}(t).$$

The functions $\eta_{L|K}$ and $\psi_{L|K}$ satisfy the following transitivity condition:

(10.8) Proposition. *If $L'|K$ is a Galois subextension of $L|K$, then*

$$\eta_{L|K} = \eta_{L'|K} \circ \eta_{L|L'} \quad \text{and} \quad \psi_{L|K} = \psi_{L|L'} \circ \psi_{L'|K}.$$

Proof: For the ramification indices of the extensions $L|K$, $L'|K$, $L|L'$ we have $e_{L|K} = e_{L'|K}e_{L|L'}$. From (10.7), we obtain $G_s/H_s = (G/H)_t$, $t = \eta_{L|L'}(s)$; thus

$$\frac{1}{e_{L|K}}\#G_s = \frac{1}{e_{L'|K}}\#(G/H)_t \frac{1}{e_{L|L'}}\#H_s.$$

This equation is equivalent to

$$\eta'_{L|K}(s) = \eta'_{L'|K}(t)\eta'_{L|L'}(s) = (\eta_{L'|K} \circ \eta_{L|L'})'(s).$$

As $\eta_{L|K}(0) = (\eta_{L'|K} \circ \eta_{L|L'})(0)$, it follows that $\eta_{L|K} = \eta_{L'|K} \circ \eta_{L|L'}$ and the formula for ψ follows. □

The advantage of the upper numbering of the ramification groups is that it is invariant when passing from $L|K$ to a Galois subextension.

(10.9) Proposition. *Let $L'|K$ be a Galois subextension of $L|K$ and $H = G(L|L')$. Then one has*

$$G^t(L|K)H/H = G^t(L'|K).$$

Proof: We put $s = \psi_{L'|K}(t)$, $G' = G(L'|K)$, apply (10.7) and (10.8), and get

$$\begin{aligned} G^t H/H &= G_{\psi_{L|K}(t)}H/H = G'_{\eta_{L|L'}(\psi_{L|K}(t))} = G'_{\eta_{L|L'}(\psi_{L|L'}(s))} \\ &= G'_s = G'^t. \end{aligned}$$ □

Exercise 1. Let $K = \mathbb{Q}_p$ and $K_n = K(\zeta)$, where ζ is a primitive p^n-th root of unity. Show that the ramification groups of $K_n|K$ are given as follows:

$$\begin{aligned} G_s &= G(K_n|K) && \text{for } s = 0, \\ G_s &= G(K_n|K_1) && \text{for } 1 \le s \le p-1, \\ G_s &= G(K_n|K_2) && \text{for } p \le s \le p^2-1, \\ &\dots \\ G_s &= 1 && \text{for } p^{n-1} \le s. \end{aligned}$$

Exercise 2. Let K' be an intermediate field of $L|K$. Describe the relation between the ramification groups of $L|K$ and of $L|K'$ in the upper numbering.

Chapter III
Riemann-Roch Theory

§ 1. Primes

Having set up the general theory of valued fields, we now return to algebraic number fields. We want to develop their basic theory from the valuation-theoretic point of view. This approach has two prominent advantages compared to the ideal-theoretic treatment given in the first chapter. The first one results from the possibility of passing to completions, thereby enacting the important "local-to-global principle". This will be done in chapter VI. The other advantage lies in the fact that the archimedean valuations bring into the picture the points at infinity, which were hitherto lacking, as the "primes at infinity". In this way a perfect analogy with the function fields is achieved. This is the task to which we now turn.

(1.1) Definition. *A* **prime** (*or* **place**) $\mathfrak{p}$ *of an algebraic number field K is a class of equivalent valuations of K. The nonarchimedean equivalence classes are called* **finite** *primes and the archimedean ones* **infinite** *primes.*

The infinite primes $\mathfrak{p}$ are obtained, according to chap. II, (8.1), from the embeddings $\tau : K \to \mathbb{C}$. There are two sorts of these: the **real primes**, which are given by embeddings $\tau : K \to \mathbb{R}$, and the **complex primes**, which are induced by the pairs of complex conjugate non-real embeddings $K \to \mathbb{C}$. $\mathfrak{p}$ is real or complex depending whether the completion $K_\mathfrak{p}$ is isomorphic to $\mathbb{R}$ or to $\mathbb{C}$. The infinite primes will be referred to by the formal notation $\mathfrak{p} \mid \infty$, the finite ones by $\mathfrak{p} \nmid \infty$.

In the case of a finite prime, the letter $\mathfrak{p}$ has a multiple meaning: it also stands for the prime ideal of the ring $\mathcal{O}$ of integers of K, or for the maximal ideal of the associated valuation ring, or even for the maximal ideal of the completion. However, this will nowhere create any risk of confusion. We write $\mathfrak{p}|p$ if p is the characteristic of the residue field $\kappa(\mathfrak{p})$ of the finite prime $\mathfrak{p}$. For an infinite prime we adopt the convention that the completion $K_\mathfrak{p}$ also serves as its own "residue field, " i.e., we put

$$\kappa(\mathfrak{p}) := K_\mathfrak{p}, \quad \text{when } \mathfrak{p}|\infty .$$

To each prime $\mathfrak{p}$ of K we now associate a canonical homomorphism

$$v_{\mathfrak{p}} : K^* \to \mathbb{R}$$

from the multiplicative group K^* of K. If $\mathfrak{p}$ is finite, then $v_{\mathfrak{p}}$ is the $\mathfrak{p}$-adic exponential valuation which is normalized by the condition $v_{\mathfrak{p}}(K^*) = \mathbb{Z}$. If $\mathfrak{p}$ is infinite, then $v_{\mathfrak{p}}$ is given by

$$v_{\mathfrak{p}}(a) = -\log|\tau a|,$$

where $\tau : K \to \mathbb{C}$ is an embedding which defines $\mathfrak{p}$.

For an arbitrary prime $\mathfrak{p}|p$ (p prime number or $p = \infty$) we put furthermore

$$f_{\mathfrak{p}} = \big[\kappa(\mathfrak{p}) : \kappa(p)\big],$$

so that $f_{\mathfrak{p}} = [K_{\mathfrak{p}} : \mathbb{R}]$ if $\mathfrak{p}|\infty$, and

$$\mathfrak{N}(\mathfrak{p}) = \begin{cases} p^{f_{\mathfrak{p}}}, & \text{if } \mathfrak{p} \nmid \infty, \\ e^{f_{\mathfrak{p}}}, & \text{if } \mathfrak{p} \mid \infty. \end{cases}$$

This convention suggests that we consider e as being an **infinite prime number**, and the extension $\mathbb{C}|\mathbb{R}$ as being *unramified* with inertia degree 2. We define the absolute value $|\ \ |_{\mathfrak{p}} : K \to \mathbb{R}$ by

$$|a|_{\mathfrak{p}} = \mathfrak{N}(\mathfrak{p})^{-v_{\mathfrak{p}}(a)}$$

for $a \neq 0$ and $|0|_{\mathfrak{p}} = 0$. If $\mathfrak{p}$ is the infinite prime associated to the embedding $\tau : K \to \mathbb{C}$, then one finds

$$|a|_{\mathfrak{p}} = |\tau a|, \quad \text{resp.} \quad |a|_{\mathfrak{p}} = |\tau a|^2,$$

if $\mathfrak{p}$ is real, resp. complex.

If $L|K$ is a finite extension of K, then we denote the primes of L by $\mathfrak{P}$ and write $\mathfrak{P}|\mathfrak{p}$ to signify that the valuations in the class $\mathfrak{P}$, when restricted to K, give those of $\mathfrak{p}$. In the case of an infinite prime $\mathfrak{P}$, we define the **inertia degree**, resp. the **ramification index**, by

$$f_{\mathfrak{P}|\mathfrak{p}} = [L_{\mathfrak{P}} : K_{\mathfrak{p}}], \quad \text{resp. } e_{\mathfrak{P}|\mathfrak{p}} = 1.$$

For arbitrary primes $\mathfrak{P}|\mathfrak{p}$ we then have the

(1.2) Proposition. (i) $\sum_{\mathfrak{P}|\mathfrak{p}} e_{\mathfrak{P}|\mathfrak{p}} f_{\mathfrak{P}|\mathfrak{p}} = \sum_{\mathfrak{P}|\mathfrak{p}} [L_{\mathfrak{P}} : K_{\mathfrak{p}}] = [L : K]$,

(ii) $\mathfrak{N}(\mathfrak{P}) = \mathfrak{N}(\mathfrak{p})^{f_{\mathfrak{P}|\mathfrak{p}}}$,

(iii) $v_{\mathfrak{P}}(a) = e_{\mathfrak{P}|\mathfrak{p}} v_{\mathfrak{p}}(a)$ for $a \in K^*$,

(iv) $v_{\mathfrak{p}}(N_{L_{\mathfrak{P}}|K_{\mathfrak{p}}}(a)) = f_{\mathfrak{P}|\mathfrak{p}} v_{\mathfrak{P}}(a)$ for $a \in L^*$,

(v) $|a|_{\mathfrak{P}} = |N_{L_{\mathfrak{P}}|K_{\mathfrak{p}}}(a)|_{\mathfrak{p}}$ for $a \in L$.

The normalized valuations $|\ |_\mathfrak{p}$ satisfy the following **product formula**, which demonstrates how important it is to include the infinite primes.

(1.3) Proposition. *Given any $a \in K^*$, one has $|a|_\mathfrak{p} = 1$ for almost all $\mathfrak{p}$, and*

$$\prod_\mathfrak{p} |a|_\mathfrak{p} = 1.$$

Proof: We have $v_\mathfrak{p}(a) = 0$ and therefore $|a|_\mathfrak{p} = 1$ for all $\mathfrak{p} \nmid \infty$ which do not occur in the prime decomposition of the principal ideal (a) (see chap. I, § 11, p. 69). This therefore holds for almost all $\mathfrak{p}$. From (1.2) and formula (8.4) of chap. II,

$$N_{K|\mathbb{Q}}(a) = \prod_{\mathfrak{p}|p} N_{K_\mathfrak{p}|\mathbb{Q}_p}(a)$$

(which includes the case $p = \infty$, $\mathbb{Q}_p = \mathbb{R}$), we obtain the product formula for K from the product formula for $\mathbb{Q}$, which was proved already in chap. II, (2.1):

$$\prod_\mathfrak{p} |a|_\mathfrak{p} = \prod_p \prod_{\mathfrak{p}|p} |a|_\mathfrak{p} = \prod_p \prod_{\mathfrak{p}|p} \big| N_{K_\mathfrak{p}|\mathbb{Q}_p}(a) \big|_p = \prod_p \big| N_{K|\mathbb{Q}}(a) \big|_p = 1. \quad \square$$

We denote by $J(\mathcal{O})$ the group of fractional ideals of K, by $P(\mathcal{O})$ the subgroup of fractional principal ideals, and by

$$Pic(\mathcal{O}) = J(\mathcal{O})/P(\mathcal{O})$$

the ideal class group Cl_K of K.

Let us now extend the notion of fractional ideal of K by taking into account also the infinite primes.

(1.4) Definition. *A* **replete ideal** *of K is an element of the group*

$$J(\bar{\mathcal{O}}) := J(\mathcal{O}) \times \prod_{\mathfrak{p}|\infty} \mathbb{R}_+^*,$$

where $\mathbb{R}_+^*$ denotes the multiplicative group of positive real numbers.

In order to unify notation, we put, for any infinite prime $\mathfrak{p}$ and any real number $\nu \in \mathbb{R}$,

$$\mathfrak{p}^\nu := e^\nu \in \mathbb{R}_+^*.$$

Given a system of real numbers $\nu_\mathfrak{p}$, $\mathfrak{p}|\infty$, let $\prod_{\mathfrak{p}|\infty}\mathfrak{p}^{\nu_\mathfrak{p}}$ always denote the vector

$$\prod_{\mathfrak{p}|\infty}\mathfrak{p}^{\nu_\mathfrak{p}} = (\ldots, e^{\nu_\mathfrak{p}}, \ldots) \in \prod_{\mathfrak{p}|\infty}\mathbb{R}_+^*,$$

and *not* the product of the quantities $e^{\nu_\mathfrak{p}}$ in $\mathbb{R}$. Then every replete ideal $\mathfrak{a} \in J(\bar{o})$ admits the unique product representation

$$\mathfrak{a} = \prod_{\mathfrak{p}\nmid\infty}\mathfrak{p}^{\nu_\mathfrak{p}} \times \prod_{\mathfrak{p}|\infty}\mathfrak{p}^{\nu_\mathfrak{p}} =: \prod_{\mathfrak{p}}\mathfrak{p}^{\nu_\mathfrak{p}},$$

where $\nu_\mathfrak{p} \in \mathbb{Z}$ for $\mathfrak{p} \nmid \infty$, and $\nu_\mathfrak{p} \in \mathbb{R}$ for $\mathfrak{p}|\infty$. Put

$$\mathfrak{a}_\mathrm{f} = \prod_{\mathfrak{p}\nmid\infty}\mathfrak{p}^{\nu_\mathfrak{p}} \quad \text{and} \quad \mathfrak{a}_\infty = \prod_{\mathfrak{p}|\infty}\mathfrak{p}^{\nu_\mathfrak{p}},$$

so that $\mathfrak{a} = \mathfrak{a}_\mathrm{f} \times \mathfrak{a}_\infty$. $\mathfrak{a}_\mathrm{f}$ is a fractional ideal of K, and $\mathfrak{a}_\infty$ is an element of $\prod_{\mathfrak{p}|\infty}\mathbb{R}_+^*$. At the same time, we view $\mathfrak{a}_\mathrm{f}$, resp. $\mathfrak{a}_\infty$, as replete ideals

$$\mathfrak{a}_\mathrm{f} \times \prod_{\mathfrak{p}|\infty} 1, \quad \text{resp. } (1) \times \mathfrak{a}_\infty.$$

Thus for all elements of $J(\bar{o})$ the decomposition

$$\mathfrak{a} = \mathfrak{a}_\mathrm{f} \cdot \mathfrak{a}_\infty$$

applies. To $a \in K^*$ we associate the **replete principal ideal**

$$[a] = \prod_{\mathfrak{p}}\mathfrak{p}^{\nu_\mathfrak{p}(a)} = (a) \times \prod_{\mathfrak{p}|\infty}\mathfrak{p}^{\nu_\mathfrak{p}(a)}.$$

These replete ideals form a subgroup $P(\bar{o})$ of $J(\bar{o})$. The factor group

$$Pic(\bar{o}) = J(\bar{o})/P(\bar{o})$$

is called the **replete ideal class group**, or **replete Picard group**.

(1.5) Definition. *The* **absolute norm** *of a replete ideal* $\mathfrak{a} = \prod_\mathfrak{p}\mathfrak{p}^{\nu_\mathfrak{p}}$ *is defined to be the positive real number*

$$\mathfrak{N}(\mathfrak{a}) = \prod_{\mathfrak{p}}\mathfrak{N}(\mathfrak{p})^{\nu_\mathfrak{p}}.$$

The absolute norm is multiplicative and induces a surjective homomorphism

$$\mathfrak{N} : J(\bar{o}) \to \mathbb{R}_+^*.$$

The absolute norm of a replete principal ideal $[a]$ is equal to 1 in view of the product formula (1.3),

$$\mathfrak{N}([a]) = \prod_{\mathfrak{p}}\mathfrak{N}(\mathfrak{p})^{\nu_\mathfrak{p}(a)} = \prod_{\mathfrak{p}}|a|_\mathfrak{p}^{-1} = 1.$$

We therefore obtain a surjective homomorphism

$$\mathfrak{N} : Pic(\bar{o}) \to \mathbb{R}_+^*$$

on the replete Picard group.

The relations between the replete ideals of a number field K and those of an extension field L are afforded by the two homomorphisms

$$J(\bar{\mathcal{O}}_K) \underset{N_{L|K}}{\overset{i_{L|K}}{\rightleftarrows}} J(\bar{\mathcal{O}}_L),$$

which are defined by

$$i_{L|K}\Big(\prod_{\mathfrak{p}} \mathfrak{p}^{\nu_{\mathfrak{p}}}\Big) = \prod_{\mathfrak{p}} \prod_{\mathfrak{P}|\mathfrak{p}} \mathfrak{P}^{e_{\mathfrak{P}|\mathfrak{p}}\nu_{\mathfrak{p}}},$$

$$N_{L|K}\Big(\prod_{\mathfrak{P}} \mathfrak{P}^{\nu_{\mathfrak{P}}}\Big) = \prod_{\mathfrak{p}} \prod_{\mathfrak{P}|\mathfrak{p}} \mathfrak{p}^{f_{\mathfrak{P}|\mathfrak{p}}\nu_{\mathfrak{P}}}.$$

Here the various product signs have to be read according to our convention. These homomorphisms satisfy the

(1.6) Proposition.

(i) *For a chain of fields $K \subseteq L \subseteq M$, one has $N_{M|K} = N_{L|K} \circ N_{M|L}$ and $i_{M|K} = i_{M|L} \circ i_{L|K}$.*

(ii) $N_{L|K}(i_{L|K}\mathfrak{a}) = \mathfrak{a}^{[L:K]}$ *for $\mathfrak{a} \in J(\bar{\mathcal{O}}_K)$.*

(iii) $\mathfrak{N}(N_{L|K}(\mathfrak{A})) = \mathfrak{N}(\mathfrak{A})$ *for $\mathfrak{A} \in J(\bar{\mathcal{O}}_L)$.*

(iv) *If $L|K$ is Galois with Galois group G, then for every prime ideal $\mathfrak{P}$ of $\mathcal{O}_L$, one has $N_{L|K}(\mathfrak{P})\mathcal{O}_L = \prod_{\sigma\in G}\sigma\mathfrak{P}$.*

(v) *For any replete principal ideal $[a]$ of K, resp. L, one has*

$$i_{L|K}([a]) = [a], \quad \text{resp. } N_{L|K}([a]) = [N_{L|K}(a)].$$

(vi) *$N_{L|K}(\mathfrak{A}_{\mathfrak{f}}) = N_{L|K}(\mathfrak{A})_{\mathfrak{f}}$ is the ideal of K generated by the norms $N_{L|K}(a)$ of all $a \in \mathfrak{A}_{\mathfrak{f}}$.*

Proof: (i) is based on the transitivity of inertia degree and ramification index. (ii) follows from (1.2) in view of the fundamental identity $\sum_{\mathfrak{P}|\mathfrak{p}} f_{\mathfrak{P}|\mathfrak{p}}e_{\mathfrak{P}|\mathfrak{p}} = [L:K]$. (iii) holds for any replete "prime ideal" $\mathfrak{P}$ of L, by (1.2):

$$\mathfrak{N}\big(N_{L|K}(\mathfrak{P})\big) = \mathfrak{N}(\mathfrak{p}^{f_{\mathfrak{P}|\mathfrak{p}}}) = \mathfrak{N}(\mathfrak{p})^{f_{\mathfrak{P}|\mathfrak{p}}} = \mathfrak{N}(\mathfrak{P}),$$

and therefore for all replete ideals of L.

(iv) The prime ideal $\mathfrak{p}$ lying below $\mathfrak{P}$ decomposes in the ring $\mathcal{O}$ of integers of L as $\mathfrak{p} = (\mathfrak{P}_1 \cdots \mathfrak{P}_r)^e$, with prime ideals $\mathfrak{P}_i = \sigma_i\mathfrak{P}$, $\sigma_i \in G/G_{\mathfrak{P}}$, which are conjugates of $\mathfrak{P}$ and thus have the same inertia degree f. Therefore

$$N_{L|K}(\mathfrak{P})\mathcal{O} = \mathfrak{p}^f\mathcal{O} = \prod_{i=1}^{r}\mathfrak{P}_i^{ef} = \prod_{i=1}^{r}\prod_{\tau\in G_{\mathfrak{P}}}\sigma_i\tau\mathfrak{P} = \prod_{\sigma\in G}\sigma\mathfrak{P}.$$

(v) For any element $a \in K^*$, (1.2) implies that $v_{\mathfrak{P}}(a) = e_{\mathfrak{P}|\mathfrak{p}} v_{\mathfrak{p}}(a)$. Hence

$$i_{L|K}([a]) = i_{L|K}\Big(\prod_{\mathfrak{p}} \mathfrak{p}^{v_{\mathfrak{p}}(a)}\Big) = \prod_{\mathfrak{p}} \prod_{\mathfrak{P}|\mathfrak{p}} \mathfrak{P}^{e_{\mathfrak{P}|\mathfrak{p}} v_{\mathfrak{p}}(a)} = \prod_{\mathfrak{P}} \mathfrak{P}^{v_{\mathfrak{P}}(a)} = [a].$$

If, on the other hand, $a \in L^*$, then (1.2) and chap. II, (8.4) imply that $v_{\mathfrak{p}}(N_{L|K}(a)) = \sum_{\mathfrak{P}|\mathfrak{p}} f_{\mathfrak{P}|\mathfrak{p}} v_{\mathfrak{P}}(a)$. Hence

$$N_{L|K}([a]) = N_{L|K}\Big(\prod_{\mathfrak{P}} \mathfrak{P}^{v_{\mathfrak{P}}(a)}\Big) = \prod_{\mathfrak{p}} \mathfrak{p}^{v_{\mathfrak{p}}(N_{L|K}(a))} = [N_{L|K}(a)].$$

(vi) Let $\mathfrak{a}_f$ be the ideal of K which is generated by all $N_{L|K}(a)$, with $a \in \mathfrak{A}_f$. If $\mathfrak{A}_f$ is a principal ideal (a), then $\mathfrak{a}_f = (N_{L|K}(a)) = N_{L|K}(\mathfrak{A}_f)$, by (v). But the argument which yielded (v) applies equally well to the localizations $\mathcal{O}_{\mathfrak{p}}|\mathcal{o}_{\mathfrak{p}}$ of the extension $\mathcal{O}|\mathcal{o}$ of maximal orders of $L|K$. $\mathcal{O}_{\mathfrak{p}}$ has only a finite number of prime ideals, and is therefore a principal ideal domain (see chap. I, § 3, exercise 4). We thus get

$$(\mathfrak{a}_f)_{\mathfrak{p}} = N_{L|K}\big((\mathfrak{A}_f)_{\mathfrak{p}}\big) = N_{L|K}(\mathfrak{A}_f)_{\mathfrak{p}}$$

for all prime ideals $\mathfrak{p}$ of $\mathcal{o}$, and consequently $\mathfrak{a}_f = N_{L|K}(\mathfrak{A}_f)$. □

Since the homomorphisms $i_{L|K}$ and $N_{L|K}$ map replete principal ideals to replete principal ideals, they induce homomorphisms of the replete Picard groups of K and L, and we obtain the

(1.7) Proposition. *For every finite extension $L|K$, the following two diagrams are commutative:*

$$\begin{array}{ccc} Pic(\bar{\mathcal{O}}_L) & \xrightarrow{\ \mathfrak{N}\ } & \mathbb{R}_+^* \\ {\scriptstyle i_{L|K}}\uparrow\downarrow{\scriptstyle N_{L|K}} & & {\scriptstyle [L:K]}\uparrow\downarrow{\scriptstyle \mathrm{id}} \\ Pic(\bar{\mathcal{O}}_K) & \xrightarrow{\ \mathfrak{N}\ } & \mathbb{R}_+^* . \end{array}$$

Let us now translate the notions we have introduced into the function-theoretic language of divisors. In chap. I, § 12, we defined the divisor group $Div(\mathcal{o})$ to consist of all formal sums

$$D = \sum_{\mathfrak{p}\nmid\infty} \nu_{\mathfrak{p}} \mathfrak{p},$$

where $\nu_{\mathfrak{p}} \in \mathbb{Z}$, and $\nu_{\mathfrak{p}} = 0$ for almost all $\mathfrak{p}$. Contained in this group is the group $\mathcal{P}(\mathcal{o})$ of principal divisors $\mathrm{div}(f) = \sum_{\mathfrak{p}\nmid\infty} v_{\mathfrak{p}}(f)\mathfrak{p}$, which allowed us to define the divisor class group

$$CH^1(\mathcal{o}) = Div(\mathcal{o})/\mathcal{P}(\mathcal{o}).$$

It follows from the main theorem of ideal theory, chap. I, (3.9), that this group is isomorphic to the ideal class group Cl_K, which is a finite group (see chap. I, (12.14)). We now extend these concepts as follows.

(1.8) Definition. *A* **replete divisor** (*or* **Arakelov divisor**) *of* K *is a formal sum*

$$D = \sum_{\mathfrak{p}} \nu_{\mathfrak{p}} \mathfrak{p},$$

where $\nu_{\mathfrak{p}} \in \mathbb{Z}$ *for* $\mathfrak{p} \nmid \infty$, $\nu_{\mathfrak{p}} \in \mathbb{R}$ *for* $\mathfrak{p}|\infty$, *and* $\nu_{\mathfrak{p}} = 0$ *for almost all* $\mathfrak{p}$.

The replete divisors form a group, which is denoted by $Div(\bar{\mathcal{O}})$. It admits a decomposition

$$Div(\bar{\mathcal{O}}) \cong Div(\mathcal{O}) \times \bigoplus_{\mathfrak{p}|\infty} \mathbb{R}\mathfrak{p}.$$

On the right-hand side, the second factor is endowed with the canonical topology, the first one with the discrete topology. On the product we have the product topology, which makes $Div(\bar{\mathcal{O}})$ into a locally compact topological group.

We now study the canonical homomorphism

$$\operatorname{div} : K^* \longrightarrow Div(\bar{\mathcal{O}}), \quad \operatorname{div}(f) = \sum_{\mathfrak{p}} v_{\mathfrak{p}}(f)\mathfrak{p}.$$

The elements of the form $\operatorname{div}(f)$ are called **replete principal divisors**.

Remark: The composite of the mapping $\operatorname{div} : K^* \to Div(\bar{\mathcal{O}})$ with the mapping

$$Div(\bar{\mathcal{O}}) \longrightarrow \prod_{\mathfrak{p}|\infty} \mathbb{R}, \quad \sum_{\mathfrak{p}} \nu_{\mathfrak{p}}\mathfrak{p} \longmapsto (\nu_{\mathfrak{p}} f_{\mathfrak{p}})_{\mathfrak{p}|\infty},$$

is equal, up to sign, to the logarithm map

$$\lambda : K^* \longrightarrow \prod_{\mathfrak{p}|\infty} \mathbb{R}, \quad \lambda(f) = \big(\ldots, \log |f|_{\mathfrak{p}}, \ldots\big),$$

of Minkowski theory (see chap. I, §7, p. 39, and chap. III, §3, p. 211). By chap. I, (7.3), it maps the unit group $\mathcal{O}^*$ onto a complete lattice $\Gamma = \lambda(\mathcal{O}^*)$ in trace-zero space $H = \{(x_{\mathfrak{p}}) \in \prod_{\mathfrak{p}|\infty} \mathbb{R} \mid \sum_{\mathfrak{p}|\infty} x_{\mathfrak{p}} = 0\}$.

(1.9) Proposition. *The kernel of* $\operatorname{div} : K^* \to Div(\bar{\mathcal{O}})$ *is the group* $\mu(K)$ *of roots of unity in* K, *and its image* $\mathcal{P}(\bar{\mathcal{O}})$ *is a discrete subgroup of* $Div(\bar{\mathcal{O}})$.

Proof: By the above remark, the composite of div with the map $Div(\bar{\mathcal{O}}) \to \prod_{\mathfrak{p}|\infty} \mathbb{R}$, $\sum_{\mathfrak{p}} \nu_{\mathfrak{p}}\mathfrak{p} \mapsto (\nu_{\mathfrak{p}} f_{\mathfrak{p}})_{\mathfrak{p}|\infty}$, yields, up to sign, the homomorphism $\lambda : K^* \to \prod_{\mathfrak{p}|\infty} \mathbb{R}$. By chap. I, (7.1), the latter fits into the exact sequence

$$1 \longrightarrow \mu(K) \longrightarrow \mathcal{O}^* \xrightarrow{\lambda} \Gamma \longrightarrow 0,$$

where Γ is a complete lattice in trace-zero space $H \subseteq \prod_{\mathfrak{p}|\infty} \mathbb{R}$. Therefore $\mu(K)$ is the kernel of div. Since Γ is a lattice, there exists a neighbourhood U of 0 in $\prod_{\mathfrak{p}|\infty} \mathbb{R}$ which contains no element of Γ except 0. Considering the isomorphism $\alpha : \prod_{\mathfrak{p}|\infty} \mathbb{R} \to \bigoplus_{\mathfrak{p}|\infty} \mathbb{R}\mathfrak{p}$, $(\nu_{\mathfrak{p}})_{\mathfrak{p}|\infty} \mapsto \sum_{\mathfrak{p}|\infty} \frac{\nu_{\mathfrak{p}}}{f_{\mathfrak{p}}}\, \mathfrak{p}$, the set $\{0\} \times \alpha U \subset Div(\mathcal{O}) \times \bigoplus_{\mathfrak{p}|\infty} \mathbb{R}\mathfrak{p} = Div(\bar{\mathcal{O}})$ is a neighbourhood of 0 in $Div(\bar{\mathcal{O}})$ which contains no replete principal divisor except 0. This shows that $\mathcal{P}(\bar{\mathcal{O}}) = \mathrm{div}(K^*)$ lies discretely in $Div(\bar{\mathcal{O}})$. □

(1.10) Definition. *The factor group*

$$CH^1(\bar{\mathcal{O}}) = Div(\bar{\mathcal{O}})/\mathcal{P}(\bar{\mathcal{O}})$$

is called the **replete divisor class group** *(or* **Arakelov class group***) of K.*

As $\mathcal{P}(\bar{\mathcal{O}})$ is discrete in $Div(\bar{\mathcal{O}})$, and is therefore in particular closed, $CH^1(\bar{\mathcal{O}})$ becomes a locally compact Hausdorff topological group with respect to the quotient topology. It is the correct analogue of the divisor class group of a function field (see chap. I, § 14). For the latter we introduced the degree map onto the group $\mathbb{Z}$; for $CH^1(\bar{\mathcal{O}})$ we obtain a **degree map** onto the group $\mathbb{R}$. It is induced by the continuous homomorphism

$$\deg : Div(\bar{\mathcal{O}}) \longrightarrow \mathbb{R}$$

which sends a replete divisor $D = \sum_{\mathfrak{p}} \nu_{\mathfrak{p}}\mathfrak{p}$ to the real number

$$\deg(D) = \sum_{\mathfrak{p}} \nu_{\mathfrak{p}} \log \mathfrak{N}(\mathfrak{p}) = \log\Big(\prod_{\mathfrak{p}} \mathfrak{N}(\mathfrak{p})^{\nu_{\mathfrak{p}}}\Big).$$

From the product formula (1.3), we find for any replete principal divisor $\mathrm{div}(f) \in \mathcal{P}(\bar{\mathcal{O}})$ that

$$\deg\big(\mathrm{div}(f)\big) = \sum_{\mathfrak{p}} \log \mathfrak{N}(\mathfrak{p})^{\nu_{\mathfrak{p}}(f)} = \log\Big(\prod_{\mathfrak{p}} |f|_{\mathfrak{p}}^{-1}\Big) = 0.$$

Thus we obtain a well-defined continuous homomorphism

$$\deg : CH^1(\bar{\mathcal{O}}) \longrightarrow \mathbb{R}.$$

The **kernel** $CH^1(\bar{\mathcal{O}})^0$ of this map is made up from the unit group $\mathcal{O}^*$ and the ideal class group $Cl_K \cong CH^1(\mathcal{O})$ of K as follows.

(1.11) Proposition. *Let $\Gamma = \lambda(\mathcal{O}^*)$ denote the complete lattice of units in trace-zero space $H = \{(x_\mathfrak{p}) \in \prod_{\mathfrak{p}|\infty} \mathbb{R} \mid \sum_{\mathfrak{p}|\infty} x_\mathfrak{p} = 0\}$. There is an exact sequence*

$$0 \longrightarrow H/\Gamma \longrightarrow CH^1(\bar{\mathcal{O}})^0 \longrightarrow CH^1(\mathcal{O}) \longrightarrow 0.$$

Proof: Let $Div(\bar{\mathcal{O}})^0$ be the kernel of $\deg : Div(\bar{\mathcal{O}}) \to \mathbb{R}$. Consider the exact sequence

$$0 \longrightarrow \prod_{\mathfrak{p}|\infty} \xrightarrow{\alpha} Div(\bar{\mathcal{O}}) \longrightarrow Div(\mathcal{O}) \longrightarrow 0,$$

where $\alpha((\nu_\mathfrak{p})) = \sum_{\mathfrak{p}|\infty} \frac{\nu_\mathfrak{p}}{f_\mathfrak{p}}\, \mathfrak{p}$. Restricting to $Div(\bar{\mathcal{O}})^0$ yields the exact commutative diagram

$$\begin{array}{ccccccccc}
0 & \longrightarrow & \lambda(\mathcal{O}^*) & \xrightarrow{\alpha} & \mathcal{P}(\bar{\mathcal{O}}) & \longrightarrow & \mathcal{P}(\mathcal{O}) & \longrightarrow & 0 \\
 & & \downarrow & & \downarrow & & \downarrow & & \\
0 & \longrightarrow & H & \xrightarrow{\alpha} & Div(\bar{\mathcal{O}})^0 & \longrightarrow & Div(\mathcal{O}) & \longrightarrow & 0.
\end{array}$$

Via the snake lemma (see [23], chap. III, § 3, (3.3)), this gives rise to the exact sequence

$$0 \longrightarrow H/\lambda(\mathcal{O}^*) \longrightarrow CH^1(\bar{\mathcal{O}})^0 \longrightarrow CH^1(\mathcal{O}) \longrightarrow 0.$$ □

The two fundamental facts of algebraic number theory, the finiteness of the class number and Dirichlet's unit theorem, now merge into (and are in fact equivalent to) the simple statement that the kernel $CH^1(\bar{\mathcal{O}})^0$ of the degree map $\deg : CH^1(\bar{\mathcal{O}}) \to \mathbb{R}$ is **compact**.

(1.12) Theorem. *The group $CH^1(\bar{\mathcal{O}})^0$ is compact.*

Proof: This follows immediately from the exact sequence

$$0 \longrightarrow H/\Gamma \longrightarrow CH^1(\bar{\mathcal{O}})^0 \longrightarrow CH^1(\mathcal{O}) \longrightarrow 0.$$

As Γ is a complete lattice in the $\mathbb{R}$-vector space H, the quotient H/Γ is a compact torus. In view of the finiteness of $CH^1(\mathcal{O})$, we obtain $CH^1(\bar{\mathcal{O}})^0$ as the union of the finitely many compact cosets of H/Γ in $CH^1(\bar{\mathcal{O}})^0$. Thus $CH^1(\bar{\mathcal{O}})^0$ itself is compact. □

The correspondence between replete ideals and replete divisors is given by the two mutually inverse mappings

$$\operatorname{div} : J(\bar{\mathcal{O}}) \longrightarrow Div(\bar{\mathcal{O}}), \qquad \operatorname{div}\Big(\prod_{\mathfrak{p}} \mathfrak{p}^{\nu_{\mathfrak{p}}}\Big) = \sum_{\mathfrak{p}} -\nu_{\mathfrak{p}}\mathfrak{p},$$

$$Div(\bar{\mathcal{O}}) \longrightarrow J(\bar{\mathcal{O}}), \qquad \sum_{\mathfrak{p}} \nu_{\mathfrak{p}}\mathfrak{p} \longmapsto \prod_{\mathfrak{p}} \mathfrak{p}^{-\nu_{\mathfrak{p}}}.$$

These are *topological isomorphisms* once we equip

$$J(\bar{\mathcal{O}}) = J(\mathcal{O}) \times \prod_{\mathfrak{p}|\infty} \mathbb{R}_+^*$$

with the product topology of the discrete topology on $J(\mathcal{O})$ and the canonical topology on $\prod_{\mathfrak{p}|\infty} \mathbb{R}_+^*$. The image of a divisor $D = \sum_{\mathfrak{p}} \nu_{\mathfrak{p}}\mathfrak{p}$ is also written as

$$\mathcal{O}(D) = \prod_{\mathfrak{p}} \mathfrak{p}^{-\nu_{\mathfrak{p}}}.$$

The minus sign here is motivated by classical usage in function theory. Replete principal ideals correspond to replete principal divisors in such a way that $P(\bar{\mathcal{O}})$ becomes a discrete subgroup of $J(\bar{\mathcal{O}})$ by (1.9), and $Pic(\bar{\mathcal{O}}) = J(\bar{\mathcal{O}})/P(\bar{\mathcal{O}})$ is a locally compact Hausdorff topological group. We obtain the following extension of chap. I, (12.14).

(1.13) Proposition. *The mapping* $\operatorname{div} : J(\bar{\mathcal{O}}) \xrightarrow{\sim} Div(\bar{\mathcal{O}})$ *induces a topological isomorphism*

$$\operatorname{div} : Pic(\bar{\mathcal{O}}) \xrightarrow{\sim} CH^1(\bar{\mathcal{O}}).$$

On the group $J(\bar{\mathcal{O}})$ we have the homomorphism $\mathfrak{N} : J(\bar{\mathcal{O}}) \to \mathbb{R}_+^*$, and on the group $Div(\bar{\mathcal{O}})$ there is the degree map $\deg : Div(\bar{\mathcal{O}}) \to \mathbb{R}$. They are obviously related by the formula

$$\deg\big(\operatorname{div}(\mathfrak{a})\big) = -\log \mathfrak{N}(\mathfrak{a}),$$

and we get a commutative diagram

$$\begin{array}{ccccccccc} 0 & \longrightarrow & Pic(\bar{\mathcal{O}})^0 & \longrightarrow & Pic(\bar{\mathcal{O}}) & \xrightarrow{\mathfrak{N}} & \mathbb{R}_+^* & \longrightarrow & 0 \\ & & \downarrow\cong & & {\scriptstyle\operatorname{div}}\downarrow\cong & & {\scriptstyle -\log}\downarrow\cong & & \\ 0 & \longrightarrow & CH^1(\bar{\mathcal{O}})^0 & \longrightarrow & CH^1(\bar{\mathcal{O}}) & \xrightarrow{\deg} & \mathbb{R} & \longrightarrow & 0 \end{array}$$

with exact rows. (1.12) now yields the

(1.14) Corollary. *The group*

$$Pic(\bar{\mathcal{O}})^0 = \{ [\mathfrak{a}] \in Pic(\bar{\mathcal{O}}) \mid \mathfrak{N}(\mathfrak{a}) = 1 \}$$

is compact.

The preceding isomorphism result (1.13) invites a philosophical reflection. Throughout the historical development of algebraic number theory, a controversy persisted between the followers of Dedekind's ideal-theoretic approach, and the divisor-theoretic method of building up the theory from the valuation-theoretic notion of primes. Both theories are equivalent in the sense of the above isomorphism result, but they are also fundamentally different in nature. The controversy has finally given way to the realization that neither approach is dominant, that each one instead emanates from its own proper world, and that the relation between these worlds is expressed by an important mathematical principle. However, all this becomes evident only in higher dimensional arithmetic algebraic geometry. There, on an algebraic $\mathbb{Z}$-scheme X, one studies on the one hand the totality of all *vector bundles*, and on the other, that of all *irreducible subschemes* of X. From the first, one constructs a series of groups $K_i(X)$ which constitute the subject of algebraic ***K*-theory**. From the second is constructed a series of groups $CH^i(X)$, the subject of **Chow theory**. Vector bundles are by definition locally free $\mathcal{O}_X$-modules. In the special case $X = \operatorname{Spec}(\mathcal{O})$ this includes the fractional ideals. The irreducible subschemes and their formal linear combinations, i.e., the **cycles** of X, are to be considered as generalizations of the primes and divisors. The isomorphism between divisor class group and ideal class group extends to the general setting as a homomorphic relation between the groups $CH^i(X)$ and $K_i(X)$. Thus the initial controversy has been resolved into a seminal mathematical theory (for further reading, see [13]).

Exercise 1 (Strong Approximation Theorem). Let S be a finite set of primes and let $\mathfrak{p}_0$ be another prime of K which does not belong to S. Let $a_\mathfrak{p} \in K$ be given numbers, for $\mathfrak{p} \in S$. Then for every $\varepsilon > 0$, there exists an $x \in K$ such that

$$|x - a_\mathfrak{p}|_\mathfrak{p} < \varepsilon \text{ for } \mathfrak{p} \in S, \text{ and } |x|_\mathfrak{p} \le 1 \text{ for } \mathfrak{p} \notin S \cup \{\mathfrak{p}_0\}.$$

Exercise 2. Let K be totally real, i.e., $K_\mathfrak{p} = \mathbb{R}$ for all $\mathfrak{p}|\infty$. Let T be a proper nonempty subset of $\operatorname{Hom}(K, \mathbb{R})$. Then there exists a unit ε of K satisfying $\tau\varepsilon > 1$ for $\tau \in S$ and $0 < \tau\varepsilon < 1$ for $\tau \notin S$.

Exercise 3. Show that the absolute norm $\mathfrak{N} : Pic(\bar{\mathbb{Z}}) \to \mathbb{R}_+^*$ is an isomorphism.

Exercise 4. Let K and L be number fields, and let $\tau : K \to L$ be a homomorphism.

Given any replete divisor $D = \sum_\mathfrak{P} \nu_\mathfrak{P} \mathfrak{P}$ of L, define a replete divisor of K by the rule

$$\tau_*(D) = \sum_\mathfrak{p} \Big(\sum_{\mathfrak{P}|\mathfrak{p}} \nu_\mathfrak{P} f_{\mathfrak{P}|\mathfrak{p}} \Big) \mathfrak{p},$$

where $f_{\mathfrak{P}|\mathfrak{p}}$ is the inertia degree of $\mathfrak{P}$ over τK and $\mathfrak{P}|\mathfrak{p}$ signifies $\tau\mathfrak{p} = \mathfrak{P}|_{\tau K}$. Show that τ_* induces a homomorphism

$$\tau_* : CH^1(\bar{o}_L) \to CH^1(\bar{o}_K).$$

Exercise 5. Given any replete divisor $D = \sum_{\mathfrak{p}} \nu_{\mathfrak{p}}\mathfrak{p}$ of K, define a replete divisor of L by the rule

$$\tau^*(D) = \sum_{\mathfrak{p}} \sum_{\mathfrak{P}|\mathfrak{p}} \nu_{\mathfrak{p}} e_{\mathfrak{P}|\mathfrak{p}} \mathfrak{P},$$

where $e_{\mathfrak{P}|\mathfrak{p}}$ denotes the ramification index of $\mathfrak{P}$ over K. Show that τ^* induces a homomorphism

$$\tau^* : CH^1(\bar{o}_K) \to CH^1(\bar{o}_L).$$

Exercise 6. Show that $\tau_* \circ \tau^* = [L : K]$ and that

$$\deg(\tau_* D) = \deg(D), \quad \deg(\tau^* D) = [L : K]\deg(D).$$

§ 2. Different and Discriminant

It is our intention to develop a framework for the theory of algebraic number fields which shows the complete analogy with the theory of function fields. This goal leads us naturally to the notions of different and discriminant, as we shall explain in § 3 and § 7. They control the ramification behaviour of an extension of valued fields.

Let $L|K$ be a finite separable field extension, $o \subseteq K$ a Dedekind domain with field of fractions K, and let $\mathcal{O} \subseteq L$ be its integral closure in L. Throughout this section we assume systematically that the residue field extensions $\lambda|\kappa$ of $\mathcal{O}|o$ are separable. The theory of the different originates from the fact that we are given a canonical nondegenerate symmetric bilinear form on the K-vector space L, *viz.*, the **trace form**

$$T(x, y) = Tr(xy)$$

(see chap. I, § 2). It allows us to associate to every fractional ideal $\mathfrak{A}$ of L the **dual** $\mathcal{O}$-module

$$^*\mathfrak{A} = \{x \in L \mid Tr(x\,\mathfrak{A}) \subseteq o\}.$$

It is again a fractional ideal. For if $\alpha_1, \dots, \alpha_n \in \mathcal{O}$ is a basis of $L|K$ and $d = \det(Tr(\alpha_i\alpha_j))$ its discriminant, then $ad^*\mathfrak{A} \subseteq \mathcal{O}$ for every nonzero $a \in \mathfrak{A} \cap o$. Indeed, if $x = x_1\alpha_1 + \dots + x_n\alpha_n \in {}^*\mathfrak{A}$, with $x_i \in K$, then the ax_i satisfy the system of linear equations $\sum_{i=1}^n ax_i\, Tr(\alpha_i\alpha_j) = Tr(xa\alpha_j) \in o$. This implies $dax_i \in o$ and thus $dax \in \mathcal{O}$.

The notion of duality is justified by the isomorphism

$$^*\mathfrak{A} \xrightarrow{\sim} \mathrm{Hom}_{\mathcal{o}}(\mathfrak{A}, \mathcal{o}), \quad x \longmapsto \big(y \mapsto Tr(xy)\big)\,.$$

Indeed, since every $\mathcal{o}$-homomorphism $f : \mathfrak{A} \to \mathcal{o}$ extends uniquely to a K-homomorphism $f : L \to K$ in view of $\mathfrak{A}K = L$, we may consider $\mathrm{Hom}_{\mathcal{o}}(\mathfrak{A}, \mathcal{o})$ as a submodule of $\mathrm{Hom}_K(L, K)$, namely, the image of $^*\mathfrak{A}$ with respect to $L \to \mathrm{Hom}_K(L, K)$, $x \mapsto (y \mapsto Tr(xy))$. The module dual to $\mathcal{O}$,

$$^*\mathcal{O} \;\cong\; \mathrm{Hom}_{\mathcal{o}}(\mathcal{O}, \mathcal{o}),$$

will obviously occupy a distinguished place in this theory.

(2.1) Definition. *The fractional ideal*

$$\mathfrak{C}_{\mathcal{O}|\mathcal{o}} = {}^*\mathcal{O} = \big\{x \in L \bigm| Tr(x\mathcal{O}) \subseteq \mathcal{o}\big\}$$

is called **Dedekind's complementary module**, *or the* **inverse different**. *Its inverse,*

$$\mathfrak{D}_{\mathcal{O}|\mathcal{o}} = \mathfrak{C}_{\mathcal{O}|\mathcal{o}}^{-1},$$

is called the **different** *of $\mathcal{O}|\mathcal{o}$.*

As $\mathfrak{C}_{\mathcal{O}|\mathcal{o}} \supseteq \mathcal{O}$, the ideal $\mathfrak{D}_{\mathcal{O}|\mathcal{o}} \subseteq \mathcal{O}$ is actually an integral ideal of L. We will frequently denote it by $\mathfrak{D}_{L|K}$, provided the intended subrings $\mathcal{o}, \mathcal{O}$ are evident from the context. In the same way, we write $\mathfrak{C}_{L|K}$ instead of $\mathfrak{C}_{\mathcal{O}|\mathcal{o}}$. The different behaves in the following manner under change of rings $\mathcal{o}$ and $\mathcal{O}$.

(2.2) Proposition.

(i) *For a tower of fields $K \subseteq L \subseteq M$, one has $\mathfrak{D}_{M|K} = \mathfrak{D}_{M|L}\mathfrak{D}_{L|K}$.*

(ii) *For any multiplicative subset S of $\mathcal{o}$, one has $\mathfrak{D}_{S^{-1}\mathcal{O}|S^{-1}\mathcal{o}} = S^{-1}\mathfrak{D}_{\mathcal{O}|\mathcal{o}}$.*

(iii) *If $\mathfrak{P}|\mathfrak{p}$ are prime ideals of $\mathcal{O}$, resp. $\mathcal{o}$, and $\mathcal{O}_{\mathfrak{P}}|\mathcal{o}_{\mathfrak{p}}$ are the associated completions, then*

$$\mathfrak{D}_{\mathcal{O}|\mathcal{o}}\mathcal{O}_{\mathfrak{P}} = \mathfrak{D}_{\mathcal{O}_{\mathfrak{P}}|\mathcal{o}_{\mathfrak{p}}}\,.$$

Proof: (i) Let $A = \mathcal{o} \subseteq K$, and let $B \subseteq L$, $C \subseteq M$ be the integral closure of $\mathcal{o}$ in L, resp. M. It then suffices to show that

$$\mathfrak{C}_{C|A} = \mathfrak{C}_{C|B}\mathfrak{C}_{B|A}\,.$$

The inclusion $\supseteq$ follows from

$$\begin{aligned} Tr_{M|K}(\mathfrak{C}_{C|B}\mathfrak{C}_{B|A}C) &= Tr_{L|K}\, Tr_{M|L}(\mathfrak{C}_{C|B}\mathfrak{C}_{B|A}C) \\ &= Tr_{L|K}\big(\mathfrak{C}_{B|A}\, Tr_{M|L}(\mathfrak{C}_{C|B}C)\big) \subseteq A\,. \end{aligned}$$

In view of $BC = C$, the inclusion $\subseteq$ is derived as follows:

$$Tr_{M|K}(\mathfrak{C}_{C|A}C) = Tr_{L|K}\big(B\, Tr_{M|L}(\mathfrak{C}_{C|A}C)\big) \subseteq A\,,$$

so that $Tr_{M|L}(\mathfrak{C}_{C|A}C) \subseteq \mathfrak{C}_{B|A}$, and thus

$$Tr_{M|L}(\mathfrak{C}_{B|A}^{-1}\mathfrak{C}_{C|A}C) = \mathfrak{C}_{B|A}^{-1}\, Tr_{M|L}(\mathfrak{C}_{C|A}C) \subseteq B\,.$$

This does indeed imply $\mathfrak{C}_{B|A}^{-1}\mathfrak{C}_{C|A} \subseteq \mathfrak{C}_{C|B}$, and so $\mathfrak{C}_{C|A} \subseteq \mathfrak{C}_{C|B}\mathfrak{C}_{B|A}$.

(ii) is trivial.

(iii) By (ii) we may assume that $\mathcal{o}$ is a discrete valuation ring. We show that $\mathfrak{C}_{\mathcal{O}|\mathcal{o}}$ is dense in $\mathfrak{C}_{\mathcal{O}_{\mathfrak{P}}|\mathcal{o}_{\mathfrak{p}}}$. In order to do this, we use the formula

$$Tr_{L|K} = \sum_{\mathfrak{P}|\mathfrak{p}} Tr_{L_{\mathfrak{P}}|K_{\mathfrak{p}}}$$

(see chap. II, (8.4)). Let $x \in \mathfrak{C}_{\mathcal{O}|\mathcal{o}}$ and $y \in \mathcal{O}_{\mathfrak{P}}$. The approximation theorem allows us to find an η in L which is close to y with respect to $v_{\mathfrak{P}}$, and close to 0 with respect to $v_{\mathfrak{P}'}$, for $\mathfrak{P}'|\mathfrak{p}$, $\mathfrak{P}' \neq \mathfrak{P}$. The left-hand side of the equation

$$Tr_{L|K}(x\eta) = Tr_{L_{\mathfrak{P}}|K_{\mathfrak{p}}}(x\eta) + \sum_{\mathfrak{P}' \neq \mathfrak{P}} Tr_{L_{\mathfrak{P}'}|K_{\mathfrak{p}}}(x\eta)$$

then belongs to $\mathcal{o}_{\mathfrak{p}}$, since $Tr_{L|K}(x\eta) \in \mathcal{o} \subseteq \mathcal{o}_{\mathfrak{p}}$, but the same is true of the elements $Tr_{L_{\mathfrak{P}'}|K_{\mathfrak{p}}}(x\eta)$ because they are close to zero with respect to $v_{\mathfrak{p}}$. Therefore $Tr_{L_{\mathfrak{P}}|K_{\mathfrak{p}}}(xy) \in \mathcal{o}_{\mathfrak{p}}$. This shows that $\mathfrak{C}_{\mathcal{O}|\mathcal{o}} \subseteq \mathfrak{C}_{\mathcal{O}_{\mathfrak{P}}|\mathcal{o}_{\mathfrak{p}}}$.

If on the other hand $x \in \mathfrak{C}_{\mathcal{O}_{\mathfrak{P}}|\mathcal{o}_{\mathfrak{p}}}$, and if $\xi \in L$ is sufficiently close to x with respect to $v_{\mathfrak{P}}$, and sufficiently close to 0 with respect to $v_{\mathfrak{P}'}$, for $\mathfrak{P}' \neq \mathfrak{P}$, then $\xi \in \mathfrak{C}_{\mathcal{O}|\mathcal{o}}$. In fact, if $y \in \mathcal{O}$, then $Tr_{L_{\mathfrak{P}}|K_{\mathfrak{p}}}(\xi y) \in \mathcal{o}_{\mathfrak{p}}$, since $Tr_{L_{\mathfrak{P}}|K_{\mathfrak{p}}}(xy) \in \mathcal{o}_{\mathfrak{p}}$. Likewise $Tr_{L_{\mathfrak{P}'}|K_{\mathfrak{p}}}(\xi y) \in \mathcal{o}_{\mathfrak{p}}$ for $\mathfrak{P}'|\mathfrak{P}$, because these elements are close to 0. Therefore $Tr_{L|K}(\xi y) \in \mathcal{o}_{\mathfrak{p}} \cap K = \mathcal{o}$, i.e., $\xi \in \mathfrak{C}_{\mathcal{O}|\mathcal{o}}$. This shows that $\mathfrak{C}_{\mathcal{O}|\mathcal{o}}$ is dense in $\mathfrak{C}_{\mathcal{O}_{\mathfrak{P}}|\mathcal{o}_{\mathfrak{p}}}$, in other words, $\mathfrak{C}_{\mathcal{O}|\mathcal{o}}\mathcal{O}_{\mathfrak{P}} = \mathfrak{C}_{\mathcal{O}_{\mathfrak{P}}|\mathcal{o}_{\mathfrak{p}}}$, and so $\mathfrak{D}_{\mathcal{O}|\mathcal{o}}\mathcal{O}_{\mathfrak{P}} = \mathfrak{D}_{\mathcal{O}_{\mathfrak{P}}|\mathcal{o}_{\mathfrak{p}}}$. □

If we put $\mathfrak{D} = \mathfrak{D}_{L|K}$ and $\mathfrak{D}_{\mathfrak{P}} = \mathfrak{D}_{L_{\mathfrak{P}}|K_{\mathfrak{p}}}$, and consider $\mathfrak{D}_{\mathfrak{P}}$ at the same time as an ideal of $\mathcal{O}$ (i.e., as the ideal $\mathcal{O} \cap \mathfrak{D}_{\mathfrak{P}}$), then (2.2), (iii) gives us the

(2.3) Corollary. $\mathfrak{D} = \prod_{\mathfrak{P}} \mathfrak{D}_{\mathfrak{P}}$.

The name "different" is explained by the following explicit description, which was Dedekind's original way to define it. Let $\alpha \in \mathcal{O}$ and let $f(X) \in o[X]$ be the minimal polynomial of α. We define the **different of the element** α by

$$\delta_{L|K}(\alpha) = \begin{cases} f'(\alpha) & \text{if } L = K(\alpha), \\ 0 & \text{if } L \neq K(\alpha). \end{cases}$$

In the special case where $\mathcal{O} = o[\alpha]$ we then obtain:

(2.4) Proposition. *If $\mathcal{O} = o[\alpha]$, then the different is the principal ideal*

$$\mathfrak{D}_{L|K} = \big(\delta_{L|K}(\alpha)\big).$$

Proof: Let $f(X) = a_0 + a_1 X + \cdots + a_n X^n$ be the minimal polynomial of α and

$$\frac{f(X)}{X-\alpha} = b_0 + b_1 X + \cdots + b_{n-1} X^{n-1}.$$

The dual basis of $1, \alpha, \ldots, \alpha^{n-1}$ with respect to $Tr(xy)$ is then given by

$$\frac{b_0}{f'(\alpha)}, \ldots, \frac{b_{n-1}}{f'(\alpha)}.$$

For if $\alpha_1, \ldots, \alpha_n$ are the roots of f, then one has

$$\sum_{i=1}^{n} \frac{f(X)}{X-\alpha_i} \frac{\alpha_i^r}{f'(\alpha_i)} = X^r, \qquad 0 \le r \le n-1,$$

as the difference of the two sides is a polynomial of degree $\le n-1$ with roots $\alpha_1, \ldots, \alpha_n$, so is identically zero. We may write this equation in the form

$$Tr\left[\frac{f(X)}{X-\alpha} \frac{\alpha^r}{f'(\alpha)}\right] = X^r.$$

Considering now the coefficient of each of the powers of X, we obtain

$$Tr\left(\alpha^i \frac{b_j}{f'(\alpha)}\right) = \delta_{ij}$$

and the claim follows.

As $\mathcal{O} = o + o\alpha + \cdots + o\alpha^{n-1}$, we get

$$\mathfrak{C}_{\mathcal{O}|o} = f'(\alpha)^{-1}(ob_0 + \cdots + ob_{n-1}).$$

From the recursive formulas

$$\begin{aligned} b_{n-1} &= 1, \\ b_{n-2} - \alpha b_{n-1} &= a_{n-1}, \\ \cdots &= \cdots \end{aligned}$$

it follows that

$$b_{n-i} = \alpha^{i-1} + a_{n-1}\alpha^{i-2} + \cdots + a_{n-i+1},$$

so that $\mathcal{O}b_0 + \cdots + \mathcal{O}b_{n-1} = \mathcal{O}[\alpha] = \mathcal{O}$; then $\mathfrak{C}_{\mathcal{O}|\mathcal{O}} = f'(\alpha)^{-1}\mathcal{O}$, and thus $\mathfrak{D}_{L|K} = (f'(\alpha))$. □

The proof shows that the module ${}^*\mathcal{O}[\alpha] = \{x \in L \mid Tr_{L|K}(x\mathcal{O}[\alpha]) \subseteq \mathcal{O}\}$, which is the dual of the $\mathcal{O}$-module $\mathcal{O}[\alpha]$, always admits the $\mathcal{O}$-basis $\alpha^i/f'(\alpha)$, $i = 0, \ldots, n-1$. We exploit this for the following characterization of the different in the general case where $\mathcal{O}$ need not be monogenous.

(2.5) Theorem. *The different $\mathfrak{D}_{L|K}$ is the ideal generated by all differents of elements $\delta_{L|K}(\alpha)$ for $\alpha \in \mathcal{O}$.*

Proof: Let $\alpha \in \mathcal{O}$ such that $L = K(\alpha)$, and let $f(X)$ be the minimal polynomial of α. In order to show that $f'(\alpha) \in \mathfrak{D}_{L|K}$, we consider the "conductor" $\mathfrak{f} = \mathfrak{f}_{\mathcal{O}[\alpha]} = \{x \in L \mid x\mathcal{O} \subseteq \mathcal{O}[\alpha]\}$ of $\mathcal{O}[\alpha]$ (see chap. I, § 12, p. 79). On putting $b = f'(\alpha)$, we have for $x \in L$:

$$\begin{aligned} x \in \mathfrak{f} \iff x\mathcal{O} \subseteq \mathcal{O}[\alpha] &\iff b^{-1}x\mathcal{O} \subseteq b^{-1}\mathcal{O}[\alpha] = {}^*\mathcal{O}[\alpha] \\ \iff Tr(b^{-1}x\mathcal{O}) \subseteq \mathcal{O} &\iff b^{-1}x \in \mathfrak{D}_{L|K}^{-1} \iff x \in b\mathfrak{D}_{L|K}^{-1}. \end{aligned}$$

Therefore $(f'(\alpha)) = \mathfrak{f}_{\mathcal{O}[\alpha]}\mathfrak{D}_{L|K}$, so in particular, $f'(\alpha) \in \mathfrak{D}_{L|K}$.

$\mathfrak{D}_{L|K}$ thus divides all the differents of elements $\delta_{L|K}(\alpha)$. In order to prove that $\mathfrak{D}_{L|K}$ is in fact the greatest common divisor of all $\delta_{L|K}(\alpha)$, it suffices to show that, for every prime ideal $\mathfrak{P}$, there exists an $\alpha \in \mathcal{O}$ such that $L = K(\alpha)$ and $v_{\mathfrak{P}}(\mathfrak{D}_{L|K}) = v_{\mathfrak{P}}(f'(\alpha))$.

We think of L as embedded into the separable closure $\overline{K}_{\mathfrak{p}}$ of $K_{\mathfrak{p}}$ in such a way that the absolute value $|\ |$ of $\overline{K}_{\mathfrak{p}}$ defines the prime $\mathfrak{P}$.

By chap. II, (10.4), we find an element β in the valuation ring $\mathcal{O}_{\mathfrak{P}}$ of the completion $L_{\mathfrak{P}}$ satisfying $\mathcal{O}_{\mathfrak{P}} = \mathcal{O}_{\mathfrak{p}}[\beta]$, and the proof *loc. cit.* shows that, for every element $\alpha \in \mathcal{O}_{\mathfrak{P}}$ which is sufficiently close to β, one also has $\mathcal{O}_{\mathfrak{P}} = \mathcal{O}_{\mathfrak{p}}[\alpha]$. From (2.2), (iii) and (2.4), it follows that

$$v_{\mathfrak{P}}(\mathfrak{D}_{L|K}) = v_{\mathfrak{P}}(\mathfrak{D}_{L_{\mathfrak{P}}|K_{\mathfrak{p}}}) = v_{\mathfrak{P}}\big(\delta_{L_{\mathfrak{P}}|K_{\mathfrak{p}}}(\alpha)\big).$$

It therefore suffices to show that we can find an element α in $\mathcal{O}$ such that $L = K(\alpha)$ and

$$v_{\mathfrak{P}}\big(\delta_{L_{\mathfrak{P}}|K_{\mathfrak{p}}}(\alpha)\big) = v_{\mathfrak{P}}\big(\delta_{L|K}(\alpha)\big).$$

For this, let $\sigma_2, \ldots, \sigma_r : L \to \overline{K}_{\mathfrak{p}}$ be K-embeddings giving the primes $\mathfrak{P}_i|\mathfrak{p}$ different from $\mathfrak{P}$. Let $a \in \mathcal{O}_{\mathfrak{p}}$ be an element such that

$$(*) \qquad |\tau\beta - a| = 1 \quad \text{for all} \quad \tau \in G_{\mathfrak{p}} = G(\overline{K}_{\mathfrak{p}}|K_{\mathfrak{p}}).$$

(Choose $a = 1$, resp. $a = 0$, according as the residue classes $\tau\beta \bmod \mathfrak{P}$ which are conjugate over $\mathcal{O}_\mathfrak{p}/\mathfrak{p}$ are zero or not.) Using the Chinese remainder theorem, we now pick an $\alpha \in \mathcal{O}$ such that $|\alpha - \beta|$ and $|\sigma_i\alpha - a|$, for $i = 2, \dots, r$, are very small. We may even assume that $L = K(\alpha)$ (if not, modify by $\alpha + \pi^\nu\gamma$, $\pi \in \mathfrak{p}$, for ν big, $\gamma \in \mathcal{O}$, $L = K(\gamma)$; for suitable $\nu \neq \mu$, one then finds $K(\alpha + \pi^\nu\gamma) = K(\alpha + \pi^\mu\gamma) = K(\gamma)$). Since α is close to β, we have $\mathcal{O}_\mathfrak{P} = \mathcal{O}_\mathfrak{p}[\alpha]$. Now

$$\delta_{L_\mathfrak{P}|K_\mathfrak{p}}(\alpha) = \prod_{\tau \neq 1}(\alpha - \tau\alpha),$$

where τ runs through the $K_\mathfrak{p}$-embeddings $L_\mathfrak{P} \to \overline{K}_\mathfrak{p}$ different from 1. Furthermore,

$$\delta_{L|K}(\alpha) = \prod_{\sigma \neq 1}(\alpha - \sigma\alpha) = \prod_{\tau \neq 1}(\alpha - \tau\alpha)\prod_{i=2}^{r}\prod_{j}(\alpha - \tau_{ij}\sigma_i\alpha),$$

where σ runs over the K-embeddings different from 1, and the τ_{ij} are certain elements in $G_\mathfrak{p}$. But now

$$|\alpha - \tau_{ij}\sigma_i\alpha| = |\tau_{ij}^{-1}\alpha - \sigma_i\alpha| = |\tau_{ij}^{-1}\alpha - a + a - \sigma_i\alpha| = 1,$$

since $|a - \sigma_i\alpha|$ is very small, and $\tau_{ij}^{-1}\alpha$ is very close to $\tau_{ij}^{-1}\beta$ (see $(*)$). Therefore $v_\mathfrak{P}(\delta_{L|K}(\alpha)) = v_\mathfrak{P}(\prod_{\tau\neq 1}(\alpha - \tau\alpha)) = v_\mathfrak{P}(\delta_{L_\mathfrak{P}|K_\mathfrak{p}}(\alpha))$, as required. □

The different characterizes the ramification behaviour of the extension $L|K$ as follows.

(2.6) Theorem. *A prime ideal $\mathfrak{P}$ of L is ramified over K if and only if $\mathfrak{P}|\mathfrak{D}_{L|K}$.*

Let $\mathfrak{P}^s$ be the maximal power of $\mathfrak{P}$ dividing $\mathfrak{D}_{L|K}$, and let e be the ramification index of $\mathfrak{P}$ over K. Then one has

$$s = e - 1, \qquad \textit{if } \mathfrak{P} \textit{ is tamely ramified,}$$

$$e \leq s \leq e - 1 + v_\mathfrak{P}(e), \qquad \textit{if } \mathfrak{P} \textit{ is wildly ramified.}$$

Proof: By (2.2), (iii), we may assume that $\mathcal{O}$ is a complete discrete valuation ring with maximal ideal $\mathfrak{p}$. Then, by chap. II, (10.4), we have $\mathcal{O} = \mathcal{o}[\alpha]$ for a suitable $\alpha \in \mathcal{O}$. Let $f(X)$ be the minimal polynomial of α. (2.4) says that $s = v_\mathfrak{P}(f'(\alpha))$. Assume $L|K$ is unramified. Then $\overline{\alpha} = \alpha \bmod \mathfrak{P}$ is a simple zero of $\overline{f}(X) = f(X) \bmod \mathfrak{p}$, so that $f'(\alpha) \in \mathcal{O}^*$ and thus $s = 0 = e - 1$.

By (2.2), (i) and chap. II, (7.5), we may now pass to the maximal unramified extension and assume that $L|K$ is totally ramified. Then α may be chosen to be a prime element of $\mathcal{O}$. In this case the minimal polynomial

$$f(X) = a_0X^e + a_1X^{e-1} + \cdots + a_e\,, \quad a_0 = 1\,,$$

is an Eisenstein polynomial. Let us look at the derivative

$$f'(\alpha) = ea_0\alpha^{e-1} + (e-1)a_1\alpha^{e-2} + \cdots + a_{e-1}\,.$$

For $i = 0, \dots, e-1$, we find

$$v_{\mathfrak{P}}\big((e-i)a_i\alpha^{e-i-1}\big) = ev_{\mathfrak{p}}(e-i) + ev_{\mathfrak{p}}(a_i) + e - i - 1 \equiv -i-1 \bmod e\,,$$

so that the individual terms of $f'(\alpha)$ have distinct valuations. Therefore

$$s = v_{\mathfrak{P}}\big(f'(\alpha)\big) = \min_{0\le i<e}\big\{v_{\mathfrak{P}}((e-i)a_i\alpha^{e-i-1})\big\}\,.$$

If now $L|K$ is tamely ramified, i.e., if $v_{\mathfrak{p}}(e) = 0$, then the minimum is obviously equal to $e-1$, and for $v_{\mathfrak{p}}(e) \ge 1$, we deduce that $e \le s \le v_{\mathfrak{P}}(e) + e - 1$. □

The geometric significance of the different, and thus also the way it fits into higher dimensional algebraic geometry, is brought out by the following characterization, which is due to *E. Kähler*. For an arbitrary extension $B|A$ of commutative rings, consider the homomorphism

$$\mu : B \otimes_A B \longrightarrow B\,, \quad x \otimes y \longmapsto xy\,,$$

whose kernel we denote by I. Then

$$\Omega^1_{B|A} := I/I^2 = I \otimes_{B\otimes B} B$$

is a $B \otimes B$-module, and hence in particular also a B-module, via the embedding $B \to B \otimes B$, $b \mapsto b \otimes 1$. It is called the **module of differentials** of $B|A$, and its elements are called **Kähler differentials**. If we put

$$dx = x \otimes 1 - 1 \otimes x \bmod I^2\,,$$

then we obtain a mapping

$$d : B \longrightarrow \Omega^1_{B|A}$$

satisfying

$$d(xy) = x\,dy + y\,dx\,,$$

$$da = 0 \quad \text{for} \quad a \in A\,.$$

Such a map is called a **derivation** of $B|A$. One can show that d is universal among all derivations of $B|A$ with values in B-modules. $\Omega^1_{B|A}$ consists of the linear combinations $\sum y_i\,dx_i$. The link with the different is now this.

(2.7) Proposition. *The different* $\mathfrak{D}_{\mathcal{O}|o}$ *is the* **annihilator** *of the* $\mathcal{O}$-*module* $\Omega^1_{\mathcal{O}|o}$, *i.e.,*

$$\mathfrak{D}_{\mathcal{O}|o} = \{x \in \mathcal{O} \mid x\,dy = 0 \quad \textit{for all} \quad y \in \mathcal{O}\}\,.$$

Proof: For greater notational clarity, let us put $\mathcal{O} = B$ and $o = A$. If A' is any commutative A-algebra and $B' = B \otimes_A A'$, then it is easy to see that $\Omega^1_{B'|A'} = \Omega^1_{B|A} \otimes_A A'$. Thus the module of differentials is preserved under localization and completion, and we may therefore assume that A is a complete discrete valuation ring. Then we find by chap. II, (10.4), that $B = A[x]$, and if $f(X) \in A[X]$ is the minimal polynomial of x, then $\Omega_{B|A}$ is generated by dx (exercise 3). The annihilator of dx is $f'(x)$. On the other hand, by (2.4) we have $\mathfrak{D}_{B|A} = (f'(x))$. This proves the claim. □

A most intimate connection holds between the different and the discriminant of $\mathcal{O}|o$. The latter is defined as follows.

(2.8) Definition. *The* **discriminant** $\mathfrak{d}_{\mathcal{O}|o}$ *is the ideal of* o *which is generated by the discriminants* $d(\alpha_1, \dots, \alpha_n)$ *of all the bases* $\alpha_1, \dots, \alpha_n$ *of* $L|K$ *which are contained in* $\mathcal{O}$.

We will frequently write $\mathfrak{d}_{L|K}$ instead of $\mathfrak{d}_{\mathcal{O}|o}$. If $\alpha_1, \dots, \alpha_n$ is an integral basis of $\mathcal{O}|o$, then $\mathfrak{d}_{L|K}$ is the principal ideal generated by $d(\alpha_1, \dots, \alpha_n) = d_{L|K}$, because all other bases contained in $\mathcal{O}$ are transforms of the given one by matrices with entries in o. The discriminant is obtained from the different by taking the norm $N_{L|K}$ (see § 1).

(2.9) Theorem. *The following relation exists between the discriminant and the different:*

$$\mathfrak{d}_{L|K} = N_{L|K}(\mathfrak{D}_{L|K})\,.$$

Proof: If S is a multiplicative subset of o, then clearly $\mathfrak{d}_{S^{-1}\mathcal{O}|S^{-1}o} = S^{-1}\mathfrak{d}_{\mathcal{O}|o}$ and $\mathfrak{D}_{S^{-1}\mathcal{O}|S^{-1}o} = S^{-1}\mathfrak{D}_{\mathcal{O}|o}$. We may therefore assume that o is a discrete valuation ring. Then, since o is a principal ideal domain, so is $\mathcal{O}$ (see chap. I, § 3, exercise 4), and it admits an integral basis $\alpha_1, \dots, \alpha_n$ (see chap. I, (2.10)). So we have $\mathfrak{d}_{L|K} = (d(\alpha_1, \dots, \alpha_n))$. Dedekind's complementary module $\mathfrak{C}_{\mathcal{O}|o}$ is generated by the dual basis $\alpha'_1, \dots, \alpha'_n$ which is characterized by $Tr_{L|K}(\alpha_i\alpha'_j) = \delta_{ij}$. On the other hand, $\mathfrak{C}_{\mathcal{O}|o}$ is a principal ideal (β) and admits the o-basis $\beta\alpha_1, \dots, \beta\alpha_n$ of discriminant

$$d(\beta\alpha_1, \dots, \beta\alpha_n) = N_{L|K}(\beta)^2 d(\alpha_1, \dots, \alpha_n)\,.$$

But $(N_{L|K}(\beta)) = N_{L|K}(\mathfrak{C}_{\mathcal{O}|o}) = N_{L|K}(\mathfrak{D}_{L|K}^{-1}) = N_{L|K}(\mathfrak{D}_{L|K})^{-1}$, and $(d(\alpha_1, \ldots, \alpha_n)) = \mathfrak{d}_{L|K}$. One has $d(\alpha_1, \ldots, \alpha_n) = \det((\sigma_i\alpha_j))^2$, $d(\alpha_1', \ldots, \alpha_n') = \det((\sigma_i\alpha_j'))^2$, for $\sigma_i \in \mathrm{Hom}_K(L, \overline{K})$, and $Tr(\alpha_i\alpha_j') = \delta_{ij}$. Then $d(\alpha_1, \ldots, \alpha_n) \cdot d(\alpha_1', \ldots, \alpha_n') = 1$. Combining these yields

$$\begin{aligned}\mathfrak{d}_{L|K}^{-1} &= (d(\alpha_1, \ldots, \alpha_n)^{-1}) = (d(\alpha_1', \ldots, \alpha_n')) = (d(\beta\alpha_1, \ldots, \beta\alpha_n)) \\ &= N_{L|K}(\mathfrak{D}_{L|K})^{-2}\mathfrak{d}_{L|K}\end{aligned}$$

and hence $N_{L|K}(\mathfrak{D}_{L|K}) = \mathfrak{d}_{L|K}$. □

(2.10) Corollary. *For a tower of fields $K \subseteq L \subseteq M$, one has*

$$\mathfrak{d}_{M|K} = \mathfrak{d}_{L|K}^{[M:L]} N_{L|K}(\mathfrak{d}_{M|L}).$$

Proof: Applying to $\mathfrak{D}_{M|K} = \mathfrak{D}_{M|L}\mathfrak{D}_{L|K}$ the norm $N_{M|K} = N_{L|K} \circ N_{M|L}$, (1.6) gives

$$\mathfrak{d}_{M|K} = N_{L|K}(\mathfrak{d}_{M|L})N_{L|K}(\mathfrak{D}_{L|K}^{[M:L]}) = N_{L|K}(\mathfrak{d}_{M|L})\mathfrak{d}_{L|K}^{[M:L]}.$$ □

Putting $\mathfrak{d} = \mathfrak{d}_{L|K}$ and $\mathfrak{d}_{\mathfrak{P}} = \mathfrak{d}_{L_{\mathfrak{P}}|K_{\mathfrak{p}}}$ and viewing $\mathfrak{d}_{\mathfrak{P}}$ also as the ideal $\mathfrak{d}_{\mathfrak{P}} \cap o$ of K, the product formula (2.3) for the different, together with theorem (2.9), yields:

(2.11) Corollary. $\mathfrak{d} = \prod_{\mathfrak{P}} \mathfrak{d}_{\mathfrak{P}}$.

The extension $L|K$ is called **unramified** if all prime ideals $\mathfrak{p}$ of K are unramified. This amounts to requiring that all primes of K be unramified. In fact, the infinite primes are always to be regarded as unramified because $e_{\mathfrak{P}|\mathfrak{p}} = 1$.

(2.12) Corollary. *A prime ideal $\mathfrak{p}$ of K is ramified in L if and only if $\mathfrak{p}|\mathfrak{d}$. In particular, the extension $L|K$ is unramified if the discriminant $\mathfrak{d} = (1)$.*

Combining this result with Minkowski theory leads to two important theorems on unramified extensions of number fields which belong to the classical body of algebraic number theory. The first of these results is the following.

(2.13) Theorem. *Let K be an algebraic number field and let S be a finite set of primes of K. Then there exist only finitely many extensions $L|K$ of given degree n which are unramified outside of S.*

Proof: If $L|K$ is an extension of degree n which is unramified outside of S, then, by (2.12) and (2.6), its discriminant $\mathfrak{d}_{L|K}$ is one of the finite number of divisors of the ideal $\mathfrak{a} = \prod_{\substack{\mathfrak{p}\in S \\ \mathfrak{p}\nmid\infty}} \mathfrak{p}^{n(1+n)}$. It therefore suffices to show that there are only finitely many extensions $L|K$ of degree n with given discriminant. We may assume without loss of generality that $K = \mathbb{Q}$. For if $L|K$ is an extension of degree n with discriminant $\mathfrak{d}$, then $L|\mathbb{Q}$ is an extension of degree $m = n[K:\mathbb{Q}]$ with discriminant $(d) = \mathfrak{d}^n_{K|\mathbb{Q}} N_{K|\mathbb{Q}}(\mathfrak{d})$. Finally, the discriminant of $L(\sqrt{-1})|\mathbb{Q}$ differs from the discriminant of $L|\mathbb{Q}$ only by a constant factor. So we are reduced to proving that there exist only finitely many fields $K|\mathbb{Q}$ of degree n containing $\sqrt{-1}$ with a given discriminant d. Such a field K has only complex embeddings $\tau : K \to \mathbb{C}$. Choose one of them: τ_0. In the Minkowski space

$$K_{\mathbb{R}} = \Big[\prod_{\tau} \mathbb{C}\Big]^+$$

(see chap. I, § 5) consider the convex, centrally symmetric subset

$$X = \Big\{ (z_\tau) \in K_{\mathbb{R}} \;\Big|\; |\mathrm{Im}(z_{\tau_0})| < C\sqrt{|d|},\quad |\mathrm{Re}(z_{\tau_0})| < 1,\ |z_\tau| < 1 \ \text{for}\ \tau \neq \tau_0, \bar{\tau}_0 \Big\},$$

where C is an arbitrarily big constant which depends only on n. For a convenient choice of C, the volume will satisfy

$$\mathrm{vol}(X) > 2^n\sqrt{|d|} = 2^n\,\mathrm{vol}(\mathcal{O}_K),$$

where $\mathrm{vol}(\mathcal{O}_K)$ is the volume of a fundamental mesh of the lattice $j\mathcal{O}_K$ in $K_{\mathbb{R}}$ – see chap. I, (5.2). By Minkowski's lattice point theorem (chap. I, (4.4)), we thus find $\alpha \in \mathcal{O}_K$, $\alpha \neq 0$, such that $j\alpha = (\tau\alpha) \in X$, that is,

$$(*) \quad \big|\mathrm{Im}(\tau_0\alpha)\big| < C\sqrt{|d|},\ \big|\mathrm{Re}(\tau_0\alpha)\big| < 1,\quad |\tau\alpha| < 1 \quad \text{for} \quad \tau \neq \tau_0, \bar{\tau}_0 .$$

This α is a primitive element of K, i.e., one has $K = \mathbb{Q}(\alpha)$. Indeed, $|N_{K|\mathbb{Q}}(\alpha)| = \prod_\tau |\tau\alpha| \geq 1$ implies $|\tau_0\alpha| > 1$; thus $\mathrm{Im}(\tau_0\alpha) \neq 0$ so that the conjugates $\tau_0\alpha$ and $\bar{\tau}_0\alpha$ of α have to be distinct. Since $|\tau\alpha| < 1$ for $\tau \neq \tau_0, \bar{\tau}_0$, one has $\tau_0\alpha \neq \tau\alpha$ for all $\tau \neq \tau_0$. This implies $K = \mathbb{Q}(\alpha)$, because if $\mathbb{Q}(\alpha) \subsetneqq K$ then the restriction $\tau_0|_{\mathbb{Q}(\alpha)}$ would admit an extension τ different from τ_0, contradicting $\tau_0\alpha \neq \tau\alpha$.

Since the conjugates $\tau\alpha$ of α are subject to the conditions $(*)$, which only depend on d and n, the coefficients of the minimal polynomial of α

are bounded once d and n are fixed. Thus every field $K\,|\,\mathbb{Q}$ of degree n with discriminant d is generated by one of the finitely many lattice points α in the bounded region X. Therefore there are only finitely many fields with given degree and discriminant. □

The second theorem alluded to above is in fact a strengthening of the first. It follows from the following *bound on the discriminant.*

(2.14) Proposition. *The discriminant of an algebraic number field K of degree n satisfies*

$$|d_K|^{1/2} \geq \frac{n^n}{n!}\Big(\frac{\pi}{4}\Big)^{n/2}.$$

Proof: In Minkowski space $K_{\mathbb{R}} = \big[\prod_\tau \mathbb{C}\big]^+$, $\tau \in \operatorname{Hom}(K,\mathbb{C})$, consider the convex, centrally symmetric subset

$$X_t = \Big\{ (z_\tau) \in K_{\mathbb{R}} \;\Big|\; \sum_\tau |z_\tau| \leq t \Big\}.$$

Its volume is

$$\operatorname{vol}(X_t) = 2^r \pi^s \frac{t^n}{n!}.$$

Leaving aside the proof of this formula for the moment (which incidentally was exercise 2 of chap. I, § 5), we deduce the proposition from Minkowski's lattice point theorem (chap. I, (4.4)) as follows. Consider in $K_{\mathbb{R}}$ the lattice $\Gamma = j\mathcal{O}$ defined by $\mathcal{O}$. By chap. I, (5.2), the volume of a fundamental mesh is $\operatorname{vol}(\Gamma) = \sqrt{|d_K|}$. The inequality

$$\operatorname{vol}(X_t) > 2^n \operatorname{vol}(\Gamma)$$

therefore holds if and only if $2^r\pi^s\frac{t^n}{n!} > 2^n\sqrt{|d_K|}$, or equivalently if

$$t^n = n!\Big(\frac{4}{\pi}\Big)^s\sqrt{|d_K|} + \varepsilon,$$

for some $\varepsilon > 0$. If this is the case, there exists an $a \in \mathcal{O}$, $a \neq 0$, such that $ja \in X_t$. As this holds for all $\varepsilon > 0$, and since X_t contains only finitely many lattice points, it continues to hold for $\varepsilon = 0$. Applying now the inequality between arithmetic and geometric means,

$$\frac{1}{n}\sum_\tau |z_\tau| \geq \Big(\prod_\tau |z_\tau|\Big)^{1/n},$$

we obtain the desired result:

$$1 \leq \big|N_{K|\mathbb{Q}}(a)\big| = \prod_\tau |\tau a| \leq \frac{1}{n^n}\Big(\sum_\tau |\tau a|\Big)^n < \frac{t^n}{n^n} = \frac{n!}{n^n}\Big(\frac{4}{\pi}\Big)^s\sqrt{|d_K|}$$

$$\leq \frac{n!}{n^n}\Big(\frac{4}{\pi}\Big)^{n/2}\sqrt{|d_K|}.$$

Given this, it remains to prove the following lemma. □

(2.15) Lemma. *In Minkowski space* $K_{\mathbb{R}} = \big[\prod_\tau \mathbb{C}\big]^+$, *the domain*

$$X_t = \Big\{ (z_\tau) \in K_{\mathbb{R}} \ \Big|\ \sum_\tau |z_\tau| < t \Big\}$$

has volume

$$\operatorname{vol}(X_t) = 2^r \pi^s \frac{t^n}{n!} \cdot$$

Proof: $\operatorname{vol}(X_t)$ is 2^s times the Lebesgue volume $\operatorname{Vol}(f(X_t))$ of the image $f(X_t)$ under the mapping chap. I, (5.1),

$$f : K_{\mathbb{R}} \longrightarrow \prod_\tau \mathbb{R}, \quad (z_\tau) \longmapsto (x_\tau),$$

where $x_\rho = z_\rho$, $x_\sigma = \operatorname{Re}(z_\sigma)$, $x_{\bar\sigma} = \operatorname{Im}(z_\sigma)$. Substituting x_i, $i = 1, \dots, r$, instead of x_ρ, and y_j, z_j, $j = 1, \dots, s$, instead of $x_\sigma, x_{\bar\sigma}$, we see that $f(X_t)$ is described by the inequality

$$|x_1| + \cdots + |x_r| + 2\sqrt{y_1^2 + z_1^2} + \cdots + 2\sqrt{y_s^2 + z_s^2} < t.$$

The factor 2 occurs because $|z_{\bar\sigma}| = |\bar z_\sigma| = |z_\sigma|$. Passing to polar coordinates $y_j = u_j \cos\theta_j$, $z_j = u_j \sin\theta_j$, where $0 \le \theta_j \le 2\pi$, $0 \le u_j$, one sees that $\operatorname{Vol}(f(X_t))$ is computed by the integral

$$I(t) = \int u_1 \cdots u_s \, dx_1 \cdots dx_r \, du_1 \cdots du_s \, d\theta_1 \cdots d\theta_s$$

extended over the domain

$$|x_1| + \cdots + |x_r| + 2u_1 + \cdots + 2u_s \le t.$$

Restricting this domain of integration to $x_i \ge 0$, the integral gets divided by 2^r. Substituting $2u_j = w_j$ gives

$$I(t) = 2^r 4^{-s} (2\pi)^s I_{r,s}(t),$$

where the integral

$$I_{r,s}(t) = \int w_1 \cdots w_s \, dx_1 \cdots dx_r \, dw_1 \cdots dw_s$$

has to be taken over the domain $x_i \ge 0$, $w_j \ge 0$ and

$$(*) \qquad x_1 + \cdots + x_r + w_1 + \cdots + w_s \le t.$$

Clearly $I_{r,s}(t) = t^{r+2s} I_{r,s}(1) = t^n I_{r,s}(1)$. Writing $x_2 + \cdots + x_r + w_1 + \cdots + w_s \le t - x_1$ instead of $(*)$, Fubini's theorem yields

$$I_{r,s}(1) = \int_0^1 I_{r-1,s}(1 - x_1)\, dx_1 = \int_0^1 (1 - x_1)^{n-1} dx_1 \cdot I_{r-1,s}(1)$$
$$= \frac{1}{n} I_{r-1,s}(1).$$

By induction, this implies that

$$I_{r,s}(1) = \frac{1}{n(n-1)\cdots(n-r+1)} I_{0,s}(1).$$

In the same way, one gets

$$I_{0,s}(1) = \int_0^1 w_1(1-w_1)^{2s-2}\,dw_1 I_{0,s-1}(1),$$

and, doing the integration, induction shows that

$$I_{0,s}(1) = \frac{1}{(2s)!}\,I_{0,0}(1) = \frac{1}{(2s)!}\,.$$

Together, this gives $I_{r,s}(1) = \frac{1}{n!}$ and therefore indeed

$$\operatorname{vol}(X_t) = 2^s \operatorname{Vol}\big(f(X_t)\big) = 2^s 2^r 4^{-s}(2\pi)^s t^n I_{r,s}(1) = \frac{2^r \pi^s}{n!} t^n. \qquad \square$$

If we combine Stirling's formula,

$$n! = \sqrt{2\pi n}\left(\frac{n}{e}\right)^n e^{\frac{\theta}{12n}}, \quad 0 < \theta < 1,$$

with the inequality (2.14), we obtain the inequality

$$|d_K| > \left(\frac{\pi}{4}\right)^{2s} \frac{1}{2\pi n}\, e^{2n - \frac{1}{6n}}.$$

This shows that the absolute value of the discriminant of an algebraic number field tends to infinity with the degree. In the proof of (2.13) we saw that there are only finitely many number fields with bounded degree and discriminant. So now, since the degree is bounded if the discriminant is, we may strengthen (2.13), obtaining

(2.16) Hermite's Theorem. *There exist only finitely many number fields with bounded discriminant.*

The expression $a_n = \frac{n^n}{n!}\left(\frac{\pi}{4}\right)^{n/2}$ satisfies

$$\frac{a_{n+1}}{a_n} = \left(\frac{\pi}{4}\right)^{1/2}\left(1 + \frac{1}{n}\right)^n > 1,$$

i.e., $a_{n+1} > a_n$. Since $a_2 = \frac{\pi}{2} > 1$, (2.14) yields

(2.17) Minkowski's Theorem. *The discriminant of a number field K different from $\mathbb{Q}$ is $\neq \pm 1$.*

Combining this result with corollary (2.12), we obtain the

(2.18) Theorem. *The field $\mathbb{Q}$ does not admit any unramified extensions.*

These last theorems are of fundamental importance for number theory. Their significance is seen especially clearly in the light of higher dimensional analogues. For instance, let us replace the finite field extensions $L|K$ of a number field K by all smooth complete (i.e., proper) algebraic curves defined over K of a fixed genus g. If $\mathfrak{p}$ is a prime ideal of K, then for any such curve X, one may define the "reduction mod $\mathfrak{p}$". This is a curve defined over the residue class field of $\mathfrak{p}$. One says that X has *good reduction* at the prime $\mathfrak{p}$ if its reduction $\bmod \mathfrak{p}$ is again a smooth curve. This corresponds to an extension $L|K$ being unramified. In analogy to Hermite's theorem, the Russian mathematician *I.S. Šafarevič* formulated the conjecture that there exist only finitely many smooth complete curves of genus g over K with good reduction outside a fixed finite set of primes S. This conjecture was proved in 1983 by the mathematician *Gerd Faltings* (see [35]). The impact of this result can be gauged by the non-expert from the fact that it was the basis for *Faltings*'s proof of the famous **Mordell Conjecture**:

Every algebraic equation

$$f(x, y) = 0$$

of genus $g > 1$ with coefficients in K admits only **finitely many solutions** in K.

A 1-dimensional analogue of Minkowski's theorem (2.18) was proved in 1985 by the French mathematician *J.-M. Fontaine*: over the field $\mathbb{Q}$, there are no smooth proper curves with good reduction $\bmod p$ for all prime numbers p (see [39]).

Exercise 1. Let $d(\alpha) = d(1, \alpha, \dots, \alpha^{n-1})$, for an element $\alpha \in \mathcal{O}$ such that $L = K(\alpha)$. Show that $\mathfrak{d}_{L|K}$ is generated by all discriminants of elements $d(\alpha)$ if o is a complete discrete valuation ring and the residue field extension $\lambda|\kappa$ is separable. In other words, $\mathfrak{d}_{L|K}$ equals the gcd of all discriminants of individual elements. This fails to hold in general. Counterexample: $K = \mathbb{Q}$, $L = \mathbb{Q}(\alpha)$, $\alpha^3 - \alpha^2 - 2\alpha - 8 = 0$. (See [60], chap. III, § 25, p. 443. The untranslatable German catch phrase for this phenomenon is: there are *"außerwesentliche Diskriminantenteiler"*.)

Exercise 2. Let $L|K$ be a Galois extension of henselian fields with separable residue field extension $\lambda|\kappa$, and let G_i, $i \geq 0$, be the i-th ramification group. Then, if $\mathfrak{D}_{L|K} = \mathfrak{P}^s$, one has

$$s = \sum_{i=0}^{\infty} (\#G_i - 1).$$

Hint: If $\mathcal{O} = o[x]$ (see chap. II, (10.4)), then $s = v_L(\delta_{L|K}(x)) = \sum_{\substack{\sigma \in G \\ \sigma \neq 1}} v_L(x - \sigma x)$.

Exercise 3. The module of differentials $\Omega^1_{\mathcal{O}|o}$ is generated by a single element dx, $x \in \mathcal{O}$, and there is an exact sequence of $\mathcal{O}$-modules

$$0 \to \mathfrak{D}_{\mathcal{O}|o} \to \mathcal{O} \to \Omega^1_{\mathcal{O}|o} \to 0.$$

Exercise 4. For a tower $M \supseteq L \supseteq K$ of algebraic number fields there is an exact sequence of o_M-modules

$$0 \to \Omega^1_{L|K} \otimes o_M \to \Omega^1_{M|K} \to \Omega^1_{M|L} \to 0.$$

Exercise 5. If ζ is a primitive p^n-th root of unity, then

$$\mathfrak{D}_{\mathbb{Q}_p(\zeta)|\mathbb{Q}_p} = \mathfrak{P}^{p^{n-1}(pn-n-1)}.$$

§ 3. Riemann-Roch

The notion of replete divisor introduced into our development of number theory in § 1 is a terminology reminiscent of the function-theoretic model. We now have to ask the question to what extent this point of view does justice to our goal to also couch the number-theoretic *results* in a geometric function-theoretic fashion, and conversely to give arithmetic significance to the classical theorems of function theory. Among the latter, the Riemann-Roch theorem stands out as the most important representative. If number theory is to proceed in a geometric manner, it must work towards finding an adequate way to incorporate this result as well. This is the task we are now going to tackle.

First recall the classical situation in function theory. There the basic data is a compact Riemann surface X with the sheaf o_X of holomorphic functions. To each divisor $D = \sum_{P \in X} \nu_P P$ on X corresponds a **line bundle** $o(D)$, i.e., an o_X-module which is locally free of rank 1. If U is an open subset of X and $K(U)$ is the ring of meromorphic functions on U, then the vector space $o(D)(U)$ of sections of the sheaf $o(D)$ over U is given as

$$o(D)(U) = \big\{ f \in K(U) \bigm| \operatorname{ord}_P(f) \geq -\nu_P \text{ for all } P \in U \big\}.$$

The Riemann-Roch problem is to calculate the dimension

$$\ell(D) = \dim H^0\big(X, o(D)\big)$$

of the vector space of global sections

$$H^0\big(X, \mathcal{O}(D)\big) = \mathcal{O}(D)(X).$$

In its first version the Riemann-Roch theorem does not provide a formula for $H^0(X, \mathcal{O}(D))$ itself, but for the **Euler-Poincaré characteristic**

$$\chi\big(\mathcal{O}(D)\big) = \dim H^0\big(X, \mathcal{O}(D)\big) - \dim H^1\big(X, \mathcal{O}(D)\big).$$

The formula reads

$$\chi\big(\mathcal{O}(D)\big) = \deg(D) + 1 - g,$$

where g is the **genus** of X. For the divisor $D = 0$, one has $\mathcal{O}(D) = \mathcal{O}_X$ and $\deg(D) = 0$, so that $\chi(\mathcal{O}_X) = 1 - g$; then this equation may also be replaced by

$$\chi(\mathcal{O}(D)) = \deg(D) + \chi(\mathcal{O}_X).$$

The classical Riemann-Roch formula

$$\ell(D) - \ell(\mathcal{K} - D) = \deg(D) + 1 - g$$

is then obtained by using *Serre* duality, which states that $H^1(X, \mathcal{O}(D))$ is dual to $H^0(X, \omega \otimes \mathcal{O}(-D))$, where $\omega = \Omega^1_X$ is the so-called **canonical module** of X, and $\mathcal{K} = \operatorname{div}(\omega)$ is the associated divisor (see for instance [51], chap. III, 7.12.1 and chap. IV, 1.1.3).

In order to mimic this state of affairs in number theory, let us recall the explanations of chap. I, § 14 and chap. III, § 1. We endow the places $\mathfrak{p}$ of an algebraic number field K with the rôle of points of a space X which should be conceived of as the analogue of a *compact* Riemann surface. The elements $f \in K^*$ will be given the rôle of "meromorphic functions" on this space X. The order of the pole, resp. zero of f at the point $\mathfrak{p} \in X$, for $\mathfrak{p} \nmid \infty$, is defined to be the integer $v_\mathfrak{p}(f)$, and for $\mathfrak{p} | \infty$ it is the real number $v_\mathfrak{p}(f) = -\log|\tau f|$. In this way we associate to each $f \in K^*$ the replete divisor

$$\operatorname{div}(f) = \sum_\mathfrak{p} v_\mathfrak{p}(f)\mathfrak{p} \in Div(\bar{\mathcal{O}}).$$

More precisely, for a given divisor $D = \sum_\mathfrak{p} \nu_\mathfrak{p}\mathfrak{p}$, we are interested in the replete ideal

$$\mathcal{O}(D) = \prod_\mathfrak{p} \mathfrak{p}^{-\nu_\mathfrak{p}}$$

and the set

$$\begin{aligned} H^0(\mathcal{O}(D)) &= \big\{ f \in K^* \mid \operatorname{div}(f) \geq -D \big\} \\ &= \big\{ f \in \mathcal{O}(D)_\mathfrak{f} \mid 0 \neq |f|_\mathfrak{p} \leq \mathfrak{N}(\mathfrak{p})^{\nu_\mathfrak{p}} \text{ for } \mathfrak{p}|\infty \big\}, \end{aligned}$$

where the relation $D' \geq D$ between divisors $D' = \sum_{\mathfrak{p}} \nu'_{\mathfrak{p}}\mathfrak{p}$ and $D = \sum_{\mathfrak{p}} \nu_{\mathfrak{p}}\mathfrak{p}$ is simply defined to mean $\nu'_{\mathfrak{p}} \geq \nu_{\mathfrak{p}}$ for all $\mathfrak{p}$. Note that $H^0(\mathcal{O}(D))$ is no longer a vector space. An analogue of $H^1(X, \mathcal{O}(D))$ is completely missing. Instead of attacking directly the problem of measuring the size of $H^0(\mathcal{O}(D))$, we proceed as in the function-theoretic model by looking at the "Euler-Poincaré characteristic" of the replete ideal $\mathcal{O}(D)$. Before defining this, we want to establish the relation between the Minkowski space $K_{\mathbb{R}} = \big[\prod_\tau \mathbb{C}\big]^+$, $\tau \in \operatorname{Hom}(K, \mathbb{C})$, and the product $\prod_{\mathfrak{p}|\infty} K_{\mathfrak{p}}$. The reader will allow us to explain this simple situation in the following sketch.

We have the correspondences

$$\rho\colon K \to \mathbb{R} \quad \longmapsto \quad \mathfrak{p} \quad \text{real prime, } \rho = \rho_{\mathfrak{p}} : K_{\mathfrak{p}} \xrightarrow{\ \sim\ } \mathbb{R}\,,$$

$$\sigma, \bar{\sigma}\colon K \to \mathbb{C} \quad \longmapsto \quad \mathfrak{p} \quad \text{complex prime, } \sigma = \sigma_{\mathfrak{p}} : K_{\mathfrak{p}} \xrightarrow{\ \sim\ } \mathbb{C}\,.$$

There are the following isomorphisms

$$K \otimes_{\mathbb{Q}} \mathbb{R} \xrightarrow{\ \sim\ } K_{\mathbb{R}}\,, \qquad a \otimes x \longmapsto \big((\tau a)x\big)_\tau\,,$$

$$K \otimes_{\mathbb{Q}} \mathbb{R} \xrightarrow{\ \sim\ } \prod_{\mathfrak{p}|\infty} K_{\mathfrak{p}}\,, \quad a \otimes x \longmapsto \big((\tau_{\mathfrak{p}} a)x\big)_{\mathfrak{p}|\infty}\,,$$

$\tau_{\mathfrak{p}}$ being the canonical embedding $K \to K_{\mathfrak{p}}$ (see chap. II, (8.3)). They fit into the commutative diagram

$$\begin{array}{ccccccc}
K \otimes \mathbb{R} & \cong & K_{\mathbb{R}} & = & \prod_\rho \mathbb{R} & \times & \prod_\sigma [\mathbb{C} \times \mathbb{C}]^+ \\
\Big\| & & \uparrow \wr & & \wr\uparrow \prod \rho_{\mathfrak{p}} & & \wr\uparrow \prod [\sigma_{\mathfrak{p}} \times \bar{\sigma}_{\mathfrak{p}}] \\
K \otimes \mathbb{R} & \cong & \prod_{\mathfrak{p}|\infty} K_{\mathfrak{p}} & = & \prod_{\mathfrak{p}\text{ real}} K_{\mathfrak{p}} & \times & \prod_{\mathfrak{p}\text{ complex}} K_{\mathfrak{p}}\,,
\end{array}$$

where the arrow on the right is given by $a \mapsto (\sigma a, \bar{\sigma} a)$. Via this isomorphism, we identify $K_{\mathbb{R}}$ with $\prod_{\mathfrak{p}|\infty} K_{\mathfrak{p}}$:

$$K_{\mathbb{R}} = \prod_{\mathfrak{p}|\infty} K_{\mathfrak{p}}\,.$$

The scalar product $\langle x, y \rangle = \sum_\tau x_\tau \bar{y}_\tau$ on $K_{\mathbb{R}}$ is then transformed into

$$\langle x, y \rangle = \sum_{\mathfrak{p}\text{ real}} x_{\mathfrak{p}} y_{\mathfrak{p}} + \sum_{\mathfrak{p}\text{ complex}} (x_{\mathfrak{p}} \bar{y}_{\mathfrak{p}} + \bar{x}_{\mathfrak{p}} y_{\mathfrak{p}})\,.$$

The Haar measure μ on $K_{\mathbb{R}}$ which is determined by $\langle x, y \rangle$ becomes the product measure

$$\mu = \prod_{\mathfrak{p}|\infty} \mu_{\mathfrak{p}}\,,$$

where

$$\mu_{\mathfrak{p}} = \text{Lebesgue measure on } K_{\mathfrak{p}} = \mathbb{R}, \text{ if } \mathfrak{p} \text{ real},$$
$$\mu_{\mathfrak{p}} = 2\,\text{Lebesgue measure on } K_{\mathfrak{p}} = \mathbb{C}, \text{ if } \mathfrak{p} \text{ complex}.$$

Indeed, the system $1/\sqrt{2}$, $i/\sqrt{2}$ is an orthonormal basis with respect to the scalar product $x\bar{y} + \bar{x}y$ on $K_{\mathfrak{p}} = \mathbb{C}$. Hence the square $Q = \{z = x + iy \mid 0 \le x, y \le 1/\sqrt{2}\}$ has volume $\mu_{\mathfrak{p}}(Q) = 1$, but Lebesgue volume 1/2.

Finally, the logarithm map

$$\ell : \Big[\prod_{\tau} \mathbb{C}^*\Big]^+ \longrightarrow \Big[\prod_{\tau} \mathbb{R}\Big]^+, \quad x \longmapsto \big(\log |x_\tau|\big)_\tau$$

studied in Minkowski theory is transformed into the mapping

$$\ell : \prod_{\mathfrak{p}|\infty} K_{\mathfrak{p}}^* \longrightarrow \prod_{\mathfrak{p}|\infty} \mathbb{R}, \quad x \longmapsto \big(\log |x_{\mathfrak{p}}|_{\mathfrak{p}}\big)\,,$$

for one has the commutative diagram

$$\begin{array}{ccc} K_{\mathbb{R}}^* & \xrightarrow{\ \ell\ } & \big[\prod_\tau \mathbb{R}\big]^+ \\ \downarrow & & \downarrow \\ \prod_{\mathfrak{p}|\infty} K_{\mathfrak{p}}^* & \xrightarrow{\ \ell\ } & \prod_{\mathfrak{p}|\infty} \mathbb{R}\,, \end{array}$$

where the arrow on the right,

$$\Big[\prod_{\tau} \mathbb{R}\Big]^+ = \prod_{\rho} \mathbb{R} \times \prod_{\sigma} [\mathbb{R} \times \mathbb{R}]^+ \longrightarrow \prod_{\mathfrak{p}|\infty} \mathbb{R},$$

is defined by $x \mapsto x$ for $\rho \leftrightarrow \mathfrak{p}$, and by $(x, x) \mapsto 2x$ for $\sigma \leftrightarrow \mathfrak{p}$. This isomorphism takes the trace map $x \mapsto \sum_\tau x_\tau$ on $\big[\prod_\tau \mathbb{R}\big]^+$ into the trace map $x \mapsto \sum_{\mathfrak{p}|\infty} x_{\mathfrak{p}}$ on $\prod_{\mathfrak{p}|\infty} \mathbb{R}$, and hence the trace-zero space $H = \big\{x \in \big[\prod_\tau \mathbb{R}\big]^+ \mid \sum_\tau x_\tau = 0\big\}$ into the trace-zero space

$$H = \Big\{x \in \prod_{\mathfrak{p}|\infty} \mathbb{R} \;\Big|\; \sum_{\mathfrak{p}|\infty} x_{\mathfrak{p}} = 0\Big\}\,.$$

In this way we have translated all necessary invariants of the Minkowski space $K_{\mathbb{R}}$ to the product $\prod_{\mathfrak{p}|\infty} K_{\mathfrak{p}}$.

To a given replete ideal

$$\mathfrak{a} = \mathfrak{a}_f \cdot \mathfrak{a}_\infty = \prod_{\mathfrak{p}\nmid\infty} \mathfrak{p}^{\nu_{\mathfrak{p}}} \times \prod_{\mathfrak{p}|\infty} \mathfrak{p}^{\nu_{\mathfrak{p}}}$$

we now associate the following complete lattice $j\mathfrak{a}$ in $K_{\mathbb{R}}$. The fractional ideal $\mathfrak{a}_f \subseteq K$ is mapped by the embedding $j : K \to K_{\mathbb{R}}$ onto a complete lattice $j\mathfrak{a}_{\mathrm{f}}$ of $K_{\mathbb{R}} = \prod_{\mathfrak{p}|\infty} K_{\mathfrak{p}}$. By componentwise multiplication, $\mathfrak{a}_\infty = \prod_{\mathfrak{p}|\infty} \mathfrak{p}^{\nu_\mathfrak{p}} = (\dots, e^{\nu_\mathfrak{p}}, \dots)_{\mathfrak{p}|\infty}$ yields an isomorphism

$$\mathfrak{a}_\infty : K_{\mathbb{R}} \to K_{\mathbb{R}}, \quad (x_\mathfrak{p})_{\mathfrak{p}|\infty} \mapsto (e^{\nu_\mathfrak{p}} x_\mathfrak{p})_{\mathfrak{p}|\infty},$$

with determinant

$$(*) \qquad \det(\mathfrak{a}_\infty) = \prod_{\mathfrak{p}|\infty} e^{\nu_\mathfrak{p} f_\mathfrak{p}} = \prod_{\mathfrak{p}|\infty} \mathfrak{N}(\mathfrak{p})^{\nu_\mathfrak{p}} = \mathfrak{N}(\mathfrak{a}_\infty).$$

The image of the lattice $j\mathfrak{a}_{\mathrm{f}}$ under this map is a complete lattice

$$j\mathfrak{a} := \mathfrak{a}_\infty j \mathfrak{a}_{\mathrm{f}}.$$

Let $\operatorname{vol}(\mathfrak{a})$ denote the **volume of a fundamental mesh** of $j\mathfrak{a}$ with respect to the canonical measure. By $(*)$, we then have

$$\operatorname{vol}(\mathfrak{a}) = \mathfrak{N}(\mathfrak{a}_\infty) \operatorname{vol}(\mathfrak{a}_{\mathrm{f}}).$$

(3.1) Definition. *If $\mathfrak{a}$ is a replete ideal of K, then the real number*

$$\chi(\mathfrak{a}) = -\log \operatorname{vol}(\mathfrak{a})$$

will be called the **Euler-Minkowski characteristic** *of $\mathfrak{a}$.*

The reason for this terminology will become clear in § 8.

(3.2) Proposition. *The Euler-Minkowski characteristic $\chi(\mathfrak{a})$ only depends on the class of $\mathfrak{a}$ in $Pic(\bar{\mathcal{O}}) = J(\bar{\mathcal{O}})/P(\bar{\mathcal{O}})$.*

Proof: Let $[a] = [a]_{\mathrm{f}} \cdot [a]_\infty = (a) \times [a]_\infty$ be a replete principal ideal. Then one has

$$[a]\mathfrak{a} = a\mathfrak{a}_{\mathrm{f}} \times [a]_\infty \mathfrak{a}_\infty.$$

The lattice $j(a\mathfrak{a}_{\mathrm{f}})$ is the image of the lattice $j\mathfrak{a}_{\mathrm{f}}$ under the linear map $ja : K_{\mathbb{R}} \to K_{\mathbb{R}}$, $(x_\mathfrak{p})_{\mathfrak{p}|\infty} \mapsto (ax_\mathfrak{p})_{\mathfrak{p}|\infty}$. The absolute value of the determinant of this mapping is obviously given by

$$\left|\det(ja)\right| = \prod_{\mathfrak{p}|\infty} |a|_\mathfrak{p} = \prod_{\mathfrak{p}|\infty} \mathfrak{N}(\mathfrak{p})^{-v_\mathfrak{p}(a)} = \mathfrak{N}([a]_\infty)^{-1}.$$

For the canonical measure, we therefore have

$$\operatorname{vol}(a\mathfrak{a}_{\mathrm{f}}) = \mathfrak{N}([a]_\infty)^{-1} \operatorname{vol}(\mathfrak{a}_{\mathrm{f}}).$$

Taken together with $(*)$, this yields

$$\operatorname{vol}([a]\mathfrak{a}) = \mathfrak{N}([a]_\infty \mathfrak{a}_\infty) \operatorname{vol}(a\mathfrak{a}_{\mathrm{f}}) = \mathfrak{N}(\mathfrak{a}_\infty) \operatorname{vol}(\mathfrak{a}_{\mathrm{f}}) = \operatorname{vol}(\mathfrak{a}),$$

so that $\chi([a]\mathfrak{a}) = \chi(\mathfrak{a})$. □

The explicit evaluation of the Euler-Minkowski characteristic results from a result of Minkowski theory, *viz.*, proposition (5.2) of chap. I.

(3.3) Proposition. *For every replete ideal* $\mathfrak{a}$ *of* K *one has*

$$\mathrm{vol}(\mathfrak{a}) = \sqrt{|d_K|}\,\mathfrak{N}(\mathfrak{a}).$$

Proof: Multiplying by a replete principal ideal $[a]$ we may assume, as $\mathrm{vol}([a]\mathfrak{a}) = \mathrm{vol}(\mathfrak{a})$ and $\mathfrak{N}([a]\mathfrak{a}) = \mathfrak{N}(\mathfrak{a})$, that $\mathfrak{a}_f$ is an integral ideal of K. By chap. I, (5.2) the volume of a fundamental mesh of $\mathfrak{a}_f$ is given by

$$\mathrm{vol}(\mathfrak{a}_f) = \sqrt{|d_K|}(\mathcal{O} : \mathfrak{a}_f).$$

Hence

$$\mathrm{vol}(\mathfrak{a}) = \mathfrak{N}(\mathfrak{a}_\infty)\,\mathrm{vol}(\mathfrak{a}_f) = \mathfrak{N}(\mathfrak{a}_\infty)\sqrt{|d_K|}\,\mathfrak{N}(\mathfrak{a}_f) = \sqrt{|d_K|}\,\mathfrak{N}(\mathfrak{a}). \qquad \square$$

In view of the commutative diagram in §1, p. 192, we will now introduce the **degree** of the replete ideal $\mathfrak{a}$ to be the real number

$$\deg(\mathfrak{a}) = -\log \mathfrak{N}(\mathfrak{a}) = \deg\big(\mathrm{div}(\mathfrak{a})\big).$$

Observing that

$$\chi(\mathcal{O}) = -\log\sqrt{|d_K|},$$

we deduce from proposition (3.3) the first version of the Riemann-Roch theorem:

(3.4) Proposition. *For every replete ideal of* K *we have the formula*

$$\chi(\mathfrak{a}) = \deg(\mathfrak{a}) + \chi(\mathcal{O}).$$

In function theory there is the following relationship between the Euler-Poincaré characteristic and the genus g of the Riemann surface X in question:

$$\chi(\mathcal{O}) = \dim H^0(X, \mathcal{O}_X) - \dim H^1(X, \mathcal{O}_X) = 1 - g.$$

There is no immediate analogue of $H^1(X, \mathcal{O}_X)$ in arithmetic. However, there is an analogue of $H^0(X, \mathcal{O}_X)$. For each replete ideal $\mathfrak{a} = \prod_{\mathfrak{p}} \mathfrak{p}^{\nu_{\mathfrak{p}}}$ of the number field K, we define

$$H^0(\mathfrak{a}) = \big\{ f \in K^* \mid v_{\mathfrak{p}}(f) \geq \nu_{\mathfrak{p}} \text{ for all } \mathfrak{p} \big\}.$$

This is a finite set because $jH^0(\mathfrak{a})$ lies in the part of the lattice $j\mathfrak{a}_f \subseteq K_{\mathbb{R}}$ which is bounded by the conditions $|f|_{\mathfrak{p}} \leq e^{-\nu_{\mathfrak{p}} f_{\mathfrak{p}}}$, $\mathfrak{p}|\infty$. As the analogue of the dimension, we put $\ell(\mathfrak{a}) = 0$ if $H^0(\mathfrak{a}) = \emptyset$, and in all other cases

$$\ell(\mathfrak{a}) := \log \frac{\#H^0(\mathfrak{a})}{\mathrm{vol}(W)},$$

where the normalizing factor $\operatorname{vol}(W)$ is the volume of the set

$$W = \Big\{ (z_\tau) \in K_{\mathbb{R}} = \big[\prod_\tau \mathbb{C}\big]^+ \;\Big|\; |z_\tau| \le 1 \Big\}.$$

This volume is given explicitly by

$$\operatorname{vol}(W) = 2^r (2\pi)^s,$$

where r, resp. s, is the number of real, resp. complex, primes of K (see the proof of chap. I, (5.3)). In particular, one has

$$H^0(\mathcal{O}) = \mu(K), \quad \text{so that} \quad \ell(\mathcal{O}) = \log \frac{\#\mu(K)}{2^r (2\pi)^s},$$

because $|f|_{\mathfrak{p}} \le 1$ for all $\mathfrak{p}$, and $\prod_{\mathfrak{p}} |f|_{\mathfrak{p}} = 1$ implies $|f|_{\mathfrak{p}} = 1$ for all $\mathfrak{p}$, so that $H^0(\mathcal{O})$ is a finite subgroup of K^* and thus must consist of all roots of unity. This normalization leads us necessarily to the following definition of the genus of a number field, which had already been proposed *ad hoc* by the French mathematician *André Weil* in 1939 (see [138]).

(3.5) Definition. *The* **genus** *of a number field K is defined to be the real number*

$$g = \ell(\mathcal{O}) - \chi(\mathcal{O}) = \log \frac{\#\mu(K)\sqrt{|d_K|}}{2^r (2\pi)^s}.$$

Observe that the genus of the field of rational numbers $\mathbb{Q}$ is 0. Using this definition, the Riemann-Roch formula (3.4) takes the following shape:

(3.6) Proposition. *For every replete ideal $\mathfrak{a}$ of K one has*

$$\chi(\mathfrak{a}) = \deg(\mathfrak{a}) + \ell(\mathcal{O}) - g.$$

The analogue of the strong Riemann-Roch formula

$$\ell(D) = \deg(D) + 1 - g + \ell(\mathcal{K} - D),$$

hinges on the following deep theorem of Minkowski theory, which is due to *Serge Lang* and which reflects an arithmetic analogue of Serre duality. As usual, let r, resp. s, denote the number of real, resp. complex, primes, and $n = [K : \mathbb{Q}]$.

(3.7) Theorem (*S. Lang*). *For replete ideals $\mathfrak{a} = \prod_{\mathfrak{p}} \mathfrak{p}^{\nu_{\mathfrak{p}}} \in J(\bar{\mathcal{O}})$, one has*

$$\#H^0(\mathfrak{a}^{-1}) = \frac{2^r (2\pi)^s}{\sqrt{|d_K|}} \mathfrak{N}(\mathfrak{a}) + O\big(\mathfrak{N}(\mathfrak{a})^{1-\frac{1}{n}}\big)$$

if $\mathfrak{N}(\mathfrak{a}) \to \infty$. Here, as usual, $O(t)$ denotes a function such that $O(t)/t$ remains bounded as $t \to \infty$.

For the proof of the theorem we need the following

(3.8) Lemma. *Let $\mathfrak{a}_1, \dots, \mathfrak{a}_h$ be fractional ideals representing the classes of the finite ideal class group $Pic(\mathcal{O})$. Let c be a positive constant and*

$$\mathfrak{A}_i = \Big\{ \mathfrak{a} = \prod_{\mathfrak{p}} \mathfrak{p}^{\nu_{\mathfrak{p}}} \;\Big|\; \mathfrak{a}_f = \mathfrak{a}_i\,,\ \mathfrak{N}(\mathfrak{p})^{\nu_{\mathfrak{p}}} \le c\,\mathfrak{N}(\mathfrak{a})^{f_{\mathfrak{p}}/n} \ \text{ for } \mathfrak{p}|\infty \Big\}\,.$$

Then the constant c can be chosen in such a way that

$$J(\bar{\mathcal{O}}) = \bigcup_{i=1}^{h} \mathfrak{A}_i P(\bar{\mathcal{O}})\,.$$

Proof: Let $\mathfrak{B}_i = \{\mathfrak{a} \in J(\bar{\mathcal{O}}) \mid \mathfrak{a}_f = \mathfrak{a}_i\}$. Multiplying by a suitable replete principal ideal $[a]$, every $\mathfrak{a} \in J(\bar{\mathcal{O}})$ may be transformed into a replete ideal $\mathfrak{a}' = \mathfrak{a}[a]$ such that $\mathfrak{a}'_f = \mathfrak{a}_i$ for some i. Consequently, one has $J(\bar{\mathcal{O}}) = \bigcup_{i=1}^{h} \mathfrak{B}_i P(\bar{\mathcal{O}})$. It therefore suffices to show that $\mathfrak{B}_i \subseteq \mathfrak{A}_i P(\bar{\mathcal{O}})$ for $i = 1, \dots, h$, if the constant c is chosen conveniently. To do this, let $\mathfrak{a} = \mathfrak{a}_i \mathfrak{a}_\infty \in \mathfrak{B}_i$, $\mathfrak{a}_\infty = \prod_{\mathfrak{p}|\infty} \mathfrak{p}^{\nu_{\mathfrak{p}}} \in \prod_{\mathfrak{p}|\infty} \mathbb{R}_+^*$. Then we find for the replete ideal

$$\mathfrak{a}'_\infty = \mathfrak{a}_\infty \mathfrak{N}(\mathfrak{a}_\infty)^{-\frac{1}{n}} = \prod_{\mathfrak{p}|\infty} \mathfrak{p}^{\nu'_{\mathfrak{p}}}\,,$$

where $\nu'_{\mathfrak{p}} = \nu_{\mathfrak{p}} - \frac{1}{n}\sum_{\mathfrak{q}|\infty} f_{\mathfrak{q}}\nu_{\mathfrak{q}}$, that $\mathfrak{N}(\mathfrak{a}'_\infty) = 1$, and thus $\sum_{\mathfrak{p}|\infty} f_{\mathfrak{p}}\nu'_{\mathfrak{p}} = 0$. The vector

$$(\dots, f_{\mathfrak{p}}\nu'_{\mathfrak{p}}, \dots) \in \prod_{\mathfrak{p}|\infty} \mathbb{R}$$

therefore lies in the trace-zero space $H = \{(x_{\mathfrak{p}}) \in \prod_{\mathfrak{p}|\infty} \mathbb{R} \mid \sum_{\mathfrak{p}|\infty} x_{\mathfrak{p}} = 0\}$. Inside it we have – see chap. I, (7.3) – the complete unit lattice $\lambda(\mathcal{O}^*)$. Thus there exists a lattice point $\lambda(u) = (\dots, -f_{\mathfrak{p}}v_{\mathfrak{p}}(u), \dots)_{\mathfrak{p}|\infty}$, $u \in \mathcal{O}^*$, such that

$$\big| f_{\mathfrak{p}}\nu'_{\mathfrak{p}} - f_{\mathfrak{p}}v_{\mathfrak{p}}(u) \big| \le f_{\mathfrak{p}} c_0$$

with a constant c_0 depending only on the lattice $\lambda(\mathcal{O}^*)$. This implies

$$\nu_{\mathfrak{p}} - v_{\mathfrak{p}}(u) = \nu'_{\mathfrak{p}} + \frac{1}{n}\sum_{\mathfrak{q}|\infty} f_{\mathfrak{q}}\nu_{\mathfrak{q}} - v_{\mathfrak{p}}(u) \le \frac{1}{n}\log\mathfrak{N}(\mathfrak{a}_\infty) + c_0 = \frac{1}{n}\log\mathfrak{N}(\mathfrak{a}) + c_1$$

with $c_1 = c_0 - \frac{1}{n}\log\mathfrak{N}(\mathfrak{a}_i)$. Putting now $\mathfrak{b} = \mathfrak{a}[u^{-1}] = \prod_{\mathfrak{p}} \mathfrak{p}^{n_{\mathfrak{p}}}$, we get $\mathfrak{b}_f = \mathfrak{a}_i$. This is because $[u]_f = (u) = (1)$ and

$$f_{\mathfrak{p}} n_{\mathfrak{p}} = f_{\mathfrak{p}}\big(\nu_{\mathfrak{p}} - v_{\mathfrak{p}}(u)\big) \le \frac{f_{\mathfrak{p}}}{n}\log\mathfrak{N}(\mathfrak{a}) + nc_1\,,$$

so that $\mathfrak{N}(\mathfrak{p})^{n_{\mathfrak{p}}} \le e^{nc_1}\mathfrak{N}(\mathfrak{a})^{f_{\mathfrak{p}}/n}$ for $\mathfrak{p}|\infty$; then $\mathfrak{b} \in \mathfrak{A}_i$, so that $\mathfrak{a} = \mathfrak{b}[u] \in \mathfrak{A}_i P(\bar{\mathcal{O}})$, where $c = e^{nc_1}$. □

Proof of (3.7): As $O(t) = O(t) - 1$, we may replace $H^0(\mathfrak{a}^{-1})$ by

$$\overline{H}^0(\mathfrak{a}^{-1}) = H^0(\mathfrak{a}^{-1}) \cup \{0\} = \left\{ f \in \mathfrak{a}_{\mathrm{f}}^{-1} \mid |f|_{\mathfrak{p}} \le \mathfrak{N}(\mathfrak{p})^{\nu_{\mathfrak{p}}} \text{ for } \mathfrak{p}|\infty \right\}.$$

We have to show that there are constants C, C' such that

$$(*) \qquad \left| \#\overline{H}^0(\mathfrak{a}^{-1}) - \frac{2^r(2\pi)^s}{\sqrt{|d_K|}} \mathfrak{N}(\mathfrak{a}) \right| \le C\mathfrak{N}(\mathfrak{a})^{1-\frac{1}{n}}$$

for all $\mathfrak{a} \in J(\bar{\mathcal{O}})$ satisfying $\mathfrak{N}(\mathfrak{a}) \ge C'$. For $a \in K^*$, the set $\overline{H}^0(\mathfrak{a}^{-1})$ is mapped bijectively via $x \mapsto ax$ onto the set $\overline{H}^0([a]\mathfrak{a}^{-1})$. The numbers $\#\overline{H}^0(\mathfrak{a}^{-1})$ and $\mathfrak{N}(\mathfrak{a})$ thus depend only on the class $\mathfrak{a}$ mod $P(\bar{\mathcal{O}})$. As by the preceding lemma $J(\bar{\mathcal{O}}) = \bigcup_{i=1}^h \mathfrak{A}_i P(\bar{\mathcal{O}})$, it suffices to show $(*)$ for $\mathfrak{a}$ ranging over the set $\mathfrak{A}_i$.

For this, we shall use the identification of Minkowski space

$$K_{\mathbb{R}} = \prod_{\mathfrak{p}|\infty} K_{\mathfrak{p}}$$

with its canonical measure. Since $\mathfrak{a}_{\mathrm{f}} = \mathfrak{a}_i$ for $\mathfrak{a} = \prod_{\mathfrak{p}} \mathfrak{p}^{\nu_{\mathfrak{p}}} \in \mathfrak{A}_i$, we have

$$\overline{H}^0(\mathfrak{a}^{-1}) = \left\{ f \in \mathfrak{a}_i^{-1} \mid |f|_{\mathfrak{p}} \le \mathfrak{N}(\mathfrak{p})^{\nu_{\mathfrak{p}}} \text{ for } \mathfrak{p}|\infty \right\}.$$

We therefore have to count the lattice points in $\Gamma = j\mathfrak{a}_i^{-1} \subseteq K_{\mathbb{R}}$ which fall into the domain

$$P_{\mathfrak{a}} = \prod_{\mathfrak{p}|\infty} D_{\mathfrak{p}}$$

where $D_{\mathfrak{p}} = \{x \in K_{\mathfrak{p}} \mid |x|_{\mathfrak{p}} \le \mathfrak{N}(\mathfrak{p})^{\nu_{\mathfrak{p}}}\}$. Let F be a fundamental mesh of Γ. We consider the sets

$$X = \left\{ \gamma \in \Gamma \mid (F+\gamma) \cap P_{\mathfrak{a}} \ne \emptyset \right\},$$

$$Y = \left\{ \gamma \in \Gamma \mid F+\gamma \subseteq \mathring{P}_{\mathfrak{a}} \right\},$$

$$X \smallsetminus Y = \left\{ \gamma \in \Gamma \mid (F+\gamma) \cap \partial P_{\mathfrak{a}} \ne \emptyset \right\}.$$

As $Y \subseteq \Gamma \cap P_{\mathfrak{a}} = \overline{H}^0(\mathfrak{a}^{-1}) \subseteq X$ and as $\bigcup_{\gamma\in Y}(F+\gamma) \subseteq P_{\mathfrak{a}} \subseteq \bigcup_{\gamma\in X}(F+\gamma)$, one has

$$\#Y \le \#\overline{H}^0(\mathfrak{a}^{-1}) \le \#X$$

as well as

$$\#Y \operatorname{vol}(F) \le \operatorname{vol}(P_{\mathfrak{a}}) \le \#X \operatorname{vol}(F).$$

This implies

$$\left| \#\overline{H}^0(\mathfrak{a}^{-1}) - \frac{\operatorname{vol}(P_{\mathfrak{a}})}{\operatorname{vol}(F)} \right| \le \#X - \#Y = \#(X \smallsetminus Y).$$

For the set $P_{\mathfrak{a}} = \prod_{\mathfrak{p}|\infty} D_{\mathfrak{p}}$, we now have

$$\operatorname{vol}(P_{\mathfrak{a}}) = \prod_{\mathfrak{p} \text{ real}} 2\mathfrak{N}(\mathfrak{p})^{\nu_{\mathfrak{p}}} \prod_{\mathfrak{p} \text{ complex}} 2\pi\mathfrak{N}(\mathfrak{p})^{\nu_{\mathfrak{p}}} = 2^r(2\pi)^s\mathfrak{N}(\mathfrak{a}_\infty)$$

(observe here that, under the identification $K_{\mathfrak{p}} = \mathbb{C}$, one has the equation $|x|_{\mathfrak{p}} = |x|^2$). For the fundamental mesh F, (3.3) yields

$$\operatorname{vol}(F) = \sqrt{|d_K|}\,\mathfrak{N}(\mathfrak{a}_f^{-1}).$$

From this we get

$$\left|\#\overline{H}^0(\mathfrak{a}^{-1}) - \frac{2^r(2\pi)^s}{\sqrt{|d_K|}}\mathfrak{N}(\mathfrak{a})\right| \le \#(X \smallsetminus Y).$$

Having obtained this inequality, it suffices to show that there exist constants C, C' such that

$$\#(X \smallsetminus Y) = \#\big\{\gamma \in \Gamma \mid (F+\gamma) \cap \partial P_{\mathfrak{a}} \neq \emptyset\big\} \le C\,\mathfrak{N}(\mathfrak{a})^{1-\frac{1}{n}},$$

for all $\mathfrak{a} \in \mathfrak{A}_i$ with $\mathfrak{N}(\mathfrak{a}) \ge C'$. We choose $C' = 1$ and find the constant C in the remainder of the proof. We parametrize the set $P_{\mathfrak{a}} = \prod_{\mathfrak{p}|\infty} D_{\mathfrak{p}}$ via the mapping

$$\varphi : I^n \to P_{\mathfrak{a}},$$

where $I = [0,1]$, which is given by

$$I \longrightarrow D_{\mathfrak{p}}, \quad t \longmapsto 2\alpha_{\mathfrak{p}}(t - \tfrac{1}{2}), \qquad \text{if } \mathfrak{p} \text{ real},$$

$$I^2 \longrightarrow D_{\mathfrak{p}}, \quad (\rho,\theta) \longmapsto \sqrt{\alpha_{\mathfrak{p}}}(\rho\cos 2\pi\theta, \rho\sin 2\pi\theta), \quad \text{if } \mathfrak{p} \text{ complex},$$

where $\alpha_{\mathfrak{p}} = \mathfrak{N}(\mathfrak{p})^{\nu_{\mathfrak{p}}}$. We bound the norm $\|d\varphi(x)\|$ of the derivative $d\varphi(x) : \mathbb{R}^n \to \mathbb{R}^n$ ($x \in I^n$). If $d\varphi(x) = (a_{ik})$, then $\|d\varphi(x)\| \le n \max |a_{ik}|$. Every partial derivative of φ is now bounded by $2\alpha_{\mathfrak{p}}$, resp. $2\pi\sqrt{\alpha_{\mathfrak{p}}}$. Since $\mathfrak{a} \in \mathfrak{A}_i$, we have that $\alpha_{\mathfrak{p}} = \mathfrak{N}(\mathfrak{p})^{\nu_{\mathfrak{p}}} \le c\mathfrak{N}(\mathfrak{a})^{f_{\mathfrak{p}}/n}$, for all $\mathfrak{p}|\infty$. It follows that

$$\big\|\, d\varphi(x)\big\| \le 2\pi n \max \alpha_{\mathfrak{p}}^{1/f_{\mathfrak{p}}} \le c_1\mathfrak{N}(\mathfrak{a})^{1/n}.$$

The mean value theorem implies that

$$(**) \qquad \big\|\varphi(x) - \varphi(y)\big\| \le c_1\mathfrak{N}(\mathfrak{a})^{1/n}\|x - y\|,$$

where $\| \;\; \|$ is the euclidean norm. The boundary of $P_{\mathfrak{a}}$,

$$\partial P_{\mathfrak{a}} = \bigcup_{\mathfrak{p}} \big[\,\partial D_{\mathfrak{p}} \times \prod_{\mathfrak{q} \neq \mathfrak{p}} D_{\mathfrak{q}}\big],$$

is parametrized by a finite number of boundary cubes I^{n-1} of I^n. We subdivide every edge of I^{n-1} into $m = [\mathfrak{N}(\mathfrak{a})^{1/n}] \ge C' = 1$ segments of equal length and obtain for I^{n-1} a decomposition into m^{n-1} small cubes of diameter $\le (n-1)^{1/2}/m$. From $(**)$, the image of such a small cube under φ has a diameter $\le \frac{(n-1)^{1/2}}{m} c_1 \mathfrak{N}(\mathfrak{a})^{1/n} \le (n-1)^{1/2} c_1 \frac{m+1}{m} \le (n-1)^{1/2} c_1 2 =: c_2$. The number of translates $F+\gamma$, $\gamma \in \Gamma$, meeting a domain of diameter $\le c_2$ is bounded by a constant c_3 which depends only on c_2 and the fundamental mesh F. The image of a small cube under φ thus meets at most c_3 translates $F+\gamma$. Since there are precisely $m^{n-1} = [\mathfrak{N}(\mathfrak{a})^{1/n}]^{n-1}$ cubes in $\varphi(I^{n-1})$, we see that $\varphi(I^{n-1})$ meets at most $c_3[\mathfrak{N}(\mathfrak{a})^{1/n}]^{n-1} \le c_3 \mathfrak{N}(\mathfrak{a})^{1-\frac{1}{n}}$ translates, and since the boundary $\partial P_{\mathfrak{a}}$ is covered by at most $2n$ such parts $\varphi(I^{n-1})$, we do indeed find that

$$\#\big\{\gamma \in \Gamma \mid (F+\gamma) \cap \partial P_{\mathfrak{a}} \neq \emptyset\big\} \le C\, \mathfrak{N}(\mathfrak{a})^{1-\frac{1}{n}}$$

for all $\mathfrak{a} \in \mathfrak{A}_i$ with $\mathfrak{N}(\mathfrak{a}) \ge 1$, where $C = 2nc_3$ is a constant which is independent of $\mathfrak{a} \in \mathfrak{A}_i$, as required. □

From the theorem we have just proved, we now obtain the strong version of the Riemann-Roch theorem. We want to formulate it in the language of divisors. Let $D = \sum_{\mathfrak{p}} \nu_{\mathfrak{p}} \mathfrak{p}$ be a replete divisor of K,

$$H^0(D) = H^0\big(\mathcal{O}(D)\big) = \big\{ f \in K^* \mid v_{\mathfrak{p}}(f) \ge -\nu_{\mathfrak{p}} \big\},$$

$$\ell(D) = \ell(\mathcal{O}(D)) = \log \frac{\#H^0(D)}{\mathrm{vol}(W)} \quad \text{and} \quad \chi(D) = \chi(\mathcal{O}(D)).$$

We call the number

$$i(D) = \ell(D) - \chi(D)$$

the **index of specialty** of D and get the

(3.9) Theorem (Riemann-Roch). *For every replete divisor $D \in Div(\overline{\mathcal{O}})$ we have the formula*

$$\ell(D) = \deg(D) + \ell(\mathcal{O}) - g + i(D).$$

The index of specialty $i(D)$ satisfies

$$i(D) = O\big(e^{-\frac{1}{n}\deg(D)}\big),$$

in particular, $i(D) \to 0$ for $\deg(D) \to \infty$.

Proof: The formula for $\ell(D)$ follows from $\chi(D) = \deg(D) + \ell(\mathcal{O}) - g$ and $\chi(D) = \ell(D) - i(D)$. Putting $\mathfrak{a}^{-1} = \mathcal{O}(D)$, we find by (3.7) that

$$\frac{\#H^0(\mathfrak{a}^{-1})}{2^r(2\pi)^s} = \frac{\mathfrak{N}(\mathfrak{a})}{\sqrt{|d_K|}}\big(1 + \varphi(\mathfrak{a})\,\mathfrak{N}(\mathfrak{a})^{-1/n}\big),$$

for some function $\varphi(\mathfrak{a})$ which remains bounded as $\mathfrak{N}(\mathfrak{a}) \to \infty$, so that $\deg(D) = -\log \mathfrak{N}(\mathfrak{a}^{-1}) = \log \mathfrak{N}(\mathfrak{a}) \to \infty$. Taking logarithms and observing that $\log(1 + O(t)) = O(t)$ and $\mathfrak{N}(\mathfrak{a})^{-1/n} = \exp(-\frac{1}{n}\deg D)$, we obtain

$$\begin{aligned}\ell(D) = \ell(\mathfrak{a}^{-1}) &= -\log\big(\sqrt{|d_K|}\,\mathfrak{N}(\mathfrak{a}^{-1})\big) + O\big(\mathfrak{N}(\mathfrak{a})^{-1/n}\big)\\ &= \chi(D) + O(e^{-\frac{1}{n}\deg D}).\end{aligned}$$

Hence $i(D) = \ell(D) - \chi(D) = O(e^{-\frac{1}{n}\deg D})$. □

To conclude this section, let us study the variation of the Euler-Minkowski characteristic and of the genus when we change the field K. Let $L|K$ be a finite extension and $\mathcal{o}$, resp. $\mathcal{O}$, the ring of integers of K, resp. L. In §2 we considered **Dedekind's complementary module**

$$\mathfrak{C}_{L|K} = \big\{x \in L \;\big|\; Tr(x\mathcal{O}) \subseteq \mathcal{o}\big\} \;\cong\; \mathrm{Hom}_{\mathcal{o}}(\mathcal{O}, \mathcal{o}).$$

It is a fractional ideal in L whose inverse is the different $\mathfrak{D}_{L|K}$. From (2.6), it is divisible only by the prime ideals of L which are ramified over K.

(3.10) Definition. *The fractional ideal*

$$\omega_K = \mathfrak{C}_{K|\mathbb{Q}} \;\cong\; \mathrm{Hom}_{\mathbb{Z}}(\mathcal{o}, \mathbb{Z})$$

is called the **canonical module** *of the number field K.*

By (2.2) we have the

(3.11) Proposition. *The canonical modules of L and K satisfy the relation*

$$\omega_L = \mathfrak{C}_{L|K}\,\omega_K\,.$$

The canonical module ω_K is related to the Euler-Minkowski characteristic $\chi(\mathcal{o})$ and the genus g of K in the following way, by formula (3.3):

$$\mathrm{vol}(\mathcal{o}) = \sqrt{|d_K|}.$$

(3.12) Proposition. $\deg \omega_K = -2\chi(\mathcal{o}) = 2g - 2\ell(\mathcal{o})$.

Proof: By (2.9) we know that $N_{K|\mathbb{Q}}(\mathfrak{D}_{K|\mathbb{Q}})$ is the discriminant ideal $\mathfrak{d}_{K|\mathbb{Q}} = (d_K)$, and therefore by (1.6), (iii):

$$\mathfrak{N}(\omega_K) = \mathfrak{N}(\mathfrak{D}_{K|\mathbb{Q}})^{-1} = \mathfrak{N}(\mathfrak{d}_{K|\mathbb{Q}})^{-1} = |d_K|^{-1},$$

so that, as $\mathrm{vol}(\mathcal{o}) = \sqrt{|d_K|}$, we have indeed

$$\deg \omega_K = -\log \mathfrak{N}(\omega_K) = \log |d_K| = 2\log \mathrm{vol}(\mathcal{o}) = -2\chi(\mathcal{o}) = 2g - 2\ell(\mathcal{o}).$$

□

As for the genus, we now obtain the following analogue of the **Riemann-Hurwitz formula** of function theory.

(3.13) Proposition. *Let $L|K$ be a finite extension and g_L, resp. g_K, the genus of L, resp. K. Then one has*

$$g_L - \ell(\mathcal{O}_L) = [L:K]\big(g_K - \ell(\mathcal{O}_K)\big) + \frac{1}{2}\deg \mathfrak{D}_{L|K}\,.$$

In particular, in the case of an unramified extension $L|K$:

$$\chi(\mathcal{O}_L) = [L:K]\chi(\mathcal{O}_K)\,.$$

Proof: Since $\omega_L = \mathfrak{D}_{L|K}\omega_K$, one has

$$\mathfrak{N}(\omega_L) = \mathfrak{N}(i_{L|K}\omega_K)\,\mathfrak{N}(\mathfrak{D}_{L|K}) = \mathfrak{N}(\omega_K)^{[L:K]}\,\mathfrak{N}(\mathfrak{D}_{L|K})\,,$$

so that

$$\deg\omega_L = [L:K]\deg\omega_K + \deg\mathfrak{D}_{L|K}\,.$$

Thus the proposition follows from (3.12). □

The Riemann-Hurwitz formula tells us in particular that, in the decision we took in § 1, we really had no choice but to consider the extension $\mathbb{C}|\mathbb{R}$ as *unramified*. In fact, in function theory the module corresponding by analogy to the ideal $\mathfrak{D}_{L|K}$ takes account of precisely the branch points of the covering of Riemann surfaces in question. In order to obtain the same phenomenon in number theory we are forced to declare all the infinite primes $\mathfrak{P}$ of L unramified, since they do not occur in the ideal $\mathfrak{D}_{L|K}$.

Thus the fact that $\mathbb{C}|\mathbb{R}$ is unramified appears to be forced by nature itself. Investigating the matter a little more closely, however, this turns out not to be the case. It is rather a consequence of a well-hidden initial choice that we made. In fact, in chap. I, § 5, we equipped the Minkowski space

$$K_{\mathbb{R}} = \Big[\prod_\tau \mathbb{C}\Big]^+$$

with the "canonical metric"

$$\langle x, y\rangle = \sum_\tau x_\tau \bar{y}_\tau\,.$$

Replacing it, for instance, by the "Minkowski metric"

$$(x, y) = \sum_\tau \alpha_\tau x_\tau \bar{y}_\tau\,,$$

$\alpha_\tau = 1$ if $\tau = \bar{\tau}$, $\alpha_\tau = \frac{1}{2}$ if $\tau \neq \bar{\tau}$, changes the whole picture. The Haar measures on $K_{\mathbb{R}}$ belonging to $\langle\, ,\rangle$ and $(\, ,)$ are related as follows:

$$\mathrm{vol}_{\text{canonical}}(X) = 2^s\, \mathrm{vol}_{\text{Minkowski}}(X)\,.$$

Distinguishing the invariants of Riemann-Roch theory with respect to the Minkowski measure by a tilde, we get the relations

$$\widetilde{\chi}(\mathfrak{a}) = \chi(\mathfrak{a}) + \log 2^s, \quad \tilde{\ell}(\mathfrak{a}) = \ell(\mathfrak{a}) + \log 2^s$$

(the latter in case that $H^0(\mathfrak{a}) \neq \emptyset$), whereas the genus remains unchanged. Substituting this into the Riemann-Hurwitz formula (3.13) preserves its shape only if one enriches $\mathfrak{C}_{L|K}$ into a replete ideal in which all infinite primes $\mathfrak{P}$ such that $L_{\mathfrak{P}} \neq K_{\mathfrak{p}}$ occur. This forces us to consider the extension $\mathbb{C}|\mathbb{R}$ as *ramified*, to put $\tilde{e}_{\mathfrak{P}|\mathfrak{p}} = [L_{\mathfrak{P}} : K_{\mathfrak{p}}]$, $\widetilde{f}_{\mathfrak{P}|\mathfrak{p}} = 1$, and in particular

$$\tilde{e}_{\mathfrak{p}} = [K_{\mathfrak{p}} : \mathbb{R}], \quad \widetilde{f}_{\mathfrak{p}} = 1\,.$$

The following modifications ensue from this. For an infinite prime $\mathfrak{p}$ one has to define

$$\tilde{v}_{\mathfrak{p}}(a) = -\tilde{e}_{\mathfrak{p}} \log|\tau a|, \quad \mathfrak{p}^\nu = e^{\nu/\tilde{e}_{\mathfrak{p}}}, \quad \widetilde{\mathfrak{N}}(\mathfrak{p}) = e.$$

The absolute norm as well as the degree of a replete ideal $\mathfrak{a}$ remain unaltered:

$$\widetilde{\mathfrak{N}}(\mathfrak{a}) = \mathfrak{N}(\mathfrak{a}), \quad \widetilde{\deg}(\mathfrak{a}) = -\log \widetilde{\mathfrak{N}}(\mathfrak{a}) = \deg(\mathfrak{a})\,.$$

The canonical module ω_K however has to be changed:

$$\tilde{\omega}_K = \omega_K \prod_{\mathfrak{p} \text{ complex}} \mathfrak{p}^{2\log 2},$$

in order for the equation

$$\deg \tilde{\omega}_K = -2\widetilde{\chi}(\mathcal{O}) = 2g - 2\tilde{\ell}(\mathcal{O})$$

to hold. By the same token, the ideal $\mathfrak{C}_{L|K}$ has to be replaced by the replete ideal

$$\tilde{\mathfrak{C}}_{L|K} = \mathfrak{C}_{L|K} \prod_{\substack{\widetilde{\mathfrak{P}}|\infty \\ e_{\widetilde{\mathfrak{P}}|\tilde{\mathfrak{p}}} \neq 1}} \mathfrak{P}^{2\log 2}$$

so that

$$\tilde{\omega}_L = \tilde{\mathfrak{C}}_{L|K}\, i_{L|K}(\tilde{\omega}_K)\,.$$

In the same way as in (3.13), this yields the Riemann-Hurwitz formula

$$g_L - \tilde{\ell}(\mathcal{O}_L) = [L : K]\big(g_K - \tilde{\ell}(\mathcal{O}_K)\big) + \frac{1}{2} \deg \tilde{\mathfrak{C}}_{L|K}\,.$$

In view of this sensitivity to the chosen metric on Minkowski space $K_{\mathbb{R}}$, the mathematician Uwe Jannsen proposes as analogues of the function fields not just number fields K by themselves but number fields equipped with a metric of the type

$$\langle x, y\rangle_K = \sum_\tau \alpha_\tau x_\tau \overline{y}_\tau ,$$

$\alpha_\tau > 0$, $\alpha_\tau = \alpha_{\overline{\tau}}$, on $K_{\mathbb{R}}$. Let these new objects be called **metrized number fields**. This idea does indeed do justice to the situation in question in a very precise manner, and it is of fundamental importance for algebraic number theory. We denote metrized number fields $(K, \langle\ ,\ \rangle_K)$ as $\widetilde{K}$ and attach to them the following invariants. Let

$$\langle x, y\rangle_K = \sum_\tau \alpha_\tau x_\tau \overline{y}_\tau .$$

Let $\mathfrak{p} = \mathfrak{p}_\tau$ be the infinite prime corresponding to $\tau : K \to \mathbb{C}$. We then put $\alpha_{\mathfrak{p}} = \alpha_\tau$. At the same time, we also use the letter $\mathfrak{p}$ for the positive real number

$$\mathfrak{p} = e^{\alpha_{\mathfrak{p}}} \in \mathbb{R}_+^* ,$$

which we interpret as the replete ideal $(1) \times (1, \cdots, 1, e^{\alpha_{\mathfrak{p}}}, 1, \cdots, 1) \in J(\mathcal{O}) \times \prod_{\mathfrak{p}|\infty} \mathbb{R}_+^*$. We put

$$e_{\mathfrak{p}} = 1/\alpha_{\mathfrak{p}} \quad \text{and} \quad f_{\mathfrak{p}} = \alpha_{\mathfrak{p}}[K_{\mathfrak{p}} : \mathbb{R}],$$

and we define the valuation $v_{\mathfrak{p}}$ of K^* associated to $\mathfrak{p}$ by

$$v_{\mathfrak{p}}(a) = -e_{\mathfrak{p}} \log |\tau a| .$$

Further, we put

$$\mathfrak{N}(\mathfrak{p}) = e^{f_{\mathfrak{p}}} \quad \text{and} \quad |a|_{\mathfrak{p}} = \mathfrak{N}(\mathfrak{p})^{-v_{\mathfrak{p}}(a)} ,$$

so that again $|a|_{\mathfrak{p}} = |\tau a|$ if $\mathfrak{p}$ is real, and $|a|_{\mathfrak{p}} = |\tau a|^2$ if $\mathfrak{p}$ is complex. For every replete ideal $\mathfrak{a}$ of K, there is a unique representation $\mathfrak{a} = \prod_{\mathfrak{p}} \mathfrak{p}^{\nu_{\mathfrak{p}}}$, which gives the absolute norm $\mathfrak{N}(\mathfrak{a}) = \prod_{\mathfrak{p}} \mathfrak{N}(\mathfrak{p})^{\nu_{\mathfrak{p}}}$, and the degree

$$\deg_{\widetilde{K}}(\mathfrak{a}) = -\log \mathfrak{N}(\mathfrak{a}) .$$

The *canonical module* of $\widetilde{K}$ is defined to be the replete ideal

$$\omega_{\widetilde{K}} = \omega_K \cdot \omega_\infty \in J(\overline{\mathcal{O}}) = J(\mathcal{O}) \times \prod_{\mathfrak{p}|\infty} \mathbb{R}_+^* ,$$

where ω_K is the inverse of the different $\mathfrak{D}_{K|\mathbb{Q}}$ of $K|\mathbb{Q}$, and

$$\omega_\infty = (\alpha_{\mathfrak{p}}^{-1})_{\mathfrak{p}|\infty} \in \prod_{\mathfrak{p}|\infty} \mathbb{R}_+^* .$$

The Riemann-Roch theory may be transferred without any problem, using the definitions given above, to metrized number fields $\widetilde{K} = (K, \langle\,,\,\rangle_K)$. Distinguishing their invariants by the suffix $\widetilde{K}$ yields the relations

$$\mathrm{vol}_{\widetilde{K}}(X) = \prod_\tau \sqrt{\alpha_\tau}\,\mathrm{vol}(X),$$

because $T_\alpha : (K_{\mathbb{R}}, \langle\,,\,\rangle_K) \to (K_{\mathbb{R}}, \langle\,,\,\rangle)$, $(x_\tau) \mapsto (\sqrt{\alpha_\tau}\, x_\tau)$, is an isometry with determinant $\prod_\tau \sqrt{\alpha_\tau}$, and therefore

$$\chi_{\widetilde{K}}(\mathcal{O}_K) = -\log \mathrm{vol}_{\widetilde{K}}(\mathcal{O}_K) \;= \chi(\mathcal{O}_K) - \log \prod_\tau \sqrt{\alpha_\tau},$$

$$\ell_{\widetilde{K}}(\mathcal{O}_K) = \log \frac{\# H^0(\mathcal{O}_K)}{\mathrm{vol}_{\widetilde{K}}(W)} = \ell(\mathcal{O}_K) - \log \prod_\tau \sqrt{\alpha_\tau}.$$

The genus

$$g_{\widetilde{K}} = \ell_{\widetilde{K}}(\mathcal{O}_K) - \chi_{\widetilde{K}}(\mathcal{O}_K) = \ell(\mathcal{O}_K) - \chi(\mathcal{O}_K) = \log \frac{\#\mu(K)\sqrt{|d_K|}}{2^r (2\pi)^s}$$

does not depend on the choice of metric.

Just as in function theory, there is then no longer one smallest field, but $\mathbb{Q}$ is replaced by the continuous family of metrized fields $(\mathbb{Q}, \alpha x y)$, $\alpha \in \mathbb{R}_+^*$, all of which have genus $g = 0$. One even has the

(3.14) Proposition. *The metrized fields $(\mathbb{Q}, \alpha x y)$ are the only metrized number fields of genus 0.*

Proof: We have

$$g_{\widetilde{K}} = \log \frac{\#\mu(K)\sqrt{|d_K|}}{2^r (2\pi)^s} = 0 \iff \#\mu(K)\sqrt{|d_K|} = 2^r (2\pi)^s .$$

Since π is transcendental, one has $s = 0$, i.e., K is totally real. Thus $\#\mu(K) = 2$ so that $|d_K| = 4^{n-1}$, where $n = r = [K : \mathbb{Q}]$. In view of the bound (2.14) on the discriminant

$$|d_K|^{1/2} \geq \frac{n^n}{n!}\left(\frac{\pi}{4}\right)^{n/2},$$

this can only happen if $n \leq 6$. But for this case one has sharper bounds, due to *Odlyzko* (see [111], table 2):

n	$=$	3	4	5	6
$\lvert d_K\rvert^{1/n}$	$\geq$	3,09	4,21	5,30	6,35

This is not compatible with $|d_K|^{1/n} = 4^{\frac{n-1}{n}}$, so we may conclude that $n \leq 2$. But there is no real quadratic field with discriminant $|d_K| = 4$ (see chap. I, § 2, exercise 4). Hence $n = 1$, so that $K = \mathbb{Q}$. $\square$

An *extension of metrized number fields* is a pair $\widetilde{K} = (K, \langle\ ,\ \rangle_K)$, $\widetilde{L} = (L, \langle\ ,\ \rangle_L)$, such that $K \subseteq L$ and the metrics

$$\langle x, y\rangle_K = \sum_\tau \alpha_\tau x_\tau \bar{y}_\tau, \quad \langle x, y\rangle_L = \sum_\sigma \beta_\sigma x_\sigma \bar{y}_\sigma$$

satisfy the relation $\alpha_\tau \geq \beta_\sigma$ whenever $\tau = \sigma|_K$. If $\mathfrak{P}|\mathfrak{p}$ are infinite primes of $L|K$, $\mathfrak{P}$ belonging to σ and $\mathfrak{p}$ to $\tau = \sigma|_K$, we define the *ramification index* and *inertia degree* by

$$e_{\mathfrak{P}|\mathfrak{p}} = \alpha_\tau/\beta_\sigma \quad \text{and} \quad f_{\mathfrak{P}|\mathfrak{p}} = \beta_\sigma/\alpha_\tau [L_\mathfrak{P} : K_\mathfrak{p}].$$

Thus the fundamental identity

$$\sum_{\mathfrak{P}|\mathfrak{p}} e_{\mathfrak{P}|\mathfrak{p}} f_{\mathfrak{P}|\mathfrak{p}} = [L : K]$$

is preserved. Also $\mathfrak{P}$ is unramified if and only if $\alpha_\tau = \beta_\sigma$. For "replete prime ideals" $\mathfrak{p} = e^{\alpha_\tau}$, $\mathfrak{P} = e^{\beta_\sigma}$, we put

$$i_{L|K}(\mathfrak{p}) = \prod_{\mathfrak{P}|\mathfrak{p}} \mathfrak{P}^{e_{\mathfrak{P}|\mathfrak{p}}}, \quad N_{L|K}(\mathfrak{P}) = \mathfrak{p}^{f_{\mathfrak{P}|\mathfrak{p}}}.$$

Finally we define the *different* of $\widetilde{L}|\widetilde{K}$ to be the replete ideal

$$\mathfrak{D}_{\widetilde{L}|\widetilde{K}} = \mathfrak{D}_{L|K} \cdot \mathfrak{D}_\infty \in J(\bar{\mathcal{O}}_L) = J(\mathcal{O}_L) \times \prod_{\mathfrak{P}|\infty} \mathbb{R}_+^*,$$

where $\mathfrak{D}_{L|K}$ is the different of $L|K$ and

$$\mathfrak{D}_\infty = (\beta_\mathfrak{P}/\alpha_\mathfrak{p})_{\mathfrak{P}|\infty} \in \prod_{\mathfrak{P}|\infty} \mathbb{R}_+^*,$$

where $\beta_\mathfrak{P} = \beta_\sigma$ and $\alpha_\mathfrak{p} = \alpha_\tau$ ($\mathfrak{P}$ belongs to σ and $\mathfrak{p}$ to $\tau = \sigma|_K$). With this convention, we obtain the general *Riemann-Hurwitz formula*

$$g_{\widetilde{L}} - \ell_{\widetilde{L}}(\mathcal{O}_L) = [L : K]\big(g_{\widetilde{K}} - \ell_{\widetilde{K}}(\mathcal{O}_K)\big) - \frac{1}{2}\deg \mathfrak{D}_{\widetilde{L}|\widetilde{K}}.$$

If we consider only number fields endowed with the Minkowski metric, then $L_\mathfrak{P} \neq K_\mathfrak{p}$ is always ramified. In this way the convention found in many textbooks is no longer incompatible with the customs introduced in the present book.

§ 4. Metrized $\mathcal{O}$-Modules

The Riemann-Roch theory which was presented in the preceding section in the case of replete ideals is embedded in a much more far-reaching theory which treats finitely generated $\mathcal{O}$-modules. It is only in this setting that the theory displays its true nature, and becomes susceptible to the most comprehensive generalization. This theory is subject to a formalism which has been constructed by *Alexander Grothendieck* for higher dimensional algebraic varieties, and which we will now develop for number fields. In doing so, our principal attention will be focused as before on the kind of compactification which is accomplished by taking into account the infinite places. The effect is that a leading rôle is claimed by linear algebra – for which we refer to [15]. Our treatment is based on a course on "Arakelov Theory and Grothendieck-Riemann-Roch" taught by *Günter Tamme*. There, however, proofs were not given directly, as we will do here, but usually deduced as special cases from the general abstract theory.

Let K be an algebraic number field and $\mathcal{O}$ the ring of integers of K. For the passage from K to $\mathbb{R}$ and $\mathbb{C}$, we start by considering the ring

$$(1) \qquad K_{\mathbb{C}} = K \otimes_{\mathbb{Q}} \mathbb{C}.$$

It admits the following two further interpretations, between which we will freely go back and forth in the sequel without further explanation. The set

$$X(\mathbb{C}) = \mathrm{Hom}(K, \mathbb{C})$$

induces a canonical decomposition of rings

$$(2) \qquad K_{\mathbb{C}} \cong \bigoplus_{\sigma \in X(\mathbb{C})} \mathbb{C}, \quad a \otimes z \longmapsto \bigoplus_{\sigma \in X(\mathbb{C})} z\sigma a.$$

Alternatively, the right-hand side may be viewed as the set $\mathbb{C}^{X(\mathbb{C})} = \mathrm{Hom}(X(\mathbb{C}), \mathbb{C})$ of all functions $x : X(\mathbb{C}) \to \mathbb{C}$, i.e.,

$$(3) \qquad K_{\mathbb{C}} \cong \mathrm{Hom}\big(X(\mathbb{C}), \mathbb{C}\big).$$

The field K is embedded in $K_{\mathbb{C}}$ via

$$K \longrightarrow K \otimes_{\mathbb{Q}} \mathbb{C}, \quad a \longmapsto a \otimes 1,$$

and we identify it with its image. In the interpretation (2), the image of $a \in K$ appears as the tuple $\bigoplus_{\sigma} \sigma a$ of conjugates of a, and in the interpretation (3) as the function $x(\sigma) = \sigma a$.

We denote the generator of the Galois group $G(\mathbb{C}|\mathbb{R})$ by F_∞, or simply by F. This underlines the fact that it has a position analogous to the Frobenius automorphism $F_p \in G(\overline{\mathbb{F}}_p|\mathbb{F}_p)$, in accordance with our decision of §1 to view the extension $\mathbb{C}|\mathbb{R}$ as unramified. F induces an involution F on $K_{\mathbb{C}}$

which, in the representation $K_{\mathbb{C}} = \mathrm{Hom}(X(\mathbb{C}), \mathbb{C})$ for $x : X(\mathbb{C}) \to \mathbb{C}$, is given by

$$(Fx)(\sigma) = \overline{x(\overline{\sigma})}.$$

F is an automorphism of the $\mathbb{R}$-algebra $K_{\mathbb{C}}$. It is called the **Frobenius correspondence**. Sometimes we also consider, besides F, the involution $z \mapsto \bar{z}$ on $K_{\mathbb{C}}$ which is given by

$$\bar{z}(\sigma) = \overline{z(\sigma)}.$$

We call it the **conjugation**. Finally, we call an element $x \in K_{\mathbb{C}}$, that is, a function $x : X(\mathbb{C}) \to \mathbb{C}$, **positive** (written $x > 0$) if it takes real values, and if $x(\sigma) > 0$ for all $\sigma \in X(\mathbb{C})$.

By convention every $\mathcal{O}$-module considered in the sequel will be supposed to be *finitely generated*. For every such $\mathcal{O}$-module M, we put

$$M_{\mathbb{C}} = M \otimes_{\mathbb{Z}} \mathbb{C}.$$

This is a module over the ring $K_{\mathbb{C}} = \mathcal{O} \otimes_{\mathbb{Z}} \mathbb{C}$, and viewing $\mathcal{O}$ as a subring of $K_{\mathbb{C}}$ – as we agreed above – we may also write

$$M_{\mathbb{C}} = M \otimes_{\mathcal{O}} K_{\mathbb{C}}$$

as $M \otimes_{\mathbb{Z}} \mathbb{C} = M \otimes_{\mathcal{O}} (\mathcal{O} \otimes_{\mathbb{Z}} \mathbb{C})$. The involution $x \mapsto Fx$ on $K_{\mathbb{C}}$ induces the involution

$$F(a \otimes x) = a \otimes Fx$$

on $M_{\mathbb{C}}$. In the representation $M_{\mathbb{C}} = M \otimes_{\mathbb{Z}} \mathbb{C}$, one clearly has

$$F(a \otimes z) = a \otimes \bar{z}.$$

(4.1) Definition. *A* **hermitian metric** *on the $K_{\mathbb{C}}$-module $M_{\mathbb{C}}$ is a sesquilinear mapping*

$$\langle\ ,\ \rangle_M : M_{\mathbb{C}} \times M_{\mathbb{C}} \longrightarrow K_{\mathbb{C}},$$

i.e., a $K_{\mathbb{C}}$-linear form $\langle x, y\rangle_M$ in the first variable satisfying

$$\overline{\langle x, y\rangle_M} = \langle y, x\rangle_M,$$

such that one has $\langle x, x\rangle_M > 0$ for $x \neq 0$.

The metric $\langle\ ,\ \rangle_M$ is called ***F*-invariant** *if we have furthermore*

$$F\langle x, y\rangle_M = \langle Fx, Fy\rangle_M.$$

This notion may be immediately reduced to the usual notion of a hermitian metric if we view the $K_{\mathbb{C}}$-module $M_{\mathbb{C}}$, by means of the decomposition $K_{\mathbb{C}} = \bigoplus_\sigma \mathbb{C}$, as a direct sum

$$M_{\mathbb{C}} = M \otimes_{\mathcal{O}} K_{\mathbb{C}} = \bigoplus_{\sigma \in X(\mathbb{C})} M_\sigma$$

of $\mathbb{C}$-vector spaces

$$M_\sigma = M \otimes_{\mathcal{O},\sigma} \mathbb{C}.$$

The hermitian metric $\langle\ ,\ \rangle_M$ then splits into the direct sum

$$\langle x, y\rangle_M = \bigoplus_{\sigma \in X(\mathbb{C})} \langle x_\sigma, y_\sigma\rangle_{M_\sigma}$$

of hermitian scalar products $\langle\ ,\ \rangle_{M_\sigma}$ on the $\mathbb{C}$-vector spaces M_σ. In this interpretation, the F-invariance of $\langle x, y\rangle_M$ amounts to the commutativity of the diagrams

$$\begin{array}{ccc} M_\sigma \times M_\sigma & \xrightarrow{\langle\ ,\ \rangle_{M_\sigma}} & \mathbb{C} \\ {\scriptstyle F\times F}\downarrow & & \downarrow{\scriptstyle F} \\ M_{\bar\sigma} \times M_{\bar\sigma} & \xrightarrow[\langle\ ,\ \rangle_{M_{\bar\sigma}}]{} & \mathbb{C}. \end{array}$$

(4.2) Definition. *A* **metrized** $\mathcal{O}$**-module** *is a finitely generated $\mathcal{O}$-module M with an F-invariant hermitian metric on $M_{\mathbb{C}}$.*

Example 1: Every fractional ideal $\mathfrak{a} \subseteq K$ of $\mathcal{O}$, in particular $\mathcal{O}$ itself, may be equipped with the **trivial metric**

$$\langle x, y\rangle = x\bar{y} = \bigoplus_{\sigma \in X(\mathbb{C})} x_\sigma \bar{y}_\sigma$$

on $\mathfrak{a} \otimes_{\mathbb{Z}} \mathbb{C} = K \otimes_{\mathbb{Q}} \mathbb{C} = K_{\mathbb{C}}$. All the F-invariant hermitian metrics on $\mathfrak{a}$ are obtained as

$$\alpha\langle x, y\rangle = \alpha x\bar{y} = \bigoplus_\sigma \alpha(\sigma) x_\sigma \bar{y}_\sigma\,,$$

where $\alpha \in K_{\mathbb{C}}$ varies over the functions $\alpha : X(\mathbb{C}) \to \mathbb{R}_+^*$ such that $\alpha(\bar\sigma) = \alpha(\sigma)$.

Example 2: Let $L|K$ be a finite extension and $\mathfrak{A}$ a fractional ideal of L, which we view as an $\mathcal{O}$-module M. If $Y(\mathbb{C}) = \operatorname{Hom}(L, \mathbb{C})$, we have the restriction map $Y(\mathbb{C}) \to X(\mathbb{C})$, $\tau \mapsto \tau|_K$, and we write $\tau|\sigma$ if $\sigma = \tau|_K$. For the complexification $M_{\mathbb{C}} = \mathfrak{A} \otimes_{\mathbb{Z}} \mathbb{C} = L_{\mathbb{C}}$, we obtain the decomposition

$$M_{\mathbb{C}} = \bigoplus_{\tau \in Y(\mathbb{C})} \mathbb{C} = \bigoplus_{\sigma \in X(\mathbb{C})} M_\sigma\,,$$

where $M_\sigma = \bigoplus_{\tau|\sigma} \mathbb{C}$. M is turned into a metrized $\mathcal{O}$-module by fixing the **standard metrics**

$$\langle x, y\rangle_{M_\sigma} = \sum_{\tau|\sigma} x_\tau \bar{y}_\tau$$

on the $[L:K]$-dimensional $\mathbb{C}$-vector spaces M_σ.

If M and M' are metrized $\mathcal{O}$-modules, then so is their direct sum $M \oplus M'$, the tensor product $M \otimes M'$, the dual $\check{M} = \operatorname{Hom}_{\mathcal{O}}(M, \mathcal{O})$ and the n-th exterior power $\bigwedge^n M$. In fact, we have that

$$(M \oplus M')_{\mathbb{C}} = M_{\mathbb{C}} \oplus M'_{\mathbb{C}}, \quad (M \otimes_{\mathcal{O}} M')_{\mathbb{C}} = M_{\mathbb{C}} \otimes_{K_{\mathbb{C}}} M'_{\mathbb{C}},$$
$$\check{M}_{\mathbb{C}} = \operatorname{Hom}_{K_{\mathbb{C}}}(M_{\mathbb{C}}, K_{\mathbb{C}}), \quad (\textstyle\bigwedge^n M)_{\mathbb{C}} = \textstyle\bigwedge^n{}_{K_{\mathbb{C}}} M_{\mathbb{C}},$$

and the metrics on these $K_{\mathbb{C}}$-modules are given by

$$\langle x \oplus x', y \oplus y'\rangle_{M \oplus M'} = \langle x, y\rangle_M + \langle x', y'\rangle_{M'}, \qquad \text{resp.}$$
$$\langle x \otimes x', y \otimes y'\rangle_{M \otimes M'} = \langle x, y\rangle_M \cdot \langle x', y'\rangle_{M'}, \qquad \text{resp.}$$
$$\langle \check{x}, \check{y}\rangle_{\check{M}} = \overline{\langle x, y\rangle_M}, \qquad \text{resp.}$$
$$\big\langle x_1 \wedge \ldots \wedge x_n, y_1 \wedge \ldots \wedge y_n\big\rangle_{\bigwedge^n M} = \det\big(\langle x_i, y_j\rangle_M\big).$$

Here $\check{x}$, in the case of the module $\check{M}_{\mathbb{C}}$, denotes the homomorphism $\check{x} = \langle\ , x\rangle_M : M_{\mathbb{C}} \to K_{\mathbb{C}}$.

Among all $\mathcal{O}$-modules M the **projective** ones play a special rôle. They are defined by the condition that for every exact sequence of $\mathcal{O}$-modules $F' \to F \to F''$ the sequence

$$\operatorname{Hom}_{\mathcal{O}}(M, F') \longrightarrow \operatorname{Hom}_{\mathcal{O}}(M, F) \longrightarrow \operatorname{Hom}_{\mathcal{O}}(M, F'')$$

is also exact. This is equivalent to any of the following conditions (the last two, because $\mathcal{O}$ is a Dedekind domain). For the proof, we refer the reader to standard textbooks of commutative algebra (see for instance [90], chap. IV, §3, or [16], chap. 7, §4).

(4.3) Proposition. *For any finitely generated $\mathcal{O}$-module M the following conditions are equivalent:*

(i) *M is projective,*

(ii) *M is a direct summand of a finitely generated free $\mathcal{O}$-module,*

(iii) *M is locally free, i.e., $M \otimes_{\mathcal{O}} \mathcal{O}_{\mathfrak{p}}$ is a free $\mathcal{O}_{\mathfrak{p}}$-module for any prime ideal $\mathfrak{p}$,*

(iv) *M is torsion free, i.e., the map $M \to M$, $x \mapsto ax$, is injective for all nonzero $a \in \mathcal{O}$,*

(v) *$M \cong \mathfrak{a} \oplus \mathcal{O}^n$ for some ideal $\mathfrak{a}$ of $\mathcal{O}$ and some integer $n \geq 0$.*

In order to distinguish them from projective $\mathcal{O}$-modules, we will henceforth call arbitrary finitely generated $\mathcal{O}$-modules *coherent*. The **rank** of a coherent $\mathcal{O}$-module M is defined to be the dimension

$$\mathrm{rk}(M) = \dim_K(M \otimes_{\mathcal{O}} K).$$

The projective $\mathcal{O}$-modules L of rank 1 are called **invertible** $\mathcal{O}$-modules, because for them $L \otimes_{\mathcal{O}} \check{L} \to \mathcal{O}$, $a \otimes \check{a} \mapsto \check{a}(a)$, is an isomorphism. The invertible $\mathcal{O}$-modules are either fractional ideals of K, or isomorphic to a fractional ideal as $\mathcal{O}$-modules. Indeed, if L is projective of rank 1 and $\alpha \in L$, $\alpha \neq 0$, then, by (4.3), (iv), mapping

$$L \longrightarrow L \otimes_{\mathcal{O}} K = K(\alpha \otimes 1), \quad x \longmapsto f(x)(\alpha \otimes 1),$$

gives an injective $\mathcal{O}$-module homomorphism $L \to K$, $x \mapsto f(x)$, onto a fractional ideal $\mathfrak{a} \subseteq K$.

To see the connection with the Riemann-Roch theory of the last section, which we are about to generalize, we observe that every replete ideal

$$\mathfrak{a} = \prod_{\mathfrak{p} \nmid \infty} \mathfrak{p}^{\nu_{\mathfrak{p}}} \prod_{\mathfrak{p} | \infty} \mathfrak{p}^{\nu_{\mathfrak{p}}} = \mathfrak{a}_f \mathfrak{a}_\infty$$

of K defines an invertible, metrized $\mathcal{O}$-module. In fact, the identity $\mathfrak{a}_\infty = \prod_{\mathfrak{p}|\infty} \mathfrak{p}^{\nu_{\mathfrak{p}}}$ yields the function

$$\alpha : X(\mathbb{C}) \longrightarrow \mathbb{R}_+, \quad \alpha(\sigma) = e^{2\nu_{\mathfrak{p}_\sigma}},$$

where $\mathfrak{p}_\sigma$ denotes as before the infinite place defined by $\sigma : K \to \mathbb{C}$. Since $\mathfrak{p}_{\overline{\sigma}} = \mathfrak{p}_\sigma$, one has $\alpha(\overline{\sigma}) = \alpha(\sigma)$, and we obtain on the complexification

$$\mathfrak{a}_{f\mathbb{C}} = \mathfrak{a}_f \otimes_{\mathbb{Z}} \mathbb{C} = K_{\mathbb{C}}$$

the F-invariant hermitian metric

$$\langle x, y\rangle_{\mathfrak{a}} = \alpha x \overline{y} = \bigoplus_{\sigma \in X(\mathbb{C})} e^{2\nu_{\mathfrak{p}_\sigma}} x_\sigma \overline{y}_\sigma$$

(see example 1, p. 227). We denote the metrized $\mathcal{O}$-module thus obtained by $L(\mathfrak{a})$.

The ordinary fractional ideals, i.e., the replete ideals $\mathfrak{a}$ such that $\mathfrak{a}_\infty = 1$, and in particular $\mathcal{O}$ itself, are thus equipped with the *trivial* metric $\langle x, y\rangle_{\mathfrak{a}} = \langle x, y\rangle = x\overline{y}$.

(4.4) Definition. *Two metrized $\mathcal{O}$-modules M and M' are called* **isometric** *if there exists an isomorphism*

$$f : M \longrightarrow M'$$

of $\mathcal{O}$-modules which induces an isometry $f_{\mathbb{C}} : M_{\mathbb{C}} \to M'_{\mathbb{C}}$.

(4.5) Proposition.

(i) *Two replete ideals* $\mathfrak{a}$ *and* $\mathfrak{b}$ *define isometric metrized* $\mathcal{O}$*-modules* $L(\mathfrak{a})$ *and* $L(\mathfrak{b})$ *if and only if they differ by a replete principal ideal* $[a]$: $\mathfrak{a} = \mathfrak{b}[a]$.

(ii) *Every invertible metrized* $\mathcal{O}$*-module is isometric to an* $\mathcal{O}$*-module* $L(\mathfrak{a})$.

(iii) $L(\mathfrak{a}\mathfrak{b}) \cong L(\mathfrak{a}) \otimes_{\mathcal{O}} L(\mathfrak{b})$, $L(\mathfrak{a}^{-1}) = \check{L}(\mathfrak{a})$.

Proof: (i) Let $\mathfrak{a} = \prod_{\mathfrak{p}} \mathfrak{p}^{\nu_{\mathfrak{p}}}$, $\mathfrak{b} = \prod_{\mathfrak{p}} \mathfrak{p}^{\mu_{\mathfrak{p}}}$, $[a] = \prod_{\mathfrak{p}} \mathfrak{p}^{v_{\mathfrak{p}}(a)}$, and let

$$\alpha(\sigma) = e^{2\nu_{\mathfrak{p}_\sigma}}, \quad \beta(\sigma) = e^{2\mu_{\mathfrak{p}_\sigma}}, \quad \gamma(\sigma) = e^{2v_{\mathfrak{p}_\sigma}(a)}.$$

If $\mathfrak{a} = \mathfrak{b}[a]$, then $\nu_{\mathfrak{p}} = \mu_{\mathfrak{p}} + v_{\mathfrak{p}}(a)$; thus $\alpha = \beta\gamma$, and $\mathfrak{a}_f = \mathfrak{b}_f(a)$. The $\mathcal{O}$-module isomorphism $\mathfrak{b}_f \to \mathfrak{a}_f$, $x \mapsto ax$, takes the form $\langle\ ,\ \rangle_{\mathfrak{b}}$ to the form $\langle\ ,\ \rangle_{\mathfrak{a}}$. Indeed, viewing a as embedded in $K_{\mathbb{C}}$, we find $a = \bigoplus_\sigma \sigma a$ and

$$a\bar{a} = \bigoplus_\sigma e^{-2v_{\mathfrak{p}_\sigma}(a)} = \gamma^{-1},$$

because $v_{\mathfrak{p}_\sigma}(a) = -\log|\sigma a|$, so that

$$\langle ax, ay\rangle_{\mathfrak{a}} = \alpha\langle ax, ay\rangle = \alpha\gamma^{-1}\langle x, y\rangle = \beta\langle x, y\rangle = \langle x, y\rangle_{\mathfrak{b}}.$$

Therefore $\mathfrak{b}_f \to \mathfrak{a}_f$, $x \mapsto ax$, gives an isometry $L(\mathfrak{a}) \cong L(\mathfrak{b})$.

Conversely, let $g : L(\mathfrak{b}) \to L(\mathfrak{a})$ be an isometry. Then the $\mathcal{O}$-module homomorphism

$$g : \mathfrak{b}_f \to \mathfrak{a}_f$$

is given as multiplication by some element $a \in \mathfrak{b}_f^{-1}\mathfrak{a}_f \cong \operatorname{Hom}_{\mathcal{O}}(\mathfrak{b}_f, \mathfrak{a}_f)$. The identity

$$\beta\langle x, y\rangle = \langle x, y\rangle_{\mathfrak{b}} = \langle g(x), g(y)\rangle_{\mathfrak{a}} = \alpha\langle ax, ay\rangle = \alpha\gamma^{-1}\langle x, y\rangle$$

then implies that $\alpha = \beta\gamma$, so that $\nu_{\mathfrak{p}} = \mu_{\mathfrak{p}} + v_{\mathfrak{p}}(a)$ for all $\mathfrak{p}|\infty$. In view of $\mathfrak{a}_f = \mathfrak{b}_f(a)$, this yields $\mathfrak{a} = \mathfrak{b}[a]$.

(ii) Let L be an invertible metrized $\mathcal{O}$-module. As mentioned before, we have an isomorphism

$$g : L \longrightarrow \mathfrak{a}_f$$

for the underlying $\mathcal{O}$-module onto a fractional ideal $\mathfrak{a}_f$. The isomorphism $g_{\mathbb{C}} : L_{\mathbb{C}} \to \mathfrak{a}_{f\mathbb{C}} = K_{\mathbb{C}}$ gives us the F-invariant hermitian metric $h(x, y) = \langle g_{\mathbb{C}}^{-1}(x), g_{\mathbb{C}}^{-1}(y)\rangle_L$ on $K_{\mathbb{C}}$. It is of the form

$$h(x, y) = \alpha x\bar{y}$$

for some function $\alpha : X(\mathbb{C}) \to \mathbb{R}_+^*$ such that $\alpha(\bar{\sigma}) = \alpha(\sigma)$. Putting now $\alpha(\sigma) = e^{2\nu_{\mathfrak{p}_\sigma}}$, with $\nu_{\mathfrak{p}_\sigma} \in \mathbb{R}$, makes $\mathfrak{a}_f$ with the metric h into the metrized

$\mathcal{O}$-module $L(\mathfrak{a})$ associated to the replete ideal $\mathfrak{a} = \mathfrak{a}_f \prod_{\mathfrak{p}|\infty} \mathfrak{p}^{\nu_\mathfrak{p}}$, and L is isometric to $L(\mathfrak{a})$.

(iii) Let $\mathfrak{a} = \prod_\mathfrak{p} \mathfrak{p}^{\nu_\mathfrak{p}}$, $\mathfrak{b} = \prod_\mathfrak{p} \mathfrak{p}^{\mu_\mathfrak{p}}$, $\alpha(\sigma) = e^{2\nu_{\mathfrak{p}\sigma}}$, $\beta(\sigma) = e^{2\mu_{\mathfrak{p}\sigma}}$. The isomorphism

$$\mathfrak{a}_f \otimes_\mathcal{O} \mathfrak{b}_f \longrightarrow \mathfrak{a}_f\mathfrak{b}_f, \quad a \otimes b \longmapsto ab,$$

between the $\mathcal{O}$-modules underlying $L(\mathfrak{a}) \otimes_\mathcal{O} L(\mathfrak{b})$ and $L(\mathfrak{ab})$ then yields, as $\langle ab, a'b'\rangle_{\mathfrak{ab}} = \alpha\beta\, ab\,\overline{a'b'} = \alpha\langle a, a'\rangle\beta\langle b, b'\rangle = \langle a, a'\rangle_\mathfrak{a}\langle b, b'\rangle_\mathfrak{b}$, an isometry $L(\mathfrak{a}) \otimes_\mathcal{O} L(\mathfrak{b}) \cong L(\mathfrak{ab})$.

The $\mathcal{O}$-module $\operatorname{Hom}_\mathcal{O}(\mathfrak{a}_f, \mathcal{O})$ underlying $\check{L}(\mathfrak{a})$ is isomorphic, via the isomorphism

$$g : \mathfrak{a}_f^{-1} \longrightarrow \operatorname{Hom}_\mathcal{O}(\mathfrak{a}_f, \mathcal{O}), \quad a \longmapsto \big(g(a) : x \mapsto ax\big),$$

to the fractional ideal $\mathfrak{a}_f^{-1}$. For the induced $K_\mathbb{C}$-isomorphism

$$g_\mathbb{C} : K_\mathbb{C} \longrightarrow \operatorname{Hom}_{K_\mathbb{C}}(K_\mathbb{C}, K_\mathbb{C})$$

we have

$$g_\mathbb{C}(x)(y) = xy = \alpha^{-1}\alpha xy = \alpha^{-1}\langle y, \bar{x}\rangle_{L(\mathfrak{a})},$$

so that $g_\mathbb{C}(x) = \alpha^{-1}\check{\bar{x}}$, and thus

$$\begin{aligned}\big\langle g_\mathbb{C}(x), g_\mathbb{C}(y)\big\rangle_{\check{L}(\mathfrak{a})} &= \alpha^{-2}\langle\check{\bar{x}}, \check{\bar{y}}\rangle_{\check{L}(\mathfrak{a})} = \alpha^{-2}\,\overline{\langle\bar{x}, \bar{y}\rangle}_{L(\mathfrak{a})} \\ &= \alpha^{-1}x\bar{y} = \langle x, y\rangle_{L(\mathfrak{a}^{-1})}.\end{aligned}$$

Thus g gives an isometry $\check{L}(\mathfrak{a}) \cong L(\mathfrak{a}^{-1})$. □

(4.6) Definition. ***A short exact sequence***

$$0 \longrightarrow M' \xrightarrow{\alpha} M \xrightarrow{\beta} M'' \longrightarrow 0$$

of metrized $\mathcal{O}$-modules is by definition a short exact sequence of the underlying $\mathcal{O}$-modules which splits isometrically, i.e., in the sequence

$$0 \longrightarrow M'_\mathbb{C} \xrightarrow{\alpha_\mathbb{C}} M_\mathbb{C} \xrightarrow{\beta_\mathbb{C}} M''_\mathbb{C} \longrightarrow 0,$$

$M'_\mathbb{C}$ is mapped isometrically onto $\alpha_\mathbb{C} M'_\mathbb{C}$, and the orthogonal complement $(\alpha_\mathbb{C} M'_\mathbb{C})^\perp$ is mapped isometrically onto $M''_\mathbb{C}$.

The homomorphisms α, β in a short exact sequence of metrized $\mathcal{O}$-modules are called an **admissible monomorphism**, resp. **epimorphism**.

To each projective metrized $\mathcal{O}$-module M is associated its **determinant** $\det M$, an invertible metrized $\mathcal{O}$-module. The determinant is the highest exterior power of M, i.e.,

$$\det M = \bigwedge\nolimits^n M, \quad n = \operatorname{rk}(M).$$

(4.7) Proposition. *If* $0 \to M' \xrightarrow{\alpha} M \xrightarrow{\beta} M'' \to 0$ *is a short exact sequence of projective metrized* $\mathcal{O}$*-modules, we have a canonical isometry*

$$\det M' \otimes_{\mathcal{O}} \det M'' \cong \det M .$$

Proof: Let $n' = \mathrm{rk}(M')$ and $n'' = \mathrm{rk}(M'')$. We obtain an isomorphism

$$\kappa : \det M' \otimes_{\mathcal{O}} \det M'' \xrightarrow{\sim} \det M$$

of projective $\mathcal{O}$-modules of rank 1 by mapping

$$(m_1' \wedge \ldots \wedge m_{n'}') \otimes (m_1'' \wedge \ldots \wedge m_{n''}'') \mapsto \alpha m_1' \wedge \ldots \wedge \alpha m_{n'}' \wedge \widetilde{m}_1'' \wedge \ldots \wedge \widetilde{m}_{n''}'' ,$$

where $\widetilde{m}_1'', \ldots, \widetilde{m}_{n''}''$ are preimages of $m_1'', \ldots, m_{n''}''$ under $\beta : M \to M''$. This mapping does not depend on the choice of the preimages, for if, say, $\widetilde{m}_1'' + \alpha m_{n'+1}'$, where $m_{n'+1}' \in M'$, is another preimage of m_1'', then

$$\begin{aligned}\alpha m_1' \wedge \ldots \wedge \alpha m_{n'}' \wedge (\widetilde{m}_1'' + \alpha m_{n'+1}') \wedge \widetilde{m}_2'' \wedge \ldots \wedge \widetilde{m}_{n''}''\\ = \alpha m_1' \wedge \ldots \wedge \alpha m_{n'}' \wedge \widetilde{m}_1'' \wedge \ldots \wedge \widetilde{m}_{n''}''\end{aligned}$$

since $\alpha m_1' \wedge \ldots \wedge \alpha m_{n'}' \wedge \alpha m_{n'+1}' = 0$. We show that the $\mathcal{O}$-module isomorphism κ is an isometry. According to the rules of multilinear algebra it induces an isomorphism

$$\kappa : \det M_{\mathbb{C}}' \otimes_{K_{\mathbb{C}}} \det M_{\mathbb{C}}'' \xrightarrow{\sim} \det M_{\mathbb{C}}$$

of $K_{\mathbb{C}}$-modules. Let $x_i', y_i' \in M_{\mathbb{C}}'$, $i = 1, \ldots, n'$, and $x_j, y_j \in \alpha M_{\mathbb{C}}'^{\perp}$, $j = 1, \ldots, n''$, and furthermore

$$x' = \textstyle\bigwedge_i x_i', \quad y' = \bigwedge_i y_i'; \qquad x = \bigwedge_j x_j, \quad y = \bigwedge_j y_j .$$

Then we have

$$\begin{aligned}\langle \kappa(x' \otimes \beta x), \kappa(y' \otimes \beta y)\rangle_{\det M} &= \langle \alpha x' \wedge x, \alpha y' \wedge y\rangle_{\det M}\\ &= \det\begin{pmatrix} \langle x_i', y_k'\rangle_{M'} & 0 \\ 0 & \langle \beta x_j, \beta y_\ell\rangle_{M''}\end{pmatrix}\\ &= \det\big(\langle x_i', y_k'\rangle_{M'}\big) \det\big(\langle \beta x_j, \beta y_\ell\rangle_{M''}\big)\\ &= \langle x', y'\rangle_{\det M'} \langle \beta x, \beta y\rangle_{\det M''}\\ &= \langle x' \otimes \beta x, y' \otimes \beta y\rangle_{\det M' \otimes \det M''} .\end{aligned}$$

Thus κ is an isometry. □

Exercise 1. If M, N, L are metrized $\mathcal{O}$-modules, then one has canonical isometries

$$M \otimes_{\mathcal{O}} N \cong N \otimes_{\mathcal{O}} M, \quad (M \otimes_{\mathcal{O}} N) \otimes_{\mathcal{O}} L \cong M \otimes_{\mathcal{O}} (N \otimes_{\mathcal{O}} L),$$

$$M \otimes_{\mathcal{O}} (N \oplus L) \cong (M \otimes_{\mathcal{O}} N) \oplus (M \otimes_{\mathcal{O}} L).$$

Exercise 2. For any two projective metrized $\mathcal{O}$-modules M, N, one has

$$\bigwedge^{n}(M \oplus N) \cong \bigoplus_{i+j=n} \bigwedge^{i} M \otimes \bigwedge^{j} N.$$

Exercise 3. For any two projective metrized $\mathcal{O}$-modules M, N, one has

$$\det(M \otimes_{\mathcal{O}} N) \cong (\det M)^{\otimes \operatorname{rk}(N)} \otimes_{\mathcal{O}} (\det N)^{\otimes \operatorname{rk}(M)}.$$

Exercise 4. If M is a projective metrized $\mathcal{O}$-module of rank n, and $p \geq 0$, then there is a canonical isometry

$$\det(\bigwedge^{p} M) \cong (\det M)^{\otimes \binom{n-1}{p-1}}.$$

§ 5. Grothendieck Groups

We will now manufacture two abelian groups from the collection of all metrized $\mathcal{O}$-modules, resp. the collection of all projective metrized $\mathcal{O}$-modules. We denote by $\{M\}$ the isometry class of a metrized $\mathcal{O}$-module M and form the free abelian group

$$F_0(\bar{\mathcal{O}}) = \bigoplus_{\{M\}} \mathbb{Z}\{M\}, \quad \text{resp.} \quad F^0(\bar{\mathcal{O}}) = \bigoplus_{\{M\}} \mathbb{Z}\{M\},$$

on the isometry classes of projective, resp. coherent, metrized $\mathcal{O}$-modules. In this group, we consider the subgroup

$$R_0(\bar{\mathcal{O}}) \subseteq F_0(\bar{\mathcal{O}}), \quad \text{resp.} \quad R^0(\bar{\mathcal{O}}) \subseteq F^0(\bar{\mathcal{O}}),$$

generated by all elements $\{M'\} - \{M\} + \{M''\}$ which arise from a short exact sequence

$$0 \longrightarrow M' \longrightarrow M \longrightarrow M'' \longrightarrow 0$$

of projective, resp. coherent, metrized $\mathcal{O}$-modules.

(5.1) Definition. *The quotient groups*

$$K_0(\bar{\mathcal{O}}) = F_0(\bar{\mathcal{O}})/R_0(\bar{\mathcal{O}}), \quad \textit{resp.} \quad K^0(\bar{\mathcal{O}}) = F^0(\bar{\mathcal{O}})/R^0(\bar{\mathcal{O}})$$

are called the **replete** *(or* **compactified***)* **Grothendieck groups** *of $\mathcal{O}$. If M is a metrized $\mathcal{O}$-module, then $[M]$ denotes the class it defines in $K_0(\bar{\mathcal{O}})$, resp. $K^0(\bar{\mathcal{O}})$.*

The construction of the Grothendieck groups is such that a short exact sequence

$$0 \longrightarrow M' \longrightarrow M \longrightarrow M'' \longrightarrow 0$$

of metrized $\mathcal{O}$-modules becomes an additive decomposition in the group:

$$[M] = [M'] + [M''].$$

In particular, one has

$$[M' \oplus M''] = [M'] + [M''].$$

The tensor product even induces a ring structure on $K_0(\bar{\mathcal{O}})$, and $K^0(\bar{\mathcal{O}})$ then becomes a $K_0(\bar{\mathcal{O}})$-module: extending the product

$$\{M\}\{M'\} := \{M \otimes_{\mathcal{O}} M'\}$$

by linearity, and observing that $N \otimes M \cong M \otimes N$ and $(M \otimes N) \otimes L \cong M \otimes (N \otimes L)$, we find right away that $F^0(\bar{\mathcal{O}})$ is a commutative ring and $F_0(\bar{\mathcal{O}})$ is a subring. Furthermore the subgroups $R_0(\bar{\mathcal{O}}) \subseteq F_0(\bar{\mathcal{O}})$ and $R^0(\bar{\mathcal{O}}) \subseteq F^0(\bar{\mathcal{O}})$ turn out to be $F_0(\bar{\mathcal{O}})$-submodules. For if

$$0 \longrightarrow M' \longrightarrow M \longrightarrow M'' \longrightarrow 0$$

is a short exact sequence of coherent metrized $\mathcal{O}$-modules, and N is a projective metrized $\mathcal{O}$-module, then it is clear that

$$0 \longrightarrow N \otimes M' \longrightarrow N \otimes M \longrightarrow N \otimes M'' \longrightarrow 0$$

is a short exact sequence of metrized $\mathcal{O}$-modules as well, so that, along with a generator $\{M'\} - \{M\} + \{M''\}$, the element

$$\{N\}\big(\{M'\} - \{M\} + \{M''\}\big) = \{N \otimes M'\} - \{N \otimes M\} + \{N \otimes M''\}$$

will also belong to $R_0(\bar{\mathcal{O}})$, resp. $R^0(\bar{\mathcal{O}})$. This is why $K_0(\bar{\mathcal{O}}) = F_0(\bar{\mathcal{O}})/R_0(\bar{\mathcal{O}})$ is a ring and $K^0(\bar{\mathcal{O}}) = F^0(\bar{\mathcal{O}})/R^0(\bar{\mathcal{O}})$ is a $K_0(\bar{\mathcal{O}})$-module.

Associating to the class $[M]$ of a projective $\mathcal{O}$-module M in $K_0(\bar{\mathcal{O}})$ its class in $K^0(\bar{\mathcal{O}})$ (which again is denoted by $[M]$), defines a homomorphism

$$K_0(\bar{\mathcal{O}}) \longrightarrow K^0(\bar{\mathcal{O}}).$$

It is called the **Poincaré homomorphism**. We will show next that the Poincaré homomorphism is an isomorphism. The proof is based on the following two lemmas.

(5.2) Lemma. *All coherent metrized $\mathcal{O}$-modules M admit a "metrized projective resolution", i.e., a short exact sequence*

$$0 \longrightarrow E \longrightarrow F \longrightarrow M \longrightarrow 0$$

of metrized $\mathcal{O}$-modules in which E and F are projective.

Proof: If $\alpha_1, \ldots, \alpha_n$ is a system of generators of M, and F is the free $\mathcal{O}$-module $F = \mathcal{O}^n$, then

$$F \longrightarrow M, \quad (x_1, \ldots, x_n) \longmapsto \sum_{i=1}^{n} x_i \alpha_i ,$$

is a surjective $\mathcal{O}$-module homomorphism. Its kernel E is torsion free, and hence a projective $\mathcal{O}$-module by (4.3). In the exact sequence

$$0 \longrightarrow E_{\mathbb{C}} \longrightarrow F_{\mathbb{C}} \xrightarrow{f} M_{\mathbb{C}} \longrightarrow 0,$$

we choose a section $s : M_{\mathbb{C}} \to F_{\mathbb{C}}$ of f, so that $F_{\mathbb{C}} = E_{\mathbb{C}} \oplus sM_{\mathbb{C}}$. We obtain a metric on $F_{\mathbb{C}}$ by transferring the metric of $M_{\mathbb{C}}$ to $sM_{\mathbb{C}}$, and by choosing any metric on $E_{\mathbb{C}}$. This makes $0 \to E \to F \to M \to 0$ into a short exact sequence of metrized $\mathcal{O}$-modules in which E and F are projective. □

In a diagram of metrized projective resolutions of M

$$\begin{array}{ccccccccc}
0 & \longrightarrow & E & \longrightarrow & F & \longrightarrow & M & \longrightarrow & 0 \\
 & & \downarrow & & \downarrow & & \| & & \\
0 & \longrightarrow & E' & \longrightarrow & F' & \longrightarrow & M & \longrightarrow & 0
\end{array}$$

the resolution in the top line will be called *dominant* if the vertical arrows are admissible epimorphisms.

(5.3) Lemma. *Let*

$$0 \longrightarrow E' \longrightarrow F' \xrightarrow{f'} M \longrightarrow 0, \qquad 0 \longrightarrow E'' \longrightarrow F'' \xrightarrow{f''} M \longrightarrow 0$$

be two metrized projective resolutions of the metrized $\mathcal{O}$-module M. Then, taking the $\mathcal{O}$-module

$$F = F' \times_M F'' = \big\{ (x', x'') \in F' \times F'' \,\big|\, f'(x') = f''(x'') \big\}$$

and the mapping $f : F \to M$, $(x', x'') \mapsto f'(x') = f''(x'')$, one obtains a third metrized projective resolution

$$0 \longrightarrow E \longrightarrow F \longrightarrow M \longrightarrow 0$$

with kernel $E = E' \times E''$ which dominates both given ones.

Proof: Since $F' \oplus F''$ is projective, so is F, being the kernel of the homomorphism $F' \oplus F'' \xrightarrow{f'-f''} M$. Thus E is also projective, being the kernel of $F \to M$. We consider the commutative diagram

$$\begin{array}{ccccccccc}
0 \longrightarrow & E'_{\mathbb{C}} & \longrightarrow & F'_{\mathbb{C}} & \underset{s'}{\overset{f'}{\rightleftarrows}} & M_{\mathbb{C}} & \longrightarrow 0 \\
& \uparrow & & \uparrow & & \| & \\
0 \longrightarrow & E_{\mathbb{C}} & \longrightarrow & F_{\mathbb{C}} & \underset{s}{\overset{f}{\rightleftarrows}} & M_{\mathbb{C}} & \longrightarrow 0 \\
& \downarrow & & \downarrow & & \| & \\
0 \longrightarrow & E''_{\mathbb{C}} & \longrightarrow & F''_{\mathbb{C}} & \underset{s''}{\overset{f''}{\rightleftarrows}} & M_{\mathbb{C}} & \longrightarrow 0,
\end{array}$$

where the vertical arrows are induced by the surjective projections

$$F \xrightarrow{\pi'} F', \quad F \xrightarrow{\pi''} F''.$$

The canonical isometries

$$s' : M_{\mathbb{C}} \longrightarrow s'M_{\mathbb{C}}, \quad s'' : M_{\mathbb{C}} \longrightarrow s''M_{\mathbb{C}}$$

give a section

$$s : M_{\mathbb{C}} \longrightarrow F_{\mathbb{C}}, \quad sx = (s'x, s''x),$$

of F which transfers the metric on $M_{\mathbb{C}}$ to a metric on $sM_{\mathbb{C}}$. $E_{\mathbb{C}} = E'_{\mathbb{C}} \times E''_{\mathbb{C}}$ carries the sum of the metrics of $E'_{\mathbb{C}}$, $E''_{\mathbb{C}}$, so that $F_{\mathbb{C}} = E_{\mathbb{C}} \oplus sM_{\mathbb{C}}$ also receives a metric, and

$$0 \longrightarrow E \longrightarrow F \longrightarrow M \longrightarrow 0$$

becomes a metrized projective resolution of M. It is trivial that the projections $E \to E'$, and $E \to E''$ are admissible epimorphisms, and it remains to show this for the projections $\pi' : F \to F'$, $\pi'' : F \to F''$. But we clearly have the exact sequence of $\mathcal{O}$-modules

$$0 \longrightarrow E'' \xrightarrow{i} F = F' \times_M F'' \xrightarrow{\pi'} F' \longrightarrow 0,$$

where $ix'' = (0, x'')$. As the restriction of the metric of F to $E = E' \times E''$ is the sum of the metrics on E' and E'', we see that $i : E''_{\mathbb{C}} \to iE''_{\mathbb{C}}$ is an isometry. The orthogonal complement of $iE''_{\mathbb{C}}$ in $F_{\mathbb{C}}$ is the space

$$F'_{\mathbb{C}} \times_{M_{\mathbb{C}}} s''M_{\mathbb{C}} = \left\{ (x', s''a) \in F'_{\mathbb{C}} \times s''M_{\mathbb{C}} \;\middle|\; f'(x') = a \right\}.$$

Indeed, on the one hand it is clearly mapped bijectively onto $F'_{\mathbb{C}}$, and on the other hand it is orthogonal to $iE''_{\mathbb{C}}$. For if we write $x' = s'a + e'$, with $e' \in E'_{\mathbb{C}}$, then

$$(x', s''a) = sa + (e', 0),$$

where $(e',0) \in E_{\mathbb{C}}$ and we find that, for all $x'' \in E''_{\mathbb{C}}$,

$$\big\langle ix'', (x', s''a)\big\rangle_F = \big\langle (0,x''), sa\big\rangle_F + \big\langle (0,x''), (e',0)\big\rangle_E = 0\,.$$

Finally, the projection $F'_{\mathbb{C}} \times_{M_{\mathbb{C}}} s''M_{\mathbb{C}} \to F'_{\mathbb{C}}$ is an isometry, for if $(x', s''a)$, $(y', s''b) \in F'_{\mathbb{C}} \times_{M_{\mathbb{C}}} s''M_{\mathbb{C}}$ and $x' = s'a + e'$, $y' = s'b + d'$, with $e', d' \in E'_{\mathbb{C}}$, then we get

$$(x', s''a) = sa + (e',0)\,, \quad (y', s''b) = sb + (d',0)$$

and

$$\begin{aligned}\big\langle (x', s''a), (y', s''b)\big\rangle_F &= \big\langle sa, sb\big\rangle_F + \big\langle sa, (d',0)\big\rangle_F + \big\langle (e',0), sb\big\rangle_F \\ &\quad + \big\langle (e',0), (d',0)\big\rangle_E \\ &= \langle a,b\rangle_M + \langle e',d'\rangle_{E'} = \langle s'a, s'b\rangle_{F'} + \langle e', d'\rangle_{E'} \\ &= \langle s'a + e', s'b + d'\rangle_{F'} = \langle x', y'\rangle_{F'}\,.\end{aligned}$$

□

(5.4) Theorem. *The Poincaré homomorphism*

$$K_0(\bar{\mathcal{O}}) \longrightarrow K^0(\bar{\mathcal{O}})$$

is an isomorphism.

Proof: We define a mapping

$$\pi : F^0(\bar{\mathcal{O}}) \longrightarrow K_0(\bar{\mathcal{O}})$$

by choosing, for every coherent metrized $\mathcal{O}$-module M, a metrized projective resolution

$$0 \longrightarrow E \longrightarrow F \longrightarrow M \longrightarrow 0$$

and associating to the class $\{M\}$ in $F^0(\bar{\mathcal{O}})$ the difference $[F] - [E]$ of the classes $[F]$ and $[E]$ in $K_0(\bar{\mathcal{O}})$. To see that this mapping is well-defined let us first consider a commutative diagram

$$\begin{array}{ccccccccc} 0 & \longrightarrow & E & \longrightarrow & F & \longrightarrow & M & \longrightarrow & 0 \\ & & \big\downarrow{\scriptstyle\alpha} & & \big\downarrow{\scriptstyle\beta} & & \big\| & & \\ 0 & \longrightarrow & E' & \longrightarrow & F' & \longrightarrow & M & \longrightarrow & 0 \end{array}$$

of two metrized projective resolutions of M, with the top one dominating the bottom one. Then $E \to F$ induces an isometry $\ker(\alpha) \xrightarrow{\sim} \ker(\beta)$, so that we have the following identity in $K_0(\bar{\mathcal{O}})$:

$$[F] - [E] = [F'] + [\ker(\beta)] - [E'] - [\ker(\alpha)] = [F'] - [E'].$$

If now $0 \to E' \to F' \to M \to 0$, and $0 \to E'' \to F'' \to M \to 0$ are two arbitrary metrized projective resolutions of M, then by (5.3) we find a third one, $0 \to E \to F \to M \to 0$, dominating both, such that

$$[F'] - [E'] = [F] - [E] = [F''] - [E''].$$

This shows that the map $\pi : F^0(\bar{\mathcal{O}}) \to K_0(\bar{\mathcal{O}})$ is well-defined. We now show that it factorizes via $K^0(\bar{\mathcal{O}}) = F^0(\bar{\mathcal{O}})/R^0(\bar{\mathcal{O}})$. Let $0 \to M' \to M \xrightarrow{\alpha} M'' \to 0$ be a short exact sequence of metrized coherent $\mathcal{O}$-modules. By (5.2), we can pick a metrized projective resolution $0 \to E \to F \xrightarrow{f} M \to 0$. Then clearly $0 \to E'' \to F \xrightarrow{f''} M'' \to 0$ is a short exact sequence of metrized $\mathcal{O}$-modules as well, where we write $f'' = \alpha \circ f$ and $E'' = \ker(f'')$. We thus get the commutative diagram

$$\begin{array}{ccccccccc}
0 & \longrightarrow & E & \longrightarrow & F & \stackrel{f}{\longrightarrow} & M & \longrightarrow & 0 \\
 & & \downarrow & & \downarrow{\scriptstyle \mathrm{id}} & & \downarrow{\scriptstyle \alpha} & & \\
0 & \longrightarrow & E'' & \longrightarrow & F & \stackrel{f''}{\longrightarrow} & M'' & \longrightarrow & 0
\end{array}$$

and the snake lemma gives the exact sequence of $\mathcal{O}$-modules

$$0 \longrightarrow E \longrightarrow E'' \stackrel{f}{\longrightarrow} M' \longrightarrow 0.$$

It is actually a short exact sequence of metrized $\mathcal{O}$-modules, for $E_{\mathbb{C}}^{\perp}$ is mapped isometrically by f onto M, so that $E''^{\perp}_{\mathbb{C}} \subseteq E_{\mathbb{C}}^{\perp}$ is mapped isometrically by f onto $M''_{\mathbb{C}}$. We therefore obtain in $K_0(\bar{\mathcal{O}})$ the identity

$$\pi\{M'\} - \pi\{M\} + \pi\{M''\} = [E''] - [E] - \big([F] - [E]\big) + [F] - [E''] = 0.$$

It shows that $\pi : F^0(\bar{\mathcal{O}}) \to K_0(\bar{\mathcal{O}})$ does indeed factorize via a homomorphism

$$K^0(\bar{\mathcal{O}}) \longrightarrow K_0(\bar{\mathcal{O}}).$$

It is the inverse of the Poincaré homomorphism because the composed maps

$$K_0(\bar{\mathcal{O}}) \longrightarrow K^0(\bar{\mathcal{O}}) \longrightarrow K_0(\bar{\mathcal{O}}) \quad \text{and} \quad K^0(\bar{\mathcal{O}}) \longrightarrow K_0(\bar{\mathcal{O}}) \longrightarrow K^0(\bar{\mathcal{O}})$$

are the identity homomorphisms. Indeed, if $0 \to E \to F \to M \to 0$ is a projective resolution of M, and M is projective, resp. coherent, then in $K_0(\bar{\mathcal{O}})$, resp. $K^0(\bar{\mathcal{O}})$, one has the identity $[M] = [F] - [E]$. □

The preceding theorem shows that the Grothendieck group $K_0(\bar{\mathcal{O}})$ does not just accommodate all projective metrized $\mathcal{O}$-modules, but in fact all coherent metrized $\mathcal{O}$-modules. This fact has fundamental significance. For when dealing with projective modules, one is led very quickly to non-projective modules, for instance, to the residue class rings $\mathcal{O}/\mathfrak{a}$. The corresponding classes in $K^0(\bar{\mathcal{O}})$, however, can act out their important rôles only inside the ring $K_0(\bar{\mathcal{O}})$, because only this ring can be immediately subjected to a more advanced theory.

The following relationship holds between the Grothendieck ring $K_0(\bar{\mathcal{O}})$ and the replete Picard group $Pic(\bar{\mathcal{O}})$, which was introduced in §1.

(5.5) Proposition. *Associating to a replete ideal $\mathfrak{a}$ of K the metrized $\mathcal{O}$-module $L(\mathfrak{a})$ yields a homomorphism*

$$Pic(\bar{\mathcal{O}}) \longrightarrow K_0(\bar{\mathcal{O}})^*, \quad [\mathfrak{a}] \longmapsto [L(\mathfrak{a})],$$

into the unit group of the ring $K_0(\bar{\mathcal{O}})$.

Proof: The correspondence $[\mathfrak{a}] \mapsto [L(\mathfrak{a})]$ is independent of the choice of a replete ideal $\mathfrak{a}$ inside the class $[\mathfrak{a}] \in Pic(\bar{\mathcal{O}})$. Indeed, if $\mathfrak{b}$ is another representative, then we have $\mathfrak{a} = \mathfrak{b}[a]$, for some replete principal ideal $[a]$, and the metrized $\mathcal{O}$-modules $L(\mathfrak{a})$ and $L(\mathfrak{b})$ are isometric by (4.5), (i), so that $[L(\mathfrak{a})] = [L(\mathfrak{b})]$. The correspondence is a multiplicative homomorphism as

$$[L(\mathfrak{a}\mathfrak{b})] = [L(\mathfrak{a}) \otimes_{\mathcal{O}} L(\mathfrak{b})] = [L(\mathfrak{a})][L(\mathfrak{b})]. \qquad \square$$

In the sequel, we simply denote the class of a metrized invertible $\mathcal{O}$-module $L(\mathfrak{a})$ in $K_0(\bar{\mathcal{O}})$ by $[\mathfrak{a}]$. In particular, to the replete ideal $\mathcal{O} = \prod_{\mathfrak{p}} \mathfrak{p}^0$ corresponds the class $\mathbf{1} = [\mathcal{O}]$ of the $\mathcal{O}$-module $\mathcal{O}$ equipped with the trivial metric.

(5.6) Proposition. *$K_0(\bar{\mathcal{O}})$ is generated as an additive group by the elements $[\mathfrak{a}]$.*

Proof: Let M be a projective metrized $\mathcal{O}$-module. By (4.3), the underlying $\mathcal{O}$-module admits as quotient a fractional ideal $\mathfrak{a}_{\mathfrak{f}}$, i.e., we have an exact sequence

$$0 \longrightarrow N \longrightarrow M \longrightarrow \mathfrak{a}_{\mathfrak{f}} \longrightarrow 0$$

of $\mathcal{O}$-modules. This becomes an exact sequence of metrized $\mathcal{O}$-modules once we restrict the metric from M to N and choose on $\mathfrak{a}_{\mathfrak{f}}$ the metric which is transferred via the isomorphism $N_{\mathbb{C}}^{\perp} \cong \mathfrak{a}_{\mathfrak{f}\mathbb{C}}$. Thus $\mathfrak{a}_{\mathfrak{f}}$ becomes the metrized $\mathcal{O}$-module $L(\mathfrak{a})$ corresponding to the replete ideal $\mathfrak{a}$ of K, so that we get the identity $[M] = [N] + [\mathfrak{a}]$ in $K_0(\bar{\mathcal{O}})$. Induction on the rank shows that for every projective metrized $\mathcal{O}$-module M, there is a decomposition

$$[M] = [\mathfrak{a}_1] + \cdots + [\mathfrak{a}_r]. \qquad \square$$

The elements $[\mathfrak{a}]$ in $K_0(\bar{\mathcal{O}})$ satisfy the following remarkable relation.

(5.7) Proposition. *For any two replete ideals $\mathfrak{a}$ and $\mathfrak{b}$ of K we have in $K_0(\bar{\mathcal{O}})$ the equation*

$$([\mathfrak{a}] - 1)([\mathfrak{b}] - 1) = 0.$$

Proof (Tamme): For every function $\alpha : X(\mathbb{C}) \to \mathbb{C}$ let us consider on the $K_{\mathbb{C}}$-module $K_{\mathbb{C}} = \bigoplus_{\sigma \in X(\mathbb{C})} \mathbb{C}$ the form

$$\alpha x \bar{y} = \bigoplus_{\sigma} \alpha(\sigma) x_\sigma \bar{y}_\sigma.$$

For every matrix $A = \begin{pmatrix} \alpha & \gamma \\ \delta & \beta \end{pmatrix}$ of such functions, we consider on the $K_{\mathbb{C}}$-module $K_{\mathbb{C}} \oplus K_{\mathbb{C}}$ the form

$$\langle x \oplus y, x' \oplus y' \rangle_A = \alpha x \bar{x}' + \gamma x \bar{y}' + \delta y \bar{x}' + \beta y \bar{y}'.$$

$\alpha x \bar{y}$, resp. $\langle\ ,\ \rangle_A$, is an F-invariant metric on $K_{\mathbb{C}}$, resp. on $K_{\mathbb{C}} \oplus K_{\mathbb{C}}$, if and only if α is F-invariant (i.e., $\overline{\alpha(\sigma)} = \alpha(\bar{\sigma})$) and $\alpha(\sigma) \in \mathbb{R}_+^*$, resp. if all the functions $\alpha, \beta, \gamma, \delta$ are F-invariant, $\alpha(\sigma), \beta(\sigma) \in \mathbb{R}_+^*$ and $\delta = \bar{\gamma}$, and if moreover $\det A = \alpha\beta - \gamma\bar{\gamma} > 0$. We now assume this in what follows.

Let $\mathfrak{a}$ and $\mathfrak{b}$ be fractional ideals of K. We have to prove the formula

$$[\mathfrak{a}] + [\mathfrak{b}] = [\mathfrak{a}\mathfrak{b}] + 1.$$

We may assume that $\mathfrak{a}_{\mathfrak{f}}$ and $\mathfrak{b}_{\mathfrak{f}}$ are *integral* ideals *relatively prime* to one another, because if necessary we may pass to replete ideals $\mathfrak{a}' = \mathfrak{a}[a]$, $\mathfrak{b}' = \mathfrak{b}[b]$ with corresponding ideals $\mathfrak{a}'_{\mathfrak{f}} = \mathfrak{a}_{\mathfrak{f}} a$, $\mathfrak{b}'_{\mathfrak{f}} = \mathfrak{b}_{\mathfrak{f}} b$ without changing the classes $[\mathfrak{a}], [\mathfrak{b}], [\mathfrak{a}\mathfrak{b}]$ in $K_0(\bar{\mathcal{O}})$. We denote the $\mathcal{O}$-module $\mathfrak{a}_{\mathfrak{f}}$, when metrized by $\alpha x \bar{y}$, by $(\mathfrak{a}_{\mathfrak{f}}, \alpha)$, and the $\mathcal{O}$-module $\mathfrak{a}_{\mathfrak{f}} \oplus \mathfrak{b}_{\mathfrak{f}}$, metrized by $\langle\ ,\ \rangle_A$, for $A = \begin{pmatrix} \alpha & \gamma \\ \bar{\gamma} & \beta \end{pmatrix}$, by $(\mathfrak{a}_{\mathfrak{f}} \oplus \mathfrak{b}_{\mathfrak{f}}, A)$. Given any two matrices $A = \begin{pmatrix} \alpha & \gamma \\ \bar{\gamma} & \beta \end{pmatrix}$ and $A' = \begin{pmatrix} \alpha' & \gamma' \\ \bar{\gamma}' & \beta' \end{pmatrix}$ we write

$$A \sim A',$$

if $[(\mathfrak{a}_{\mathfrak{f}} \oplus \mathfrak{b}_{\mathfrak{f}}), A] = [(\mathfrak{a}_{\mathfrak{f}} \oplus \mathfrak{b}_{\mathfrak{f}}), A']$ in $K_0(\bar{\mathcal{O}})$. We now consider the canonical exact sequence

$$0 \longrightarrow \mathfrak{a}_{\mathfrak{f}} \longrightarrow \mathfrak{a}_{\mathfrak{f}} \oplus \mathfrak{b}_{\mathfrak{f}} \longrightarrow \mathfrak{b}_{\mathfrak{f}} \longrightarrow 0.$$

Once we equip $\mathfrak{a}_\mathfrak{f} \oplus \mathfrak{b}_\mathfrak{f}$ with the metric $\langle\ ,\ \rangle_A$ which is given by $A = \begin{pmatrix} \alpha & \gamma \\ \overline{\gamma} & \beta \end{pmatrix}$, we obtain the following exact sequence of metrized $\mathcal{O}$-modules:

$$(*) \qquad 0 \longrightarrow (\mathfrak{a}_\mathfrak{f}, \alpha) \longrightarrow (\mathfrak{a}_\mathfrak{f} \oplus \mathfrak{b}_\mathfrak{f}, A) \longrightarrow \left(\mathfrak{b}_\mathfrak{f}, \beta - \frac{\gamma\overline{\gamma}}{\alpha}\right) \longrightarrow 0.$$

Indeed, in the exact sequence

$$0 \longrightarrow K_\mathbb{C} \longrightarrow K_\mathbb{C} \oplus K_\mathbb{C} \longrightarrow K_\mathbb{C} \longrightarrow 0,$$

the restriction of $\langle\ ,\ \rangle_A$ to $K_\mathbb{C} \oplus \{0\}$ yields the metric $\alpha x \overline{y}$ on $K_\mathbb{C}$, and the orthogonal complement V of $K_\mathbb{C} \oplus \{0\}$ consists of all elements $a + b \in K_\mathbb{C} \oplus K_\mathbb{C}$ such that

$$\langle x \oplus 0, a \oplus b\rangle = \alpha x \overline{a} + \gamma x \overline{b} = 0,$$

for all $x \in K_\mathbb{C}$, so that

$$V = \left\{(-\overline{\gamma}/\alpha)b \oplus b \,\middle|\, b \in K_\mathbb{C}\right\}.$$

The isomorphism $V \xrightarrow{\pi} K_\mathbb{C}$, $(-\overline{\gamma}/\alpha)b \oplus b \longmapsto b$, transfers the metric $\langle\ ,\ \rangle_A$ on V into the metric $\delta x \overline{y}$, where δ is determined by the rule

$$\begin{aligned}\delta &= \left\langle \pi^{-1}(1), \pi^{-1}(1)\right\rangle_A = \left\langle (-\overline{\gamma}/\alpha)1 \oplus 1, (-\overline{\gamma}/\alpha)1 \oplus 1\right\rangle_A \\ &= \alpha\frac{\overline{\gamma}\gamma}{\alpha^2} - \gamma\frac{\overline{\gamma}}{\alpha} - \overline{\gamma}\frac{\gamma}{\alpha} + \beta = \beta - \frac{\overline{\gamma}\gamma}{\alpha}.\end{aligned}$$

This shows that $(*)$ is a short exact sequence of metrized $\mathcal{O}$-modules, i.e.,

$$\begin{pmatrix} \alpha & \gamma \\ \overline{\gamma} & \beta \end{pmatrix} \sim \begin{pmatrix} \alpha & 0 \\ 0 & \beta - \frac{\gamma\overline{\gamma}}{\alpha} \end{pmatrix}.$$

Replacing β by $\beta + \frac{\gamma\overline{\gamma}}{\alpha}$, we get

$$\begin{pmatrix} \alpha & \gamma \\ \overline{\gamma} & \beta + \frac{\gamma\overline{\gamma}}{\alpha} \end{pmatrix} \sim \begin{pmatrix} \alpha & 0 \\ 0 & \beta \end{pmatrix}.$$

Applying the same procedure to the exact sequence $0 \to \mathfrak{b}_\mathfrak{f} \to \mathfrak{a}_\mathfrak{f} \oplus \mathfrak{b}_\mathfrak{f} \to \mathfrak{a}_\mathfrak{f} \to 0$ and the metric $\begin{pmatrix} \alpha' & \gamma \\ \overline{\gamma} & \beta' \end{pmatrix}$ on $\mathfrak{a}_\mathfrak{f} \oplus \mathfrak{b}_\mathfrak{f}$, we obtain

$$\begin{pmatrix} \alpha' + \frac{\gamma\overline{\gamma}}{\beta'} & \gamma \\ \overline{\gamma} & \beta' \end{pmatrix} \sim \begin{pmatrix} \alpha' & 0 \\ 0 & \beta' \end{pmatrix}.$$

Choosing

$$\beta' = \beta + \frac{\gamma\overline{\gamma}}{\alpha}, \quad \alpha' = \frac{\alpha\beta}{\beta + \frac{\gamma\overline{\gamma}}{\alpha}},$$

makes the matrices on the left equal, and yields

$$\begin{pmatrix} \alpha & 0 \\ 0 & \beta \end{pmatrix} \sim \begin{pmatrix} \frac{\alpha\beta}{\beta+\frac{\gamma\bar{\gamma}}{\alpha}} & 0 \\ 0 & \beta+\frac{\gamma\bar{\gamma}}{\alpha} \end{pmatrix},$$

or, if we put $\delta = \beta + \frac{\gamma\bar{\gamma}}{\alpha}$,

$$(**) \qquad \begin{pmatrix} \alpha & 0 \\ 0 & \beta \end{pmatrix} \sim \begin{pmatrix} \frac{\alpha\beta}{\delta} & 0 \\ 0 & \delta \end{pmatrix},$$

which is valid for any F-invariant function $\delta : X(\mathbb{C}) \to \mathbb{R}$ such that $\delta \geq \beta$. This implies furthermore

$$(***) \qquad \begin{pmatrix} \frac{\alpha\beta}{\delta} & 0 \\ 0 & \delta \end{pmatrix} \sim \begin{pmatrix} \frac{\alpha\beta}{\varepsilon} & 0 \\ 0 & \varepsilon \end{pmatrix}$$

for any two F-invariant functions $\delta, \varepsilon : X(\mathbb{C}) \to \mathbb{R}_+^*$. For if $\kappa : X(\mathbb{C}) \to \mathbb{R}$ is an F-invariant function such that $\kappa \geq \delta, \kappa \geq \varepsilon$, then $(**)$ gives

$$\begin{pmatrix} \frac{\alpha\beta}{\delta} & 0 \\ 0 & \delta \end{pmatrix} \sim \begin{pmatrix} \frac{\alpha\beta}{\kappa} & 0 \\ 0 & \kappa \end{pmatrix} \sim \begin{pmatrix} \frac{\alpha\beta}{\varepsilon} & 0 \\ 0 & \varepsilon \end{pmatrix}.$$

Now putting $\delta = \beta$ and $\varepsilon = 1$ in $(***)$, we find

$$[(\mathfrak{a}_f, \alpha)] + [(\mathfrak{b}_f, \beta)] = [(\mathfrak{a}_f, \alpha\beta)] + [\mathfrak{b}_f].$$

For the replete ideals $\mathfrak{a} = \prod_\mathfrak{p} \mathfrak{p}^{\nu_\mathfrak{p}}$, $\mathfrak{b} = \prod_\mathfrak{p} \mathfrak{p}^{\nu_\mathfrak{p}}$, this means

$$(1) \qquad [\mathfrak{a}] + [\mathfrak{b}] = [\mathfrak{a}\mathfrak{b}_\infty] + [\mathfrak{b}_f],$$

for if we put $\alpha(\sigma) = e^{2\nu_{\mathfrak{p}\sigma}}$, $\beta(\sigma) = e^{2\mu_{\mathfrak{p}\sigma}}$, then we have

$$(\mathfrak{a}_f, \alpha) = L(\mathfrak{a}), \quad (\mathfrak{b}_f, \beta) = L(\mathfrak{b}), \quad (\mathfrak{a}_f, \alpha\beta) = L(\mathfrak{a}\mathfrak{b}_\infty).$$

On the other hand, we obtain the formula

$$(2) \qquad [\mathfrak{a}] + [\mathfrak{b}_f] = [\mathfrak{a}\mathfrak{b}_f] + 1$$

in the following manner. We have two exact sequences of *coherent* metrized $\mathcal{O}$-modules:

$$0 \longrightarrow (\mathfrak{a}_f\mathfrak{b}_f, \alpha) \longrightarrow (\mathfrak{a}_f, \alpha) \longrightarrow \mathfrak{a}_f/\mathfrak{a}_f\mathfrak{b}_f \longrightarrow 0,$$

$$0 \longrightarrow (\mathfrak{b}_f, 1) \longrightarrow (\mathcal{O}, 1) \longrightarrow \mathcal{O}/\mathfrak{b}_f \longrightarrow 0.$$

As $\mathfrak{a}_f$ and $\mathfrak{b}_f$ are relatively prime, i.e., $\mathfrak{a}_f + \mathfrak{b}_f = \mathcal{O}$, it follows that

$$\mathfrak{a}_f/\mathfrak{a}_f\mathfrak{b}_f \longrightarrow \mathcal{O}/\mathfrak{b}_f$$

is an isomorphism, so that in the group $K^0(\bar{\mathcal{O}})$ one has the identity $[\mathfrak{a}_f/\mathfrak{a}_f\mathfrak{b}_f] = [\mathcal{O}/\mathfrak{b}_f]$, and therefore

$$[(\mathfrak{a}_f, \alpha)] - [(\mathfrak{a}_f\mathfrak{b}_f, \alpha)] = [(\mathcal{O}, 1)] - [(\mathfrak{b}_f, 1)],$$

and so

$$[\mathfrak{a}] - [\mathfrak{a}\mathfrak{b}_f] = 1 - [\mathfrak{b}_f].$$

From (1) and (2) it now follows that

$$[\mathfrak{a}] + [\mathfrak{b}] = [\mathfrak{a}\mathfrak{b}_\infty] + [\mathfrak{b}_f] = [\mathfrak{a}\mathfrak{b}_\infty\mathfrak{b}_f] + 1 = [\mathfrak{a}\mathfrak{b}] + 1.$$

In view of the isomorphism $K_0(\bar{\mathcal{O}}) \cong K^0(\bar{\mathcal{O}})$, this is indeed an identity in $K_0(\bar{\mathcal{O}})$. □

§6. The Chern Character

The Grothendieck ring $K_0(\bar{\mathcal{O}})$ is equipped with a canonical surjective homomorphism

$$\mathrm{rk} : K_0(\bar{\mathcal{O}}) \longrightarrow \mathbb{Z}.$$

Indeed, the rule which associates to every isometry class $\{M\}$ of projective metrized $\mathcal{O}$-modules the rank

$$\mathrm{rk}\{M\} = \dim_K(M \otimes_{\mathcal{O}} K)$$

extends by linearity to a ring homomorphism $F_0(\bar{\mathcal{O}}) \to \mathbb{Z}$. For a short exact sequence $0 \to M' \to M \to M'' \to 0$ of metrized $\mathcal{O}$-modules one has $\mathrm{rk}(M) = \mathrm{rk}(M') + \mathrm{rk}(M'')$, and so $\mathrm{rk}(\{M'\} - \{M\} + \{M''\}) = 0$. Thus rk is zero on the ideal $R_0(\bar{\mathcal{O}})$ and induces therefore a homomorphism $K_0(\bar{\mathcal{O}}) \to \mathbb{Z}$. It is called the **augmentation** of $K_0(\bar{\mathcal{O}})$ and its kernel $I = \ker(\mathrm{rk})$ is called the **augmentation ideal**.

(6.1) Proposition. *The ideal I, resp. I^2, is generated as an additive group by the elements $[\mathfrak{a}] - 1$, resp. $([\mathfrak{a}] - 1)([\mathfrak{b}] - 1)$, where $\mathfrak{a}, \mathfrak{b}$ vary over the replete ideals of K.*

Proof: By (5.6), every element $\xi \in K_0(\bar{\mathcal{O}})$ is of the form

$$\xi = \sum_{i=1}^{r} n_i[\mathfrak{a}_i].$$

If $\xi \in I$, then $\operatorname{rk}(\xi) = \sum_{i=1}^{r} n_i = 0$, and thus

$$\xi = \sum_i n_i[\mathfrak{a}_i] - \sum_i n_i = \sum_i n_i([\mathfrak{a}_i] - 1).$$

The ideal I^2 is therefore generated by the elements $([\mathfrak{a}] - 1)([\mathfrak{b}] - 1)$. As

$$[\mathfrak{c}]([\mathfrak{a}] - 1)([\mathfrak{b}] - 1) = \big(([\mathfrak{c}\mathfrak{a}] - 1) - ([\mathfrak{c}] - 1)\big)([\mathfrak{b}] - 1),$$

these elements already form a system of generators of the abelian group I^2. □

By (5.7), this gives us the

(6.2) Corollary. $I^2 = 0$.

We now define

$$\operatorname{gr} K_0(\bar{\mathcal{O}}) = \mathbb{Z} \oplus I$$

and turn this additive group into a ring by putting $xy = 0$ for $x, y \in I$.

(6.3) Definition. *The additive homomorphism*

$$c_1 : K_0(\bar{\mathcal{O}}) \longrightarrow I, \quad c_1(\xi) = \xi - \operatorname{rk}(\xi)$$

is called the **first Chern class**. *The mapping*

$$\operatorname{ch} : K_0(\bar{\mathcal{O}}) \to \operatorname{gr} K_0(\bar{\mathcal{O}}), \quad \operatorname{ch}(\xi) = \operatorname{rk}(\xi) + c_1(\xi),$$

is called the **Chern character** *of $K_0(\bar{\mathcal{O}})$.*

(6.4) Proposition. *The Chern character*

$$\operatorname{ch} : K_0(\bar{\mathcal{O}}) \longrightarrow \operatorname{gr} K_0(\bar{\mathcal{O}})$$

is an isomorphism of rings.

Proof: The mappings rk and c_1 are homomorphisms of additive groups, and both are also multiplicative. For rk this is clear, and for c_1 it is enough to check it on the generators $x = [\mathfrak{a}]$, $y = [\mathfrak{b}]$. This works because

$$c_1(xy) = xy - 1 = (x-1) + (y-1) + (x-1)(y-1) = c_1(x) + c_1(y),$$

because $(x-1)(y-1) = 0$ by (5.7). Therefore ch is a ring homomorphism. The mapping

$$\mathbb{Z} \oplus I \longrightarrow K_0(\bar{\mathcal{O}}), \quad n \oplus \xi \longmapsto \xi + n,$$

is obviously an inverse mapping, so that ch is even an isomorphism. □

We obtain a complete and explicit description of the Chern character by taking into account another homomorphism, as well as the homomorphism $\mathrm{rk} : K_0(\bar{\mathcal{O}}) \to \mathbb{Z}$, namely

$$\det : K_0(\bar{\mathcal{O}}) \longrightarrow Pic(\bar{\mathcal{O}})$$

which is induced by taking determinants $\det M$ of projective $\mathcal{O}$-modules M as follows (see §4). $\det M$ is an invertible metrized $\mathcal{O}$-module, and therefore of the form $L(\mathfrak{a})$ for some replete ideal $\mathfrak{a}$, which is well determined up to isomorphism. Denoting by $[\det M]$ the class of $\mathfrak{a}$ in $Pic(\bar{\mathcal{O}})$, the linear extension of the map $\{M\} \mapsto [\det M]$ gives a homomorphism

$$\det : F_0(\bar{\mathcal{O}}) \longrightarrow Pic(\bar{\mathcal{O}}).$$

It maps the subgroup $R_0(\bar{\mathcal{O}})$ to 1, because it is generated by the elements $\{M'\} - \{M\} + \{M''\}$ which arise from short exact sequences

$$0 \longrightarrow M' \longrightarrow M \longrightarrow M'' \longrightarrow 0$$

of projective metrized $\mathcal{O}$-modules and which, by (4.7), satisfy

$$\begin{aligned}\det\{M\} &= [\det M] = [\det M' \otimes \det M''] \\ &= [\det M'][\det M''] = \det\{M'\}\det\{M''\}.\end{aligned}$$

Thus we get an induced homomorphism $\det : K_0(\bar{\mathcal{O}}) \to Pic(\bar{\mathcal{O}})$. It satisfies the following proposition.

(6.5) Proposition. (i) *The canonical homomorphism*

$$Pic(\bar{\mathcal{O}}) \longrightarrow K_0(\bar{\mathcal{O}})^*$$

is injective.

(ii) *The restriction of* det *to* I,

$$\det : I \longrightarrow Pic(\bar{\mathcal{O}}),$$

is an isomorphism.

Proof: (i) The composite of both mappings

$$Pic(\bar{\mathcal{O}}) \longrightarrow K_0(\bar{\mathcal{O}})^* \xrightarrow{\det} Pic(\bar{\mathcal{O}})$$

is the identity, since for an invertible metrized $\mathcal{O}$-module M, one clearly has $\det M = M$. This gives (i).

(ii) Next, viewing the elements of $Pic(\bar{\mathcal{O}})$ as elements of $K_0(\bar{\mathcal{O}})$,

$$\delta : Pic(\bar{\mathcal{O}}) \longrightarrow I, \quad \delta(x) = x - 1,$$

gives us an inverse mapping to $\det : I \to Pic(\bar{\mathcal{O}})$. In fact, one has $\det \circ \delta = \mathrm{id}$ since $\det([\mathfrak{a}] - 1) = \det[\mathfrak{a}] = [\mathfrak{a}]$, and $\delta \circ \det = \mathrm{id}$ since $\delta(\det([\mathfrak{a}] - 1)) = \delta(\det[\mathfrak{a}]) = \delta([\mathfrak{a}]) = [\mathfrak{a}] - 1$ and because of the fact that I is generated by elements of the form $[\mathfrak{a}] - 1$ (see (6.1)). □

From the isomorphism $\det : I \xrightarrow{\sim} Pic(\bar{\mathcal{O}})$, we now obtain an isomorphism

$$\operatorname{gr} K_0(\bar{\mathcal{O}}) \xrightarrow{\sim} \mathbb{Z} \oplus Pic(\bar{\mathcal{O}})$$

and the composite

$$K_0(\bar{\mathcal{O}}) \xrightarrow{\text{ch}} \operatorname{gr} K_0(\bar{\mathcal{O}}) \xrightarrow{\text{id}\oplus\det} \mathbb{Z} \oplus Pic(\bar{\mathcal{O}})$$

will again be called the Chern character of $K_0(\bar{\mathcal{O}})$. Observing that $\det(c_1(\xi)) = \det(\xi - \operatorname{rk}(\xi)\cdot 1) = \det(\xi)$, this yields the explicit description of the Grothendieck group $K_0(\bar{\mathcal{O}})$:

(6.6) Theorem. *The Chern character gives an isomorphism*

$$\text{ch} : K_0(\bar{\mathcal{O}}) \xrightarrow{\sim} \mathbb{Z} \oplus Pic(\bar{\mathcal{O}}), \quad \text{ch}(\xi) = \operatorname{rk}(\xi) \oplus \det(\xi)\,.$$

The expert should note that this homomorphism is a realization map from K-theory into Chow-theory. Identifying $Pic(\bar{\mathcal{O}})$ with the divisor class group $CH^1(\bar{\mathcal{O}})$, we have to view $\mathbb{Z} \oplus Pic(\bar{\mathcal{O}})$ as the "replete" *Chow ring* $CH(\bar{\mathcal{O}})$.

§ 7. Grothendieck-Riemann-Roch

We now consider a finite extension $L|K$ of algebraic number fields and study the relations between the Grothendieck groups of L and K. Let $\mathcal{O}$, resp. $\mathcal{O}$, be the ring of integers of K, resp. L, and write $X(\mathbb{C}) = \operatorname{Hom}(K,\mathbb{C})$, $Y(\mathbb{C}) = \operatorname{Hom}(L,\mathbb{C})$. The inclusion $i : \mathcal{O} \to \mathcal{O}$ and the surjection $Y(\mathbb{C}) \to X(\mathbb{C})$, $\sigma \mapsto \sigma|_K$, give two canonical homomorphisms

$$i^* : K_0(\bar{\mathcal{O}}) \longrightarrow K_0(\bar{\mathcal{O}}) \quad \text{and} \quad i_* : K_0(\bar{\mathcal{O}}) \longrightarrow K_0(\bar{\mathcal{O}}),$$

defined as follows.

If M is a projective metrized $\mathcal{O}$-module, then $M \otimes_{\mathcal{O}} \mathcal{O}$ is a projective $\mathcal{O}$-module. As

$$(M \otimes_{\mathcal{O}} \mathcal{O})_{\mathbb{C}} = M \otimes_{\mathcal{O}} \mathcal{O} \otimes_{\mathbb{Z}} \mathbb{C} = M_{\mathbb{C}} \otimes_{K_{\mathbb{C}}} L_{\mathbb{C}},$$

the hermitian metric on the $K_{\mathbb{C}}$-module $M_{\mathbb{C}}$ extends canonically to an F-invariant metric of the $L_{\mathbb{C}}$-module $(M \otimes_{\mathcal{O}} \mathcal{O})_{\mathbb{C}}$. Therefore $M \otimes_{\mathcal{O}} \mathcal{O}$ is automatically a metrized $\mathcal{O}$-module, which we denote by i^*M. If

$$0 \longrightarrow M' \longrightarrow M \longrightarrow M'' \longrightarrow 0$$

is a short exact sequence of projective metrized $\mathfrak{o}$-modules, then

$$0 \longrightarrow M' \otimes_{\mathfrak{o}} \mathcal{O} \longrightarrow M \otimes_{\mathfrak{o}} \mathcal{O} \longrightarrow M'' \otimes_{\mathfrak{o}} \mathcal{O} \longrightarrow 0$$

is a short exact sequence of metrized $\mathcal{O}$-modules, because $\mathcal{O}$ is a projective $\mathfrak{o}$-module and the metrics in the sequence

$$0 \longrightarrow M'_{\mathbb{C}} \longrightarrow M_{\mathbb{C}} \longrightarrow M''_{\mathbb{C}} \longrightarrow 0$$

simply extend $L_{\mathbb{C}}$-sesquilinearly to metrics in the sequence of $L_{\mathbb{C}}$-modules

$$0 \longrightarrow M'_{\mathbb{C}} \otimes_{K_{\mathbb{C}}} L_{\mathbb{C}} \longrightarrow M_{\mathbb{C}} \otimes_{K_{\mathbb{C}}} L_{\mathbb{C}} \longrightarrow M''_{\mathbb{C}} \otimes_{K_{\mathbb{C}}} L_{\mathbb{C}} \longrightarrow 0.$$

This is why mapping, in the usual way (i.e., via the representation $K_0(\bar{\mathfrak{o}}) = F_0(\bar{\mathfrak{o}})/R_0(\bar{\mathfrak{o}})$),

$$M \longmapsto [i^*M] = [M \otimes_{\mathfrak{o}} \mathcal{O}]$$

gives a well-defined homomorphism

$$i^* : K_0(\bar{\mathfrak{o}}) \longrightarrow K_0(\bar{\mathcal{O}}).$$

The reader may verify for himself that this is in fact a ring homomorphism.

On the other hand, if M is a projective metrized $\mathcal{O}$-module, then M is automatically also a projective $\mathfrak{o}$-module. For the complexification $M_{\mathbb{C}} = M \otimes_{\mathbb{Z}} \mathbb{C}$ we have the decomposition

$$M_{\mathbb{C}} = \bigoplus_{\tau \in Y(\mathbb{C})} M_\tau = \bigoplus_{\sigma \in X(\mathbb{C})} \bigoplus_{\tau | \sigma} M_\tau = \bigoplus_{\sigma \in X(\mathbb{C})} M_\sigma ,$$

where $M_\tau = M \otimes_{\mathcal{O},\tau} \mathbb{C}$ and

$$M_\sigma = M \otimes_{\mathfrak{o},\sigma} \mathbb{C} = \bigoplus_{\tau | \sigma} M_\tau .$$

The $\mathbb{C}$-vector spaces M_τ carry hermitian metrics $\langle\ ,\ \rangle_{M_\tau}$, and we define the metric $\langle\ ,\ \rangle_{M_\sigma}$ on the $\mathbb{C}$-vector space M_σ to be the orthogonal sum

$$\langle x, y\rangle_{M_\sigma} = \sum_{\tau | \sigma} \langle x_\tau, y_\tau \rangle_{M_\tau} .$$

This gives a hermitian metric on the $K_{\mathbb{C}}$-module $M_{\mathbb{C}}$, whose F-invariance is clearly guaranteed by the F-invariance of the original metric $\langle\ ,\ \rangle_M$. We denote the metrized $\mathfrak{o}$-module M thus constructed by i_*M.

If $0 \to M' \to M \to M'' \to 0$ is a short exact sequence of projective metrized $\mathcal{O}$-modules, then

$$0 \longrightarrow i_*M' \longrightarrow i_*M \longrightarrow i_*M'' \longrightarrow 0$$

is clearly an exact sequence of projective metrized $\mathfrak{o}$-modules. As before, this is why the correspondence

$$M \longmapsto [i_*M]$$

gives us a well-defined (additive) homomorphism

$$i_* : K_0(\bar{\mathcal{O}}) \longrightarrow K_0(\bar{\mathfrak{o}}).$$

(7.1) Proposition (Projection Formula). *The diagram*

$$\begin{array}{ccccc} K_0(\bar{\mathcal{O}}) & \times & K_0(\bar{\mathcal{O}}) & \stackrel{\cdot}{\longrightarrow} & K_0(\bar{\mathcal{O}}) \\ {\scriptstyle i_*}\downarrow & & \uparrow{\scriptstyle i^*} & & \downarrow{\scriptstyle i_*} \\ K_0(\bar{o}) & \times & K_0(\bar{o}) & \stackrel{\cdot}{\longrightarrow} & K_0(\bar{o}) \end{array}$$

is commutative, where the horizontal arrows are multiplication.

Proof: If M, resp. N, is a projective metrized $\mathcal{O}$-module, resp. o-module, there is an isometry

$$i_*(M \otimes_{\mathcal{O}} i^*N) \cong i_*M \otimes_o N$$

of projective metrized o-modules. Indeed, we have an isomorphism of the underlying o-modules

$$M \otimes_{\mathcal{O}} (N \otimes_o \mathcal{O}) \cong M \otimes_o N, \quad a \otimes (b \otimes c) \mapsto ca \otimes b.$$

Tensoring with $\mathbb{C}$, it induces an isomorphism

$$M_{\mathbb{C}} \otimes_{L_{\mathbb{C}}} (N_{\mathbb{C}} \otimes_{K_{\mathbb{C}}} L_{\mathbb{C}}) \cong M_{\mathbb{C}} \otimes_{K_{\mathbb{C}}} N_{\mathbb{C}}.$$

That this is an isometry of metrized $K_{\mathbb{C}}$-modules results from the distributivity

$$\sum_{\tau|\sigma} \langle\ ,\ \rangle_{M_\tau} \langle\ ,\ \rangle_{N_\sigma} = \Big(\sum_{\tau|\sigma} \langle\ ,\ \rangle_{M_\tau}\Big) \langle\ ,\ \rangle_{N_\sigma}$$

by applying mathematical grammar. □

The Riemann-Roch problem in Grothendieck's perspective is the task of computing the Chern character $\mathrm{ch}(i_*M)$ for a projective metrized $\mathcal{O}$-module M in terms of $\mathrm{ch}(M)$. By (6.6), this amounts to computing $\det(i_*M)$ in terms of $\det M$. But $\det M$ is an invertible metrized $\mathcal{O}$-module and is therefore isometric by (4.5) to the metrized $\mathcal{O}$-module $L(\mathfrak{A})$ of a replete ideal $\mathfrak{A}$ of L. $N_{L|K}(\mathfrak{A})$ is then a replete ideal of K, and we put

$$N_{L|K}(\det M) := L\big(N_{L|K}(\mathfrak{A})\big).$$

This is an invertible metrized o-module which is well determined by M up to isometry. With this notation we first establish the following theorem.

(7.2) Theorem. *For any projective metrized $\mathcal{O}$-module M one has:*

$$\mathrm{rk}(i_*M) = \mathrm{rk}(M)\,\mathrm{rk}(\mathcal{O}),$$

$$\det(i_*M) \cong N_{L|K}(\det M) \otimes_o (\det i_*\mathcal{O})^{\mathrm{rk}(M)}.$$

Here we have $\mathrm{rk}(\mathcal{O}) = [L : K]$.

Proof: One has $M_K := M \otimes_{\mathcal{o}} K = M \otimes_{\mathcal{O}} \mathcal{O} \otimes_{\mathcal{o}} K = M \otimes_{\mathcal{O}} L =: M_L$ and therefore

$$\mathrm{rk}(i_*M) = \dim_K(M_K) = \dim_K(M_L) = \dim_L(M_L)[L:K] = \mathrm{rk}(M)\,\mathrm{rk}(\mathcal{O}).$$

In order to prove the second equation, we first reduce to a special case. Let

$$\lambda(M) = \det(i_*M) \quad \text{and} \quad \rho(M) = N_{L|K}(\det M) \otimes_{\mathcal{o}} (\det i_*\mathcal{O})^{\mathrm{rk}(M)}.$$

If $0 \to M' \to M \to M'' \to 0$ is a short exact sequence of projective metrized $\mathcal{O}$-modules, one has

$$(*) \qquad \lambda(M) \cong \lambda(M') \otimes_{\mathcal{o}} \lambda(M'') \quad \text{and} \quad \rho(M) \cong \rho(M') \otimes_{\mathcal{o}} \rho(M'').$$

The isomorphism on the left follows from the exact sequence $0 \to i_*M' \to i_*M \to i_*M'' \to 0$ by (4.7), and the one on the right from (4.7) also, from the multiplicativity of the norm $N_{L|K}$ and the additivity of the rank rk. As in the proof of (5.6), we now make use of the fact that every projective metrized $\mathcal{O}$-module M projects via an admissible epimorphism onto a suitable $\mathcal{O}$-module of the form $L(\mathfrak{A})$ for some replete ideal $\mathfrak{A}$. Thus $(*)$ allows us to reduce by induction on $\mathrm{rk}(M)$ to the case $M = L(\mathfrak{A})$. Here $\mathrm{rk}(M) = 1$, so we have to establish the isomorphism

$$\det\big(i_*L(\mathfrak{A})\big) = L\big(N_{L|K}(\mathfrak{A})\big) \otimes_{\mathcal{o}} \det_{\mathcal{o}} \mathcal{O}.$$

For the underlying $\mathcal{o}$-modules this amounts to the identity

$$(**) \qquad \det_{\mathcal{o}} \mathfrak{A}_{\mathfrak{f}} = N_{L|K}(\mathfrak{A}_{\mathfrak{f}}) \det_{\mathcal{o}} \mathcal{O},$$

which has to be viewed as inside $\det_K L$ and which is proved as follows. If $\mathcal{O}$ and $\mathcal{o}$ were principal ideal domains, it would be obvious. In fact, in that case we could choose a generator α of $\mathfrak{A}_{\mathfrak{f}}$ and an integral basis $\omega_1, \dots, \omega_n$ of $\mathcal{O}$ over $\mathcal{o}$. Since $N_{L|K}(\alpha)$ is by definition the determinant $\det(T_\alpha)$ of the transformation $T_\alpha : L \to L$, $x \mapsto \alpha x$, we would get the equation

$$\alpha\omega_1 \wedge \ldots \wedge \alpha\omega_n = N_{L|K}(\alpha)(\omega_1 \wedge \ldots \wedge \omega_n),$$

the left-hand side, resp. right-hand side, of which would, by (1.6), generate the left-hand side, resp. right-hand side, of $(**)$. But we may always produce a principal ideal domain as desired by passing from $\mathcal{O}|\mathcal{o}$ to the localization $\mathcal{O}_{\mathfrak{p}}|\mathcal{o}_{\mathfrak{p}}$ for every prime ideal $\mathfrak{p}$ of $\mathcal{o}$ (see chap. I, § 11 and § 3, exercise 4). The preceding argument then shows that

$$(\det_{\mathcal{o}} \mathfrak{A}_{\mathfrak{f}})_{\mathfrak{p}} = \det_{\mathcal{o}_{\mathfrak{p}}} \mathfrak{A}_{\mathfrak{f}_{\mathfrak{p}}} = N_{L|K}(\mathfrak{A}_{\mathfrak{f}_{\mathfrak{p}}}) \det_{\mathcal{o}_{\mathfrak{p}}} \mathcal{O}_{\mathfrak{p}} = (N_{L|K}(\mathfrak{A}_{\mathfrak{f}}) \det_{\mathcal{o}} \mathcal{O})_{\mathfrak{p}},$$

and since this identity is valid for all prime ideals $\mathfrak{p}$ of $\mathcal{o}$, we deduce the equality $(**)$.

In order to prove that the metrics agree on both sides of $(**)$, we put $M = L(\mathfrak{A})$, $N = L(\mathcal{O})$, $\mathfrak{a} = N_{L|K}(\mathfrak{A})$ and we view $M, N, \mathfrak{a}$ as metrized $\mathcal{O}$-modules. One has $M_{\mathbb{C}} = N_{\mathbb{C}} = L_{\mathbb{C}}$ and $\mathfrak{a}_{\mathbb{C}} = K_{\mathbb{C}}$, and we consider the metrics on the components

$$M_\sigma = \bigoplus_{\tau|\sigma} \mathbb{C}, \quad N_\sigma = \bigoplus_{\tau|\sigma} \mathbb{C}, \quad \mathfrak{a}_\sigma = \mathbb{C},$$

where $\sigma \in \operatorname{Hom}(K, \mathbb{C})$ and $\tau \in \operatorname{Hom}(L, \mathbb{C})$ is such that $\tau|\sigma$. We have to show that, for $\xi, \eta \in \det_{\mathbb{C}} M_\sigma$ and $a, b \in \mathbb{C}$, one has the identity

$$\langle a\xi, b\eta \rangle_{\det M_\sigma} = \langle a, b \rangle_{\mathfrak{a}_\sigma} \langle \xi, \eta \rangle_{\det N_\sigma}.$$

For this, let $\mathfrak{A}_\infty = \prod_{\mathfrak{P}|\infty} \mathfrak{P}^{\nu_{\mathfrak{P}}}$, so that one gets

$$\mathfrak{a}_\infty = N_{L|K}(\mathfrak{A}_\infty) = \prod_{\mathfrak{p}|\infty} \mathfrak{p}^{\nu_{\mathfrak{p}}}$$

with $\nu_{\mathfrak{p}} = \sum_{\mathfrak{P}|\mathfrak{p}} f_{\mathfrak{P}|\mathfrak{p}} \nu_{\mathfrak{P}}$. Then

$$\langle x, y \rangle_{N_\sigma} = \sum_{\tau|\sigma} x_\tau \bar{y}_\tau, \qquad \langle x, y \rangle_{M_\sigma} = \sum_{\tau|\sigma} e^{2\nu_{\mathfrak{P}_\tau}} x_\tau \bar{y}_\tau,$$

$$\langle a, b \rangle_{\mathfrak{a}_\sigma} = e^{2\nu_{\mathfrak{p}_\sigma}} a\bar{b}, \qquad \nu_{\mathfrak{p}_\sigma} = \sum_{\mathfrak{P}|\mathfrak{p}_\sigma} f_{\mathfrak{P}|\mathfrak{p}_\sigma} \nu_{\mathfrak{P}} = \sum_{\tau|\sigma} \nu_{\mathfrak{P}_\tau}.$$

Let $\xi = x_1 \wedge \ldots \wedge x_n$, $\eta = y_1 \wedge \ldots \wedge y_n$. We number the embeddings $\tau|\sigma$, $\tau_1, \ldots, \tau_n$, put $\nu_k = \nu_{\mathfrak{P}_{\tau_k}}$ and form the matrices

$$A = (x_{i\tau_k}), \quad B = (\bar{y}_{i\tau_k}), \quad D = \begin{pmatrix} e^{\nu_1} & & 0 \\ & \ddots & \\ 0 & & e^{\nu_n} \end{pmatrix}.$$

Then, observing that

$$\det(D) = \prod_\tau e^{\nu_{\mathfrak{P}_\tau}} = \prod_{\mathfrak{P}|\mathfrak{p}_\sigma} e^{f_{\mathfrak{P}|\mathfrak{p}_\sigma} \nu_{\mathfrak{P}}} = e^{\nu_{\mathfrak{p}_\sigma}},$$

we do indeed get

$$\begin{aligned} \langle a\xi, b\eta \rangle_{\det M_\sigma} &= a\bar{b}\, \langle \xi, \eta \rangle_{\det M_\sigma} \\ &= a\bar{b} \det\big((AD)(BD)^t\big) = a\bar{b}\,(\det D)^2 \det(AB^t) \\ &= e^{2\nu_{\mathfrak{p}_\sigma}} a\bar{b}\, \langle \xi, \eta \rangle_{\det N_\sigma} = \langle a, b \rangle_{\mathfrak{a}_\sigma} \langle \xi, \eta \rangle_{\det N_\sigma}. \end{aligned}$$

This proves our theorem. □

Extending the formulas of (7.2) to the free abelian group

$$F_0(\bar{\mathcal{O}}) = \bigoplus_{\{M\}} \mathbb{Z}\{M\}$$

by linearity, and passing to the quotient group $K_0(\bar{\mathcal{O}}) = F_0(\bar{\mathcal{O}})/R_0(\bar{\mathcal{O}})$ yields the following corollary.

(7.3) Corollary. *For every class* $\xi \in K_0(\overline{\mathcal{O}})$, *one has the formulas*

$$\operatorname{rk}(i_*\xi) = [L:K]\operatorname{rk}(\xi),$$
$$\det(i_*\xi) = [\det i_*\mathcal{O}]^{\operatorname{rk}(\xi)} N_{L|K}(\det \xi).$$

The square of the metrized $\mathcal{o}$-module $\det i_*\mathcal{O}$ appearing in the second formula can be computed to be the *discriminant* $\mathfrak{d}_{L|K}$ of the extension $L|K$, which we view as a metrized $\mathcal{o}$-module with the *trivial* metric.

(7.4) Proposition. *There is a canonical isomorphism*

$$(\det i_*\mathcal{O})^{\otimes 2} \cong \mathfrak{d}_{L|K}$$

of metrized $\mathcal{o}$*-modules.*

Proof: Consider on $\mathcal{O}$ the bilinear trace map

$$T : \mathcal{O}\times\mathcal{O} \longrightarrow \mathcal{o}, \quad (x,y) \longmapsto Tr_{L|K}(xy).$$

It induces an $\mathcal{o}$-module homomorphism

$$T : \det\mathcal{O} \otimes \det\mathcal{O} \longrightarrow \mathcal{o},$$

given by

$$T\big((\alpha_1\wedge\ldots\wedge\alpha_n)\otimes(\beta_1\wedge\ldots\wedge\beta_n)\big) = \det\big(Tr_{L|K}(\alpha_i\beta_j)\big).$$

The image of T is the discriminant ideal $\mathfrak{d}_{L|K}$, which, by definition, is generated by the discriminants

$$d(\omega_1,\ldots,\omega_n) = \det\big(Tr_{L|K}(\omega_i\omega_j)\big)$$

of all bases of $L|K$ which are contained in $\mathcal{O}$. This is clear if $\mathcal{O}$ admits an integral basis over $\mathcal{o}$, since the α_i and β_i can be written in terms of such a basis with coefficients in $\mathcal{o}$. If there is no such integral basis, it will exist after localizing $\mathcal{O}_\mathfrak{p}|\mathcal{o}_\mathfrak{p}$ at every prime ideal $\mathfrak{p}$ (see chap. I, (2.10)). The image of

$$T_\mathfrak{p} : (\det\mathcal{O}_\mathfrak{p})\otimes(\det\mathcal{O}_\mathfrak{p}) \longrightarrow \mathcal{o}_\mathfrak{p}$$

is therefore the discriminant ideal of $\mathcal{O}_\mathfrak{p}|\mathcal{o}_\mathfrak{p}$ and at the same time the localization of the image of T. Since two ideals are equal when their localizations are, we find $\operatorname{image}(T) = \mathfrak{d}_{L|K}$. Furthermore, T has to be injective since $(\det\mathcal{O})^{\otimes 2}$ is an invertible $\mathcal{o}$-module. Therefore T is an $\mathcal{o}$-module isomorphism.

We now check that

$$T_{\mathbb{C}} : \left(\det(\mathcal{O})^{\otimes 2}\right)_{\mathbb{C}} \longrightarrow (\mathfrak{d}_{L|K})_{\mathbb{C}}$$

is indeed an isometry. For $\mathcal{O}_{\mathbb{C}} = \mathcal{O} \otimes_{\mathbb{Z}} \mathbb{C}$, we obtain the $K_{\mathbb{C}}$-module decomposition

$$\mathcal{O}_{\mathbb{C}} = \bigoplus_{\sigma} \mathcal{O}_{\sigma},$$

where σ varies over the set $\mathrm{Hom}(K, \mathbb{C})$, and the direct sum

$$\mathcal{O}_{\sigma} = \bigoplus_{\tau|\sigma} (\mathcal{O} \otimes_{o,\tau} \mathbb{C}) = \bigoplus_{\tau|\sigma} \mathbb{C}$$

is taken over all $\tau \in \mathrm{Hom}(L, \mathbb{C})$ such that $\tau|_K = \sigma$. The mapping $\mathcal{O}_{\mathbb{C}} \to K_{\mathbb{C}}$ induced by $Tr_{L|K} : \mathcal{O} \to o$ is given, for $x = \bigoplus_{\sigma} x_{\sigma}$, $x_{\sigma} \in \mathcal{O}_{\sigma}$, by

$$Tr_{L|K}(x) = \sum_{\sigma} Tr_{\sigma}(x_{\sigma}),$$

where $Tr_{\sigma}(x_{\sigma}) = \sum_{\tau|\sigma} x_{\sigma,\tau}$, the $x_{\sigma,\tau} \in \mathbb{C}$ being the components of x_{σ}. The metric on $(i_*\mathcal{O})_{\mathbb{C}} = \mathcal{O}_{\mathbb{C}}$ is the orthogonal sum of the standard metrics

$$\langle x, y\rangle_{\sigma} = \sum_{\tau|\sigma} x_{\tau}\overline{y}_{\tau} = Tr_{\sigma}(x\overline{y})$$

on the $\mathbb{C}$-vector spaces $(i_*\mathcal{O})_{\sigma} = \mathcal{O}_{\sigma} = \bigoplus_{\tau|\sigma} \mathbb{C}$. Now let $x_i, y_i \in \mathcal{O}_{\sigma}$, $i = 1, \dots, n$, and write $x = x_1 \wedge \dots \wedge x_n$, $y = y_1 \wedge \dots \wedge y_n \in \det(\mathcal{O}_{\sigma})$. The map $T_{\mathbb{C}}$ splits into the direct sum $T_{\mathbb{C}} = \bigoplus_{\sigma} T_{\sigma}$ of the maps

$$T_{\sigma} : \det(\mathcal{O}_{\sigma}) \otimes_{\mathbb{C}} \det(\mathcal{O}_{\sigma}) \to (\mathfrak{d}_{L|K})_{\sigma} = \mathbb{C}$$

which are given by

$$T_{\sigma}(x \otimes y) = \det\big(Tr_{\sigma}(x_i y_j)\big).$$

For any two n-tuples x_i', $y_i' \in \mathcal{O}_{\sigma}$ we form the matrices

$$A = \big(Tr_{\sigma}(x_i y_j)\big),\ A' = \big(Tr_{\sigma}(\overline{x}_i'\overline{y}_j')\big),\ B = \big(Tr_{\sigma}(x_i\overline{x}_j')\big),\ B' = \big(Tr_{\sigma}(y_i\overline{y}_j')\big).$$

Then one has $AA' = BB'$, and we obtain

$$\begin{aligned}
\big\langle T_{\sigma}(x \otimes y), T_{\sigma}(x' \otimes y')\big\rangle_{(\mathfrak{d}_{L|K})_{\sigma}} &= T_{\sigma}(x \otimes y)\,\overline{T_{\sigma}(x' \otimes y')} \\
&= \det\big(Tr_{\sigma}(x_i y_j)\big)\det\big(Tr_{\sigma}(\overline{x}_i'\overline{y}_j')\big) = \det(AA') = \det(BB') \\
&= \det\big(Tr_{\sigma}(x_i\overline{x}_j')\big)\det\big(Tr_{\sigma}(y_i\overline{y}_j')\big) = \det\big(\langle x_i, x_j'\rangle_{\sigma}\big)\det\big(\langle y_i, y_j'\rangle_{\sigma}\big) \\
&= \langle x, x'\rangle_{\det \mathcal{O}_{\sigma}} \langle y, y'\rangle_{\det \mathcal{O}_{\sigma}} = \langle x \otimes y, x' \otimes y'\rangle_{(\det \mathcal{O}_{\sigma})^{\otimes 2}}.
\end{aligned}$$

This shows that $T_{\mathbb{C}}$ is an isometry. □

We now set out to rewrite the results obtained in (7.2) and (7.4) in the language of GROTHENDIECK's general formalism. For the homomorphism i_* there is the commutative diagram

$$\begin{array}{ccc} K_0(\overline{\mathcal{O}}) & \xrightarrow{\text{rk}} & \mathbb{Z} \\ {\scriptstyle i_*}\downarrow & & \downarrow{\scriptstyle [L:K]} \\ K_0(\bar{\mathcal{o}}) & \xrightarrow{\text{rk}} & \mathbb{Z}, \end{array}$$

because $[L:K]$ times the rank of an $\mathcal{O}$-module M is its rank as $\mathcal{o}$-module. Therefore i_* induces a homomorphism

$$i_* : I(\overline{\mathcal{O}}) \longrightarrow I(\bar{\mathcal{o}})$$

between the kernels of both rank homomorphisms, so that there is a homomorphism

$$i_* : \operatorname{gr} K_0(\overline{\mathcal{O}}) \longrightarrow \operatorname{gr} K_0(\bar{\mathcal{o}}).$$

It is called the **Gysin map**. (7.3) immediately gives the following explicit description of it.

(7.5) Corollary. *The diagram*

$$\begin{array}{ccc} \operatorname{gr} K_0(\overline{\mathcal{O}}) & \xrightarrow[\sim]{\text{id}\oplus\det} & \mathbb{Z} \oplus Pic(\overline{\mathcal{O}}) \\ {\scriptstyle i_*}\downarrow & & \downarrow{\scriptstyle [L:K]\oplus N_{L|K}} \\ \operatorname{gr} K_0(\bar{\mathcal{o}}) & \xrightarrow[\sim]{\text{id}\oplus\det} & \mathbb{Z} \oplus Pic(\bar{\mathcal{o}}) \end{array}$$

is commutative.

We now consider the following diagram

$$\begin{array}{ccc} K_0(\overline{\mathcal{O}}) & \xrightarrow{\text{ch}} & \operatorname{gr} K_0(\overline{\mathcal{O}}) \\ {\scriptstyle i_*}\downarrow & & \downarrow{\scriptstyle i_*} \\ K_0(\bar{\mathcal{o}}) & \xrightarrow{\text{ch}} & \operatorname{gr} K_0(\bar{\mathcal{o}}) \end{array}$$

where the Gysin map i_* on the right is explicitly given by (7.5), whereas the determination of the composite $\text{ch} \circ i_*$ is precisely the Riemann-Roch problem. The difficulty that confronts us here lies in the fact that the diagram is *not commutative*. In order to make it commute, we need a correction, which will be provided via the module of differentials (with trivial metric), by the *Todd class*, which is defined as follows.

The module $\Omega^1_{\mathcal{O}|o}$ of differentials is only a coherent, and not a projective $\mathcal{O}$-module. But its class $[\Omega^1_{\mathcal{O}|o}]$ is viewed as an element of $K_0(\overline{\mathcal{O}})$ via the Poincaré isomorphism

$$K_0(\overline{\mathcal{O}}) \xrightarrow{\sim} K^0(\overline{\mathcal{O}}),$$

and since $\mathrm{rk}_{\mathcal{O}}(\Omega^1_{\mathcal{O}|o}) = 0$, it lies in $I(\overline{\mathcal{O}})$.

(7.6) Definition. *The* **Todd class** *of $\mathcal{O}|o$ is defined to be the element*

$$\mathrm{Td}(\mathcal{O}|o) = 1 - \frac{1}{2}c_1\big([\Omega^1_{\mathcal{O}|o}]\big) = 1 - \frac{1}{2}[\Omega^1_{\mathcal{O}|o}] \in \mathrm{gr}\, K_0(\overline{\mathcal{O}}) \otimes \mathbb{Z}[\tfrac{1}{2}].$$

Because of the factor $\frac{1}{2}$, the Todd class does not belong to the ring $\mathrm{gr}\, K_0(\overline{\mathcal{O}})$ itself, but is only an element of $\mathrm{gr}\, K_0(\overline{\mathcal{O}}) \otimes \mathbb{Z}[\frac{1}{2}]$. The module of differentials $\Omega^1_{\mathcal{O}|o}$ is connected with the **different** $\mathfrak{D}_{L|K}$ of the extension $L|K$ by the exact sequence

$$0 \longrightarrow \mathfrak{D}_{L|K} \longrightarrow \mathcal{O} \longrightarrow \Omega^1_{\mathcal{O}|o} \longrightarrow 0$$

of $\mathcal{O}$-modules (with trivial metrics) (see §2, exercise 3). This implies that $[\Omega^1_{\mathcal{O}|o}] = 1 - [\mathfrak{D}_{L|K}]$. We may therefore describe the Todd class also by the different:

$$\mathrm{Td}(\mathcal{O}|o) = 1 + \frac{1}{2}([\mathfrak{D}_{L|K}] - 1) = \frac{1}{2}\big(1 + [\mathfrak{D}_{L|K}]\big)\,.$$

The main result now follows from (7.3) using the Todd class.

(7.7) Theorem (Grothendieck-Riemann-Roch). *The diagram*

$$\begin{array}{ccc} K_0(\overline{\mathcal{O}}) & \xrightarrow{\mathrm{Td}(\mathcal{O}|o)\mathrm{ch}} & \mathrm{gr}\, K_0(\overline{\mathcal{O}}) \\ {\scriptstyle i_*}\downarrow & & \downarrow{\scriptstyle i_*} \\ K_0(\overline{o}) & \xrightarrow{\mathrm{ch}} & \mathrm{gr}\, K_0(\overline{o}) \end{array}$$

is commutative.

Proof: For $\xi \in K_0(\overline{\mathcal{O}})$, we have to show the identity

$$\mathrm{ch}(i_*\xi) = i_*\big(\mathrm{Td}(\mathcal{O}|o)\,\mathrm{ch}(\xi)\big)\,.$$

Decomposing $\operatorname{ch}(i_*\xi) = \operatorname{rk}(i_*\xi) \oplus c_1(i_*\xi)$ and $\operatorname{ch}(\xi) = \operatorname{rk}(\xi) \oplus c_1(\xi)$ and observing that

$$\begin{aligned}\operatorname{Td}(\mathcal{O}|\mathcal{o})\operatorname{ch}(\xi) &= \big(1 + \tfrac{1}{2}\big([\mathfrak{D}_{L|K}] - 1\big)\big)(\operatorname{rk}(\xi) + c_1(\xi)) \\ &= \operatorname{rk}(\xi) + \big[c_1(\xi) + \tfrac{1}{2}\operatorname{rk}(\xi)([\mathfrak{D}_{L|K}] - 1)\big],\end{aligned}$$

it suffices to check the equations

(a) $$\operatorname{rk}(i_*\xi) = \operatorname{rk}(\xi)\operatorname{rk}(i_*[\mathcal{O}]),$$

(b) $$c_1(i_*\xi) = i_*(c_1(\xi)) + \operatorname{rk}(\xi)c_1(i_*[\mathcal{O}])$$

and

(c) $$\operatorname{rk}\big(i_*[\mathcal{O}]\big) = i_*(1),$$

(d) $$2c_1\big(i_*[\mathcal{O}]\big) = i_*\big([\mathfrak{D}_{L|K}] - 1\big)$$

in $\operatorname{gr} K_0(\bar{\mathcal{o}})$. The equations (a) and (c) are clear because of $\operatorname{rk}(i_*[\mathcal{O}]) = \operatorname{rk}(i_*\mathcal{O}) = [L : K]$. To show (b) and (d), we apply det to both sides and are reduced by the commutative diagram (7.5) to the equations

(e) $$\det(i_*\xi) = N_{L|K}\big((\det\xi)\big)[\det i_*\mathcal{O}]^{\operatorname{rk}(\xi)},$$

(f) $$(\det i_*\mathcal{O})^{\otimes 2} = N_{L|K}(\det \mathfrak{D}_{L|K}).$$

But (e) is the second identity of (7.3), and (f) follows from (7.4) and (2.9). □

With this final theorem, the theory of algebraic integers can be integrated completely into a general programme of algebraic geometry as a special case. What is needed is the use of the geometric language for the objects considered. Thus the ring $\mathcal{o}$ is interpreted as the scheme $X = \operatorname{Spec}(\mathcal{o})$, the projective metrized $\mathcal{o}$-modules as metrized *vector bundles*, the invertible $\mathcal{o}$-modules as *line bundles*, the inclusion $i : \mathcal{o} \to \mathcal{O}$ as morphism $f : Y = \operatorname{Spec}(\mathcal{O}) \to X$ of schemes, the class $\Omega^1_{\mathcal{O}|\mathcal{o}}$ as the *cotangent element*, etc. In this way one realizes in the present context the old idea of viewing number theory as part of geometry.

§ 8. The Euler-Minkowski Characteristic

Considering the theorem of Grothendieck-Riemann-Roch in the special case of an extension $K|\mathbb{Q}$, amounts to revisiting the Riemann-Roch theory of §3 from our new point of view. At the center of that theory was the Euler-Minkowski characteristic

$$\chi(\mathfrak{a}) = -\log \operatorname{vol}(\mathfrak{a})$$

of replete ideals $\mathfrak{a}$ of K. Here, $\operatorname{vol}(\mathfrak{a})$ was the *canonical measure* of a fundamental mesh of the lattice in Minkowski space $K_{\mathbb{R}} = \mathfrak{a} \otimes_{\mathbb{Z}} \mathbb{R}$ defined by $\mathfrak{a}$. This definition is properly explained in the theory of metrized modules of higher rank. More precisely, instead of considering $\mathfrak{a}$ as a metrized $\mathcal{O}$-module of rank 1, it should be viewed as a metrized $\mathbb{Z}$-module of rank $[K : \mathbb{Q}]$. This point of view leads us necessarily to the following definition of the Euler-Minkowski characteristic.

(8.1) Proposition. *The degree map*

$$\deg_K : Pic(\bar{\mathcal{O}}) \longrightarrow \mathbb{R}, \quad \deg_K([\mathfrak{a}]) = -\log \mathfrak{N}(\mathfrak{a}),$$

extends uniquely to a homomorphism

$$\chi_K : K_0(\bar{\mathcal{O}}) \longrightarrow \mathbb{R}$$

on $K_0(\bar{\mathcal{O}})$, and thereby on $K^0(\bar{\mathcal{O}})$. It is given by

$$\chi_K = \deg \circ \det$$

and called the **Euler-Minkowski characteristic** *over K.*

Proof: Since, by (5.6), $K_0(\bar{\mathcal{O}})$ is generated as an additive group by the elements $[\mathfrak{a}] \in Pic(\bar{\mathcal{O}})$, the map $\deg_K$ on $Pic(\bar{\mathcal{O}})$ determines a unique homomorphism $K_0(\bar{\mathcal{O}}) \to \mathbb{R}$ which extends $\deg_K$. But such a homomorphism is given by the composite of the homomorphisms

$$K_0(\bar{\mathcal{O}}) \xrightarrow{\det} Pic(\bar{\mathcal{O}}) \xrightarrow{\deg} \mathbb{R},$$

as the composite $Pic(\bar{\mathcal{O}}) \hookrightarrow K_0(\bar{\mathcal{O}}) \xrightarrow{\det} Pic(\bar{\mathcal{O}})$ is the identity. □

Via the Poincaré isomorphism $K_0(\bar{\mathcal{O}}) \xrightarrow{\sim} K^0(\bar{\mathcal{O}})$, we transfer the maps det and χ_K to the Grothendieck group $K^0(\bar{\mathcal{O}})$ of coherent metrized $\mathcal{O}$-modules. Then proposition (8.1) is equally valid for $K^0(\bar{\mathcal{O}})$ as for $K_0(\bar{\mathcal{O}})$. We define in what follows $\chi_K(M) = \chi_K([M])$ for a metrized $\mathcal{O}$-module M. If $L|K$ is an extension of algebraic number fields and $i : \mathcal{O} \to \mathcal{O}$ the inclusion of the maximal orders of K, resp. L, then applying $\deg_K$ to the formula (7.2) and using

$$\deg_L(\mathfrak{A}) = -\log \mathfrak{N}(\mathfrak{A}) = -\log \mathfrak{N}\big(N_{L|K}(\mathfrak{A})\big) = \deg_K\big(N_{L|K}(\mathfrak{A})\big)$$

(see (1.6), (iii)) gives the

(8.2) Theorem. *For every coherent $\mathcal{O}$-module M, the Riemann-Roch formula*

$$\chi_K(i_*M) = \deg_L(\det M) + \mathrm{rk}(M)\,\chi_K(i_*\mathcal{O})$$

is valid, and in particular, for an invertible metrized $\mathcal{O}$-module M, we have

$$\chi_K(i_*M) = \deg_L(M) + \chi_K(i_*\mathcal{O}).$$

We now specialize to the case of the base field $K = \mathbb{Q}$, that is, we consider metrized $\mathbb{Z}$-modules. Such a module is simply a finitely generated abelian group M together with a euclidean metric on the real vector space

$$M_{\mathbb{R}} = M \otimes_{\mathbb{Z}} \mathbb{R}.$$

Indeed, since $\mathbb{Q}$ has only a single embedding into $\mathbb{C}$, i.e., $\mathbb{Q}_{\mathbb{C}} = \mathbb{C}$, a metric on M is simply given by a hermitian scalar product on the $\mathbb{C}$-vector space $M_{\mathbb{C}} = M_{\mathbb{R}} \otimes \mathbb{C}$. Restricting this to $M_{\mathbb{R}}$ gives a euclidean metric the sesquilinear extension of which reproduces the original metric.

If M is a *projective* metrized $\mathbb{Z}$-module, then the underlying $\mathbb{Z}$-module is a finitely generated free abelian group. The canonical map $M \to M \otimes \mathbb{R}$, $a \mapsto a \otimes 1$, identifies M with a complete lattice in $M_{\mathbb{R}}$. If $\alpha_1, \dots, \alpha_n$ is a $\mathbb{Z}$-basis of M, then the set

$$\Phi = \left\{ x_1\alpha_1 + \cdots + x_n\alpha_n \;\middle|\; x_i \in \mathbb{R},\ 0 \le x_i < 1 \right\}$$

is a *fundamental mesh* of the lattice M. The euclidean metric $\langle\ ,\ \rangle_M$ defines a Haar measure on $M_{\mathbb{R}}$. Once we choose an orthonormal basis $e_1, \dots, e_n$ of $M_{\mathbb{R}}$, this Haar measure can be expressed, via the isomorphism $M_{\mathbb{R}} \xrightarrow{\sim} \mathbb{R}^n$, $x_1e_1 + \cdots + x_ne_n \longmapsto (x_1, \dots, x_n)$, by the Lebesgue measure on $\mathbb{R}^n$. With respect to this measure, the volume of the fundamental mesh Φ is given by

$$\mathrm{vol}(\Phi) = \left|\det(\langle\alpha_i, \alpha_j\rangle)\right|^{1/2}.$$

It will be denoted by $\mathrm{vol}(M)$ for short. It does not depend on the choice of $\mathbb{Z}$-basis $\alpha_1, \dots, \alpha_n$ because a different choice is linked to the original one by a matrix with integer coefficients which also has an inverse with integer coefficients, hence has determinant of absolute value 1.

A more elegant definition of $\mathrm{vol}(M)$ can be given in terms of the invertible metrized $\mathbb{Z}$-module $\det M$. $\det M_{\mathbb{R}}$ is a one-dimensional $\mathbb{R}$-vector space with metric $\langle\ ,\ \rangle_{\det M}$, and with the lattice $\det M$ isomorphic to $\mathbb{Z}$. If $x \in \det M$ is a generator (for instance, $x = \alpha_1 \wedge \ldots \wedge \alpha_n$), then

$$\mathrm{vol}(M) = \|x\|_{\det M} = \sqrt{\langle x, x\rangle_{\det M}}\,.$$

In the present case, where the base field is $\mathbb{Q}$, the degree map

$$\deg : Pic(\overline{\mathbb{Z}}) \longrightarrow \mathbb{R}$$

is an isomorphism (see § 1, exercise 3), and we call the unique homomorphism arising from this,

$$\chi = \deg \circ \det : K^0(\overline{\mathbb{Z}}) \longrightarrow \mathbb{R},$$

the Euler-Minkowski characteristic. It is computed explicitly as follows.

(8.3) Proposition. *For a coherent metrized $\mathbb{Z}$-module M, one has*

$$\chi(M) = \log \#M_{\rm tor} - \log \mathrm{vol}(M/M_{\rm tor})\,.$$

In this formula $M_{\rm tor}$ denotes the torsion subgroup of M and $M/M_{\rm tor}$ the projective metrized $\mathbb{Z}$-module which receives its metric from M via $M \otimes \mathbb{R} = M/M_{\rm tor} \otimes \mathbb{R}$.

Proof of (8.3): If M is a finite $\mathbb{Z}$-module, then the determinant of the class $[M] \in K^0(\overline{\mathbb{Z}})$ is computed from a free resolution

$$0 \longrightarrow E \longrightarrow F \xrightarrow{\alpha} M \longrightarrow 0,$$

where $F = \mathbb{Z}^n$ and $E = \ker(\alpha) \cong \mathbb{Z}^n$. If we equip $F \otimes \mathbb{R} = E \otimes \mathbb{R} = \mathbb{R}^n$ with the standard metric, the sequence becomes a short exact sequence of metrized $\mathbb{Z}$-modules, because $M \otimes \mathbb{R} = 0$. We therefore have in $K^0(\overline{\mathbb{Z}})$:

$$[M] = [F] - [E].$$

Let A be the matrix corresponding to the change of basis from the standard basis $e_1, \dots, e_n$ of F to a $\mathbb{Z}$-basis $e_1', \dots, e_n'$ of E. Then $x = e_1 \wedge \dots \wedge e_n$, resp. $x' = e_1' \wedge \dots \wedge e_n'$, is a generator of $\det F$, resp. $\det E$, and

$$x' = \det A \cdot x = (F : E) \cdot x = \#M \cdot x\,.$$

The metric $\| \ \|$ on $\det E$ is the same as that on $\det F$, so that

$$\chi(E) = \deg(\det E) = -\log \|x'\| = -\log\big(\#M \|x\|\big) = -\log \#M + \chi(F),$$

and then

$$\chi(M) = \chi\big([F] - [E]\big) = \chi(F) - \chi(E) = \log \#M.$$

For an arbitrary coherent metrized $\mathbb{Z}$-module M we have the direct sum decomposition

$$M = M_{\rm tor} \bigoplus M/M_{\rm tor}$$

into metrized $\mathbb{Z}$-modules. If $\alpha_1, \dots, \alpha_n$ is a basis of the lattice M/M_{tor}, then $x = \alpha_1 \wedge \dots \wedge \alpha_n$ is a generator of $\det M/M_{\text{tor}}$; then $\chi(M/M_{\text{tor}}) = \deg(\det M/M_{\text{tor}}) = -\log \|x\| = -\log \operatorname{vol}(M/M_{\text{tor}})$. We therefore conclude that

$$\chi(M) = \chi(M_{\text{tor}}) + \chi(M/M_{\text{tor}}) = \log \# M_{\text{tor}} - \log \operatorname{vol}(M/M_{\text{tor}}). \qquad \square$$

The Euler-Minkowski characteristic of a replete ideal $\mathfrak{a}$,

$$\chi(\mathfrak{a}) = -\log \operatorname{vol}(\mathfrak{a}),$$

which we defined *ad hoc* in §3 via the Minkowski measure $\operatorname{vol}(\mathfrak{a})$ now appears as a simple special case of the Euler-Minkowski characteristic for metrized $\mathbb{Z}$-modules to which the detailed development of the theory has led us. Indeed, viewing the metrized $\mathcal{O}$-module $L(\mathfrak{a})$ of rank 1 associated to $\mathfrak{a}$ as the metrized $\mathbb{Z}$-module $i_*L(\mathfrak{a})$ of rank $[K : \mathbb{Q}]$, we get the

(8.4) Proposition. $\chi(\mathfrak{a}) = \chi\big(i_*L(\mathfrak{a})\big)$.

Proof: Let $\mathfrak{a} = \mathfrak{a}_f \mathfrak{a}_\infty = \mathfrak{a}_f \prod_{\mathfrak{p}|\infty} \mathfrak{p}^{\nu_\mathfrak{p}}$. The metric $\langle\ ,\ \rangle_{i_*L(\mathfrak{a})}$ on the $\mathbb{C}$-vector space $K_\mathbb{C} = \prod_{\tau \in X(\mathbb{C})} \mathbb{C}$ is then given by

$$\langle x, y \rangle_{i_*L(\mathfrak{a})} = \sum_\tau e^{2\nu_{\mathfrak{p}_\tau}} x_\tau \bar{y}_\tau ,$$

where $\mathfrak{p}_\tau$ is the infinite place of K corresponding to the embedding $\tau : K \to \mathbb{C}$. It results from the standard metric $\langle\ ,\ \rangle$ via the F-invariant transformation

$$T : K_\mathbb{C} \longrightarrow K_\mathbb{C}, \quad (x_\tau)_{\tau \in X(\mathbb{C})} \longmapsto (e^{\nu_{\mathfrak{p}_\tau}} x_\tau)_{\tau \in X(\mathbb{C})} .$$

Equivalently,

$$\langle x, y \rangle_{i_*L(\mathfrak{a})} = \langle Tx, Ty \rangle .$$

The volume $\operatorname{vol}(i_*L(\mathfrak{a}))$ of a fundamental mesh of the lattice $\mathfrak{a}_f$ in $K_\mathbb{R}$ with respect to the Haar measure defined by the euclidean metric on $K_\mathbb{R}$ is then the volume of a fundamental mesh of the lattice $T\mathfrak{a}_f$ with respect to the canonical measure defined by $\langle\ ,\ \rangle$. Thus

$$\operatorname{vol}\big(i_*L(\mathfrak{a})\big) = \operatorname{vol}(T\mathfrak{a}_f) .$$

In the representation $K_\mathbb{R} = \prod_{\mathfrak{p}|\infty} K_\mathfrak{p}$, the canonical embedding

$$K_\mathbb{R} = K \otimes_\mathbb{Q} \mathbb{R} \longrightarrow K_\mathbb{C} = K \otimes_\mathbb{Q} \mathbb{C}$$

maps an element $(x_\mathfrak{p})_{\mathfrak{p}|\infty}$ to the element $(x_\tau)_{\tau \in X(\mathbb{C})}$ with $x_\tau = \tau x_{\mathfrak{p}_\tau}$. Here we extend τ to $K_{\mathfrak{p}_\tau}$. The restriction of the transformation $T : (x_\tau) \mapsto (e^{\nu_{\mathfrak{p}_\tau}} x_\tau)$ to $K_\mathbb{R} = \prod_{\mathfrak{p}|\infty} K_\mathfrak{p}$ is therefore given by $(x_\mathfrak{p}) \mapsto (e^{\nu_\mathfrak{p}} x_\mathfrak{p})$. The lattice $T\mathfrak{a}_f$ is then the same lattice which was denoted $\mathfrak{a}$ in §3. So we obtain

$$\operatorname{vol}\big(i_*L(\mathfrak{a})\big) = \operatorname{vol}(\mathfrak{a}),$$

i.e., $\chi(i_*L(\mathfrak{a})) = \chi(\mathfrak{a})$. □

Given this identification, the Riemann-Roch theorem (3.4) proven in §3 for replete ideals $\mathfrak{a}$,

$$\chi(\mathfrak{a}) = \deg(\mathfrak{a}) + \chi(\mathcal{O}),$$

now appears as a special case of theorem (8.2), which says that

$$\chi\big(i_*L(\mathfrak{a})\big) = \deg\big(L(\mathfrak{a})\big) + \chi(i_*\mathcal{O}).$$

Chapter IV
Abstract Class Field Theory

§ 1. Infinite Galois Theory

Every field k is equipped with a distinguished Galois extension: the separable closure $\bar{k}|k$. Its Galois group $G_k = G(\bar{k}|k)$ is called the **absolute Galois group** of k. As a rule, this extension will have infinite degree. It does, however, have the advantage of collecting all finite Galois extensions of k. This is why it is reasonable to try to give it a prominent place in Galois theory. But such an attempt faces the difficulty that the main theorem of Galois theory does not remain true for infinite extensions. Let us explain this in the following

Example: The absolute Galois group $G_{\mathbb{F}_p} = G(\overline{\mathbb{F}}_p|\mathbb{F}_p)$ of the field $\mathbb{F}_p$ with p elements contains the Frobenius automorphism φ which is given by

$$x^\varphi = x^p \quad \text{for all} \quad x \in \overline{\mathbb{F}}_p.$$

The subgroup $(\varphi) = \{\varphi^n \,|\, n \in \mathbb{Z}\}$ has the same fixed field $\mathbb{F}_p$ as the whole of $G_{\mathbb{F}_p}$. But contrary to what we are used to in finite Galois theory, we find $(\varphi) \neq G_{\mathbb{F}_p}$. In order to check this, let us construct an element $\psi \in G_{\mathbb{F}_p}$ which does not belong to (φ). We choose a sequence $\{a_n\}_{n\in\mathbb{N}}$ of integers satisfying

$$a_n \equiv a_m \bmod m$$

whenever $m|n$, but such that there is no integer a satisfying $a_n \equiv a \bmod n$ for all $n \in \mathbb{N}$. An example of such a sequence is given by $a_n = n'x_n$, where we write $n = n'p^{v_p(n)}$, $(n', p) = 1$, and $1 = n'x_n + p^{v_p(n)}y_n$. Now put

$$\psi_n = \varphi^{a_n}|_{\mathbb{F}_{p^n}} \in G(\mathbb{F}_{p^n}|\mathbb{F}_p).$$

If $\mathbb{F}_{p^m} \subseteq \mathbb{F}_{p^n}$, then $m|n$, so that $a_n \equiv a_m \bmod m$, and therefore

$$\psi_n|_{\mathbb{F}_{p^m}} = \varphi^{a_n}|_{\mathbb{F}_{p^m}} = \varphi^{a_m}|_{\mathbb{F}_{p^m}} = \psi_m.$$

Observe that $\varphi|_{\mathbb{F}_{p^m}}$ has order m. Therefore the ψ_n define an automorphism ψ of $\overline{\mathbb{F}}_p = \bigcup_{n=1}^\infty \mathbb{F}_{p^n}$. Now ψ cannot belong to (φ) because $\psi = \varphi^a$, for $a \in \mathbb{Z}$, would imply $\psi|_{\mathbb{F}_{p^n}} = \varphi^{a_n}|_{\mathbb{F}_{p^n}} = \varphi^a|_{\mathbb{F}_{p^n}}$ and hence $a_n \equiv a \bmod n$ for all n, which is what we ruled out by construction.

The example does not mean, however, that we have to chuck the main theorem of Galois theory altogether in the case of infinite extensions. We just have to amend it using the observation that the Galois group $G = G(\Omega|k)$ of any Galois extension $\Omega|k$ carries a canonical topology. This topology is called the **Krull topology** and is obtained as follows. For every $\sigma \in G$ we take the cosets

$$\sigma G(\Omega|K)$$

as a basis of neighbourhoods of σ, with $K|k$ ranging over *finite* Galois subextensions of $\Omega|k$. The multiplication and the inverse map

$$G \times G \longrightarrow G,\ (\sigma,\tau) \longmapsto \sigma\tau, \quad \text{and} \quad G \longrightarrow G,\ \sigma \longmapsto \sigma^{-1},$$

are continuous maps, since the preimage of a fundamental open neighbourhood $\sigma\tau G(\Omega|K)$, resp. $\sigma^{-1}G(\Omega|K)$, contains the open neighbourhood $\sigma G(\Omega|K) \times \tau G(\Omega|K)$, resp. $\sigma G(\Omega|K)$. Thus G is a topological group which satisfies the following

(1.1) Proposition. *For every (finite or infinite) Galois extension $\Omega|k$ the Galois group $G = G(\Omega|k)$ is compact Hausdorff with respect to the Krull topology.*

Proof: If $\sigma,\tau \in G$ and $\sigma \neq \tau$, then there exists a finite Galois subextension $K|k$ of $\Omega|k$ such that $\sigma|_K \neq \tau|_K$, so that $\sigma G(\Omega|K) \neq \tau G(\Omega|K)$ and thus $\sigma G(\Omega|K) \cap \tau G(\Omega|K) = \emptyset$. This shows that G is Hausdorff. In order to prove compactness, consider the mapping

$$h : G \longrightarrow \prod_K G(K|k), \quad \sigma \longmapsto \prod_K \sigma|_K,$$

where $K|k$ varies over the finite Galois subextensions. We view the finite groups $G(K|k)$ as discrete compact topological groups. Their product is therefore a compact topological space, by Tykhonov's theorem (see [98]). The homomorphism h is injective, because $\sigma|_K = 1$ for all K is equivalent to $\sigma = 1$. The sets $U = \prod_{K \neq K_0} G(K|k) \times \{\bar\sigma\}$ form a subbasis of open sets of the product $\prod_K G(K|k)$, where $K_0|k$ varies over the finite subextensions of $\Omega|k$ and $\bar\sigma \in G(K_0|k)$. If $\sigma \in G$ is a preimage of $\bar\sigma$, then $h^{-1}(U) = \sigma G(\Omega|K_0)$. Thus h is continuous. Moreover $h(\sigma G(\Omega|K_0)) = h(G) \cap U$, so $h : G \mapsto h(G)$ is open, and thus a homeomorphism. It therefore suffices to show that $h(G)$ is closed in the compact set $\prod_K G(K|k)$. To see this we consider, for each pair $L' \supseteq L$ of finite Galois subextensions of $\Omega|k$, the set

$$M_{L'|L} = \Big\{ \prod_K \sigma_K \in \prod_K G(K|k) \;\Big|\; \sigma_{L'}|_L = \sigma_L \Big\}.$$

One clearly has $h(G) = \bigcap_{L'\supseteq L} M_{L'|L}$. So it suffices to show that $M_{L'|L}$ is closed. But if $G(L|k) = \{\sigma_1, \dots, \sigma_n\}$, and $S_i \subseteq G(L'|k)$ is the set of extensions of σ_i to L', then

$$M_{L'|L} = \bigcup_{i=1}^{n} \Big(\prod_{K\neq L, L'} G(K|k) \times S_i \times \{\sigma_i\}\Big),$$

i.e., $M_{L'|L}$ is indeed closed. □

The main theorem of Galois theory for infinite extensions can now be formulated as follows.

(1.2) Theorem. *Let $\Omega|k$ be a (finite or infinite) Galois extension. Then the assignment*

$$K \longmapsto G(\Omega|K)$$

is a 1–1-correspondence between the subextensions $K|k$ of $\Omega|k$ and the closed subgroups of $G(\Omega|k)$. The open subgroups of $G(\Omega|k)$ correspond precisely to the finite subextensions of $\Omega|k$.

Proof: Every open subgroup of $G(\Omega|k)$ is also closed, because it is the complement of the union of its open cosets. If $K|k$ is a finite subextension, then $G(\Omega|K)$ is open, because each $\sigma \in G(\Omega|K)$ admits the open neighbourhood $\sigma G(\Omega|N) \subseteq G(\Omega|K)$, where $N|k$ is the normal closure of $K|k$. If $K|k$ is an arbitrary subextension, then

$$G(\Omega|K) = \bigcap_i G(\Omega|K_i),$$

where $K_i|k$ varies over the finite subextensions of $K|k$. Therefore $G(\Omega|K)$ is closed.

The assignment $K \mapsto G(\Omega|K)$ is injective, since K is the fixed field of $G(\Omega|K)$. To prove surjectivity, we have to show that, given an arbitrary closed subgroup H of $G(\Omega|k)$, we always have

$$H = G(\Omega|K),$$

where K is the fixed field of H. The inclusion $H \subseteq G(\Omega|K)$ is trivial. Conversely, let $\sigma \in G(\Omega|K)$. If $L|K$ is a finite Galois subextension of $\Omega|K$, then $\sigma G(\Omega|L)$ is a fundamental open neighbourhood of σ in $G(\Omega|K)$. The map $H \to G(L|K)$ is certainly surjective, because the image $\overline{H}$ has fixed field K and is therefore equal to $G(L|K)$, by the main theorem of Galois theory for finite extensions. Thus we may choose a $\tau \in H$ such that

$\tau|_L = \sigma|_L$, i.e., $\tau \in H \cap \sigma G(\Omega|L)$. This shows that σ belongs to the closure of H in $G(\Omega|K)$, and thus to H itself, so that $H = G(\Omega|K)$.

If H is an open subgroup of $G(\Omega|k)$, then it is also closed, and therefore of the form $H = G(\Omega|K)$. But $G(\Omega|k)$ is the disjoint union of the open cosets of H. Since $G(\Omega|k)$ is compact, a finite number of cosets suffices to cover the group. Thus there is only a finite number of them; $H = G(\Omega|K)$ has finite index in $G(\Omega|k)$, and this implies that $K|k$ has finite degree. □

The topological Galois groups $G = G(\Omega|k)$ have the special property that there is a fundamental system of neighbourhoods of the neutral element $1 \in G$ which consists of normal subgroups. This property leads us to the abstract, purely group-theoretical notion of a profinite group.

(1.3) Definition. *A* **profinite group** *is a topological group G which is Hausdorff and compact, and which admits a basis of neighbourhoods of $1 \in G$ consisting of normal subgroups.*

It can be shown that the last condition is tantamount to G being totally disconnected, i.e., to the condition that each element of G is equal to its own connected component. Every closed subgroup H of G is obviously again a profinite group. The disjoint coset decomposition

$$G = \dot{\bigcup_i} \sigma_i H$$

shows immediately that H is open if and only if the index $(G : H)$ is finite.

Profinite groups are fairly close relatives of finite groups. They can be reconstituted rather easily from their finite quotients. For the precise description of this we need the notion of *projective limit*, which naturally occurs in various places in number theory and which we will introduce next.

Exercise 1. Let $L|k$ be a Galois extension and $K|k$ an arbitrary extension, both contained in a common extension $\Omega|k$. If $L \cap K = k$, then the mapping

$$G(LK|K) \to G(L|k), \quad \sigma \mapsto \sigma|_L,$$

is a topological isomorphism, that is, an isomorphism of groups and a homeomorphism of topological spaces.

Exercise 2. Given a family of Galois extensions $K_i|k$ in $\Omega|k$, let $K|k$ be the composite of all $K_i|k$, and $K_i'|k$ the composite of the extensions $K_j|k$ such that $j \neq i$. If $K_i \cap K_i' = k$ for all i, then one has a topological isomorphism

$$G(K|k) \cong \prod_i G(K_i|k).$$

Exercise 3. A compact Hausdorff group is totally disconnected if and only if its neutral element admits a basis of neighbourhoods consisting only of normal subgroups.

Exercise 4. Every quotient G/H of a profinite group G by a closed normal subgroup H is a profinite group.

Exercise 5. Let G' be the closure of the commutator subgroup of a profinite group, and $G^{ab} = G/G'$. Show that every continuous homomorphism $G \to A$ into an abelian profinite group factorizes through G^{ab}.

§ 2. Projective and Inductive Limits

The notions of projective, resp. inductive limit generalize the operations of intersection, resp. union. If $\{X_i\}_{i\in I}$ is a family of subsets of a topological space X which for any two sets X_i, X_j also contains the set $X_i \cap X_j$ (resp. $X_i \cup X_j$), then the projective (resp. inductive) limit of this family is simply de ined by

$$\varprojlim_{i\in I} X_i = \bigcap_{i\in I} X_i \quad \text{(resp.} \quad \varinjlim_{i\in I} X_i = \bigcup_{i\in I} X_i).$$

Writing $i \leq j$ if $X_j \subseteq X_i$ (resp. $X_i \subseteq X_j$) makes the indexing set I into a *directed system*, i.e., an ordered set in which, for every pair i, j, there exists a k such that $i \leq k$ and $j \leq k$. In the case at hand, such a k is given by $X_k = X_i \cap X_j$ (resp. $X_k = X_i \cup X_j$). For $i \leq j$ we denote the inclusion $X_j \hookrightarrow X_i$ (resp. $X_i \hookrightarrow X_j$) by f_{ij} and obtain a system $\{X_i, f_{ij}\}$ of sets and maps. The operations of intersection and union are now generalized by replacing the inclusions f_{ij} with arbitrary maps.

(2.1) Definition. *Let I be a directed system. A* **projective**, *resp.* **inductive system** *over I is a family $\{X_i, f_{ij} \mid i, j \in I,\ i \leq j\}$ of topological spaces X_i and continuous maps*

$$f_{ij} : X_j \longrightarrow X_i, \quad \text{resp.} \quad f_{ij} : X_i \longrightarrow X_j,$$

such that one has $f_{ii} = \mathrm{id}_{X_i}$ and

$$f_{ik} = f_{ij} \circ f_{jk}, \quad \text{resp.} \quad f_{ik} = f_{jk} \circ f_{ij},$$

when $i \leq j \leq k$.

In order to define the projective, resp. inductive limit of a projective, resp. inductive system $\{X_i, f_{ij}\}$, we make use of the direct product $\prod_{i\in I} X_i$, resp. the disjoint union $\coprod_{i\in I} X_i$.

(2.2) Definition. *The* **projective limit**

$$X = \varprojlim_{i \in I} X_i$$

of the projective system $\{X_i, f_{ij}\}$ *is defined to be the subset*

$$X = \Big\{ (x_i)_{i\in I} \in \prod_{i\in I} X_i \mid f_{ij}(x_j) = x_i \quad \text{for} \quad i \le j \Big\}$$

of the product $\prod_{i\in I} X_i$.

The product $\prod_{i\in I} X_i$ is equipped with the product topology. If the X_i are Hausdorff, then so is the product, and it contains in this case X as a closed subspace. Indeed, one has

$$X = \bigcap_{i\le j} X_{ij},$$

where $X_{ij} = \big\{ (x_k)_{k\in I} \in \prod_k X_k \mid f_{ij}(x_j) = x_i \big\}$, so that it suffices to show the closedness of the sets X_{ij}. Writing $p_i : \prod_{k\in I} X_k \to X_i$ for the i-th projection, the two maps $g = p_i$, $f = f_{ij} \circ p_j : \prod_{k\in I} X_k \to X_i$ are continuous, and we may write $X_{ij} = \big\{ x \in \prod_k X_k \mid g(x) = f(x) \big\}$. But in the Hausdorff case the equation $g(x) = f(x)$ defines a closed subset. This representation $X = \bigcap_{i\le j} X_{ij}$ also gives the following

(2.3) Proposition. *The projective limit* $X = \varprojlim_i X_i$ *of nonempty compact spaces* X_i *is itself nonempty and compact.*

Proof: If all the X_i are compact, then so is the product $\prod_{i\in I} X_i$, by Tykhonov's theorem, and thus also the closed subset X. Furthermore, $X = \bigcap_{i\le j} X_{ij}$ cannot be the empty set if the X_i are nonempty. In fact, as the product $\prod_i X_i$ is compact, there would have to be an intersection of finitely many X_{ij} which is empty. But this is impossible: if all indices entering into this finite intersection satisfy $i, j \le n$, and if $x_n \in X_n$, then the element $(x_i)_{i\in I}$ belongs to this intersection, where we choose $x_i = f_{in}(x_n)$, for $i \le n$, and arbitrarily for all other i. □

(2.4) Definition. *The* **inductive limit**

$$X = \varinjlim_{i\in I} X_i$$

of an inductive system $\{X_i, f_{ij}\}$ *is defined to be the quotient*

$$X = \big(\textstyle\coprod_{i\in I} X_i\big) / \sim$$

of the disjoint union $\coprod_{i\in I} X_i$, *where we consider two elements* $x_i \in X_i$ *and* $x_j \in X_j$ *equivalent if there exists a* $k \ge i, j$ *such that*

$$f_{ik}(x_i) = f_{jk}(x_j).$$

In the applications, the projective and inductive systems $\{X_i, f_{ij}\}$ that occur will not just be systems of topological spaces and continuous maps, but the X_i will usually be topological groups, rings or modules, etc., and the f_{ij} will be continuous homomorphisms. In what follows, we will deal explicitly only with projective and inductive systems $\{G_i, g_{ij}\}$ of topological groups. But since everything works exactly the same way for systems of rings or modules, these cases may be thought of tacitly as being treated as well.

Let $\{G_i, g_{ij}\}$ be a projective, resp. inductive system of topological groups. Then the projective, resp. inductive limit

$$G = \varprojlim_{i\in I} G_i, \quad \text{resp.} \quad G = \varinjlim_{i\in I} G_i$$

is a topological group as well. The multiplication in the projective limit is induced by the componentwise multiplication in the product $\prod_{i\in I} G_i$. In the case of the inductive limit, given two equivalence classes $x, y \in G = \varinjlim_{i\in I} G_i$, one has to choose representatives x_k and y_k in the same G_k in order to define

$$xy = \text{equivalence class of } x_k y_k .$$

We leave it to the reader to check that this definition is independent of the choice of representatives, and that the operation thus defined makes G into a group.

The projections $p_i : \prod_{i\in I} G_i \to G_i$, resp. the inclusions $\iota_i : G_i \to \coprod_{i\in I} G_i$, induce a family of continuous homomorphisms

$$g_i : G \longrightarrow G_i, \quad \text{resp.} \quad g_i : G_i \longrightarrow G$$

such that $g_i = g_{ij} \circ g_j$, resp. $g_i = g_j \circ g_{ij}$, for $i \le j$. This family has the following universal property.

(2.5) Proposition. *If H is a topological group and*

$$h_i : H \longrightarrow G_i, \quad \textit{resp.} \quad h_i : G_i \longrightarrow H$$

is a family of continuous homomorphisms such that

$$h_i = g_{ij} \circ h_j, \quad \textit{resp.} \quad h_i = h_j \circ g_{ij}$$

for $i \le j$, then there exists a unique continuous homomorphism

$$h : H \longrightarrow G = \varprojlim_{i} G_i, \quad \textit{resp.} \quad h : G = \varinjlim_{i} G_i \longrightarrow H$$

satisfying $h_i = g_i \circ h$, resp. $h_i = h \circ g_i$, for all $i \in I$.

The easy proof is left to the reader. A **morphism** between two projective, resp. inductive systems $\{G_i, g_{ij}\}$ and $\{G'_i, g'_{ij}\}$ of topological groups is a family of continuous homomorphisms $f_i : G_i \to G'_i$, $i \in I$, such that the diagrams

$$\begin{array}{ccc} G_j & \xrightarrow{f_j} & G'_j \\ {\scriptstyle g_{ij}}\downarrow & & \downarrow{\scriptstyle g'_{ij}}, \\ G_i & \xrightarrow{f_i} & G'_i \end{array} \qquad \text{resp.} \qquad \begin{array}{ccc} G_j & \xrightarrow{f_j} & G'_j \\ {\scriptstyle g_{ij}}\uparrow & & \uparrow{\scriptstyle g'_{ij}} \\ G_i & \xrightarrow{f_i} & G'_i \end{array}$$

commute for $i \le j$. Such a family $(f_i)_{i \in I}$ defines a mapping

$$f : \prod_{i \in I} G_i \longrightarrow \prod_{i \in I} G'_i, \quad \text{resp.} \quad f : \coprod_{i \in I} G_i \longrightarrow \coprod_{i \in I} G'_i,$$

which induces a homomorphism

$$f : \varprojlim_{i \in I} G_i \longrightarrow \varprojlim_{i \in I} G'_i, \quad \text{resp.} \quad f : \varinjlim_{i \in I} G_i \longrightarrow \varinjlim_{i \in I} G'_i.$$

In this way $\varprojlim$, resp. $\varinjlim$, becomes a functor. A particularly important property of this functor is its so-called "exactness". For the inductive limit $\varinjlim$, exactness holds without restrictions. In other words, one has the

(2.6) Proposition. *Let $\alpha : \{G'_i, g'_{ij}\} \to \{G_i, g_{ij}\}$ and $\beta : \{G_i, g_{ij}\} \to \{G''_i, g''_{ij}\}$ be morphisms between inductive systems of topological groups such that the sequence*

$$G'_i \xrightarrow{\alpha_i} G_i \xrightarrow{\beta_i} G''_i$$

is exact for every $i \in I$. Then the induced sequence

$$\varinjlim_{i \in I} G'_i \xrightarrow{\alpha} \varinjlim_{i \in I} G_i \xrightarrow{\beta} \varinjlim_{i \in I} G''_i$$

is also exact.

Proof: Let $G' = \varinjlim_{i} G'_i$, $G = \varinjlim_{i} G_i$, $G'' = \varinjlim_{i} G''_i$. We consider the commutative diagram

$$\begin{array}{ccccc} G'_i & \xrightarrow{\alpha_i} & G_i & \xrightarrow{\beta_i} & G''_i \\ \downarrow{\scriptstyle g'_i} & & \downarrow{\scriptstyle g_i} & & \downarrow{\scriptstyle g''_i} \\ G' & \xrightarrow{\alpha} & G & \xrightarrow{\beta} & G''. \end{array}$$

Let $x \in G$ be such that $\beta(x) = 1$. Then there exists an i and an $x_i \in G_i$ such that $g_i(x_i) = x$. As

$$g_i'' \beta_i(x_i) = \beta g_i(x_i) = \beta(x) = 1,$$

there exists $j \geq i$ such that $\beta_i(x_i)$ equals 1 in G_j''. Changing notation, we may therefore assume that $\beta_i(x_i) = 1$, so that there exists $y_i \in G_i'$ such that $\alpha_i(y_i) = x_i$. Putting $y = g_i'(y_i)$, we have $\alpha(y) = x$. □

The projective limit is not exact in complete generality, but only for **compact** groups, so that we have the

(2.7) Proposition. *Let $\alpha : \{G_i', g_{ij}'\} \to \{G_i, g_{ij}\}$ and $\beta : \{G_i, g_{ij}\} \to \{G_i'', g_{ij}''\}$ be morphisms between projective systems of* **compact** *topological groups such that the sequence*

$$G_i' \xrightarrow{\alpha_i} G_i \xrightarrow{\beta_i} G_i''$$

is exact for every $i \in I$. Then

$$\varprojlim_i G_i' \xrightarrow{\alpha} \varprojlim_i G_i \xrightarrow{\beta} \varprojlim_i G_i''$$

is again an exact sequence of compact topological groups.

Proof: Let $x = (x_i)_{i \in I} \in \varprojlim_i G_i$ and $\beta(x) = 1$, so that $\beta_i(x_i) = 1$ for all $i \in I$. The preimages $Y_i = \alpha_i^{-1}(x_i) \subseteq G_i'$ then form a projective system of nonempty closed, and hence compact subsets of the G_i'. By (2.3), this means that the projective limit $Y = \varprojlim_i Y_i \subseteq \varprojlim_i G_i'$ is nonempty, and α maps every element $y \in Y$ to x. □

Now that we have at our disposal the notion of projective limit, we return to our starting point, the profinite groups. Recall that these are the topological groups which are Hausdorff, compact and totally disconnected, i.e., they admit a basis of neighbourhoods of the neutral element consisting of normal subgroups. The next proposition shows that they are precisely the projective limits of finite groups (which we view as compact topological groups with respect to the discrete topology).

(2.8) Proposition. *If G is a profinite group, and if N varies over the open normal subgroups of G, then one has, algebraically as well as topologically, that*

$$G \cong \varprojlim_{N} G/N\,.$$

If conversely $\{G_i, g_{ij}\}$ is a projective system of finite (or even profinite) groups, then

$$G = \varprojlim_{i} G_i$$

is a profinite group.

Proof: Let G be a profinite group and let $\{N_i \mid i \in I\}$ be the family of its open normal subgroups. We make I into a directed system by defining $i \le j$ if $N_i \supseteq N_j$. The groups $G_i = G/N_i$ are finite since the cosets of N_i in G form a disjoint open covering of G, which must be finite because G is compact. For $i \le j$ we have the projections $g_{ij} : G_j \to G_i$ and obtain a projective system $\{G_i, g_{ij}\}$ of finite, and hence discrete, compact groups. We show that the homomorphism

$$f : G \longrightarrow \varprojlim_{i\in I} G_i\,, \quad \sigma \longmapsto \prod_{i\in I} \sigma_i\,, \quad \sigma_i = \sigma \bmod N_i\,,$$

is an isomorphism and a homeomorphism. f is injective because its kernel is the intersection $\bigcap_{i\in I} N_i$, which equals $\{1\}$ because G is Hausdorff and the N_i form a basis of neighbourhoods of 1. The groups

$$U_S = \prod_{i\notin S} G_i \times \prod_{i\in S} \{1_{G_i}\},$$

with S varying over the finite subsets of I, form a basis of neighbourhoods of the neutral element in $\prod_{i\in I} G_i$. As $f^{-1}(U_S \cap \varprojlim G_i) = \bigcap_{i\in S} N_i$, we see that f is continuous. Moreover, as G is compact, the image $f(G)$ is closed in $\varprojlim G_i$. On the other hand it is also dense. For if $x = (x_i)_{i\in I} \in \varprojlim G_i$, and $x(U_S \cap \varprojlim G_i)$ is a fundamental neighbourhood of x, then we may choose a $y \in G$ which is mapped to x_k under the projection $G \to G/N_k$, where we put $N_k = \bigcap_{i\in S} N_i$. Then $y \bmod N_i = x_i$ for all $i \in S$, so that $f(y)$ belongs to the neighbourhood $x(U_S \cap \varprojlim G_i)$. Therefore the closed set $f(G)$ is indeed dense in $\varprojlim G_i$, and so $f(G) = \varprojlim G_i$. Since G is compact, f maps closed sets into closed sets, and thus also open sets into open sets. This shows that $f : G \to \varprojlim G_i$ is an isomorphism and a homeomorphism.

Conversely, let $\{G_i, g_{ij}\}$ be a projective system of profinite groups. As the G_i are Hausdorff and compact, so is the projective limit $G = \varprojlim G_i$, by (2.3). If N_i varies over a basis of neighbourhoods of the neutral element in G_i which consists of normal subgroups, then the groups

$$U_S = \prod_{i \notin S} G_i \times \prod_{i \in S} N_i \,,$$

with S varying over the finite subsets of I, make up a basis of neighbourhoods of the neutral element in $\prod_{i \in I} G_i$ consisting of normal subgroups. The normal subgroups $U_S \cap \varprojlim G_i$ therefore form a basis of neighbourhoods of the neutral element in $\varprojlim G_i$; thus $\varprojlim G_i$ is a profinite group. □

Let us now illustrate the notions of profinite group and projective limit by a few concrete examples.

Example 1: The Galois group $G = G(\Omega|k)$ of a Galois extension $\Omega|k$ is a profinite group with respect to the Krull topology. This was already stated in § 1. If $K|k$ varies over the finite Galois subextensions of $\Omega|k$, then, by definition of the Krull topology, $G(\Omega|K)$ varies over the open normal subgroups of G. In view of the identity $G(K|k) = G(\Omega|k)/G(\Omega|K)$ and of (2.8), we therefore obtain the Galois group $G(\Omega|k)$ as the projective limit

$$G(\Omega|k) \cong \varprojlim G(K|k)$$

of the finite Galois groups $G(K|k)$.

Example 2: If p is a prime number, then the rings $\mathbb{Z}/p^n\mathbb{Z}$, $n \in \mathbb{N}$, form a projective system with respect to the projections $\mathbb{Z}/p^n\mathbb{Z} \to \mathbb{Z}/p^m\mathbb{Z}$, for $n \geq m$. The projective limit

$$\mathbb{Z}_p = \varprojlim_n \mathbb{Z}/p^n\mathbb{Z}$$

is the ring of p-adic integers (see chap. II, § 1).

Example 3: Let $\mathcal{O}$ be the valuation ring in a $\mathfrak{p}$-adic number field K and $\mathfrak{p}$ its maximal ideal. The ideals $\mathfrak{p}^n$, $n \in \mathbb{N}$, make up a basis of neighbourhoods of the zero element 0 in $\mathcal{O}$. $\mathcal{O}$ is Hausdorff and compact, and so is a profinite ring. The rings $\mathcal{O}/\mathfrak{p}^n$, $n \in \mathbb{N}$, are finite and we have a topological isomorphism

$$\mathcal{O} \cong \varprojlim_n \mathcal{O}/\mathfrak{p}^n \,, \quad a \longmapsto \prod_{n \in \mathbb{N}} (a \bmod \mathfrak{p}^n) \,.$$

The group of units $U = \mathcal{O}^*$ is closed in $\mathcal{O}$, hence Hausdorff and compact, and the subgroups $U^{(n)} = 1 + \mathfrak{p}^n$ form a basis of neighbourhoods of $1 \in U$. Thus

$$U \cong \varprojlim_n U/U^{(n)}$$

is also a profinite group. In fact, we have seen all this already in chap. II, § 4.

Example 4: The rings $\mathbb{Z}/n\mathbb{Z}$, $n \in \mathbb{N}$, form a projective system with respect to the projections $\mathbb{Z}/n\mathbb{Z} \to \mathbb{Z}/m\mathbb{Z}$, $n|m$, where the ordering on $\mathbb{N}$ is now given by divisibility, $n|m$. The projective limit

$$\widehat{\mathbb{Z}} = \varprojlim_{n} \mathbb{Z}/n\mathbb{Z}$$

was originally called the **Prüfer ring**, whereas nowadays it has become customary to refer to it by the somewhat curt abbreviation "zed-hat" (or "zee-hat"). This ring is going to occupy quite an important position in what follows. It contains $\mathbb{Z}$ as a dense subring. The groups $n\widehat{\mathbb{Z}}$, $n \in \mathbb{N}$, are precisely the open subgroups of $\widehat{\mathbb{Z}}$, and it is easy to verify that

$$\widehat{\mathbb{Z}}/n\widehat{\mathbb{Z}} \cong \mathbb{Z}/n\mathbb{Z}.$$

Taking, for each natural number n, the prime factorization $n = \prod_p p^{\nu_p}$, the Chinese remainder theorem implies the decomposition

$$\mathbb{Z}/n\mathbb{Z} \cong \prod_p \mathbb{Z}/p^{\nu_p}\mathbb{Z},$$

and passing to the projective limit,

$$\widehat{\mathbb{Z}} \cong \prod_p \mathbb{Z}_p.$$

This takes the natural embedding of $\mathbb{Z}$ into $\widehat{\mathbb{Z}}$ to the diagonal embedding $\mathbb{Z} \to \prod_p \mathbb{Z}_p$, $a \mapsto (a, a, a, \ldots)$.

Example 5: For the field $\mathbb{F}_q$ with q elements, we get isomorphisms

$$G(\mathbb{F}_{q^n}|\mathbb{F}_q) \cong \mathbb{Z}/n\mathbb{Z},$$

one for every $n \in \mathbb{N}$, by mapping the Frobenius automorphism φ_n to $1 \bmod n\mathbb{Z}$. Passing to the projective limit gives an isomorphism

$$G(\overline{\mathbb{F}}_q|\mathbb{F}_q) \cong \widehat{\mathbb{Z}}$$

which sends the Frobenius automorphism $\varphi \in G(\overline{\mathbb{F}}_q|\mathbb{F}_q)$ to $1 \in \widehat{\mathbb{Z}}$, and the subgroup $(\varphi) = \{\varphi^n \mid n \in \mathbb{Z}\}$ onto the dense (but not closed) subgroup $\mathbb{Z}$ of $\widehat{\mathbb{Z}}$. Given this, it is now clear, in the example at the beginning of this chapter, how we were able to construct an element $\psi \in G(\overline{\mathbb{F}}_q|\mathbb{F}_q)$ which did not belong to (φ). In fact, looking at it via the isomorphism $G(\overline{\mathbb{F}}_q|\mathbb{F}_q) \cong \widehat{\mathbb{Z}}$, what we did amounted to writing down the element

$$(\ldots, 0, 0, 1_p, 0, 0, \ldots) \in \prod_{\ell} \mathbb{Z}_\ell = \widehat{\mathbb{Z}},$$

which does not belong to $\mathbb{Z}$.

Example 6: Let $\widetilde{\mathbb{Q}}|\mathbb{Q}$ be the extension obtained by adjoining all roots of unity. Its Galois group $G(\widetilde{\mathbb{Q}}|\mathbb{Q})$ is then canonically isomorphic (as a topological group) to the group of units $\widehat{\mathbb{Z}}^* \cong \prod_p \mathbb{Z}_p^*$ of $\widehat{\mathbb{Z}}$,

$$G(\widetilde{\mathbb{Q}}|\mathbb{Q}) \cong \widehat{\mathbb{Z}}^* .$$

This isomorphism is obtained by passing to the projective limit from the canonical isomorphisms

$$G(\mathbb{Q}(\mu_n)|\mathbb{Q}) \cong (\mathbb{Z}/n\mathbb{Z})^* ,$$

where μ_n denotes the group of n-th roots of unity.

Example 7: The groups $\mathbb{Z}_p$ and $\widehat{\mathbb{Z}}$ are (additive) special cases of the class of **procyclic** groups. These are profinite groups G which are topologically generated by a single element σ; i.e., G is the closure $\overline{(\sigma)}$ of the subgroup $(\sigma) = \{\sigma^n \mid n \in \mathbb{Z}\}$. The open subgroups of a procyclic group $G = \overline{(\sigma)}$ are all of the form G^n. Indeed, G^n is closed, being the image of the continuous map $G \to G$, $x \mapsto x^n$, and the quotient group G/G^n is finite, because it contains the finite group $\{\sigma^\nu \bmod G^n \mid 0 \le \nu < n\}$ as a dense subgroup, and is therefore equal to it. Conversely, if H is a subgroup of G of index n, then $G^n \subseteq H \subseteq G$ and $n = (G : H) \le (G : G^n) \le n$, so that $H = G^n$.

Every procyclic group G is a quotient of the group $\widehat{\mathbb{Z}}$. In fact, if $G = \overline{(\sigma)}$, then we have for every n the surjective homomorphism

$$\mathbb{Z}/n\mathbb{Z} \to G/G^n , \quad 1 \bmod n\mathbb{Z} \mapsto \sigma \bmod G^n ,$$

and in view of (2.7), passing to the projective limit yields a continuous surjection $\widehat{\mathbb{Z}} \to G$.

Example 8: Let A be an abelian torsion group. Then the **Pontryagin dual**

$$\chi(A) = \mathrm{Hom}(A, \mathbb{Q}/\mathbb{Z})$$

is a profinite group. For one has

$$A = \bigcup_i A_i ,$$

where A_i varies over the finite subgroups of A, and thus

$$\chi(A) = \varprojlim_i \chi(A_i)$$

with finite groups $\chi(A_i)$. If for instance,

$$A = \mathbb{Q}/\mathbb{Z} = \bigcup_{n\in\mathbb{N}} \tfrac{1}{n}\mathbb{Z}/\mathbb{Z},$$

then $\chi(\frac{1}{n}\mathbb{Z}/\mathbb{Z}) = \mathbb{Z}/n\mathbb{Z}$, so that

$$\chi(\mathbb{Q}/\mathbb{Z}) \cong \varprojlim_n \mathbb{Z}/n\mathbb{Z} = \widehat{\mathbb{Z}}.$$

Example 9: If G is any group and N varies over all normal subgroups of finite index, then the profinite group

$$\widehat{G} = \varprojlim_N G/N$$

is called the *profinite completion* of G. The profinite completion of $\mathbb{Z}$, for example, is the group $\widehat{\mathbb{Z}} = \varprojlim_n \mathbb{Z}/n\mathbb{Z}$.

Exercise 1. Show that, for a profinite group G, the power map $G \times \mathbb{Z} \to G$, $(\sigma, n) \mapsto \sigma^n$, extends to a continuous map

$$G \times \widehat{\mathbb{Z}} \to G, \quad (\sigma, a) \mapsto \sigma^a,$$

and that one has $(\sigma^a)^b = \sigma^{ab}$ and $\sigma^{a+b} = \sigma^a\sigma^b$ if G is abelian.

Exercise 2. If $\sigma \in G$ and $a = \lim_{i\to\infty} a_i \in \widehat{\mathbb{Z}}$ with $a_i \in \mathbb{Z}$, then $\sigma^a = \lim_{i\to\infty} \sigma^{a_i}$ is in G.

Exercise 3. A *pro-p-group* is a profinite group G whose quotients G/N, modulo all open normal subgroups N, are finite p-groups. Imitating exercise 1, make sense of the powers σ^a, for all $\sigma \in G$ and $a \in \mathbb{Z}_p$.

Exercise 4. A closed subgroup H of a profinite group G is called a ***p*-Sylow subgroup** of G if, for every open normal subgroup N of G, the group HN/N is a p-Sylow subgroup of G/N. Show:

(i) For every prime number p, there exists a p-Sylow subgroup of G.

(ii) Every pro-p-subgroup of G is contained in a p-Sylow subgroup.

(iii) Every two p-Sylow subgroups of G are conjugate.

Exercise 5. What is the p-Sylow subgroup of $\widehat{\mathbb{Z}}$ and of $\mathbb{Z}_p^*$?

Exercise 6. If $\{G_i\}$ is a projective system of profinite groups and $G = \varprojlim_i G_i$, then $G^{ab} = \varprojlim_i G_i^{ab}$ (see § 1, exercise 5).

§ 3. Abstract Galois Theory

Class field theory is the final outcome of a long development of algebraic number theory the beginning of which was Gauss's reciprocity law

$$\left(\frac{a}{b}\right)\left(\frac{b}{a}\right) = (-1)^{\frac{a-1}{2}\frac{b-1}{2}}.$$

The endeavours to generalize this law finally produced a theory of the abelian extensions of algebraic and $\mathfrak{p}$-adic number fields. These extensions $L|K$ are classified by certain subgroups $\mathcal{N}_L = N_{L|K}A_L$ of a group A_K attached to the base field. In the local case, A_K is the multiplicative group K^* and in the global case it is a modification of the ideal class group. At the heart of this theory there is a mysterious canonical isomorphism

$$G(L|K) \cong A_K/N_{L|K}A_L,$$

which – if we view things in the right way – encapsulates the reciprocity law in its most general form. Now, this map can be abstracted completely from the field-theoretic situation and treated on a purely group theoretical basis. In this way, class field theory can be given an abstract, but elementary foundation, to which we will now turn.

We begin our considerations by giving ourselves a profinite group G. The theory we are about to develop is purely group theoretical in nature. However, the only applications we have in mind are field theoretical, and the language of field theory allows immediate insights into the group theoretical relations. We will therefore formally interpret the profinite group G as a Galois group in the following way. (Let us remark in passing that every profinite group is indeed the Galois group $G = G(\bar{k}|k)$ of a Galois field extension $\bar{k}|k$; this will allow the reader to rely on his standard knowledge of Galois theory whenever the formal development in terms of group theory alone would seem odd.)

We denote the closed subgroups of G by G_K, and call these indices K "fields"; K will be called the fixed field of G_K. The field k such that $G_k = G$ is called the base field, and $\bar{k}$ denotes the field satisfying $G_{\bar{k}} = \{1\}$. The field belonging to the closure $\overline{(\sigma)}$ of the cyclic group $(\sigma) = \{\sigma^k \mid k \in \mathbb{Z}\}$ generated by an element $\sigma \in G$ is simply called the fixed field of σ.

We write formally $K \subseteq L$ or $L|K$ if $G_L \subseteq G_K$, and we call the pair $L|K$ a field extension. $L|K$ is called a finite extension, if G_L is open, i.e., of finite index in G_K, and this index

$$[L : K] := (G_K : G_L)$$

will be called the degree of $L|K$. $L|K$ is said to be normal or Galois if G_L is a normal subgroup of G_K. If this is the case, we define the Galois group of $L|K$ by

$$G(L|K) = G_K/G_L .$$

If $N \supseteq L \supseteq K$ are Galois extensions of K, we define the restriction of an element $\sigma \in G(N|K)$ to L by

$$\sigma|_L = \sigma \bmod G(N|L) \in G(L|K).$$

This gives a homomorphism

$$G(N|K) \longrightarrow G(L|K), \quad \sigma \longmapsto \sigma|_L ,$$

with kernel $G(N|L)$. The extension $L|K$ is called cyclic, abelian, solvable, etc., if the Galois group $G(L|K)$ has these properties. We put

$$K = \bigcap_i K_i \qquad \text{("intersection")}$$

if G_K is topologically generated by the subgroups G_{K_i}, and

$$K = \prod_i K_i \quad \text{("composite")}$$

if $G_K = \bigcap_i G_{K_i}$. If $G_{K'} = \sigma^{-1} G_K \sigma$ for $\sigma \in G$, we write $K' = K^\sigma$.

Now let A be a *continuous multiplicative* G-module. By this we mean a multiplicative abelian group A on which the elements $\sigma \in G$ operate as automorphisms on the right, $\sigma : A \to A$, $a \mapsto a^\sigma$. This action must satisfy

(i) $a^1 = a$,

(ii) $(ab)^\sigma = a^\sigma b^\sigma$,

(iii) $a^{\sigma\tau} = (a^\sigma)^\tau$,

(iv) $A = \bigcup_{[K:k]<\infty} A_K$,

where A_K in the last condition denotes the fixed module A^{G_K} under G_K, so that

$$A_K = \{ a \in A \mid a^\sigma = a \ \text{ for all } \sigma \in G_K \} ,$$

and where K varies over all extensions that are finite over k. The condition (iv) says that G operates continuously on A, i.e., the map

$$G \times A \longrightarrow A, \quad (\sigma, a) \longmapsto a^\sigma ,$$

is continuous, where A is equipped with the discrete topology. Indeed, this continuity is equivalent to the fact that, for every element $(\sigma, a) \in G \times A$, there exists an open subgroup $U = G_K$ of G such that the neighbourhood $\sigma U \times \{a\}$ of (σ, a) is mapped to the open set $\{a^\sigma\}$, and this means simply that $a^\sigma \in A^U = A_K$.

Remark: In the exponential notation a^σ, the operation of G on A appears as an action on the right. This notation is adequate for many computations in the case of multiplicative G-modules A. For instance, the notation $a^{\sigma-1} := a^\sigma a^{-1}$ is to be preferred to writing $(\sigma - 1)a = \sigma a \cdot a^{-1}$. On the other hand, classical usage often calls for an operation on the left. Thus in the case of a Galois extension $L|K$ of actual fields, the Galois group $G(L|K)$ acts as the automorphism group on L from the left, and therefore also in the same way on the multiplicative group L^*. This occasional switch from the left to the right should not confuse the reader.

For every extension $L|K$ we have $A_K \subseteq A_L$, and if $L|K$ is finite, then we have the norm map

$$N_{L|K} : A_L \longrightarrow A_K, \quad N_{L|K}(a) = \prod_\sigma a^\sigma,$$

where σ varies over a system of representatives of $G_L \backslash G_K$. If $L|K$ is Galois, then A_L is a $G(L|K)$-module and one has

$$A_L^{G(L|K)} = A_K.$$

At the center of class field theory there is the **norm residue group**

$$H^0\big(G(L|K), A_L\big) = A_K / N_{L|K} A_L.$$

We also consider the group

$$H^{-1}\big(G(L|K), A_L\big) = {}_{N_{L|K}}A_L / I_{G(L|K)} A_L,$$

where

$${}_{N_{L|K}}A_L = \big\{a \in A_L \mid N_{L|K}(a) = 1\big\}$$

is the "norm-one group" and $I_{G(L|K)}A_L$ is the subgroup of ${}_{N_{L|K}}A_L$ which is generated by all elements

$$a^{\sigma-1} := a^\sigma a^{-1},$$

with $a \in A_L$, and $\sigma \in G(L|K)$. If $G(L|K)$ is cyclic and σ is a generator, then $I_{G(L|K)}A_L$ is simply the group

$$A_L^{\sigma-1} = \big\{a^{\sigma-1} \mid a \in A_L\big\}.$$

In fact, the formal identity $\sigma^k - 1 = (1 + \sigma + \cdots + \sigma^{k-1})(\sigma - 1)$ implies $a^{\sigma^k - 1} = b^{\sigma-1}$ with $b = \prod_{i=0}^{k-1} a^{\sigma^i}$.

Let us now apply the notions introduced so far to the example of **Kummer theory**. For this, we impose on the G-module A the following axiomatic condition.

(3.1) Axiom. *One has* $H^{-1}(G(L|K), A_L) = 1$ *for all finite cyclic extensions* $L|K$.

The theory we are about to develop makes reference to a surjective G-homomorphism

$$\wp : A \longrightarrow A, \quad a \longmapsto a^{\wp},$$

with finite *cyclic* kernel $\mu_{\wp}$. The order $n = \#\mu_{\wp}$ is called the *exponent* of the operator $\wp$. The case of prime interest to us is when $\wp$ is the n-th power map $a \mapsto a^n$, and $\mu_{\wp} = \mu_n = \{\xi \in A \mid \xi^n = 1\}$ is the group of "n-th roots of unity" in A.

We now fix a field K such that $\mu_{\wp} \subseteq A_K$. For every subset $B \subseteq A$, let $K(B)$ denote the fixed field of the closed subgroup

$$H = \big\{\sigma \in G_K \bigm| b^{\sigma} = b \text{ for all } b \in B\big\}$$

of G_K. If B is G_K-invariant, then $K(B)|K$ is obviously Galois. A **Kummer extension** (with respect to $\wp$) is by definition an extension of the form

$$K\big(\wp^{-1}(\Delta)\big) \mid K,$$

where $\Delta \subseteq A_K$. A Kummer extension $K(\wp^{-1}(\Delta))|K$ is always Galois, and its Galois group is abelian of exponent n. Indeed, for an extension $K(\wp^{-1}(a))|K$, we have the injective homomorphism

$$G\big(K(\wp^{-1}(a))|K\big) \hookrightarrow \mu_{\wp}, \quad \sigma \longmapsto \alpha^{\sigma-1},$$

where $\alpha \in \wp^{-1}(a)$. Since $\mu_{\wp} \subseteq A_K$, this definition does not depend on the choice of α. Thus, for a Kummer extension $L = K(\wp^{-1}(\Delta)) = \prod_{a\in\Delta} K(\wp^{-1}(a))$, the composite map

$$G(L|K) \longrightarrow \prod_{a\in\Delta} G\big(K(\wp^{-1}(a))|K\big) \longrightarrow \mu_{\wp}^{\Delta}$$

is an injective homomorphism.

The following proposition says that conversely, any abelian extension $L|K$ of exponent n is a Kummer extension.

(3.2) Proposition. *If* $L|K$ *is an abelian extension of exponent* n, *then*

$$L = K\big(\wp^{-1}(\Delta)\big) \quad \textit{with} \quad \Delta = A_L^{\wp} \cap A_K.$$

If in particular, $L|K$ *is cyclic, then we find* $L = K(\alpha)$ *with* $\alpha^{\wp} = a \in A_K$.

Proof: We have $\wp^{-1}(\Delta) \subseteq A_L$, for if $x \in A$ and $x^{\wp} = \alpha^{\wp} = a \in A_K$, $\alpha \in A_L$, then $x = \xi\alpha \in A_L$ for some $\xi \in \mu_{\wp} \subseteq A_K$. Therefore $K(\wp^{-1}(\Delta)) \subseteq L$. On the other hand, the extension $L|K$ is the composite of its cyclic subextensions. For it is the composite of its *finite* subextensions, and the Galois group of a finite subextension is the product of cyclic groups, which may be interpreted as Galois groups of cyclic subextensions. Let now $M|K$ be a cyclic subextension of $L|K$. It suffices to show that $M \subseteq K(\wp^{-1}(\Delta))$. Let σ be a generator of $G(M|K)$ and ζ a generator of $\mu_{\wp}$. Let $d = [M : K]$, $d' = n/d$ and $\xi = \zeta^{d'}$. Since $N_{M|K}(\xi) = \xi^d = 1$, (3.1) shows that $\xi = \alpha^{\sigma-1}$ for some $\alpha \in A_M$. Thus $K \subseteq K(\alpha) \subseteq M$. But $\alpha^{\sigma^i} = \xi^i\alpha$. Thus $\alpha^{\sigma^i} = \alpha$ is equivalent to $i \equiv 0 \bmod d$, so that $K(\alpha) = M$. But $(\alpha^{\wp})^{\sigma-1} = (\alpha^{\sigma-1})^{\wp} = \xi^{\wp} = 1$, so that $a = \alpha^{\wp} \in A_K$; then $\alpha \in \wp^{-1}(\Delta)$, and therefore $M \subseteq K(\wp^{-1}(\Delta))$. □

As the main result of general Kummer theory, we now obtain the following

(3.3) Theorem. *The correspondence*

$$\Delta \longmapsto L = K\big(\wp^{-1}(\Delta)\big)$$

is a 1-1-correspondence between the groups Δ such that $A_K^{\wp} \subseteq \Delta \subseteq A_K$ and the abelian extensions $L|K$ of exponent n.

If Δ and L correspond to each other, then $A_L^{\wp} \cap A_K = \Delta$, and we have a canonical isomorphism

$$\Delta/A_K^{\wp} \cong \operatorname{Hom}\big(G(L|K), \mu_{\wp}\big), \quad a \bmod A_K^{\wp} \longmapsto \chi_a,$$

where the character $\chi_a : G(L|K) \to \mu_{\wp}$ is given by $\chi_a(\sigma) = \alpha^{\sigma-1}$, for $\alpha \in \wp^{-1}(a)$.

Proof: Let $L|K$ be an abelian extension of exponent n. By (3.2), we then find $L = K(\wp^{-1}(\Delta))$ with $\Delta = A_L^{\wp} \cap A_K$. We consider the homomorphism

$$\Delta \longrightarrow \operatorname{Hom}\big(G(L|K), \mu_{\wp}\big), \quad a \longmapsto \chi_a,$$

where $\chi_a(\sigma) = \alpha^{\sigma-1}$, $\alpha \in \wp^{-1}(a)$. Since

$$\begin{aligned} \chi_a = 1 &\iff \alpha^{\sigma-1} = 1 \ \text{ for all } \sigma \in G(L|K) \\ &\iff \alpha \in A_K \iff a = \alpha^{\wp} \in A_K^{\wp}, \end{aligned}$$

it has the kernel $A_K^{\wp}$. To prove the surjectivity, we let $\chi \in \operatorname{Hom}(G(L|K), \mu_{\wp})$. χ defines a cyclic extension $M|K$ and is the composite of homomorphisms $G(L|K) \to G(M|K) \xrightarrow{\bar{\chi}} \mu_{\wp}$. Let σ be a generator of $G(M|K)$. Since

$N_{M|K}(\overline{\chi}(\sigma)) = \overline{\chi}(\sigma)^{[M:K]} = 1$, we deduce from (3.1) that $\overline{\chi}(\sigma) = \alpha^{\sigma-1}$ for some $\alpha \in A_M$. Now, $(\alpha^{\wp})^{\sigma-1} = (\alpha^{\sigma-1})^{\wp} = \overline{\chi}(\sigma)^{\wp} = 1$, so that $a = \alpha^{\wp} \in A_L^{\wp} \cap A_K = \Delta$. For $\tau \in G(L|K)$, one has $\chi(\tau) = \overline{\chi}(\tau|_M) = \alpha^{\tau-1} = \chi_a(\tau)$, so that $\chi = \chi_a$. This proves the surjectivity, and we obtain an isomorphism

$$\Delta/A_K^{\wp} \cong \mathrm{Hom}\big(G(L|K), \mu_{\wp}\big).$$

If Δ is any group between $A_K^{\wp}$ and A_K and if $L = K(\wp^{-1}(\Delta))$, then $\Delta = A_L^{\wp} \cap A_K$. In fact, putting $\Delta' = A_L^{\wp} \cap A_K$, we have just seen that one has

$$\Delta'/A_K^{\wp} \cong \mathrm{Hom}\big(G(L|K), \mu_{\wp}\big).$$

The subgroup $\Delta/A_K^{\wp}$ corresponds under Pontryagin duality to the subgroup $\mathrm{Hom}(G(L|K)/H, \mu_{\wp})$, where

$$H = \big\{\sigma \in G(L|K) \,\big|\, \chi_a(\sigma) = 1 \quad \text{for all } a \in \Delta\big\}.$$

As $\alpha^{\sigma-1} = \chi_a(\sigma)$ for $\alpha \in \wp^{-1}(a)$, H leaves fixed the elements of $\wp^{-1}(\Delta)$, and as $K(\wp^{-1}(\Delta)) = L$, we find that $H = 1$, so that $\mathrm{Hom}(G(L|K)/H, \mu_{\wp}) = \mathrm{Hom}(G(L|K), \mu_{\wp})$. It follows that $\Delta/A_K^{\wp} = \Delta'/A_K^{\wp}$, i.e., $\Delta = \Delta'$.

It is therefore clear that the correspondence $\Delta \mapsto L = K(\wp^{-1}(\Delta))$ is a $1-1$-correspondence, as claimed. This finishes the proof of the theorem. □

Remarks and Examples: 1) If $L|K$ is infinite, then $\mathrm{Hom}(G(L|K), \mu_{\wp})$ has to be interpreted as the group of all *continuous* homomorphisms $\chi : G(L|K) \to \mu_{\wp}$, i.e., as the character group of the *topological* group $G(L|K)$.

2) The composite of two abelian extensions of K of exponent n is again of the same type, and all of them lie in the *maximal* abelian extension of exponent n. It is given by $\widehat{K} = K(\wp^{-1}(A_K))$, and for the Pontryagin dual

$$G(\widehat{K}|K)^* = \mathrm{Hom}\big(G(\widehat{K}|K), \mathbb{Q}/\mathbb{Z}\big) = \mathrm{Hom}\big(G(\widehat{K}|K), \mu_{\wp}\big)$$

we have by (3.3) that

$$G(\widehat{K}|K)^* \cong A_K/A_K^{\wp}.$$

3) If k is an actual field of positive characteristic p and $\bar{k}$ is the separable closure of k, then A may be chosen to be the additive group $\bar{k}$ and $\wp$ to be the operator

$$\wp : \bar{k} \longrightarrow \bar{k}, \quad a \longmapsto \wp a = a^p - a.$$

Then axiom (3.1) is indeed satisfied, for we have, in complete generality:

(3.4) Proposition. *For every cyclic finite field extension $L|K$, one has $H^{-1}(G(L|K), L) = 1$.*

Proof: The extension $L|K$ always admits a normal basis $\{\sigma c \mid \sigma \in G(L|K)\}$, so that $L = \bigoplus_\sigma K\sigma c$. This means that L is a $G(L|K)$-induced module in the sense of §7, and then $H^{-1}(G(L|K), L) = 1$, by (7.4). □

The Kummer theory with respect to the operator $\wp a = a^p - a$ is usually called **Artin-Schreier theory**.

4) The chief application of the theory developed above is to the case where G is the absolute Galois group $G(\bar{k}|k)$ of an actual field k, A is the multiplicative group $\bar{k}^*$ of the algebraic closure, and $\wp$ is the n-th power map $a \mapsto a^n$, for some natural number n which is relatively prime to the characteristic of k (in particular, n is arbitrary if $\text{char}(k) = 0$). Axiom (3.1) is always satisfied in this case and is called **Hilbert 90** because this statement occurs as *Satz* number 90 among the 169 theorems in Hilbert's famous "Zahlbericht" [72]. Thus we have the

(3.5) Theorem (Hilbert 90). *For a cyclic field extension $L|K$ one always has*

$$H^{-1}\big(G(L|K), L^*\big) = 1.$$

In other words:

An element $\alpha \in L^*$ of norm $N_{L|K}(\alpha) = 1$ is of the form $\alpha = \beta^{\sigma-1}$, where $\beta \in L^*$ and σ is a generator of $G(L|K)$.

Proof: Let $n = [L:K]$. By virtue of the linear independence of the automorphisms $1, \sigma, \ldots, \sigma^{n-1}$ (see [15], chap. 5, §7, no. 5), there exists an element $\gamma \in L^*$ such that

$$\beta = \gamma + \alpha\gamma^\sigma + \alpha^{1+\sigma}\gamma^{\sigma^2} + \cdots + \alpha^{1+\sigma+\cdots+\sigma^{n-2}}\gamma^{\sigma^{n-1}} \neq 0.$$

As $N_{L|K}(\alpha) = 1$, one gets $\alpha\beta^\sigma = \beta$, and thus $\alpha = \beta^{1-\sigma}$. □

If now the field K contains the group μ_n of n-th roots of unity, the operator $\wp(a) = a^n$ has exponent n, and we obtain the following corollary, which is the most important special case of theorem (3.3).

(3.6) Corollary. *Let n be a natural number which is relatively prime to the characteristic of the field K, and assume that $\mu_n \subseteq K$.*

Then the abelian extensions $L|K$ of exponent n correspond 1–1 to the subgroups $\Delta \subseteq K^*$ which contain K^{*n}, via the rule

$$\Delta \mapsto L = K(\sqrt[n]{\Delta}),$$

and we have

$$G(L|K) \cong \operatorname{Hom}(\Delta/K^{*n}, \mu_n).$$

Hilbert's theorem 90, which is the main basis of this corollary, admits the following generalization to arbitrary Galois extensions $L|K$, which goes back to the mathematician *Emmy Noether* (1882–1935). Let G be a finite group and A a multiplicative G-module. A **1-cocycle**, or **crossed homomorphism**, of G with values in A is a function $f : G \to A$ satisfying

$$f(\sigma\tau) = f(\sigma)^\tau f(\tau)$$

for all $\sigma, \tau \in G$. The 1-cocycles form an abelian group $Z^1(G, A)$. For every $a \in A$, the function

$$f_a(\sigma) = a^{\sigma-1}$$

is a 1-cocycle, for one has

$$f_a(\sigma\tau) = a^{\sigma\tau-1} = (a^{\sigma-1})^\tau a^{\tau-1} = f_a(\sigma)^\tau f_a(\tau).$$

The functions f_a are called **1-coboundaries** and form a subgroup $B^1(G, A)$ of $Z^1(G, A)$. We define

$$H^1(G, A) = Z^1(G, A)/B^1(G, A)$$

and obtain as a first result about this group the

(3.7) Proposition. *If G is cyclic, then $H^1(G, A) \cong H^{-1}(G, A)$.*

Proof: Let $G = (\sigma)$. If $f \in Z^1(G, A)$, then for $k \geq 1$

$$f(\sigma^k) = f(\sigma^{k-1})^\sigma f(\sigma) = f(\sigma^{k-2})^{\sigma^2} f(\sigma)^\sigma f(\sigma) = \cdots = \prod_{i=0}^{k-1} f(\sigma)^{\sigma^i},$$

and $f(1) = 1$ because $f(1) = f(1)f(1)$. If $n = \#G$, then

$$N_G f(\sigma) = \prod_{i=0}^{n-1} f(\sigma)^{\sigma^i} = f(\sigma^n) = f(1) = 1,$$

so that $f(\sigma) \in {}_{N_G}A = \{a \in A \mid N_G a = \prod_{i=0}^{n-1} a^{\sigma^i} = 1\}$. Conversely we obtain, for every $a \in A$ such that $N_G a = 1$, a 1-cocycle by putting $f(\sigma) = a$ and

$$f(\sigma^k) = \prod_{i=0}^{k-1} a^{\sigma^i}.$$

The reader is invited to check this. The map $f \mapsto f(\sigma)$ therefore is an isomorphism between $Z^1(G, A)$ and ${}_{N_G}A$. This isomorphism maps $B^1(G, A)$ onto $I_G A$, because $f \in B^1(G, A) \iff f(\sigma^k) = a^{\sigma^k - 1}$ for some fixed $a \iff f(\sigma) = a^{\sigma - 1} \iff f(\sigma) \in I_G A$. □

Noether's generalization of Hilbert's theorem 90 now reads:

(3.8) Proposition. *For a finite Galois field extension $L|K$, one has that*

$$H^1\big(G(L|K), L^*\big) = 1.$$

Proof: Let $f : G \to L^*$ be a 1-cocycle. For $c \in L^*$, we put

$$\alpha = \sum_{\sigma \in G(L|K)} f(\sigma) c^\sigma.$$

Since the automorphisms σ are linearly independent (see [15], chap. 5, §7, no. 5), we can choose $c \in L^*$ such that $\alpha \neq 0$. For $\tau \in G(L|K)$, we obtain

$$\alpha^\tau = \sum_\sigma f(\sigma)^\tau c^{\sigma\tau} = \sum_\sigma f(\tau)^{-1} f(\sigma\tau) c^{\sigma\tau} = f(\tau)^{-1}\alpha,$$

i.e., $f(\tau) = \beta^{\tau - 1}$ with $\beta = \alpha^{-1}$. □

This proposition will only be applied once in this book (see chap. VI, (2.5)).

Exercise 1. Show that Hilbert 90 in Noether's formulation also holds for the additive group L of a Galois extension $L|K$.

Hint: Use the normal basis theorem.

Exercise 2. Let k be a field of characteristic p and $\bar{k}$ its separable closure. For fixed $n \geq 1$, consider in the ring of Witt vectors $W(\bar{k})$ (see chap. II, §4, exercise 2–6) the additive group $W_n(\bar{k})$ of truncated Witt vectors $a = (a_0, a_1, \ldots, a_{n-1})$. Show that axiom (3.1) holds for the $G(\bar{k}|k)$-module $A = W_n(\bar{k})$.

Exercise 3. Show that the operator

$$\wp : W_n(\bar{k}) \to W_n(\bar{k}), \quad \wp a = Fa - a,$$

is a homomorphism with cyclic kernel $\mu_\wp$ of order p^n. Discuss the corresponding Kummer theory for the abelian extensions of exponent p^n.

Exercise 4. Let G be a profinite group and A a continuous G-module. Put

$$H^1(G,A) = Z^1(G,A)/B^1(G,A),$$

where $Z^1(G,A)$ consists of all continuous maps $f : G \to A$ (with respect to the discrete topology on A) such that $f(\sigma\tau) = f(\sigma)^\tau f(\tau)$, and $B^1(G,A)$ consists of all functions of the form $f_a(\sigma) = a^{\sigma-1}$, $a \in A$. Show that if g is a closed normal subgroup of G, then one has an exact sequence

$$1 \to H^1(G/g, A^g) \to H^1(G,A) \to H^1(g,A).$$

Exercise 5. Show that $H^1(G,A) = \varinjlim H^1(G/N, A^N)$, where N varies over all the open normal subgroups of G.

Exercise 6. If $1 \to A \to B \to C \to 1$ is an exact sequence of continuous G-modules, then one has an exact sequence

$$1 \to A^G \to B^G \to C^G \to H^1(G,A) \to H^1(G,B) \to H^1(G,C).$$

Remark: The group $H^1(G,A)$ is only the first term of a whole series of groups $H^i(G,A)$, $i = 1,2,3,\dots$, which are the objects of **group cohomology** (see [145]). Class field theory can also be built upon this theory (see [10], [108]).

Exercise 7. Even for infinite Galois extensions $L|K$, one has Hilbert's theorem 90: $H^1(G(L|K), L^*) = 1$.

Exercise 8. If n is not divisible by the characteristic of the field K and if μ_n denotes the group of n-th roots of unity in the separable closure $\overline{K}$, then

$$H^1(G_K, \mu_n) \cong K^*/K^{*n}.$$

§ 4. Abstract Valuation Theory

The further development will now be based on a fixed choice of a surjective continuous homomorphism

$$d : G \longrightarrow \widehat{\mathbb{Z}}$$

from the profinite group G onto the procyclic group $\widehat{\mathbb{Z}} = \varprojlim \mathbb{Z}/n\mathbb{Z}$ (see §2, example 4). This homomorphism will produce a theory which is an abstract reflection of the ramification theory of $\mathfrak{p}$-adic number fields. Indeed, in the case where G is the absolute Galois group $G_k = G(\bar{k}|k)$ of a $\mathfrak{p}$-adic number field k, such a surjective homomorphism $d : G \to \widehat{\mathbb{Z}}$ arises via the maximal unramified extension $\tilde{k}|k$: if $\mathbb{F}_q$ is the residue class field of k, then, by chap. II, §9, p. 173 and example 5 in §2, we have canonical isomorphisms

$$G(\tilde{k}|k) \cong G(\overline{\mathbb{F}}_q|\mathbb{F}_q) \cong \widehat{\mathbb{Z}}$$

which associate to the element $1 \in \widehat{\mathbb{Z}}$ the Frobenius automorphism $\varphi \in G(\tilde{k}|k)$. It is defined by

$$a^{\varphi} \equiv a^q \bmod \tilde{\mathfrak{p}} \quad \text{for} \quad a \in \tilde{\mathcal{O}},$$

where $\tilde{\mathcal{O}}$, resp. $\tilde{\mathfrak{p}}$, denote the valuation ring of $\tilde{k}$, resp. its maximal ideal. The homomorphism $d : G \to \widehat{\mathbb{Z}}$ in question is then given, in this concrete case, as the composite

$$G \longrightarrow G(\tilde{k}|k) \xrightarrow{\sim} \widehat{\mathbb{Z}}.$$

In the abstract situation, the initial choice of a surjective homomorphism $d : G \to \widehat{\mathbb{Z}}$ mimics the p-adic case, but the applications of the theory are by no means confined to $\mathfrak{p}$-adic number fields. The kernel I of d has a certain fixed field $\tilde{k}|k$, and d induces an isomorphism $G(\tilde{k}|k) \cong \widehat{\mathbb{Z}}$.

More generally, for any field K we denote by I_K the kernel of the restriction $d : G_K \to \widehat{\mathbb{Z}}$, and call it the **inertia group** over K. Since

$$I_K = G_K \cap I = G_K \cap G_{\tilde{k}} = G_{K\tilde{k}},$$

the fixed field $\widetilde{K}$ of I_K is the composite

$$\widetilde{K} = K\tilde{k}.$$

We call $\widetilde{K}|K$ the **maximal unramified extension** of K. We put

$$f_K = \big(\widehat{\mathbb{Z}} : d(G_K)\big), \quad e_K = (I : I_K)$$

and obtain, when f_K is finite, a surjective homomorphism

$$d_K = \frac{1}{f_K} d : G_K \longrightarrow \widehat{\mathbb{Z}}$$

with kernel I_K, and an isomorphism

$$d_K : G(\widetilde{K}|K) \xrightarrow{\sim} \widehat{\mathbb{Z}}.$$

(4.1) Definition. *The element $\varphi_K \in G(\widetilde{K}|K)$ such that $d_K(\varphi_K) = 1$ is called the* **Frobenius** *over K.*

For a field extension $L|K$ we define the **inertia degree** $f_{L|K}$ and the **ramification index** $e_{L|K}$ by

$$f_{L|K} = \big(d(G_K) : d(G_L)\big) \quad \text{and} \quad e_{L|K} = (I_K : I_L).$$

For a tower of fields $K \subseteq L \subseteq M$ this definition obviously implies that

$$f_{M|K} = f_{L|K}\, f_{M|L} \quad \text{and} \quad e_{M|K} = e_{L|K}\, e_{M|L}.$$

(4.2) Proposition. *For every extension $L|K$ we have the "fundamental identity"*

$$[L : K] = f_{L|K}\, e_{L|K}\,.$$

Proof: The exact commutative diagram

$$\begin{array}{ccccccccc} 1 & \longrightarrow & I_L & \longrightarrow & G_L & \longrightarrow & d(G_L) & \longrightarrow & 1 \\ & & \downarrow & & \downarrow & & \downarrow & & \\ 1 & \longrightarrow & I_K & \longrightarrow & G_K & \longrightarrow & d(G_K) & \longrightarrow & 1 \end{array}$$

immediately yields, if $L|K$ is Galois, the exact sequence

$$1 \longrightarrow I_K/I_L \longrightarrow G(L|K) \longrightarrow d(G_K)/d(G_L) \longrightarrow 1\,.$$

If $L|K$ is not Galois, we pass to a Galois extension $M|K$ containing L, and get the result from the above transitivity rules for e and f. □

$L|K$ is called **unramified** if $e_{L|K} = 1$, i.e., if $L \subseteq \widetilde{K}$. $L|K$ is called **totally ramified** if $f_{L|K} = 1$, i.e., if $L \cap \widetilde{K} = K$. In the unramified case, we have the surjective homomorphism

$$G(\widetilde{K}|K) \to G(L|K)$$

and, if $f_K < \infty$, we call the image $\varphi_{L|K}$ of φ_K the **Frobenius automorphism** of $L|K$.

For an arbitrary extension $L|K$ one has

$$\widetilde{L} = L\widetilde{K}\,,$$

since $L\widetilde{K} = LK\tilde{k} = L\tilde{k} = \widetilde{L}$, and $L \cap \widetilde{K}|K$ is the maximal unramified subextension of $L|K$. It clearly has degree

$$f_{L|K} = [L \cap \widetilde{K} : K]\,.$$

Equally obvious is the

(4.3) Proposition. *If f_K and f_L are finite, then $f_{L|K} = f_L/f_K$, and we have the commutative diagram*

$$\begin{array}{ccc} G_L & \xrightarrow{d_L} & \widehat{\mathbb{Z}} \\ \downarrow & & \downarrow {\scriptstyle f_{L|K}} \\ G_K & \xrightarrow{d_K} & \widehat{\mathbb{Z}}\,. \end{array}$$

In particular, one has $\varphi_L|_{\widetilde{K}} = \varphi_K^{f_{L|K}}$.

The Frobenius automorphism governs the entire class field theory like a king. It is therefore most remarkable that in the case of a finite Galois extension $L|K$, every $\sigma \in G(L|K)$ becomes a Frobenius automorphism once it is manœuvered into the right position. This is achieved in the following manner. For what follows, let us assume systematically that $f_K < \infty$. We pass from the Galois extension $L|K$ to the extension $\widetilde{L}|K$ and consider in the Galois group $G(\widetilde{L}|K)$ the semigroup

$$\mathrm{Frob}(\widetilde{L}|K) = \left\{\sigma \in G(\widetilde{L}|K) \mid d_K(\sigma) \in \mathbb{N}\right\}.$$

Observe here that $d_K : G_K \longrightarrow \widehat{\mathbb{Z}}$ factorizes through $G(\widetilde{L}|K)$ because $G_{\widetilde{L}} = I_L \subseteq I_K$; recall also that $0 \notin \mathbb{N}$. Firstly, we have the

(4.4) Proposition. *For a finite Galois extension $L|K$ the mapping*

$$\mathrm{Frob}(\widetilde{L}|K) \longrightarrow G(L|K), \quad \sigma \longmapsto \sigma|_L,$$

is surjective.

Proof: Let $\sigma \in G(L|K)$ and let $\varphi \in G(\widetilde{L}|K)$ be an element such that $d_K(\varphi) = 1$. Then $\varphi|_{\widetilde{K}} = \varphi_K$ and $\varphi|_{L\cap\widetilde{K}} = \varphi_{L\cap\widetilde{K}|K}$. Restricting σ to the maximal unramified subextension $L \cap \widetilde{K}|K$, it becomes a power of the Frobenius automorphism, $\sigma|_{L\cap\widetilde{K}} = \varphi^n_{L\cap\widetilde{K}|K}$, so we may choose n in $\mathbb{N}$. As $\widetilde{L} = L\widetilde{K}$, we have

$$G(\widetilde{L}|\widetilde{K}) \cong G(L|L \cap \widetilde{K}).$$

If now $\tau \in G(\widetilde{L}|\widetilde{K})$ is mapped to $\sigma\varphi^{-n}|_L$ under this isomorphism, then $\tilde{\sigma} = \tau\varphi^n$ is an element satisfying $\tilde{\sigma}|_L = \tau\varphi^n|_L = \sigma\varphi^{-n}\varphi^n|_L = \sigma$ and $\tilde{\sigma}|_{\widetilde{K}} = \varphi^n_K$. Hence $d_K(\tilde{\sigma}) = n$, and so $\tilde{\sigma} \in \mathrm{Frob}(\widetilde{L}|K)$. □

Thus every element $\sigma \in G(L|K)$ may be lifted to an element $\tilde{\sigma} \in \mathrm{Frob}(\widetilde{L}|K)$. The following proposition shows that this lifting, considered over its fixed field, is actually the Frobenius automorphism.

(4.5) Proposition. *Let $\tilde{\sigma} \in \mathrm{Frob}(\widetilde{L}|K)$, and let Σ be the fixed field of $\tilde{\sigma}$. Then we have:*

(i) $f_{\Sigma|K} = d_K(\tilde{\sigma})$, (ii) $[\Sigma : K] < \infty$, (iii) $\widetilde{\Sigma} = \widetilde{L}$, (iv) $\tilde{\sigma} = \varphi_\Sigma$.

Proof: (i) $\Sigma \cap \widetilde{K}$ is the fixed field of $\tilde{\sigma}|_{\widetilde{K}} = \varphi_K^{d_K(\tilde{\sigma})}$, so that

$$f_{\Sigma|K} = [\Sigma \cap \widetilde{K} : K] = d_K(\tilde{\sigma}).$$

(ii) One has $\widetilde{K} \subseteq \Sigma\widetilde{K} = \widetilde{\Sigma} \subseteq \widetilde{L}$; thus

$$e_{\Sigma|K} = (I_K : I_\Sigma) = \#G(\widetilde{\Sigma}|\widetilde{K}) \le \#G(\widetilde{L}|\widetilde{K})$$

is finite. Therefore $[\Sigma : K] = f_{\Sigma|K} e_{\Sigma|K}$ is finite as well.

(iii) The canonical surjection $\Gamma = G(\widetilde{L}|\Sigma) \to G(\widetilde{\Sigma}|\Sigma) \cong \widehat{\mathbb{Z}}$ has to be bijective. For since $\Gamma = \overline{(\tilde{\sigma})}$ is procyclic, one finds $(\Gamma : \Gamma^n) \le n$ for every $n \in \mathbb{N}$ (see §2, p. 273). Thus the induced maps $\Gamma/\Gamma^n \cong \widehat{\mathbb{Z}}/n\widehat{\mathbb{Z}}$ are bijective and so is $\Gamma \to \widehat{\mathbb{Z}}$. But $G(\widetilde{L}|\Sigma) = G(\widetilde{\Sigma}|\Sigma)$ implies that $\widetilde{L} = \widetilde{\Sigma}$.

(iv) $f_{\Sigma|K} d_\Sigma(\tilde{\sigma}) = d_K(\tilde{\sigma}) = f_{\Sigma|K}$; thus $d_\Sigma(\tilde{\sigma}) = 1$, and so $\tilde{\sigma} = \varphi_\Sigma$. □

Let us illustrate the situation described in the last proposition by a diagram, which one should keep in mind for the sequel.

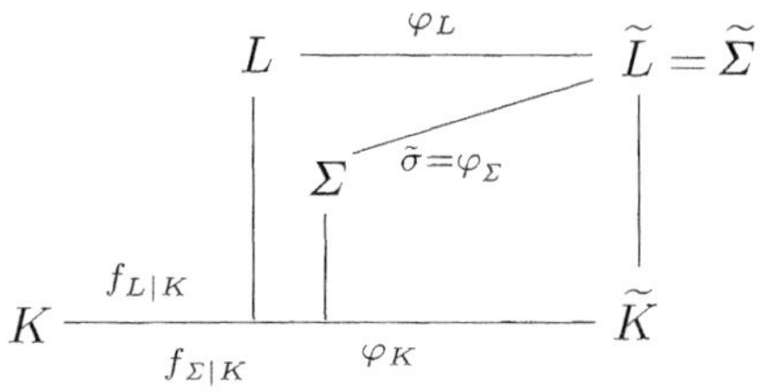

All the preceding discussions arose entirely from the initial datum of the homomorphism $d : G \to \widehat{\mathbb{Z}}$. We now add to the data a multiplicative G-module A, which we equip with a homomorphism that is to play the rôle of a henselian valuation.

(4.6) Definition. *A* **henselian valuation** *of A_k with respect to $d : G \to \widehat{\mathbb{Z}}$ is a homomorphism*

$$v : A_k \to \widehat{\mathbb{Z}}$$

satisfying the following properties:

(i) $v(A_k) = Z \supseteq \mathbb{Z}$ *and* $Z/nZ \cong \mathbb{Z}/n\mathbb{Z}$ *for all* $n \in \mathbb{N}$,

(ii) $v(N_{K|k} A_K) = f_K Z$ *for all finite extensions* $K|k$.

Exactly like the original homomorphism $d : G_k \to \widehat{\mathbb{Z}}$, the henselian valuation $v : A_k \to \widehat{\mathbb{Z}}$ has the property of reproducing itself over every finite extension K of k.

(4.7) Proposition. *For every field K which is finite over k, the formula*

$$v_K = \frac{1}{f_K} v \circ N_{K|k} : A_K \longrightarrow Z$$

defines a surjective homomorphism satisfying the following properties:

(i) $v_K = v_{K^\sigma} \circ \sigma$ *for all* $\sigma \in G$.

(ii) *For every finite extension $L|K$, one has the commutative diagram*

$$\begin{array}{ccc} A_L & \xrightarrow{v_L} & \widehat{\mathbb{Z}} \\ {\scriptstyle N_{L|K}}\downarrow & & \downarrow {\scriptstyle f_{L|K}} \\ A_K & \xrightarrow{v_K} & \widehat{\mathbb{Z}}\,. \end{array}$$

Proof: (i) If τ runs through a system of representatives of G_k/G_K, then $\sigma^{-1}\tau\sigma$ sweeps across a system of representatives of $G_k/\sigma^{-1}G_K\sigma = G_k/G_{K^\sigma}$. Hence we have, for $a \in A_K$,

$$\begin{aligned} v_{K^\sigma}(a^\sigma) &= \frac{1}{f_{K^\sigma}} v\big(\prod_\tau a^{\sigma\sigma^{-1}\tau\sigma}\big) = \frac{1}{f_K} v\big(\big(\prod_\tau a^\tau\big)^\sigma\big) = \frac{1}{f_K} v\big(N_{K|k}(a)\big) \\ &= v_K(a)\,. \end{aligned}$$

(ii) For $a \in A_L$ one has:

$$\begin{aligned} f_{L|K}\, v_L(a) &= f_{L|K} \frac{1}{f_L} v\big(N_{L|k}(a)\big) = \frac{1}{f_K} v\big(N_{K|k}(N_{L|K}(a))\big) \\ &= v_K\big(N_{L|K}(a)\big)\,. \end{aligned}$$

□

(4.8) Definition. *A* **prime element** *of A_K is an element $\pi_K \in A_K$ such that $v_K(\pi_K) = 1$. We put*

$$U_K = \big\{u \in A_K \bigm| v_K(u) = 0\big\}\,.$$

For an unramified extension $L|K$, that is, an extension such that $f_{L|K} = [L:K]$, we have from (4.7), (ii) that $v_L|_{A_K} = v_K$. In particular, a prime element of A_K is itself also a prime element of A_L. If on the other hand, $L|K$ is totally ramified, i.e., $f_{L|K} = 1$, and if π_L is a prime element of A_L, then $\pi_K = N_{L|K}(\pi_L)$ is a prime element of A_K.

Exercise 1. Assume that every closed abelian subgroup of G is procyclic. Let $K|k$ be a finite extension. A **microprime** $\mathfrak{p}$ of K is by definition a conjugacy class $\langle\sigma\rangle \subseteq G_K$ of some Frobenius element $\sigma \in \mathrm{Frob}(\bar{k}|K)$ which is not a proper power σ'^n, $n > 1$, of some other Frobenius element $\sigma' \in \mathrm{Frob}(\bar{k}|K)$. Let $\mathrm{spec}(K)$ be the set of all microprimes of K. Show that if $L|K$ is a finite extension, then there is a canonical mapping

$$\pi : \mathrm{spec}(L) \to \mathrm{spec}(K).$$

Above any microprime $\mathfrak{p}$ there are only finitely many microprimes $\mathfrak{P}$ of L, i.e., the set $\pi^{-1}(\mathfrak{p})$ is finite. We write $\mathfrak{P}|\mathfrak{p}$ to mean $\mathfrak{P} \in \pi^{-1}(\mathfrak{p})$.

Exercise 2. For a finite extension $L|K$ and a microprime $\mathfrak{P}|\mathfrak{p}$ of L, let $f_{\mathfrak{P}|\mathfrak{p}} = d(\mathfrak{P})/d(\mathfrak{p})$. Show that

$$\sum_{\mathfrak{P}|\mathfrak{p}} f_{\mathfrak{P}|\mathfrak{p}} = [L:K].$$

Exercise 3. For an infinite extension $L|K$, let

$$\mathrm{spec}(L) = \varprojlim_{\alpha} \mathrm{spec}(L_\alpha),$$

where $L_\alpha|K$ varies over the finite subextensions of $L|K$. What are the microprimes of $\bar{k}$?

Exercise 4. Show that if $L|K$ is Galois, then the Galois group $G(L|K)$ operates transitively on $\mathrm{spec}(L)$. The "decomposition group"

$$G_{\mathfrak{P}}(L|K) = \{\sigma \in G(L|K) \mid \mathfrak{P}^\sigma = \mathfrak{P}\}$$

is cyclic, and if $Z_{\mathfrak{P}} = L^{G_{\mathfrak{P}}(L|K)}$ is the "decomposition field" of $\mathfrak{P} \in \mathrm{spec}(L)$, then $L|Z_{\mathfrak{P}}$ is unramified.

§ 5. The Reciprocity Map

Continuing with the notation of the previous section, we consider again a profinite group G, a continuous G-module A, and a pair of homomorphisms

$$d : G \longrightarrow \widehat{\mathbb{Z}}, \quad v : A_k \longrightarrow \widehat{\mathbb{Z}},$$

such that d is continuous and surjective and v is a henselian valuation with respect to d. In the following we introduce the convention that the letter K, whenever it occurs without embellishments or commentary to the contrary, will always denote a field of *finite* degree over k. We furthermore impose the following axiomatic condition, which will be systematically assumed in the sequel.

(5.1) Axiom. *For every unramified finite extension $L|K$ one has*

$$H^i\big(G(L|K), U_L\big) = 1 \quad \text{for} \quad i = 0, -1.$$

For an infinite extension $L|K$ we set

$$N_{L|K}A_L = \bigcap_M N_{M|K}A_M\,,$$

with $M|K$ varying over the finite subextensions of $L|K$.

Our goal is to define a canonical homomorphism

$$r_{L|K} : G(L|K) \longrightarrow A_K/N_{L|K}A_L$$

for every finite Galois extension $L|K$. To this end, we pass from $L|K$ to the extension $\widetilde{L}|K$ and define first a mapping on the semigroup

$$\mathrm{Frob}(\widetilde{L}|K) = \big\{\sigma \in G(\widetilde{L}|K) \mid d_K(\sigma) \in \mathbb{N}\big\}\,.$$

(5.2) Definition. *The* **reciprocity map**

$$r_{\widetilde{L}|K} : \mathrm{Frob}(\widetilde{L}|K) \longrightarrow A_K/N_{\widetilde{L}|K}A_{\widetilde{L}}$$

is defined by

$$r_{\widetilde{L}|K}(\sigma) = N_{\Sigma|K}(\pi_\Sigma) \bmod N_{\widetilde{L}|K}A_{\widetilde{L}}\,,$$

where Σ is the fixed field of σ and $\pi_\Sigma \in A_\Sigma$ is a prime element.

Observe that Σ is of finite degree over K by (4.5), and σ becomes the Frobenius automorphism φ_Σ over Σ. The definition of $r_{\widetilde{L}|K}(\sigma)$ does not depend on the choice of the element π_Σ. For another one differs from π_Σ only by an element $u \in U_\Sigma$, and for this we have $N_{\Sigma|K}(u) \in N_{\widetilde{L}|K}A_{\widetilde{L}}$, so that $N_{\Sigma|K}(u) \in N_{M|K}A_M$ for every finite Galois subextension $M|K$ of $\widetilde{L}|K$. To see this, we may clearly assume that $\Sigma \subseteq M$. Applying (5.1) to the unramified extension $M|\Sigma$, one finds $u = N_{M|\Sigma}(\varepsilon)$, $\varepsilon \in U_M$, and thus

$$N_{\Sigma|K}(u) = N_{\Sigma|K}\big(N_{M|\Sigma}(\varepsilon)\big) = N_{M|K}(\varepsilon) \in N_{M|K}A_M\,.$$

Next we want to show that the reciprocity map $r_{\widetilde{L}|K}$ is multiplicative. To do this, we consider for every $\sigma \in G(\widetilde{L}|K)$ and every $n \in \mathbb{N}$ the endomorphisms

$$\sigma - 1 \colon A_{\widetilde{L}} \longrightarrow A_{\widetilde{L}}, \qquad a \longmapsto a^{\sigma-1} = a^\sigma/a,$$

$$\sigma_n \colon A_{\widetilde{L}} \longrightarrow A_{\widetilde{L}}, \qquad a \longmapsto a^{\sigma_n} = \prod_{i=0}^{n-1} a^{\sigma^i}\,.$$

In formal notation, this gives $\sigma_n = \dfrac{\sigma^n - 1}{\sigma - 1}$, and we find that

$$(\sigma - 1) \circ \sigma_n = \sigma_n \circ (\sigma - 1) = \sigma^n - 1\,.$$

Now we introduce the homomorphism

$$N = N_{\widetilde{L}|\widetilde{K}} : A_{\widetilde{L}} \longrightarrow A_{\widetilde{K}}$$

and prove two lemmas for it.

(5.3) Lemma. *Let $\varphi, \sigma \in \mathrm{Frob}(\widetilde{L}|K)$ with $d_K(\varphi) = 1, d_K(\sigma) = n$. If Σ is the fixed field of σ and $a \in A_\Sigma$, then*

$$N_{\Sigma|K}(a) = (N \circ \varphi_n)(a) = (\varphi_n \circ N)(a).$$

Proof: The maximal unramified subextension $\Sigma^0 = \Sigma \cap \widetilde{K}|K$ is of degree n, and its Galois group $G(\Sigma^0|K)$ is generated by the Frobenius automorphism $\varphi_{\Sigma^0|K} = \varphi_K|_{\Sigma^0} = \varphi|_{\widetilde{K}}|_{\Sigma^0} = \varphi|_{\Sigma^0}$. Consequently, $N_{\Sigma^0|K} = \varphi_n|_{A_{\Sigma^0}}$. On the other hand, one has $\Sigma\widetilde{K} = \widetilde{L}$ and $\Sigma \cap \widetilde{K} = \Sigma^0$, and therefore $N_{\Sigma|\Sigma^0} = N|_{A_\Sigma}$. For $a \in A_\Sigma$ we thus get

$$N_{\Sigma|K}(a) = N_{\Sigma^0|K}\big(N_{\Sigma|\Sigma^0}(a)\big) = N(a)^{\varphi_n} = N(a^{\varphi_n}).$$

The last equation follows from $\varphi G(\widetilde{L}|\widetilde{K}) = G(\widetilde{L}|\widetilde{K})\varphi$. □

The subgroup $I_{G(\widetilde{L}|\widetilde{K})}U_{\widetilde{L}}$, which is generated by all elements of the form $u^{\tau-1}$, $u \in U_{\widetilde{L}}$, $\tau \in G(\widetilde{L}|\widetilde{K})$, is mapped to 1 by the homomorphism $N = N_{\widetilde{L}|\widetilde{K}} : U_{\widetilde{L}} \to U_{\widetilde{K}}$. We therefore obtain an induced homomorphism

$$N : H_0(G(\widetilde{L}|\widetilde{K}), U_{\widetilde{L}}) \to U_{\widetilde{K}}$$

on the quotient group $H_0(G(\widetilde{L}|\widetilde{K}), U_{\widetilde{L}}) = U_{\widetilde{L}}/I_{G(\widetilde{L}|\widetilde{K})}U_{\widetilde{L}}$. For this group, we have the following lemma.

(5.4) Lemma. *If $x \in H_0(G(\widetilde{L}|\widetilde{K}), U_{\widetilde{L}})$ is fixed by an element $\varphi \in G(\widetilde{L}|K)$ such that $d_K(\varphi) = 1$, i.e., $x^\varphi = x$, then*

$$N(x) \in N_{\widetilde{L}|K}U_{\widetilde{L}}.$$

Proof: Let $x = u \bmod I_{G(\widetilde{L}|\widetilde{K})}U_{\widetilde{L}}$, with $x^{\varphi-1} = 1$, so that

$$(*) \qquad u^{\varphi-1} = \prod_{i=1}^{r} u_i^{\tau_i-1}, \quad u_i \in U_{\widetilde{L}}, \quad \tau_i \in G(\widetilde{L}|\widetilde{K}).$$

Let $M|K$ be a finite Galois subextension of $\widetilde{L}|K$. In order to prove that $N(u) \in N_{M|K}U_M$, we may assume that $u, u_i \in U_M$ and $L \subseteq M$. Let $n = [M : K]$, $\sigma = \varphi^n$ and let $\Sigma \supseteq M$ be the fixed field of σ. Further, let $\Sigma_n|\Sigma$ be the unramified extension of degree n, i.e., the fixed field of $\sigma^n = \varphi_\Sigma^n$. By (5.1), we can then find elements $\tilde{u}, \tilde{u}_i \in U_{\Sigma_n}$ such that

$$u = N_{\Sigma_n|\Sigma}(\tilde{u}) = \tilde{u}^{\sigma_n}, \quad u_i = N_{\Sigma_n|\Sigma}(\tilde{u}_i) = \tilde{u}_i^{\sigma_n}.$$

By $(*)$, the elements $\tilde{u}^{\varphi-1}$ and $\prod_i \tilde{u}_i^{\tau_i-1}$ only differ by an element $\tilde{x} \in U_{\Sigma_n}$ such that $N_{\Sigma_n|\Sigma}(\tilde{x}) = 1$. Hence – again by (5.1) – they differ by an element of the form $\tilde{y}^{\sigma-1}$, with $\tilde{y} \in U_{\Sigma_n}$. We may thus write

$$\tilde{u}^{\varphi-1} = \tilde{y}^{\varphi^n-1} \prod_i \tilde{u}_i^{\tau_i-1} = (\tilde{y}^{\varphi_n})^{\varphi-1} \prod_i \tilde{u}_i^{\tau_i-1}.$$

Applying N gives $N(\tilde{u})^{\varphi-1} = N(\tilde{y}^{\varphi_n})^{\varphi-1}$, so that

$$N(\tilde{u}) = N(\tilde{y}^{\varphi_n}) \cdot z,$$

for some $z \in U_{\tilde{K}}$ such that $z^{\varphi-1} = 1$; therefore $z^{\varphi} = z$, and $z \in U_K$. Finally, applying σ_n and putting $y = \tilde{y}^{\sigma_n} = N_{\Sigma_n|\Sigma}(\tilde{y}) \in U_\Sigma$, we obtain, observing $n = [M:K]$ and using (5.3), that

$$\begin{aligned} N(u) = N(\tilde{u})^{\sigma_n} &= N(\tilde{y}^{\varphi_n})^{\sigma_n} z^{\sigma_n} = N(y^{\varphi_n}) z^n \\ &= N_{\Sigma|K}(y) N_{M|K}(z) \in N_{M|K} U_M . \end{aligned}$$

□

(5.5) Proposition. *The reciprocity map*

$$r_{\tilde{L}|K} : \mathrm{Frob}(\tilde{L}|K) \longrightarrow A_K / N_{\tilde{L}|K} A_{\tilde{L}}$$

is multiplicative.

Proof: Let $\sigma_1\sigma_2 = \sigma_3$ be an equation in $\mathrm{Frob}(\tilde{L}|K)$, $n_i = d_K(\sigma_i)$, Σ_i the fixed field of σ_i and $\pi_i \in A_{\Sigma_i}$ a prime element, for $i = 1, 2, 3$. We have to show that

$$N_{\Sigma_1|K}(\pi_1) N_{\Sigma_2|K}(\pi_2) \equiv N_{\Sigma_3|K}(\pi_3) \bmod N_{\tilde{L}|K} A_{\tilde{L}} .$$

Choose a fixed $\varphi \in G(\tilde{L}|K)$ such that $d_K(\varphi) = 1$ and put

$$\tau_i = \sigma_i^{-1} \varphi^{n_i} \in G(\tilde{L}|\tilde{K}).$$

From $\sigma_1\sigma_2 = \sigma_3$ and $n_1 + n_2 = n_3$, we then deduce that

$$\tau_3 = \sigma_2^{-1}\sigma_1^{-1}\varphi^{n_2+n_1} = \sigma_2^{-1}\varphi^{n_2}(\varphi^{-n_2}\sigma_1\varphi^{n_2})^{-1}\varphi^{n_1} .$$

Putting $\sigma_4 = \varphi^{-n_2}\sigma_1\varphi^{n_2}$, $n_4 = d_K(\sigma_4) = n_1$, $\Sigma_4 = \Sigma_1^{\varphi^{n_2}}$, $\pi_4 = \pi_1^{\varphi^{n_2}} \in A_{\Sigma_4}$ and $\tau_4 = \sigma_4^{-1}\varphi^{n_4}$, we find $\tau_3 = \tau_2\tau_4$ and

$$N_{\Sigma_4|K}(\pi_4) = N_{\Sigma_1|K}(\pi_1) .$$

We may therefore pass to the congruence

$$N_{\Sigma_3|K}(\pi_3) \equiv N_{\Sigma_2|K}(\pi_2) N_{\Sigma_4|K}(\pi_4) \bmod N_{\tilde{L}|K} A_{\tilde{L}}$$

the proof of which uses the identity $\tau_3 = \tau_2\tau_4$. From (5.3), we have $N_{\Sigma_i|K}(\pi_i) = N(\pi_i^{\varphi_{n_i}})$. Thus, if we put

$$u = \pi_3^{\varphi_{n_3}} \pi_4^{-\varphi_{n_4}} \pi_2^{-\varphi_{n_2}},$$

then the congruence amounts simply to the relation $N(u) \in N_{\tilde{L}|K} A_{\tilde{L}}$. For this, however, lemma (5.4) gives us all that we need.

Since $\varphi_{n_i} \circ (\varphi - 1) = \varphi^{n_i} - 1$ and $\pi_i^{\varphi^{n_i}-1} = \pi_i^{\sigma_i^{-1}\varphi^{n_i}-1} = \pi_i^{\tau_i - 1}$, we have

$$u^{\varphi-1} = \pi_3^{\tau_3-1} \pi_4^{1-\tau_4} \pi_2^{1-\tau_2}.$$

From the identity $\tau_3 = \tau_2\tau_4$, it follows that $(\tau_3 - 1) + (1 - \tau_2) + (1 - \tau_4) = (1 - \tau_2)(1 - \tau_4)$. Putting now

$$\pi_3 = u_3\pi_4, \quad \pi_2 = u_2^{-1}\pi_4, \quad \pi_4^{\tau_2} = u_4\pi_4, \quad u_i \in U_{\tilde{L}},$$

we obtain

$$u^{\varphi-1} = \prod_{i=2}^{4} u_i^{\tau_i - 1}.$$

For the element $x = u \bmod I_{G(\tilde{L}|\tilde{K})} U_{\tilde{L}} \in H_0(G(\tilde{L}|\tilde{K}), U_{\tilde{L}})$, this means that $x^{\varphi-1} = 1$, and so $x^{\varphi} = x$; then by (5.4), we do get $N(u) = N(x) \in N_{\tilde{L}|K} A_{\tilde{L}}$. □

From the surjectivity of the mapping

$$\mathrm{Frob}(\tilde{L}|K) \longrightarrow G(L|K)$$

and the fact that $N_{\tilde{L}|K} A_{\tilde{L}} \subseteq N_{L|K} A_L$, we now have the

(5.6) Proposition. *For every finite Galois extension $L|K$, there is a canonical homomorphism*

$$r_{L|K} : G(L|K) \longrightarrow A_K / N_{L|K} A_L$$

given by

$$r_{L|K}(\sigma) = N_{\Sigma|K}(\pi_\Sigma) \bmod N_{L|K} A_L,$$

where Σ is the fixed field of a preimage $\tilde{\sigma} \in \mathrm{Frob}(\tilde{L}|K)$ of $\sigma \in G(L|K)$ and $\pi_\Sigma \in A_\Sigma$ is a prime element. It is called the **reciprocity homomorphism** *of $L|K$.*

Proof: We first show that the definition of $r_{L|K}(\sigma)$ is independent of the choice of the preimage $\tilde{\sigma} \in \mathrm{Frob}(\widetilde{L}|K)$ of σ. For this, let $\tilde{\sigma}' \in \mathrm{Frob}(\widetilde{L}|K)$ be another preimage, Σ' its fixed field and $\pi_{\Sigma'} \in A_{\Sigma'}$ a prime element. If $d_K(\tilde{\sigma}) = d_K(\tilde{\sigma}')$, then $\tilde{\sigma}|_{\widetilde{K}} = \tilde{\sigma}'|_{\widetilde{K}}$ and $\tilde{\sigma}|_L = \tilde{\sigma}'|_L$, so that $\tilde{\sigma} = \tilde{\sigma}'$, and there is nothing to show. However, if we have, say, $d_K(\tilde{\sigma}) < d_K(\tilde{\sigma}')$, then $\tilde{\sigma}' = \tilde{\sigma}\tilde{\tau}$ for some $\tilde{\tau} \in \mathrm{Frob}(\widetilde{L}|K)$, and $\tilde{\tau}|_L = 1$. The fixed field Σ'' of $\tilde{\tau}$ contains L, so $r_{\widetilde{L}|K}(\tilde{\tau}) \equiv N_{\Sigma''|K}(\pi_{\Sigma''}) \equiv 1 \bmod N_{L|K}A_L$. It follows therefore that $r_{\widetilde{L}|K}(\tilde{\sigma}') = r_{\widetilde{L}|K}(\tilde{\sigma})\, r_{\widetilde{L}|K}(\tilde{\tau}) = r_{\widetilde{L}|K}(\tilde{\sigma})$.

The fact that the mapping is a homomorphism now follows directly from (5.5): if $\tilde{\sigma}_1, \tilde{\sigma}_2 \in \mathrm{Frob}(\widetilde{L}|K)$ are preimages of $\sigma_1, \sigma_2 \in G(L|K)$, then $\tilde{\sigma}_3 = \tilde{\sigma}_1\tilde{\sigma}_2$ is a preimage of $\sigma_3 = \sigma_1\sigma_2$. □

The definition of the reciprocity map expresses the fundamental principle of class field theory to the effect that Frobenius automorphisms correspond to prime elements: the element $\sigma = \varphi_\Sigma \in G(\widetilde{L}|\Sigma)$ is mapped to $\pi_\Sigma \in A_\Sigma$; for reasons of functoriality, the inclusion $G(\widetilde{L}|\Sigma) \hookrightarrow G(\widetilde{L}|K)$ corresponds to the norm map $N_{\Sigma|K} : A_\Sigma \to A_K$. So the definition of $r_{L|K}(\sigma)$ is already forced upon us by these requirements. This principle appears at its purest in the

(5.7) Proposition. *If $L|K$ is an unramified extension, then the reciprocity map*

$$r_{L|K} : G(L|K) \longrightarrow A_K/N_{L|K}A_L$$

is given by

$$r_{L|K}(\varphi_{L|K}) = \pi_K \bmod N_{L|K}A_L,$$

and is an isomorphism.

Proof: In this case one has $\widetilde{L} = \widetilde{K}$ and $\varphi_K \in G(\widetilde{K}|K)$ is a preimage of $\varphi_{L|K}$ with fixed field K, i.e., $r_{L|K}(\varphi_{L|K}) = \pi_K \bmod N_{L|K}A_L$. The fact that we have an isomorphism is seen from the composite

$$G(L|K) \longrightarrow A_K/N_{L|K}A_L \longrightarrow Z/nZ \cong \mathbb{Z}/n\mathbb{Z},$$

with $n = [L:K]$, where the second map is induced by the valuation $v_K : A_K \to Z$ because $v_K(N_{L|K}A_L) \subseteq nZ$. It is an isomorphism, for if $v_K(a) \equiv 0 \bmod nZ$, then $a = u\pi_K^{dn}$, and since $u = N_{L|K}(\varepsilon)$ for some $\varepsilon \in U_L$, by (5.1), we find $a = N_{L|K}(\varepsilon\pi_K^d) \equiv 1 \bmod N_{L|K}A_L$. On the side of the homomorphisms, the generators $\varphi_{L|K}$, $\pi_K \bmod N_{L|K}A_L$, and $1 \bmod nZ$ correspond to each other, and everything is proved. □

The reciprocity homomorphism $r_{L|K}$ exhibits the following functorial behaviour.

(5.8) Proposition. *Let $L|K$ and $L'|K'$ be finite Galois extensions, so that $K \subseteq K'$ and $L \subseteq L'$, and let $\sigma \in G$. Then we have the commutative diagrams*

$$\begin{array}{ccc} G(L'|K') & \xrightarrow{r_{L'|K'}} & A_{K'}/N_{L'|K'}A_{L'} \\ \downarrow & & \downarrow N_{K'|K} \\ G(L|K) & \xrightarrow{r_{L|K}} & A_K/N_{L|K}A_L \end{array} \qquad \begin{array}{ccc} G(L|K) & \xrightarrow{r_{L|K}} & A_K/N_{L|K}A_L \\ \downarrow \sigma^* & & \downarrow \sigma \\ G(L^\sigma|K^\sigma) & \xrightarrow{r_{L^\sigma|K^\sigma}} & A_{K^\sigma}/N_{L^\sigma|K^\sigma}A_{L^\sigma} \end{array}$$

where the vertical arrows on the left are given by $\sigma' \mapsto \sigma'|_L$, resp. by the conjugation $\tau \mapsto \sigma^{-1}\tau\sigma$.

Proof: Let $\sigma' \in G(L'|K')$ and $\sigma = \sigma'|_L \in G(L|K)$. If $\tilde\sigma' \in \mathrm{Frob}(\widetilde{L}'|K')$ is a preimage of σ', then $\tilde\sigma = \tilde\sigma'|_{\widetilde{L}} \in \mathrm{Frob}(\widetilde{L}|K)$ is a preimage of σ such that $d_K(\tilde\sigma) = f_{K'|K}d_{K'}(\tilde\sigma') \in \mathbb{N}$. Let Σ' be the fixed field of $\tilde\sigma'$. Then $\Sigma = \Sigma' \cap \widetilde{L} = \Sigma' \cap \widetilde{\Sigma}$ is the fixed field of $\tilde\sigma$ and $f_{\Sigma'|\Sigma} = 1$. If now $\pi_{\Sigma'} \in A_{\Sigma'}$ is a prime element of Σ', then $\pi_\Sigma = N_{\Sigma'|\Sigma}(\pi_\Sigma) \in A_\Sigma$ is a prime element of Σ. The commutativity of the diagram on the left therefore follows from the equality of norms

$$N_{\Sigma|K}(\pi_\Sigma) = N_{\Sigma|K}\big(N_{\Sigma'|\Sigma}(\pi_{\Sigma'})\big) = N_{\Sigma'|K}(\pi_{\Sigma'}) = N_{K'|K}\big(N_{\Sigma'|K'}(\pi_{\Sigma'})\big)\,.$$

On the other hand, let $\tau \in G(L|K)$, and let $\tilde\tau$ be a preimage in $\mathrm{Frob}(\widetilde{L}|K)$ with fixed field Σ, and $\hat\tau \in G$ a lifting of $\tilde\tau$ to $\bar{k}$. Then Σ^σ is the fixed field of $\sigma^{-1}\hat\tau\sigma|_{\widetilde{L}^\sigma}$, and if $\pi \in A_\Sigma$ is a prime element of Σ, then $\pi^\sigma \in A_{\Sigma^\sigma}$ is a prime element of Σ^σ. The commutativity of the diagram on the right therefore follows from the equality of norms

$$N_{\Sigma^\sigma|K^\sigma}(\pi^\sigma) = N_{\Sigma|K}(\pi)^\sigma\,. \qquad \square$$

Another very interesting functorial property of the reciprocity map is obtained via the *transfer* (*Verlagerung* in German). For an arbitrary group G, let G' denote the commutator subgroup and write

$$G^{ab} = G/G'$$

for the maximal abelian quotient group. If then $H \subseteq G$ is a subgroup of finite index, we have a canonical homomorphism

$$\mathrm{Ver} : G^{ab} \longrightarrow H^{ab}\,,$$

which is called **transfer from G to H**. This homomorphism is defined as follows (see [75], chap. IV, § 1).

Let R be a system of representatives for the left cosets of H in G, $G = RH$, $1 \in R$. If $\sigma \in G$ we write, for every $\rho \in R$,

$$\sigma\rho = \rho'\sigma_\rho\,, \quad \sigma_\rho \in H\,, \quad \rho' \in R\,,$$

and we define

$$\mathrm{Ver}(\sigma \bmod G') = \prod_{\rho\in R} \sigma_\rho \bmod H'\,.$$

Another description of the transfer results from the double coset decomposition

$$G = \dot{\bigcup_\tau} (\sigma)\tau H$$

of G in terms of the subgroups (σ) and H. Letting $f(\tau)$ denote the smallest natural number such that $\sigma_\tau = \tau^{-1}\sigma^{f(\tau)}\tau \in H$, one has $H \cap (\tau^{-1}\sigma\tau) = (\sigma_\tau)$, and we find that

$$\mathrm{Ver}(\sigma \bmod G') = \prod_\tau \sigma_\tau \bmod H'\,.$$

This formula is obtained from the one above by choosing for R the set $\{\sigma^i\tau \mid i = 1, \dots, f(\tau)\}$. Applying this to the reciprocity homomorphism

$$r_{L|K} : G(L|K)^{ab} \longrightarrow A_K/N_{L|K}A_L$$

we get the

(5.9) Proposition. *Let $L|K$ be a finite Galois extension and K' an intermediate field. Then we have the commutative diagram*

$$\begin{array}{ccc} G(L|K')^{ab} & \xrightarrow{r_{L|K'}} & A_{K'}/N_{L|K'}A_L \\ {\scriptstyle \mathrm{Ver}}\uparrow & & \uparrow \\ G(L|K)^{ab} & \xrightarrow{r_{L|K}} & A_K/N_{L|K}A_L\,, \end{array}$$

where the arrow on the right is induced by inclusion.

Proof: Let us write temporarily $G = G(\tilde{L}|K)$ and $H = G(\tilde{L}|K')$. Let $\sigma \in G(L|K)$, and let $\tilde{\sigma}$ be a preimage in $\mathrm{Frob}(\tilde{L}|K)$ with fixed field Σ and $S = G(\tilde{L}|\Sigma) = \overline{(\tilde{\sigma})}$. We consider the double coset decomposition $G = \dot{\bigcup}_\tau S\tau H$ and put $S_\tau = \tau^{-1}S\tau \cap H$ and $\tilde{\sigma}_\tau = \tau^{-1}\tilde{\sigma}^{f(\tau)}\tau$ as above. Let
$\overline{G} = G(L|K)$, $\overline{H} = G(L|K')$, $\overline{S} = (\sigma)$, $\overline{\tau} = \tau|_L$ and $\sigma_\tau = \tilde{\sigma}_\tau|_L$.
Then we obviously have

$$\overline{G} = \bigcup_\tau \overline{S}\,\overline{\tau}\,\overline{H}\,,$$

and therefore

$$\mathrm{Ver}\big(\sigma \bmod G(L|K)'\big) = \prod_\tau \sigma_\tau \bmod G(L|K')'.$$

For every τ, let ω_τ vary over a system of right coset representatives of H/S_τ. Then one has

$$H = \dot{\bigcup_{\omega_\tau}} S_\tau \omega_\tau \quad \text{and} \quad G = \dot{\bigcup_{\tau,\omega_\tau}} S\tau\omega_\tau .$$

Let Σ_τ be the fixed field of $\tilde{\sigma}_\tau$, i.e., the fixed field of S_τ. Σ^τ is the fixed field of $\tau^{-1}\tilde{\sigma}\tau$ so that $\Sigma_\tau|\Sigma^\tau$ is the unramified subextension of degree $f(\tau)$ in $\widetilde{L}|\Sigma^\tau$. If now $\pi \in A_\Sigma$ is a prime element of Σ, then $\pi^\tau \in A_{\Sigma^\tau}$ is a prime element of Σ^τ, and thus also of Σ_τ. In view of the above double coset decomposition, we therefore find

$$N_{\Sigma|K}(\pi) = \prod_{\tau,\omega_\tau} \pi^{\tau\omega_\tau} = \prod_\tau\Big(\prod_{\omega_\tau}(\pi^\tau)^{\omega_\tau}\Big) = \prod_\tau N_{\Sigma_\tau|K'}(\pi^\tau),$$

and since $\tilde{\sigma}_\tau \in \mathrm{Frob}(\widetilde{L}|K')$ is a preimage of $\sigma_\tau \in G(L|K')$, it follows that

$$r_{L|K}(\sigma) \equiv \prod_\tau r_{L|K'}(\sigma_\tau) \equiv r_{L|K'}\Big(\prod_\tau \sigma_\tau\Big) \equiv r_{L|K'}\big(\mathrm{Ver}(\sigma \bmod G(L|K)')\big).$$

□

Exercise 1. Let $L|K$ be abelian and totally ramified, and let π be a prime element of A_L. If then $\sigma \in G(L|K)$ and

$$y^{\varphi-1} = \pi^{\sigma-1}$$

with $y \in U_{\widetilde{L}}$, then $r_{L|K}(\sigma) = N(y) \bmod N_{L|K}A_L$, where $N = N_{\widetilde{L}|\widetilde{K}}$ (*B. Dwork*, see [122], chap. XIII, § 5).

Exercise 2. Generalize the theory developed so far in the following way. Let P be a set of prime numbers and let G be a pro-P-group, i.e., a profinite group all of whose quotients G/N by open normal subgroups N have order divisible only by primes in P.

Let $d : G \to \mathbb{Z}_P$ be a surjective homomorphism onto the group $\mathbb{Z}_P = \prod_{p\in P}\mathbb{Z}_p$, and let A be a G-module. A **henselian P-valuation** with respect to d is by definition a homomorphism

$$v : A_k \to \mathbb{Z}_P$$

which satisfies the following properties:

(i) $v(A_K) = Z \supseteq \mathbb{Z}$ and $Z/nZ \cong \mathbb{Z}/n\mathbb{Z}$ for all natural numbers n which are divisible only by primes in P,

(ii) $v(N_{K|k}A_K) = f_K Z$ for all finite extensions $K|k$, where $f_K = (d(G) : d(G_K))$.

Under the hypothesis that $H^i(G(L|K), U_L) = 1$ for $i = 0, -1$, for all unramified extensions $L|K$, prove the existence of a canonical reciprocity homomorphism $r_{L|K} : G(L|K)^{ab} \to A_K/N_{L|K}A_L$ for every finite Galois extension $L|K$.

Exercise 3. Let $d : G \to \widehat{\mathbb{Z}}$ be a surjective homomorphism, A a G-module which satisfies axiom (5.1), and let $v : A_k \to \widehat{\mathbb{Z}}$ be a henselian valuation with respect to d.

Let $K|k$ be a finite extension and let $\operatorname{spec}(K)$ be the set of **microprimes** of K (see § 4, exercise 1–5). Define a canonical mapping

$$r_K : \operatorname{spec}(K) \to A_K/N_{\tilde{k}|K} A_{\tilde{k}},$$

and show that, for a finite extension, the diagram

$$\begin{array}{ccc} \operatorname{spec}(L) & \xrightarrow{r_L} & A_L/N_{\tilde{k}|L} A_{\tilde{k}} \\ {\scriptstyle \pi}\downarrow & & \downarrow{\scriptstyle N_{L|K}} \\ \operatorname{spec}(K) & \xrightarrow{r_K} & A_K/N_{\tilde{k}|K} A_{\tilde{k}} \end{array}$$

commutes. Show furthermore that, for every finite Galois extension $L|K$, r_K induces the reciprocity isomorphism

$$r_{L|K} : G(L|K) \to A_K/N_{L|K} A_L .$$

Hint: Let $\varphi \in G_K$ be an element such that $d_K(\varphi) \in \mathbb{N}$. Let Σ be the fixed field of φ and

$$\widehat{A}_\Sigma = \varprojlim_{\alpha} A_{K_\alpha},$$

where $K_\alpha|K$ varies over the finite subextensions of $\Sigma|K$, and where the projective limit is taken with respect to the norm maps $N_{K_\beta|K_\alpha} : A_{K_\beta} \to A_{K_\alpha}$. Then there is a surjective homomorphism $v_\Sigma : \widehat{A}_\Sigma \to Z$ and a homomorphism $N_{\Sigma|K} : \widehat{A}_\Sigma \to A_K$.

§ 6. The General Reciprocity Law

We now impose on the continuous G-module A the following condition.

(6.1) Class Field Axiom. *For every cyclic extension $L|K$ one has*

$$\#H^i\big(G(L|K), A_L\big) = \begin{cases} [L:K] & \text{for } i = 0, \\ 1 & \text{for } i = -1. \end{cases}$$

Among the cyclic extensions there are in particular the unramified ones. For them the above condition amounts precisely to requiring axiom (5.1), so that one has

(6.2) Proposition. *For a finite unramified extension $L|K$, one has*

$$H^i\big(G(L|K), U_L\big) = 1 \quad \text{for} \quad i = 0, -1 .$$

Proof: Since $L|K$ is unramified, a prime element π_K of A_K is also a prime element of A_L. As $H^{-1}(G(L|K), A_L) = 1$, every element $u \in U_L$ such that $N_{L|K}(u) = 1$ is of the form $u = a^{\sigma-1}$, with $a \in A_L$, $\sigma = \varphi_{L|K}$. So writing $a = \varepsilon\pi_K^m$, $\varepsilon \in U_L$, we obtain $u = \varepsilon^{\sigma-1}$. This shows that $H^{-1}(G(L|K), U_L) = 1$.

On the other hand, the homomorphism $v_K : A_K \to Z$ gives rise to a homomorphism

$$v_K : A_K/N_{L|K}A_L \longrightarrow Z/nZ = \mathbb{Z}/n\mathbb{Z},$$

where $n = [L:K] = f_{L|K}$, because $v_K(N_{L|K}A_L) = f_{L|K}Z = nZ$. This homomorphism is surjective as $v_K(\pi_K \bmod N_{L|K}A_L) = 1 \bmod nZ$, and it is bijective as $\#A_K/N_{L|K}A_L = n$. If now $u \in U_K$, then we have $u = N_{L|K}(a)$, with $a \in A_L$, since $v_K(u) = 0$. But $0 = v_K(u) = v_K(N_{L|K}(a)) = nv_L(a)$, so we get in fact $a \in U_L$. This proves that $H^0(G(L|K), U_L) = 1$. □

By definition, a **class field theory** is a pair of homomorphisms

$$\big(d : G \to \widehat{\mathbb{Z}},\ v : A \to \widehat{\mathbb{Z}}\big),$$

where A is a G-module which satisfies axiom (6.1), d is a surjective continuous homomorphism, and v is a henselian valuation. From proposition (6.2) and § 5, we obtain for every finite Galois extension $L|K$, the reciprocity homomorphism

$$r_{L|K} : G(L|K)^{ab} \longrightarrow A_K/N_{L|K}A_L.$$

But the class field axiom yields moreover the following theorem, which represents the main theorem of class field theory, and which we will call the **general reciprocity law**.

(6.3) Theorem. *For every finite Galois extension $L|K$, the homomorphism*

$$r_{L|K} : G(L|K)^{ab} \longrightarrow A_K/N_{L|K}A_L$$

is an isomorphism.

Proof: If $M|K$ is a Galois subextension of $L|K$, we get from (5.8) the commutative exact diagram

$$\begin{array}{ccccccccc}
1 \longrightarrow & G(L|M) & \longrightarrow & G(L|K) & \longrightarrow & G(M|K) & \longrightarrow 1 \\
& \downarrow{\scriptstyle r_{L|M}} & & \downarrow{\scriptstyle r_{L|K}} & & \downarrow{\scriptstyle r_{M|K}} & \\
& A_M/N_{L|M}A_L & \xrightarrow{N_{M|K}} & A_K/N_{L|K}A_L & \longrightarrow & A_K/N_{M|K}A_M & \longrightarrow 1.
\end{array}$$

We use this diagram to make three reductions.

First reduction. We may assume that $G(L|K)$ is abelian. For if the theorem is proved in this case, then, putting $M = L^{ab}$ the maximal abelian subextension of $L|K$, we find $G(L|K)^{ab} = G(M|K)$, and the commutator subgroup $G(L|M)$ of $G(L|K)$ is precisely the kernel of $r_{L|K}$, i.e., $G(L|K)^{ab} \to A_K/N_{L|K}A_L$ is injective. The surjectivity follows by induction on the degree. Indeed, in the case where $G(L|K)$ is solvable, one has either $M = L$ or $[L:M] < [L:K]$, and if $r_{M|K}$ and $r_{L|M}$ are surjective, then so is $r_{L|K}$. In the general case, let M be the fixed field of a p-Sylow subgroup. $M|K$ need not be Galois, but we may use the left part of the diagram, where $r_{L|M}$ is surjective. It then suffices to show that the image of $N_{M|K}$ is the p-Sylow subgroup S_p of $A_K/N_{L|K}A_L$. That this holds true for all p amounts to saying that $r_{L|K}$ is surjective. Now the inclusion $A_K \subseteq A_M$ induces a homomorphism

$$i : A_K/N_{L|K}A_L \longrightarrow A_M/N_{L|M}A_L\,,$$

such that $N_{M|K} \circ i = [M:K]$. As $([M:K], p) = 1$, $S_p \xrightarrow{[M:K]} S_p$ is surjective, so S_p lies in the image of $N_{M|K}$, and therefore of $r_{L|K}$.

Second reduction. We may assume that $L|K$ is cyclic. For if $M|K$ varies over all cyclic subextensions of $L|K$, then the diagram shows that the kernel of $r_{L|K}$ lies in the kernel of the map $G(L|K) \to \prod_M G(M|K)$. Since $G(L|K)$ is abelian, this map is injective and hence the same is true of $r_{L|K}$. Choosing a proper cyclic subextension $M|K$ of $L|K$, surjectivity follows by induction on the degree as in the first reduction for solvable extensions.

Third reduction. Let $L|K$ be cyclic. We may assume that $f_{L|K} = 1$. To see this, let $M = L \cap \widetilde{K}$ be the maximal unramified subextension of $L|K$. Then $f_{L|M} = 1$ and $r_{M|K}$ is an isomorphism by (5.7). In the bottom sequence of our diagram, the map $N_{M|K}$ is injective because the groups in this sequence have the respective orders $[L:M]$, $[L:K]$, $[M:K]$ by axiom (6.1). Therefore $r_{L|K}$ is an isomorphism if $r_{L|M}$ is.

Now let $L|K$ be cyclic and totally ramified, i.e., $f_{L|K} = 1$. Let σ be a generator of $G(L|K)$. We view σ via the isomorphism $G(L|K) \cong G(\widetilde{L}|\widetilde{K})$ as an element of $G(\widetilde{L}|\widetilde{K})$, and obtain the element $\tilde{\sigma} = \sigma\varphi_L \in \mathrm{Frob}(\widetilde{L}|K)$, which is a preimage of $\sigma \in G(L|K)$ such that $d_K(\tilde{\sigma}) = d_K(\varphi_L) = f_{L|K} = 1$. We thus find for the fixed field $\Sigma|K$ of $\tilde{\sigma}$ that $f_{\Sigma|K} = 1$, and so $\Sigma \cap \widetilde{K} = K$. Let $M|K$ be a finite Galois subextension of $\widetilde{L}|K$ containing Σ and L, let $M^0 = M \cap \widetilde{K}$ be the maximal unramified subextension of $M|K$, and put $N = N_{M|M^0}$. As $f_{\Sigma|K} = f_{L|K} = 1$, one finds $N|_{A_\Sigma} = N_{\Sigma|K}$, $N|_{A_L} = N_{L|K}$ (see the proof of (5.3)).

For the injectivity of $r_{L|K}$, we have to prove this: if $r_{L|K}(\sigma^k) = 1$, where $0 \le k < n = [L:K]$, then $k = 0$.

In order to do this, let $\pi_\Sigma \in A_\Sigma$, $\pi_L \in A_L$ be prime elements. Since $\Sigma, L \subseteq M$, π_Σ and π_L are both prime elements of M. Putting $\pi_\Sigma^k = u\pi_L^k$, $u \in U_M$, we obtain

$$r_{L|K}(\sigma^k) \equiv N(\pi_\Sigma^k) \equiv N(u) \cdot N(\pi_L^k) \equiv N(u) \bmod N_{L|K} A_L .$$

From $r_{L|K}(\sigma^k) = 1$, it thus follows that $N(u) = N(v)$ for some $v \in U_L$, so that $N(u^{-1}v) = 1$. From axiom (6.1), we may write $u^{-1}v = a^{\sigma-1}$ for some $a \in A_M$, and find in A_M the equation

$$(\pi_L^k v)^{\sigma-1} = (\pi_L^k v)^{\tilde{\sigma}-1} = (\pi_\Sigma^k u^{-1} v)^{\tilde{\sigma}-1} = (a^{\sigma-1})^{\tilde{\sigma}-1} = (a^{\tilde{\sigma}-1})^{\sigma-1},$$

and so $x = \pi_L^k v a^{1-\tilde{\sigma}} \in A_{M^0}$. Now $v_{M^0}(x) \in \widehat{\mathbb{Z}}$ and $n v_{M^0}(x) = v_M(x) = k$ imply that one has $k = 0$, and so $r_{L|K}$ is injective. The surjectivity then follows from (6.1). □

The inverse of the mapping $r_{L|K} : G(L|K)^{ab} \to A_K / N_{L|K} A_L$ gives, for every finite Galois extension $L|K$, a surjective homomorphism

$$(\ ,L|K) : A_K \longrightarrow G(L|K)^{ab}$$

with kernel $N_{L|K} A_L$. This map is called the **norm residue symbol** of $L|K$. From (5.8) and (5.9) we have the

(6.4) Proposition. *Let $L|K$ and $L'|K'$ be finite Galois extensions such that $K \subseteq K'$ and $L \subseteq L'$, and let $\sigma \in G$. Then we have the commutative diagrams*

$$\begin{array}{ccc} A_{K'} & \xrightarrow{(\ ,L'|K')} & G(L'|K')^{ab} \\ {\scriptstyle N_{K'|K}}\downarrow & & \downarrow \\ A_K & \xrightarrow{(\ ,L|K)} & G(L|K)^{ab}, \end{array} \qquad \begin{array}{ccc} A_K & \xrightarrow{(\ ,L|K)} & G(L|K)^{ab} \\ {\scriptstyle \sigma}\downarrow & & \downarrow{\scriptstyle \sigma^*} \\ A_{K^\sigma} & \xrightarrow{(\ ,L^\sigma|K^\sigma)} & G(L^\sigma|K^\sigma)^{ab}, \end{array}$$

and if $K' \subseteq L$, we have the commutative diagram

$$\begin{array}{ccc} A_{K'} & \xrightarrow{(\ ,L|K')} & G(L|K')^{ab} \\ \uparrow & & \uparrow{\scriptstyle \mathrm{Ver}} \\ A_K & \xrightarrow{(\ ,L|K)} & G(L|K)^{ab}. \end{array}$$

The definition of the norm residue symbol automatically extends to infinite Galois extensions $L|K$. For if $L_i|K$ varies over the finite Galois subextensions, then

$$G(L|K)^{ab} = \varprojlim_i G(L_i|K)^{ab}$$

(see §2, exercise 6). As $(a, L_{i'}|K)|_{L_i^{ab}} = (a, L_i|K)$ for $L_{i'} \supseteq L_i$, the individual norm residue symbols $(a, L_i|K)$, $a \in A_K$, determine an element

$$(a, L|K) \in G(L|K)^{ab}.$$

In the special case of the extension $\widetilde{K}|K$, we find the following intimate connection between the maps d_K, v_K, and $(\ ,\widetilde{K}|K)$.

(6.5) Proposition. *One has*

$$(a, \widetilde{K}|K) = \varphi_K^{v_K(a)}, \quad \text{and thus} \quad d_K \circ (\ ,\widetilde{K}|K) = v_K.$$

Proof: Let $L|K$ be the subextension of $\widetilde{K}|K$ of degree f. As $Z/fZ = \mathbb{Z}/f\mathbb{Z}$, we have $v_K(a) = n + fz$, with $n \in \mathbb{Z}$, $z \in Z$; that is, $a = u\pi_K^n b^f$, with $u \in U_K$, $b \in A_K$. From (5.7), we obtain

$$(a, \widetilde{K}|K)|_L = (a, L|K) = (\pi_K, L|K)^n (b, L|K)^f = \varphi_{L|K}^n = \varphi_K^{v_K(a)}|_L.$$

Thus we must have $(a, \widetilde{K}|K) = \varphi_K^{v_K(a)}$. □

The main goal of field theory is to classify all algebraic extensions of a given field K. The law governing the constitution of extensions of K is hidden in the inner structure of the base field K itself, and should therefore be expressed in terms of entities directly associated with it. Class field theory solves this problem as far as the abelian extensions of K are concerned. It establishes a 1–1-correspondence between these extensions and certain subgroups of A_K. More precisely, this is done as follows.

For every field K, we equip the group A_K with a topology by declaring the cosets $aN_{L|K}A_L$ to be a basis of neighbourhoods of $a \in A_K$, where $L|K$ varies over all finite Galois extensions of K. We call this topology the **norm topology** of A_K.

(6.6) Proposition. (i) *The open subgroups of A_K are precisely the closed subgroups of finite index.*

(ii) *The valuation $v_K : A_K \to \widehat{\mathbb{Z}}$ is continuous.*

(iii) *If $L|K$ is a finite extension, then $N_{L|K} : A_L \to A_K$ is continuous.*

(iv) *A_K is Hausdorff if and only if the group*

$$A_K^0 = \bigcap_L N_{L|K} A_L$$

of **universal norms** *is trivial.*

Proof: (i) If $\mathcal{N}$ is a subgroup of A_K, then

$$\mathcal{N} = A_K \setminus \bigcup_{a\mathcal{N} \neq \mathcal{N}} a\mathcal{N}.$$

Now, if $\mathcal{N}$ is open, so are all cosets $a\mathcal{N}$, so that $\mathcal{N}$ is closed, and since $\mathcal{N}$ has to contain one of the neighbourhoods $N_{L|K} A_L$ of the basis of neighbourhoods of 1, $\mathcal{N}$ is also of finite index. If, conversely, $\mathcal{N}$ is closed and of finite index, then the union of the finitely many cosets $a\mathcal{N} \neq \mathcal{N}$ is closed, and so $\mathcal{N}$ is open.

(ii) The groups $f\widehat{\mathbb{Z}}$, $f \in \mathbb{N}$, form a basis of neighbourhoods of $0 \in \widehat{\mathbb{Z}}$ (see § 2), and if $L|K$ is the unramified extension of degree f, then it follows from (4.7) that

$$v_K(N_{L|K} A_L) = f v_L(A_L) \subseteq f\widehat{\mathbb{Z}},$$

which shows the continuity of v_K.

(iii) Let $N_{M|K} A_M$ be an open neighbourhood of $1 \in A_K$. Then

$$N_{L|K}(N_{ML|L} A_{ML}) = N_{ML|K} A_{ML} \subseteq N_{M|K} A_M,$$

which shows the continuity of $N_{L|K}$.

(iv) is self-evident. □

The finite abelian extensions $L|K$ are now classified as follows.

(6.7) Theorem. *Associating*

$$L \longmapsto \mathcal{N}_L = N_{L|K} A_L$$

sets up a 1–1-correspondence between the finite abelian extensions $L|K$ and the open subgroups $\mathcal{N}$ of A_K. Furthermore, one has

$$L_1 \subseteq L_2 \Longleftrightarrow \mathcal{N}_{L_1} \supseteq \mathcal{N}_{L_2}, \quad \mathcal{N}_{L_1L_2} = \mathcal{N}_{L_1} \cap \mathcal{N}_{L_2}, \quad \mathcal{N}_{L_1 \cap L_2} = \mathcal{N}_{L_1}\mathcal{N}_{L_2}.$$

The field L corresponding to the subgroup $\mathcal{N}$ of A_K is called the **class field** associated with $\mathcal{N}$. By (6.3), it satisfies

$$G(L|K) \cong A_K/\mathcal{N}.$$

Proof of (6.7): If L_1 and L_2 are abelian extensions of K, then the transitivity of the norm implies $\mathcal{N}_{L_1L_2} \subseteq \mathcal{N}_{L_1} \cap \mathcal{N}_{L_2}$. If, conversely, $a \in \mathcal{N}_{L_1} \cap \mathcal{N}_{L_2}$, then the element $(a, L_1L_2|K) \in G(L_1L_2|K)$ projects trivially onto $G(L_i|K)$, that is, $(a, L_i|K) = 1$ for $i = 1, 2$. Thus $(a, L_1L_2|K) = 1$, i.e., $a \in \mathcal{N}_{L_1L_2}$. We therefore have $\mathcal{N}_{L_1L_2} = \mathcal{N}_{L_1} \cap \mathcal{N}_{L_2}$, and so

$$\mathcal{N}_{L_1} \supseteq \mathcal{N}_{L_2} \iff \mathcal{N}_{L_1} \cap \mathcal{N}_{L_2} = \mathcal{N}_{L_1L_2} = \mathcal{N}_{L_2} \iff [L_1L_2 : K] = [L_2 : K] \iff L_1 \subseteq L_2 .$$

This shows the injectivity of the correspondence $L \mapsto \mathcal{N}_L$.

If $\mathcal{N}$ is any open subgroup, then it contains the norm group $\mathcal{N}_L = N_{L|K}A_L$ of some Galois extension $L|K$. (6.3) implies that $\mathcal{N}_L = \mathcal{N}_{L^{ab}}$, so we may assume $L|K$ to be abelian. But $(\mathcal{N}, L|K) = G(L|L')$ for some intermediate field L' of $L|K$. Since $\mathcal{N} \supseteq \mathcal{N}_L$, the group $\mathcal{N}$ is the full preimage of $G(L|L')$ under the map $(\ ,L|K) : A_K \to G(L|K)$, and thus it is the full kernel of $(\ ,L'|K) : A_K \to G(L'|K)$. Thus $\mathcal{N} = \mathcal{N}_{L'}$. This shows that the correspondence $L \mapsto \mathcal{N}_L$ is surjective.

Finally, the equality $\mathcal{N}_{L_1 \cap L_2} = \mathcal{N}_{L_1}\mathcal{N}_{L_2}$ is obtained as follows. $L_1 \cap L_2 \subseteq L_i$ implies that $\mathcal{N}_{L_1 \cap L_2} \supseteq \mathcal{N}_{L_i}$, $i = 1, 2$, and thus

$$\mathcal{N}_{L_1 \cap L_2} \supseteq \mathcal{N}_{L_1}\mathcal{N}_{L_2} .$$

As $\mathcal{N}_{L_1}\mathcal{N}_{L_2}$ is open, we have just shown that $\mathcal{N}_{L_1}\mathcal{N}_{L_2} = \mathcal{N}_L$ for some finite abelian extension $L|K$. But $\mathcal{N}_{L_i} \subseteq \mathcal{N}_L$ implies $L \subseteq L_1 \cap L_2$, so that

$$\mathcal{N}_{L_1}\mathcal{N}_{L_2} = \mathcal{N}_L \supseteq \mathcal{N}_{L_1 \cap L_2} .$$ □

Exercise 1. Let n be a natural number, and assume the group $\mu_n = \{\xi \in A \mid \xi^n = 1\}$ is cyclic of order n, and $A \subseteq A^n$. Let K be a field such that $\mu_n \subseteq A_K$, and let $L|K$ be the maximal abelian extension of exponent n. If $L|K$ is finite, then one has $N_{L|K}A_L = A_K^n$.

Exercise 2. Under the hypotheses of exercise 1, Kummer theory and class field theory yield, via Pontryagin duality $G(L|K) \times \mathrm{Hom}(G(L|K), \mu_n) \to \mu_n$, a nondegenerate bilinear mapping (the abstract "Hilbert symbol")

$$(\ ,\ ) : A_K/A_K^n \times A_K/A_K^n \longrightarrow \mu_n .$$

Exercise 3. Let p be a prime number and $(d : G \to \mathbb{Z}_p,\ v : A_k \to \mathbb{Z}_p)$ a p-class field theory in the sense of §5, exercise 2. Let $d' : G \to \mathbb{Z}_p$ be another surjective homomorphism, and $\widehat{K}'|K$ the $\mathbb{Z}_p$-extension defined by d'. Let $v' : A_k \to \mathbb{Z}_p$ be the composite of

$$A_k \xrightarrow{(\ ,\widehat{K}'|K)} G(\widehat{K}'|K) \xrightarrow{d'} \mathbb{Z}_p .$$

Then (d', v') is also a p-class field theory. The norm residue symbols associated to (d, v) and (d', v') coincide. (No analogous statement holds in the case of $\widehat{\mathbb{Z}}$-class field theories $(d : G \to \widehat{\mathbb{Z}},\ v : A_k \to \widehat{\mathbb{Z}})$.)

Exercise 4. (Generalization to infinite extensions.) Let $(d : G \to \widehat{\mathbb{Z}},\ v : A_k \to \widehat{\mathbb{Z}})$ be a class field theory. We assume that the kernel U_k of $v_k : A_k \to \widehat{\mathbb{Z}}$ is compact for every finite extension $K|k$. For an infinite extension $K|k$, put

$$\widehat{A}_K = \varprojlim A_{K_\alpha},$$

where $K_\alpha|k$ varies over the finite subextensions of $K|k$ and the projective limit is taken with respect to the norm maps $N_{K_\beta|K_\alpha} : A_{K_\beta} \to A_{K_\alpha}$. Show:

1) For every (finite or infinite) extension $L|K$, one has a norm map

$$N_{L|K} : \widehat{A}_L \to \widehat{A}_K,$$

and if $L|K$ is finite, there is an injection $i_{L|K} : \widehat{A}_K \to \widehat{A}_L$. If furthermore $L|K$ is Galois, then one has $\widehat{A}_K \cong \widehat{A}_L^{G(L|K)}$.

2) For every extension $K|k$ with finite inertia degree $f_K = [K \cap \tilde{k} : k]$, (d, v) induces a class field theory $(d_K : G_K \to \widehat{\mathbb{Z}},\ v_K : \widehat{A}_K \to \widehat{\mathbb{Z}})$.

3) If $K \subseteq K'$ are extensions of k with $f_K,\ f_{K'} < \infty$, and $L|K$ and $L'|K'$ are (finite or infinite) Galois extensions with $L \subseteq L'$, then one has a commutative diagram

$$\begin{array}{ccc} \widehat{A}_{K'} & \xrightarrow{(\ ,L'|K')} & G(L'|K')^{ab} \\ {\scriptstyle N_{K'|K}}\downarrow & & \downarrow \\ \widehat{A}_K & \xrightarrow{(\ ,L|K)} & G(L|K)^{ab}. \end{array}$$

Exercise 5. If $L|K$ is a finite Galois extension, then G_L^{ab} is a $G(L|K)$-module in a canonical way, and the transfer from G_K to G_L is a homomorphism

$$\mathrm{Ver} : G_K^{ab} \to (G_L^{ab})^{G(L|K)}.$$

Exercise 6. (Tautological class field theory.) Assume that the profinite group G satisfies the condition: for every finite Galois extension,

$$\mathrm{Ver} : G_K^{ab} \to (G_L^{ab})^{G(L|K)}$$

is an isomorphism. (These are the profinite groups of "strict cohomological dimension 2" (see [145], chap. III, th. (3.6.4)).) Put $A_K = G_K^{ab}$ and form the direct limit $A = \varinjlim A_K$ via the transfer. Then A_K is identified with A^{G_K}.

Show that for every cyclic extension $L|K$ one has

$$\#H^i(G(L|K), A_L) = \begin{cases} [L : K] & \text{for } i = 0, \\ 1 & \text{for } i = -1, \end{cases}$$

and that for every surjective homomorphism $d : G \to \widehat{\mathbb{Z}}$, the induced map $v : A_k = G^{ab} \to \widehat{\mathbb{Z}}$ is a henselian valuation with respect to d. The corresponding reciprocity map $r_{L|K} : G(L|K) \to A_K/N_{L|K}A_L$ is essentially the identity.

Abstract class field theory acquires a much broader range of applications if it is generalized as follows.

Exercise 7. Let G be a profinite group and $B(G)$ the category of finite G-sets, i.e., of finite sets X with a continuous G-operation. Show that the *connected*, i.e., transitive G-sets in $B(G)$ are, up to isomorphism, the sets G/G_K, where G_K is an open subgroup of G, and G operates via multiplication on the left.

If X is a finite G-set and $x \in X$, then

$$\pi_1(X,x) = G_x = \{\sigma \in G \mid \sigma x = x\}$$

is called the **fundamental group** of X with base point x. For a map $f : X \to Y$ in $B(G)$, we put

$$G(X|Y) = \mathrm{Aut}_Y(X).$$

f is called Galois if X and Y are connected and $G(X|Y)$ operates transitively on the fibres $f^{-1}(y)$.

Exercise 8. Let $f : X \to Y$ be a map of connected finite G-sets, and let $x \in X$, $y = f(x) \in Y$. Show that f is Galois if and only if $\pi_1(X,x)$ is a normal subgroup of $\pi_1(Y,y)$. In this case, one has a canonical isomorphism

$$G(X|Y) \cong \pi_1(Y,y)/\pi_1(X,x).$$

A pair of functors

$$A = (A^*, A_*) : B(G) \to (ab),$$

consisting of a contravariant functor A^* and a covariant functor A_* from $B(G)$ to the category (ab) of abelian groups is called a **double functor** if

$$A^*(X) = A_*(X) =: A(X)$$

for all $X \in B(G)$. We define

$$A_K = A(G/G_K).$$

If $f : X \to Y$ is a morphism in $B(G)$, then we put

$$A^*(f) = f^* \quad \text{and} \quad A_*(f) = f_*.$$

A homomorphism $h : A \to B$ of double functors is a family of homomorphisms $h(X) : A(X) \to B(X)$ representing natural transformations $A^* \to B^*$ and $A_* \to B_*$.

A G-**modulation** is defined to be a double functor A such that

(i) $A(X \cup Y) = A(X) \times A(Y)$.

(ii) If among the two diagrams

$$\begin{array}{ccc} X & \xleftarrow{g'} & X' \\ {\scriptstyle f}\downarrow & & \downarrow{\scriptstyle f'} \\ Y & \xleftarrow{g} & Y' \end{array} \qquad \text{and} \qquad \begin{array}{ccc} A(X) & \xrightarrow{g'^*} & A(X') \\ {\scriptstyle f_*}\downarrow & & \downarrow{\scriptstyle f'_*} \\ A(Y) & \xrightarrow{g^*} & A(Y') \end{array}$$

in $B(G)$, resp. (ab), the one on the left is cartesian, then the one on the right is commutative.

Remark: G-modulations were introduced in a general context by *A. Dress* under the name of **Mackey functors** (see [32]).

Exercise 9. G-modulations form an abelian category.

Exercise 10. If A is a G-module, then the function $\mathbf{A}(G/G_K) = A^{G_K}$ extends to a G-modulation $\mathbf{A}$ in such a way that, for an extension $L|K$, the map $f^* : A_K \to A_L$, resp. $f_* : A_L \to A_K$, induced by $f : G/G_L \to G/G_K$, is the inclusion, resp. the norm $N_{L|K}$.

The rule $A \mapsto \boldsymbol{A}$ is an equivalence between the category of G-modules and the category of G-modulations with "Galois descent", i.e., of those G-modulations $\boldsymbol{A}$ such that

$$f^* : \boldsymbol{A}(Y) \to \boldsymbol{A}(X)^{G(X|Y)},$$

for every Galois mapping $f : X \to Y$, is an isomorphism.

Exercise 11. G-modulations are explicitly given by the following data. Let $B_0(G)$ be the category whose objects are the G-sets G/U, where U varies over the open subgroups of G, and whose morphisms are just the projections $\pi : G/U \to G/V$ for $U \subseteq V$, as well as the maps $c(\sigma) : G/U \to G/\sigma U \sigma^{-1}$, $\tau U \mapsto \tau U \sigma^{-1} = \tau\sigma^{-1}(\sigma U \sigma^{-1})$, for $\sigma \in G$.

Let $A = (A^*, A_*) : B_0(G) \to (ab)$ be a double functor and for $\pi : G/U \to G/V$ $(U \subseteq V)$, resp. $c(\sigma) : G/U \to G/\sigma U \sigma^{-1}$ $(\sigma \in G)$, define

$$\begin{aligned} \mathrm{Ind}_V^U &= A_*(\pi)\colon A(G/U) \to A(G/V), \\ \mathrm{Res}_U^V &= A^*(\pi)\colon A(G/V) \to A(G/U), \\ c(\sigma)_* &= A_*(c(\sigma))\colon A(G/U) \to A(G/\sigma U \sigma^{-1}). \end{aligned}$$

If for any three open subgroups $U, V \subseteq W$ of G, one has the *induction formula*

$$\mathrm{Res}_U^W \circ \mathrm{Ind}_W^V = \sum_{U\backslash W/V} \mathrm{Ind}_U^{U\cap\sigma V\sigma^{-1}} \circ c(\sigma)_* \circ \mathrm{Res}^V_{V\cap\sigma^{-1}U\sigma},$$

then A extends uniquely to a G-modulation $A : B(G) \to (ab)$.

Hint: If X is an arbitrary finite G-set, then the disjoint union

$$A_X = \dot{\bigcup_{x\in X}} A(G/G_x)$$

is again a G-set, because $c(\sigma)_* A(G/G_x) = A(G/G_{\sigma x})$. Define $A(X)$ to be the group

$$A(X) = \mathrm{Hom}_X(X, A_X)$$

of all G-equivariant sections $X \to A_X$ of the projection $A_X \to X$.

Exercise 12. The function $\pi^{ab}(G/G_K) = G_K^{ab}$ extends to a G-modulation

$$\pi^{ab} : B(G) \to (pro\text{-}ab)$$

into the category of pro-abelian groups. Thus, for an extension $L|K$, the maps $f^* : G_K^{ab} \to G_L^{ab}$, resp. $f_* : G_L^{ab} \to G_K^{ab}$, induced by $f : G/G_L \to G/G_K$ are given by the transfer, resp. the inclusion $G_L \to G_K$.

Exercise 13. Let A be a G-modulation. For every connected finite G-set X, let

$$NA(X) = \bigcap f_* A(Y),$$

where the intersection is taken over all Galois maps $f : Y \to X$. Show that the function $NA(X)$ defines a G-submodulation NA of A, the **modulation of universal norms**.

Exercise 14. If A is a G-modulation, then the **completion** $\widehat{A}$ is again a G-modulation which, for connected X, is given by

$$\widehat{A}(X) = \varprojlim A(X)/f_* A(Y),$$

where the projective limit is taken over all Galois maps $f : Y \to X$.

For the following, let $d : G \to \widehat{\mathbb{Z}}$ be a fixed surjective homomorphism. Let $f : X \to Y$ be a map of connected finite G-sets and $x \in X$, $y = f(x) \in Y$. The **inertia degree**, resp. the **ramification index**, of f is defined by

$$f_{X|Y} = (d(G_y) : d(G_x)), \quad \text{resp.} \quad e_{X|Y} = (I_y : I_x),$$

where I_y, resp. I_x, is the kernel of $d : G_y \to \widehat{\mathbb{Z}}$, resp. $d : G_x \to \widehat{\mathbb{Z}}$. f is called **unramified** if $e_{X|Y} = 1$.

Exercise 15. d defines a G-modulation $\widehat{\mathbf{Z}}$ such that the maps f^*, f_*, corresponding to a mapping $f : X \to Y$ of connected G-sets, are given by

$$\widehat{\mathbf{Z}}(Y) = \widehat{\mathbf{Z}} \underset{f_{X|Y}}{\overset{e_{X|Y}}{\rightleftarrows}} \widehat{\mathbf{Z}} = \widehat{\mathbf{Z}}(X).$$

This gives a homomorphism of G-modulations

$$d : \pi^{ab} \longrightarrow \widehat{\mathbf{Z}}.$$

Exercise 16. An unramified map $f : X \to Y$ of connected finite G-sets is Galois, and d induces an isomorphism

$$G(X|Y) \cong \mathbb{Z}/f_{X|Y}\mathbb{Z}.$$

Let $\varphi_{X|Y} \in G(X|Y)$ be the element which is mapped to 1 mod $f_{X|Y}\mathbb{Z}$.

Let A be a G-modulation. We define a *henselian valuation* of A to be a homomorphism

$$v : A \to \widehat{\mathbf{Z}}$$

such that the submodulation $v(A)$ of $\widehat{\mathbf{Z}}$ comes from a subgroup $Z \subseteq \widehat{\mathbf{Z}}$ which contains $\mathbb{Z}$ and satisfies $Z/nZ = \mathbb{Z}/n\mathbb{Z}$ for all $n \in \mathbb{N}$. Let U denote the kernel of A.

Exercise 17. Compare this definition with the definition (4.6) of a henselian valuation of a G-module A.

Exercise 18. Assume that for every unramified map $f : X \to Y$ of connected finite G-sets, the sequence

$$0 \to U(Y) \xrightarrow{f^*} U(X) \xrightarrow{\varphi^*_{X|Y} - 1} U(X) \xrightarrow{f_*} U(Y) \to 0$$

is exact, and that $A(Y)^{[X:Y]} \subseteq f_*A(X)$ for every Galois mapping $f : X \to Y$ (the latter is a consequence of the condition which will be imposed in exercise 19). Then the pair (d, v) gives, for every Galois mapping $f : X \to Y$, a canonical "reciprocity homomorphism"

$$r_{X|Y} : G(X|Y) \to A(Y)/f_*A(X).$$

Exercise 19. Assume, beyond the condition required in exercise 18, that for every Galois mapping $f : X \to Y$ with cyclic Galois group $G(X|Y)$, one has

$$(A(Y) : f_*A(X)) = [X : Y] \quad \text{and} \quad \ker f_* = \operatorname{im}(\sigma^* - 1),$$

where $[X : Y] = \#f^{-1}(y)$, with $y \in Y$, and σ is a generator of $G(X|Y)$. Then if $r_{X|Y}$ is an isomorphism for every Galois mapping $f : X \to Y$ of prime degree $[X : Y]$, so is

$$r_{X|Y} : G(X|Y)^{ab} \to A(Y)/f_*A(X),$$

for every Galois mapping $f : X \to Y$.

Exercise 20. Under the hypotheses of exercise 18 and 19 one obtains a canonical homomorphism of G-modulations

$$A \longrightarrow \pi^{ab}$$

whose kernel is the G-modulation NA of universal norms (see exercise 13). It induces an isomorphism

$$\widehat{A} \xrightarrow{\sim} \pi^{ab}$$

of the completion $\widehat{A}$ of A (see exercise 14).

Remark: The theory sketched above and contained in the exercises has a very interesting application to **higher dimensional class field theory**. In chap. V, (1.3), we will show that, for a Galois extension $L|K$ of local fields, there is a reciprocity isomorphism

$$G(L|K)^{ab} \cong K^*/N_{L|K}L^*.$$

The multiplicative group K^* may be interpreted in K-*theory* as the group $K_1(K)$ of the field K. The group $K_2(K)$ is defined to be the quotient group

$$K_2(K) = (K^* \otimes K^*)/R,$$

where R is generated by all elements of the form $x \otimes (1-x)$. Treating Galois extensions $L|K$ of "2-local fields" – these are discretely valued complete fields with residue class field a local field (e.g., $\mathbb{Q}_p((x))$, $\mathbb{F}_p((x))((y))$) – the Japanese mathematician *Kazuya Kato* (see [83]) has established a canonical isomorphism

$$G(L|K)^{ab} \cong K_2(K)/N_{L|K}K_2(L).$$

Kato's proof is intricate and needs heavy machinery. It was simplified by the Russian mathematician *I. Fesenko* (see [36], [37], [38]). His proof may be viewed as a special case of the theory sketched above. The basic idea is the following. The correspondence $K \mapsto K_2(K)$ may be extended to a G-modulation K_2. It does not satisfy the hypothesis of exercise 15, so that one may not apply the abstract theory directly to K_2. But *Fesenko* considers on K_2 the finest topology for which the canonical map $(\ ,\ ): K^* \times K^* \to K_2(K)$ is sequentially continuous, and for which one has $x_n + y_n \to 0$, $-x_n \to 0$ whenever $x_n \to 0$, $y_n \to 0$. He puts

$$K_2^{\mathrm{top}}(K) = K_2(K)/\Lambda_2(K),$$

where $\Lambda_2(K)$ is the intersection of all open neighbourhoods of 1 in $K_2(K)$, and he shows that

$$K_2^{\mathrm{top}}(K)/N_{L|K}K_2^{\mathrm{top}}(L) \cong K_2(K)/N_{L|K}K_2(L)$$

for every Galois extension $L|K$, and that $K_2^{\mathrm{top}}(K)$ satisfies properties which imply the hypothesis of exercise 18 and 19 when viewing K_2^{top} as a G-modulation. This makes *Kato*'s theorem into a special case of the theory developed above.

§ 7. The Herbrand Quotient

The preceding section concluded abstract class field theory. In order to be able to apply it to the concrete situations encountered in number theory, it is all important to verify the *class field axiom* (6.1) in these contexts. An excellent tool for this is the **Herbrand quotient**. It is a group-theoretic formalism, which we develop here for future use.

Let G be a finite cyclic group of order n, let σ be a generator, and A a G-module. As before, we form the two groups

$$H^0(G,A) = A^G/N_GA \quad \text{and} \quad H^{-1}(G,A) = {}_{N_G}A/I_GA\,,$$

where

$$A^G = \{a \in A \mid a^\sigma = a\}\,, \qquad N_GA = \Big\{N_Ga = \prod_{i=0}^{n-1} a^{\sigma^i} \mid a \in A\Big\}\,,$$

$${}_{N_G}A = \{a \in A \mid N_Ga = 1\}\,, \qquad I_GA = \{a^{\sigma-1} \mid a \in A\}\,.$$

(7.1) Proposition. *If* $1 \to A \to B \to C \to 1$ *is an exact sequence of G-modules, then we obtain an exact hexagon*

$$\begin{array}{ccccc}
 & H^0(G,A) & \xrightarrow{f_1} & H^0(G,B) & \\
{\scriptstyle f_6}\nearrow & & & & \searrow{\scriptstyle f_2} \\
H^{-1}(G,C) & & & & H^0(G,C) \\
{\scriptstyle f_5}\nwarrow & & & & \swarrow{\scriptstyle f_3} \\
 & H^{-1}(G,B) & \xleftarrow{f_4} & H^{-1}(G,A) &
\end{array}\,.$$

Proof: The homomorphisms f_1, f_4 and f_2, f_5 are induced by $A \xrightarrow{i} B$ and $B \xrightarrow{j} C$. We identify A with its image in B so that i becomes the inclusion. Then f_3 is defined as follows. Let $c \in C^G$ and let $b \in B$ be an element such that $j(b) = c$. Then we have $j(b^{\sigma-1}) = c^{\sigma-1} = 1$ and $N_G(b^{\sigma-1}) = N_G(b^\sigma)/N_G(b) = 1$, so that $b^{\sigma-1} \in {}_{N_G}A$. f_3 is thus defined by $c \bmod N_GC \mapsto b^{\sigma-1} \bmod I_GA$. In order to define f_6, let $c \in {}_{N_G}C$, and let $b \in B$ be an element such that $j(b) = c$. Then $j(N_Gb) = N_Gc = 1$, so that $N_Gb \in A$. The map f_6 is now given by $c \bmod I_GA \mapsto N_Gb \bmod N_GA$.

We now prove exactness at the place $H^0(G,A)$. Let $a \in A^G$ such that $f_1(a \bmod N_GA) = 1$; in other words, $a = N_Gb$ for some $b \in B$. Writing $c = j(b)$, we find $f_6(c \bmod I_GC) = a \bmod N_GA$. Exactness at $H^{-1}(G,A)$ is deduced as follows: let $a \in {}_{N_G}A$ and $f_4(a \bmod I_GA) = 1$, i.e., $a = b^{\sigma-1}$, with $b \in B$. Writing $c = j(b)$, we find $f_3(c \bmod N_GC) = a \bmod I_GA$. The exactness at all other places is seen even more easily. $\square$

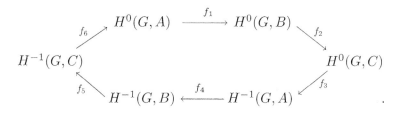

(7.2) Definition. *The* **Herbrand quotient** *of the G-module A is defined to be*

$$h(G,A) = \frac{\#H^0(G,A)}{\#H^{-1}(G,A)},$$

provided that both orders are finite.

The salient property of the Herbrand quotient is its *multiplicativity*.

(7.3) Proposition. *If $1 \to A \to B \to C \to 1$ is an exact sequence of G-modules, then one has*

$$h(G,B) = h(G,A)\,h(G,C)$$

in the sense that, whenever two of these quotients are defined, so is the third and the identity holds.

For a finite G-module A, one has $h(G,A) = 1$.

Proof: We consider the exact hexagon (7.1). Calling n_i the order of the image of f_i, we find

$$\begin{aligned} &\#H^0(G,A) = n_6 n_1, \quad &&\#H^0(G,B) = n_1 n_2, \quad &&\#H^0(G,C) = n_2 n_3, \\ &\#H^{-1}(G,A) = n_3 n_4, \quad &&\#H^{-1}(G,B) = n_4 n_5, \quad &&\#H^{-1}(G,C) = n_5 n_6, \end{aligned}$$

and thus

$$\begin{aligned} &\#H^0(G,A)\cdot\#H^0(G,C)\cdot\#H^{-1}(G,B) \\ &\qquad = \#H^0(G,B)\cdot\#H^{-1}(G,A)\cdot\#H^{-1}(G,C). \end{aligned}$$

At the same time, we see that if any two of the quotients are well-defined, then so is the third. And from the last equation, we obtain $h(G,B) = h(G,A)h(G,C)$. Finally, if A is a finite G-module, then the exact sequences

$$1 \longrightarrow A^G \longrightarrow A \xrightarrow{\sigma-1} I_G A \longrightarrow 1, \quad 1 \longrightarrow {}_{N_G}A \longrightarrow A \xrightarrow{N_G} N_G A \longrightarrow 1,$$

show that $\#A = \#A^G \cdot \#I_G A = \#_{N_G}A \cdot \#N_G A$, and $h(G,A) = 1$. □

If G is an arbitrary group and g a subgroup, then to any g-module B, we may associate the so-called **induced G-module**

$$A = \operatorname{Ind}_G^g(B).$$

It consists of all functions $f : G \to B$ such that $f(x\tau) = f(x)^\tau$ for all $\tau \in g$. The operation of $\sigma \in G$ is given by

$$f^\sigma(x) = f(\sigma x).$$

If $g = \{1\}$, we write $\mathrm{Ind}_G(B)$ instead of $\mathrm{Ind}_G^g(B)$. We have a canonical g-homomorphism

$$\pi : \mathrm{Ind}_G^g(B) \longrightarrow B, \quad f \longmapsto f(1),$$

which maps the g-submodule

$$B' = \left\{ f \in \mathrm{Ind}_G^g(B) \mid f(x) = 1 \ \text{ for } x \notin g \right\}$$

isomorphically onto B. We identify B' with B. If g is of finite index, we find

$$\mathrm{Ind}_G^g(B) = \prod_{\sigma \in G/g} B^\sigma,$$

where the notation $\sigma \in G/g$ signifies that σ varies over a system of left coset representatives of G/g.

Indeed, for any $f \in \mathrm{Ind}_G^g(B)$ we have a unique factorization $f = \prod_\sigma f_\sigma^\sigma$, where f_σ denotes the function in B' which is determined by $f_\sigma(1) = f(\sigma^{-1})$.

If conversely A is a G-module with a g-submodule B such that A is the direct product

$$A = \prod_{\sigma \in G/g} B^\sigma,$$

then $A \cong \mathrm{Ind}_G^g(B)$ via $B \cong B'$.

(7.4) Proposition. *Let G be a finite cyclic group, g a subgroup and B a g-module. Then we have canonically*

$$H^i\big(G, \mathrm{Ind}_G^g(B)\big) \cong H^i(g, B) \quad \textit{for} \quad i = 0, -1.$$

Proof: Let $A = \mathrm{Ind}_G^g(B)$ and let R be a system of right coset representatives for G/g with $1 \in R$. We consider the g-homomorphisms

$$\pi : A \longrightarrow B, \quad f \longmapsto f(1); \qquad \nu : A \longrightarrow B, \quad f \longmapsto \prod_{\rho \in R} f(\rho).$$

Both admit the g-homomorphism

$$s : B \longrightarrow A, \quad b \longmapsto f_b(\sigma) = \begin{cases} b^\sigma & \text{for } \sigma \in g, \\ 1 & \text{for } \sigma \notin g, \end{cases}$$

as a section, i.e., $\pi \circ s = \nu \circ s = \mathrm{id}$, and we have

$$\pi \circ N_G = N_g \circ \nu,$$

because one finds that, for $f \in A$,

$$(N_G f)(1) = \prod_{\tau \in g} \prod_{\rho \in R} f^{\rho\tau}(1) = \prod_{\tau} \prod_{\rho} f(\rho\tau) = \prod_{\tau} \big(\prod_{\rho} f(\rho) \big)^{\tau} = N_g\big(\nu(f)\big) .$$

If $f \in A^G$, then $f(\sigma) = f(1)$ for all $\sigma \in G$, and $f(1) = f(\tau) = f(1)^\tau$ for all $\tau \in g$. The map π therefore induces an isomorphism

$$\pi : A^G \longrightarrow B^g .$$

It sends $N_G A$ onto $N_g B$, for one has $\pi(N_G A) = N_g(\nu A) \subseteq N_g B$ on the one hand, and on the other, $N_g(B) = N_g(\nu s B) = \pi(N_G(sB)) \subseteq \pi(N_G A)$. Therefore $H^0(G, A) = H^0(g, B)$.

As $N_g \circ \nu = \pi \circ N_G$, the g-homomorphism $\nu : A \to B$ induces a g-homomorphism

$$\nu : {}_{N_G}A \longrightarrow {}_{N_g}B .$$

It is surjective since $\nu \circ s = \mathrm{id}$. We show that $I_G A$ is the preimage of $I_g B$. $I_G A$ consists of all elements $f^{\sigma - 1}$, $f \in A$, $\sigma \in G$. For if $G = (\sigma_0)$ and $\sigma = \sigma_0^i$, then $f^{\sigma-1} = f^{(1+\sigma_0+\cdots+\sigma_0^{i-1})(\sigma_0 - 1)} \in I_G A$. In the same way, one has $I_g B = \{ b^{\tau - 1} \mid b \in B,\ \tau \in g\}$. Writing now $\sigma\rho = \rho'\tau_\rho$, with $\rho, \rho' \in R$, $\tau_\rho \in g$, we obtain

$$\nu(f^{\sigma-1}) = \prod_{\rho \in R} \frac{f(\sigma\rho)}{f(\rho)} = \prod_{\rho'} \frac{f(\rho')^{\tau_\rho}}{f(\rho')} = \prod_{\rho} b_\rho^{\tau_\rho - 1} \in I_g B .$$

On the other hand, for $b^{\tau-1} \in I_g B$, the function $f^{\tau-1}$, with $f = sb$, is a preimage as $\nu(f^{\tau-1}) = \nu s(b)^{\tau-1} = b^{\tau-1}$. After this it remains to show $\ker(\nu) \subseteq I_G A$. Let $G = (\varphi)$, $n = (G : g)$, $R = \{1, \varphi, \dots, \varphi^{n-1}\}$. Let $f \in_{N_G} A$ be such that $\nu(f) = \prod_{i=0}^{n-1} \varphi^i = 1$. Define the function $h \in A$ by $h(1) = 1$, $h(\varphi^k) = \prod_{i=1}^{k-1} f(\varphi^i)$. Then $f(\varphi^k) = h(\varphi^k)/h(\varphi^{k-1}) = h(\varphi^{k-1})^{1-\varphi^{-1}}$ for $0 < k < n$, and $f(1) h^{\varphi^{-1}-1}(1) = \prod_{i=0}^{n-1} f(\varphi^i) = 1$. Hence $f = h^{1-\varphi^{-1}} \in I_G A$. Thus we finally get $H^{-1}(G, A) = H^{-1}(g, B)$. □

Exercise 1. Let f, g be endomorphisms of an abelian group A such that $f \circ g = g \circ f = 0$. Make sense of the following statement. The quotient

$$q_{f,g}(A) = \frac{(\ker f : \operatorname{im} g)}{(\ker g : \operatorname{im} f)}$$

is multiplicative.

Exercise 2. Let f, g be two commuting endomorphisms of an abelian group A. Show that

$$q_{0,gf}(A) = q_{0,g}(A) q_{0,f}(A) ,$$

provided all quotients are defined.

Exercise 3. Let G be a cyclic group of prime order p, and let A be a G-module such that $q_{0,p}(A)$ is defined. Show that

$$h(G,A)^{p-1} = q_{0,p}(A^G)^p / q_{0,p}(A).$$

Hint: Use the exact sequence

$$0 \to A^G \longrightarrow A \xrightarrow{\sigma-1} A^{\sigma-1} \to 0.$$

Let $N = 1 + \sigma + \cdots + \sigma^{p-1}$ in the group ring $\mathbb{Z}[G]$. Show that the ring $\mathbb{Z}[G]/\mathbb{Z}N$ is isomorphic to $\mathbb{Z}[\zeta]$, for ζ a primitive p-th root of unity, and that in this ring one has

$$p = (\sigma - 1)^{p-1}\varepsilon,$$

where ε is a unit in $\mathbb{Z}[G]/\mathbb{Z}N$.

Exercise 4. Let $L|K$ be a cyclic extension of prime degree. Using exercise 3, compute the Herbrand quotient of the group of units $\mathcal{O}_L^*$ of L, viewed as a $G(L|K)$-module.

Exercise 5. If G is a group, g a normal subgroup and A a g-module, then $H^1(G, \operatorname{Ind}_G^g(A)) \cong H^1(g, A)$.

Chapter V

Local Class Field Theory

§ 1. The Local Reciprocity Law

The abstract class field theory that we have developed in the last chapter is now going to be applied to the case of a *local field*, i.e., to a field which is complete with respect to a discrete valuation, and which has a finite residue class field. By chap. II, (5.2), these are precisely the finite extensions K of the fields $\mathbb{Q}_p$ or $\mathbb{F}_p((t))$. We will use the following notation. Let

v_K be the discrete valuation normalized by $v_K(K^*) = \mathbb{Z}$,

$\mathcal{O}_K = \{a \in K \mid v_K(a) \geq 0\}$ the valuation ring,

$\mathfrak{p}_K = \{a \in K \mid v_K(a) > 0\}$ the maximal ideal,

$\kappa = \mathcal{O}_K/\mathfrak{p}_K$ the residue class field,

$U_K = \{a \in K^* \mid v_K(a) = 0\}$ the unit group,

$U_K^{(n)} = 1 + \mathfrak{p}_K^n$ the group of n-th higher units, $n = 1, 2, \ldots,$

$q = q_K = \#\kappa,$

$|a|_\mathfrak{p} = q^{-v_K(a)}$ the normalized $\mathfrak{p}$-adic absolute value,

μ_n the group of n-th roots of unity, and $\mu_n(K) = \mu_n \cap K^*$.

π_K, or simply π, denotes a prime element of K, i.e., $\mathfrak{p}_K = \pi\mathcal{O}_K$.

In local class field theory, the rôle of the profinite group G of abstract class field theory is taken by the absolute Galois group $G(\bar{k}|k)$ of a fixed local field k, and that of the G-module A by the multiplicative group $\bar{k}^*$ of the separable closure $\bar{k}$ of k. For a finite extension $K|k$ we thus have $A_K = K^*$, and the crucial point is to verify for the multiplicative group of a local field the class field axiom:

(1.1) Theorem. *For a cyclic extension $L|K$ of local fields, one has*

$$\#H^i\big(G(L|K), L^*\big) = \begin{cases} [L:K] & \text{for } i = 0, \\ 1 & \text{for } i = -1. \end{cases}$$

Proof: For $i = -1$ this is the claim of proposition (3.5) ("Hilbert 90") in chap. IV. So all we have to show is that the Herbrand quotient is $h(G,L^*) = \#H^0(G,L^*) = [L:K]$, where we have put $G = G(L|K)$. The exact sequence

$$1 \longrightarrow U_L \longrightarrow L^* \xrightarrow{v_L} \mathbb{Z} \longrightarrow 0,$$

in which $\mathbb{Z}$ has to be viewed as the trivial G-module, yields, by chap. IV, (7.3),

$$h(G,L^*) = h(G,\mathbb{Z})\,h(G,U_L) = [L:K]\,h(G,U_L).$$

Hence we have to show that $h(G,U_L) = 1$. For this we choose a normal basis $\{\alpha^\sigma \mid \sigma \in G\}$ of $L|K$ (see [93], chap. VIII, § 12, th. 20), $\alpha \in \mathcal{O}_L$, and consider in $\mathcal{O}_L$ the open (and closed) G-module $M = \sum_{\sigma\in G} \mathcal{O}_K\alpha^\sigma$. Then the open sets

$$V^n = 1 + \pi_K^n M, \quad n = 1,2,\dots,$$

form a basis of open neighbourhoods of 1 in U_L. Since M is open, we have $\pi_K^N \mathcal{O}_L \subseteq M$ for suitable N, and for $n \geq N$ the V^n are even subgroups (of finite index) of U_L, because we have

$$(\pi_K^n M)(\pi_K^n M) = \pi_K^{2n} MM \subseteq \pi_K^{2n}\mathcal{O}_L \subseteq \pi_K^{2n-N} M \subseteq \pi_K^n M.$$

Hence $V^nV^n \subseteq V^n$, and since $1 - \pi_K^n\mu$, for $\mu \in M$, lies in V^n, so does $(1-\pi_K^n\mu)^{-1} = 1 + \pi_K^n(\sum_{i=1}^\infty \mu^i \pi_K^{n(i-1)})$. Via the correspondence $1 + \pi_K^n\alpha \mapsto \alpha \bmod \pi_K M$, we obtain G-isomorphisms as in II, (3.10),

$$V^n/V^{n+1} \cong M/\pi_K M = \bigoplus_{\sigma\in G} (\mathcal{O}_K/\mathfrak{p}_K)\alpha^\sigma = \operatorname{Ind}_G(\mathcal{O}_K/\mathfrak{p}_K).$$

So by chap. IV, (7.4), we have $H^i(G, V^n/V^{n+1}) = 1$ for $i = 0, -1$ and $n \geq N$. This in turn implies that $H^i(G,V^n) = 1$ for $i = 0, -1$ and $n \geq N$. Indeed, if for instance $i = 0$ and $a \in (V^n)^G$, then $a = (N_G b_0)a_1$, with $b_0 \in V^n$, $a_1 \in (V^{n+1})^G$, and thus $a_1 = (N_G b_1)a_2$, for some $b_1 \in V^{n+1}$, $a_2 \in (V^{n+2})^G$, etc.; in general,

$$a_i = (N_G b_i)a_{i+1}, \quad b_i \in V^{n+i}, \quad a_{i+1} \in (V^{n+i+1})^G.$$

This yields $a = N_G b$, with the convergent product $b = \prod_{i=0}^\infty b_i \in V^n$, so that $H^0(G,V^n) = 1$. In the same way we have for $a \in V^n$ such that $N_G a = 1$, that $a = b^{\sigma-1}$, for some $b \in V^n$, where σ is a generator of G. Thus $H^{-1}(G,V^n) = 1$. We now obtain

$$h(G,U_L) = h(G,U_L/V^n)\,h(G,V^n) = 1$$

because U_L/V^n is finite. □

(1.2) Corollary. *If $L|K$ is an unramified extension of local fields, then for $i = 0, -1$, one has*

$$H^i\big(G(L|K), U_L\big) = 1 \quad \text{and} \quad H^i\big(G(L|K), U_L^{(n)}\big) = 1 \quad \text{for} \quad n = 1, 2, \dots$$

In particular,

$$N_{L|K} U_L = U_K \quad \text{and} \quad N_{L|K} U_L^{(n)} = U_K^{(n)}.$$

Proof: Let $G = G(L|K)$. We have already seen that $H^i(G, U_L) = 1$ in chap. IV, (6.2). In order to prove $H^i(G, U_L^{(n)}) = 1$, we first show that

$$H^i(G, \lambda^*) = 1 \quad \text{and} \quad H^i(G, \lambda) = 1,$$

for the residue class field λ of L. It is enough to prove this for $i = -1$, as λ is finite, and so $h(G, \lambda^*) = h(G, \lambda) = 1$. We have $H^{-1}(G, \lambda^*) = 1$ by Hilbert 90 (see chap. IV, (3.5)). Let $f = [\lambda : \kappa]$ be the degree of λ over the residue class field κ of K, and let φ be the Frobenius automorphism of $\lambda|\kappa$. Then we have

$$\#_{N_G}\lambda = \#\Big\{x \in \lambda \;\Big|\; \sum_{i=0}^{f-1} x^{\varphi^i} = \sum_{i=0}^{f-1} x^{q^i} = 0\Big\} \leq q^{f-1}$$

and

$$\#(\varphi - 1)\lambda = q^{f-1},$$

since the map $\lambda \xrightarrow{\varphi-1} \lambda$ has kernel κ. Therefore $H^{-1}(G, \lambda) = {}_{N_G}\lambda/(\varphi - 1)\lambda = 1$.

Applying now the exact hexagon of chap. IV, (7.1), to the exact sequence of G-modules

$$1 \longrightarrow U_L^{(1)} \longrightarrow U_L \longrightarrow \lambda^* \longrightarrow 1,$$

we obtain $H^i(G, U_L^{(1)}) = H^i(G, U_L) = 1$, because $H^i(G, \lambda^*) = 1$. If π is a prime element of K, then π is also a prime element of L, so the map $U_L^{(n)} \to \lambda$ given by $1 + a\pi^n \mapsto a \bmod \mathfrak{p}_L$ is a G-homomorphism. From the exact sequence

$$1 \longrightarrow U_L^{(n+1)} \longrightarrow U_L^{(n)} \longrightarrow \lambda \longrightarrow 1,$$

we now deduce by induction just as above, because $H^i(G, \lambda) = 0$, that

$$H^i(G, U_L^{(n+1)}) = H^i(G, U_L^{(n)}) = 1,$$

since $H^i(G, U_L^{(1)}) = 1$. □

We now consider the maximal unramified extension $\tilde{k}|k$ over the ground field k. By chap. II, §9, the residue class field of $\tilde{k}$ is the algebraic closure $\overline{\kappa}$ of the residue class field κ of k. By chap. II, (9.9), we get a canonical isomorphism

$$G(\tilde{k}|k) \cong G(\overline{\kappa}|\kappa) \cong \widehat{\mathbb{Z}}.$$

It associates to the element $1 \in \widehat{\mathbb{Z}}$ the Frobenius automorphism $x \mapsto x^q$ in $G(\overline{\kappa}|\kappa)$, and the Frobenius automorphism φ_k in $G(\tilde{k}|k)$ which is given by

$$a^{\varphi_k} \equiv a^q \bmod \mathfrak{p}_{\tilde{k}}, \quad a \in \mathcal{O}_{\tilde{k}}.$$

For the absolute Galois group $G = G(\bar{k}|k)$ we therefore obtain the continuous and surjective homomorphism

$$d : G \longrightarrow \widehat{\mathbb{Z}}.$$

Thus the abstract notions of chap. IV, §4, based on this homomorphism, like "unramified", "ramification index", "inertia degree", etc., do agree, in the case at hand, with the corresponding concrete notions defined in chap. II.

As stated above we choose $A = \bar{k}^*$ to be our G-module. Hence $A_K = K^*$, for every finite extension $K|k$. The usual normalized exponential valuation $v_k : k^* \to \mathbb{Z}$ is then henselian with respect to d, in the sense of chap. IV, (4.6). For, given any finite extension $K|k$, $\frac{1}{e_K} v_K$ is the extension of v_k to K^*, and by chap. II, (4.8),

$$\frac{1}{e_K} v_K(K^*) = \frac{1}{[K:k]} v_k(N_{K|k}K^*) = \frac{1}{e_K f_K} v_k(N_{K|k}K^*),$$

i.e., $v_k(N_{K|k}K^*) = f_K v_K(K^*) = f_K\mathbb{Z}$. The pair of homomorphisms

$$(d : G \to \widehat{\mathbb{Z}},\ v_k : k^* \to \mathbb{Z})$$

therefore satisfies all the properties of a class field theory, and we obtain the **Local Reciprocity Law**:

(1.3) Theorem. *For every finite Galois extension $L|K$ of local fields we have a canonical isomorphism*

$$r_{L|K} : G(L|K)^{ab} \xrightarrow{\sim} K^*/N_{L|K}L^*.$$

The general definition of the reciprocity map in chap. IV, (5.6), was actually inspired by the case of local class field theory. This is why it is especially transparent in this case: let $\sigma \in G(L|K)$, and let $\tilde{\sigma}$ be an extension of σ to the maximal unramified extension $\widetilde{L}|K$ of L such that $d_K(\tilde{\sigma}) \in \mathbb{N}$

or, in other words, $\tilde{\sigma}\,|_{\tilde{K}} = \varphi_K^n$, for some $n \in \mathbb{N}$. If Σ is the fixed field of $\tilde{\sigma}$ and $\pi_\Sigma \in \Sigma$ is a prime element, then

$$r_{L|K}(\sigma) = N_{\Sigma|K}(\pi_\Sigma) \bmod N_{L|K}L^*.$$

Inverting $r_{L|K}$ gives us the **local norm residue symbol**

$$(\quad, L|K) : K^* \longrightarrow G(L|K)^{ab}.$$

It is surjective and has kernel $N_{L|K}L^*$.

In global class field theory we will have to take into account the field $\mathbb{R} = \mathbb{Q}_\infty$ along with the p-adic number fields $\mathbb{Q}_p$. It also admits a reciprocity law: for the unique non-trivial Galois extension $\mathbb{C}|\mathbb{R}$, we define the norm residue symbol

$$(\quad, \mathbb{C}|\mathbb{R}) : \mathbb{R}^* \longrightarrow G(\mathbb{C}|\mathbb{R})$$

by

$$(a, \mathbb{C}|\mathbb{R})\sqrt{-1} = \sqrt{-1}^{\,\mathrm{sgn}(a)}.$$

The kernel of $(\ , \mathbb{C}|\mathbb{R})$ is the group $\mathbb{R}_+^*$ of all positive real numbers, which is again the group of norms $N_{\mathbb{C}|\mathbb{R}}\mathbb{C}^* = \{z\bar{z} \mid z \in \mathbb{C}^*\}$.

The reciprocity law gives us a very simple classification of the abelian extensions of a local field K. It is formulated in the following

(1.4) Theorem. *The rule*

$$L \longmapsto \mathcal{N}_L = N_{L|K}L^*$$

gives a 1–1-correspondence between the finite abelian extensions of a local field K and the open subgroups $\mathcal{N}$ of finite index in K^*. Furthermore,

$$L_1 \subseteq L_2 \Longleftrightarrow \mathcal{N}_{L_1} \supseteq \mathcal{N}_{L_2}, \quad \mathcal{N}_{L_1L_2} = \mathcal{N}_{L_1} \cap \mathcal{N}_{L_2}, \quad \mathcal{N}_{L_1 \cap L_2} = \mathcal{N}_{L_1}\mathcal{N}_{L_2}.$$

Proof: By chap. IV, (6.7), all we have to show is that the subgroups $\mathcal{N}$ of K^* which are open in the norm topology are precisely the subgroups of finite index which are open in the valuation topology. A subgroup $\mathcal{N}$ which is open in the norm topology contains by definition a group of norms $N_{L|K}L^*$. By (1.3), this has finite index in K^*. It is also open because it contains the subgroup $N_{L|K}U_L$ which itself is open, for it is closed, being the image of the compact group U_L, and has finite index in U_K. We prove the converse first in

The case $\mathrm{char}(K) \nmid n$. Let $\mathcal{N}$ be a subgroup of finite index $n = (K^* : \mathcal{N})$. Then $K^{*n} \subseteq \mathcal{N}$, and it is enough to show that K^{*n} contains a group of norms. For this we use Kummer theory (see chap. IV, §3). We may assume that K^* contains the group μ_n of n-th roots of unity. For if it does not, we put $K_1 = K(\mu_n)$. If K_1^{*n} contains a group of norms $N_{L_1|K_1}L_1^*$, and $L|K$ is a Galois extension containing L_1, then

$$\begin{aligned} N_{L|K}L^* = N_{K_1|K}(N_{L|K_1}L^*) &\subseteq N_{K_1|K}(N_{L_1|K_1}L_1^*) \\ &\subseteq N_{K_1|K}(K_1^{*n}) \subseteq K^{*n}. \end{aligned}$$

So let $\mu_n \subseteq K$, and let $L = K(\sqrt[n]{K^*})$ be the maximal abelian extension of exponent n. Then by chap. IV, §3, we have

$$(*) \qquad \mathrm{Hom}(G(L|K), \mu_n) \cong K^*/K^{*n}.$$

By chap. II, (5.8), K^*/K^{*n} is finite, and then so is $G(L|K)$. Since $K^*/N_{L|K}L^*$ is isomorphic to $G(L|K)$ and has exponent n, we have that $K^{*n} \subseteq N_{L|K}L^*$, and $(*)$ yields

$$\#K^*/K^{*n} = \#G(L|K) = \#K^*/N_{L|K}L^*,$$

and therefore $K^{*n} = N_{L|K}L^*$.

The case $\mathrm{char}(K) = p|n$. In this case the proof will follow from Lubin-Tate theory which we will develop in §4. But it is also possible to do without this theory, at the expense of *ad hoc* arguments which turn out to be somewhat elaborate. Since the result has no further use in the remainder of this book, we simply refer the reader to the beautiful treatment in [122], chap. XI, §5, and chap. XIV, §6. □

The proof also shows the following

(1.5) Proposition. *If K contains the n-th roots of unity, and if the characteristic of K does not divide n, then the extension $L = K(\sqrt[n]{K^*})|K$ is finite, and one has*

$$N_{L|K}L^* = K^{*n} \quad \textit{and} \quad G(L|K) \cong K^*/K^{*n}.$$

Theorem (1.4) is called the **existence theorem**, because its essential statement is that, for every open subgroup $\mathcal{N}$ of finite index in K^*, there exists an abelian extension $L|K$ such that $N_{L|K}L^* = \mathcal{N}$. This is the "class field" of $\mathcal{N}$. (Incidentally, when $\mathrm{char}(K) = 0$, every subgroup of finite index is automatically open – see chap. II, (5.7).) Every open subgroup of K^* contains some higher unit group $U_K^{(n)}$, as these form a basis of neighbourhoods of 1 in K^*. We put $U_K^{(0)} = U_K$ and define:

(1.6) Definition. *Let $L|K$ be a finite abelian extension, and n the smallest number ≥ 0 such that $U_K^{(n)} \subseteq N_{L|K}L^*$. Then the ideal*

$$\mathfrak{f} = \mathfrak{p}_K^n$$

is called the **conductor** *of $L|K$.*

(1.7) Proposition. *A finite abelian extension $L|K$ is unramified if and only if its conductor is $\mathfrak{f} = 1$.*

Proof: If $L|K$ is unramified, then $U_K = N_{L|K}U_L$ by (1.2), so that $\mathfrak{f} = 1$. If conversely $\mathfrak{f} = 1$, then $U_K \subseteq N_{L|K}U_L$ and $\pi_K^n \in N_{L|K}L^*$, for $n = (K^* : N_{L|K}L^*)$. If $M|K$ is the unramified extension of degree n, then $N_{M|K}M^* = (\pi_K^n) \times U_K \subseteq N_{L|K}L^*$, and then $M \supseteq L$, i.e., $L|K$ is unramified. □

Every open subgroup $\mathcal{N}$ of finite index in K^* contains a group of the form $(\pi^f) \times U_K^{(n)}$. This is again open and of finite index. Hence every finite abelian extension $L|K$ is contained in the class field of such a group $(\pi^f) \times U_K^{(n)}$. Therefore the class fields for the groups $(\pi^f) \times U_K^{(n)}$ are particularly important. We will characterize them explicitly in §5, as immediate analogues of the cyclotomic fields over $\mathbb{Q}_p$. In the case of the ground field $K = \mathbb{Q}_p$, the class field of the group $(p) \times U_K^{(n)}$ is precisely the field $\mathbb{Q}_p(\mu_{p^n})$ of p^n-th roots of unity:

(1.8) Proposition. *The group of norms of the extension $\mathbb{Q}_p(\mu_{p^n})|\mathbb{Q}_p$ is the group $(p) \times U_{\mathbb{Q}_p}^{(n)}$.*

Proof: Let $K = \mathbb{Q}_p$ and $L = \mathbb{Q}_p(\mu_{p^n})$. By chap. II, (7.13), the extension $L|K$ is totally ramified of degree $p^{n-1}(p-1)$, and if ζ is a primitive p^n-th root of unity, then $1-\zeta$ is a prime element of L of norm $N_{L|K}(1-\zeta) = p$. We now consider the exponential map of $\mathbb{Q}_p$. By chap. II, (5.5), it gives an isomorphism

$$\exp : \mathfrak{p}_K^\nu \longrightarrow U_K^{(\nu)}$$

for $\nu \geq 1$, provided $p \neq 2$, and for $\nu \geq 2$, even if $p = 2$. It transforms the isomorphism $\mathfrak{p}_K^\nu \to \mathfrak{p}_K^{\nu+s-1}$ given by $a \mapsto p^{s-1}(p-1)a$, into the isomorphism $U_K^{(\nu)} \to U_K^{(\nu+s-1)}$ given by $x \mapsto x^{p^{s-1}(p-1)}$, so that $(U_K^{(1)})^{p^{n-1}(p-1)} = U_K^{(n)}$ if $p \neq 2$, and $(U_K^{(2)})^{2^{n-2}} = U_K^{(n)}$ if $p = 2$, $n > 1$

(the case $p = 2$, $n = 1$ is trivial). Consequently, we have $U_K^{(n)} \subseteq N_{L|K}L^*$ if $p \neq 2$. For $p = 2$ we note that

$$U_K^{(2)} = U_K^{(3)} \cup 5U_K^{(3)} = \big(U_K^{(2)}\big)^2 \cup 5\big(U_K^{(2)}\big)^2,$$

because a number that is congruent to 1 mod 4 is congruent to 1 or 5 mod 8. Hence

$$U_K^{(n)} = \big(U_K^{(2)}\big)^{2^{n-1}} \cup 5^{2^{n-2}}\big(U_K^{(2)}\big)^{2^{n-1}}.$$

It is easy to show that $5^{2^{n-2}} = N_{L|K}(2+i)$, so $U_K^{(n)} \subseteq N_{L|K}L^*$ holds also in case $p = 2$. Since $p = N_{L|K}(1-\zeta)$, we have $(p) \times U_K^{(n)} \subseteq N_{L|K}L^*$, and since both groups have index $p^{n-1}(p-1)$ in K^*, we do find that $N_{L|K}L^* = (p) \times U_K^{(n)}$ as claimed. □

As an immediate consequence of this last proposition, we obtain a local version of the famous theorem of *Kronecker-Weber*, to the effect that every finite abelian extension of $\mathbb{Q}$ is contained in a cyclotomic field.

(1.9) Corollary. *Every finite abelian extension of $L|\mathbb{Q}_p$ is contained in a field $\mathbb{Q}_p(\zeta)$, where ζ is a root of unity. In other words:*

The maximal abelian extension $\mathbb{Q}_p^{ab}|\mathbb{Q}_p$ is generated by adjoining all roots of unity.

Proof: For suitable f and n, we have $(p^f) \times U_{\mathbb{Q}_p}^{(n)} \subseteq N_{L|K}L^*$. Therefore L is contained in the class field M of the group

$$(p^f) \times U_{\mathbb{Q}_p}^{(n)} = \big((p^f) \times U_{\mathbb{Q}_p}\big) \cap \big((p) \times U_{\mathbb{Q}_p}^{(n)}\big).$$

By (1.4), M is the composite of the class field for $(p^f) \times U_{\mathbb{Q}_p}$ – this being the unramified extension of degree f – and the class field for $(p) \times U_{\mathbb{Q}_p}^{(n)}$. M is therefore generated by the $(p^f - 1)p^n$-th roots of unity. □

From the local Kronecker-Weber theorem, one may readily deduce the global, classical **Theorem of Kronecker-Weber**.

(1.10) Theorem. *Every finite abelian extension $L|\mathbb{Q}$ is contained in a field $\mathbb{Q}(\zeta)$ generated by a root of unity ζ.*

Proof: Let S be the set of all prime numbers p that are ramified in L, and let L_p be the completion of L with respect to some prime lying above p. Then $L_p|\mathbb{Q}_p$ is abelian, and therefore $L_p \subseteq \mathbb{Q}_p(\mu_{n_p})$, for a suitable n_p. Let p^{e_p} be the precise power of p dividing n_p, and let

$$n = \prod_{p \in S} p^{e_p}.$$

We will show that $L \subseteq \mathbb{Q}(\mu_n)$. For this let $M = L(\mu_n)$. Then $M|\mathbb{Q}$ is abelian, and if p is ramified in $M|\mathbb{Q}$, then p must lie in S. If M_p is the completion with respect to a prime of M above p whose restriction to L gives the completion L_p, then

$$M_p = L_p(\mu_n) = \mathbb{Q}_p(\mu_{p^{e_p}n'}) = \mathbb{Q}_p(\mu_{p^{e_p}})\,\mathbb{Q}_p(\mu_{n'}),$$

with $(n', p) = 1$. $\mathbb{Q}_p(\mu_{n'})|\mathbb{Q}_p$ is the maximal unramified subextension of $\mathbb{Q}_p(\mu_{p^{e_p}n'})|\mathbb{Q}_p$. The inertia group I_p of $M_p|\mathbb{Q}_p$ is therefore isomorphic to the group $G(\mathbb{Q}_p(\mu_{p^{e_p}})|\mathbb{Q}_p)$, and consequently has order $\varphi(p^{e_p})$, where φ is Euler's function. Let I be the subgroup of $G(M|\mathbb{Q})$ generated by all I_p, $p \in S$. The fixed field of I is then unramified, and hence by Minkowski's theorem from chap. III, (2.18), it equals $\mathbb{Q}$, i.e., $I = G(M|\mathbb{Q})$. On the other hand we have

$$\#I \le \prod_{p \in S} \#I_p = \prod_{p \in S} \varphi(p^{e_p}) = \varphi(n) = \big[\mathbb{Q}(\mu_n) : \mathbb{Q}\big],$$

and therefore $[M : \mathbb{Q}] = [\mathbb{Q}(\mu_n) : \mathbb{Q}]$, so that $M = \mathbb{Q}(\mu_n)$. This shows that $L \subseteq \mathbb{Q}(\mu_n)$. □

The following exercises 1–3 presuppose exercises 4–8 of chap. IV, § 3.

Exercise 1. For the Galois group $\Gamma = G(\widetilde{K}|K)$, one has canonically

$$H^1(\Gamma, \mathbb{Z}/n\mathbb{Z}) \cong \mathbb{Z}/n\mathbb{Z} \quad \text{and} \quad H^1(\Gamma, \mu_n) \cong U_K K^{*n}/K^{*n},$$

the latter provided that n is not divisible by the residue characteristic.

Exercise 2. For an arbitrary field K and a G_K-module A, put

$$H^1(K, A) = H^1(G_K, A).$$

If K is a $\mathfrak{p}$-adic number field and n a natural number, then there exists a nondegenerate pairing

$$H^1(K, \mathbb{Z}/n\mathbb{Z}) \times H^1(K, \mu_n) \longrightarrow \mathbb{Z}/n\mathbb{Z}$$

of finite groups given by

$$(\chi, a) \mapsto \chi((a, \widetilde{K}|K)).$$

If n is not divisible by the residue characteristic p, then the orthogonal complement of

$$H^1_{nr}(K, \mathbb{Z}/n) := H^1(G(\widetilde{K}|K), \mathbb{Z}/n\mathbb{Z}) \subseteq H^1(K, \mathbb{Z}/n\mathbb{Z})$$

is the group

$$H^1_{nr}(K, \mu_n) := H^1(G(\widetilde{K}|K), \mu_n) \subseteq H^1(K, \mu_n).$$

Exercise 3. If $L|K$ is a finite extension of $\mathfrak{p}$-adic number fields, then one has a commutative diagram

$$\begin{array}{ccccc} H^1(L,\mathbb{Z}/n\mathbb{Z}) & \times & H^1(L,\mu_n) & \to & \mathbb{Z}/n\mathbb{Z} \\ \uparrow & & \downarrow N_{L|K} & & \| \\ H^1(K,\mathbb{Z}/n\mathbb{Z}) & \times & H^1(K,\mu_n) & \to & \mathbb{Z}/n\mathbb{Z}. \end{array}$$

Exercise 4 (Local Tate Duality). Show that the statements of exercises 2 and 3 generalize to an arbitrary finite G_K-module A instead of $\mathbb{Z}/n\mathbb{Z}$, and $A' = \operatorname{Hom}(A, \bar{K}^*)$ instead of μ_n.

Hint: Use exercises 4–8 of chap. IV, § 3.

Exercise 5. Let $L|K$ be the composite of all $\mathbb{Z}_p$-extensions of a $\mathfrak{p}$-adic number field K (i.e., extensions with Galois group isomorphic to $\mathbb{Z}_p$). Show that the Galois group $G(L|K)$ is a free, finitely generated $\mathbb{Z}_p$-module and determine its rank.

Hint: Use chap. II, (5.7).

Exercise 6. There is only one unramified $\mathbb{Z}_p$-extension of K. Generate it by roots of unity.

Exercise 7. Let p be the residue characteristic of K, and let L be the field generated by all roots of unity of p-power order. The fixed field of the torsion subgroup of $G(L|K)$ is a $\mathbb{Z}_p$-extension. It is called the **cyclotomic** $\mathbb{Z}_p$-extension.

Exercise 8. Let $\widehat{\mathbb{Q}}_p|\mathbb{Q}_p$ be the cyclotomic $\mathbb{Z}_p$-extension of $\mathbb{Q}_p$, let $G(\widehat{\mathbb{Q}}_p|\mathbb{Q}_p) \cong \mathbb{Z}_p$ be a chosen isomorphism, and let $\hat{d} : G_{\mathbb{Q}_p} \to \mathbb{Z}_p$ be the induced homomorphism of the absolute Galois group. Show:

For a suitable topological generator u of the group of principal units of $\mathbb{Q}_p$,

$$\hat{v} : \mathbb{Q}_p^* \to \mathbb{Z}_p, \quad \hat{v}(a) = \frac{\log a}{\log u},$$

defines a henselian valuation with respect to $\hat{d}$, in the sense of abstract p-class field theory (see chap. IV, § 5, exercise 2).

Exercise 9. Determine all p-class field theories $(d : G_K \to \mathbb{Z}_p,\ v : K^* \to \mathbb{Z}_p)$ over a p-adic number field K.

Exercise 10. Determine all class field theories $(d : G_K \to \widehat{\mathbb{Z}},\ v : K^* \to \widehat{\mathbb{Z}})$ over a $\mathfrak{p}$-adic number field K.

Exercise 11. The **Weil group** of a local field K is the preimage W_K of $\mathbb{Z}$ under the mapping $d_K : G_K \to \widehat{\mathbb{Z}}$. Show:

The norm residue symbol $(\ ,K^{ab}|K)$ of the maximal abelian extension $K^{ab}|K$ yields an isomorphism

$$(\ ,K^{ab}|K) : K^* \xrightarrow{\sim} W_K^{ab},$$

which maps the unit group U_K onto the inertia group $I(K^{ab}|K)$, and the group of principal units $U_K^{(1)}$ onto the ramification group $R(K^{ab}|K)$.

§ 2. The Norm Residue Symbol over $\mathbb{Q}_p$

If ζ is a primitive m-th root of unity, with $(m, p) = 1$, then $\mathbb{Q}_p(\zeta)|\mathbb{Q}_p$ is unramified, and the norm residue symbol is obviously given by

$$\big(a, \mathbb{Q}_p(\zeta)|\mathbb{Q}_p\big)\zeta = \zeta^{p^{v_p(a)}}.$$

But if ζ is a primitive p^n-th root of unity, then we obtain the norm residue symbol for the extension $\mathbb{Q}_p(\zeta)|\mathbb{Q}_p$ explicitly in the simple form

$$\big(a, \mathbb{Q}_p(\zeta)|\mathbb{Q}_p\big)\zeta = \zeta^{u^{-1}},$$

where $a = up^{v_p(a)}$, and $\zeta^{u^{-1}}$ is the power ζ^r with any rational integer $r \equiv u^{-1} \bmod p^n$. This result is important, not only in the local situation, but it will play an essential rôle when we develop global class field theory (see chap. VI, § 5). Unfortunately, there is no direct algebraic proof of this fact known to date. We have to invoke a transcendental method which makes use of the *completion* $\widehat{K}$ of the maximal unramified extension $\widetilde{K}$ of a local field K. We extend the Frobenius $\varphi \in G(\widetilde{K}|K)$ to $\widehat{K}$ by continuity. First we prove the

(2.1) Lemma. *For every $c \in \mathcal{O}_{\widehat{K}}$, resp. every $c \in U_{\widehat{K}}$, the equation*

$$x^{\varphi} - x = c, \quad \text{resp.} \quad x^{\varphi-1} = c,$$

admits a solution in $\mathcal{O}_{\widehat{K}}$, resp. in $U_{\widehat{K}}$. If $x^{\varphi} = x$ for $x \in \mathcal{O}_{\widehat{K}}$, then $x \in \mathcal{O}_K$.

Proof: Let π be a prime element of K. Then π is also a prime element of $\widehat{K}$, and we have the φ-invariant isomorphisms

$$U_{\widehat{K}}/U_{\widehat{K}}^{(1)} \cong \bar{\kappa}^*, \quad U_{\widehat{K}}^{(n)}/U_{\widehat{K}}^{(n+1)} \cong \bar{\kappa}$$

(see chap. II, (3.10)). Let $c \in U_{\widehat{K}}$ and $\bar{c} = c \bmod \mathfrak{p}_{\widehat{K}}$. Since the residue class field $\bar{\kappa}$ of $\widehat{K}$ is algebraically closed, the equation $\bar{x}^{\varphi} = \bar{x}^q = \bar{x}\cdot\bar{c}$ $(q = q_K)$ has a solution $\neq 0$ in $\bar{\kappa} = \mathcal{O}_{\widehat{K}}/\mathfrak{p}_{\widehat{K}}$, i.e.,

$$c = x_1^{\varphi-1}a_1, \quad x_1 \in U_{\widehat{K}}, \quad a_1 \in U_{\widehat{K}}^{(1)}.$$

For similar reasons, we find that $a_1 = x_2^{\varphi-1}a_2$, for some $x_2 \in U_{\widehat{K}}^{(1)}$ and $a_2 \in U_{\widehat{K}}^{(2)}$, so that $c = (x_1x_2)^{\varphi-1}a_2$. Indeed, putting $a_1 = 1 + b_1\pi$, $x_2 = 1 + y_2\pi$, gives $a_1x_2^{1-\varphi} \equiv 1 - (y_2^{\varphi} - y_2 - b_1)\pi \bmod \pi^2$, i.e., we have to solve the congruence $y_2^{\varphi} - y_2 - b_1 \equiv 0 \bmod \pi$, or equivalently the equation $\bar{y}_2^q - \bar{y}_2 - \bar{b}_1 = 0$ in $\bar{\kappa}$. This is possible because $\bar{\kappa}$ is algebraically closed. Continuing in this way, we get

$$c = (x_1 x_2 \cdots x_n)^{\varphi-1} a_n, \quad x_n \in U_{\widehat{K}}^{(n-1)}, \quad a_n \in U_{\widehat{K}}^{(n)},$$

and passing to the limit finally gives $c = x^{\varphi-1}$, where $x = \prod_{n=1}^{\infty} x_n \in U_{\widehat{K}}$. The solvability of the equation $x^{\varphi} - x = c$ follows analogously, using the isomorphisms $\mathfrak{p}_{\widehat{K}}^n / \mathfrak{p}_{\widehat{K}}^{n+1} \cong \bar{\kappa}$.

Now let $x \in \mathcal{O}_{\widehat{K}}$ and $x^{\varphi} = x$. Then, for every $n \geq 1$, one has

$$(*) \qquad x = x_n + \pi^n y_n \quad \text{with } x_n \in \mathcal{O}_K \text{ and } y_n \in \mathcal{O}_{\widehat{K}}.$$

Indeed, for $n = 1$ we have $x = a + \pi b$, with $a \in \mathcal{O}_{\tilde{K}}$, $b \in \mathcal{O}_{\widehat{K}}$, and $x^{\varphi} = x$ implies $a^{\varphi} \equiv a \bmod \pi$. Hence $a = x_1 + \pi c$, with $x_1 \in \mathcal{O}_K$, $c \in \mathcal{O}_{\tilde{K}}$, and therefore $x = x_1 + \pi(b+c) = x_1 + \pi y_1$, $y_1 \in \mathcal{O}_{\widehat{K}}$. The equation $x = x_n + \pi^n y_n$ implies furthermore that $y_n^{\varphi} = y_n$, so that we get as above $y_n = c_n + \pi d_n$, with $c_n \in \mathcal{O}_K$, $d_n \in \mathcal{O}_{\widehat{K}}$, and therefore $x = (x_n + c_n \pi^n) + \pi^{n+1} d_n = x_{n+1} + \pi^{n+1} y_{n+1}$, for some $x_{n+1} \in \mathcal{O}_K$, $y_{n+1} \in \mathcal{O}_{\widehat{K}}$. Now passing to the limit in the equation $(*)$ gives $x = \lim_{n\to\infty} x_n \in \mathcal{O}_K$, because K is complete. □

For a power series $F(X_1, \ldots, X_n) \in \mathcal{O}_{\widehat{K}}[[X_1, \ldots, X_n]]$, let F^{φ} be the power series in $\mathcal{O}_{\widehat{K}}[[X_1, \ldots, X_n]]$ which arises from F by applying φ to the coefficients of F. A **Lubin-Tate series** for a prime element π of K is by definition a power series $e(X) \in \mathcal{O}_K[[X]]$ with the properties

$$e(X) \equiv \pi X \bmod \deg 2 \quad \text{and} \quad e(X) \equiv X^q \bmod \pi,$$

where $q = q_K$ denotes, as always, the number of elements in the residue class field of K. The totality of all Lubin-Tate series is denoted by $\mathcal{E}_{\pi}$. In $\mathcal{E}_{\pi}$ there are in particular the polynomials

$$e(X) = uX^q + \pi(a_{q-1}X^{q-1} + \cdots + a_2 X^2) + \pi X,$$

where $u, a_i \in \mathcal{O}_K$ and $u \equiv 1 \bmod \pi$. These are called the **Lubin-Tate polynomials**. The simplest one among them is the polynomial $X^q + \pi X$. In the case $K = \mathbb{Q}_p$ for example, $e(X) = (1+X)^p - 1$ is a Lubin-Tate polynomial for the prime element p.

(2.2) Proposition. *Let π and $\bar{\pi}$ be prime elements of $\widehat{K}$, and let $e(X) \in \mathcal{E}_{\pi}$, $\bar{e}(X) \in \mathcal{E}_{\bar{\pi}}$ be Lubin-Tate series. Let $L(X_1, \ldots, X_n) = \sum_{i=1}^{n} a_i X_i$ be a linear form with coefficients $a_i \in \mathcal{O}_{\widehat{K}}$ such that*

$$\pi L(X_1, \ldots, X_n) = \bar{\pi} L^{\varphi}(X_1, \ldots, X_n).$$

Then there is a uniquely determined power series $F(X_1, \dots, X_n) \in \mathcal{O}_{\widehat{K}}[[X_1, \dots, X_n]]$ *satisfying*

$$F(X_1, \dots, X_n) \equiv L(X_1, \dots, X_n) \bmod \deg 2,$$

$$e\big(F(X_1, \dots, X_n)\big) = F^{\varphi}\big(\bar{e}(X_1), \dots, \bar{e}(X_n)\big).$$

If the coefficients of $e, \bar{e}, L$ *lie in a complete subring* $\mathcal{O}$ *of* $\mathcal{O}_{\widehat{K}}$ *such that* $\mathcal{O}^{\varphi} = \mathcal{O}$, *then* F *has coefficients in* $\mathcal{O}$ *as well.*

Proof: Let $\mathcal{O}$ be a complete subring of $\mathcal{O}_{\widehat{K}}$ such that $\mathcal{O}^{\varphi} = \mathcal{O}$, which contains the coefficients of $e, \bar{e}, L$. We put $X = (X_1, \dots, X_n)$ and $e(X) = (e(X_1), \dots, e(X_n))$. Let

$$F(X) = \sum_{\nu=1}^{\infty} E_{\nu}(X) \in \mathcal{O}[[X]]$$

be a power series, $E_{\nu}(X)$ its homogeneous part of degree ν, and let

$$F_r(X) = \sum_{\nu=1}^{r} E_{\nu}(X).$$

Clearly, $F(X)$ is a solution of the above problem if and only if $F_1(X) = L(X)$ and

$$(1) \qquad e\big(F_r(X)\big) \equiv F_r^{\varphi}\big(\bar{e}(X)\big) \bmod \deg(r+1)$$

for every $r \geq 1$. We determine the polynomials $E_{\nu}(X)$ inductively. For $\nu = 1$ we are forced to take $E_1(X) = L(X)$. Condition (1) is then satisfied for $r = 1$ by hypothesis. Assume that the $E_{\nu}(X)$, for $\nu = 1, \dots, r$, have already been found, and that they are uniquely determined by condition (1). We then put $F_{r+1}(X) = F_r(X) + E_{r+1}(X)$ with a homogeneous polynomial $E_{r+1}(X) \in \mathcal{O}[X]$ of degree $r+1$ which has yet to be determined. The congruences

$$e\big(F_{r+1}(X)\big) \equiv e\big(F_r(X)\big) + \pi E_{r+1}(X) \bmod \deg(r+2),$$

$$F_{r+1}^{\varphi}\big(\bar{e}(X)\big) \equiv F_r^{\varphi}\big(\bar{e}(X)\big) + \bar{\pi}^{r+1} E_{r+1}^{\varphi}(X) \bmod \deg(r+2)$$

show that $E_{r+1}(X)$ has to satisfy the congruence

$$(2) \qquad G_{r+1}(X) + \pi E_{r+1}(X) - \bar{\pi}^{r+1} E_{r+1}^{\varphi}(X) \equiv 0 \bmod \deg(r+2)$$

with $G_{r+1}(X) = e(F_r(X)) - F_r^{\varphi}(\bar{e}(X)) \in \mathcal{O}[[X]]$. We have $G_{r+1}(X) \equiv 0 \bmod \deg(r+1)$ and

$$(3) \qquad G_{r+1}(X) \equiv F_r(X)^q - F_r^{\varphi}(X^q) \equiv 0 \bmod \pi$$

because $e(X) \equiv \bar{e}(X) \equiv X^q \bmod \pi$ and $\alpha^\varphi \equiv \alpha^q \bmod \pi$ for $\alpha \in \mathcal{O}$. Now let $X^i = X_1^{i_1} \cdots X_n^{i_n}$ be a monomial of degree $r+1$ in $\mathcal{O}[X]$. By (3), the coefficient of X^i in G_{r+1} is of the form $-\pi\beta$, with $\beta \in \mathcal{O}$. Let α be the coefficient of the same monomial X^i in E_{r+1}. Then $\pi\alpha - \bar{\pi}\alpha^\varphi$ is the coefficient of X^i in $\pi E_{r+1} - \bar{\pi} E_{r+1}^\varphi$. Since $G_{r+1}(X) \equiv 0 \bmod \deg(r+1)$, (2) holds if and only if the coefficient α of X^i in E_{r+1} satisfies the equation

$$(4) \qquad -\pi\beta + \pi\alpha - \bar{\pi}^{r+1}\alpha^\varphi = 0$$

for every monomial X^i of degree $r+1$. This equation has a unique solution α in $\mathcal{O}_{\widehat{K}}$, which actually belongs to $\mathcal{O}$. For if we put $\gamma = \pi^{-1}\bar{\pi}^{r+1}$, we obtain the equation

$$\alpha - \gamma\alpha^\varphi = \beta\,,$$

which is clearly solved by the series

$$\alpha = \beta + \gamma\beta^\varphi + \gamma^{1+\varphi}\beta^{\varphi^2} + \cdots \in \mathcal{O}$$

(the series converges because $v_{\widehat{K}}(\gamma) \ge 1$). If α' is another solution, then $\alpha - \alpha' = \gamma(\alpha^\varphi - \alpha'^\varphi)$, hence $v_{\widehat{K}}(\alpha - \alpha') = v_{\widehat{K}}(\gamma) + v_{\widehat{K}}((\alpha - \alpha')^\varphi) = v_{\widehat{K}}(\gamma) + v_{\widehat{K}}(\alpha - \alpha')$, i.e., $v_{\widehat{K}}(\alpha - \alpha') = \infty$ because $v_{\widehat{K}}(\gamma) \ge 1$, and therefore $\alpha = \alpha'$. As a consequence, for every monomial X^i of degree $r+1$, equation (4) has a unique solution α in $\mathcal{O}$, i.e., there exists a unique $E_{r+1}(X) \in \mathcal{O}[X]$ satisfying (2). This finishes the proof. □

(2.3) Corollary. *Let π and $\bar{\pi}$ be prime elements of K, and let $e \in \mathcal{E}_\pi$, $\bar{e} \in \mathcal{E}_{\bar{\pi}}$ be Lubin-Tate series with coefficients in $\mathcal{O}_K$. Let $\pi = u\bar{\pi}$, $u \in U_K$, and $u = \varepsilon^{\varphi-1}$, $\varepsilon \in U_{\widehat{K}}$. Then there is a uniquely determined power series $\theta(X) \in \mathcal{O}_{\widehat{K}}[[X]]$ such that $\theta(X) \equiv \varepsilon X \bmod \deg 2$ and*

$$e \circ \theta = \theta^\varphi \circ \bar{e}\,.$$

Furthermore, there is a uniquely determined power series $[u](X) \in \mathcal{O}_K[[X]]$ such that $[u](X) \equiv uX \bmod \deg 2$ and

$$\bar{e} \circ [u] = [u] \circ \bar{e}\,.$$

They satisfy

$$\theta^\varphi = \theta \circ [u]\,.$$

Proof: Putting $L(X) = \varepsilon X$, we have $\pi L(X) = \bar{\pi} L^\varphi(X)$ and the first claim follows immediately from (2.2). In the same way, with the linear form $L(X) = uX$, one obtains the existence and uniqueness of the power series $[u](X) \in \mathcal{O}_K[[X]]$. Finally, defining $\theta_1 = \theta^{\varphi^{-1}} \circ [u]$, we get

$$e \circ \theta_1 = (e \circ \theta)^{\varphi^{-1}} \circ [u] = (\theta^\varphi \circ \bar{e})^{\varphi^{-1}} \circ [u] = (\theta^{\varphi^{-1}} \circ [u])^\varphi \circ \bar{e} = \theta_1^\varphi \circ \bar{e},$$

and thus $\theta_1 = \theta$ because of uniqueness. Hence $\theta^\varphi = \theta \circ [u]$. □

(2.4) Theorem. *Let $a = up^{v_p(a)} \in \mathbb{Q}_p^*$, and let ζ be a primitive p^n-th root of unity. Then one has*

$$\big(a, \mathbb{Q}_p(\zeta)|\mathbb{Q}_p\big)\,\zeta = \zeta^{u^{-1}}.$$

Proof: As $\mathbb{N}$ is dense in $\mathbb{Z}_p$, we may assume that $u \in \mathbb{N}$, $(u,p) = 1$. Let $K = \mathbb{Q}_p$, $L = \mathbb{Q}_p(\zeta)$, and let $\sigma \in G(L|K)$ be the automorphism defined by

$$\zeta^\sigma = \zeta^{u^{-1}}.$$

Since $\mathbb{Q}_p(\zeta)|\mathbb{Q}_p$ is totally ramified, we have $G(L|K) \cong G(\widetilde{L}|\widetilde{K})$, and we view σ as an element of $G(\widetilde{L}|K)$. Then $\tilde{\sigma} = \sigma\varphi_L \in \mathrm{Frob}(\widetilde{L}|K)$ is an element such that $d_K(\tilde{\sigma}) = 1$ and $\tilde{\sigma}|_L = \sigma$. The fixed field Σ of $\tilde{\sigma}$ is totally ramified because $f_{\Sigma|K} = d_K(\tilde{\sigma}) = 1$ by chap. IV, (4.5). The proof of the theorem is based on the fact that the field Σ can be explicitly generated by a prime element π_Σ which is given by the power series θ of (2.3).

In order to do this, assume $\tilde{\sigma}$ and $\varphi = \varphi_L$ have been extended continuously to the completion $\widehat{L}$ of $\widetilde{L}$, and consider the two Lubin-Tate polynomials

$$e(X) = upX + X^p \quad \text{and} \quad \bar{e}(X) = (1+X)^p - 1$$

as well as the polynomial $[u](X) = (1+X)^u - 1$. Then $\bar{e}([u](X)) = (1+X)^{up} - 1 = [u](\bar{e}(X))$. By (2.3), there is a power series $\theta(X) \in \mathcal{O}_{\widehat{K}}[[X]]$ such that

$$e \circ \theta = \theta^\varphi \circ \bar{e} \quad \text{and} \quad \theta^\varphi = \theta \circ [u].$$

Substituting the prime element $\lambda = \zeta - 1$ of L, we obtain a prime element of Σ by

$$\pi_\Sigma = \theta(\lambda).$$

Indeed, $[u](\lambda^\sigma) = (1+\lambda^\sigma)^u - 1 = \zeta^{\sigma u} - 1 = \zeta - 1 = \lambda$, and therefore

$$\pi_\Sigma^{\tilde{\sigma}} = \theta^\varphi(\lambda^\sigma) = \theta\big([u](\lambda^\sigma)\big) = \theta(\lambda) = \pi_\Sigma,$$

i.e., $\pi_\Sigma \in \Sigma$. We will show that

$$P(X) = e^{n-1}(X)^{p-1} + up \in \mathbb{Z}_p[X]$$

is the minimal polynomial of π_Σ, where $e^i(X)$ is defined by $e^0(X) = X$ and $e^i(X) = e(e^{i-1}(X))$. $P(X)$ is monic of degree $p^{n-1}(p-1)$ and irreducible by Eisenstein's criterion, as $e(X) \equiv X^p \bmod p$, and so $e^{n-1}(X)^{p-1} \equiv X^{p^{n-1}(p-1)} \bmod p$. Finally, $e^n(X) = e^{n-1}(X)\cdot(up + e^{n-1}(X)^{p-1}) = e^{n-1}(X)P(X)$, so that

$$P(\pi_\Sigma)e^{n-1}(\pi_\Sigma) = e^n(\pi_\Sigma).$$

Since $e^i(\pi_\Sigma) = e^i(\theta(\lambda)) = \theta^{\varphi^i}(\bar{e}^i(\lambda)) = \theta^{\varphi^i}((1+\lambda)^{p^i} - 1) = \theta^{\varphi^i}(\zeta^{p^i} - 1)$, we have $e^n(\pi_\Sigma) = 0$, $e^{n-1}(\pi_\Sigma) \neq 0$, and thus $P(\pi_\Sigma) = 0$.

Observing that $N_{L|K}(\zeta - 1) = (-1)^d p$, $d = [L:K]$ (see chap. II, (7.13)), we obtain

$$N_{\Sigma|K}(\pi_\Sigma) = (-1)^d P(0) = (-1)^d pu \equiv u \bmod N_{L|K}L^*$$

and therefore $r_{L|K}(\sigma) = u \bmod N_{L|K}L^*$, i.e., $(u, L|K) = (a, L|K) = \sigma$, as required. □

In order to really understand this proof of theorem (2.4), one has to read §4. Let us note that one would get a direct, purely algebraic proof, if one could show without using the power series θ that the splitting field of the polynomial $e^n(X)$ is abelian, and that its elements are all fixed under $\tilde{\sigma} = \sigma\varphi_L$. This splitting field would then have to be equal to the field Σ and every zero of $P(X) = e^n(X)/e^{n-1}(X)$ would have to be a prime element $\pi_\Sigma \in \Sigma$ such that $N_{\Sigma|K}(\pi_\Sigma) \equiv u \bmod N_{L|K}L^*$, in which case $r_{L|K}(\sigma) \equiv u \bmod N_{L|K}L^*$, and so $(u, L|K) = \sigma$.

Exercise 1. The p-class field theory $(d : G_{\mathbb{Q}_p} \to \mathbb{Z}_p, v : \mathbb{Q}_p^* \to \mathbb{Z})$ for the unramified $\mathbb{Z}_p$-extension of $\mathbb{Q}_p$, and the p-class field theory $(\hat{d} : G_{\mathbb{Q}_p} \to \mathbb{Z}_p, \hat{v} : \mathbb{Q}_p^* \to \mathbb{Z}_p)$ for the cyclotomic $\mathbb{Z}_p$-extension of $\mathbb{Q}_p$ (see § 1, exercise 7) yield the same norm residue symbol $(\ , L|K)$.

Hint: Show that this statement is equivalent to formula (2.4): $(u, \mathbb{Q}_p(\zeta)|\mathbb{Q}_p)\zeta = \zeta^{u^{-1}}$.

Exercise 2. Let $L|K$ be a totally ramified Galois extension, and let $\widehat{L}$ (resp. $\widehat{K}$) be the completion of the maximal unramified extension $\widetilde{L}$ (resp. $\widetilde{K}$) of L (resp. K). Show that $N_{\hat{L}|\hat{K}}\widehat{L}^* = \widehat{K}^*$, and that every $y \in \widehat{L}^*$ with $N_{\hat{L}|\hat{K}}(y) = 1$ is of the form $y = \prod_i z_i^{\sigma_i - 1}$, $\sigma_i \in G(L|K)$.

Exercise 3 (Theorem of Dwork). Let $L|K$ be a totally ramified abelian extension of $\mathfrak{p}$-adic number fields. Let $x \in K^*$ and $y \in \widehat{L}^*$ such that $N_{\hat{L}|\hat{K}}(y) = x$. Let $z_i \in \widehat{L}^*$ and choose $\sigma_i \in G(L|K)$ such that

$$y^{\varphi_K - 1} = \prod_i z_i^{\sigma_i - 1}.$$

Putting $\sigma = \prod_i \sigma_i$, one has $(x, L|K) = \sigma^{-1}$.

Hint: See chap. IV, § 5, exercise 1.

Exercise 4. Deduce from exercises 2 and 3 the formula $(u, \mathbb{Q}_p(\zeta)|\mathbb{Q}_p)\zeta = \zeta^{u^{-1}}$, for some p^n-th root of unity ζ.

§ 3. The Hilbert Symbol

Let K be a local field, or $K = \mathbb{R}$, $K = \mathbb{C}$. We assume that K contains the group μ_n of n-th roots of unity, where n is a natural number which is relatively prime to the characteristic of K (i.e., n can be arbitrary if $\mathrm{char}(K) = 0$). Over such a field K we then have at our disposal, on the one hand, Kummer theory (see chap. IV, § 3), and on the other, class field theory. It is the interplay between both theories, which gives rise to the "Hilbert symbol". This is a highly remarkable phenomenon which will lead us to a generalization of the classical reciprocity law of Gauss, to n-th power residues.

Let $L = K(\sqrt[n]{K^*})$ be the maximal abelian extension of exponent n. By (1.5), we then have

$$N_{L|K}L^* = K^{*n},$$

and class field theory gives us the canonical isomorphism

$$G(L|K) \cong K^*/K^{*n}.$$

On the other hand, Kummer theory gives the canonical isomorphism

$$\mathrm{Hom}(G(L|K),\mu_n) \cong K^*/K^{*n}.$$

The bilinear map

$$G(L|K) \times \mathrm{Hom}(G(L|K),\mu_n) \longrightarrow \mu_n, \quad (\sigma,\chi) \longmapsto \chi(\sigma),$$

therefore defines a nondegenerate bilinear pairing

$$\left(\frac{\ ,\ }{\mathfrak{p}}\right) : K^*/K^{*n} \times K^*/K^{*n} \longrightarrow \mu_n$$

(bilinear in the multiplicative sense). This pairing is called the **Hilbert symbol**. Its relation to the norm residue symbol is described explicitly in the following proposition.

(3.1) Proposition. *For $a,b \in K^*$, the Hilbert symbol $\left(\frac{a,b}{\mathfrak{p}}\right) \in \mu_n$ is given by*

$$\left(a, K(\sqrt[n]{b})|K\right)\sqrt[n]{b} = \left(\frac{a,b}{\mathfrak{p}}\right)\sqrt[n]{b}.$$

Proof: The image of a under the isomorphism $K^*/K^{*n} \cong G(L|K)$ of class field theory is the norm residue symbol $\sigma = (a, L|K)$. The image of b under the isomorphism $K^*/K^{*n} \cong \mathrm{Hom}(G(L|K),\mu_n)$ of Kummer theory is the character $\chi_b : G(L|K) \to \mu_n$ given by $\chi_b(\tau) = \tau\sqrt[n]{b}/\sqrt[n]{b}$. By definition of the Hilbert symbol, we have

$$\left(\frac{a,b}{\mathfrak{p}}\right) = \chi_b(\sigma) = \sigma\sqrt[n]{b}\,/\sqrt[n]{b},$$

hence $(a, K(\sqrt[n]{b})|K)\sqrt[n]{b} = (a, L|K)\sqrt[n]{b} = \left(\frac{a,b}{\mathfrak{p}}\right)\sqrt[n]{b}$. □

The Hilbert symbol has the following fundamental properties:

(3.2) Proposition.

(i) $\left(\frac{aa',b}{\mathfrak{p}}\right) = \left(\frac{a,b}{\mathfrak{p}}\right)\left(\frac{a',b}{\mathfrak{p}}\right)$,

(ii) $\left(\frac{a,bb'}{\mathfrak{p}}\right) = \left(\frac{a,b}{\mathfrak{p}}\right)\left(\frac{a,b'}{\mathfrak{p}}\right)$,

(iii) $\left(\frac{a,b}{\mathfrak{p}}\right) = 1 \iff$ *a is a norm from the extension $K(\sqrt[n]{b})|K$,*

(iv) $\left(\frac{a,b}{\mathfrak{p}}\right) = \left(\frac{b,a}{\mathfrak{p}}\right)^{-1}$,

(v) $\left(\frac{a,1-a}{\mathfrak{p}}\right) = 1$ *and* $\left(\frac{a,-a}{\mathfrak{p}}\right) = 1$,

(vi) *If* $\left(\frac{a,b}{\mathfrak{p}}\right) = 1$ *for all $b \in K^*$, then $a \in K^{*n}$.*

Proof: (i) and (ii) are clear from the definition, (iii) follows from (3.1), and (vi) reformulates the nondegenerateness of the Hilbert symbol.

If $b \in K^*$ and $x \in K$ such that $x^n - b \neq 0$, then

$$x^n - b = \prod_{i=0}^{n-1}(x - \zeta^i\beta), \quad \beta^n = b,$$

for some primitive n-th root of unity ζ. Let d be the greatest divisor of n such that $y^d = b$ has a solution in K, and let $n = dm$. Then the extension $K(\beta)|K$ is cyclic of degree m, and the conjugates of $x - \zeta^i\beta$ are the elements $x - \zeta^j\beta$ such that $j \equiv i \bmod d$. We may therefore write

$$x^n - b = \prod_{i=0}^{d-1} N_{K(\beta)|K}(x - \zeta^i\beta).$$

Hence $x^n - b$ is a norm from $K(\sqrt[n]{b})|K$, i.e.,

$$\left(\frac{x^n - b, b}{\mathfrak{p}}\right) = 1.$$

Choosing $x = 1$, $b = 1 - a$, and $x = 0$, $b = -a$ then yield (v). (iv) finally follows from

$$\begin{aligned}\left(\frac{a,b}{\mathfrak{p}}\right)\left(\frac{b,a}{\mathfrak{p}}\right) &= \left(\frac{a,-a}{\mathfrak{p}}\right)\left(\frac{a,b}{\mathfrak{p}}\right)\left(\frac{b,a}{\mathfrak{p}}\right)\left(\frac{b,-b}{\mathfrak{p}}\right)\\ &= \left(\frac{a,-ab}{\mathfrak{p}}\right)\left(\frac{b,-ab}{\mathfrak{p}}\right) = \left(\frac{ab,-ab}{\mathfrak{p}}\right) = 1.\end{aligned}$$ □

In the case $K = \mathbb{R}$ we have $n = 1$ or $n = 2$. For $n = 1$ one finds, of course, $\left(\frac{a,b}{\mathfrak{p}}\right) = 1$, and for $n = 2$ we have

$$\left(\frac{a,b}{\mathfrak{p}}\right) = (-1)^{\frac{\operatorname{sgn} a - 1}{2} \cdot \frac{\operatorname{sgn} b - 1}{2}},$$

because $(a, \mathbb{R}(\sqrt{b})|\mathbb{R}) = 1$ for $b > 0$, and $= (-1)^{\frac{\operatorname{sgn} a - 1}{2}}$ for $b < 0$. Here the letter $\mathfrak{p}$ symbolically stands for an infinite place.

Next we determine the Hilbert symbol explicitly in the case where K is a local field ($\neq \mathbb{R}, \mathbb{C}$) whose residue characteristic p does not divide n. We call this the case of the **tame Hilbert symbol**. Since $\mu_n \subseteq \mu_{q-1}$ one has $n \,|\, q - 1$ in that case. First we establish the

(3.3) Lemma. *Let $(n, p) = 1$ and $x \in K^*$. The extension $K(\sqrt[n]{x})|K$ is unramified if and only if $x \in U_K K^{*n}$.*

Proof: Let $x = uy^n$ with $u \in U_K$, $y \in K^*$, so that $K(\sqrt[n]{x}) = K(\sqrt[n]{u})$. Let κ' be the splitting field of the polynomial $X^n - u \bmod \mathfrak{p}$ over the residue class field κ, and let $K'|K$ be the unramified extension with residue class field κ' (see chap. II, § 9, p. 173). By Hensel's lemma, $X^n - u$ splits over K' into linear factors, so $K(\sqrt[n]{u}) \subseteq K'$ is unramified. Assume conversely that $L = K(\sqrt[n]{x})$ is unramified over K, and let $x = u\pi^r$, where $u \in U_K$ and π is a prime element of K. Then $v_L(\sqrt[n]{u\pi^r}) = \frac{1}{n} v_L(\pi^r) = \frac{r}{n} \in \mathbb{Z}$, hence $n|r$, i.e., $\pi^r \in K^{*n}$, and thus $x \in U_K K^{*n}$. □

Since $U_K = \mu_{q-1} \times U_K^{(1)}$, every unit $u \in U_K$ has a unique decomposition

$$u = \omega(u)\langle u \rangle$$

with $\omega(u) \in \mu_{q-1}$ and $\langle u \rangle \in U_K^{(1)}$, $u \equiv \omega(u) \bmod \mathfrak{p}$. With this notation we will now prove the

(3.4) Proposition. *If $(n, p) = 1$ and $a, b \in K^*$, then*

$$\left(\frac{a,b}{\mathfrak{p}}\right) = \omega\left((-1)^{\alpha\beta} \frac{b^\alpha}{a^\beta}\right)^{(q-1)/n},$$

where $\alpha = v_K(a)$, $\beta = v_K(b)$.

Proof: The function

$$\langle a,b\rangle := \omega\Big((-1)^{\alpha\beta}\frac{b^\alpha}{a^\beta}\Big)^{(q-1)/n}$$

is obviously bilinear (in the multiplicative sense). We may therefore assume that a and b are prime elements: $a=\pi$, $b=-\pi u$, $u\in U_K$. Since clearly $\langle\pi,-\pi\rangle = \big(\frac{\pi,\,-\pi}{\mathfrak p}\big)=1$, we may restrict to the case $a=\pi$, $b=u$. Let $y=\sqrt[n]{u}$ and $K'=K(y)$. Then we have

$$\langle\pi,u\rangle = \omega(u)^{(q-1)/n} \quad\text{and}\quad (\pi,K'|K)y = \Big(\frac{\pi,u}{\mathfrak p}\Big)y\,.$$

By (3.3), we see that $K'|K$ is unramified and by chap. IV, (5.7), $(\pi,K'|K)$ is the Frobenius automorphism $\varphi=\varphi_{K'|K}$. Consequently,

$$\Big(\frac{\pi,u}{\mathfrak p}\Big) = \frac{\varphi y}{y} \equiv y^{q-1}\equiv u^{(q-1)/n}\equiv \omega(u)^{(q-1)/n}\equiv\langle\pi,u\rangle \bmod \mathfrak p\,,$$

hence $\big(\frac{\pi,u}{\mathfrak p}\big)=\langle\pi,u\rangle$, because μ_{q-1} is mapped isomorphically onto κ^* by $U_K\to\kappa^*$. □

The proposition shows in particular that the Hilbert symbol

$$\Big(\frac{\pi,u}{\mathfrak p}\Big) = \omega(u)^{(q-1)/n}$$

(in the case $(n,p)=1$) is independent of the choice of the prime element π. We may therefore put

$$\Big(\frac{u}{\mathfrak p}\Big) := \Big(\frac{\pi,u}{\mathfrak p}\Big) \quad\text{for}\quad u\in U_K\,.$$

$\big(\frac{u}{\mathfrak p}\big)$ is the root of unity determined by

$$\Big(\frac{u}{\mathfrak p}\Big)\equiv u^{(q-1)/n} \bmod \mathfrak p_K\,.$$

We call it the **Legendre symbol**, or the n-th **power residue symbol**. Both names are justified by the

(3.5) Proposition. *Let $(n,p)=1$ and $u\in U_K$. Then one has*

$$\Big(\frac{u}{\mathfrak p}\Big)=1 \iff u \text{ is an } n\text{-th power} \bmod \mathfrak p_K\,.$$

Proof: Let ζ be a primitive $(q-1)$-th root of unity, and let $m = \frac{q-1}{n}$. Then ζ^n is a primitive m-th root of unity, and

$$\left(\frac{u}{\mathfrak{p}}\right) = \omega(u)^m = 1 \Longleftrightarrow \omega(u) \in \mu_m \Longleftrightarrow \omega(u) = (\zeta^n)^i$$
$$\Longleftrightarrow u \equiv \omega(u) = (\zeta^i)^n \bmod \mathfrak{p}_K . \qquad \square$$

It is an important, but in general difficult, problem to find explicit formulae for the Hilbert symbol $\left(\frac{a,b}{\mathfrak{p}}\right)$ also in the case $p|n$. Let us look at the case where $n = 2$ and $K = \mathbb{Q}_p$. If $a \in \mathbb{Z}_2$, then $(-1)^a$ means

$$(-1)^a = (-1)^r ,$$

where r is a rational integer $\equiv a \bmod 2$.

(3.6) Theorem. *Let $n = 2$. For $a, b \in \mathbb{Q}_p^*$ we write*

$$a = p^\alpha a', \quad b = p^\beta b', \quad a', b' \in U_{\mathbb{Q}_p} .$$

If $p \neq 2$, then

$$\left(\frac{a,b}{p}\right) = (-1)^{\frac{p-1}{2}\alpha\beta}\left(\frac{a'}{p}\right)^\beta \left(\frac{b'}{p}\right)^\alpha .$$

In particular, one has $\left(\frac{p,p}{p}\right) = (-1)^{(p-1)/2}$ and $\left(\frac{p,u}{p}\right) = \left(\frac{u}{p}\right)$, if u is a unit. If $p = 2$ and $a, b \in U_{\mathbb{Q}_2}$, then

$$\left(\frac{2,a}{2}\right) = (-1)^{(a^2-1)/8} ,$$
$$\left(\frac{a,b}{2}\right) = \left(\frac{b,a}{2}\right) = (-1)^{\frac{a-1}{2}\frac{b-1}{2}} .$$

Proof: The claim for the case $p \neq 2$ is an immediate consequence of (3.4), and will be left to the reader. So let $p = 2$. We put $\eta(a) = \frac{a^2-1}{8}$ and $\varepsilon(a) = \frac{a-1}{2}$. An elementary computation shows that

$$\eta(a_1 a_2) \equiv \eta(a_1) + \eta(a_2) \bmod 2 \quad \text{and} \quad \varepsilon(a_1 a_2) \equiv \varepsilon(a_1) + \varepsilon(a_2) \bmod 2 .$$

Thus both sides of the equations we have to prove are multiplicative and it is enough to check the claim for a set of generators of $U_{\mathbb{Q}_2}/U_{\mathbb{Q}_2}^2$. $\{5, -1\}$ is such a set. We postpone this for the moment and define $(a,b) = \left(\frac{a,b}{2}\right)$.

We have $(-1,x) = 1$ if and only if x is a norm from $\mathbb{Q}_2(\sqrt{-1})|\mathbb{Q}_2$, i.e., $x = y^2 + z^2$, $y, z \in \mathbb{Q}_2$. Since $5 = 4 + 1$ and $2 = 1 + 1$, we find that $(-1,2) = (-1,5) = 1$. If we had $(-1,-1) = 1$, then it would follow that $(-1,x) = 1$ for all x, i.e., -1 would be a square in $\mathbb{Q}_2^*$, which is not the case. Therefore we have $(-1,-1) = -1$.

We have $(2,2) = (2,-1) = 1$ and $(5,5) = (5,-1) = 1$. It remains therefore to determine $(2,5)$. $(2,5) = 1$ would imply $(2,x) = 1$ for all x, i.e., 2 would be a square in $\mathbb{Q}_2^*$, which is not the case. Hence $(2,5) = -1$.

By direct verification one sees that the values we just found coincide with those of $(-1)^{\eta(a)}$, resp. $(-1)^{\varepsilon(a)\varepsilon(b)}$, in the respective cases.

It remains to show that $U_{\mathbb{Q}_2}/U_{\mathbb{Q}_2}^2$ is generated by $\{5,-1\}$. We set $U = U_{\mathbb{Q}_2}$, $U^{(n)} = U_{\mathbb{Q}_2}^{(n)}$. By chap. II, (5.5), $\exp : 2^n\mathbb{Z}_2 \to U^{(n)}$ is an isomorphism for $n > 1$. Since $a \mapsto 2a$ defines an isomorphism $2^2\mathbb{Z}_2 \to 2^3\mathbb{Z}_2$, $x \mapsto x^2$ defines an isomorphism $U^{(2)} \to U^{(3)}$. It follows that $U^{(3)} \subseteq U^2$. Since $\{1,-1,5,-5\}$ is a system of representatives of $U/U^{(3)}$, U/U^2 is generated by -1 and 5. □

It is much more difficult to determine the n-th Hilbert symbol in the general case. It was discovered only in 1964 by the mathematician *Helmut Brückner*. Since the result has not previously been published in an easily accessible place, we state it here without proof for the case $n = p^\nu$ of odd residue characteristic p of K.

So let $\mu_{p^\nu} \subseteq K$, choose a prime element π of K, and let W be the ring of integers of the maximal unramified subextension T of $K|\mathbb{Q}_p$ (i.e., the ring of Witt vectors over the residue class field of K). Then every element $x \in K$ can be written in the form

$$x = f(\pi),$$

with a Laurent series $f(X) \in W((X))$.

For an arbitrary Laurent series $f(X) = \sum_{i \geq -m} a_i X^i \in W((X))$, let $f^{\mathcal{P}}(X)$ denote the series

$$f^{\mathcal{P}}(X) = \sum_i a_i^\varphi X^{ip},$$

where φ is the Frobenius automorphism of W. Further, let $\operatorname{Res}(f\,dX) \in W$ denote the residue of the differential $f\,dX$,

$$d\log f := \frac{f'}{f}\,dX,$$

and

$$\log f := \sum_{i=1}^{\infty} (-1)^{i+1} \frac{(f-1)^i}{i},$$

if $f \in 1 + pW[[X]]$.

Now let ζ be a primitive p^ν-th root of unity. Then $1-\zeta$ is a prime element of $\mathbb{Q}_p(\zeta)$, and thus

$$1-\zeta = \pi^e \varepsilon,$$

for some unit ε of K, where e is the ramification index of $K|\mathbb{Q}_p(\zeta)$. Let $\eta(X) \in W[[X]]$ be a power series such that

$$\varepsilon = \eta(\pi),$$

and let $h(X)$ be the series

$$h(X) = \frac{1}{2}\,\frac{1+(1-X^e\eta(X))^{p^\nu}}{1-(1-X^e\eta(X))^{p^\nu}} = \sum_{i=-\infty}^{\infty} a_i X^i\,,\quad a_i \in W\,,\quad \lim_{i\to-\infty} a_i = 0.$$

With this notation we can now state *Brückner*'s formula for the p^ν-th Hilbert symbol $\left(\frac{x,y}{\mathfrak{p}}\right)$, $p = \operatorname{char}(\kappa) \neq 2$.

(3.7) Theorem. *If $x, y \in K^*$ and $f, g \in W((X))^*$ such that $f(\pi) = x$ and $g(\pi) = y$, then*

$$\left(\frac{x,y}{\mathfrak{p}}\right) = \zeta^{w(x,y)}$$

where

$$w(x,y) = Tr_{W|\mathbb{Z}_p} \operatorname{Res} h\cdot\Big(\frac{1}{p}\log\frac{f^p}{f^{\mathcal{P}}}\,d\log g - \frac{1}{p}\log\frac{g^p}{g^{\mathcal{P}}}\,\frac{1}{p}\,d\log f^{\mathcal{P}}\Big) \bmod p^\nu.$$

For the proof of this theorem, we have to refer to [20] (see also [69] and [135]). *Brückner* has also deduced an explicit formula for the case $n = 2^\nu$, but it is much more complicated. A more recent treatment of the theorem, which also includes the case $n = 2^\nu$, has been given by *G. Henniart* [69].

It would be interesting to deduce from these formulae the following classical result of *Iwasawa* [80], *Artin* and *Hasse* (see [9]) relative to the field

$$\Phi_\nu = \mathbb{Q}_p(\zeta),$$

where ζ is a primitive p^ν-th root of unity ($p \neq 2$). Putting $\pi = 1-\zeta$ and denoting by S the trace map from Φ_ν to $\mathbb{Q}_p$, we obtain for the p^ν-th Hilbert symbol $\left(\frac{x,y}{\mathfrak{p}}\right)$ of the field Φ_ν the

(3.8) Proposition. *For $a \in U^{(2p^{\nu-1})}_{\Phi_\nu}$ and $b \in \Phi_\nu^*$ one has*

$$\left(\frac{a,b}{\mathfrak{p}}\right) = \zeta^{S(\zeta \log a \; D\log b)/p^\nu}, \tag{1}$$

where $D\log b$ denotes the formal logarithmic derivative in π of an arbitrary representation of b as an integral power series in π with coefficients in $\mathbb{Z}_p$.

For $a \in U^{(1)}_{\Phi_\nu}$ one has furthermore the two **supplementary theorems**

$$\left(\frac{\zeta,a}{\mathfrak{p}}\right) = \zeta^{S(\log a)/p^\nu}, \tag{2}$$

$$\left(\frac{a,\pi}{\mathfrak{p}}\right) = \zeta^{S((\zeta/\pi)\log a)/p^\nu}. \tag{3}$$

The supplementary theorems (2) and (3) go back to *Artin* and *Hasse* [9]. The formula (1) was proved independently by *Artin* [10] and *Hasse* [61] in the case $\nu = 1$, and by *Iwasawa* [80] in general. In the case $\nu = 1$, for instance, one can indeed obtain the formulae from *Brückner*'s theorem (3.7). Since

$$\frac{1}{p}S(\zeta\pi^i) \equiv \begin{cases} 1 \bmod p, & i = p-1, \\ 0 \bmod p, & i \neq p-1, \end{cases} \qquad \text{and} \quad \log a \equiv 0 \bmod \mathfrak{p}^2,$$

one may also interpret the ζ-exponent in the formulae (1)–(3) as the $(p-1)$-st coefficient of a π-adic expansion of $\log a \; D\log b$. In this way it appears as a formal residue $\operatorname{Res}_\pi \frac{1}{\pi^p}\log a \; D\log b$. As to the supplementary theorems, one has to define also $D\log\zeta = -\zeta^{-1}$, $D\log\pi = \pi^{-1}$.

Exercise 1. For $n = 2$ the Hilbert symbol has the following concrete meaning:

$$\left(\frac{a,b}{\mathfrak{p}}\right) = 1 \iff ax^2 + by^2 - z^2 = 0 \quad \text{has a nontrivial solution in } K.$$

Exercise 2. Deduce proposition (3.8) from theorem (3.7).

Exercise 3. Let K be a local field of characteristic p, let $\overline{K}$ be its separable closure, and let $W_n(\overline{K})$ be the ring of Witt vectors of length n, with the operator $\wp : W_n(\overline{K}) \to W_n(\overline{K})$, $\wp a = Fa - a$ (see chap. IV, §3, exercises 2 and 3). Show that one has $\ker(\wp) = W_n(\mathbb{F}_p)$.

Exercise 4. Abstract Kummer theory (chap. IV, (3.3)) yields for the maximal abelian extension $L|K$ of exponent n a surjective homomorphism

$$W_n(K) \to \operatorname{Hom}(G(L|K), W_n(\mathbb{F}_p)), \quad x \mapsto \chi_x,$$

where one has $\chi_x(\sigma) = \sigma\xi - \xi$ for all $\sigma \in G(L|K)$, with an arbitrary $\xi \in W_n(L)$ such that $\wp\xi = x$. Show that $x \mapsto \chi_x$ has kernel $\wp W_n(K)$.

Exercise 5. Define, for $x \in W_n(K)$ and $a \in K^*$, the symbol $[x,a) \in W_n(\mathbb{F}_p)$ by

$$[x,a) := \chi_x((a, L|K)),$$

where $(\ , L|K)$ is the norm residue symbol. Show:

(i) $[x,a) = (a, K(\xi)|K)\xi - \xi$, if $\xi \in W_n(\bar{K})$ with $\wp\xi = x$.

(ii) $[x+y,a) = [x,a) + [y,a)$.

(iii) $[x,ab) = [x,a) + [x,b)$.

(iv) $[x,a) = 0 \iff a \in N_{K(\xi)|K}K(\xi)^*$, where $\xi \in W_n(\bar{K})$ is an element such that $\wp\xi = x$.

(v) $[x,a) = 0$ for all $a \in K^* \iff x \in \wp W_n(K)$.

(vi) $[x,a) = 0$ for all $x \in W_n(K) \iff a \in K^{*p^n}$.

Exercise 6. Let κ be the residue class field of K and π a prime element such that $K = \kappa((\pi))$. Let

$$d : K \to \Omega^1_{K|\kappa}, \quad f \mapsto df,$$

be the canonical map to the differential module of $K|\kappa$ (see chap. III, § 2, p. 200). For every $f \in K$ one has

$$df = f'_\pi d\pi,$$

where f'_π is the formal derivative of f in the expansion according to powers of π with coefficients in κ. Show that for $\omega = (\sum_{i>-\infty} a_i\pi^i)d\pi$, the *residue*

$$\operatorname{Res}\omega := a_{-1}$$

does not depend on the choice of the prime element π.

Exercise 7. Show that in the case $n = 1$ the symbol $[x,a)$ is given by

$$[x,a) = Tr_{\kappa|\mathbb{F}_p}\operatorname{Res}\left(x\frac{da}{a}\right).$$

Remark: Such a formula can also be given for $n \geq 1$ (*P.* KÖLCZE [88]).

§ 4. Formal Groups

The most explicit realization of local class field theory we have encountered for the case of cyclotomic fields over the field $\mathbb{Q}_p$, i.e., with the extensions $\mathbb{Q}_p(\zeta)|\mathbb{Q}_p$, where ζ is a p^n-th root of unity. The notion of formal group allows us to construct such an explicit cyclotomic theory over an arbitrary local field K by introducing a new kind of roots of unity which are "division points" that do the same for the field K as the p^n-th roots of unity do for the field $\mathbb{Q}_p$.

(4.1) Definiton. *A (1-dimensional, commutative)* **formal group** *over a ring* $\mathcal{O}$ *is a formal power series* $F(X,Y) \in \mathcal{O}[[X,Y]]$ *with the following properties:*

(i) $F(X,Y) \equiv X + Y \bmod \deg 2$,

(ii) $F(X,Y) = F(Y,X)$ *"commutativity"*,

(iii) $F(X,F(Y,Z)) = F(F(X,Y),Z)$ *"associativity"*.

From a formal group one gets an ordinary group by evaluating in a domain where the power series converge. If for instance $\mathcal{O}$ is a complete valuation ring and $\mathfrak{p}$ its maximal ideal, then the operation

$$x \underset{F}{+} y := F(x,y)$$

defines a new structure of abelian group on the set $\mathfrak{p}$.

Examples:

1. $\mathbb{G}_a(X,Y) = X + Y$ (the formal additive group).

2. $\mathbb{G}_m(X,Y) = X + Y + XY$ (the formal multiplicative group). Since

$$X + Y + XY = (1+X)(1+Y) - 1,$$

we have

$$(x \underset{\mathbb{G}_m}{+} y) + 1 = (x+1) \cdot (y+1)\,.$$

So the new operation $\underset{\mathbb{G}_m}{+}$ is obtained from multiplication $\cdot$ via the translation $x \mapsto x + 1$.

3. A power series $f(X) = a_1X + a_2X^2 + \cdots \in \mathcal{O}[[X]]$ whose first coefficient a_1 is a unit admits an "inverse", i.e., there exists a power series

$$f^{-1}(X) = a_1^{-1}X + \cdots \in \mathcal{O}[[X]],$$

such that $f^{-1}(f(X)) = f(f^{-1}(X)) = X$. For every such power series,

$$F(X,Y) = f^{-1}\big(f(X) + f(Y)\big)$$

is a formal group.

(4.2) Definition. *A* **homomorphism** $f : F \to G$ *between two formal groups is a power series* $f(X) = a_1X + a_2X^2 + \cdots \in \mathcal{O}[[X]]$ *such that*

$$f(F(X,Y)) = G\big(f(X), f(Y)\big)\,.$$

In example 3, for instance, the power series f is a homomorphism of the formal group F to the additive group $\mathbb{G}_a$. It is called the *logarithm* of F.

A homomorphism $f : F \to G$ is an *isomorphism* if $a_1 = f'(0)$ is a unit, i.e., if there is a homomorphism $g = f^{-1} : G \to F$ such that

$$f\big(g(X)\big) = g\big(f(X)\big) = X .$$

If the power series $f(X) = a_1X + a_2X^2 + \cdots$ satisfies the equation $f(F(X,Y)) = G(f(X), f(Y))$, but its coefficients belong to an extension ring $\mathcal{O}'$, then we call this a homomorphism *defined over* $\mathcal{O}'$. The following proposition is immediately evident.

(4.3) Proposition. *The homomorphisms $f : F \to F$ of a formal group F over $\mathcal{O}$ form a ring $\operatorname{End}_{\mathcal{O}}(F)$ in which addition and multiplication are defined by*

$$(f \underset{F}{+} g)(X) = F\big(f(X), g(X)\big), \quad (f \circ g)(X) = f\big(g(X)\big).$$

(4.4) Definition. *A* **formal** $\mathcal{O}$**-module** *is a formal group F over $\mathcal{O}$ together with a ring homomorphism*

$$\mathcal{O} \longrightarrow \operatorname{End}_{\mathcal{O}}(F), \quad a \longmapsto [a]_F(X),$$

such that $[a]_F(X) \equiv aX \bmod \deg 2$.

A homomorphism (over $\mathcal{O}' \supseteq \mathcal{O}$) *between formal $\mathcal{O}$-modules F, G is a homomorphism $f : F \to G$ of formal groups* (over $\mathcal{O}'$) *in the sense of (4.2) such that*

$$f\big([a]_F(X)\big) = [a]_G\big(f(X)\big) \quad \textit{for all} \quad a \in \mathcal{O}.$$

Now let $\mathcal{O} = \mathcal{O}_K$ be the valuation ring of a local field K, and write $q = (\mathcal{O}_K : \mathfrak{p}_K)$. We consider the following special formal $\mathcal{O}_K$-modules.

(4.5) Definition. *A* **Lubin-Tate module** *over $\mathcal{O}_K$ for the prime element π is a formal $\mathcal{O}_K$-module F such that*

$$[\pi]_F(X) \equiv X^q \bmod \pi .$$

This definition reflects once more the dominating principle of class field theory, to the effect that prime elements correspond to Frobenius elements. In fact, if we reduce the coefficients of some formal $\mathcal{O}$-module F modulo π, we obtain a formal group $\overline{F}(X,Y)$ over the residue class field $\mathbb{F}_q$. The reduction mod π of $[\pi]_F(X)$ is an endomorphism of $\overline{F}$. But on the other hand, $f(X) = X^q$ is clearly an endomorphism of $\overline{F}$, its *Frobenius endomorphism*. Thus F is a Lubin-Tate module if the endomorphism defined by a prime element π gives via reduction the Frobenius endomorphism of $\overline{F}$.

Example: The formal multiplicative group $\mathbb{G}_m$ is a formal $\mathbb{Z}_p$-module with respect to the mapping

$$\mathbb{Z}_p \to \operatorname{End}_{\mathbb{Z}_p}(\mathbb{G}_m), \quad a \mapsto [a]_{\mathbb{G}_m}(X) = (1+X)^a - 1 = \sum_{\nu=1}^{\infty} \binom{a}{\nu} X^\nu .$$

$\mathbb{G}_m$ is a Lubin-Tate module for the prime element p because

$$[p]_{\mathbb{G}_m}(X) = (1+X)^p - 1 \equiv X^p \bmod p .$$

The following theorem gives a complete and explicit overall view of the totality of all Lubin-Tate modules. Let $e(X), \bar{e}(X) \in \mathcal{O}_K[[X]]$ be Lubin-Tate series for the prime element π of K, and let

$$F_e(X,Y) \in \mathcal{O}_K[[X,Y]] \quad \text{and} \quad [a]_{e,\bar{e}}(X) \in \mathcal{O}_K[[X]]$$

($a \in \mathcal{O}_K$) be the power series (uniquely determined according to (2.2)) such that

$$F_e(X,Y) \equiv X + Y \bmod \deg 2, \qquad e\big(F_e(X,Y)\big) = F_e\big(e(X), e(Y)\big),$$

$$[a]_{e,\bar{e}}(X) \equiv aX \bmod \deg 2, \qquad e\big([a]_{e,\bar{e}}(X)\big) = [a]_{e,\bar{e}}\big(\bar{e}(X)\big).$$

If $e(X) = \bar{e}(X)$ we simply write $[a]_{e,\bar{e}}(X) = [a]_e(X)$.

(4.6) Theorem. (i) *The Lubin-Tate modules for π are precisely the series $F_e(X,Y)$, with the formal $\mathcal{O}_K$-module structure given by*

$$\mathcal{O}_K \longrightarrow \operatorname{End}_{\mathcal{O}_K}(F_e), \quad a \longmapsto [a]_e(X).$$

(ii) *For every $a \in \mathcal{O}_K$ the power series $[a]_{e,\bar{e}}(X)$ is a homomorphism*

$$[a]_{e,\bar{e}} : F_{\bar{e}} \longrightarrow F_e$$

of formal $\mathcal{O}$-modules, and it is an isomorphism if a is a unit.

Proof: If F is any Lubin-Tate module, then $e(X) := [\pi]_F(X) \in \mathcal{E}_\pi$ and $F = F_e$ because $e(F(X,Y)) = F(e(X),e(Y))$, and because of the uniqueness statement of (2.2). For the other claims of the theorem one has to show the following formulae.

(1) $F_e(X,Y) = F_e(Y,X)$,

(2) $F_e(X,F_e(Y,Z)) = F_e(F_e(X,Y),Z)$,

(3) $[a]_{e,\bar{e}}(F_{\bar{e}}(X,Y)) = F_e([a]_{e,\bar{e}}(X),[a]_{e,\bar{e}}(Y))$,

(4) $[a+b]_{e,\bar{e}}(X) = F_e([a]_{e,\bar{e}}(X),[b]_{e,\bar{e}}(X))$,

(5) $[ab]_{e,\bar{\bar{e}}}(X) = [a]_{e,\bar{e}}([b]_{\bar{e},\bar{\bar{e}}}(X))$,

(6) $[\pi]_e(X) = e(X)$.

(1) and (2) show that F_e is a formal group. (3), (4), and (5) show that

$$\mathcal{O}_K \longrightarrow \operatorname{End}_{\mathcal{O}_K}(F_e), \quad a \longmapsto [a]_e,$$

is a homomorphism of rings, i.e., that F_e is a formal $\mathcal{O}_K$-module, and that $[a]_{e,\bar{e}}$ is a homomorphism of formal $\mathcal{O}_K$-modules from $F_{\bar{e}}$ to F_e. Finally, (6) shows that F_e is a Lubin-Tate module.

The proofs of these formulae all follow the same pattern. One checks that both sides of each formula are solutions of the same problem of (2.2), and then deduces their equality from the uniqueness statement. In (6) for instance, both power series commence with the linear form πX and satisfy the condition $e([\pi]_e(X)) = [\pi]_e(e(X))$, resp. $e(e(X)) = e(e(X))$. □

Exercise 1. $\operatorname{End}_{\mathcal{O}}(\mathbb{G}_a)$ consists of all aX such that $a \in \mathcal{O}$.

Exercise 2. Let R be a commutative $\mathbb{Q}$-algebra. Then for every formal group $F(X,Y)$ over R, there exists a unique isomorphism

$$\log_F : F \xrightarrow{\sim} \mathbb{G}_a,$$

such that $\log_F(X) \equiv X \bmod \deg 2$, the **logarithm** of F.

Hint: Let $F_1 = \partial F/\partial Y$. Differentiating $F(F(X,Y),Z) = F(X,F(Y,Z))$ yields $F_1(X,0) \equiv 1 \bmod \deg 1$. Let $\psi(X) = 1 + \sum_{n=1}^\infty a_n X^n \in R[[X]]$ be the power series such that $\psi(X)F_1(X,0) = 1$. Then $\log_F(X) = X + \sum_{n=1}^\infty \frac{a_n}{n} X^n$ does what we want.

Exercise 3. $\log_{\mathbb{G}_m}(X) = \sum\limits_{n=1}^{\infty}(-1)^{n+1}\dfrac{X^n}{n} = \log(1+X)$.

Exercise 4. Let π be a prime element of the local field K, and let $f(X) = X + \pi^{-1}X^q + \pi^{-2}X^{q^2} + \cdots$ Then

$$F(X,Y) = f^{-1}(f(X)+f(Y)), \quad [a]_F(X) = f^{-1}(af(X)), \quad a \in \mathcal{O}_K,$$

defines a Lubin-Tate module with logarithm $\log_F = f$.

Exercise 5. Two Lubin-Tate modules over the valuation ring $\mathcal{O}_K$ of a local field K, but for different prime elements π and $\bar{\pi}$, are never isomorphic.

Exercise 6. Two Lubin-Tate modules F_e and $F_{\bar{e}}$ for prime elements π and $\bar{\pi}$ always become isomorphic over $\mathcal{O}_{\widehat{K}}$, where $\widehat{K}$ is the completion of the maximal unramified extension $\widetilde{K}|K$.

Hint: The power series θ of (2.3) yields an isomorphism $\theta : F_{\bar{e}} \to F_e$.

§ 5. Generalized Cyclotomic Theory

Formal groups are relevant for local class field theory in that they allow us to construct a perfect analogue of the theory of the p^n-th cyclotomic field $\mathbb{Q}_p(\zeta)$ over $\mathbb{Q}_p$, with its fundamental isomorphism

$$G(\mathbb{Q}_p(\zeta)|\mathbb{Q}_p) \xrightarrow{\sim} (\mathbb{Z}/p^n\mathbb{Z})^*$$

(see chap. II (7.13)), replacing $\mathbb{Q}_p$ by an arbitrary local ground field K. The formal groups furnish a generalization of the notion of p^n-th root of unity, and provide an explicit version of the local reciprocity law in the corresponding extensions.

A formal $\mathcal{O}_K$-module gives rise to an ordinary $\mathcal{O}_K$-module if we read the power series over a domain in which they converge. We now choose for this the maximal ideal $\bar{\mathfrak{p}}$ of the valuation ring of the algebraic closure $\bar{K}$ of the given local field K. If $G(X_1, \ldots, X_n) \in \mathcal{O}_K[[X_1, \ldots, X_n]]$ is a power series with constant coefficient 0, and if $x_1, \ldots, x_n \in \bar{\mathfrak{p}}$, then the series $G(x_1, \ldots, x_n)$ converges in the complete field $K(x_1, \ldots, x_n)$ to an element in $\bar{\mathfrak{p}}$. From the definition of the formal $\mathcal{O}$-modules and their homomorphisms we therefore obtain immediately the

(5.1) Proposition. *Let F be a formal $\mathcal{O}_K$-module. Then the set $\bar{\mathfrak{p}}$ with the operations*

$$x \underset{F}{+} y = F(x, y) \quad \text{and} \quad a \cdot x = [a]_F(x),$$

$x, y \in \bar{\mathfrak{p}}$, $a \in \mathcal{O}_K$, is an $\mathcal{O}_K$-module in the usual sense. We denote it by $\bar{\mathfrak{p}}_F$.

If $f : F \to G$ is a homomorphism (isomorphism) of formal $\mathcal{O}_K$-modules, then

$$f : \overline{\mathfrak{p}}_F \longrightarrow \overline{\mathfrak{p}}_G, \quad x \longmapsto f(x),$$

is a homomorphism (isomorphism) of ordinary $\mathcal{O}_K$-modules.

The operations in $\overline{\mathfrak{p}}_F$, and particularly scalar multiplication $a \bullet x = [a]_F(x)$, must of course not be confused with the usual operations in the field $\overline{K}$.

We now consider a Lubin-Tate module F for the prime element π of $\mathcal{O}_K$. We define the group of π^n-**division points** by

$$F(n) = \{\lambda \in \overline{\mathfrak{p}}_F \mid \pi^n \bullet \lambda = 0\} = \ker([\pi^n]_F).$$

This is an $\mathcal{O}_K$-module, and an $\mathcal{O}_K/\pi^n\mathcal{O}_K$-module because it is killed by $\pi^n\mathcal{O}_K$.

(5.2) Proposition. *$F(n)$ is a free $\mathcal{O}_K/\pi^n\mathcal{O}_K$-module of rank* 1.

Proof: An isomorphism $f : F \to G$ of Lubin-Tate modules obviously induces isomorphisms $f : \overline{\mathfrak{p}}_F \to \overline{\mathfrak{p}}_G$ and $f : F(n) \to G(n)$ of $\mathcal{O}_K$-modules. By (4.6), Lubin-Tate modules for the same prime element π are all isomorphic. We may therefore assume that $F = F_e$, with $e(X) = X^q + \pi X = [\pi]_F(X)$. $F(n)$ then consists of the q^n zeroes of the iterated polynomial $e^n(X) = (e \circ \cdots \circ e)(X) = [\pi^n]_F(X)$, which is easily shown, by induction on n, to be separable. Now if $\lambda_n \in F(n) \smallsetminus F(n-1)$, then

$$\mathcal{O}_K \longrightarrow F(n), \quad a \longmapsto a \bullet \lambda_n,$$

is a homomorphism of $\mathcal{O}_K$-modules with kernel $\pi^n\mathcal{O}_K$. It induces a bijective homomorphism $\mathcal{O}_K/\pi^n\mathcal{O}_K \to F(n)$ because both sides are of order q^n. □

(5.3) Corollary. *Associating $a \mapsto [a]_F$ we obtain canonical isomorphisms*

$$\mathcal{O}_K/\pi^n\mathcal{O}_K \longrightarrow \mathrm{End}_{\mathcal{O}_K}(F(n)) \quad \textit{and} \quad U_K/U_K^{(n)} \longrightarrow \mathrm{Aut}_{\mathcal{O}_K}(F(n)).$$

Proof: The map on the left is an isomorphism since $\mathcal{O}_K/\pi^n\mathcal{O}_K \cong F(n)$ and $\mathrm{End}_{\mathcal{O}_K}(\mathcal{O}_K/\pi^n\mathcal{O}_K) = \mathcal{O}_K/\pi^n\mathcal{O}_K$. The one on the right is obtained by taking the unit groups of these rings. □

Given a Lubin-Tate module F for the prime element π, we now define the **field of π^n-division points** by

$$L_n = K\big(F(n)\big).$$

Since $F(n) \subseteq F(n+1)$ we get a tower of fields

$$K \subseteq L_1 \subseteq L_2 \subseteq \ldots \subseteq L_\pi := \bigcup_{n=1}^{\infty} L_n.$$

These fields are also called the **Lubin-Tate extensions**. They only depend on the prime element π, not on the Lubin-Tate module F. For if G is another Lubin-Tate module for π, then by (4.6), there is an isomorphism $f : F \to G$, $f \in \mathcal{O}_K[[X]]$ such that $G(n) = f(F(n)) \subseteq K(F(n))$, and hence $K(G(n)) = K(F(n))$. If F is the Lubin-Tate module F_e belonging to a Lubin-Tate *polynomial* $e(X) \in \mathcal{E}_\pi$, then $e(X) = [\pi]_F(X)$ and $L_n|K$ is the splitting field of the n-fold iteration

$$e^n(X) = (e \circ \cdots \circ e)(X) = [\pi^n]_F(X).$$

Example: If $\mathcal{O}_K = \mathbb{Z}_p$ and F is the Lubin-Tate module $\mathbb{G}_m$, then

$$e^n(X) = [\,p^n]_{\mathbb{G}_m}(X) = (1+X)^{p^n} - 1.$$

So $\mathbb{G}_m(n)$ consists of the elements $\zeta - 1$, where ζ varies over the p^n-th roots of unity. $L_n|K$ is therefore the p^n-th cyclotomic extension $\mathbb{Q}_p(\mu_{p^n})|\mathbb{Q}_p$. The following theorem shows the complete analogy of Lubin-Tate extensions with cyclotomic fields.

(5.4) Theorem. *$L_n|K$ is a totally ramified abelian extension of degree $q^{n-1}(q-1)$ with Galois group*

$$G(L_n|K) \cong \mathrm{Aut}_{\mathcal{O}_K}\big(F(n)\big) \cong U_K/U_K^{(n)},$$

i.e., for every $\sigma \in G(L_n|K)$ there is a unique class u mod $U_K^{(n)}$, with $u \in U_K$ such that

$$\lambda^\sigma = [u]_F(\lambda) \quad \text{for} \quad \lambda \in F(n).$$

Furthermore the following is true: let F be the Lubin-Tate module F_e associated to the polynomial $e(X) \in \mathcal{E}_\pi$, and let $\lambda_n \in F(n) \setminus F(n-1)$. Then λ_n is a prime element of L_n, i.e., $L_n = K(\lambda_n)$, and

$$\phi_n(X) = \frac{e^n(X)}{e^{n-1}(X)} = X^{q^{n-1}(q-1)} + \cdots + \pi \in \mathcal{O}_K[X]$$

is its minimal polynomial. In particular one has $N_{L_n|K}(-\lambda_n) = \pi$.

Proof: If

$$e(X) = X^q + \pi(a_{q-1}X^{q-1} + \cdots + a_2X^2) + \pi X$$

is a Lubin-Tate polynomial, then

$$\begin{aligned}\phi_n(X) &= \frac{e^n(X)}{e^{n-1}(X)}\\ &= e^{n-1}(X)^{q-1} + \pi\big(a_{q-1}e^{n-1}(X)^{q-2} + \cdots + a_2 e^{n-1}(X)\big) + \pi\end{aligned}$$

is an Eisenstein polynomial of degree $q^{n-1}(q-1)$. If F is the Lubin-Tate module associated to e, and $\lambda_n \in F(n) \smallsetminus F(n-1)$, then λ_n is clearly a zero of this Eisenstein polynomial, and is therefore a prime element of the totally ramified extension $K(\lambda_n)|K$ of degree $q^{n-1}(q-1)$. Each $\sigma \in G(L|K)$ induces an automorphism of $F(n)$. We therefore obtain a homomorphism

$$G(L_n|K) \longrightarrow \mathrm{Aut}_{\mathcal{O}_K}\big(F(n)\big) \cong U_K/U_K^{(n)}.$$

It is injective because L_n is generated by $F(n)$, and it is surjective because

$$\#G(L_n|K) \geq \big[K(\lambda_n) : K\big] = q^{n-1}(q-1) = \#U_K/U_K^{(n)}.$$

This proves the theorem. □

Generalizing the explicit norm residue symbol of the cyclotomic fields $\mathbb{Q}_p(\mu_{p^n})|\mathbb{Q}_p$ (see (2.4)), we obtain the following explicit formula for the norm residue symbol of the Lubin-Tate extensions.

(5.5) Theorem. *For the field $L_n|K$ of π^n-division points and for $a = u\pi^{v_K(a)} \in K^*$, $u \in U_K$, one has*

$$(a, L_n|K)\lambda = [u^{-1}]_F(\lambda), \quad \lambda \in F(n).$$

Proof: The proof is the same as that of (2.4). Let $\sigma \in G(L_n|K)$ be the automorphism such that

$$\lambda^\sigma = [u^{-1}]_F(\lambda), \quad \lambda \in F(n).$$

Let $\tilde{\sigma}$ be an element in $\mathrm{Frob}(\widetilde{L}_n|K)$ such that $\sigma = \tilde{\sigma}|_{L_n}$ and $d_{\widetilde{K}}(\tilde{\sigma}) = 1$. We view $\tilde{\sigma}$ as an automorphism of the completion $\widehat{L}_n = L_n\widehat{K}$ of $\widetilde{L}_n$. Let Σ be the fixed field of $\tilde{\sigma}$. Since $f_{\Sigma|K} = d_K(\tilde{\sigma}) = 1$, $\Sigma|K$ is totally ramified. It has degree $q^{n-1}(q-1)$ because $\Sigma \cap \widetilde{K} = K$ and $\widetilde{\Sigma} = \Sigma\widetilde{K} = \widetilde{L}_n$. Consequently $[\Sigma : K] = [\widetilde{L}_n : \widetilde{K}] = [L_n : K]$.

Now let $e \in \mathcal{E}_\pi$, $\bar{e} \in \mathcal{E}_{\overline{\pi}}$ be Lubin-Tate series over $\mathcal{O}_K$, where $\pi = u\overline{\pi}$, and let $F = F_{\bar{e}}$. By (2.3), there exists a power series $\theta(X) = \varepsilon X + \cdots \in \mathcal{O}_{\widehat{K}}[[X]]$, with $\varepsilon \in U_{\widehat{K}}$, such that

$$\theta^\varphi = \theta \circ [u]_F \quad \text{and} \quad \theta^\varphi \circ \bar{e} = e \circ \theta \quad (\varphi = \varphi_K).$$

Let $\lambda_n \in F(n) \smallsetminus F(n-1)$. λ_n is a prime element of L_n, and

$$\pi_\Sigma = \theta(\lambda_n)$$

is a prime element of Σ because

$$\pi_\Sigma^{\tilde{\sigma}} = \theta^\varphi(\lambda_n^\sigma) = \theta^\varphi\big([u^{-1}]_F(\lambda_n)\big) = \theta(\lambda_n) = \pi_\Sigma .$$

Since $e^i(\theta(\lambda_n)) = \theta^{\varphi^i}(\bar{e}^i(\lambda_n)) = 0$ for $i = n$, and $\neq 0$ for $i = n-1$, we have $\pi_\Sigma \in F_e(n) \smallsetminus F_e(n-1)$. Hence $\Sigma = K(\pi_\Sigma)$ is the field of π^n-division points of F_e, and $N_{\Sigma|K}(-\pi_\Sigma) = \pi = u\overline{\pi}$ by (5.4). Since $\overline{\pi} = N_{L_n|K}(-\lambda_n) \in N_{L_n|K}L_n^*$, we get

$$r_{L_n|K}(\sigma) = N_{\Sigma|K}(-\pi_\Sigma) = \pi \equiv u \bmod N_{L_n|K}L_n^*,$$

and thus

$$(a, L_n|K) = (\pi^{v_K(a)}, L_n|K)(u, L_n|K) = (u, L_n|K) = \sigma . \qquad \square$$

(5.6) Corollary. *The field $L_n|K$ of π^n-division points is the class field relative to the group $(\pi) \times U_K^{(n)} \subseteq K^*$.*

Proof: For $a = u\pi^{v_K(a)}$ we have

$$a \in N_{L_n|K}L_n^* \iff (a, L_n|K) = 1 \iff [u^{-1}]_F(\lambda) = \lambda \quad \text{for all } \lambda \in F(n)$$
$$\iff [u^{-1}]_F = \mathrm{id}_{F(n)} \iff u^{-1} \in U_K^{(n)} \iff a \in (\pi) \times U_K^{(n)}.$$

$\square$

For the maximal abelian extension $K^{ab}|K$, this gives the following generalization of the local Kronecker-Weber theorem (1.9):

(5.7) Corollary. *The maximal abelian extension of K is the composite*

$$K^{ab} = \widetilde{K} L_\pi ,$$

where L_π is the union $\bigcup_{n=1}^\infty L_n$ of the fields L_n of π^n-division points.

Proof: Let $L|K$ be a finite abelian extension. Then we have $\pi^f \in N_{L|K}L^*$ for suitable f. Since $N_{L|K}L^*$ is open in K^*, and since the $U_K^{(n)}$ form a basis of neighbourhoods of 1, we have $(\pi^f) \times U_K^{(n)} \subseteq N_{L|K}L^*$ for a suitable n. Hence L is contained in the class field of the group $(\pi^f) \times U_K^{(n)} = ((\pi) \times U_K^{(n)}) \cap ((\pi^f) \times U_K)$. The class field of $(\pi) \times U_K^{(n)}$ is L_n, and that of $(\pi^f) \times U_K$ is the unramified extension K_f of degree f. It follows that $L \subseteq K_f L_n \subseteq \widetilde{K} L_\pi = K^{ab}$. □

Exercise 1. Let $F = F_e$ be the Lubin-Tate module for the Lubin-Tate series $e \in \mathcal{E}_\pi$, with the endomorphisms $[a] = [a]_e$. Let $S = \mathcal{O}_K[[X]]$ and $S^* = \{g \in S \mid g(0) \in U_K\}$. Show:

(i) If $g \in S$ is a power series such that $g(F(1)) = 0$, then g is divisible by $[\pi]$, i.e., $g(X) = [\pi](X)h(X)$, $h(X) \in S$.

(ii) Let $g \in S$ be a power series such that

$$g(X \underset{F}{+} \lambda) = g(X) \quad \text{for all} \quad \lambda \in F(1),$$

where we write $X \underset{F}{+} \lambda = F(X, \lambda)$. Then there exists a unique power series $h(X)$ in S such that

$$g = h \circ \pi .$$

Exercise 2. If $h(X)$ is a power series in S, then the power series

$$h_1(X) = \prod_{\lambda \in F(1)} h(X \underset{F}{+} \lambda)$$

also belongs to S, and one has $h_1(X \underset{F}{+} \lambda) = h_1(X)$ for all $\lambda \in F(1)$.

Exercise 3. Let $N(h) \in S$ be the power series (uniquely determined by exercise 1 and exercise 2) such that

$$N(h) \circ [\pi] = \prod_{\lambda \in F(1)} h(X \underset{F}{+} \lambda).$$

The mapping $N : S \to S$ is called **Coleman's norm operator**. Show:

(i) $N(h_1 h_2) = N(h_1)N(h_2)$.

(ii) $N(h) \equiv h \bmod \mathfrak{p}$.

(iii) $h \in X^i S^*$ for $i \geq 0 \Rightarrow N(h) \in X^i S^*$.

(iv) $h \equiv 1 \bmod \mathfrak{p}^i$ for $i \geq 1 \Rightarrow N(h) \equiv 1 \bmod \mathfrak{p}^{i+1}$.

(v) For the operators $N^0(h) = h$, $N^n(h) = N(N^{n-1}(h))$, one has

$$N^n(h) \circ [\pi^n] = \prod_{\lambda \in F(n)} h(X \underset{F}{+} \lambda), \quad n \geq 0.$$

(vi) If $h \in X^i S^*$, $i \geq 0$, then $N^{n+1}(h)/N^n(h) \in S^*$ and

$$N^{n+1}(h) \equiv N^n(h) \bmod \mathfrak{p}^{n+1}, \quad n \geq 0.$$

Exercise 4. Let $\lambda \in F(n+1) \smallsetminus F(n)$, $n \geq 0$, and $\lambda_i = [\pi^{n-i}](\lambda) \in F(i+1)$ for $0 \leq i \leq n$. Then λ_i is a prime element of the Lubin-Tate extension $L_{i+1} = K(F(i+1))$, and $\mathcal{O}_{i+1} = \mathcal{O}_K[\lambda_i]$ is the valuation ring of L_{i+1}, with maximal ideal $\mathfrak{p}_{i+1} = \lambda_i \mathcal{O}_{i+1}$. Show:

let $\beta_i \in \pi^{n-i}\mathfrak{p}_1\mathcal{O}_{i+1}$, $0 \leq i \leq n$. Then there exists a power series $h(X) \in S$ such that

$$h(\lambda_i) = \beta_i \quad \text{for} \quad 0 \leq i \leq n.$$

Hint: Write $\beta_i = \pi^{n-i}\lambda_0 h_i(\lambda_i)$, with $h_i(X) \in \mathcal{O}[X]$ and put, for $0 \leq i \leq n$: $g_i(X) = [\pi^{n+1}][\pi^i]/[\pi^{i+1}]$. Then $h = \sum_{i=0}^{n} h_i g_i$ is a solution.

Exercise 5. Let $\lambda \in F(n+1) \smallsetminus F(n)$ and $\lambda_i = [\pi^{n-i}](\lambda)$, $0 \leq i \leq n$. For every $u \in U_{L_{n+1}}$, there exists a power series $h(X) \in \mathcal{O}[[X]]$ such that

$$N_{n,i}(u) = h(\lambda_i) \quad \text{for} \quad 0 \leq i \leq n,$$

where $N_{n,i}$ is the norm from L_{n+1} to L_{i+1}.

Hint: Write $u = h_1(\lambda)$, $h_1(X) \in \mathcal{O}[X]$, and put $h_2 = N^n(h_1) \in S^*$. Show that $\beta_i = N_{n,i}(u) - h_2(\lambda_i) \in \pi^{n-i}\mathfrak{p}_1\mathcal{O}_{i+1}$. Then by exercise 4 there is a power series $h_3(X) \in \mathcal{O}[[X]]$ such that $\beta_i = h_3(\lambda_i)$, $0 \leq i \leq n$. Show that $h = h_2 + h_3$ works.

Remark: The solutions of these exercises are discussed in detail in [79], 5.2.

§ 6. Higher Ramification Groups

Considering the homomorphism

$$(\quad, L|K) : K^* \longrightarrow G(L|K)$$

defined for an abelian extension $L|K$ of local fields by the norm residue symbol, it is striking that both groups are equipped with a canonical filtration: in the group K^* on the left we have the descending chain

$$(*) \qquad K^* \supseteq U_K = U_K^{(0)} \supseteq U_K^{(1)} \supseteq U_K^{(2)} \supseteq \cdots$$

of *higher unit groups* $U_K^{(i)}$, and on the right there is the descending chain

$$(**) \qquad G(L|K) \supseteq G^0(L|K) \supseteq G^1(L|K) \supseteq G^2(L|K) \supseteq \cdots$$

of *ramification groups* $G^i(L|K)$ in the upper numbering (see chap. II, § 10). The latter arose from the ramification groups in the lower numbering

$$G_i(L|K) = \left\{\sigma \in G(L|K) \,\middle|\, v_L(\sigma a - a) \geq i+1 \quad \text{for all} \quad a \in \mathcal{O}_L\right\}$$

via the strictly increasing function

$$\eta_{L|K}(s) = \int_0^s \frac{dx}{(G_0 : G_x)}$$

by the rule

$$G^i(L|K) = G_{\psi_{L|K}(i)}(L|K),$$

where ψ is the inverse function of η. We will now prove the remarkable arithmetic fact that the norm residue symbol $(\ ,L|K)$ relates both filtrations $(*)$ and $(**)$ in a precise way. To this end we determine (generalizing chap. II, § 10, exercise 1) the higher ramification groups of the Lubin-Tate extensions.

(6.1) Proposition. *Let $L_n|K$ be the field of π^n-division points of a Lubin-Tate module for the prime element π. Then*

$$G_i(L_n|K) = G(L_n|L_k) \quad \text{for} \quad q^{k-1} \le i \le q^k - 1.$$

Proof: By (5.4) and (5.5), the norm residue symbol gives an isomorphism $U_K/U_K^{(k)} \to G(L_k|K)$ for every k. Hence $G(L_n|L_k) = (U_K^{(k)}, L_n|K)$. We therefore have to show that

$$G_i(L_n|K) = \big(U_K^{(k)}, L_n|K\big) \quad \text{for} \quad q^{k-1} \le i \le q^k - 1.$$

Let $\sigma \in G_1(L_n|K)$ and $\sigma = (u^{-1}, L_n|K)$. Then we have necessarily $u \in U_K^{(1)}$ because $(,L_n|K) : U_K/U_K^{(n)} \xrightarrow{\sim} G(L_n|K)$ maps the p-Sylow subgroup $U_K^{(1)}/U_K^{(n)}$ onto the p-Sylow subgroup $G_1(L_n|K)$ of $G(L_n|K)$. Let $u = 1 + \varepsilon\pi^m$, $\varepsilon \in U_K$, and $\lambda \in F(n) \setminus F(n-1)$. Then λ is a prime element of L_n and from (5.4) we get that

$$\lambda^\sigma = [u]_F(\lambda) = F\big(\lambda, [\varepsilon\pi^m]_F(\lambda)\big).$$

If $m \ge n$, then $\sigma = 1$, so that $v_{L_n}(\lambda^\sigma - \lambda) = \infty$. If $m < n$, then $\lambda_{n-m} = [\pi^m]_F(\lambda)$ is a prime element of L_{n-m} and therefore also $[\varepsilon\pi^m]_F(\lambda) = [\varepsilon]_F(\lambda_{n-m})$. As $L_n|L_{n-m}$ is totally ramified of degree q^m we may write $[\varepsilon\pi^m]_F(\lambda) = \varepsilon_0\lambda^{q^m}$ for some $\varepsilon_0 \in U_{L_n}$. Since $F(X,0) = X$, $F(0,Y) = Y$, we have $F(X,Y) = X + Y + XYG(X,Y)$ with $G(X,Y) \in \mathcal{O}_K[[X,Y]]$. Thus

$$\lambda^\sigma - \lambda = F(\lambda, \varepsilon_0\lambda^{q^m}) - \lambda = \varepsilon_0\lambda^{q^m} + a\lambda^{q^m+1}, \quad a \in \mathcal{O}_{L_n},$$

i.e.,

$$i_{L_n|K}(\sigma) := v_{L_n}(\lambda^\sigma - \lambda) = \begin{cases} q^m, & \text{if } m < n, \\ \infty, & \text{if } m \ge n. \end{cases}$$

By chap. II, § 10, we have $G_i(L_n|K) = \{\sigma \in G(L_n|K) \mid i_{L_n|K}(\sigma) \ge i + 1\}$. Now let $q^{k-1} \le i \le q^k - 1$. If $u \in U_K^{(k)}$, then $m \ge k$, i.e., $i_{L_n|K}(\sigma) \ge q^k \ge i + 1$, and so $\sigma \in G_i(L_n|K)$. This proves the inclusion $(U_K^{(k)}, L_n|K) \subseteq G_i(L_n|K)$. If conversely $\sigma \in G_i(L_n|K)$ and $\sigma \ne 1$, then $i_{L_n|K}(\sigma) = q^m > i \ge q^{k-1}$, i.e., $m \ge k$. Consequently $u \in U_K^{(k)}$, and this shows the inclusion $G_i(L_n|K) \subseteq (U_K^{(k)}, L_n|K)$. □

From this proposition we get the following result, which may be considered the main theorem of higher ramification theory.

(6.2) Theorem. *If $L|K$ is a finite abelian extension, then the norm residue symbol*

$$(\quad, L|K) : K^* \longrightarrow G(L|K)$$

maps the group $U_K^{(n)}$ onto the group $G^n(L|K)$, for $n \geq 0$.

Proof: We may assume that $L|K$ is totally ramified. For if $L^0|K$ is the maximal unramified subextension of $L|K$, then we have on the one hand $G^n(L|K) = G^n(L|L^0)$ because $\psi_{L^0|K}(s) = s$ and $\psi_{L|K}(s) = \psi_{L|L^0}(\psi_{L^0|K}(s)) = \psi_{L|L^0}(s)$ (see chap. II, (10.8)). On the other hand, by chap. IV, (6.4), and chap. V, (1.2), we have

$$\big(U_{L^0}^{(n)}, L|L^0\big) = \big(N_{L^0|K}U_{L^0}^{(n)}, L|K\big) = \big(U_K^{(n)}, L|K\big),$$

so we may replace $L|K$ by $L|L^0$.

If now $L|K$ is totally ramified and π_L is a prime element of L, then $\pi = N_{L|K}(\pi_L)$ is a prime element of K and $(\pi) \times U_K^{(m)} \subseteq N_{L|K}L^*$ for m sufficiently big. Therefore $L|K$ is contained in the class field of $(\pi) \times U_K^{(m)}$, which, by (5.6), is equal to the field L_m of π^m-division points of some Lubin-Tate module for π. In view of chap. II, (10.9), and chap. IV, (6.4), we may even assume that $L = L_m$. By (6.1), the norm residue symbol maps the group $U_K^{(n)}$ onto the group

$$G(L_m|L_n) = G_i(L_m|K) \quad \text{for} \quad q^{n-1} \leq i \leq q^n - 1.$$

But we have (see chap. II, § 10)

$$\eta_{L|K}(q^n - 1) = \frac{1}{g_0}(g_1 + \cdots + g_{q^n-1})$$

with $g_i = \#G_i(L|K) = \#G(L_m|L_n) = (q^{m-1} - q^{n-1})(q-1)$ for $q^{n-1} \leq i \leq q^n - 1$. This yields $\eta_{L|K}(q^n - 1) = n$ and thus $(U_K^{(n)}, L|K) = G_{q^n-1}(L|K) = G^n(L|K)$. □

Higher ramification groups $G^t(L|K)$ were introduced for arbitrary real numbers $t \geq -1$. Thus we may ask for which numbers they change. We call these numbers the *jumps* of the filtration $\{G^t(L|K)\}_{t \geq -1}$ of $G(L|K)$. In other words, t is a jump if for all $\varepsilon > 0$, one has

$$G^t(L|K) \neq G^{t+\varepsilon}(L|K).$$

(6.3) Proposition (*Hasse-Arf*). *For a finite abelian extension $L|K$, the jumps of the filtration $\{G^t(L|K)\}_{t\geq -1}$ of $G(L|K)$ are rational integers.*

Proof: As in the proof of (6.2), we may assume (since $G^t(L|K) = G^t(L|L^0)$) that $L|K$ is totally ramified and contained in a Lubin-Tate extension $L_m|K$. If now t is a jump of $\{G^t(L|K)\}$, then by chap. II (10.9), t is also a jump of $\{G^t(L_m|K)\}$. Since by (6.1), the jumps of $\{G_s(L_m|K)\}$ are the numbers $q^n - 1$, for $n = 0, \dots, m-1$ ($q = 2$ is an exception: 0 is not a jump), the jumps of $\{G^t(L_m|K)\}$ are the numbers $\eta_{L_m|K}(q^n - 1) = n$, for $n = 0, \dots, m-1$. □

The theorem of *Hasse-Arf* has an important application to *Artin L-series*, which we will study in chap. VII (see chap. VII, (11.4)).

Chapter VI

Global Class Field Theory

§ 1. Idèles and Idèle Classes

The rôle held in local class field theory by the multiplicative group of the base field is taken in global class field theory by the idèle class group. The notion of idèle is a modification of the notion of ideal. It was introduced by the French mathematician *Claude Chevalley* (1909–1984) with a view to providing a suitable basis for the important local-to-global principle, i.e., for the principle which reduces problems concerning a number field K to analogous problems for the various completions $K_{\mathfrak{p}}$. *Chevalley* used the term "ideal element", which was abbreviated as id. el.

An **adèle** of K – this curious expression, which has the stress on the second syllable, is derived from the original term "additive idèle" – is a family

$$\alpha = (\alpha_{\mathfrak{p}})$$

of elements $\alpha_{\mathfrak{p}} \in K_{\mathfrak{p}}$ where $\mathfrak{p}$ runs through all primes of K, and $\alpha_{\mathfrak{p}}$ is integral in $K_{\mathfrak{p}}$ for almost all $\mathfrak{p}$. The adèles form a ring, which is denoted by

$$\mathbb{A}_K = \prod_{\mathfrak{p}} K_{\mathfrak{p}}.$$

Addition and multiplication are defined componentwise. This kind of product is called the "restricted product" of the $K_{\mathfrak{p}}$ with respect to the subrings $\mathcal{O}_{\mathfrak{p}} \subseteq K_{\mathfrak{p}}$.

The **idèle group** of K is defined to be the unit group

$$I_K = \mathbb{A}_K^*.$$

Thus an **idèle** is a family

$$\alpha = (\alpha_{\mathfrak{p}})$$

of elements $\alpha_{\mathfrak{p}} \in K_{\mathfrak{p}}^*$ where $\alpha_{\mathfrak{p}}$ is a unit in the ring $\mathcal{O}_{\mathfrak{p}}$ of integers of $K_{\mathfrak{p}}$, for almost all $\mathfrak{p}$. In analogy with $\mathbb{A}_K$, we write the idèle group as the restricted product

$$I_K = \prod_{\mathfrak{p}} K_{\mathfrak{p}}^*$$

with respect to the unit groups $\mathcal{O}_\mathfrak{p}^*$. For every finite set of primes S, I_K contains the subgroup

$$I_K^S = \prod_{\mathfrak{p}\in S} K_\mathfrak{p}^* \times \prod_{\mathfrak{p}\notin S} U_\mathfrak{p}$$

of **S-idèles**, where $U_\mathfrak{p} = K_\mathfrak{p}^*$ for $\mathfrak{p}$ infinite complex, and $U_\mathfrak{p} = \mathbb{R}_+^*$ for $\mathfrak{p}$ infinite real. One clearly has

$$I_K = \bigcup_S I_K^S\,,$$

if S varies over all finite sets of primes of K.

The inclusions $K \subseteq K_\mathfrak{p}$ allow us to define the diagonal embedding

$$K^* \longrightarrow I_K\,,$$

which associates to $a \in K^*$ the idèle $\alpha \in I_K$ whose $\mathfrak{p}$-th component is the element a in $K_\mathfrak{p}$. We thus view K^* as a subgroup of I_K and we call the elements of K^* in I_K **principal idèles**. The intersection

$$K^S = K^* \cap I_K^S$$

consists of the numbers $a \in K^*$ which are units at all primes $\mathfrak{p} \notin S$, $\mathfrak{p} \nmid \infty$, and which are positive in $K_\mathfrak{p} = \mathbb{R}$ for all real infinite places $\mathfrak{p} \notin S$. They are called **S-units**. In particular, for the set S_∞ of infinite places, K^{S_∞} is the unit group $\mathcal{O}_K^*$ of $\mathcal{O}_K$. We get the following generalization of Dirichlet's unit theorem.

(1.1) Proposition. *If S contains all infinite places, then the homomorphism*

$$\lambda : K^S \longrightarrow \prod_{\mathfrak{p}\in S} \mathbb{R}\,, \quad \lambda(a) = \big(\log|a|_\mathfrak{p}\big)_{\mathfrak{p}\in S}\,,$$

has kernel $\mu(K)$, and its image is a complete lattice in the $(s-1)$-dimensional trace-zero space $H = \big\{(x_\mathfrak{p}) \in \prod_{\mathfrak{p}\in S}\mathbb{R} \;\big|\; \sum_{\mathfrak{p}\in S} x_\mathfrak{p} = 0\big\}$, $s = \#S$.

Proof: For the set $S_\infty = \{\mathfrak{p} \mid \infty\}$, this is the claim of chap. I, (7.1) and (7.3). Let $S_f = S \setminus S_\infty$, and let $J(S_f)$ be the subgroup of J_K generated by the prime ideals $\mathfrak{p} \in S_f$. Associating to every $a \in K^S$ the principal ideal $ia = (a) \in J(S_f)$, we obtain the commutative diagram

$$\begin{array}{ccccccc}
1 \longrightarrow & \mathcal{O}_K^* & \longrightarrow & K^S & \xrightarrow{\;i\;} & J(S_f) \\
& \big\downarrow{\scriptstyle\lambda'} & & \big\downarrow{\scriptstyle\lambda} & & \big\downarrow{\scriptstyle\lambda''} \\
0 \longrightarrow & \prod\limits_{\mathfrak{p}\in S_\infty}\mathbb{R} & \longrightarrow & \prod\limits_{\mathfrak{p}\in S}\mathbb{R} & \xrightarrow{\;i\;} & \prod\limits_{\mathfrak{p}\in S_f}\mathbb{R}
\end{array}$$

with exact rows. The map λ'' on the right is given by

$$\lambda''\Big(\prod_{\mathfrak{p}\in S_f} \mathfrak{p}^{\nu_\mathfrak{p}}\Big) = -\prod_{\mathfrak{p}\in S_f} \nu_\mathfrak{p} \log \mathfrak{N}(\mathfrak{p})$$

(observe that $|a|_\mathfrak{p} = \mathfrak{N}(\mathfrak{p})^{-v_\mathfrak{p}(a)}$), and maps $J(S_f)$ isomorphically onto the complete lattice spanned by the vectors

$$e_\mathfrak{p} = \big(0, \dots, 0, \log \mathfrak{N}(\mathfrak{p}), 0, \dots, 0\big),$$

for $\mathfrak{p} \in S_f$. It follows that $\ker(\lambda) = \ker(\lambda') = \mu(K)$, and we obtain the exact sequence

$$(*) \qquad 0 \longrightarrow \operatorname{im}(\lambda') \longrightarrow \operatorname{im}(\lambda) \xrightarrow{\ i\ } \operatorname{im}(\lambda''),$$

where the groups on the left and on the right are lattices. This implies that the group in the middle is also a lattice. For if $x \in \operatorname{im}(\lambda)$, and U is a neighbourhood of $i(x)$ which contains no other point of $\operatorname{im}(\lambda'')$, then $i^{-1}(U)$ contains the coset $x + \operatorname{im}(\lambda')$, and no other. It is discrete since $\operatorname{im}(\lambda')$ is discrete.

For every $\mathfrak{p} \in S_f$, if h is the class number of K, then $\mathfrak{p}^h$ belongs to $i(K^S)$, i.e.,

$$J(S_f)^h \subseteq i(K^S) \subseteq J(S_f).$$

The groups on the left and on the right have rank $\#S_f$, hence so does $i(K^S)$. In the sequence $(*)$, the image of i therefore has rank $\#S_f$, and the kernel has rank $\#S_\infty - 1$. Hence $\operatorname{im}(\lambda)$ is a lattice of rank $\#S_\infty - 1 + \#S_f = \#S - 1$. It lies in the $(\#S-1)$-dimensional trace-zero space H, since $\prod_{\mathfrak{p}\in S} |a|_\mathfrak{p} = \prod_\mathfrak{p} |a|_\mathfrak{p} = 1$ for $a \in K^S$. □

(1.2) Definition. *The elements of the subgroup K^* of I_K are called* **principal idèles** *and the quotient group*

$$C_K = I_K/K^*$$

is called the **idèle class group** *of K.*

The relation between the ideal class group Cl_K and the idèle class group C_K is as follows. There is a surjective homomorphism

$$(\ ) : I_K \longrightarrow J_K, \quad \alpha \longmapsto (\alpha) = \prod_{\mathfrak{p}\nmid\infty} \mathfrak{p}^{v_\mathfrak{p}(\alpha_\mathfrak{p})},$$

from the idèle group I_K to the ideal group J_K. Its kernel is

$$I_K^{S_\infty} = \prod_{\mathfrak{p}|\infty} K_\mathfrak{p}^* \times \prod_{\mathfrak{p}\nmid\infty} U_\mathfrak{p}.$$

It induces a surjective homomorphism

$$C_K \longrightarrow Cl_K$$

with kernel $I_K^{S_\infty} K^*/K^*$. We may also consider the surjective homomorphism

$$I_K \longrightarrow J(\bar{\mathcal{O}}), \quad \alpha \longmapsto \prod_{\mathfrak{p}} \mathfrak{p}^{v_\mathfrak{p}(\alpha_\mathfrak{p})},$$

onto the *replete* ideal group $J(\bar{\mathcal{O}})$. Its kernel is

$$I_K^0 = \{ (\alpha_\mathfrak{p}) \in I_K \mid |\alpha_\mathfrak{p}|_\mathfrak{p} = 1 \text{ for all } \mathfrak{p} \}$$

(see chap. III, § 1). It takes principal idèles to replete principal ideals and induces a surjective homomorphism

$$C_K \longrightarrow Pic(\bar{\mathcal{O}})$$

onto the replete ideal class group, with kernel $I_K^0 K^*/K^*$. We therefore have the

(1.3) Proposition. *$Cl_K \cong I_K/I_K^{S_\infty} K^*$, and $Pic(\bar{\mathcal{O}}) \cong I_K/I_K^0 K^*$.*

In contrast to the ideal class group, the idèle class group is not finite. But the finiteness of the former is reflected in terms of the latter as follows.

(1.4) Proposition. *$I_K = I_K^S K^*$, i.e., $C_K = I_K^S K^*/K^*$, if S is a sufficiently big finite set of places of K.*

Proof: Let $\mathfrak{a}_1, \dots, \mathfrak{a}_h$ be ideals representing the h classes of J_K/P_K. They are composed of a finite number of prime ideals $\mathfrak{p}_1, \dots, \mathfrak{p}_n$. Now if S is any finite set of places containing these primes and the places at infinity, then one has $I_K = I_K^S K^*$.

In order to see this, we use the isomorphism $I_K/I_K^{S_\infty} \cong J_K$. If $\alpha \in I_K$, then the corresponding ideal $(\alpha) = \prod_{\mathfrak{p}\nmid\infty} \mathfrak{p}^{v_\mathfrak{p}(\alpha_\mathfrak{p})}$ belongs to some class $\mathfrak{a}_i P_K$, i.e., $(\alpha) = \mathfrak{a}_i(a)$ for some principal ideal (a). The idèle $\alpha' = \alpha a^{-1}$ is mapped by $I_K \to J_K$ to the ideal $\mathfrak{a}_i = \prod_{\mathfrak{p}\nmid\infty} \mathfrak{p}^{v_\mathfrak{p}(\alpha'_\mathfrak{p})}$. Since the prime ideals occurring in $\mathfrak{a}_i$ lie in S, we have $v_\mathfrak{p}(\alpha'_\mathfrak{p}) = 0$, i.e., $\alpha'_\mathfrak{p} \in U_\mathfrak{p}$ for all $\mathfrak{p} \notin S$. Hence $\alpha' = \alpha a^{-1} \in I_K^S$, and thus $\alpha \in I_K^S K^*$. □

The idèle group comes equipped with a canonical topology. A basic system of neighbourhoods of $1 \in I_K$ is given by the sets

$$\prod_{\mathfrak{p}\in S} W_{\mathfrak{p}} \times \prod_{\mathfrak{p}\notin S} U_{\mathfrak{p}} \subseteq I_K ,$$

where S runs through the finite sets of places of K which contain all $\mathfrak{p}|\infty$, and $W_{\mathfrak{p}} \subseteq K_{\mathfrak{p}}^*$ is a basic system of neighbourhoods of $1 \in K_{\mathfrak{p}}^*$. The groups $U_{\mathfrak{p}}$ are compact for $\mathfrak{p} \notin S$. Therefore the same is true of the group $\prod_{\mathfrak{p}\notin S} U_{\mathfrak{p}}$. If the $W_{\mathfrak{p}}$, for $\mathfrak{p}|\infty$, are bounded, then $\prod_{\mathfrak{p}\in S} W_{\mathfrak{p}} \times \prod_{\mathfrak{p}\notin S} U_{\mathfrak{p}}$ is a neighbourhood of 1 in I_K whose closure is compact. Therefore I_K is a **locally compact topological group**.

(1.5) Proposition. *K^* is a discrete, and therefore closed, subgroup of I_K.*

Proof: It is enough to show that $1 \in I_K$ has a neighbourhood which contains no other principal idèle besides 1.

$$\mathfrak{U} = \left\{\alpha \in I_K \;\middle|\; |\alpha_{\mathfrak{p}}|_{\mathfrak{p}} = 1 \text{ for } \mathfrak{p} \nmid \infty,\ |\alpha_{\mathfrak{p}} - 1|_{\mathfrak{p}} < 1 \text{ for } \mathfrak{p}|\infty\right\},$$

is such a neighbourhood. For if we had a principal idèle $x \in \mathfrak{U}$ different from 1, then we get the contradiction

$$\begin{aligned} 1 = \prod_{\mathfrak{p}} |x-1|_{\mathfrak{p}} &= \prod_{\mathfrak{p}\nmid\infty} |x-1|_{\mathfrak{p}} \cdot \prod_{\mathfrak{p}|\infty} |x-1|_{\mathfrak{p}} \\ &< \prod_{\mathfrak{p}\nmid\infty} |x-1|_{\mathfrak{p}} \le \prod_{\mathfrak{p}\nmid\infty} \max\{|x|_{\mathfrak{p}}, 1\} = 1 . \end{aligned}$$

That the subgroup is closed follows for a completely general reason: since $(x,y) \mapsto xy^{-1}$ is continuous, there is a neighbourhood V of 1 such that $VV^{-1} \subseteq \mathfrak{U}$. For every $y \in I_K$, the neighbourhood yV then contains at most one $x \in K^*$. Indeed, from $x_1 = yv_1$, $x_2 = yv_2 \in K^*$, with $x_1 \neq x_2$, one deduces $x_1x_2^{-1} = v_1v_2^{-1} \in \mathfrak{U}$, a contradiction. □

As K^* is closed in I_K, the fact that I_K is a locally compact Hausdorff topological group carries over to the idèle class group $C_K = I_K/K^*$. For any idèle $\alpha = (\alpha_{\mathfrak{p}}) \in I_K$, its class in C_K will be denoted by $[\alpha]$. We define the **absolute norm** of α to be the real number

$$\mathfrak{N}(\alpha) = \prod_{\mathfrak{p}} \mathfrak{N}(\mathfrak{p})^{v_{\mathfrak{p}}(\alpha_{\mathfrak{p}})} = \prod_{\mathfrak{p}} |\alpha_{\mathfrak{p}}|_{\mathfrak{p}}^{-1} .$$

If $x \in K^*$ is a principal idèle, then we find by chap. III, (1.3), that $\mathfrak{N}(x) = \prod_{\mathfrak{p}} |x|_{\mathfrak{p}}^{-1} = 1$. We thus have a continuous homomorphism

$$\mathfrak{N} : C_K \longrightarrow \mathbb{R}_+^* .$$

It is related to the absolute norm on the replete Picard group $Pic(\bar{\mathcal{O}})$ via the commutative diagram

$$\begin{array}{ccc} C_K & \xrightarrow{\mathfrak{N}} & \mathbb{R}_+^* \\ \downarrow & & \| \\ Pic(\bar{\mathcal{O}}) & \xrightarrow{\mathfrak{N}} & \mathbb{R}_+^* . \end{array}$$

Here the arrow

$$C_K \longrightarrow Pic(\bar{\mathcal{O}})$$

is induced by the continuous surjective homomorphism

$$I_K \longrightarrow J(\bar{\mathcal{O}}), \quad (\alpha_\mathfrak{p}) \longmapsto \prod_\mathfrak{p} \mathfrak{p}^{v_\mathfrak{p}(\alpha_\mathfrak{p})},$$

with kernel

$$I_K^0 = \left\{ (\alpha_\mathfrak{p}) \in I_K \;\middle|\; |\alpha_\mathfrak{p}|_\mathfrak{p} = 1 \text{ for all } \mathfrak{p} \right\}.$$

As to the kernel C_K^0 of $\mathfrak{N} : C_K \to \mathbb{R}_+$, we obtain, in analogy with chap. III, (1.14), the following important theorem. It reflects the finiteness of the unit rank of K as well as the finiteness of the class number.

(1.6) Theorem. *The group* $C_K^0 = \{[\alpha] \in C_K \mid \mathfrak{N}([\alpha]) = 1\}$ *is compact.*

Proof: The claim concerning the commutative exact diagram

$$\begin{array}{ccccccccc} 1 & \longrightarrow & C_K^0 & \longrightarrow & C_K & \longrightarrow & \mathbb{R}_+^* & \longrightarrow & 1 \\ & & \downarrow & & \downarrow & & \| & & \\ 1 & \rightarrow & Pic(\bar{\mathcal{O}})^0 & \longrightarrow & Pic(\bar{\mathcal{O}}) & \longrightarrow & \mathbb{R}_+^* & \longrightarrow & 1 \end{array}$$

will be reduced to the compactness of the group $Pic(\bar{\mathcal{O}})^0$, which was proved in chap. III, (1.14). The kernel of the vertical arrow in the middle is the group $I_K^0 K^*/K^* = I_K^0/I_K^0 \cap K^*$, where we have $I_K^0 = \prod_\mathfrak{p} I_\mathfrak{p}^0$, $I_\mathfrak{p}^0 = \{\alpha_\mathfrak{p} \in K_\mathfrak{p} \mid |\alpha_\mathfrak{p}|_\mathfrak{p} = 1\}$, and $I_K^0 \cap K^* = \mu(K)$ by chap. III, (1.9). This kernel is clearly compact. We obtain an exact sequence

$$1 \longrightarrow I_K^0 K^*/K^* \longrightarrow C_K^0 \longrightarrow Pic(\bar{\mathcal{O}})^0 \longrightarrow 1$$

of continuous homomorphisms. Since $Pic(\bar{\mathcal{O}})^0$ is compact, and the same is true for the fibres of the mapping $C_K^0 \to Pic(\bar{\mathcal{O}})^0$ (they are cosets, all homeomorphic to $I_K^0 K^*/K^*$), hence so is C_K^0. □

The idèle class group C_K plays a similar rôle for the algebraic number field K as the multiplicative group $K_\mathfrak{p}^*$ does for a $\mathfrak{p}$-adic number field $K_\mathfrak{p}$. It comes equipped with a collection of canonical subgroups which are to be viewed as analogues of the higher unit groups $U_\mathfrak{p}^{(n)} = 1 + \mathfrak{p}^n$ of a $\mathfrak{p}$-adic number field $K_\mathfrak{p}$. Instead of $\mathfrak{p}^n$, we take any integral ideal $\mathfrak{m} = \prod_{\mathfrak{p}\nmid\infty} \mathfrak{p}^{n_\mathfrak{p}}$. We may also write it as a replete ideal

$$\mathfrak{m} = \prod_\mathfrak{p} \mathfrak{p}^{n_\mathfrak{p}}$$

with $n_\mathfrak{p} = 0$ for $\mathfrak{p}|\infty$, and we treat it in what follows as a **module** of K. For every place $\mathfrak{p}$ of K we put $U_\mathfrak{p}^{(0)} = U_\mathfrak{p}$, and

$$U_\mathfrak{p}^{(n_\mathfrak{p})} := \begin{cases} 1 + \mathfrak{p}^{n_\mathfrak{p}}, & \text{if } \mathfrak{p} \nmid \infty, \\ \mathbb{R}_+^* \subset K_\mathfrak{p}^*, & \text{if } \mathfrak{p} \text{ is real}, \\ \mathbb{C}^* = K_\mathfrak{p}^*, & \text{if } \mathfrak{p} \text{ is complex}, \end{cases}$$

for $n_\mathfrak{p} > 0$. Given $\alpha_\mathfrak{p} \in K_\mathfrak{p}^*$ we write

$$\alpha_\mathfrak{p} \equiv 1 \bmod \mathfrak{p}^{n_\mathfrak{p}} \iff \alpha_\mathfrak{p} \in U_\mathfrak{p}^{(n_\mathfrak{p})}.$$

For a finite prime $\mathfrak{p}$ and $n_\mathfrak{p} > 0$ this means the usual congruence; for a real place, it symbolizes positivity, and for a complex place it is the empty condition.

(1.7) Definition. *The group*

$$C_K^\mathfrak{m} = I_K^\mathfrak{m} K^*/K^*,$$

formed from the idèle group

$$I_K^\mathfrak{m} = \prod_\mathfrak{p} U_\mathfrak{p}^{(n_\mathfrak{p})},$$

is called the **congruence subgroup** mod $\mathfrak{m}$, *and the quotient group* $C_K/C_K^\mathfrak{m}$ *is called the* **ray class group** mod $\mathfrak{m}$.

Remark: This definition of the ray class group does correspond to the classical one, as given (in the ideal-theoretic version) for instance in Hasse's "Zahlbericht" [53]. It differs from those found in modern textbooks, and also from that given in [107] by the author: in the present book, the components $\alpha_\mathfrak{p}$ of idèles α in $I_K^\mathfrak{m}$ are always positive at all real places $\mathfrak{p}$, so we have here fewer congruence subgroups than in the other texts. This choice does not only simplify matters. Most of all, it was made substantially because of the choice of the canonical metric $\langle\ ,\ \rangle$ on the Minkowski space $K_{\mathbb{R}}$ (see chap. I, § 5). In fact, we saw in chap. III, § 3, that this choice forces the extension $\mathbb{C}|\mathbb{R}$ to be unramified. We will explain in § 6 below how to interpret this situation, and how to reconcile it with the definition of ray classes in other texts.

The significance of the congruence subgroups lies in that they provide an overview over all closed subgroups of finite index in C_K. More precisely, we have the

(1.8) Proposition. *The closed subgroups of finite index of C_K are precisely those subgroups that contain a congruence subgroup $C_K^{\mathfrak{m}}$.*

Proof: $C_K^{\mathfrak{m}}$ is open in C_K because $I_K^{\mathfrak{m}} = \prod_{\mathfrak{p}} U_{\mathfrak{p}}^{(n_{\mathfrak{p}})}$ is open in I_K. $I_K^{\mathfrak{m}}$ is contained in the group $I_K^{S_\infty} = \prod_{\mathfrak{p}|\infty} K_{\mathfrak{p}}^* \times \prod_{\mathfrak{p}\nmid\infty} U_{\mathfrak{p}}$, and since $(C_K : I_K^{S_\infty} K^*/K^*) = \# Cl_K = h < \infty$, the index

$$(C_K : C_K^{\mathfrak{m}}) = h(I_K^{S_\infty} K^* : I_K^{\mathfrak{m}} K^*) \le h(I_K^{S_\infty} : I_K^{\mathfrak{m}})$$
$$= h \prod_{\mathfrak{p}\nmid\infty} (U_{\mathfrak{p}} : U_{\mathfrak{p}}^{(n_{\mathfrak{p}})}) \prod_{\mathfrak{p}|\infty} (K_{\mathfrak{p}}^* : U_{\mathfrak{p}}^{(n_{\mathfrak{p}})})$$

is finite. Being the complement of the nontrivial open cosets, which are finite in number, $C_K^{\mathfrak{m}}$ is closed of finite index. Consequently, every group containing $C_K^{\mathfrak{m}}$ is also closed of finite index, for it is the union of finitely many cosets of $C_K^{\mathfrak{m}}$.

Conversely, let $\mathcal{N}$ be an arbitrary closed subgroup of finite index. Then $\mathcal{N}$ is also open, being the complement of a finite number of closed cosets. Thus the preimage J of $\mathcal{N}$ in I_K is also open, and it thus contains a subset of the form

$$W = \prod_{\mathfrak{p}\in S} W_{\mathfrak{p}} \times \prod_{\mathfrak{p}\notin S} U_{\mathfrak{p}},$$

where S is a finite set of places of K containing the infinite ones, and $W_{\mathfrak{p}}$ is an open neighbourhood of $1 \in K_{\mathfrak{p}}^*$. If $\mathfrak{p} \in S$ is finite, we are liable to choose $W_{\mathfrak{p}} = U_{\mathfrak{p}}^{(n_{\mathfrak{p}})}$, because the groups $U_{\mathfrak{p}}^{(n_{\mathfrak{p}})} \subseteq K_{\mathfrak{p}}^*$ form a basic system of neighbourhoods of $1 \in K_{\mathfrak{p}}^*$. If $\mathfrak{p} \in S$ is real, we may choose $W_{\mathfrak{p}} \subseteq \mathbb{R}_+^*$. The open set $W_{\mathfrak{p}}$ will then generate the group $\mathbb{R}_+^*$, resp. $K_{\mathfrak{p}}^*$ in the case of a complex place $\mathfrak{p}$. The subgroup of J generated by W is therefore of the form $I_K^{\mathfrak{m}}$, so $\mathcal{N}$ contains the congruence subgroup $C_K^{\mathfrak{m}}$. □

The ray class groups can be given the following purely ideal-theoretic description. Let $J_K^{\mathfrak{m}}$ be the group of all fractional ideals relatively prime to $\mathfrak{m}$, and let $P_K^{\mathfrak{m}}$ be the group of all principal ideals $(a) \in P_K$ such that

$$a \equiv 1 \bmod \mathfrak{m} \quad \text{and} \quad a \text{ totally positive.}$$

The latter condition means that, for every real embedding $K \to \mathbb{R}$, a turns out to be positive. The congruence $a \equiv 1 \bmod \mathfrak{m}$ means that a is the quotient b/c of two *integers* relatively prime to $\mathfrak{m}$ such that $b \equiv c \bmod \mathfrak{m}$. This is tantamount to saying that $a \equiv 1 \bmod \mathfrak{p}^{n_\mathfrak{p}}$ in $K_\mathfrak{p}$, i.e., $a \in U_\mathfrak{p}^{(n_\mathfrak{p})}$ for all $\mathfrak{p}|\mathfrak{m} = \prod_{\mathfrak{p}\nmid\infty} \mathfrak{p}^{n_\mathfrak{p}}$. We put

$$Cl_K^{\mathfrak{m}} = J_K^{\mathfrak{m}}/P_K^{\mathfrak{m}}.$$

We then have the

(1.9) Proposition. *The homomorphism*

$$(\ ) : I_K \longrightarrow J_K, \quad \alpha \longmapsto (\alpha) = \prod_{\mathfrak{p}\nmid\infty} \mathfrak{p}^{v_\mathfrak{p}(\alpha_\mathfrak{p})},$$

induces an isomorphism

$$C_K/C_K^{\mathfrak{m}} \cong Cl_K^{\mathfrak{m}}.$$

Proof: Let $\mathfrak{m} = \prod_\mathfrak{p} \mathfrak{p}^{n_\mathfrak{p}}$, and let

$$I_K^{(\mathfrak{m})} = \left\{\alpha \in I_K \;\middle|\; \alpha_\mathfrak{p} \in U_\mathfrak{p}^{(n_\mathfrak{p})} \text{ for } \mathfrak{p}|\mathfrak{m}\infty\right\}.$$

Then $I_K = I_K^{(\mathfrak{m})}K^*$, because for every $\alpha \in I_K$, by the approximation theorem, there exists an $a \in K^*$ such that $\alpha_\mathfrak{p} a \equiv 1 \bmod \mathfrak{p}^{n_\mathfrak{p}}$ for $\mathfrak{p}|\mathfrak{m}$, and $\alpha_\mathfrak{p} a > 0$ for $\mathfrak{p}$ real. Thus $\beta = (\alpha_\mathfrak{p} a) \in I_K^{(\mathfrak{m})}$, so that $\alpha = \beta a^{-1} \in I_K^{(\mathfrak{m})}K^*$. The elements $a \in I_K^{(\mathfrak{m})} \cap K^*$ are precisely those generating principal ideals in $P_K^{\mathfrak{m}}$. Therefore the correspondence $\alpha \mapsto (\alpha) = \prod_{\mathfrak{p}\nmid\infty} \mathfrak{p}^{v_\mathfrak{p}(\alpha_\mathfrak{p})}$ defines a surjective homomorphism

$$C_K = I_K^{(\mathfrak{m})}K^*/K^* = I_K^{(\mathfrak{m})}/I_K^{(\mathfrak{m})} \cap K^* \longrightarrow J_K^{\mathfrak{m}}/P_K^{\mathfrak{m}}.$$

Since $(\alpha) = 1$ for $\alpha \in I_K^{\mathfrak{m}}$, the group $C_K^{\mathfrak{m}} = I_K^{\mathfrak{m}}K^*/K^*$ is certainly contained in the kernel. Conversely, if the class $[\alpha]$ represented by $\alpha \in I_K^{(\mathfrak{m})}$ belongs to the kernel, then there is an $(a) \in P_K^{\mathfrak{m}}$, with $a \in I_K^{(\mathfrak{m})} \cap K^*$, such that $(\alpha) = (a)$. The components of the idèle $\beta = \alpha a^{-1}$ satisfy $\beta_\mathfrak{p} \in U_\mathfrak{p}$ for $\mathfrak{p} \nmid \mathfrak{m}\infty$, and $\beta_\mathfrak{p} \in U_\mathfrak{p}^{(n_\mathfrak{p})}$ for $\mathfrak{p}|\mathfrak{m}\infty$, in other words, $\beta \in I_K^{\mathfrak{m}}$, and hence $[\alpha] = [\beta] \in I_K^{\mathfrak{m}}K^*/K^* = C_K^{\mathfrak{m}}$. Therefore $C_K^{\mathfrak{m}}$ is the kernel of the above mapping, and the proposition is proved. □

The ray class groups in the ideal-theoretic version $Cl_K^{\mathfrak{m}} = J_K^{\mathfrak{m}}/P_K^{\mathfrak{m}}$ were introduced by *Heinrich Weber* (1842–1913) as a common generalization of ideal class groups on the one hand, and the groups $(\mathbb{Z}/m\mathbb{Z})^*$ on the other. These latter groups may be viewed as the ray class groups of the field $\mathbb{Q}$:

(1.10) Proposition. *For any module* $\mathfrak{m} = (m)$ *of the field* $\mathbb{Q}$, *one has*

$$C_{\mathbb{Q}}/C_{\mathbb{Q}}^{\mathfrak{m}} \cong Cl_{\mathbb{Q}}^{\mathfrak{m}} \cong (\mathbb{Z}/m\mathbb{Z})^* .$$

Proof: Every ideal $(a) \in J_{\mathbb{Q}}^{\mathfrak{m}}$ has two generators, a and $-a$. Mapping the *positive* generator onto the residue class mod m, we get a surjective homomorphism $J_{\mathbb{Q}}^{\mathfrak{m}} \to (\mathbb{Z}/m\mathbb{Z})^*$ whose kernel consists of all ideals (a) which have a positive generator $\equiv 1 \bmod m$. But these are precisely the ideals (a) such that $a \equiv 1 \bmod p^{n_p}$ for $p|\mathfrak{m}\infty$, i.e., the kernel of $P_{\mathbb{Q}}^{\mathfrak{m}}$. □

The group $(\mathbb{Z}/m\mathbb{Z})^*$ is canonically isomorphic to the Galois group $G(\mathbb{Q}(\mu_m)|\mathbb{Q})$ of the m-th cyclotomic field $\mathbb{Q}(\mu_m)$. We therefore obtain a canonical isomorphism

$$G(\mathbb{Q}(\mu_m)|\mathbb{Q}) \cong C_{\mathbb{Q}}/C_{\mathbb{Q}}^{\mathfrak{m}} .$$

It is class field theory, which provides a far-reaching generalization of this important fact. For all modules $\mathfrak{m}$ of an arbitrary number field K, there will be Galois extensions $K^{\mathfrak{m}}|K$ generalizing the cyclotomic fields: the so-called **ray class fields**, which satisfy canonically

$$G(K^{\mathfrak{m}}|K) \cong C_K/C_K^{\mathfrak{m}}$$

(see § 6). The ray class group mod 1 is of particular interest here. It is related to the ideal class group Cl_K – which according to our definition here, is in general not a ray class group – as follows.

(1.11) Proposition. *There is an exact sequence*

$$1 \longrightarrow \mathcal{O}^*/\mathcal{O}_+^* \longrightarrow \prod_{\mathfrak{p}\ \mathrm{real}} \mathbb{R}^*/\mathbb{R}_+^* \longrightarrow Cl_K^1 \longrightarrow Cl_K \longrightarrow 1,$$

where $\mathcal{O}_+^*$ *is the group of totally positive units of* K.

Proof: One has $Cl_K^1 \cong C_K/C_K^1 = I_K/I_K^1K^*$ and, by (1.3), $Cl_K \cong I_K/I_K^{S_\infty}K^*$, where $I_K^1 = \prod_{\mathfrak{p}} U_{\mathfrak{p}}$ and $I_K^{S_\infty} = \prod_{\mathfrak{p}\nmid\infty} U_{\mathfrak{p}} \times \prod_{\mathfrak{p}|\infty} K_{\mathfrak{p}}^*$. We therefore obtain an exact sequence

$$1 \longrightarrow I_K^{S_\infty}K^*/I_K^1K^* \longrightarrow C_K/C_K^1 \longrightarrow Cl_K \longrightarrow 1.$$

For the group on the left we have the exact sequence

$$1 \longrightarrow I_K^{S_\infty} \cap K^*/I_K^1 \cap K^* \longrightarrow I_K^{S_\infty}/I_K^1 \longrightarrow I_K^{S_\infty}K^*/I_K^1K^* \longrightarrow 1.$$

But $I_K^{S_\infty} \cap K^* = \mathcal{O}^*$, $I_K^1 \cap K^* = \mathcal{O}_+^*$, and $I_K^{S_\infty}/I_K^1 = \prod_{\mathfrak{p}|\infty} K_\mathfrak{p}^*/U_\mathfrak{p} = \prod_{\mathfrak{p}\ \text{real}} \mathbb{R}^*/\mathbb{R}_+^*$. □

Exercise 1. (i) $\mathbb{A}_\mathbb{Q} = (\widehat{\mathbb{Z}} \otimes_\mathbb{Z} \mathbb{Q}) \times \mathbb{R}$.

(ii) The quotient group $\mathbb{A}_\mathbb{Q}/\mathbb{Z}$ is compact and connected.

(iii) $\mathbb{A}_\mathbb{Q}/\mathbb{Z}$ is arbitrarily and uniquely divisible, i.e., the equation $nx = y$ has a unique solution, for every $n \in \mathbb{N}$ and $y \in \mathbb{A}_\mathbb{Q}/\mathbb{Z}$.

Exercise 2. Let K be a number field, $m = 2^\nu m'$ (m' odd), and let S be a finite set of primes. Let $a \in K^*$ and $a \in K_\mathfrak{p}^{*m}$, for all $\mathfrak{p} \notin S$. Show:

(i) If $K(\zeta_{2^\nu})|K$ is cyclic, where ζ_{2^ν} is a primitive 2^ν-th root of unity, then $a \in K^{*m}$.

(ii) Otherwise one has at least that $a \in K^{*m/2}$.

Hint: Use the following fact, proved in (3.8): if $L|K$ is a finite extension in which almost all prime ideals split completely, then $L = K$.

Exercise 3. Write $I_K^1 = I_\mathrm{f}^1 \times I_\infty^1$, with $I_\mathrm{f}^1 = \prod_{\mathfrak{p}\nmid\infty} U_\mathfrak{p}$, $I_\infty^1 = \prod_{\mathfrak{p}|\infty} U_\mathfrak{p}$. Show that taking integer powers of idèles $\alpha \in I_\mathrm{f}^1$ extends by continuity to exponentiation α^x with $x \in \widehat{\mathbb{Z}}$.

Exercise 4. Let $\varepsilon_1, \dots, \varepsilon_t \in \mathcal{O}_K^*$ be independent units. The images $\bar\varepsilon_1, \dots, \bar\varepsilon_t$ in I_f^1 are then independent units with respect to the exponentiation with elements of $\widehat{\mathbb{Z}}$, i.e., any relation

$$\bar\varepsilon_1^{x_1} \cdots \bar\varepsilon_t^{x_r} = 1, \quad x_i \in \widehat{\mathbb{Z}},$$

implies $x_i = 0$, $i = 1, \dots, t$.

Exercise 5. Let $\varepsilon \in \mathcal{O}_K^*$ be totally positive, i.e., $\varepsilon \in I_K^1$. Extend the exponentiation $\mathbb{Z} \to I_K^1$, $n \mapsto \varepsilon^n$, by continuity to an exponentiation $\widehat{\mathbb{Z}} \times \mathbb{R} \to I_K^1 = I_\mathrm{f}^1 \times I_\infty^1$, $\varepsilon \mapsto \varepsilon^\lambda$, in such a way that $\mathfrak{N}(\varepsilon^\lambda) = 1$.

Exercise 6. Let $\mathfrak{p}_1, \dots, \mathfrak{p}_s$ be the complex primes of K. For $y \in \mathbb{R}$, let $\phi_k(y)$ be the idèle having component $e^{2\pi i y}$ at $\mathfrak{p}_k$, and components 1 at all other places. Let $\varepsilon_1, \dots, \varepsilon_t$ be a $\mathbb{Z}$-basis of the group of totally positive units of K.

(i) The idèles of the form

$$\alpha = \varepsilon_1^{\lambda_1} \cdots \varepsilon_t^{\lambda_t}\, \phi_1(y_1) \cdots \phi_s(y_s), \quad \lambda_i \in \widehat{\mathbb{Z}} \times \mathbb{R},\ y_i \in \mathbb{R},$$

form a group, and have absolute norm $\mathfrak{N}(\alpha) = 1$.

(ii) α is a principal ideal if and only if $\lambda_i \in \mathbb{Z} \subseteq \widehat{\mathbb{Z}} \times \mathbb{R}$ and $y_i \in \mathbb{Z} \subseteq \mathbb{R}$.

Exercise 7. Sending

$$(\lambda_1, \dots, \lambda_t, y_1, \dots, y_s) \mapsto \varepsilon_1^{\lambda_1} \cdots \varepsilon_t^{\lambda_t} \phi_1(y_1) \cdots \phi_s(y_s)$$

defines a continuous homomorphism

$$f : (\widehat{\mathbb{Z}} \times \mathbb{R})^t \times \mathbb{R}^s \longrightarrow C_K^0$$

into the group $C_K^0 = \{[\alpha] \in C_K \mid \mathfrak{N}([\alpha]) = 1\}$, with kernel $\mathbb{Z}^t \times \mathbb{Z}^s$.

Exercise 8. (i) The image D_K^0 of f is compact, connected and arbitrarily divisible.
(ii) f yields a topological isomorphism

$$f : ((\widehat{\mathbb{Z}} \times \mathbb{R})/\mathbb{Z})^t \times (\mathbb{R}/\mathbb{Z})^s \xrightarrow{\sim} D_K^0 .$$

Exercise 9. The group D_K^0 is the intersection of all closed subgroups of finite index in C_K^0, and it is the connected component of 1 in C_K^0.

Exercise 10. The connected component D_K of 1 in the idèle class group C_K is the direct product of t copies of the "solenoid" $(\widehat{\mathbb{Z}} \times \mathbb{R})/\mathbb{Z}$, s circles $\mathbb{R}/\mathbb{Z}$, and a real line.

Exercise 11. Every ideal class of the ray class group $Cl_K^{\mathfrak{m}}$ can be represented by an integral ideal which is prime to an arbitrary fixed ideal.

Exercise 12. Let $\mathcal{O} = \mathcal{O}_K$. Every class in $(\mathcal{O}/\mathfrak{m})^*$ can be represented by a totally positive number in $\mathcal{O}$ which is prime to an arbitrary fixed ideal.

Exercise 13. For every module $\mathfrak{m}$, one has an exact sequence

$$1 \to \mathcal{O}_+^*/\mathcal{O}_+^{\mathfrak{m}} \longrightarrow (\mathcal{O}/\mathfrak{m})^* \longrightarrow Cl_K^{\mathfrak{m}} \longrightarrow Cl_K^1 \to 1,$$

where $\mathcal{O}_+^*$, resp. $\mathcal{O}_+^{\mathfrak{m}}$, is the group of totally positive units of $\mathcal{O}$, resp. of totally positive units $\equiv 1 \bmod \mathfrak{m}$.

Exercise 14. Compute the kernels of $Cl_K^{\mathfrak{m}} \to Cl_K$ and $Cl_K^{\mathfrak{m}} \to Cl_K^{\mathfrak{m}'}$ for $\mathfrak{m}'|\mathfrak{m}$.

§ 2. Idèles in Field Extensions

We shall now study the behaviour of idèles and idèle classes when we pass from a field K to an extension L. So let $L|K$ be a finite extension of algebraic number fields. We embed the idèle group I_K of K into the idèle group I_L of L by sending an idèle $\alpha = (\alpha_{\mathfrak{p}}) \in I_K$ to the idèle $\alpha' = (\alpha'_{\mathfrak{P}}) \in I_L$ whose components $\alpha'_{\mathfrak{P}}$ are given by

$$\alpha'_{\mathfrak{P}} = \alpha_{\mathfrak{p}} \in K_{\mathfrak{p}}^* \subseteq L_{\mathfrak{P}}^* \quad \text{for} \quad \mathfrak{P}|\mathfrak{p}.$$

In this way we obtain an injective homomorphism

$$I_K \longrightarrow I_L ,$$

which will always be tacitly used to consider I_K as a subgroup of I_L. An element $\alpha = (\alpha_{\mathfrak{P}}) \in I_L$ therefore belongs to the group I_K if and only if its components $\alpha_{\mathfrak{P}}$ belong to $K_{\mathfrak{p}}$ $(\mathfrak{P}|\mathfrak{p})$, and if one has furthermore $\alpha_{\mathfrak{P}} = \alpha_{\mathfrak{P}'}$ whenever $\mathfrak{P}$ and $\mathfrak{P}'$ lie above the same place $\mathfrak{p}$ of K.

Every isomorphism $\sigma : L \to \sigma L$ induces an isomorphism

$$\sigma : I_L \longrightarrow I_{\sigma L}$$

like this. For each place $\mathfrak{P}$ of L, σ induces an isomorphism

$$\sigma : L_{\mathfrak{P}} \longrightarrow (\sigma L)_{\sigma\mathfrak{P}}.$$

For if we have $\alpha = \mathfrak{P}\text{-}\lim \alpha_i$, for some sequence $\alpha_i \in L$, then the sequence $\sigma\alpha_i \in \sigma L$ converges with respect to $|\ \ |_{\sigma\mathfrak{P}}$ in $(\sigma L)_{\sigma\mathfrak{P}}$, and the isomorphism is given by

$$\alpha = \mathfrak{P}\text{-}\lim \alpha_i \mapsto \sigma\alpha = \sigma\mathfrak{P}\text{-}\lim \sigma\alpha_i.$$

For an idèle $\alpha \in I_L$, we then define $\sigma\alpha \in I_L$ to be the idèle with components

$$(\sigma\alpha)_{\sigma\mathfrak{P}} = \sigma\alpha_{\mathfrak{P}} \in (\sigma L)_{\sigma\mathfrak{P}}.$$

If $L|K$ is a Galois extension with Galois group $G = G(L|K)$, then every $\sigma \in G$ yields an automorphism $\sigma : I_L \to I_L$, i.e., I_L is turned into an G-module. As to the fixed module $I_L^G = \{\alpha \in I_L \mid \sigma\alpha = \alpha \text{ for all } \sigma \in G\}$, we have the

(2.1) Proposition. *If $L|K$ is a Galois extension with Galois group G, then*

$$I_L^G = I_K.$$

Proof: Let $\alpha \in I_K \subseteq I_L$. For $\sigma \in G$, the induced map $\sigma : L_{\mathfrak{P}} \to L_{\sigma\mathfrak{P}}$ is a $K_{\mathfrak{p}}$-isomorphism, if $\mathfrak{P}|\mathfrak{p}$. Therefore

$$(\sigma\alpha)_{\sigma\mathfrak{P}} = \sigma\alpha_{\mathfrak{P}} = \alpha_{\mathfrak{P}} = \alpha_{\sigma\mathfrak{P}},$$

so that $\sigma\alpha = \alpha$, and therefore $\alpha \in I_L^G$. If conversely $\alpha = (\alpha_{\mathfrak{P}}) \in I_L^G$, then

$$(\sigma\alpha)_{\sigma\mathfrak{P}} = \sigma\alpha_{\mathfrak{P}} = \alpha_{\sigma\mathfrak{P}}$$

for all $\sigma \in G$. In particular, if σ belongs to the decomposition group $G_{\mathfrak{P}} = G(L_{\mathfrak{P}}|K_{\mathfrak{p}})$, then $\sigma\mathfrak{P} = \mathfrak{P}$ and $\sigma\alpha_{\mathfrak{P}} = \alpha_{\mathfrak{P}}$ so that $\alpha_{\mathfrak{P}} \in K_{\mathfrak{p}}^*$. If $\sigma \in G$ is arbitrary, then $\sigma : L_{\mathfrak{P}} \to L_{\sigma\mathfrak{P}}$ induces the identity on $K_{\mathfrak{p}}$, and we get $\alpha_{\mathfrak{P}} = \sigma\alpha_{\mathfrak{P}} = \alpha_{\sigma\mathfrak{P}}$ for any two places $\mathfrak{P}$ and $\sigma\mathfrak{P}$ above $\mathfrak{p}$. This shows that $\alpha \in I_K$. □

The idèle group I_L is the unit group of the ring of adèles $\mathbb{A}_L$ of L. It is convenient to write this ring as

$$\mathbb{A}_L = \prod_{\mathfrak{p}} L_{\mathfrak{p}},$$

where

$$L_{\mathfrak{p}} = \prod_{\mathfrak{P}|\mathfrak{p}} L_{\mathfrak{P}}.$$

The restricted product $\prod_{\mathfrak{p}} L_{\mathfrak{p}}$ consists of all families $(\alpha_{\mathfrak{p}})$ of elements $\alpha_{\mathfrak{p}} \in L_{\mathfrak{p}}$ such that $\alpha_{\mathfrak{p}} \in \mathcal{O}_{\mathfrak{p}} = \prod_{\mathfrak{P}|\mathfrak{p}} \mathcal{O}_{\mathfrak{P}}$ for almost all $\mathfrak{p}$. Via the diagonal embedding

$$K_{\mathfrak{p}} \longrightarrow L_{\mathfrak{p}},$$

the factor $L_{\mathfrak{p}}$ is a commutative $K_{\mathfrak{p}}$-algebra of degree $\sum_{\mathfrak{P}|\mathfrak{p}}[L_{\mathfrak{P}} : K_{\mathfrak{p}}] = [L : K]$. These embeddings yield the embedding

$$\mathbb{A}_K \longrightarrow \mathbb{A}_L,$$

whose restriction

$$I_K = \mathbb{A}_K^* \hookrightarrow \mathbb{A}_L^* = I_L$$

turns out to be the inclusion considered above.

Every $\alpha_{\mathfrak{p}} \in L_{\mathfrak{p}}^*$ defines an automorphism

$$\alpha_{\mathfrak{p}} : L_{\mathfrak{p}} \longrightarrow L_{\mathfrak{p}}, \quad x \longmapsto \alpha_{\mathfrak{p}} x,$$

of the $K_{\mathfrak{p}}$-vector space $L_{\mathfrak{p}}$, and as in the case of a field extension, we define the norm of $\alpha_{\mathfrak{p}}$ by

$$N_{L_{\mathfrak{p}}|K_{\mathfrak{p}}}(\alpha_{\mathfrak{p}}) = \det(\alpha_{\mathfrak{p}}).$$

In this way we obtain a homomorphism

$$N_{L_{\mathfrak{p}}|K_{\mathfrak{p}}} : L_{\mathfrak{p}}^* \longrightarrow K_{\mathfrak{p}}^*.$$

It induces a norm homomorphism

$$N_{L|K} : I_L \longrightarrow I_K$$

between the idèle groups $I_L = \prod_{\mathfrak{p}} L_{\mathfrak{p}}^*$ and $I_K = \prod_{\mathfrak{p}} K_{\mathfrak{p}}^*$. Explicitly the norm of an idèle is given by the following proposition.

(2.2) Proposition. *If $L|K$ is a finite extension and $\alpha = (\alpha_{\mathfrak{P}}) \in I_L$, the local components of the idèle $N_{L|K}(\alpha)$ are given by*

$$N_{L|K}(\alpha)_{\mathfrak{p}} = \prod_{\mathfrak{P}|\mathfrak{p}} N_{L_{\mathfrak{P}}|K_{\mathfrak{p}}}(\alpha_{\mathfrak{P}}).$$

Proof: Putting $\alpha_{\mathfrak{p}} = (\alpha_{\mathfrak{P}})_{\mathfrak{P}|\mathfrak{p}} \in L_{\mathfrak{p}}$, the $K_{\mathfrak{p}}$-automorphism $\alpha_{\mathfrak{p}} : L_{\mathfrak{p}} \to L_{\mathfrak{p}}$ is the direct product of the $K_{\mathfrak{p}}$-automorphisms $\alpha_{\mathfrak{P}} : L_{\mathfrak{P}} \to L_{\mathfrak{P}}$. Therefore

$$N_{L_{\mathfrak{p}}|K_{\mathfrak{p}}}(\alpha_{\mathfrak{p}}) = \det(\alpha_{\mathfrak{p}}) = \prod_{\mathfrak{P}|\mathfrak{p}} \det(\alpha_{\mathfrak{P}}) = \prod_{\mathfrak{P}|\mathfrak{p}} N_{L_{\mathfrak{P}}|K_{\mathfrak{p}}}(\alpha_{\mathfrak{P}}). \qquad \square$$

The idèle norm enjoys the following properties.

(2.3) Proposition. (i) *For a tower of fields* $K \subseteq L \subseteq M$ *we have* $N_{M|K} = N_{L|K} \circ N_{M|L}$.

(ii) *If* $L|K$ *is embedded into the Galois extension* $M|K$ *and if* $G = G(M|K)$ *and* $H = G(M|L)$, *then one has for* $\alpha \in I_L$: $N_{L|K}(\alpha) = \prod_{\sigma \in G/H} \sigma\alpha$.

(iii) $N_{L|K}(\alpha) = \alpha^{[L:K]}$ for $\alpha \in I_K$.

(iv) *The norm of the principal idèle* $x \in L^*$ *is the principal idèle of* K *defined by the usual norm* $N_{L|K}(x)$.

The proofs of (i), (ii), (iii) are literally the same as for the norm in a field extension (see chap. I, §2). (iv) follows from the fact that, once we identify $L_\mathfrak{p} = L \otimes_K K_\mathfrak{p}$ (see chap. II, (8.3)), the $K_\mathfrak{p}$-automorphism $f_x : L_\mathfrak{p} \to L_\mathfrak{p}$, $y \mapsto xy$, arises from the K-automorphism $x : L \to L$ by tensoring with $K_\mathfrak{p}$. Hence $\det(f_x) = \det(x)$.

Remark: For fundamental as well as practical reasons, it is convenient to adopt a formal point of view for the above considerations which allows us to avoid the constant back and forth between idèles and their components. This point of view is based on identifying the ring of adèles $\mathbb{A}_L$ of L as

$$\mathbb{A}_L = \mathbb{A}_K \otimes_K L,$$

which results from the canonical isomorphisms (see chap. II, (8.3))

$$K_\mathfrak{p} \otimes_K L \xrightarrow{\sim} L_\mathfrak{p} = \prod_{\mathfrak{P}|\mathfrak{p}} L_\mathfrak{P}, \quad \alpha_\mathfrak{p} \otimes a \longmapsto \alpha_\mathfrak{p} \cdot (\tau_\mathfrak{P} a).$$

Here $\tau_\mathfrak{P}$ denotes the canonical embedding $\tau_\mathfrak{P} : L \to L_\mathfrak{P}$.

In this way the inclusion by components $I_K \subseteq I_L$ is simply given by the embedding $\mathbb{A}_K \hookrightarrow \mathbb{A}_L$, $\alpha \mapsto \alpha \otimes 1$, induced by $K \subseteq L$. An isomorphism $L \to \sigma L$ then yields the isomorphism

$$\sigma : \mathbb{A}_L = \mathbb{A}_K \otimes_K L \longrightarrow \mathbb{A}_K \otimes_K \sigma L = \mathbb{A}_{\sigma L}$$

via $\sigma(\alpha \otimes a) = \alpha \otimes \sigma a$, and the norm of an L-idèle $\alpha \in \mathbb{A}_L^*$ is simply the determinant

$$N_{L|K}(\alpha) = \det_{\mathbb{A}_K}(\alpha)$$

of the endomorphism $\alpha : \mathbb{A}_L \to \mathbb{A}_L$ which α induces on the finite $\mathbb{A}_K$-algebra $\mathbb{A}_L = \mathbb{A}_K \otimes_K L$.

Here are consequences of the preceding investigations for the **idèle class groups**.

(2.4) Proposition. *If $L|K$ is a finite extension, then the homomorphism $I_K \to I_L$ induces an injection of idèle class groups*

$$C_K \longrightarrow C_L\,, \quad \alpha K^* \longmapsto \alpha L^*\,.$$

Proof: The injection $I_K \to I_L$ clearly maps K^* into L^*. For the injectivity, we have to show that $I_K \cap L^* = K^*$. Let $M|K$ be a finite Galois extension with Galois group G containing L. Then we have $I_K \subseteq I_L \subseteq I_M$, and

$$I_K \cap L^* \subseteq I_K \cap M^* \subseteq (I_K \cap M^*)^G = I_K \cap M^{*G} = I_K \cap K^* = K^*. \quad \square$$

Via the embedding $C_K \to C_L$, the idèle class group C_K becomes a subgroup of C_L: an element $\alpha L^* \in C_L$ ($\alpha \in I_L$) lies in C_K if and only if the class αL^* has a representative α' in I_K. It is important to know that we have **Galois descent** for the idèle class group:

(2.5) Proposition. *If $L|K$ is a Galois extension and $G = G(L|K)$, then C_L is canonically a G-module and $C_L^G = C_K$.*

Proof: The G-module I_L contains L^* as a G-submodule. Hence every $\sigma \in G$ induces an automorphism

$$C_L \xrightarrow{\sigma} C_L\,, \quad \alpha L^* \longmapsto (\sigma\alpha)L^*\,.$$

This gives us an exact sequence of G-modules

$$1 \longrightarrow L^* \longrightarrow I_L \longrightarrow C_L \longrightarrow 1\,.$$

We claim that the sequence

$$1 \longrightarrow L^{*G} \longrightarrow I_L^G \longrightarrow C_L^G \longrightarrow 1$$

deduced from the first is still exact. The injectivity of $L^{*G} \to I_L^G$ is trivial. The kernel of $I_L^G \to C_L^G$ is $I_L^G \cap L^* = I_K \cap L^* = K^* = L^{*G}$. The surjectivity of $I_L^G \to C_L^G$ is not altogether straightforward. To prove it, let $\alpha L^* \in C_L^G$. For every $\sigma \in G$, one then has $\sigma(\alpha L^*) = \alpha L^*$, i.e., $\sigma\alpha = \alpha x_\sigma$ for some $x_\sigma \in L^*$. This x_σ is a "crossed homomorphism", i.e., we have

$$x_{\sigma\tau} = x_\sigma \cdot \sigma x_\tau\,.$$

Indeed, $x_{\sigma\tau} = \frac{\sigma\tau\alpha}{\alpha} = \frac{\sigma\tau\alpha}{\sigma\alpha}\cdot\frac{\sigma\alpha}{\alpha} = \sigma(\frac{\tau\alpha}{\alpha})\frac{\sigma\alpha}{\alpha} = \sigma x_\tau x_\sigma$. By Hilbert 90 in Noether's version (see chap. IV, (3.8)) such a crossed homomorphism is of the form $x_\sigma = \sigma y/y$ for some $y \in L^*$. Putting $\alpha' = \alpha y^{-1}$ yields $\alpha' L^* = \alpha L^*$ and $\sigma\alpha' = \sigma\alpha\sigma y^{-1} = \alpha x_\sigma \sigma y^{-1} = \alpha y^{-1} = \alpha'$, hence $\alpha' \in I_L^G$. This proves surjectivity. $\square$

The norm map $N_{L|K} : I_L \to I_K$ sends principal idèles to principal idèles by (2.3). Hence we get a norm map also for the idèle class group C_L,

$$N_{L|K} : C_L \longrightarrow C_K .$$

It enjoys the same properties (2.3), (i), (ii), (iii), as the norm map on the idèle group.

Exercise 1. Let $\omega_1, \dots, \omega_n$ be a basis of $L|K$. Then the isomorphism $L \otimes_K K_\mathfrak{p} \cong \prod_{\mathfrak{P}|\mathfrak{p}} L_\mathfrak{P}$ induces, for almost all prime ideals $\mathfrak{p}$ of K, an isomorphism

$$\omega_1 \mathcal{O}_\mathfrak{p} \oplus \cdots \oplus \omega_n \mathcal{O}_\mathfrak{p} \cong \prod_{\mathfrak{P}|\mathfrak{p}} \mathcal{O}_\mathfrak{P},$$

where $\mathcal{O}_\mathfrak{p}$, resp. $\mathcal{O}_\mathfrak{P}$, is the valuation ring of $K_\mathfrak{p}$, resp. $L_\mathfrak{P}$.

Exercise 2. Let $L|K$ be a finite extension. The absolute norm $\mathfrak{N}$ of idèles of K, resp. L, behaves as follows under the inclusion $i_{L|K} : I_K \to I_L$, resp. under the norm $N_{L|K} : I_L \to I_K$:

$$\mathfrak{N}(i_{L|K}(\alpha)) = \mathfrak{N}(\alpha)^{[L:K]} \quad \text{for} \quad \alpha \in I_K ,$$
$$\mathfrak{N}(N_{L|K}(\alpha)) = \mathfrak{N}(\alpha) \quad \text{for} \quad \alpha \in I_L .$$

Exercise 3. The correspondence between idèles and ideals, $\alpha \mapsto (\alpha)$, satisfies the following rule, in the case of a Galois extension $L|K$,

$$(N_{L|K}(\alpha)) = N_{L|K}((\alpha)) .$$

(For the norm on ideals, see chap. III, § 1.)

Exercise 4. The ideal class group, unlike the idèle class group, does not have Galois descent. More precisely, for a Galois extension $L|K$, the homomorphism $Cl_K \to Cl_L^{G(L|K)}$ is in general neither injective nor surjective.

Exercise 5. Define the trace $Tr_{L|K} : \mathbb{A}_L \to \mathbb{A}_K$ by $Tr_{L|K}(\alpha)$ = trace of the endomorphism $x \mapsto \alpha x$ of the $\mathbb{A}_K$-algebra $\mathbb{A}_L$, and show:

(i) $Tr_{L|K}(\alpha)_\mathfrak{p} = \sum_{\mathfrak{P}|\mathfrak{p}} Tr_{L_\mathfrak{P}|K_\mathfrak{p}}(\alpha_\mathfrak{P})$.

(ii) For a tower of fields $K \subseteq L \subseteq M$, one has $Tr_{M|K} = Tr_{L|K} \circ Tr_{M|L}$.

(iii) If $L|K$ is embedded into the Galois extension $M|K$, and if $G = G(M|K)$ and $H = G(M|L)$, then one has for $\alpha \in \mathbb{A}_L$, $Tr_{L|K}(\alpha) = \sum_{\sigma \in G/H} \sigma\alpha$.

(iv) $Tr_{L|K}(\alpha) = [L : K]\alpha$ for $\alpha \in \mathbb{A}_K$.

(v) The trace of a principal adèle $x \in L$ is the principal adèle in $\mathbb{A}_K$ defined by the usual trace $Tr_{L|K}(x)$.

§ 3. The Herbrand Quotient of the Idèle Class Group

Our goal now is to show that the idèle class group satisfies the class field axiom of chap. IV, (6.1). To do this we will first compute its Herbrand quotient. It is constituted on the one hand by the Herbrand quotient of the idèle group, and by that of the unit group on the other. We study the idèle group first.

Let $L|K$ be a finite Galois extension with Galois group G. The G-module I_L may be described in the following simple manner, which immediately reduces us to local fields. For every place $\mathfrak{p}$ of K we put

$$L_{\mathfrak{p}}^* = \prod_{\mathfrak{P}|\mathfrak{p}} L_{\mathfrak{P}}^* \quad \text{and} \quad U_{L,\mathfrak{p}} = \prod_{\mathfrak{P}|\mathfrak{p}} U_{\mathfrak{P}}.$$

Since the automorphisms $\sigma \in G$ permute the places of L above $\mathfrak{p}$, the groups $L_{\mathfrak{p}}^*$ and $U_{L,\mathfrak{p}}$ are G-modules, and we have for the G-module I_L the decomposition

$$I_L = \coprod_{\mathfrak{p}} L_{\mathfrak{p}}^*,$$

where the restricted product is taken with respect to the subgroups $U_{L,\mathfrak{p}} \subseteq L_{\mathfrak{p}}^*$. Choose a place $\mathfrak{P}$ of L above $\mathfrak{p}$, and let $G_{\mathfrak{P}} = G(L_{\mathfrak{P}}|K_{\mathfrak{p}}) \subseteq G$ be its decomposition group. As σ varies over a system of representatives of $G/G_{\mathfrak{P}}$, $\sigma\mathfrak{P}$ runs through the various places of L above $\mathfrak{p}$, and we get

$$L_{\mathfrak{p}}^* = \prod_{\sigma} L_{\sigma\mathfrak{P}}^* = \prod_{\sigma} \sigma(L_{\mathfrak{P}}^*), \quad U_{L,\mathfrak{p}} = \prod_{\sigma} U_{\sigma\mathfrak{P}} = \prod_{\sigma} \sigma(U_{\mathfrak{P}}).$$

In terms of the notion of *induced module* introduced in chap. IV, § 7, we thus get the following

(3.1) Proposition. *$L_{\mathfrak{p}}^*$ and $U_{L,\mathfrak{p}}$ are the induced G-modules*

$$L_{\mathfrak{p}}^* = \operatorname{Ind}_G^{G_{\mathfrak{P}}}(L_{\mathfrak{P}}^*), \quad U_{L,\mathfrak{p}} = \operatorname{Ind}_G^{G_{\mathfrak{P}}}(U_{\mathfrak{P}}).$$

Now let S be a finite set of places of K containing the infinite places. We then define $I_L^S = I_L^{\bar{S}}$, where $\bar{S}$ denotes the set of all places of L which lie above the places of S. For I_L^S we have the G-module decomposition

$$I_L^S = \prod_{\mathfrak{p}\in S} L_{\mathfrak{p}}^* \times \prod_{\mathfrak{p}\notin S} U_{L,\mathfrak{p}},$$

and (3.1) gives the

(3.2) Proposition. *If $L|K$ is a cyclic extension, and if S contains all primes ramified in L, then we have for $i = 0, -1$ that*

$$H^i(G, I_L^S) \cong \bigoplus_{\mathfrak{p}\in S} H^i(G_{\mathfrak{P}}, L_{\mathfrak{P}}^*) \quad \text{and} \quad H^i(G, I_L) \cong \bigoplus_{\mathfrak{p}} H^i(G_{\mathfrak{P}}, L_{\mathfrak{P}}^*),$$

where for each $\mathfrak{p}$, $\mathfrak{P}$ is a chosen prime of L above $\mathfrak{p}$.

Proof: The decomposition $I_L^S = (\bigoplus_{\mathfrak{p}\in S} L_\mathfrak{p}^*) \oplus V$, $V = \prod_{\mathfrak{p}\notin S} U_{L,\mathfrak{p}}$, gives us an isomorphism

$$H^i(G, I_L^S) = \bigoplus_{\mathfrak{p}\in S} H^i(G, L_\mathfrak{p}^*) \ \oplus\ H^i(G, V),$$

and an injection $H^i(G,V) \to \prod_{\mathfrak{p}\notin S} H^i(G, U_{L,\mathfrak{p}})$. By (3.1) and chap. IV, (7.4), we have the isomorphisms $H^i(G, L_\mathfrak{p}^*) \cong H^i(G_\mathfrak{P}, L_\mathfrak{P}^*)$ and $H^i(G, U_{L,\mathfrak{p}}) \cong H^i(G_\mathfrak{P}, U_\mathfrak{P})$. For $\mathfrak{p} \notin S$, $L_\mathfrak{P}|K_\mathfrak{p}$ is unramified. Hence $H^i(G_\mathfrak{P}, U_\mathfrak{P}) = 1$, by chap. V, (1.2). This shows the first claim of the proposition. The second is an immediate consequence:

$$H^i(G, I_L) = \varinjlim_S H^i(G, I_L^S) \cong \varinjlim_S \bigoplus_{\mathfrak{p}\in S} H^i(G_\mathfrak{P}, L_\mathfrak{P}^*) = \bigoplus_\mathfrak{p} H^i(G_\mathfrak{P}, L_\mathfrak{P}^*).$$

□

The proposition says that one has $H^{-1}(G, I_L) = \{1\}$, because $H^{-1}(G_\mathfrak{P}, L_\mathfrak{P}^*) = \{1\}$ by Hilbert 90. Further it says that

$$I_K / N_{L|K} I_L = \bigoplus_\mathfrak{p} K_\mathfrak{p}^* / N_{L_\mathfrak{P}|K_\mathfrak{p}} L_\mathfrak{P}^*,$$

where $\mathfrak{P}$ is a chosen place above $\mathfrak{p}$. In other words:

An idèle $\alpha \in I_K$ is a norm of an idèle of L if and only if it is *a norm locally everywhere*, i.e., if every component $\alpha_\mathfrak{p}$ is the norm of an element of $L_\mathfrak{P}^*$.

As for the Herbrand quotient $h(G, I_L^S)$ we obtain the result:

(3.3) Proposition. *If $L|K$ is a cyclic extension and if S contains all ramified primes, then*

$$h(G, I_L^S) = \prod_{\mathfrak{p}\in S} n_\mathfrak{p},$$

where $n_\mathfrak{p} = [L_\mathfrak{P} : K_\mathfrak{p}]$.

Proof: We have $H^{-1}(G, I_L^S) = \prod_{\mathfrak{p}\in S} H^{-1}(G_\mathfrak{P}, L_\mathfrak{P}^*) = 1$ and

$$H^0(G, I_L^S) = \prod_{\mathfrak{p}\in S} H^0(G_\mathfrak{P}, L_\mathfrak{P}^*).$$

By local class field theory, we find $\# H^0(G_\mathfrak{P}, L_\mathfrak{P}^*) = (K_\mathfrak{p}^* : N_{L_\mathfrak{P}|K_\mathfrak{p}} L_\mathfrak{P}^*) = n_\mathfrak{p}$. Hence

$$h(G, I_L^S) = \frac{\# H^0(G, I_L^S)}{\# H^{-1}(G, I_L^S)} = \prod_{\mathfrak{p}\in S} n_\mathfrak{p}.$$

□

Next we determine the Herbrand quotient of the G-module $L^S = L \cap I_L^S$. For this we need the following general

(3.4) Lemma. *Let V be an s-dimensional $\mathbb{R}$-vector space, and let G be a finite group of automorphisms of V which operates as a permutation group on the elements of a basis $v_1, \dots, v_s$: $\sigma v_i = v_{\sigma(i)}$.*

If Γ is a G-invariant complete lattice in V, i.e., $\sigma\Gamma \subseteq \Gamma$ for all σ, then there exists a complete sublattice in Γ,

$$\Gamma' = \mathbb{Z}w_1 + \cdots + \mathbb{Z}w_s,$$

such that $\sigma w_i = w_{\sigma(i)}$ for all $\sigma \in G$.

Proof: Let $|\ |$ be the sup-norm with respect to the coordinates of the basis $v_1, \dots, v_s$. Since Γ is a lattice, there exists a number b such that for every $x \in V$, there is a $\gamma \in \Gamma$ satisfying

$$|x - \gamma| < b.$$

Choose a large positive number $t \in \mathbb{R}$, and a $\gamma \in \Gamma$ such that

$$|tv_1 - \gamma| < b,$$

and define

$$w_i = \sum_{\sigma(1)=i} \sigma\gamma, \quad i = 1, \dots, s,$$

i.e., the summation is over all $\sigma \in G$ such that $\sigma(1) = i$. For every $\tau \in G$ we then have

$$\tau w_i = \sum_{\sigma(1)=i} \tau\sigma\gamma = \sum_{\rho(1)=\tau(i)} \rho\gamma = w_{\tau(i)}.$$

It is therefore enough to check the linear independence of the w_i. To do this, let

$$\sum_{i=1}^{s} c_i w_i = 0, \quad c_i \in \mathbb{R}.$$

If not all of the $c_i = 0$, then we may assume $|c_i| \le 1$ and $c_j = 1$ for some j. Let

$$\gamma = tv_1 - y,$$

for some vector y of absolute value $|y| < b$. Then

$$w_i = \sum_{\sigma(1)=i} \sigma\gamma = t \sum_{\sigma(1)=i} v_{\sigma(1)} - y_i = tn_i v_i - y_i,$$

where $|y_i| \le gb$, for $g = \#G$, and $n_i = \#\{\sigma \in G \mid \sigma(1) = i\}$. We therefore get

$$0 = \sum_{i=1}^{s} c_i w_i = t \sum_{i=1}^{s} c_i n_i v_i - z,$$

with $|z| \le sgb$, i.e.,

$$z = tn_j v_j + \sum_{i \ne j} tc_i n_i v_i.$$

If t was chosen sufficiently large, then z cannot be written in this way. This contradiction proves the lemma. □

Now let $L|K$ be a cyclic extension of degree n with Galois group $G = G(L|K)$, let S be a finite set of places containing the infinite places, and let $\bar{S}$ be the set of places of L that lie above the places of S. We denote the group $L^{\bar{S}}$ of $\bar{S}$-units simply by L^S.

(3.5) Proposition. *The Herbrand quotient of the G-module L^S satisfies*

$$h(G, L^S) = \frac{1}{n} \prod_{\mathfrak{p} \in S} n_{\mathfrak{p}},$$

where $n_{\mathfrak{p}} = [L_{\mathfrak{P}} : K_{\mathfrak{p}}]$.

Proof: Let $\{\mathfrak{e}_{\mathfrak{P}} \mid \mathfrak{P} \in \bar{S}\}$ be the standard basis of the vector space $V = \prod_{\mathfrak{P} \in \bar{S}} \mathbb{R}$. By (1.1), the homomorphism

$$\lambda : L^S \longrightarrow V, \quad \lambda(a) = \sum_{\mathfrak{P} \in \bar{S}} \log |a|_{\mathfrak{P}} \mathfrak{e}_{\mathfrak{P}},$$

has kernel $\mu(L)$ and its image is an $(\bar{s}-1)$-dimensional lattice, $\bar{s} = \#\bar{S}$. We make G operate on V via

$$\sigma \mathfrak{e}_{\mathfrak{P}} = \mathfrak{e}_{\sigma\mathfrak{P}}.$$

Then λ is a G-homomorphism because we have, for $\sigma \in G$,

$$\begin{aligned} \lambda(\sigma a) &= \sum_{\mathfrak{P}} \log |\sigma a|_{\mathfrak{P}} \mathfrak{e}_{\mathfrak{P}} = \sum_{\mathfrak{P}} \log |a|_{\sigma^{-1}\mathfrak{P}} \sigma\, \mathfrak{e}_{\sigma^{-1}\mathfrak{P}} \\ &= \sigma\Big(\sum_{\mathfrak{P}} \log |a|_{\sigma^{-1}\mathfrak{P}} \mathfrak{e}_{\sigma^{-1}\mathfrak{P}}\Big) = \sigma\lambda(a). \end{aligned}$$

Therefore $\mathfrak{e}_0 = \sum_{\mathfrak{P} \in \bar{S}} \mathfrak{e}_{\mathfrak{P}}$ and $\lambda(L^S)$ generate a G-invariant complete lattice Γ in V. Since $\mathbb{Z}\mathfrak{e}_0$ is G-isomorphic to $\mathbb{Z}$, the exact sequence

$$0 \longrightarrow \mathbb{Z}\mathfrak{e}_0 \longrightarrow \Gamma \longrightarrow \Gamma/\mathbb{Z}\mathfrak{e}_0 \longrightarrow 0,$$

together with the fact that $\Gamma/\mathbb{Z}\mathfrak{e}_0 = \lambda(L^S)$, yields the identities

$$h(G, L^S) = h(G, \lambda(L^S)) = h(G, \mathbb{Z})^{-1} h(G, \Gamma) = \frac{1}{n} h(G, \Gamma).$$

We now choose in Γ a sublattice Γ', in accordance with lemma (3.4). Then we have

$$\Gamma' = \bigoplus_{\mathfrak{P}} \mathbb{Z} w_{\mathfrak{P}} = \bigoplus_{\mathfrak{p} \in S} \bigoplus_{\mathfrak{P}|\mathfrak{p}} \mathbb{Z} w_{\mathfrak{P}} = \bigoplus_{\mathfrak{p} \in S} \Gamma'_{\mathfrak{p}}$$

and $\sigma w_{\mathfrak{P}} = w_{\sigma\mathfrak{P}}$. This identifies $\Gamma'_{\mathfrak{p}}$ as the induced G-module

$$\Gamma'_{\mathfrak{p}} = \bigoplus_{\mathfrak{P}|\mathfrak{p}} \mathbb{Z} w_{\mathfrak{P}} = \bigoplus_{\sigma \in G/G_{\mathfrak{p}}} \sigma(\mathbb{Z} w_{\mathfrak{P}_0}) = \operatorname{Ind}_G^{G_{\mathfrak{p}}}(\mathbb{Z} w_{\mathfrak{P}_0}),$$

where $\mathfrak{P}_0$ is a chosen place above $\mathfrak{p}$, and $G_{\mathfrak{p}}$ is its decomposition group. The lattice Γ' has the same rank as Γ, so is therefore of finite index in Γ. From chap. IV, (7.4), we conclude that

$$h(G,L^S) = \frac{1}{n}h(G,\Gamma') = \frac{1}{n}\prod_{\mathfrak{p}\in S} h(G,\Gamma'_{\mathfrak{p}}) = \frac{1}{n}\prod_{\mathfrak{p}\in S} h(G_{\mathfrak{p}},\mathbb{Z}w_{\mathfrak{P}_o})$$
$$= \frac{1}{n}\prod_{\mathfrak{p}\in S} h(G_{\mathfrak{p}},\mathbb{Z}).$$

Thus we do find that $h(G,L^S) = \frac{1}{n}\prod_{\mathfrak{p}\in S} n_{\mathfrak{p}}$, where $n_{\mathfrak{p}} = \#G_{\mathfrak{p}} = [L_{\mathfrak{P}} : K_{\mathfrak{p}}]$. □

From the Herbrand quotient of I_L^S and L^S we immediately get the Herbrand quotient of the idèle class group C_L. To do it choose a finite set of places S containing all infinite ones and all primes ramified in L, such that $I_L = I_L^S L^*$. Such a set exists by (1.4). From the exact sequence

$$1 \longrightarrow L^S \longrightarrow I_L^S \longrightarrow I_L^S L^*/L^* \longrightarrow 1$$

arises the identity

$$h(G,C_L) = h(G,I_L^S)h(G,L^S)^{-1},$$

and from (3.3) and (3.5) we obtain the

(3.6) Theorem. *If $L|K$ is a cyclic extension of degree n with Galois group $G = G(L|K)$, then*

$$h(G,C_L) = \frac{\#H^0(G,C_L)}{\#H^{-1}(G,C_L)} = n.$$

In particular $(C_K : N_{L|K}C_L) \geq n$.

From this result we deduce the following interesting consequence.

(3.7) Corollary. *If $L|K$ is cyclic of prime power degree $n = p^\nu$ $(\nu > 0)$, then there are infinitely many places of K which do not split in L.*

Proof: Assume that the set S of nonsplit primes were finite. Let $M|K$ be the subextension of $L|K$ of degree p. For every $\mathfrak{p} \notin S$, the decomposition group $G_{\mathfrak{p}}$ of $L|K$ is different from $G(L|K)$. Hence $G_{\mathfrak{p}} \subseteq G(L|M)$. Therefore every $\mathfrak{p} \notin S$ splits completely in M. We deduce from this that $N_{M|K}C_M = C_K$, thus contradicting (3.6).

Indeed, let $\alpha \in I_K$. By the approximation theorem of chap. II, (3.4), there exists an $a \in K^*$ such that $\alpha_{\mathfrak{p}} a^{-1}$ is contained in the open subgroup $N_{M_{\mathfrak{P}}|K_{\mathfrak{p}}} M_{\mathfrak{P}}^*$, for all $\mathfrak{p} \in S$. If $\mathfrak{p} \notin S$, then $\alpha_{\mathfrak{p}} a^{-1}$ is automatically contained in $N_{M_{\mathfrak{P}}|K_{\mathfrak{p}}} M_{\mathfrak{P}}^*$ because $M_{\mathfrak{P}} = K_{\mathfrak{p}}$. Since

$$I_K / N_{M|K} I_M = \bigoplus_{\mathfrak{p}} K_{\mathfrak{p}}^* / N_{M_{\mathfrak{P}}|K_{\mathfrak{p}}} M_{\mathfrak{P}}^*,$$

the idèle αa^{-1} is a norm of some idèle β of I_M, i.e., $\alpha = (N_{M|K}\beta)a \in N_{M|K} I_M K^*$. This shows that the class of α belongs to $N_{M|K} C_M$, so that $C_K = N_{M|K} C_M$. □

(3.8) Corollary. *Let $L|K$ be a finite extension of algebraic number fields. If almost all primes of K split completely in L, then $L = K$.*

Proof: We may assume without loss of generality that $L|K$ is Galois. In fact, let $M|K$ be the normal closure of $L|K$, and write $G = G(M|K)$ and $H = G(M|L)$. Also let $\mathfrak{p}$ be a place of K, $\mathfrak{P}$ a place of M above $\mathfrak{p}$, and let $G_{\mathfrak{P}}$ be its decomposition group. Then the number of places of L above $\mathfrak{p}$ equals the number $\# H \backslash G / G_{\mathfrak{P}}$ of double cosets $H\sigma G_{\mathfrak{P}}$ in G (see chap. I, §9). Hence $\mathfrak{p}$ splits completely in L if $\# H \backslash G / G_{\mathfrak{P}} = [L:K] = \# H \backslash G$. But this is tantamount to $G_{\mathfrak{P}} = 1$, and hence to the fact that $\mathfrak{p}$ splits completely in M.

So assume $L|K$ is Galois, $L \neq K$, and let $\sigma \in G(L|K)$ be an element of prime order, with fixed field K'. If almost all primes $\mathfrak{p}$ of K were completely split in L, then the same would hold for the primes $\mathfrak{p}'$ of K'. This contradicts (3.7). □

Exercise 1. If the Galois extension $L|K$ is not cyclic, then there are at most finitely many primes of K which do not split in L.

Exercise 2. If $L|K$ is a finite Galois extension, then the Galois group $G(L|K)$ is generated by the Frobenius automorphisms $\varphi_{\mathfrak{P}}$ of all prime ideals $\mathfrak{P}$ of L which are unramified over K.

Exercise 3. Let $L|K$ be a finite abelian extension, and let D be a subgroup of I_K such that K^*D is dense in I_K and $D \subseteq N_{L|K} L^*$. Then $L = K$.

Exercise 4. Let $L_1, \dots, L_r|K$ be cyclic extensions of prime degree p such that $L_i \cap L_j = K$ for $i \neq j$. Then there are infinitely many primes $\mathfrak{p}$ of K which split completely in L_i, for $i > 1$, but which are nonsplit in L_1.

§ 4. The Class Field Axiom

Having determined the Herbrand quotient $h(G,C_L)$ to be the degree $n=[L:K]$ of the cyclic extension $L|K$, it will now be enough to show either $H^{-1}(G,C_L)=1$ or $H^0(G,C_L)=(C_K:N_{L|K}C_L)=n$. The first identity is curiously inaccessible by way of direct attack. We are thus stuck with the second. We will reduce the problem to the case of a *Kummer extension*. For such an extension the norm group $N_{L|K}C_L$ can be written down explicitly, and this allows us to compute the index $(C_K:N_{L|K}C_L)$.

So let K be a number field that contains the n-th roots of unity, where n is a fixed *prime power*, and let $L|K$ be a Galois extension with a Galois group of the form

$$G(L|K)\;\cong\;(\mathbb{Z}/n\mathbb{Z})^r\,.$$

We choose a finite set of places S containing the ramified places, those that divide n, and the infinite ones, and which is such that $I_K=I_K^SK^*$. We write again $K^S=I_K^S\cap K^*$ for the group of S-units, and we put $s=\#S$.

(4.1) Proposition. *One has $s\ge r$, and there exists a set T of $s-r$ primes of K that do not belong to S such that*

$$L=K(\sqrt[n]{\Delta}\,),$$

where Δ is the kernel of the map $K^S\to\prod_{\mathfrak p\in T}K_{\mathfrak p}^*/K_{\mathfrak p}^{*n}$.

Proof: We show first that $L=K(\sqrt[n]{\Delta}\,)$ if $\Delta=L^{*n}\cap K^S$, and then that Δ is the said kernel. By chap. IV, (3.6), we certainly have that $L=K(\sqrt[n]{D}\,)$, with $D=L^{*n}\cap K$. If $x\in D$, then $K_{\mathfrak p}(\sqrt[n]{x}\,)|K_{\mathfrak p}$ is unramified for all $\mathfrak p\notin S$ because S contains the places ramified in L. By chap. V, (3.3), we may therefore write $x=u_{\mathfrak p}y_{\mathfrak p}^n$, with $u_{\mathfrak p}\in U_{\mathfrak p}$, $y_{\mathfrak p}\in K_{\mathfrak p}^*$. Putting $y_{\mathfrak p}=1$ for $\mathfrak p\in S$, we get an idèle $y=(y_{\mathfrak p})$ which can be written as a product $y=\alpha z$ with $\alpha\in I_K^S$, $z\in K^*$. Then $xz^{-n}=u_{\mathfrak p}\alpha_{\mathfrak p}^n\in U_{\mathfrak p}$ for all $\mathfrak p\notin S$, i.e., $xz^{-n}\in I_K^S\cap K^*=K^S$, so that $xz^{-n}\in\Delta$. This shows that $D=\Delta K^{*n}$, and thus $L=K(\sqrt[n]{\Delta}\,)$.

The field $N=K(\sqrt[n]{K^S}\,)$ contains the field L because $\Delta=L^{*n}\cap K^S\subseteq K^S$. By Kummer theory, chap. IV, (3.6), we have

$$G(N|K)\;\cong\;\operatorname{Hom}\bigl(K^S/(K^S)^n,\mathbb{Z}/n\mathbb{Z}\bigr)\,.$$

By (1.1), K^S is the product of a free group of rank $s-1$ and of the cyclic group $\mu(K)$ whose order is divisible by n. Therefore $K^S/(K^S)^n$

is a free $(\mathbb{Z}/n\mathbb{Z})$-module of rank s, and so is $G(N|K)$. Moreover, $G(N|K)/G(N|L) \cong G(L|K) \cong (\mathbb{Z}/n\mathbb{Z})^r$ is a free $(\mathbb{Z}/n\mathbb{Z})$-module of rank r so that $r \le s$, and $G(N|L)$ is a free $(\mathbb{Z}/n\mathbb{Z})$-module of rank $s-r$. Let $\sigma_1, \dots, \sigma_{s-r}$ be a $\mathbb{Z}/n\mathbb{Z}$-basis of $G(N|L)$, and let N_i be the fixed field of σ_i, $i = 1, \dots, s-r$. Then $L = \bigcap_{i=1}^{s-r} N_i$. For every $i = 1, \dots, s-r$ we choose a prime $\mathfrak{P}_i$ of N_i which is nonsplit in N such that the primes $\mathfrak{p}_1, \dots, \mathfrak{p}_{s-r}$ of K lying below $\mathfrak{P}_1, \dots, \mathfrak{P}_{s-r}$ are all distinct, and do not belong to S. This is possible by (3.7). We now show that the set $T = \{\mathfrak{p}_1, \dots, \mathfrak{p}_{s-r}\}$ realizes the group $\Delta = L^{*n} \cap K^S$ as the kernel of $K^S \to \prod_{\mathfrak{p}\in T} K_\mathfrak{p}^*/K_\mathfrak{p}^{*n}$.

N_i is the decomposition field of $N|K$ at the unique prime $\mathfrak{P}_i'$ above $\mathfrak{P}_i$, for $i = 1, \dots, s-r$. Indeed, this decomposition field Z_i is contained in N_i because $\mathfrak{P}_i$ is nonsplit in N. On the other hand, the prime $\mathfrak{p}_i$ is unramified in N, because by chap. V, (3.3), it is unramified in every extension $K(\sqrt[n]{u})$, $u \in K^S$. The decomposition group $G(N|Z_i) \supseteq G(N|N_i)$ is therefore cyclic, and necessarily of order n since each element of $G(N|K)$ has order dividing n. This shows that $N_i = Z_i$.

From $L = \bigcap_{i=1}^{s-r} N_i$ it follows that $L|K$ is the maximal subextension of $N|K$ in which the primes $\mathfrak{p}_1, \dots, \mathfrak{p}_{s-r}$ split completely. For $x \in K^S$ we therefore have

$$x \in \Delta \iff K(\sqrt[n]{x}) \subseteq L \iff K_{\mathfrak{p}_i}(\sqrt[n]{x}) = K_{\mathfrak{p}_i},\ i = 1, \dots, s-r,$$
$$\iff x \in K_{\mathfrak{p}_i}^{*n},\ i = 1, \dots, s-r.$$

This shows that Δ is the kernel of the map $K^S \to \prod_{i=1}^{s-r} K_{\mathfrak{p}_i}^*/K_{\mathfrak{p}_i}^{*n}$. □

(4.2) Theorem. *Let T be a set of places as in (4.1), and let*

$$C_K(S,T) = I_K(S,T)K^*/K^*,$$

where

$$I_K(S,T) = \prod_{\mathfrak{p}\in S} K_\mathfrak{p}^{*n} \times \prod_{\mathfrak{p}\in T} K_\mathfrak{p}^* \times \prod_{\mathfrak{p}\notin S\cup T} U_\mathfrak{p}.$$

Then one has

$$N_{L|K}C_L \supseteq C_K(S,T) \quad \textit{and} \quad \big(C_K : C_K(S,T)\big) = [L:K].$$

In particular, if $L|K$ is cyclic, then $N_{L|K}C_L = C_K(S,T)$.

Remark: It will follow from (5.5) that $N_{L|K}C_L = C_K(S,T)$ also holds in general.

For the proof of the theorem we need the following

(4.3) Lemma. $I_K(S,T) \cap K^* = (K^{S\cup T})^n$.

Proof: The inclusion $(K^{S\cup T})^n \subseteq I_K(S,T) \cap K^*$ is trivial. Let $y \in I_K(S,T) \cap K^*$, and $M = K(\sqrt[n]{y})$. It suffices to show that $N_{M|K}C_M = C_K$, for then (3.6) implies $M = K$, hence $y \in K^{*n} \cap I_K(S,T) \subseteq (K^{S\cup T})^n$. Let $[\alpha] \in C_K = I_K^S K^*/K^*$, and let $\alpha \in I_K^S$ be a representative of the class $[\alpha]$. The map

$$K^S \longrightarrow \prod_{\mathfrak{p}\in T} U_\mathfrak{p}/U_\mathfrak{p}^n$$

is surjective. For if Δ denotes its kernel, then obviously $K^{*n} \cap \Delta = (K^S)^n$, and $\Delta K^{*n}/K^{*n} = \Delta/(K^S)^n$. From (1.1) and Kummer theory, we therefore get

$$\#(K^S/\Delta) = \frac{\#K^S/(K^S)^n}{\#\Delta/(K^S)^n} = \frac{n^s}{\#G(L|K)} = n^{s-r}\,.$$

This is also the order of the product because by chap. II, (5.8), we have $\#U_\mathfrak{p}/U_\mathfrak{p}^n = n$ since $\mathfrak{p} \nmid n$. We thus find an element $x \in K^S$ such that $\alpha_\mathfrak{p} = x u_\mathfrak{p}^n$, $u_\mathfrak{p} \in U_\mathfrak{p}$, for $\mathfrak{p} \in T$. The idèle $\alpha' = \alpha x^{-1}$ belongs to the same class as α, and we show that $\alpha' \in N_{M|K} I_M$. By (3.2), this amounts to checking that every component $\alpha'_\mathfrak{p}$ is a norm from $M_\mathfrak{P}|K_\mathfrak{p}$. For $\mathfrak{p} \in S$ this holds because $y \in K_\mathfrak{p}^{*n}$. Hence we have $M_\mathfrak{P} = K_\mathfrak{p}$ for $\mathfrak{p} \in T$ since $\alpha'_\mathfrak{p} = u_\mathfrak{p}^n$ is a n-th power. For $\mathfrak{p} \notin S \cup T$ it holds because $\alpha'_\mathfrak{p}$ is a unit and $M_\mathfrak{P}|K_\mathfrak{p}$ is unramified (see chap. V, (3.3)). This is why $[\alpha] \in N_{M|K}C_M$, q.e.d. $\square$

Proof of theorem (4.2): The identity $(C_K : C_K(S,T)) = [L : K]$ follows from the exact sequence

$$\begin{aligned} 1 \longrightarrow I_K^{S\cup T} \cap K^*/I_K(S,T) \cap K^* &\longrightarrow I_K^{S\cup T}/I_K(S,T) \\ &\longrightarrow I_K^{S\cup T} K^*/I_K(S,T)K^* \longrightarrow 1\,. \end{aligned}$$

Since $I_K^{S\cup T} K^* = I_K$, the order of the group on the right is

$$\begin{aligned} \big(I_K^{S\cup T} K^* : I_K(S,T)K^*\big) &= \big(I_K K^*/K^* : I_K(S,T)K^*/K^*\big) \\ &= \big(C_K : C_K(S,T)\big)\,. \end{aligned}$$

The order of the group on the left is

$$\big(I_K^{S\cup T} \cap K^* : I_K(S,T) \cap K^*\big) = \big(K^{S\cup T} : (K^{S\cup T})^n\big) = n^{2s-r}$$

because $\#(S \cup T) = 2s - r$, and $\mu_n \subseteq K^{S\cup T}$. In view of chap. II, (5.8), the order of the group in the middle is

$$\big(I_K^{S\cup T} : I_K(S,T)\big) = \prod_{\mathfrak{p}\in S} (K_\mathfrak{p}^* : K_\mathfrak{p}^{*n}) = \prod_{\mathfrak{p}\in S} \frac{n^2}{|n|_\mathfrak{p}} = n^{2s} \prod_{\mathfrak{p}} |n|_\mathfrak{p}^{-1} = n^{2s}\,.$$

Altogether this gives

$$\big(C_K : C_K(S,T)\big) = \frac{n^{2s}}{n^{2s-r}} = n^r = [L:K].$$

We now show the inclusion $C_K(S,T) \subseteq N_{L|K}C_L$. Let $\alpha \in I_K(S,T)$. In order to show that $\alpha \in N_{L|K}I_L$ all we have to check, by (3.2), is again that every component $\alpha_\mathfrak{p}$ is a norm from $L_\mathfrak{P}|K_\mathfrak{p}$. For $\mathfrak{p} \in S$ this is true because $\alpha_\mathfrak{p} \in K_\mathfrak{p}^{*n}$ is an n-th power, hence a norm from $K_\mathfrak{p}(\sqrt[n]{K_\mathfrak{p}^*})$ (see chap. V, (1.5)), so in particular also from $L_\mathfrak{P}|K_\mathfrak{p}$. For $\mathfrak{p} \in T$ it holds because (4.1) gives $\Delta \subseteq K_\mathfrak{p}^{*n}$, and thus $L_\mathfrak{P} = K_\mathfrak{p}$. Finally, it holds for $\mathfrak{p} \notin S \cup T$ since $\alpha_\mathfrak{p}$ is a unit and $L_\mathfrak{P}|K_\mathfrak{p}$ is unramified (see chap. V, (3.3)). We therefore have $I_K(S,T) \subseteq N_{L|K}I_L$, i.e., $C_K(S,T) \subseteq N_{L|K}C_L$.

Now if $L|K$ is cyclic, i.e., if $r=1$, then from (3.6),

$$[L:K] \le (C_K : N_{L|K}C_L) \le \big(C_K : C_K(S,T)\big) = [L:K],$$

hence $N_{L|K}C_L = C_K(S,T)$. □

Now that we have an explicit picture in the case of a Kummer field, the result we want follows also in complete generality:

(4.4) Theorem (Global Class Field Axiom). *If $L|K$ is a cyclic extension of algebraic number fields, then*

$$\#H^i\big(G(L|K), C_L\big) = \begin{cases} [L:K] & \textit{for } i=0,\\ 1 & \textit{for } i=-1.\end{cases}$$

Proof: Since $h(G(L|K),C_L) = [L:K]$, it is clearly enough to show that $\#H^0(G(L|K),C_L) \mid [L:K]$. We will prove this by induction on the degree $n=[L:K]$. We write for short $H^0(L|K)$ instead of $H^0(G(L|K),C_L)$. Let $M|K$ be a subextension of prime degree p. We consider the exact sequence

$$C_M/N_{L|M}C_L \xrightarrow{N_{M|K}} C_K/N_{L|K}C_L \longrightarrow C_K/N_{M|K}C_M \longrightarrow 1,$$

i.e., the exact sequence

$$H^0(L|M) \longrightarrow H^0(L|K) \longrightarrow H^0(M|K) \longrightarrow 1.$$

If $p<n$, then $\#H^0(L|M) \mid [L:M]$, $\#H^0(M|K) \mid [M:K]$ by the induction hypothesis, hence $\#H^0(L|K) \mid [L:M][M:K] = [L:K]$.

Now let $p=n$. We put $K' = K(\mu_p)$ and $L' = L(\mu_p)$. Since $d = [K':K] \mid p-1$, we have $G(L|K) \cong G(L'|K')$. $L'|K'$ is a cyclic

Kummer extension, so by (4.2), $\#H^0(L'|K') = [L' : K'] = p$. It therefore suffices to show that the homomorphism

$$(*) \qquad H^0(L|K) \longrightarrow H^0(L'|K')$$

induced by the inclusion $C_L \to C_{L'}$ is injective. $H^0(L|K)$ has exponent p, because for $x \in C_K$ we always have $x^p = N_{L|K}(x)$. Taking $d = [K' : K]$-th powers on $H^0(L|K)$ is therefore an isomorphism. Now let $\bar{x} = x \bmod N_{L|K}C_L$ belong to the kernel of $(*)$. We write $\bar{x} = \bar{y}^d$, for some $\bar{y} = y \bmod N_{L|K}C_L$. Then $\bar{y}$ also is in the kernel of $(*)$, i.e., $y = N_{L'|K'}(z')$, $z' \in C_{L'}$, and we find:

$$y^d = N_{K'|K}(y) = N_{L'|K}(z') = N_{L|K}(N_{L'|L}(z')) \in N_{L|K}C_L\,.$$

Hence $\bar{x} = \bar{y}^d = 1$. □

An immediate consequence of the theorem we have just proved is the famous **Hasse Norm Theorem**:

(4.5) Corollary. *Let $L|K$ be a cyclic extension. An element $x \in K^*$ is a norm if and only if it is a norm locally everywhere, i.e., a norm in every completion $L_{\mathfrak{P}}|K_{\mathfrak{p}}$ ($\mathfrak{P}|\mathfrak{p}$).*

Proof: Let $G = G(L|K)$ and $G_{\mathfrak{P}} = G(L_{\mathfrak{P}}|K_{\mathfrak{p}})$. The exact sequence

$$1 \longrightarrow L^* \longrightarrow I_L \longrightarrow C_L \longrightarrow 1$$

of G-modules gives, by chap. IV, (7.1), an exact sequence

$$H^{-1}(G, C_L) \longrightarrow H^0(G, L^*) \longrightarrow H^0(G, I_L)\,.$$

By (4.4), we have $H^{-1}(G, C_L) = 1$, and from (3.2) it follows that $H^0(G, I_L) = \bigoplus_{\mathfrak{p}} H^0(G_{\mathfrak{P}}, L_{\mathfrak{P}}^*)$. Therefore the homomorphism

$$K^*/N_{L|K}L^* \longrightarrow \bigoplus_{\mathfrak{p}} K_{\mathfrak{p}}^*/N_{L_{\mathfrak{P}}|K_{\mathfrak{p}}}L_{\mathfrak{P}}^*$$

is injective. But this is the claim of the corollary. □

It should be noted that cyclicity is crucial for Hasse's norm theorem. In fact, whereas it is true by (3.2) that an element $x \in K^*$ which is everywhere locally a norm, is always the norm of some idèle α of L, this need not be by any means a principal idèle, not even in the case of arbitrary abelian extensions.

Exercise 1. Determine the norm group $N_{L|K}C_L$ for an arbitrary Kummer extension in a way analogous to the case treated in (4.2) where $G(L|K) \cong (\mathbb{Z}/p^a\mathbb{Z})^r$.

Exercise 2. Let ζ be a primitive m-th root of unity. Show that the norm group $N_{\mathbb{Q}(\zeta)|\mathbb{Q}}C_{\mathbb{Q}}$ equals the ray class group $\bmod\ \mathfrak{m} = (m)$ in $C_{\mathbb{Q}}$.

Exercise 3. An equation $x^2 - ay^2 = b$, $a, b \in K^*$, has a solution in K if and only if it is solvable everywhere locally, i.e., in each completion $K_{\mathfrak{p}}$.

Hint: $x^2 - ay^2 = N_{K(\sqrt{a})|K}(x - \sqrt{a}y)$ if $a \notin K^{*2}$.

Exercise 4. If a quadratic form $a_1x_1^2 + \cdots + a_nx_n^2$ represents zero over a field K with more than five elements (i.e., $a_1x_1^2 + \cdots + a_nx_n^2 = 0$ has a nontrivial solution in K), then there is a representation of zero in which all $x_i \neq 0$.

Hint: If $a\xi^2 = \lambda \neq 0$, $b \neq 0$, then there are non-zero elements α and β such that $a\alpha^2 + b\beta^2 = \lambda$. To prove this, multiply the identity

$$\frac{(t-1)^2}{(t+1)^2} + \frac{4t}{(t+1)^2} = 1$$

by $a\xi^2 = \lambda$ and insert $t = b\gamma^2/a$, for some element $\gamma \neq 0$ such that $t \neq \pm 1$. Use this to prove the claim by induction.

Exercise 5. A quadratic form $ax^2 + by^2 + cz^2$, $a, b, c \in K^*$, represents zero if and only if it represents zero everywhere locally.

Remark: In complete generality, one has the following "local-to-global principle":

Theorem of Minkowski-Hasse: A quadratic form over a number field K represents zero if and only if it represents zero over every completion $K_{\mathfrak{p}}$.

The proof follows from the result stated in exercise 5 by pure algebra (see [113]).

§ 5. The Global Reciprocity Law

Now that we know that the idèle class group satisfies the class field axiom, we proceed to determine a pair of homomorphisms

$$(G_{\mathbb{Q}} \xrightarrow{d} \widehat{\mathbb{Z}},\ C_{\mathbb{Q}} \xrightarrow{v} \widehat{\mathbb{Z}})$$

obeying the rules of abstract class field theory as developed in chap. IV, §4. For the $\widehat{\mathbb{Z}}$-extension of $\mathbb{Q}$ given by d, we have only one choice. It is described in the following

(5.1) Proposition. *Let $\Omega|\mathbb{Q}$ be the field obtained by adjoining all roots of unity, and let T be the torsion subgroup of $G(\Omega|K)$ (i.e., the group of all elements of finite order). Then the fixed field $\widetilde{\mathbb{Q}}|\mathbb{Q}$ of T is a $\widehat{\mathbb{Z}}$-extension.*

Proof: Since $\Omega = \bigcup_{n\ge 1}\mathbb{Q}(\mu_n)$, we find

$$G(\Omega|\mathbb{Q}) = \varprojlim_n G\big(\mathbb{Q}(\mu_n)|\mathbb{Q}\big) = \varprojlim_n (\mathbb{Z}/n\mathbb{Z})^* = \widehat{\mathbb{Z}}^*.$$

But $\widehat{\mathbb{Z}} = \prod_p \mathbb{Z}_p$, and $\mathbb{Z}_p^* \cong \mathbb{Z}_p \times \mathbb{Z}/(p-1)\mathbb{Z}$ for $p \neq 2$ and $\mathbb{Z}_2^* \cong \mathbb{Z}_2 \times \mathbb{Z}/2\mathbb{Z}$. Consequently,

$$G(\Omega|\mathbb{Q}) \cong \widehat{\mathbb{Z}}^* \cong \widehat{\mathbb{Z}} \times \widehat{T}, \quad \text{where} \quad \widehat{T} = \prod_{p\neq 2} \mathbb{Z}/(p-1)\mathbb{Z} \times \mathbb{Z}/2\mathbb{Z}.$$

This shows that the torsion subgroup T of $G(\Omega|\mathbb{Q})$ is isomorphic to the torsion subgroup of $\widehat{T}$. Since the latter contains the group $\bigoplus_{p\neq 2}\mathbb{Z}/(p-1)\mathbb{Z} \oplus \mathbb{Z}/2\mathbb{Z}$, we see that the closure $\overline{T}$ of T is isomorphic to $\widehat{T}$. Now, if $\widetilde{\mathbb{Q}}$ is the fixed field of T, this implies that $G(\widetilde{\mathbb{Q}}|\mathbb{Q}) = G(\Omega|\mathbb{Q})/\overline{T} \cong \widehat{\mathbb{Z}}$. □

Another description of the $\widehat{\mathbb{Z}}$-extension $\widetilde{\mathbb{Q}}|\mathbb{Q}$ is obtained in the following manner. For every prime number p, let $\Omega_p|\mathbb{Q}$ be the field obtained by adjoining all roots of unity of p-power order. Then

$$G(\Omega_p|\mathbb{Q}) = \varprojlim_\nu G\big(\mathbb{Q}(\mu_{p^\nu})|\mathbb{Q}\big) = \varprojlim_\nu (\mathbb{Z}/p^\nu\mathbb{Z})^* = \mathbb{Z}_p^*,$$

and $\mathbb{Z}_p^* \cong \mathbb{Z}_p \times \mathbb{Z}/(p-1)\mathbb{Z}$ for $p \neq 2$ and $\mathbb{Z}_2^* \cong \mathbb{Z}_2 \times \mathbb{Z}/2\mathbb{Z}$. The torsion subgroup of $\mathbb{Z}_p^*$ is isomorphic to $\mathbb{Z}/(p-1)\mathbb{Z}$, resp. $\mathbb{Z}/2\mathbb{Z}$, and taking its fixed field gives an extension $\widetilde{\mathbb{Q}}^{(p)}|\mathbb{Q}$ with Galois group

$$G(\widetilde{\mathbb{Q}}^{(p)}|\mathbb{Q}) \cong \mathbb{Z}_p.$$

The $\widehat{\mathbb{Z}}$-extension $\widetilde{\mathbb{Q}}|\mathbb{Q}$ is then the composite $\widetilde{\mathbb{Q}} = \prod_p \widetilde{\mathbb{Q}}^{(p)}$.

We fix an isomorphism $G(\widetilde{\mathbb{Q}}|\mathbb{Q}) \cong \widehat{\mathbb{Z}}$. There is no canonical choice as in the case of local fields. However, the reciprocity law will not depend on the choice. Via $G(\widetilde{\mathbb{Q}}|\mathbb{Q}) \cong \widehat{\mathbb{Z}}$, we obtain a continuous surjective homomorphism

$$d : G_{\mathbb{Q}} \longrightarrow \widehat{\mathbb{Z}}$$

of the absolute Galois group $G_{\mathbb{Q}} = G(\overline{\mathbb{Q}}|\mathbb{Q})$. With this we continue as in chap. IV, §4, choosing $k = \mathbb{Q}$ as our base field. If $K|\mathbb{Q}$ is a finite extension, then we put $f_K = [K \cap \widetilde{\mathbb{Q}} : \mathbb{Q}]$ and get a surjective homomorphism

$$d_K = \frac{1}{f_K} d : G_K \longrightarrow \widehat{\mathbb{Z}},$$

which defines the $\widehat{\mathbb{Z}}$-extension $\widetilde{K} = K\widetilde{\mathbb{Q}}$ of K. $\widetilde{K}|K$ is called the **cyclotomic $\widehat{\mathbb{Z}}$-extension** of K. We denote again by φ_K the element of $G(\widetilde{K}|K)$ which is mapped to 1 by the isomorphism $G(\widetilde{K}|K) \cong \widehat{\mathbb{Z}}$, and by $\varphi_{L|K}$ the restriction $\varphi_K|_L$ if $L|K$ is a subextension of $\widetilde{K}|K$. The automorphism $\varphi_{L|K}$ must not be confused with the Frobenius automorphism corresponding to a prime ideal of L (see §7).

For the $G_{\mathbb{Q}}$-module A, we choose the union of the idèle class groups C_K of all finite extensions $K|\mathbb{Q}$. Thus $A_K = C_K$. The henselian valuation $v: C_{\mathbb{Q}} \to \widehat{\mathbb{Z}}$ will be obtained as the composite

$$C_{\mathbb{Q}} \xrightarrow{[\ ,\widetilde{\mathbb{Q}}|\mathbb{Q}]} G(\widetilde{\mathbb{Q}}|\mathbb{Q}) \xrightarrow{d} \widehat{\mathbb{Z}},$$

where the mapping $[\ ,\widetilde{\mathbb{Q}}|\mathbb{Q}]$ will later turn out to be the norm residue symbol $(\ ,\widetilde{\mathbb{Q}}|\mathbb{Q})$ of global class field theory (see (5.7)). For the moment we merely define it as follows.

For an arbitrary finite abelian extension $L|K$, we define the homomorphism

$$[\ ,L|K] : I_K \longrightarrow G(L|K)$$

by

$$[\alpha, L|K] = \prod_{\mathfrak{p}} (\alpha_{\mathfrak{p}}, L_{\mathfrak{p}}|K_{\mathfrak{p}}),$$

where $L_{\mathfrak{p}}$ denotes the completion of L with respect to a place $\mathfrak{P}|\mathfrak{p}$, and $(\alpha_{\mathfrak{p}}, L_{\mathfrak{p}}|K_{\mathfrak{p}})$ is the norm residue symbol of local class field theory. Note that almost all factors in the product are 1 because almost all extensions $L_{\mathfrak{p}}|K_{\mathfrak{p}}$ are unramified and almost all $\alpha_{\mathfrak{p}}$ are units.

(5.2) Proposition. *If $L|K$ and $L'|K'$ are two abelian extensions of finite algebraic number fields such that $K \subseteq K'$ and $L \subseteq L'$, then we have the commutative diagram*

$$\begin{array}{ccc} I_{K'} & \xrightarrow{[\ ,L'|K']} & G(L'|K') \\ {\scriptstyle N_{K'|K}}\downarrow & & \downarrow \\ I_K & \xrightarrow{[\ ,L|K]} & G(L|K). \end{array}$$

Proof: For an idèle $\alpha = (\alpha_{\mathfrak{P}}) \in I_{K'}$ of K', we find by chap. IV, (6.4), that

$$(\alpha_{\mathfrak{P}}, L'_{\mathfrak{P}}\,|\,K'_{\mathfrak{P}})|_{L_{\mathfrak{p}}} = \big(N_{K'_{\mathfrak{P}}|K_{\mathfrak{p}}}(\alpha_{\mathfrak{P}}), L_{\mathfrak{p}}|K_{\mathfrak{p}}\big), \quad (\mathfrak{P}|\mathfrak{p}),$$

and (2.2) implies

$$[N_{K'|K}(\alpha), L|K] = \prod_{\mathfrak{p}} (N_{K'|K}(\alpha)_{\mathfrak{p}}, L_{\mathfrak{p}}|K_{\mathfrak{p}}) = \prod_{\mathfrak{p}} \prod_{\mathfrak{P}|\mathfrak{p}} (N_{K'_{\mathfrak{P}}|K_{\mathfrak{p}}}(\alpha_{\mathfrak{P}}), L_{\mathfrak{p}}|K_{\mathfrak{p}})$$

$$= \prod_{\mathfrak{P}} (\alpha_{\mathfrak{P}}, L'_{\mathfrak{P}}|K'_{\mathfrak{P}})|_{L} = [\alpha, L'|K']|_{L}. \qquad \square$$

If $L|K$ is an abelian extension of infinite degree, then we define the homomorphism

$$[\ \ , L|K] : I_K \longrightarrow G(L|K)$$

by its restrictions $[\ \ , L|K]|_{L'} := [\ \ , L'|K]$ to the finite subextensions L' of $L|K$. In other words, if $\alpha \in I_K$, then the elements $[\alpha, L'|K]$ define, by (5.2), an element of the projective limit $\varprojlim_{L'} G(L'|K)$, and $[\alpha, L|K]$ is precisely this element, once we identify $G(L|K) = \varprojlim G(L'|K)$. Again one has the equation

$$[\alpha, L|K] = \prod_{\mathfrak{p}} (\alpha_{\mathfrak{p}}, L_{\mathfrak{p}}|K_{\mathfrak{p}}),$$

where $L_{\mathfrak{p}}$ does not denote the completion of L with respect to a place above $\mathfrak{p}$, but rather the *localization*, i.e., the union of the completions $L'_{\mathfrak{p}}|K_{\mathfrak{p}}$ of all finite subextensions (see chap. II, § 8). Then $L_{\mathfrak{p}}|K_{\mathfrak{p}}$ is Galois, $G(L_{\mathfrak{p}}|K_{\mathfrak{p}}) \subseteq G(L|K)$, and the product $\prod_{\mathfrak{p}} (\alpha_{\mathfrak{p}}, L_{\mathfrak{p}}|K_{\mathfrak{p}})$ converges in the profinite group to the element $[\alpha, L|K]$. Indeed, if $L'|K$ varies over the finite subextensions of $L|K$, then the sets $S_{L'} = \{\mathfrak{p} \mid (\alpha_{\mathfrak{p}}, L'_{\mathfrak{p}}|K_{\mathfrak{p}}) \neq 1\}$ are all finite, so that we may write down the finite products

$$\sigma_{L'} = \prod_{\mathfrak{p} \in S_{L'}} (\alpha_{\mathfrak{p}}, L_{\mathfrak{p}}|K_{\mathfrak{p}}) \in G(L|K).$$

They converge to $[\alpha, L|K]$, for if $[\alpha, L|K]G(L|N)$ is one of the fundamental neighbourhoods (i.e., $N|K$ is one of the finite subextensions of $L|K$), then

$$\sigma_{L'} \in [\alpha, L|K]G(L|N)$$

for all $L' \supseteq N$ because

$$\sigma_{L'}|_N = \prod_{\mathfrak{p}} (\alpha_{\mathfrak{p}}, N_{\mathfrak{p}}|K_{\mathfrak{p}}) = [\alpha, N|K] = [\alpha, L|K]|_N.$$

This shows that $[\alpha, L|K]$ is the only accumulation point of the family $\{\sigma_{L'}\}$.

It is clear that proposition (5.2) remains true for infinite extensions L and L' of finite algebraic number fields K and K'.

(5.3) Proposition. *For every root of unity ζ and every principal idèle $a \in K^*$ one has*

$$[a,\, K(\zeta)|K] = 1\,.$$

Proof: By (5.2), we have $[N_{K|\mathbb{Q}}(a), \mathbb{Q}(\zeta)|\mathbb{Q}] = [a, K(\zeta)|K]|_{\mathbb{Q}(\zeta)}$. Hence we may assume that $K = \mathbb{Q}$. Likewise we may assume that ζ has prime power order $\ell^m \neq 2$. Now let $a \in \mathbb{Q}^*$, let v_p be the normalized exponential valuation of $\mathbb{Q}$ for $p \neq \infty$ and write $a = u_p p^{v_p(a)}$. For $p \neq \ell, \infty$, $\mathbb{Q}_p(\zeta)|\mathbb{Q}_p$ is unramified and $(p, \mathbb{Q}_p(\zeta)|\mathbb{Q}_p)$ is the Frobenius automorphism $\varphi_p : \zeta \to \zeta^p$. From chap. V, (2.4), we thus get

$$\big(a, \mathbb{Q}_p(\zeta)|\mathbb{Q}_p\big)\,\zeta = \zeta^{n_p} \quad \text{with} \quad n_p = \begin{cases} p^{v_p(a)} & \text{for } p \neq \ell, \infty\,, \\ u_p^{-1} & \text{for } p = \ell\,, \\ \operatorname{sgn}(a) & \text{for } p = \infty\,. \end{cases}$$

Hence

$$[a, \mathbb{Q}(\zeta)|\mathbb{Q}]\zeta = \prod_p \big(a, \mathbb{Q}_p(\zeta)|\mathbb{Q}_p\big)\,\zeta = \zeta^{\alpha}$$

where $\alpha = \prod\limits_p n_p = \operatorname{sgn}(a) \prod\limits_{p \neq \ell, \infty} p^{v_p(a)} u_\ell^{-1} = \operatorname{sgn}(a) \prod\limits_{p \neq \infty} p^{v_p(a)} a^{-1} = 1.$ □

Since the extension $\widetilde{K}|K$ is contained in the field of all roots of unity over K, the proposition implies

$$[a, \widetilde{K}|K] = 1$$

for all $a \in K^*$. The homomorphism $[\;\;, \widetilde{K}|K] : I_K \to G(\widetilde{K}|K)$ therefore induces a homomorphism

$$[\;\;, \widetilde{K}|K] : C_K \longrightarrow G(\widetilde{K}|K)\,,$$

and we consider its composite

$$v_K : C_K \longrightarrow \widehat{\mathbb{Z}}$$

with $d_K : G(\widetilde{K}|K) \to \widehat{\mathbb{Z}}$. The pair (d_K, v_K) is then a class field theory, for we have the

(5.4) Proposition. *The map $v_K : C_K \to \widehat{\mathbb{Z}}$ is surjective and is a henselian valuation with respect to d_K.*

Proof: We first show surjectivity. If $L|K$ is a finite subextension of $\widetilde{K}|K$, then the map

$$[\ ,L|K] = \prod_{\mathfrak{p}} (\ ,L_{\mathfrak{p}}|K_{\mathfrak{p}}) : I_K \longrightarrow G(L|K)$$

is surjective. Indeed, since $(\ ,L_{\mathfrak{p}}|K_{\mathfrak{p}}) : K_{\mathfrak{p}}^* \to G(L_{\mathfrak{p}}|K_{\mathfrak{p}})$ is surjective, $[I_K, L|K]$ contains all decomposition groups $G(L_{\mathfrak{p}}|K_{\mathfrak{p}})$. Thus all $\mathfrak{p}$ split completely in the fixed field M of $[I_K, L|K]$. By (3.8), this implies that $M = K$, and so $[I_K, L|K] = G(L|K)$. This yields furthermore that $[I_K, \widetilde{K}|K] = [C_K, \widetilde{K}|K]$ is dense in $G(\widetilde{K}|K)$. In the exact sequence

$$1 \longrightarrow C_K^0 \longrightarrow C_K \xrightarrow{\mathfrak{N}} \mathbb{R}_+^* \longrightarrow 1$$

(see §1) the group C_K^0 is compact by (1.6), and we obtain a splitting, if we identify $\mathbb{R}_+^*$ with the group of positive real numbers in any infinite completion $K_{\mathfrak{p}}$. Thus $C_K = C_K^0 \times \mathbb{R}_+^*$. Now, $[\mathbb{R}_+^*, \widetilde{K}|K] = 1$, for if $x \in \mathbb{R}_+^*$, then $[x, \widetilde{K}|K]|_L = [x, L|K] = 1$ for every finite subextension $L|K$ of $\widetilde{K}|K$, because we may always write $x = y^n$ with $y \in \mathbb{R}_+^*$ and $n = [L:K]$. Therefore $[C_K, \widetilde{K}|K] = [C_K^0, \widetilde{K}|K]$ is a closed, dense subgroup of $G(\widetilde{K}|K)$ and therefore equal to $G(\widetilde{K}|K)$. This proves the surjectivity of $v_K = d_K \circ [\ , \widetilde{K}|K]$.

In the definition of a henselian valuation given in chap. IV, (4.6), condition (i) is satisfied because $v_K(C_K) = \widehat{\mathbb{Z}}$, and condition (ii) follows from (5.2) because for every finite extension $L|K$ we have the identity

$$\begin{aligned} v_K(N_{L|K}C_L) = v_K(N_{L|K}I_L) &= d_K[N_{L|K}I_L, \widetilde{K}|K] \\ &= f_{L|K} d_L[I_L, \widetilde{L}|L] = f_{L|K} v_L(C_L) = f_{L|K}\widehat{\mathbb{Z}}. \end{aligned}$$ □

In view of the fact that the idèle class group C_K satisfies the class field axiom, the pair

$$(d_{\mathbb{Q}} : G_{\mathbb{Q}} \to \widehat{\mathbb{Z}},\ v_{\mathbb{Q}} : C_{\mathbb{Q}} \to \widehat{\mathbb{Z}})$$

constitutes a class field theory, the "global class field theory". The above homomorphism $v_K = d_K \circ [\ , \widetilde{K}|K] : C_K \to \widehat{\mathbb{Z}}$, for finite extensions $K|\mathbb{Q}$, satisfies the formula

$$v_K = \frac{1}{f_K} d_{\mathbb{Q}} \circ [\ , \widetilde{\mathbb{Q}}|\mathbb{Q}] \circ N_{K|\mathbb{Q}} = \frac{1}{f_K} v_{\mathbb{Q}} \circ N_{K|\mathbb{Q}}$$

and is therefore precisely the induced homomorphism in the sense of the abstract theory in chap. IV, (4.7).

As the main result of global class field theory we now obtain the **Artin reciprocity law:**

(5.5) Theorem. *For every Galois extension $L|K$ of finite algebraic number fields we have a canonical isomorphism*

$$r_{L|K} : G(L|K)^{ab} \xrightarrow{\sim} C_K/N_{L|K}C_L .$$

The inverse map of $r_{L|K}$ yields a surjective homomorphism

$$(\ \ , L|K) : C_K \to G(L|K)^{ab}$$

with kernel $N_{L|K}C_L$. The map $(\ \ ,L|K)$ is called the **global norm residue symbol**. We view it also as a homomorphism $I_K \to G(L|K)^{ab}$.

For every place $\mathfrak{p}$ of K, we have on the one hand the embedding $G(L_\mathfrak{p}|K_\mathfrak{p}) \hookrightarrow G(L|K)$, and on the other the canonical injection

$$\langle\ \ \rangle : K_\mathfrak{p}^* \to C_K ,$$

which sends $a_\mathfrak{p} \in K_\mathfrak{p}^*$ to the class of the idèle

$$\langle a_\mathfrak{p}\rangle = (\dots, 1, 1, 1, a_\mathfrak{p}, 1, 1, 1, \dots).$$

These homomorphisms express the compatibility of local and global class field theory, as follows.

(5.6) Proposition. *If $L|K$ is an abelian extension and $\mathfrak{p}$ is a place of K, then the diagram*

$$\begin{array}{ccc} K_\mathfrak{p}^* & \xrightarrow{(\ ,L_\mathfrak{p}|K_\mathfrak{p})} & G(L_\mathfrak{p}|K_\mathfrak{p}) \\ {\scriptstyle\langle\ \rangle}\downarrow & & \downarrow \\ C_K & \xrightarrow{(\ ,L|K)} & G(L|K) \end{array}$$

is commutative.

Proof: We first show that the proposition holds if $L|K$ is a subextension of $\widetilde{K}|K$, or if $L = K(i)$, $i = \sqrt{-1}$, and $\mathfrak{p}|\infty$. Indeed, the two maps $[\ \ ,\widetilde{K}|K]$, $(\ \ ,\widetilde{K}|K) : I_K \to G(\widetilde{K}|K)$ agree because from chap. IV, (6.5), we have

$$d_K \circ (\ \ ,\widetilde{K}|K) = v_K = d_K \circ [\ \ ,\widetilde{K}|K] .$$

Thus, if $L|K$ is a subextension of $\widetilde{K}|K$ and $\alpha = (\alpha_\mathfrak{p}) \in I_K$, then

$$(\alpha, L|K) = [\alpha, L|K] = \prod_\mathfrak{p} (\alpha_\mathfrak{p}, L_\mathfrak{p}|K_\mathfrak{p}) .$$

In particular, for $a_\mathfrak{p} \in K_\mathfrak{p}^*$ we have the identity

$$(\langle a_\mathfrak{p}\rangle, L|K) = (a_\mathfrak{p}, L_\mathfrak{p}|K_\mathfrak{p}) ,$$

which shows that the diagram is commutative when restricted to the finite subextensions of $\widetilde{K}|K$.

On the other hand, let $L = K(i)$, $\mathfrak{p}|\infty$, and $L_{\mathfrak{p}} \neq K_{\mathfrak{p}}$. Then $K_{\mathfrak{p}}^* = \mathbb{R}^*$, $\mathbb{R}_+^*$ is the kernel of $(\ ,L_{\mathfrak{p}}|K_{\mathfrak{p}})$, and $(-1, L_{\mathfrak{p}}|K_{\mathfrak{p}})$ is complex conjugation in $G(L_{\mathfrak{p}}|K_{\mathfrak{p}}) = G(\mathbb{C}|\mathbb{R})$. Thus, all we have to show is that $(\langle -1\rangle, L|K) \neq 1$. If we had $(\langle -1\rangle, L|K) = 1$, then the class of $\langle -1\rangle$ would be the norm of a class of C_L, i.e., $\langle -1\rangle a = N_{L|K}(\alpha)$ for some $a \in K^*$ and an idèle $\alpha \in I_L$. This would mean that $a = N_{L_{\mathfrak{q}}|K_{\mathfrak{q}}}(\alpha_{\mathfrak{q}})$ for $\mathfrak{q} \neq \mathfrak{p}$ and $-a = N_{L_{\mathfrak{p}}|K_{\mathfrak{p}}}(\alpha_{\mathfrak{p}})$, i.e., $(a, L_{\mathfrak{q}}|K_{\mathfrak{q}}) = 1$ for $\mathfrak{q} \neq \mathfrak{p}$ and $(-a, L_{\mathfrak{p}}|K_{\mathfrak{p}}) = 1$. By (5.3), we would have $1 = [a, L|K] = \prod_{\mathfrak{q}}(a, L_{\mathfrak{q}}|K_{\mathfrak{q}}) = (a, L_{\mathfrak{p}}|K_{\mathfrak{p}})$, so that $(-1, L_{\mathfrak{p}}|K_{\mathfrak{p}}) = 1$, and therefore $-1 \in N_{L_{\mathfrak{p}}|K_{\mathfrak{p}}}(L_{\mathfrak{p}}^*) = N_{\mathbb{C}|\mathbb{R}}\mathbb{C}^* = \mathbb{R}_+^*$, a contradiction.

We now reduce the general case to these special cases as follows. Let $L'|K'$ be an abelian extension, so that $K \subseteq K'$, $L \subseteq L'$. We then consider the diagram

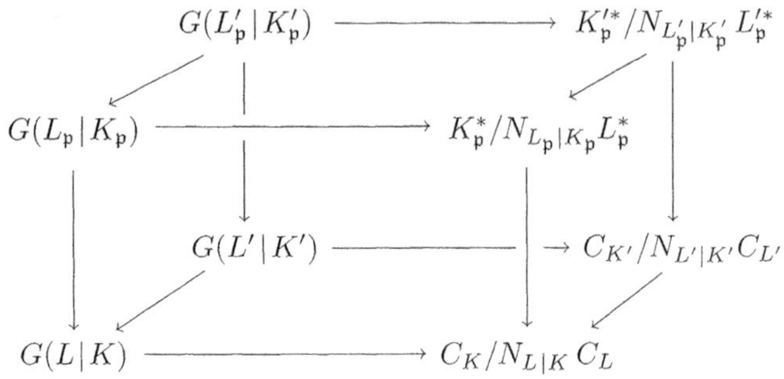

where $L_{\mathfrak{p}} = K_{\mathfrak{p}}L$, $K'_{\mathfrak{p}} = K_{\mathfrak{p}}K'$, $L'_{\mathfrak{p}} = K_{\mathfrak{p}}L'$. In this diagram, the top and bottom are commutative by chap. IV, (6.4), and the sides are commutative for trivial reasons. If now $L'|K'$ is one of the special extensions for which the proposition is already established, then the back diagram is commutative, and hence also the front one, for all elements of $G(L_{\mathfrak{p}}|K_{\mathfrak{p}})$ in the image of $G(L'_{\mathfrak{p}}|K'_{\mathfrak{p}}) \to G(L_{\mathfrak{p}}|K_{\mathfrak{p}})$. This makes it clear that it is enough to find, for every $\sigma \in G(L_{\mathfrak{p}}|K_{\mathfrak{p}})$, some special extension $L'|K'$ such that σ lies in the image of $G(L'_{\mathfrak{p}}|K'_{\mathfrak{p}})$. It is even sufficient to do this only for all σ of prime power order, because they generate the group. Passing to the fixed field of σ we may assume moreover that $G(L|K)$ is generated by σ.

When $\mathfrak{p}|\infty$ and $L_{\mathfrak{p}} \neq K_{\mathfrak{p}}$, i.e., $K_{\mathfrak{p}} = \mathbb{R}$, $L_{\mathfrak{p}} = \mathbb{C}$, we put $L' = L(i) \subseteq \mathbb{C}$, and choose for K' the fixed field of the restriction of complex conjugation to L'. Then $L' = K'(i)$ and $K'_{\mathfrak{p}} = \mathbb{R}$, $L'_{\mathfrak{p}} = \mathbb{C}$, so the mapping $G(L'_{\mathfrak{p}}|K'_{\mathfrak{p}}) \to G(L_{\mathfrak{p}}|K_{\mathfrak{p}})$ is surjective.

When $\mathfrak{p} \nmid \infty$, we find the extension $L'|K'$ as follows. Let σ be of p-power order. We denote by $\widehat{K}|K$, resp. $\widehat{L}|L$, the $\mathbb{Z}_p$-extension contained in $\widetilde{K}|K$, resp. $\widetilde{L}|L$, and consider the field diagram

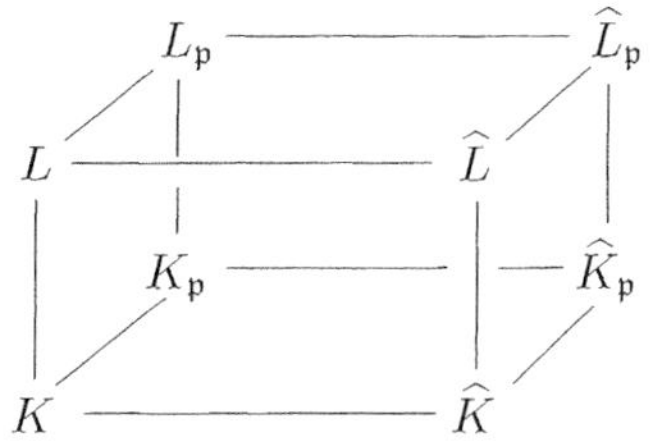

with localizations $\widehat{K}_\mathfrak{p} = K_\mathfrak{p}\widehat{K}$, $\widehat{L}_\mathfrak{p} = L_\mathfrak{p}\widehat{L}$ (all fields are considered to lie in a common bigger field). We may now lift $\sigma \in G(L_\mathfrak{p}|K_\mathfrak{p}) = G(L|K)$ to an automorphism $\hat{\sigma}$ of $\widehat{L}_\mathfrak{p}$ such that

(1) $\hat{\sigma} \in G(\widehat{L}_\mathfrak{p}|K_\mathfrak{p})$,

(2) $\hat{\sigma}|_{\widehat{K}} = \varphi^n_{\widehat{K}|K}$ for some $n \in \mathbb{N}$.

Indeed, since $\widehat{K}_\mathfrak{p} = K_\mathfrak{p}\widehat{K} \neq K_\mathfrak{p}$, the group $G(\widehat{K}_\mathfrak{p}|K_\mathfrak{p}) \neq 1$, and thus is of finite index if viewed as subgroup of $G(\widehat{K}|K) \cong \mathbb{Z}_p$. It is therefore generated by a natural power $\psi = \varphi^k_{\widehat{K}|K}$ of Frobenius $\varphi_{\widehat{K}|K} \in G(\widehat{K}|K)$. As in the proof of chap. IV, (4.4), we may then lift σ to a $\hat{\sigma} \in G(\widehat{L}_\mathfrak{p}|K_\mathfrak{p})$ such that $\hat{\sigma}|_{\widehat{K}_\mathfrak{p}} = \psi^m$, $m \in \mathbb{N}$, so that $\hat{\sigma}|_{\widehat{K}} = \varphi^{km}_{\widehat{K}|K}$.

We now take the fixed field K' of $\hat{\sigma}|_{\widehat{L}}$, and the extension $L' = K'L$. As in chap. IV, (4.5), conditions (ii) and (iii), it then follows that $[K' : K] < \infty$ and $\widehat{K}' = \widehat{L}$. $L'|K'$ is therefore a subextension of $\widehat{K}'|K'$, and σ is the image of $\hat{\sigma}|_{L'_\mathfrak{p}}$ under $G(L'_\mathfrak{p}|K'_\mathfrak{p}) \to G(L_\mathfrak{p}|K_\mathfrak{p})$. This finishes the proof. □

(5.7) Corollary. *If $L|K$ is an abelian extension and $\alpha = (\alpha_\mathfrak{p}) \in I_K$, then*

$$(\alpha, L|K) = \prod_\mathfrak{p} (\alpha_\mathfrak{p}, L_\mathfrak{p}|K_\mathfrak{p}).$$

In particular, for a principal idèle $a \in K^*$ we have the product formula

$$\prod_\mathfrak{p} (a, L_\mathfrak{p}|K_\mathfrak{p}) = 1.$$

Proof: Since I_K is topologically generated by the idèles of the form $\alpha = \langle a_\mathfrak{p} \rangle$, $a_\mathfrak{p} \in K^*_\mathfrak{p}$, it is enough to prove the first formula for these idèles. But this is exactly the statement of (5.6):

$$(\alpha, L|K) = (\langle a_\mathfrak{p} \rangle, L|K) = (a_\mathfrak{p}, L_\mathfrak{p}|K_\mathfrak{p}) = \prod_\mathfrak{q} (\alpha_\mathfrak{q}, L_\mathfrak{q}|K_\mathfrak{q}).$$

The product formula is a consequence of the fact that $(\alpha, L|K)$ depends only on the idèle class $\alpha \bmod K^*$. □

Identifying $K_{\mathfrak{p}}^*$ with its image in C_K under the map $a_{\mathfrak{p}} \mapsto \langle a_{\mathfrak{p}} \rangle$, we obtain the following further corollary, where we use the abbreviations $N = N_{L|K}$ and $N_{\mathfrak{p}} = N_{L_{\mathfrak{p}}|K_{\mathfrak{p}}}$.

(5.8) Corollary. *For every finite abelian extension one has*

$$N C_L \cap K_{\mathfrak{p}}^* = N_{\mathfrak{p}} L_{\mathfrak{p}}^* .$$

Proof: For $x_{\mathfrak{p}} \in N_{\mathfrak{p}} L_{\mathfrak{p}}^*$ we see from (5.6) that $(\langle x_{\mathfrak{p}} \rangle, L|K) = (x_{\mathfrak{p}}, L_{\mathfrak{p}}|K_{\mathfrak{p}}) = 1$. Thus the class of $\langle x_{\mathfrak{p}} \rangle$ is contained in NC_L. Therefore $N_{\mathfrak{p}} L_{\mathfrak{p}}^* \subseteq NC_L$. Conversely, let $\bar{\alpha} \in NC_L \cap K_{\mathfrak{p}}^*$. Then $\bar{\alpha}$ is represented on the one hand by a norm idèle $\alpha = N\beta$, $\beta \in I_L$, and on the other hand by an idèle $\langle x_{\mathfrak{p}} \rangle$, $x_{\mathfrak{p}} \in K_{\mathfrak{p}}^*$. This gives $\langle x_{\mathfrak{p}} \rangle a = N\beta$ with $a \in K^*$. Passing to components shows that a is a norm from $L_{\mathfrak{q}}|K_{\mathfrak{q}}$ for every $\mathfrak{q} \neq \mathfrak{p}$, and the product formula (5.7) shows that a is also a norm from $L_{\mathfrak{p}}|K_{\mathfrak{p}}$. Therefore $x_{\mathfrak{p}} \in N_{\mathfrak{p}} L_{\mathfrak{p}}^*$, and this proves the inclusion $NC_L \cap K^* \subseteq N_{\mathfrak{p}} L_{\mathfrak{p}}^*$. □

Exercise 1. If D_K is the connected component of the unit element of C_K, and if $K^{ab}|K$ is the maximal abelian extension of K, then $C_K/D_K \cong G(K^{ab}|K)$.

Exercise 2. For every place $\mathfrak{p}$ of K one has $K_{\mathfrak{p}}^{ab} = K^{ab} K_{\mathfrak{p}}$.

Hint: Use (5.6) and (5.8).

Exercise 3. Let p be a prime number, and let $M_p|K$ be the maximal abelian p-extension unramified outside of $\{\mathfrak{p}|p\}$. Further, let $H|K$ be the maximal unramified subextension of $M_p|K$ in which the infinite places split completely. Then there is an exact sequence

$$1 \to G(M_p|H) \to G(M_p|K) \to Cl_K(p) \to 1,$$

where $Cl_K(p)$ is the p-Sylow subgroup of the ideal class group Cl_K, and there is a canonical isomorphism

$$G(M_p|H) \cong \prod_{\mathfrak{p}|p} U_{\mathfrak{p}}^{(1)} / \Big(\prod_{\mathfrak{p}|p} U_{\mathfrak{p}}^{(1)} \cap \bar{E}\Big),$$

where $\bar{E}$ is the closure of the (diagonally embedded) unit group $E = \mathcal{O}_K^*$ in $\prod_{\mathfrak{p}|p} U_{\mathfrak{p}}$.

Exercise 4. The group $\bar{E}(p) := \bar{E} \cap \prod_{\mathfrak{p}|p} U_{\mathfrak{p}}^{(1)}$ is a $\mathbb{Z}_p$-module of rank $r_p(E) := \operatorname{rank}_{\mathbb{Z}_p}(\bar{E}(p)) = [K : \mathbb{Q}] - \operatorname{rank}_{\mathbb{Z}_p} G(M_p|K)$. $r_p(E)$ is called the ***p*-adic unit rank**.

Problem: For the p-adic unit rank, one has the famous **Leopoldt conjecture**:

$$r_p(E) = r + s - 1,$$

where r, resp. s, is the number of all real, resp. complex, places; in other words,

$$\operatorname{rank}_{\mathbb{Z}_p} G(M_p|K) = s + 1.$$

The Leopoldt conjecture was proved for abelian number fields $K|\mathbb{Q}$ by the American mathematician ARMAND BRUMER [22]. The general case is still open to date.

§ 6. Global Class Fields

As in local class field theory, the reciprocity law provides also in global class field theory a complete classification of all abelian extensions of a finite algebraic number field K. For this it is necessary to view the idèle class group C_K as a topological group, equipped with its natural topology which the valuations of the various completions $K_\mathfrak{p}$ impress upon it (see § 1).

(6.1) Theorem. *The map*

$$L \longmapsto \mathcal{N}_L = N_{L|K}C_L$$

is a 1−1*-correspondence between the finite abelian extensions $L|K$ and the closed subgroups of finite index in C_K. Moreover one has:*

$$L_1 \subseteq L_2 \Longleftrightarrow \mathcal{N}_{L_1} \supseteq \mathcal{N}_{L_2}, \quad \mathcal{N}_{L_1 L_2} = \mathcal{N}_{L_1} \cap \mathcal{N}_{L_2}, \quad \mathcal{N}_{L_1 \cap L_2} = \mathcal{N}_{L_1}\mathcal{N}_{L_2}.$$

The field $L|K$ corresponding to the subgroup $\mathcal{N}$ of C_K is called the **class field** *of $\mathcal{N}$. It satisfies*

$$G(L|K) \cong C_K/\mathcal{N}.$$

Proof: By chap. IV, (6.7), all we have to show is that the subgroups $\mathcal{N}$ of C_K which are open in the norm topology are precisely the closed subgroups of finite index for the natural topology.

If the subgroup $\mathcal{N}$ is open in the norm topology, then it contains a norm group $N_{L|K}C_L$ and is therefore of finite index, because from (5.5), $(C_K : N_{L|K}C_L) = \#G(L|K)^{ab}$. To show that $\mathcal{N}$ is closed it is enough to show that $N_{L|K}C_L$ is. For this, we choose an infinite place $\mathfrak{p}$ of K and denote by Γ_K the image of the subgroup of positive real numbers in $K_\mathfrak{p}$ under the mapping $\langle\ \rangle : K_\mathfrak{p}^* \to C_K$. Then Γ_K is a group of representatives for the homomorphism $\mathfrak{N} : C_K \to \mathbb{R}_+^*$ with kernel C_K^0 (see § 1), i.e., $C_K = C_K^0 \times \Gamma_K$. By the same token, Γ_K is a group of representatives for the homomorphism $\mathfrak{N} : C_L \to \mathbb{R}_+^*$. We therefore get

$$N_{L|K}C_L = N_{L|K}C_L^0 \times N_{L|K}\Gamma_K = N_{L|K}C_L^0 \times \Gamma_K^n = N_{L|K}C_L^0 \times \Gamma_K.$$

The norm map is continuous, and C_L^0 is compact by (1.6). Hence $N_{L|K}C_L^0$ is closed. Since Γ_K is clearly also closed in C_K, we get that $N_{L|K}C_L$ is closed.

Conversely let $\mathcal{N}$ be a closed subgroup of C_K of finite index. We have to show that $\mathcal{N}$ is open in the norm topology, i.e., contains a norm group $N_{L|K}C_L$. For this we may assume that the index n is a prime power. For if $n = p_1^{\nu_1} \cdots p_r^{\nu_r}$, and $\mathcal{N}_i \subseteq C_K$ is the group containing $\mathcal{N}$ of index $p_i^{\nu_i}$, then $\mathcal{N} = \bigcap_{i=1}^r \mathcal{N}_i$, and if the $\mathcal{N}_i$ are open in the norm topology, then so is $\mathcal{N}$.

Now let J be the preimage of $\mathcal{N}$ with respect to the projection $I_K \to C_K$. Then J is open in I_K because $\mathcal{N}$ is open in C_K (with respect to the natural topology). Therefore J contains a group

$$U_K^S = \prod_{\mathfrak{p}\in S} \{1\} \times \prod_{\mathfrak{p}\notin S} U_\mathfrak{p},$$

where S is a sufficiently big finite set of places of K containing the infinite ones and those primes that divide n, such that $I_K = I_K^S K^*$. Since $(I_K : J) = n$, J also contains the group $\prod_{\mathfrak{p}\in S} K_\mathfrak{p}^{*n} \times \prod_{\mathfrak{p}\notin S}\{1\}$, and hence the group

$$I_K(S) = \prod_{\mathfrak{p}\in S} K_\mathfrak{p}^{*n} \times \prod_{\mathfrak{p}\notin S} U_\mathfrak{p}.$$

Thus it is enough to show that $C_K(S) = I_K(S)K^*/K^* \subseteq \mathcal{N}$ contains a norm group. If the n-th roots of unity belong to K, then $C_K(S) = N_{L|K}C_L$ with $L = K(\sqrt[n]{K^S})$, because of the remark following (4.2). If they do not belong to K, then we adjoin them and obtain an extension $K'|K$. Let S' be the set of primes of K' lying above primes in S. If S was chosen sufficiently large, then $I_{K'} = I_{K'}^{S'}K'^*$ and $C_{K'}(S') = N_{L'|K'}C_{L'}$, with $L' = K'(\sqrt[n]{K'^{S'}})$, by the above argument. Using chap. V, (1.5), this gives on the other hand that $N_{K'|K}(I_{K'}(S')) \subseteq I_K(S)$, so that

$$N_{L'|K}C_{L'} = N_{K'|K}(N_{L'|K'}C_{L'}) = N_{K'|K}(C_{K'}(S')) \subseteq C_K(S).$$

This finishes the proof. □

The above theorem is called the "existence theorem" of global class field theory because its main assertion is the existence, for any given closed subgroup $\mathcal{N}$ of finite index in C_K, of an abelian extension $L|K$ such that $N_{L|K}C_L = \mathcal{N}$. This extension L is the class field for $\mathcal{N}$. The existence theorem gives a clear overview of all the abelian extensions of K once we bring in the *congruence subgroups* $C_K^\mathfrak{m}$ of C_K corresponding to the **modules** $\mathfrak{m} = \prod_{\mathfrak{p}\nmid\infty} \mathfrak{p}^{n_\mathfrak{p}}$ (see (1.7)). They are closed of finite index by (1.8), and they prompt the following definition.

(6.2) Definition. *The class field $K^\mathfrak{m}|K$ for the congruence subgroup $C_K^\mathfrak{m}$ is called the* **ray class field** mod $\mathfrak{m}$.

The Galois group of the ray class field is canonically isomorphic to the ray class group mod $\mathfrak{m}$:

$$G(K^\mathfrak{m}|K) \cong C_K/C_K^\mathfrak{m}.$$

One has

$$\mathfrak{m}|\mathfrak{m}' \Longrightarrow K^{\mathfrak{m}} \subseteq K^{\mathfrak{m}'},$$

because clearly $C_K^{\mathfrak{m}} \supseteq C_K^{\mathfrak{m}'}$. Since the closed subgroups of finite index in C_K are by (1.8) precisely those subgroups containing a congruence subgroup $C_K^{\mathfrak{m}}$, we get from (6.1) the

(6.3) Corollary. *Every finite abelian extension $L|K$ is contained in a ray class field $K^{\mathfrak{m}}|K$.*

(6.4) Definition. *Let $L|K$ be a finite abelian extension, and let $\mathcal{N}_L = N_{L|K}C_L$. The* **conductor** *$\mathfrak{f}$ of $L|K$ (or of $\mathcal{N}_L$) is the gcd of all modules $\mathfrak{m}$ such that $L \subseteq K^{\mathfrak{m}}$ (i.e., $C_K^{\mathfrak{m}} \subseteq \mathcal{N}_L$).*

$K^{\mathfrak{f}}|K$ is therefore the smallest ray class field containing $L|K$. But it is not true in general that $\mathfrak{m}$ is the conductor of $K^{\mathfrak{m}}|K$. In chap. V, (1.6), we defined the conductor $\mathfrak{f}_{\mathfrak{p}}$ of a $\mathfrak{p}$-adic extension $L_{\mathfrak{p}}|K_{\mathfrak{p}}$ for a finite place $\mathfrak{p}$, to be the smallest power $\mathfrak{f}_{\mathfrak{p}} = \mathfrak{p}^n$ such that $U_K^{(n)} \subseteq N_{L_{\mathfrak{p}}|K_{\mathfrak{p}}}L_{\mathfrak{p}}^*$. For an infinite place $\mathfrak{p}$ we define $\mathfrak{f}_{\mathfrak{p}} = 1$. Then we view $\mathfrak{f}$ as the replete ideal $\mathfrak{f}\prod_{\mathfrak{p}|\infty}\mathfrak{p}^0$ and obtain the

(6.5) Proposition. *If $\mathfrak{f}$ is the conductor of the abelian extension $L|K$, and $\mathfrak{f}_{\mathfrak{p}}$ is the conductor of the local extension $L_{\mathfrak{p}}|K_{\mathfrak{p}}$, then*

$$\mathfrak{f} = \prod_{\mathfrak{p}} \mathfrak{f}_{\mathfrak{p}}.$$

Proof: Let $\mathcal{N} = N_{L|K}C_L$, and let $\mathfrak{m} = \prod_{\mathfrak{p}}\mathfrak{p}^{n_{\mathfrak{p}}}$ be a module ($n_{\mathfrak{p}} = 0$ for $\mathfrak{p}|\infty$). One then has

$$C_K^{\mathfrak{m}} \subseteq \mathcal{N} \Longleftrightarrow \mathfrak{f}|\mathfrak{m} \quad \text{and} \quad \prod_{\mathfrak{p}} \mathfrak{f}_{\mathfrak{p}}|\mathfrak{m} \Longleftrightarrow \mathfrak{f}_{\mathfrak{p}}|\mathfrak{p}^{n_{\mathfrak{p}}} \quad \text{for all } \mathfrak{p}.$$

So to prove $\mathfrak{f} = \prod_{\mathfrak{p}}\mathfrak{f}_{\mathfrak{p}}$, we have to show the equivalence

$$C_K^{\mathfrak{m}} \subseteq \mathcal{N} \Longleftrightarrow \mathfrak{f}_{\mathfrak{p}}|\mathfrak{p}^{n_{\mathfrak{p}}} \quad \text{for all } \mathfrak{p}.$$

It follows from the identity $\mathcal{N} \cap K_{\mathfrak{p}}^* = N_{\mathfrak{p}}L_{\mathfrak{p}}^*$ (see (5.8)):

$$\begin{aligned} C_K^{\mathfrak{m}} \subseteq \mathcal{N} &\Longleftrightarrow (\alpha \in I_K^{\mathfrak{m}} \Rightarrow \overline{\alpha} \in \mathcal{N}) \quad \text{for } \alpha \in I_K \\ &\Longleftrightarrow (\alpha_{\mathfrak{p}} \equiv 1 \bmod \mathfrak{p}^{n_{\mathfrak{p}}} \Rightarrow (\alpha_{\mathfrak{p}}) \in \mathcal{N} \cap K_{\mathfrak{p}}^* = N_{\mathfrak{p}}L_{\mathfrak{p}}^*) \quad \text{for all } \mathfrak{p} \\ &\Longleftrightarrow (\alpha_{\mathfrak{p}} \in U_{\mathfrak{p}}^{(n_{\mathfrak{p}})} \Rightarrow \alpha_{\mathfrak{p}} \in N_{\mathfrak{p}}L_{\mathfrak{p}}^*) \Longleftrightarrow U_{\mathfrak{p}}^{(n_{\mathfrak{p}})} \subseteq N_{\mathfrak{p}}L_{\mathfrak{p}}^* \Longleftrightarrow \mathfrak{f}_{\mathfrak{p}}|\mathfrak{p}^{n_{\mathfrak{p}}}. \end{aligned}$$

□

By chap. V, (1.7), the local extension $L_\mathfrak{p}|K_\mathfrak{p}$, for a finite prime $\mathfrak{p}$, is ramified if and only if its conductor $\mathfrak{f}_\mathfrak{p}$ is $\neq 1$. This continues to hold also for an infinite place $\mathfrak{p}$, provided we call the extension $L_\mathfrak{p}|K_\mathfrak{p}$ *unramified* in this case, as we did in chap. III. Then (6.5) yields the

(6.6) Corollary. *Let $L|K$ be a finite abelian extension and $\mathfrak{f}$ its conductor. Then:*

$$\mathfrak{p} \text{ is ramified in } L \iff \mathfrak{p}|\mathfrak{f}.$$

In the case of the base field $\mathbb{Q}$, the ray class fields are nothing but the familiar cyclotomic fields:

(6.7) Proposition. *Let m be a natural number and $\mathfrak{m} = (m)$. Then the ray class field* mod $\mathfrak{m}$ *of $\mathbb{Q}$ is the field*

$$\mathbb{Q}^\mathfrak{m} = \mathbb{Q}(\mu_m)$$

of m-th roots of unity.

Proof: Let $m = \prod_{p \neq p_\infty} p^{n_p}$. Then $I_\mathbb{Q}^\mathfrak{m} = \prod_{p \neq p_\infty} U_p^{(n_p)} \times \mathbb{R}_+^*$. Let $m = m' p^{n_p}$. Then $U_p^{(n_p)}$ is certainly contained in the norm group of the unramified extension $\mathbb{Q}_p(\mu_{m'})|\mathbb{Q}_p$, but also in the norm group of $\mathbb{Q}_p(\mu_{p^{n_p}})|\mathbb{Q}_p$, according to chap. V, (1.8). This means, by § 3, that every idèle in $I_\mathbb{Q}^\mathfrak{m}$ is a norm of some idèle of $\mathbb{Q}(\mu_m)$. Thus $C_\mathbb{Q}^\mathfrak{m} \subseteq NC_{\mathbb{Q}(\mu_m)}$. On the other hand, $C_\mathbb{Q}/C_\mathbb{Q}^\mathfrak{m} \cong (\mathbb{Z}/m\mathbb{Z})^*$ by (1.10), and therefore

$$\left(C_\mathbb{Q} : C_\mathbb{Q}^\mathfrak{m}\right) = \left[\mathbb{Q}(\mu_m) : \mathbb{Q}\right] = \left(C_\mathbb{Q} : NC_{\mathbb{Q}(\mu_m)}\right),$$

so that $C_\mathbb{Q}^\mathfrak{m} = NC_{\mathbb{Q}(\mu_m)}$, and this proves the claim. □

According to this proposition, one may view the general ray class fields $K^\mathfrak{m}|K$ as analogues of the cyclotomic fields $\mathbb{Q}(\mu_m)|\mathbb{Q}$. Nonetheless, they are not made to take over the important rôle of the latter because all we know about them is that they exist, but not how to generate them. In the case of local fields things were different. There the analogues of the ray class fields were the Lubin-Tate extensions which could be generated by the division points of formal groups – a fact that carries a long way (see chap. V, § 5). This local discovery does, however, originate from the problem of generating global class fields, which will be discussed at the end of this section.

Note in passing that the above proposition gives another proof of the theorem of Kronecker and Weber (see chap. V, (1.10)) to the effect that every finite abelian extension $L|\mathbb{Q}$ is contained in a field $\mathbb{Q}(\mu_m)|\mathbb{Q}$, because by (1.8) the norm group $N_{L|\mathbb{Q}}C_L$ lies in some congruence subgroup $C_{\mathbb{Q}}^{\mathfrak{m}}$, $\mathfrak{m} = (m)$, so that $L \subseteq \mathbb{Q}(\mu_m)$.

Among all abelian extensions of K, the ray class field mod 1 occupies a special place. It is called the **big Hilbert class field** and has Galois group

$$G(K^1|K) \cong Cl_K^1 .$$

By (1.11), the group Cl_K^1 is linked to the ordinary ideal class group by the exact sequence

$$1 \longrightarrow \mathcal{O}^*/\mathcal{O}_+^* \longrightarrow \prod_{\mathfrak{p} \text{ real}} \mathbb{R}^*/\mathbb{R}_+^* \longrightarrow Cl_K^1 \longrightarrow Cl_K \longrightarrow 1.$$

The big Hilbert class field has conductor $\mathfrak{f} = 1$ and may therefore be characterized by (6.6) in the following way.

(6.8) Proposition. *The big Hilbert class field is the maximal unramified abelian extension of K.*

Since the infinite places are always unramified, this means that all prime ideals are unramified. The **Hilbert class field**, or more precisely, the "small Hilbert class field", is defined to be the maximal unramified abelian extension $H|K$ in which all infinite places split completely, i.e., the real places stay real. It satisfies the

(6.9) Proposition. *The Galois group of the small Hilbert class field $H|K$ is canonically isomorphic to the ideal class group:*

$$G(H|K) \cong Cl_K .$$

In particular, the degree $[H : K]$ is the class number h_K of K.

Proof: We consider the big Hilbert class field $K^1|K$ and, for every infinite place $\mathfrak{p}$, the commutative diagram (see (5.6))

$$\begin{array}{ccc} K_{\mathfrak{p}}^* & \xrightarrow{(\ ,\,K_{\mathfrak{p}}^1|K_{\mathfrak{p}})} & G(K_{\mathfrak{p}}^1|K_{\mathfrak{p}}) \\ {\scriptstyle (\ )}\downarrow & & \downarrow \\ I_K/I_K^1K^* & \xrightarrow{(\ ,\,K^1|K)} & G(K^1|K). \end{array}$$

The small Hilbert class field $H|K$ is the fixed field of the subgroup G_∞ generated by all $G(K_\mathfrak{p}^1|K_\mathfrak{p})$, $\mathfrak{p}|\infty$. Under $(\ ,K^1|K)$ this is the image of

$$\big(\prod_{\mathfrak{p}|\infty} K_\mathfrak{p}^*\big) I_K^1 K^*/I_K^1 K^* = I_K^{S_\infty} K^*/I_K^1 K^*,$$

where $I_K^{S_\infty} = \prod_{\mathfrak{p}|\infty} K_\mathfrak{p}^* \times \prod_{\mathfrak{p}\nmid\infty} U_\mathfrak{p}$. Therefore by (1.3),

$$G(H|K) = G(K^1|K)/G_\infty \cong I_K/I_K^{S_\infty} K^* \cong Cl_K\,. \qquad \square$$

Remark: The small Hilbert class field is in general not a ray class field in terms of the theory developed here. But it is in many other textbooks where ray class groups and ray class fields are defined differently (see for instance [107]). This other theory is obtained by equipping all number fields with the Minkowski metric

$$\langle x,y\rangle_K = \sum_\tau \alpha_\tau x_\tau \bar{y}_\tau \qquad \big(\tau \in \mathrm{Hom}(K,\mathbb{C})\big)\,,$$

$\alpha_\tau = 1$ if $\tau = \bar\tau$, $\alpha_\tau = \frac{1}{2}$ if $\tau \neq \bar\tau$. A ray class group can then be attached to any *replete* module

$$\mathfrak{m} = \prod_\mathfrak{p} \mathfrak{p}^{n_\mathfrak{p}},$$

where $n_\mathfrak{p} \in \mathbb{Z}$, $n_\mathfrak{p} \geq 0$, and $n_\mathfrak{p} = 0$ or $= 1$ if $\mathfrak{p}|\infty$. The groups $U_\mathfrak{p}^{(n_\mathfrak{p})}$ attached to the metrized number field $(K,\langle\ ,\ \rangle_K)$ are defined by

$$U_\mathfrak{p}^{(n_\mathfrak{p})} = \begin{cases} 1+\mathfrak{p}^{n_\mathfrak{p}}, & \text{for } n_\mathfrak{p} > 0, \text{ and } U_\mathfrak{p} \text{ for } n_\mathfrak{p} = 0, \text{ if } \mathfrak{p}\nmid\infty,\\ \mathbb{R}^*, & \text{if } \mathfrak{p} \text{ is real and } n_\mathfrak{p} = 0,\\ \mathbb{R}_+^*, & \text{if } \mathfrak{p} \text{ is real and } n_\mathfrak{p} = 1,\\ \mathbb{C}^* = K_\mathfrak{p}^*, & \text{if } \mathfrak{p} \text{ is complex.}\end{cases}$$

The *congruence subgroup* mod $\mathfrak{m}$ of $(K,\langle\ ,\ \rangle_K)$ is then the subgroup $C_K^\mathfrak{m} = I_K^\mathfrak{m} K^*/K^*$ of C_K formed with the group

$$I_K^\mathfrak{m} = \prod_\mathfrak{p} U_\mathfrak{p}^{(n_\mathfrak{p})},$$

and the factor group $C_K/C_K^\mathfrak{m}$ is the *ray class group* mod $\mathfrak{m}$. The *ray class field* mod $\mathfrak{m}$ of $(K,\langle\ ,\ \rangle_K)$ is again the class field of K corresponding to the group $C_K^\mathfrak{m} \subseteq C_K$. As explained in chap. III, § 3, the infinite places $\mathfrak{p}$ have to be considered as *ramified* in an extension $L|K$ if $L_\mathfrak{P} \neq K_\mathfrak{p}$. Likewise, the *conductor* of an abelian extension $L|K$, i.e., the gcd of all modules $\mathfrak{m} = \prod_{\mathfrak{p}} \mathfrak{p}^{n_\mathfrak{p}}$ such that $C_K^{\mathfrak{m}} \subseteq N_{L|K}C_L$, is the replete ideal

$$\mathfrak{f} = \prod_{\mathfrak{p}} \mathfrak{f}_\mathfrak{p},$$

where now for an infinite place $\mathfrak{p}$, we have $\mathfrak{f}_\mathfrak{p} = \mathfrak{p}^{n_\mathfrak{p}}$ with $n_\mathfrak{p} = 0$ if $L_\mathfrak{p} = K_\mathfrak{p}$, and $n_\mathfrak{p} = 1$ if $L_\mathfrak{p} \neq K_\mathfrak{p}$. Corollary (6.6) then continues to hold: a place $\mathfrak{p}$ is ramified in L if and only if $\mathfrak{p}$ occurs in the conductor $\mathfrak{f}$.

This entails the following modifications of the above theory, as far as ray class fields are concerned. The ray class field mod 1 is the *small* Hilbert class field. It is now the maximal abelian extension of K which is unramified at *all* places. The big Hilbert class field is the ray class field for the module $\mathfrak{m} = \prod_{\mathfrak{p}|\infty} \mathfrak{p}$. In the case of the base field $\mathbb{Q}$, the field $\mathbb{Q}(\zeta)$ of m-th roots of unity is the ray class field mod mp_∞, where p_∞ is the infinite place. The ray class field for the module m becomes the maximal real subextension $\mathbb{Q}(\zeta + \zeta^{-1})$, which was not a ray class field before. This is the theory one finds in the textbooks alluded to above. It corresponds to the number fields with the Minkowski metric. The theory of ray class fields according to the treatment of this book is forced upon us already by the choice of the standard metric $\langle x, y\rangle = \sum_\tau x_\tau \bar{y}_\tau$ on $K_\mathbb{R}$ taken in chap. I, § 5. It is compatible with the Riemann-Roch theory of chap. III, and has the advantage of being simpler.

Over the field $\mathbb{Q}$, the ray class field mod (m) can be generated, according to (6.7), by the m-th roots of unity, i.e., by special values of the exponential function $e^{2\pi i z}$. The question suggested by this observation is whether one may construct the abelian extensions of an arbitrary number field in a similarly concrete way, via special values of analytic functions. This was the historic origin of the notion of class field. A completely satisfactory answer to this question has been given only in the case of an imaginary quadratic field K. The results for this case are subsumed under the name of **Kronecker's Jugendtraum** (Kronecker's dream of his youth). We will briefly describe them here. For the proofs, which presuppose an in-depth knowledge of the theory of *elliptic curves*, we have to refer to [96] and [28].

An elliptic curve is given as the quotient $E = \mathbb{C}/\Gamma$ of $\mathbb{C}$ by a complete lattice $\Gamma = \mathbb{Z}\omega_1 + \mathbb{Z}\omega_2$ in $\mathbb{C}$. This is a torus which receives the structure of an algebraic curve via the **Weierstrass $\wp$-function**

$$\wp(z) = \wp_\Gamma(z) = \frac{1}{z^2} + \sum_{\omega\in\Gamma'} \left[\frac{1}{(z-\omega)^2} - \frac{1}{\omega^2}\right],$$

where $\Gamma' = \Gamma \smallsetminus \{0\}$. $\wp(z)$ is a meromorphic doubly periodic function, i.e.,

$$\wp(z+\omega) = \wp(z) \quad \text{for all} \quad \omega \in \Gamma,$$

and it satisfies, along with its derivative $\wp'(z)$, an identity

$$\wp'(z)^2 = 4\wp(z)^3 - g_2\wp(z) - g_3 .$$

The constants g_2, g_3 only depend on the lattice Γ, and are given by $g_2 = g_2(\Gamma) = 60\sum_{\omega\neq 0}\frac{1}{\omega^4}$, $g_3 = g_3(\Gamma) = 140\sum_{\omega\neq 0}\frac{1}{\omega^6}$. $\wp$ and $\wp'$ may thus be interpreted as functions on $\mathbb{C}/\Gamma$. If one takes away the finite set $S \subseteq \mathbb{C}/\Gamma$ of poles, one gets a bijection

$$\mathbb{C}/\Gamma \smallsetminus S \xrightarrow{\sim} \{(x,y) \in \mathbb{C}^2 \mid y^2 = 4x^3 - g_2x - g_3\}, \quad z \longmapsto (\wp(z), \wp'(z)),$$

onto the affine algebraic curve in $\mathbb{C}^2$ given by the equation $y^2 = 4x^3 - g_2x - g_3$. This gives the torus $\mathbb{C}/\Gamma$ the structure of an algebraic curve E over $\mathbb{C}$ of genus 1. An important rôle is played by the **j-invariant**

$$j(E) = j(\Gamma) = \frac{2^6 3^3 g_2^3}{\Delta} \qquad \text{with} \quad \Delta = g_2^3 - 27g_3^2 .$$

It determines the elliptic curve E up to isomorphism. Writing generators ω_1, ω_2 of Γ in such an order that $\tau = \omega_1/\omega_2$ lies in the upper half-plane $\mathbb{H}$, then $j(E)$ becomes the value $j(\tau)$ of a **modular function**, i.e., of a holomorphic function j on $\mathbb{H}$ which is invariant under the substitution $\tau \mapsto \dfrac{a\tau+b}{c\tau+d}$ for every matrix $\begin{pmatrix} a & b \\ c & d \end{pmatrix} \in SL_2(\mathbb{Z})$.

Now let $K \subseteq \mathbb{C}$ be an imaginary quadratic number field. Then the ring $\mathcal{O}_K$ of integers forms a lattice in $\mathbb{C}$, and more generally, any ideal $\mathfrak{a}$ of $\mathcal{O}_K$ does as well. The tori $\mathbb{C}/\mathfrak{a}$ constructed in this way are elliptic curves with *complex multiplication*. This means the following. An endomorphism of an elliptic curve $E = \mathbb{C}/\Gamma$ is given as multiplication by a complex number z such that $z\Gamma \subseteq \Gamma$. Generically, one has $\mathrm{End}(E) = \mathbb{Z}$. If this is not the case, then $\mathrm{End}(E) \otimes \mathbb{Q}$ is necessarily an imaginary quadratic number field K, and one says that this is an elliptic curve with complex multiplication. The curves $\mathbb{C}/\mathfrak{a}$ are obviously of this kind.

The consequences of these analytic investigations for class field theory are the following.

(6.10) Theorem. *Let K be an imaginary quadratic number field and $\mathfrak{a}$ an ideal of $\mathcal{O}_K$. Then one has:*

(i) *The j-invariant $j(\mathfrak{a})$ of $\mathbb{C}/\mathfrak{a}$ is an algebraic integer which depends only on the ideal class $\mathfrak{K}$ of $\mathfrak{a}$, and will therefore be denoted by $j(\mathfrak{K})$.*

(ii) *Every $j(\mathfrak{a})$ generates the Hilbert class field over K.*

(iii) *If* $\mathfrak{a}_1, \dots, \mathfrak{a}_h$ *are representatives of the ideal class group* Cl_K, *then the numbers* $j(\mathfrak{a}_i)$ *are conjugate to one another over* K.

(iv) *For almost all prime ideals* $\mathfrak{p}$ *of* K *one has*

$$\varphi_{\mathfrak{p}} j(\mathfrak{a}) = j(\mathfrak{p}^{-1}\mathfrak{a}),$$

where $\varphi_{\mathfrak{p}} \in G(K(j(\mathfrak{a}))|K)$ *is the Frobenius automorphism of a prime ideal* $\mathfrak{P}$ *of* $K(j(\mathfrak{a}))$ *above* $\mathfrak{p}$.

Note that for a totally imaginary field K there is no difference between big and small Hilbert class field. In order to go beyond the Hilbert class field, i.e., the ray class field mod 1, to the ray class fields for arbitrary modules $\mathfrak{m} \neq 1$, we form, for any lattice $\Gamma \subseteq \mathbb{C}$, the **Weber function**

$$\tau_{\Gamma}(z) = \begin{cases} -2^7 3^5 \frac{g_2 g_3}{\Delta} \wp_{\Gamma}(z), & \text{if } g_2 g_3 \neq 0, \\ -2^9 3^6 \frac{g_3}{\Delta} \wp_{\Gamma}^3(z), & \text{if } g_2 = 0, \\ 2^8 3^4 \frac{g_2^2}{\Delta} \wp_{\Gamma}^2(z), & \text{if } g_3 = 0. \end{cases}$$

Let $\mathfrak{K} \in Cl_K$ be an ideal class chosen once and for all. We denote by $\mathfrak{K}^*$ the classes in the ray class group $Cl_K^{\mathfrak{m}} = J_K^{\mathfrak{m}}/P_K^{\mathfrak{m}}$ which under the homomorphism

$$Cl_K^{\mathfrak{m}} \longrightarrow Cl_K$$

are sent to the ideal class $(\mathfrak{m})\mathfrak{K}^{-1}$. Let $\mathfrak{a}$ be an ideal in $\mathfrak{K}$, and let $\mathfrak{b}$ be an integral ideal in $\mathfrak{K}^*$. Then $\mathfrak{a}\mathfrak{b}\mathfrak{m}^{-1} = (a)$ is a principal ideal. The value $\tau_{\mathfrak{a}}(a)$ only depends on the class $\mathfrak{K}^*$, not on the choice of $\mathfrak{a}, \mathfrak{b}$ and a. It will be denoted by

$$\tau(\mathfrak{K}^*) = \tau_{\mathfrak{a}}(a).$$

With these conventions we then have the

(6.11) Theorem. (i) *The invariants* $\tau(\mathfrak{K}_1^*)$, $\tau(\mathfrak{K}_2^*)$, ..., *for a fixed ideal class* $\mathfrak{K}$, *are distinct algebraic numbers which are conjugate over the Hilbert class field* $K^1 = K(j(\mathfrak{K}))$.

(ii) *For an arbitrary* $\mathfrak{K}^*$, *the field* $K(j(\mathfrak{K}), \tau(\mathfrak{K}^*))$ *is the ray class field* mod $\mathfrak{m}$ *over* K:

$$K^{\mathfrak{m}} = K\big(j(\mathfrak{K}), \tau(\mathfrak{K}^*)\big).$$

Exercise 1. Let $K^1|K$ be the big, and $H|K$ the small Hilbert class field. Then $G(K^1|H) \cong (\mathbb{Z}/2\mathbb{Z})^{r-t}$, where r is the number of real places, and $2^t = (\mathcal{O}^* : \mathcal{O}^*_+)$.

Exercise 2. Let $d > 0$ be squarefree, and $K = \mathbb{Q}(\sqrt{d})$. Let ε be a totally positive fundamental unit of K. Then one has $[K^1 : H] = 1$ or $= 2$, according as $N_{K|\mathbb{Q}}(\varepsilon) = -1$ or $= 1$.

Exercise 3. The group $(C_K)^n = (I_K)^n K^*/K^*$ is the intersection of the norm groups $N_{L|K}C_L$ of all abelian extensions $L|K$ of exponent n.

Exercise 4. (i) For a number field K, local Tate duality (see chap. V, § 1, exercise 2) yields a non-degenerate pairing

$$(*) \qquad \prod_{\mathfrak{p}} H^1(K_{\mathfrak{p}}, \mathbb{Z}/n\mathbb{Z}) \times \prod_{\mathfrak{p}} H^1(K_{\mathfrak{p}}, \mu_n) \to \mathbb{Z}/n\mathbb{Z}$$

of locally compact groups, where the restricted products are taken with respect to the subgroups $H^1_{nr}(K_{\mathfrak{p}}, \mathbb{Z}/n\mathbb{Z})$, resp. $H^1_{nr}(K_{\mathfrak{p}}, \mu_n)$. For $\chi = (\chi_{\mathfrak{p}})$ in the first and $\alpha = (\alpha_{\mathfrak{p}})$ in the second product, it is given by

$$(\chi, \alpha) = \sum_{\mathfrak{p}} \chi_{\mathfrak{p}}(\alpha_{\mathfrak{p}}, \overline{K}_{\mathfrak{p}}|K_{\mathfrak{p}}).$$

(ii) If $L|K$ is a finite extension, then one has a commutative diagram

$$\begin{array}{ccccc} \prod_{\mathfrak{P}} H^1(L_{\mathfrak{P}}, \mathbb{Z}/n\mathbb{Z}) & \times & \prod_{\mathfrak{P}} H^1(L_{\mathfrak{P}}, \mu_n) & \longrightarrow & \mathbb{Z}/n\mathbb{Z} \\ \uparrow & & \downarrow N_{L|K} & & \| \\ \prod_{\mathfrak{p}} H^1(K_{\mathfrak{p}}, \mathbb{Z}/n\mathbb{Z}) & \times & \prod_{\mathfrak{p}} H^1(K_{\mathfrak{p}}, \mu_n) & \longrightarrow & \mathbb{Z}/n\mathbb{Z}. \end{array}$$

(iii) The images of

$$H^1(K, \mathbb{Z}/n\mathbb{Z}) \to \prod_{\mathfrak{p}} H^1(K_{\mathfrak{p}}, \mathbb{Z}/n\mathbb{Z})$$

and

$$H^1(K, \mu_n) \to \prod_{\mathfrak{p}} H^1(K_{\mathfrak{p}}, \mu_n)$$

are mutual orthogonal complements with respect to the pairing $(*)$.

Hint for (iii): The cokernel of the second map is $C_K/(C_K)^n$, and one has $H^1(K, \mathbb{Z}/n\mathbb{Z}) = \mathrm{Hom}(G(L|K), \mathbb{Z}/n\mathbb{Z})$, where $L|K$ is the maximal abelian extension of exponent n.

Exercise 5 (Global Tate Duality). Show that the statements of exercise 4 extend to an arbitrary finite G_K-module A instead of $\mathbb{Z}/n\mathbb{Z}$, and $A' = \mathrm{Hom}(A, \overline{K}^*)$ instead of μ_n.

Hint: Use exercises 4–8 of chap. IV, § 3, and exercise 4 of chap. V, § 1.

Exercise 6. If S is a finite set of places of K, then the map

$$H^1(K, \mathbb{Z}/n\mathbb{Z}) \to \prod_{\mathfrak{p} \in S} H^1(K_{\mathfrak{p}}, \mathbb{Z}/n\mathbb{Z})$$

is surjective if and only if the map

$$H^1(K, \mu_n) \to \prod_{\mathfrak{p} \notin S} H^1(K_{\mathfrak{p}}, \mu_n)$$

is injective. This is the case in particular if either the extension $K(\mu_{2^\nu})|K$ is cyclic, $n = 2^\nu m$, $(m, 2) = 1$, or if S does not contain all places $\mathfrak{p}|2$ which are nonsplit in $K(\mu_{2^\nu})$ (see § 1, exercise 2).

Exercise 7 (Theorem of *GRUNWALD***).** If the last condition of exercise 6 is satisfied for the triple (K, n, S), then, given cyclic extensions $L_\mathfrak{p}|K_\mathfrak{p}$ for $\mathfrak{p} \in S$, there always exists a cyclic extension $L|K$ which has $L_\mathfrak{p}|K_\mathfrak{p}$ as a completion for $\mathfrak{p} \in S$, and which satisfies the identity of degrees

$$[L : K] = \mathrm{scm}\{[L_\mathfrak{p} : K_\mathfrak{p}]\}$$

(see also [10], chap. X, § 2).

Note: Let G be a finite group of order prime to $\#\mu(K)$, let S be a finite set of places, and let $L_\mathfrak{p}|K_\mathfrak{p}$, $\mathfrak{p} \in S$, be given Galois extensions whose Galois groups $G_\mathfrak{p}$ can be embedded into G. Then there exists a Galois extension $L|K$ which on the one hand has Galois group isomorphic to G, and which on the other hand has the given extensions $L_\mathfrak{p}|K_\mathfrak{p}$ as completions (see [109]).

§ 7. The Ideal-Theoretic Version of Class Field Theory

Class field theory has found its idèle-theoretic formulation only after it had been completed in the language of ideals. From the very start, it was guided by the desire to classify all abelian extensions of a number field K. But at first, instead of the idèle class group C_K, there was only the ideal class group Cl_K at hand to do this, along with its subgroups. In terms of the insights that we have gained in the preceding section, this means the restriction to the subfields of the Hilbert class field, i.e., to the *unramified* abelian extensions of K. If the base field is $\mathbb{Q}$, this restriction is of course radical, for $\mathbb{Q}$ has no unramified extensions at all by Minkowski's theorem. But over $\mathbb{Q}$, we naturally encounter the cyclotomic fields $\mathbb{Q}(\mu_m)|\mathbb{Q}$ with their familiar isomorphisms $G(\mathbb{Q}(\mu_m)|\mathbb{Q}) \cong (\mathbb{Z}/m\mathbb{Z})^*$. *HEINRICH WEBER* realized, as was already mentioned, that the groups Cl_K and $(\mathbb{Z}/m\mathbb{Z})^*$ are – with a grain of salt – only different instances of a common concept, that of a ray class group, which he defined in an ideal-theoretic way as the quotient group

$$Cl_K^\mathfrak{m} = J_K^\mathfrak{m}/P_K^\mathfrak{m}$$

of all ideals relatively prime to a given module $\mathfrak{m}$, by the principal ideals (α) with $\alpha \equiv 1 \bmod \mathfrak{m}$, and α totally positive. He conjectured that this group $Cl_K^\mathfrak{m}$, along with its subgroups, would do the same for the subextensions of a "ray class field" $K^\mathfrak{m}|K$ (which at first was only postulated to exist) as the ideal class group Cl_K and its subgroups did for the subfields of the Hilbert class field. Moreover, he stated the hypothesis that every abelian extension ought to be captured by such a ray class field, as was suggested by the case where the base field is $\mathbb{Q}$, whose abelian extensions are all contained in cyclotomic fields $\mathbb{Q}(\mu_m)|\mathbb{Q}$ by the Kronecker-Weber theorem. After the seminal work of the Austrian mathematician *Philipp Furtwängler* [44], these conjectures were confirmed by the Japanese arithmetician *Teiji Takagi* (1875–1960), and cast by *Emil Artin* (1898–1962) into a definite, canonical form.

The idèle-theoretic language introduced by *Chevalley* brought the simplification that the idèle class group C_K encapsulated all abelian extensions of $L|K$ at once, avoiding choosing a module $\mathfrak{m}$ every time such an extension was given, in order to accommodate it into the ray class field $K^{\mathfrak{m}}|K$, and thereby make it amenable to class field theory. The classical point of view can be vindicated in terms of the idèle-theoretic version by looking at congruence subgroups $C_K^{\mathfrak{m}}$ in C_K, which define the ray class fields $K^{\mathfrak{m}}|K$. Their subfields correspond, according to the new point of view, to the groups between $C_K^{\mathfrak{m}}$ and C_K, and hence, in view of the isomorphism

$$C_K/C_K^{\mathfrak{m}} \cong Cl_K^{\mathfrak{m}},$$

to the subgroups of the ray class group $Cl_K^{\mathfrak{m}}$.

In what follows, we want to deduce the classical, ideal-theoretic version of global class field theory from the idèle-theoretic one. This is not only an obligation towards history, but a factual necessity that is forced upon us by the numerous applications of the more elementary and more immediately accessible ideal groups.

Let $L|K$ be an abelian extension, and let $\mathfrak{p}$ be an unramified prime ideal of K and $\mathfrak{P}$ a prime ideal of L lying above $\mathfrak{p}$. The decomposition group $G(L_{\mathfrak{P}}|K_{\mathfrak{p}}) \subseteq G(L|K)$ is then generated by the classical **Frobenius automorphism**

$$\varphi_{\mathfrak{p}} = (\pi_{\mathfrak{p}}, L_{\mathfrak{P}}|K_{\mathfrak{p}}),$$

where $\pi_{\mathfrak{p}}$ is a prime element of $K_{\mathfrak{p}}$. As an automorphism of L, $\varphi_{\mathfrak{p}}$ is obviously characterized by the congruence

$$\varphi_{\mathfrak{p}} a \equiv a^q \bmod \mathfrak{P} \quad \text{for all} \quad a \in \mathcal{O}_L$$

where q is the number of elements in the residue class field of $\mathfrak{p}$. We put

$$\varphi_{\mathfrak{p}} =: \left(\frac{L|K}{\mathfrak{p}}\right).$$

Now let $\mathfrak{m}$ be a module of K such that L lies in the ray class field mod $\mathfrak{m}$. Such a module is called an **module of definition** for L. Since by (6.6) each prime ideal $\mathfrak{p} \nmid \mathfrak{m}$ is unramified in L, we get a canonical homomorphism

$$\left(\frac{L|K}{\quad}\right) : J_K^{\mathfrak{m}} \longrightarrow G(L|K)$$

from the group $J_K^{\mathfrak{m}}$ of all ideals of K which are relatively prime to $\mathfrak{m}$ by putting, for any ideal $\mathfrak{a} = \prod_{\mathfrak{p}} \mathfrak{p}^{\nu_{\mathfrak{p}}}$:

$$\left(\frac{L|K}{\mathfrak{a}}\right) = \prod_{\mathfrak{p}} \left(\frac{L|K}{\mathfrak{p}}\right)^{\nu_{\mathfrak{p}}}.$$

$\left(\frac{L|K}{\mathfrak{a}}\right)$ is called the **Artin symbol**. If $\mathfrak{p} \in J_K^{\mathfrak{m}}$ is a prime ideal and $\pi_{\mathfrak{p}}$ a prime element of $K_{\mathfrak{p}}$, then clearly

$$\left(\frac{L|K}{\mathfrak{p}}\right) = \left(\langle \pi_{\mathfrak{p}} \rangle, L|K\right),$$

if $\langle \pi_{\mathfrak{p}} \rangle \in C_K$ denotes the class of the idèle $(\dots, 1, 1, \pi_{\mathfrak{p}}, 1, 1, \dots)$.

The relation between the idèle-theoretic and the ideal-theoretic formulation of the Artin reciprocity law is now provided by the following theorem.

(7.1) Theorem. *Let $L|K$ be an abelian extension, and let $\mathfrak{m}$ be a module of definition for it. Then the Artin symbol induces a surjective homomorphism*

$$\left(\frac{L|K}{\quad}\right) : Cl_K^{\mathfrak{m}} \longrightarrow G(L|K)$$

with kernel $H^{\mathfrak{m}}/P_K^{\mathfrak{m}}$, where $H^{\mathfrak{m}} = (N_{L|K} J_L^{\mathfrak{m}}) P_K^{\mathfrak{m}}$, and we have an exact commutative diagram

$$\begin{array}{ccccccccc}
1 \longrightarrow & N_{L|K}C_L & \longrightarrow & C_K & \xrightarrow{(\ ,L|K)} & G(L|K) & \longrightarrow 1 \\
& \downarrow & & \downarrow & & \downarrow \text{id} & \\
1 \longrightarrow & H^{\mathfrak{m}}/P_K^{\mathfrak{m}} & \longrightarrow & Cl_K^{\mathfrak{m}} & \xrightarrow{\left(\frac{L|K}{\ }\right)} & G(L|K) & \longrightarrow 1.
\end{array}$$

Proof: In § 1, we obtained the isomorphism $(\ ) : C_K/C_K^{\mathfrak{m}} \to Cl_K^{\mathfrak{m}} = J_K^{\mathfrak{m}}/P_K^{\mathfrak{m}}$ by sending an idèle $\alpha = (\alpha_{\mathfrak{p}})$ to the ideal $(\alpha) = \prod_{\mathfrak{p}\nmid\infty} \mathfrak{p}^{v_{\mathfrak{p}}(\alpha_{\mathfrak{p}})}$. This isomorphism yields a commutative diagram

$$\begin{array}{ccc}
C_K/C_K^{\mathfrak{m}} & \xrightarrow{(\ ,L|K)} & G(L|K) \\
(\ )\downarrow & & \downarrow \text{id} \\
Cl_K^{\mathfrak{m}} & \xrightarrow{f} & G(L|K),
\end{array}$$

and we show that f is given by the Artin symbol.

Let $\mathfrak{p}$ be a prime ideal not dividing $\mathfrak{m}$, $\pi_{\mathfrak{p}}$ a prime element of $K_{\mathfrak{p}}$, and $c \in C_K/C_K^{\mathfrak{m}}$ the class of the idèle $\langle \pi_{\mathfrak{p}} \rangle = (\dots, 1, 1, \pi_{\mathfrak{p}}, 1, 1, \dots)$. Then $(c) = \mathfrak{p} \bmod P_K^{\mathfrak{m}}$ and

$$f\big((c)\big) = (c, L|K) = \big(\langle \pi_{\mathfrak{p}} \rangle, L|K\big) = \Big(\frac{L|K}{\mathfrak{p}}\Big).$$

This shows that $f : J_K^{\mathfrak{m}}/P_K^{\mathfrak{m}} \to G(L|K)$ is induced by the Artin symbol $\big(\frac{L|K}{\ }\big) : J_K^{\mathfrak{m}} \to G(L|K)$, and that it is surjective.

It remains to show that the image of $N_{L|K}C_L$ under the map $(\ ) : C_K \to J_K^{\mathfrak{m}}/P_K^{\mathfrak{m}}$ is the group $H^{\mathfrak{m}}/P_K^{\mathfrak{m}}$. We view the module $\mathfrak{m} = \prod_{\mathfrak{p}\nmid\infty} \mathfrak{p}^{n_{\mathfrak{p}}}$ as a module of L by substituting for each prime ideal $\mathfrak{p}$ of K the product $\mathfrak{p} = \prod_{\mathfrak{P}|\mathfrak{p}} \mathfrak{P}^{e_{\mathfrak{P}|\mathfrak{p}}}$. As in the proof of (1.9), we then get $C_L = I_L^{(\mathfrak{m})}L^*/L^*$, where $I_L^{(\mathfrak{m})} = \{\alpha \in I_L \mid \alpha_{\mathfrak{P}} \in U_{\mathfrak{P}}^{(e_{\mathfrak{P}|\mathfrak{p}}n_{\mathfrak{p}})} \text{ for } \mathfrak{P}|\mathfrak{m}\infty\}$. The elements of

$$N_{L|K}C_L = N_{L|K}(I_L^{(\mathfrak{m})})K^*/K^*$$

are the classes of norm idèles $N_{L|K}(\alpha)$, for $\alpha \in I_L^{(\mathfrak{m})}$. As

$$N_{L|K}(\alpha)_{\mathfrak{p}} = \prod_{\mathfrak{P}|\mathfrak{p}} N_{L_{\mathfrak{P}}|K_{\mathfrak{p}}}(\alpha_{\mathfrak{P}})$$

(see (2.2)), and since $v_{\mathfrak{p}}(N_{L_{\mathfrak{P}}|K_{\mathfrak{p}}}(\alpha_{\mathfrak{P}})) = f_{\mathfrak{P}|\mathfrak{p}}v_{\mathfrak{P}}(\alpha_{\mathfrak{P}})$ (see chap. III, (1.2)), the idèle $N_{L|K}(\alpha)$ is mapped by $(\ )$ to the ideal

$$(N_{L|K}(\alpha)) = \prod_{\mathfrak{p}\nmid\infty}\prod_{\mathfrak{P}|\mathfrak{p}} \mathfrak{p}^{f_{\mathfrak{P}|\mathfrak{p}}v_{\mathfrak{P}}(\alpha_{\mathfrak{P}})} = N_{L|K}\Big(\prod_{\mathfrak{P}\nmid\infty} \mathfrak{P}^{v_{\mathfrak{P}}(\alpha_{\mathfrak{P}})}\Big).$$

Therefore the image of $N_{L|K}C_L$ under the homomorphism $(\ ) : C_K \to J_K^{\mathfrak{m}}/P_K^{\mathfrak{m}}$ is precisely the group $(N_{L|K}J_L^{\mathfrak{m}})P_K^{\mathfrak{m}}/P_K^{\mathfrak{m}}$, q.e.d. □

(7.2) Corollary. *The Artin symbol* $\big(\frac{L|K}{\mathfrak{a}}\big)$, *for* $\mathfrak{a} \in J_K^{\mathfrak{m}}$, *only depends on the class* $\mathfrak{a} \bmod P_K^{\mathfrak{m}}$. *It defines an isomorphism*

$$\Big(\frac{L|K}{\ }\Big) : J_K^{\mathfrak{m}}/H^{\mathfrak{m}} \xrightarrow{\ \sim\ } G(L|K).$$

The group $H^{\mathfrak{m}} = (N_{L|K}J_L^{\mathfrak{m}})P_K^{\mathfrak{m}}$ is called the "*ideal group defined* mod $\mathfrak{m}$" belonging to the extension $L|K$. From the existence theorem (6.1), we see that the correspondence $L \mapsto H^{\mathfrak{m}}$ is 1–1 between subextensions of the ray class field mod $\mathfrak{m}$ and subgroups of $J_K^{\mathfrak{m}}$ containing $P_K^{\mathfrak{m}}$.

The most important consequence of theorem (7.1) is a precise analysis of the kind of decomposition of any unramified prime ideal $\mathfrak{p}$ in an abelian extension $L|K$. It can be immediately read off the ideal group $H^{\mathfrak{m}} \subseteq J_K^{\mathfrak{m}}$ which determines the field L as class field.

(7.3) Theorem (Decomposition Law). *Let $L|K$ be an abelian extension of degree n, and let $\mathfrak{p}$ be an unramified prime ideal. Let $\mathfrak{m}$ be a module of definition for $L|K$ that is not divisible by $\mathfrak{p}$ (for instance the conductor), and let $H^{\mathfrak{m}}$ be the corresponding ideal group.*

If f is the order of $\mathfrak{p}$ mod $H^{\mathfrak{m}}$ in the class group $J_K^{\mathfrak{m}}/H^{\mathfrak{m}}$, i.e., the smallest positive integer such that

$$\mathfrak{p}^f \in H^{\mathfrak{m}},$$

then $\mathfrak{p}$ decomposes in L into a product

$$\mathfrak{p} = \mathfrak{P}_1 \cdots \mathfrak{P}_r$$

of $r = n/f$ distinct prime ideals of degree f over $\mathfrak{p}$.

Proof: Let $\mathfrak{p} = \mathfrak{P}_1 \cdots \mathfrak{P}_r$ be the prime decomposition of $\mathfrak{p}$ in L. Since $\mathfrak{p}$ is unramified, the $\mathfrak{P}_i$ are all distinct and have the same degree f. This degree is the order of the decomposition group of $\mathfrak{P}_i$ over K, i.e., the order of the Frobenius automorphism $\varphi_{\mathfrak{p}} = \left(\frac{L|K}{\mathfrak{p}}\right)$. In view of the isomorphism $J_K^{\mathfrak{m}}/H^{\mathfrak{m}} \cong G(L|K)$, this is also the order of $\mathfrak{p}$ mod $H^{\mathfrak{m}}$ in $J_K^{\mathfrak{m}}/H^{\mathfrak{m}}$. This finishes the proof. □

The theorem shows in particular that the prime ideals which split completely are precisely those contained in the ideal group $H^{\mathfrak{f}}$, if $\mathfrak{f}$ is the conductor of $L|K$.

Let us highlight two special cases. If the base field is $K = \mathbb{Q}$ and we look at the cyclotomic field $\mathbb{Q}(\mu_m)|\mathbb{Q}$, the conductor is the module $\mathfrak{m} = (m)$, and the ideal group corresponding to $\mathbb{Q}(\mu_m)$ in $J_{\mathbb{Q}}^{\mathfrak{m}}$ is the group $P_{\mathbb{Q}}^{\mathfrak{m}}$. As $J_{\mathbb{Q}}^{\mathfrak{m}}/P_{\mathbb{Q}}^{\mathfrak{m}} \cong (\mathbb{Z}/m\mathbb{Z})^*$ (see (1.10)), we obtain for the decomposition of rational primes $p \nmid m$, the law which we had already deduced in chap. I, (10.4), and in particular the fact that the prime numbers which split completely are characterized by

$$p \equiv 1 \bmod m.$$

In the case of the Hilbert class field $L|K$, i.e., of the field inside the ray class field mod 1 in which the infinite places split completely, the corresponding ideal group $H \subseteq J_K^1 = J_K$ is the group P_K of principal ideals (see (6.9)). This gives us the strikingly simple

(7.4) Corollary. *The prime ideals of K which split completely in the Hilbert class field are precisely the principal prime ideals.*

Another highly remarkable property of the Hilbert class field is expressed by the following theorem, known as the **principal ideal theorem**.

(7.5) Theorem. *In the Hilbert class field every ideal* $\mathfrak{a}$ *of* K *becomes a principal ideal.*

Proof: Let $K_1|K$ be the Hilbert class field of K and let $K_2|K_1$ be the Hilbert class field of K_1. We have to show that the canonical homomorphism

$$J_K/P_K \longrightarrow J_{K_1}/P_{K_1}$$

is trivial. By chap. IV, (5.9), we have a commutative diagram

$$\begin{array}{ccccc} J_{K_1}/P_{K_1} & \cong & C_{K_1}/N_{K_2|K_1}C_{K_2} & \cong & G(K_2|K_1) \\ \uparrow & & \uparrow{\scriptstyle i} & & \uparrow{\scriptstyle \mathrm{Ver}} \\ J_K/P_K & \cong & C_K/N_{K_1|K}C_{K_1} & \cong & G(K_1|K), \end{array}$$

where i is induced by the inclusion $C_K \subseteq C_{K_1}$. It is therefore enough to show that the transfer

$$\mathrm{Ver} : G(K_1|K) \longrightarrow G(K_2|K_1)$$

is the trivial homomorphism. Since $K_1|K$ is the maximal unramified abelian extension of K in which the infinite places split completely, i.e., the maximal abelian subextension of $K_2|K$, we see that $G(K_2|K_1)$ is the commutator subgroup of $G(K_2|K)$. The proof of the principal ideal theorem is thus reduced to the following purely group-theoretic result. □

(7.6) Theorem. *Let* G *be a finitely generated group,* G' *its commutator subgroup, and* G'' *the commutator subgroup of* G'*. If* $(G:G') < \infty$*, then the transfer*

$$\mathrm{Ver} : G/G' \longrightarrow G'/G''$$

is the trivial homomorphism.

We give a proof of this theorem which is due to *Ernst Witt* [141]. In the group ring $\mathbb{Z}[G] = \{\sum_{\sigma\in G} n_\sigma\sigma \mid n_\sigma \in \mathbb{Z}\}$, we consider the **augmentation ideal** I_G, which is by definition the kernel of the ring homomorphism

$$\mathbb{Z}[G] \longrightarrow \mathbb{Z}, \quad \sum_\sigma n_\sigma\sigma \longmapsto \sum_\sigma n_\sigma .$$

For every subgroup H of G, we have $I_H \subseteq I_G$, and $\{\tau - 1 \mid \tau \in H, \tau \neq 1\}$ is a $\mathbb{Z}$-basis of I_H. We first establish the following lemma, which also has independent interest in that it gives an additive interpretation of the transfer.

(7.7) Lemma. *For every subgroup H of finite index in G, one has a commutative diagram*

$$\begin{array}{ccc} G/G' & \xrightarrow{\text{Ver}} & H/H' \\ \delta\downarrow\cong & & \delta\downarrow\cong \\ I_G/I_G^2 & \xrightarrow{S} & (I_H+I_GI_H)/I_GI_H, \end{array}$$

where the homomorphisms δ are induced by $\sigma\mapsto\delta\sigma=\sigma-1$, and the homomorphism S is given by

$$S(x \bmod I_G^2) = x\sum_{\rho\in R}\rho \bmod I_GI_H,$$

for a system of representatives of the left cosets $R\ni 1$ of G/H.

Proof: We first show that the homomorphism

$$(*)\qquad H/H' \xrightarrow{\delta} (I_H+I_GI_H)/I_GI_H$$

induced by $\tau\mapsto\delta\tau=\tau-1$ has an inverse. The elements $\rho\delta\tau$, $\tau\in H$, $\tau\neq 1$, $\rho\in R$, form a $\mathbb{Z}$-basis of $I_H+I_GI_H$. Indeed, it follows from

$$\rho\delta\tau=\delta\tau+\delta\rho\delta\tau$$

that they generate $I_H+I_GI_H$, and if

$$0=\sum_{\rho,\tau}n_{\rho,\tau}\rho\delta\tau=\sum_{\rho,\tau}n_{\rho,\tau}(\rho\tau-\rho)=\sum_{\rho,\tau}n_{\rho,\tau}\rho\tau-\sum_{\rho}\Big(\sum_{\tau}n_{\rho,\tau}\Big)\rho,$$

then we conclude that $n_{\rho,\tau}=0$ because the $\rho\tau,\rho$ are pairwise distinct. Mapping $\rho\delta\tau$ to $\tau \bmod H'$, we now have a surjective homomorphism

$$I_H+I_GI_H \longrightarrow H/H'.$$

It sends $\delta(\rho\tau')\delta\tau\in I_GI_H$ to $\tau'\tau\tau'^{-1}\tau^{-1}\equiv 1 \bmod H'$ because $\delta(\rho\tau')\delta\tau = \rho\delta(\tau'\tau)-\rho\delta\tau'-\delta\tau$. It thus induces a homomorphism which is inverse to $(*)$. In particular, if $H=G$, we obtain the isomorphism $G/G'\xrightarrow{\delta} I_G/I_G^2$.

The transfer is now obtained as

$$\text{Ver}(\sigma \bmod G') = \prod_{\rho\in R}\sigma_\rho \bmod H',$$

where $\sigma_\rho\in H$ is defined by $\sigma\rho=\rho'\sigma_\rho$, $\rho'\in R$. Ver thus induces the homomorphism

$$S: I_G/I_G^2 \longrightarrow (I_H+I_GI_H)/I_GI_H$$

given by $S(\delta\sigma \bmod I_G^2) = \sum_{\rho\in R} \delta\sigma_\rho \bmod I_G I_H$. From $\sigma\rho = \rho'\sigma_\rho$ follows the identity

$$\delta\rho + (\delta\sigma)\rho = \delta\sigma_\rho + \delta\rho' + \delta\rho'\delta\sigma_\rho .$$

Since ρ' runs through the set R if ρ does, we get as claimed

$$S(\delta\rho \bmod I_G^2) \equiv \sum_{\rho\in R} \delta\sigma_\rho \equiv \sum_{\rho\in R} (\delta\sigma)\rho \equiv \delta\sigma \sum_{\rho\in R} \rho \bmod I_G I_H . \qquad \square$$

Proof of theorem (7.6): Replacing G by G/G'', we may assume that $G'' = \{1\}$, i.e., that G' is abelian. Let $R \ni 1$ be a system of representatives of left cosets of G/G', and let $\sigma_1, \dots, \sigma_n$ be generators of G. Mapping $e_i = (0, \dots, 0, 1, 0, \dots, 0) \in \mathbb{Z}^n$ to σ_i, we get an exact sequence

$$0 \longrightarrow \mathbb{Z}^n \xrightarrow{f} \mathbb{Z}^n \longrightarrow G/G' \longrightarrow 1,$$

where f is given by an $n \times n$-matrix (m_{ik}) with $\det(m_{ik}) = (G : G')$. Consequently,

$$\prod_{i=1}^{n} \sigma_i^{m_{ik}} \tau_k = 1 \quad \text{with} \quad \tau_k \in G' .$$

The formulae $\delta(xy) = \delta x + \delta y + \delta x \delta y$, $\delta(x^{-1}) = -(\delta x)x^{-1}$ yield by iteration that

$$\delta\Big(\prod_{i=1}^{n} \sigma_i^{m_{ik}} \tau_k\Big) = \sum_{i=1}^{n} (\delta\sigma_i)\mu_{ik} = 0,$$

where $\mu_{ik} \equiv m_{ik} \bmod I_G$. In fact, the τ_k are products of commutators of the σ_i and σ_i^{-1}. We view (μ_{ik}) as a matrix over the commutative ring

$$\mathbb{Z}[G/G'] \cong \mathbb{Z}[G]/\mathbb{Z}[G]I_{G'},$$

which gives a meaning to the determinant $\mu = \det(\mu_{ik}) \in \mathbb{Z}[G/G']$. Let (λ_{kj}) be the adjoint matrix of (μ_{ik}). Then

$$(\delta\sigma_j)\mu = \sum_{i,k} (\delta\sigma_i)\mu_{ik}\lambda_{kj} \equiv 0 \bmod I_G \mathbb{Z}[G] I_{G'},$$

so that $(\delta\sigma)\mu \equiv 0 \bmod I_G\mathbb{Z}[G]I_{G'} = I_G I_{G'}$ for all σ. This yields

$$\mu \equiv \sum_{\rho\in R} \rho \bmod \mathbb{Z}[G]I_{G'} .$$

For if we put $\mu = \sum_{\rho\in R} n_\rho \bar\rho$, where $\bar\rho = \rho \bmod G'$, then for all $\bar\sigma \in G/G'$,

$$\bar\sigma\mu = \sum_\rho n_\rho \bar\sigma\bar\rho = \sum_\rho n_\rho \bar\rho .$$

This implies that all n_ρ are equal, hence $\mu \equiv m\sum_{\rho\in R}\rho \bmod \mathbb{Z}[G]I_{G'}$, and as

$$\mu \equiv \det(m_{ik}) \equiv (G:G') \equiv m(G:G') \bmod I_G,$$

we even have $m = 1$. Applying now lemma (7.7), we see that the transfer is the trivial homomorphism since

$$S(\delta\sigma \bmod I_G^2) \equiv \delta\sigma \sum_{\rho\in R}\rho \equiv (\delta\sigma)\mu \equiv 0 \bmod I_G I_{G'}. \qquad \square$$

A problem which is closely related to the principal ideal theorem and which was first put forward by *Philipp Furtwängler* is the **problem of the class field tower**. This is the question whether the class field tower

$$K = K_0 \subseteq K_1 \subseteq K_2 \subseteq K_3 \subseteq \dots,$$

where K_{i+1} is the Hilbert class field of K_i, stops after a finite number of steps. A positive answer would have the implication that the last field in the tower had class number 1 so that in it not only the ideals of K, but in fact all its ideals become principal. This perspective naturally generated the greatest interest. But the problem, after withstanding for a long time all attempts to solve it, was finally decided in the negative by the Russian mathematicians *E.S. Golod* and *I.R. Šafarevič* in 1964 (see [48], [24]).

Exercise 1. The decomposition law for the prime ideals $\mathfrak{p}$ which are *ramified* in an abelian extension $L|K$ can be formulated like this. Let $\mathfrak{f}$ be the conductor of $L|K$, $H^{\mathfrak{f}} \subseteq J_K^{\mathfrak{f}}$ the ideal group for L, and $H_{\mathfrak{p}}$ the smallest ideal group containing $H^{\mathfrak{f}}$ of conductor prime to $\mathfrak{p}$.

If $e = (H_{\mathfrak{p}} : H^{\mathfrak{f}})$ and $\mathfrak{p}^f$ is the smallest power of $\mathfrak{p}$ which belongs to $H_{\mathfrak{p}}$, then

$$\mathfrak{p} = (\mathfrak{P}_1 \cdots \mathfrak{P}_r)^e,$$

where the $\mathfrak{P}_i$ are of degree f over K, and $r = \frac{n}{ef}$, $n = [L:K]$.

Hint: The class field for $H_{\mathfrak{p}}$ is the inertia field above $\mathfrak{p}$.

The following exercises 2–6 concern a non-abelian example of *E. Artin*.

Exercise 2. The polynomial $f(X) = X^5 - X + 1$ is irreducible. The discriminant of a root α (i.e., the discriminant of $\mathbb{Z}[\alpha]$) is $d = 19 \cdot 151$.

Hint: The discriminant of a root of $X^5 + aX + b$ is $5^5b^4 + 2^8a^5$.

Exercise 3. Let $k = \mathbb{Q}(\alpha)$. Then $\mathbb{Z}[\alpha]$ is the ring $\mathcal{O}_k$ of integers of k.

Hint: The discriminant of $\mathbb{Z}[\alpha]$ equals the discriminant of $\mathcal{O}_k$ because on the one hand, both differ only by a square, and on the other hand, it is squarefree. The transition matrix from $1, \alpha, \dots, \alpha^{n-1}$ to an integral basis $\omega_1, \dots, \omega_n$ of $\mathcal{O}_k$ is therefore invertible over $\mathbb{Z}$.

Exercise 4. The decomposition field $K|\mathbb{Q}$ of $f(X)$ has as Galois group the symmetric group $\mathfrak{S}_5$, i.e., it is of degree 120.

Exercise 5. K has class number 1.

Hint: Show, using chap. I, § 6, exercise 3, that every ideal class of K contains an ideal $\mathfrak{a}$ with $\mathfrak{N}(\mathfrak{a}) < 4$. If $\mathfrak{N}(\mathfrak{a}) \neq 1$, then $\mathfrak{a}$ has to be a prime ideal $\mathfrak{p}$ such that $\mathfrak{N}(\mathfrak{p}) = 2$ or 3. Hence $\mathcal{O}_k/\mathfrak{p} = \mathbb{Z}/2\mathbb{Z}$ or $= \mathbb{Z}/3\mathbb{Z}$, so f has a root $\bmod 2$ or 3, which is not the case.

Exercise 6. Show that $K \mid \mathbb{Q}(\sqrt{19 \cdot 151}\,)$ is a (non-abelian!) unramified extension.

Exercise 7. For every Galois extension $L|K$ of finite algebraic number fields, there exist infinitely many finite extensions K' such that $L \cap K' = K$, and such that $LK'|K'$ is unramified.

Hint: Let S be the set of places ramified in $L|K$, and let $L_\mathfrak{p} = K_\mathfrak{p}(\alpha_\mathfrak{p})$. By the approximation theorem, choose an algebraic number α which, for every $\mathfrak{p} \in S$, is close to $\alpha_\mathfrak{p}$ when embedded into $\overline{K}_\mathfrak{p}$. Then $K_\mathfrak{p}(\alpha_\mathfrak{p}) \subseteq K_\mathfrak{p}(\alpha)$ by Krasner's lemma, chap. II, § 6, exercise 2. Put $K' = K(\alpha)$ and show that $LK'|K'$ is unramified. To show that α can be chosen such that $L \cap K' = K$ use (3.7), and the fact that $G(L|K)$ is generated by elements of prime power order.

§ 8. The Reciprocity Law of the Power Residues

In class field theory Gauss's reciprocity law meets its most general and definite formulation. Let n be a positive integer ≥ 2 and K a number field containing the group μ_n of n-th roots of unity. In chap. V, § 3, we introduced, for every place $\mathfrak{p}$ of K, the n-th Hilbert symbol

$$\Big(\frac{\;,\;}{\mathfrak{p}}\Big) : K_\mathfrak{p}^* \times K_\mathfrak{p}^* \longrightarrow \mu_n .$$

It is given via the norm residue symbol by

$$\big(a, K_\mathfrak{p}(\sqrt[n]{b}\,)|K_\mathfrak{p}\big)\sqrt[n]{b} = \Big(\frac{a,b}{\mathfrak{p}}\Big)\sqrt[n]{b}.$$

These symbols all fit together in the following *product formula*.

(8.1) Theorem. *For $a, b \in K^*$ one has*

$$\prod_\mathfrak{p} \Big(\frac{a,b}{\mathfrak{p}}\Big) = 1 .$$

Proof: From (5.7), we find

$$\Big[\prod_{\mathfrak p}\Big(\frac{a,b}{\mathfrak p}\Big)\Big]\sqrt[n]{b} = \Big[\prod_{\mathfrak p}(a, K_{\mathfrak p}(\sqrt[n]{b})|K_{\mathfrak p})\Big]\sqrt[n]{b} = \big(a, K(\sqrt[n]{b})|K\big)\sqrt[n]{b} = \sqrt[n]{b},$$

and hence the theorem. □

In chap. V, § 3, we defined the n-th power residue symbol in terms of the Hilbert symbol:

$$\Big(\frac{a}{\mathfrak p}\Big) = \Big(\frac{\pi, a}{\mathfrak p}\Big),$$

where $\mathfrak p$ is a prime ideal of K not dividing n, $a \in U_{\mathfrak p}$, and π is a prime element of $K_{\mathfrak p}$. We have seen that this definition does not depend on the choice of the prime element π and that one has

$$\Big(\frac{a}{\mathfrak p}\Big) = 1 \iff a \equiv \alpha^n \bmod \mathfrak p,$$

and more generally

$$\Big(\frac{a}{\mathfrak p}\Big) \equiv a^{(q-1)/n} \bmod \mathfrak p\,, \quad q = \mathfrak N(\mathfrak p)\,.$$

(8.2) Definition. *For every ideal $\mathfrak b = \prod_{\mathfrak p\nmid n}\mathfrak p^{\nu_{\mathfrak p}}$ prime to n, and every number a prime to $\mathfrak b$, we define the* **n-th power residue symbol** *by*

$$\Big(\frac{a}{\mathfrak b}\Big) = \prod_{\mathfrak p\nmid n}\Big(\frac{a}{\mathfrak p}\Big)^{\nu_{\mathfrak p}}.$$

Here $\Big(\frac{a}{\mathfrak p}\Big)^{\nu_{\mathfrak p}} = 1$ *when* $\nu_{\mathfrak p} = 0$.

The power residue symbol $\big(\frac{a}{\mathfrak b}\big)$ is obviously multiplicative in both arguments. If $\mathfrak b$ is a principal ideal (b), we write for short $\big(\frac{a}{\mathfrak b}\big) = \big(\frac{a}{b}\big)$. We now prove the **general reciprocity law for the n-th power residues**.

(8.3) Theorem. *If $a, b \in K^*$ are prime to each other and to n, then*

$$\Big(\frac{a}{b}\Big)\Big(\frac{b}{a}\Big)^{-1} = \prod_{\mathfrak p|n\infty}\Big(\frac{a,b}{\mathfrak p}\Big).$$

Proof: If $\mathfrak{p}$ is prime to $bn\infty$, then we have

$$\left(\frac{b}{\mathfrak{p}}\right)^{v_\mathfrak{p}(a)} = \left(\frac{\pi,b}{\mathfrak{p}}\right)^{v_\mathfrak{p}(a)} = \left(\frac{a,b}{\mathfrak{p}}\right),$$

where π is a prime element of $K_\mathfrak{p}$. For if we put $a = u\pi^{v_\mathfrak{p}(a)}$, then $(\frac{u,b}{\mathfrak{p}}) = 1$ because $u, b \in U_\mathfrak{p}$. For the same reason, we find

$$\left(\frac{a,b}{\mathfrak{p}}\right) = 1 \quad \text{for } \mathfrak{p} \text{ prime to } abn\infty.$$

(8.1) then gives

$$\left(\frac{a}{b}\right)\left(\frac{b}{a}\right)^{-1} = \prod_{\mathfrak{p}|(b)}\left(\frac{a}{\mathfrak{p}}\right)^{v_\mathfrak{p}(b)} \prod_{\mathfrak{p}|(a)}\left(\frac{b}{\mathfrak{p}}\right)^{-v_\mathfrak{p}(a)} = \prod_{\mathfrak{p}|(b)}\left(\frac{b,a}{\mathfrak{p}}\right) \prod_{\mathfrak{p}|(a)}\left(\frac{a,b}{\mathfrak{p}}\right)^{-1}$$
$$= \prod_{\mathfrak{p}|(ab)}\left(\frac{b,a}{\mathfrak{p}}\right) = \prod_{\mathfrak{p}\nmid n\infty}\left(\frac{b,a}{\mathfrak{p}}\right) = \prod_{\mathfrak{p}|n\infty}\left(\frac{a,b}{\mathfrak{p}}\right).$$

Here $\mathfrak{p}|(b)$ means that $\mathfrak{p}$ occurs in the prime decomposition of (b). □

Gauss's reciprocity law, for which we gave an elementary proof using the theory of Gauss sums in chap. I, (8.6), in the case of two odd prime numbers p, l, is contained in the general reciprocity law (8.3) as a special case. For if we substitute, in the case $K = \mathbb{Q}$, $n = 2$, into formula (8.3) the explicit description (chap. V, (3.6)) of the Hilbert symbol $\left(\frac{a,b}{p}\right)$ for $p = 2$ and $p = \infty$, we obtain the following theorem, which is more general than chap. I, (8.6).

(8.4) Gauss's Reciprocity Law. *Let $K = \mathbb{Q}$, $n = 2$, and let a and b be odd, relatively prime integers. Then one has*

$$\left(\frac{a}{b}\right)\left(\frac{b}{a}\right) = (-1)^{\frac{a-1}{2}\frac{b-1}{2}}(-1)^{\frac{\operatorname{sgn} a-1}{2}\frac{\operatorname{sgn} b-1}{2}},$$

and for positive odd integers b, we have the two "supplementary theorems"

$$\left(\frac{-1}{b}\right) = (-1)^{\frac{b-1}{2}}, \quad \left(\frac{2}{b}\right) = (-1)^{\frac{b^2-1}{8}}.$$

For the last equation we need again the product formula:

$$\left(\frac{2}{b}\right) = \prod_{p\neq 2,\infty}\left(\frac{p,2}{p}\right)^{v_p(b)} = \prod_{p\neq 2,\infty}\left(\frac{b,2}{p}\right) = \left(\frac{2,b}{2}\right)\left(\frac{2,b}{\infty}\right) = (-1)^{\frac{b^2-1}{2}}.$$

The symbol $\left(\frac{a}{b}\right)$ is called the **Jacobi symbol**, or also the **quadratic residue symbol** (although, for b not a prime number, the condition that the symbol $\left(\frac{a}{b}\right) = 1$ is no longer equivalent to the condition that a is a quadratic residue modulo b).

In the above formulation, the reciprocity law allows us to compute simply by iteration the quadratic residue symbol $\left(\frac{a}{b}\right)$, as is shown in the following example:

$$\left(\frac{40077}{65537}\right) = \left(\frac{65537}{40077}\right) = \left(\frac{25460}{40077}\right) = \left(\frac{2^2}{40077}\right)\left(\frac{6365}{40077}\right) = \left(\frac{40077}{6365}\right) =$$
$$\left(\frac{1887}{6365}\right) = \left(\frac{6365}{1887}\right) = \left(\frac{704}{1887}\right) = \left(\frac{4^3}{1887}\right)\left(\frac{11}{1887}\right) = -\left(\frac{1887}{11}\right) =$$
$$-\left(\frac{6}{11}\right) = -\left(\frac{2}{11}\right)\left(\frac{3}{11}\right) = \left(\frac{3}{11}\right) = -\left(\frac{11}{3}\right) = -\left(\frac{2}{3}\right) = 1.$$

Class field theory originated from Gauss's reciprocity law. The quest for a similar law for the n-th power residues dominated number theory for a long time, and the all-embracing answer was finally found in Artin's reciprocity law. The above reciprocity law (8.3) of the power residues now appears as a simple and special consequence of Artin's reciprocity law. But to really settle the original problem, class field theory was still lacking the explicit computation of the Hilbert symbols $\left(\frac{a,b}{\mathfrak{p}}\right)$ for $\mathfrak{p} | n\infty$. This was finally completed in the 1960s by the mathematician *Helmut Brückner*, see chap. V, (3.7).

Chapter VII

Zeta Functions and L-series

§ 1. The Riemann Zeta Function

One of the most astounding phenomena in number theory consists in the fact that a great number of deep arithmetic properties of a number field are hidden within a single analytic function, its **zeta function**. This function has a simple shape, but it is unwilling to yield its mysteries. Each time, however, that we succeed in stealing one of these well-guarded truths, we may expect to be rewarded by the revelation of some surprising and significant relationship. This is why zeta functions, as well as their generalizations, the L-series, have increasingly moved to the foreground of the arithmetic scene, and today are more than ever the focus of number-theoretic research. The fundamental prototype of such a function is **Riemann's zeta function**

$$\zeta(s) = \sum_{n=1}^{\infty} \frac{1}{n^s},$$

where s is a complex variable. It is to this important function that we turn first.

(1.1) Proposition. *The series* $\zeta(s) = \sum_{n=1}^{\infty} \frac{1}{n^s}$ *is absolutely and uniformly convergent in the domain* $\mathrm{Re}(s) \geq 1 + \delta$, *for every* $\delta > 0$. *It therefore represents an analytic function in the half-plane* $\mathrm{Re}(s) > 1$. *One has* **Euler's identity**

$$\zeta(s) = \prod_p \frac{1}{1 - p^{-s}},$$

where p runs through the prime numbers.

Proof: For $\mathrm{Re}(s) = \sigma \geq 1+\delta$, the series $\sum_{n=1}^{\infty} |1/n^s| = \sum_{n=1}^{\infty} 1/n^{\sigma}$ admits the convergent majorant $\sum_{n=1}^{\infty} 1/n^{1+\delta}$, i.e., $\zeta(s)$ is absolutely and uniformly convergent in this domain. In order to prove Euler's identity, we remind ourselves that an infinite product $\prod_{n=1}^{\infty} a_n$ of complex numbers a_n is said to converge if the sequence of partial products $P_n = a_1 \cdots a_n$ has a nonzero limit. This is the case if and only if the series $\sum_{n=1}^{\infty} \log a_n$ converges, where log denotes the principal branch of the logarithm (see [2], chap. V, 2.2). The product is called absolutely convergent if the series converges absolutely. In this case the product converges to the same limit even after a reordering of its terms a_n.

Let us now formally take the logarithm of the product

$$E(s) = \prod_p \frac{1}{1-p^{-s}}\,.$$

We obtain the series

$$\log E(s) = \sum_p \sum_{n=1}^{\infty} \frac{1}{np^{ns}}\,.$$

It converges absolutely for $\operatorname{Re}(s) = \sigma \geq 1+\delta$. In fact, since $|p^{ns}| = p^{n\sigma} \geq p^{(1+\delta)n}$, one has the convergent majorant

$$\sum_p \sum_{n=1}^{\infty} \Big(\frac{1}{p^{1+\delta}}\Big)^n = \sum_p \frac{1}{p^{1+\delta}-1} \leq 2\sum_p \frac{1}{p^{1+\delta}}\,.$$

This implies the absolute convergence of the product

$$E(s) = \prod_p \frac{1}{1-p^{-s}} = \exp\Big(\sum_p\Big(\sum_{n=1}^{\infty} \frac{1}{np^{ns}}\Big)\Big)\,.$$

In this product we now expand the product of the factors

$$\frac{1}{1-p^{-s}} = 1 + \frac{1}{p^s} + \frac{1}{p^{2s}} + \cdots$$

for all prime numbers $p_1, \ldots, p_r \leq N$, and obtain the equality

$$(*) \qquad \prod_{p\leq N} \frac{1}{1-p^{-s}} = \sum_{\nu_1,\ldots,\nu_r=0}^{\infty} \frac{1}{(p_1^{\nu_1}\ldots p_r^{\nu_r})^s} = {\sum_n}' \frac{1}{n^s}\,,$$

where $\sum'$ denotes the sum over all natural numbers which are divisible only by prime numbers $p \leq N$. Since the sum $\sum'$ contains in particular the terms corresponding to all $n \leq N$, we may also write

$$\prod_{p\leq N} \frac{1}{1-p^{-s}} = \sum_{n\leq N} \frac{1}{n^s} + {\sum_{n>N}}' \frac{1}{n^s}\,.$$

Comparing now in $(*)$ the sum $\sum'$ with the series $\zeta(s)$, we get

$$\Big|\prod_{p\leq N} \frac{1}{1-p^{-s}} - \zeta(s)\Big| \leq \Big|\sum_{\substack{n>N\\ p_i\nmid n}} \frac{1}{n^s}\Big| \leq \sum_{n>N} \frac{1}{n^{1+\delta}}\,,$$

where the right hand side goes to zero as $N \to \infty$ because it is the remainder of a convergent series. This proves Euler's identity. □

Euler's identity expresses the law of unique prime factorization of natural numbers in a single equation. This already demonstrates the number-theoretic significance of the zeta function. It challenges us to study its properties more closely. By its definition, the function is only given on the half-plane $\operatorname{Re}(s) > 1$. It does, however, admit an analytic continuation to the whole complex plane, with the point $s = 1$ removed, and it satisfies a functional equation which relates the argument s to the argument $1 - s$. These crucial facts will be proved next. The proof hinges on an **integral formula** for the zeta function $\zeta(s)$ which arises from the well-known **gamma function**. This latter is defined for $\operatorname{Re}(s) > 0$ by the absolutely convergent integral

$$\Gamma(s) = \int_0^\infty e^{-y} y^s \frac{dy}{y}$$

and obeys the following rules (see [34], vol. I, chap. I).

(1.2) Proposition. (i) *The gamma function is analytic and admits a meromorphic continuation to all of* $\mathbb{C}$.

(ii) *It is nowhere zero and has simple poles at* $s = -n$, $n = 0, 1, 2, \ldots$, *with residues* $(-1)^n/n!$. *There are no poles anywhere else.*

(iii) *It satisfies the functional equations*

1) $\Gamma(s+1) = s\Gamma(s)$,

2) $\Gamma(s)\Gamma(1-s) = \dfrac{\pi}{\sin \pi s}$,

3) $\Gamma(s)\Gamma(s+\frac{1}{2}) = \dfrac{2\sqrt{\pi}}{2^{2s}}\Gamma(2s)$ (**Legendre's duplication formula**).

(iv) *It has the special values* $\Gamma(1/2) = \sqrt{\pi}$, $\Gamma(1) = 1$, $\Gamma(k+1) = k!$, $k = 0, 1, 2, \ldots$.

To relate the gamma function to the zeta function, start with the substitution $y \mapsto \pi n^2 y$, which gives the equation

$$\pi^{-s}\Gamma(s)\frac{1}{n^{2s}} = \int_0^\infty e^{-\pi n^2 y} y^s \frac{dy}{y}.$$

Now sum over all $n \in \mathbb{N}$ and get

$$\pi^{-s}\Gamma(s)\zeta(2s) = \int_0^\infty \sum_{n=1}^\infty e^{-\pi n^2 y} y^s \frac{dy}{y}.$$

Observe that it is legal to interchange the sum and the integral because

$$\sum_{n=1}^{\infty}\int_0^{\infty}|e^{-\pi n^2 y}y^s|\frac{dy}{y} = \sum_{n=1}^{\infty}\int_0^{\infty}e^{-\pi n^2 y}y^{\mathrm{Re}(s)}\frac{dy}{y}$$
$$= \pi^{-\mathrm{Re}(s)}\Gamma\big(\mathrm{Re}(s)\big)\zeta\big(2\,\mathrm{Re}(s)\big) < \infty.$$

Now the series under the integral,

$$g(y) = \sum_{n=1}^{\infty} e^{-\pi n^2 y},$$

arises from **Jacobi's** classical **theta series**

$$\theta(z) = \sum_{n\in\mathbb{Z}} e^{\pi i n^2 z} = 1 + 2\sum_{n=1}^{\infty} e^{\pi i n^2 z},$$

i.e., we have $g(y) = \frac{1}{2}(\theta(iy) - 1)$. The function

$$Z(s) = \pi^{-s/2}\Gamma(s/2)\zeta(s)$$

is called the **completed zeta function**. We obtain the

(1.3) Proposition. *The completed zeta function $Z(s)$ admits the integral representation*

$$Z(s) = \frac{1}{2}\int_0^{\infty}\big(\theta(iy) - 1\big)\,y^{s/2}\frac{dy}{y}.$$

The proof of the functional equation for the function $Z(s)$ is based on the following general principle. For a continuous function $f : \mathbb{R}_+^* \to \mathbb{C}$ on the group $\mathbb{R}_+^*$ of positive real numbers, we define the **Mellin transform** to be the improper integral

$$L(f,s) = \int_0^{\infty}\big(f(y) - f(\infty)\big)\,y^s\frac{dy}{y},$$

provided the limit $f(\infty) = \lim_{y\to\infty} f(y)$ and the integral exist. The following theorem is of pivotal importance, also for later applications. We will often refer to it as the **Mellin principle**.

(1.4) Theorem. *Let $f, g : \mathbb{R}_+^* \to \mathbb{C}$ be continuous functions such that*

$$f(y) = a_0 + O(e^{-cy^\alpha}), \quad g(y) = b_0 + O(e^{-cy^\alpha}),$$

for $y \to \infty$, with positive constants c, α. If these functions satisfy the equation

$$f\Big(\frac{1}{y}\Big) = Cy^k g(y),$$

for some real number $k > 0$ and some complex number $C \neq 0$, then one has:

(i) *The integrals $L(f, s)$ and $L(g, s)$ converge absolutely and uniformly if s varies in an arbitrary compact domain contained in $\{s \in \mathbb{C} \mid \mathrm{Re}(s) > k\}$. They are therefore holomorphic functions on $\{s \in \mathbb{C} \mid \mathrm{Re}(s) > k\}$. They admit holomorphic continuations to $\mathbb{C} \smallsetminus \{0, k\}$.*

(ii) *They have simple poles at $s = 0$ and $s = k$ with residues*

$$\mathrm{Res}_{s=0} L(f, s) = -a_0, \quad \mathrm{Res}_{s=k} L(f, s) = Cb_0, \quad \textit{resp.}$$
$$\mathrm{Res}_{s=0} L(g, s) = -b_0, \quad \mathrm{Res}_{s=k} L(g, s) = C^{-1}a_0.$$

(iii) *They satisfy the functional equation*

$$L(f, s) = CL(g, k - s).$$

Remark 1: The symbol $\varphi(y) = O(\psi(y))$ means, as usual, that one has $\varphi(y) = c(y)\psi(y)$, for some function $c(y)$ which stays bounded under the limit in question, so in our case, as $y \to \infty$.

Remark 2: Condition (ii) is to be understood to say that there is no pole if $a_0 = 0$, resp. $b_0 = 0$. But there is a pole, which is simple, if $a_0 \neq 0$, resp. $b_0 \neq 0$.

Proof: If s varies over a compact subset of $\mathbb{C}$, then the function $e^{-cy^\alpha} y^\sigma$, $\sigma = \mathrm{Re}(s)$, is bounded for $y \geq 1$ by a constant which is independent of σ. Therefore the condition $f(y) = a_0 + O(e^{-cy^\alpha})$ gives the following upper bound for the integrand of the Mellin integral $L(f, s)$.

$$\big|(f(y) - a_0)y^{s-1}\big| \leq B\, e^{-cy^\alpha} y^{\sigma+1} y^{-2} \leq B' \frac{1}{y^2},$$

for all $y \geq 1$, with constants B, B'. The integral $\int_1^\infty (f(y) - a_0)y^{s-1}\, dy$ therefore admits the convergent majorant $\int_1^\infty \frac{B'}{y^2}\, dy$ which is independent of s. It therefore converges absolutely and uniformly, for all s in the compact subset. The same holds for $\int_1^\infty (g(y) - b_0)y^{s-1}\, dy$.

Now let $\mathrm{Re}(s) > k$. We cut the interval of integration $(0,\infty)$ into $(0,1]$ and $(1,\infty)$ and write

$$L(f,s) = \int_1^\infty \big(f(y) - a_0\big)\, y^s \frac{dy}{y} + \int_0^1 \big(f(y) - a_0\big)\, y^s \frac{dy}{y}.$$

For the second integral, the substitution $y \mapsto 1/y$ and the equation $f(1/y) = Cy^k g(y)$ give:

$$\int_0^1 \big(f(y) - a_0\big)\, y^s \frac{dy}{y} = -a_0 \frac{y^s}{s}\Big|_0^1 + \int_1^\infty f\Big(\frac{1}{y}\Big) y^{-s} \frac{dy}{y}$$

$$= -\frac{a_0}{s} + C \int_1^\infty (g(y) - b_0) y^{k-s-1}\, dy - \frac{Cb_0}{k-s}.$$

By the above, it also converges absolutely and uniformly for $\mathrm{Re}(s) > k$. We therefore obtain

$$L(f,s) = -\frac{a_0}{s} + \frac{Cb_0}{s-k} + F(s),$$

where

$$F(s) = \int_1^\infty \big[(f(y) - a_0) y^s + C(g(y) - b_0) y^{k-s}\big] \frac{dy}{y}.$$

Swapping f and g, we see from $g(1/y) = C^{-1} y^k f(y)$ that:

$$L(g,s) = -\frac{b_0}{s} + \frac{C^{-1}a_0}{s-k} + G(s)$$

where

$$G(s) = \int_1^\infty \big[(g(y) - b_0) y^s + C^{-1}(f(y) - a_0) y^{k-s}\big] \frac{dy}{y}.$$

The integrals $F(s)$ and $G(s)$ converge absolutely and locally uniformly on the whole complex plane, as we saw above. So they represent holomorphic functions, and one obviously has $F(s) = CG(k-s)$. Thus $L(f,s)$ and $L(g,s)$ have been continued to all of $\mathbb{C} \smallsetminus \{0,k\}$ and we have $L(f,s) = CL(g,k-s)$. This finishes the proof of the theorem. □

The result can now be applied to the integral (1.3) representing the function $Z(s)$. In fact, Jacobi's theta function $\theta(z)$ is characterized by the following property.

(1.5) Proposition. *The series*

$$\theta(z) = \sum_{n\in\mathbb{Z}} e^{\pi i n^2 z}$$

converges absolutely and uniformly in the domain $\{z \in \mathbb{C} \mid \mathrm{Im}(z) \geq \delta\}$, *for every* $\delta > 0$. *It therefore represents an analytic function on the upper half-plane* $\mathbb{H} = \{z \in \mathbb{C} \mid \mathrm{Im}(z) > 0\}$, *and satisfies the transformation formula*

$$\theta(-1/z) = \sqrt{z/i}\,\theta(z).$$

We will prove this proposition in much greater generality in § 3 (see (3.6)), so we take it for granted here. Observe that if z lies in $\mathbb{H}$ then so does $-1/z$. The square root $\sqrt{z/i}$ is understood to be the holomorphic function

$$h(z) = e^{\frac{1}{2}\log z/i},$$

where log indicates the principal branch of the logarithm. It is determined uniquely by the conditions

$$h(z)^2 = z/i \quad \text{and} \quad h(iy) = \sqrt{y} > 0 \;\text{ for } y \in \mathbb{R}_+^*.$$

(1.6) Theorem. *The completed zeta function*

$$Z(s) = \pi^{-s/2}\Gamma(s/2)\zeta(s)$$

admits an analytic continuation to $\mathbb{C} \smallsetminus \{0, 1\}$, *has simple poles at* $s = 0$ *and* $s = 1$ *with residues* -1 *and* 1, *respectively, and satisfies the functional equation*

$$Z(s) = Z(1-s).$$

Proof: By (1.3), we have

$$Z(2s) = \frac{1}{2}\int_0^\infty \big(\theta(iy) - 1\big)\, y^s \frac{dy}{y},$$

i.e., $Z(2s)$ is the Mellin transform

$$Z(2s) = L(f, s)$$

of the function $f(y) = \frac{1}{2}\theta(iy)$. Since

$$\theta(iy) = 1 + 2e^{-\pi y}\Big(1 + \sum_{n=2}^{\infty} e^{-\pi(n^2-1)y}\Big),$$

one has $f(y) = \frac{1}{2} + O(e^{-\pi y})$. From (1.5), we get the transformation formula

$$f(1/y) = \frac{1}{2}\theta(-1/iy) = \frac{1}{2}y^{1/2}\theta(iy) = y^{1/2}f(y).$$

By (1.4), $L(f,s)$ has a holomorphic continuation to $\mathbb{C} \smallsetminus \{0, 1/2\}$ and simple poles at $s = 0, 1/2$ with residues $-1/2$ and $1/2$, respectively, and it satisfies the functional equation

$$L(f,s) = L\Big(f, \frac{1}{2} - s\Big).$$

Accordingly, $Z(s) = L(f, s/2)$ has a holomorphic continuation to $\mathbb{C} \smallsetminus \{0, 1\}$ and simple poles at $s = 0, 1$ with residues -1 and 1, respectively, and satisfies the functional equation

$$Z(s) = L\Big(f, \frac{s}{2}\Big) = L\Big(f, \frac{1}{2} - \frac{s}{2}\Big) = Z(1-s). \qquad \square$$

For the Riemann zeta function itself, the theorem gives the

(1.7) Corollary. *The Riemann zeta function $\zeta(s)$ admits an analytic continuation to $\mathbb{C} \smallsetminus \{1\}$, has a simple pole at $s = 1$ with residue 1 and satisfies the functional equation*

$$\zeta(1-s) = 2(2\pi)^{-s}\Gamma(s)\cos\Big(\frac{\pi s}{2}\Big)\zeta(s).$$

Proof: $Z(s) = \pi^{-s/2}\Gamma(s/2)\zeta(s)$ has a simple pole at $s = 0$, but so does $\Gamma(s/2)$. Hence $\zeta(s)$ has no pole. At $s = 1$, however, $Z(s)$ has a simple pole, and so does $\zeta(s)$, as $\Gamma(1/2) = \sqrt{\pi}$. The residue comes out to be

$$\operatorname{Res}_{s=1}\zeta(s) = \pi^{1/2}\Gamma(1/2)^{-1}\operatorname{Res}_{s=1}Z(s) = 1.$$

The equation $Z(1-s) = Z(s)$ translates into

$$(*) \qquad \zeta(1-s) = \pi^{\frac{1}{2}-s}\frac{\Gamma(\frac{s}{2})}{\Gamma(\frac{1-s}{2})}\zeta(s).$$

Substituting $(1-s)/2$, resp. $s/2$, into the formulae (1.2), (iii), 2) and 3) gives

$$\Gamma\Big(\frac{s}{2}\Big)\Gamma\Big(\frac{1+s}{2}\Big) = \frac{2\sqrt{\pi}}{2^s}\Gamma(s),$$

$$\Gamma\Big(\frac{1-s}{2}\Big)\Gamma\Big(\frac{1+s}{2}\Big) = \frac{\pi}{\cos(\pi s/2)},$$

and after taking the quotient,

$$\Gamma\Big(\frac{s}{2}\Big)\Big/\Gamma\Big(\frac{1-s}{2}\Big) = \frac{2}{2^s\sqrt{\pi}}\cos\frac{\pi s}{2}\,\Gamma(s)\,.$$

Inserting this into $(*)$ now yields the functional equation claimed. □

At some point during the first months of studies every mathematics student has the suprise to discover the remarkable formula

$$\sum_{n=1}^{\infty}\frac{1}{n^2} = \frac{1}{6}\pi^2.$$

It is only the beginning of a sequence:

$$\sum_{n=1}^{\infty}\frac{1}{n^4} = \frac{1}{90}\pi^4, \quad \sum_{n=1}^{\infty}\frac{1}{n^6} = \frac{1}{945}\pi^6, \text{ etc.}$$

These are explicit evaluations of the special values of the Riemann zeta function at the points $s = 2k$, $k \in \mathbb{N}$. The phenomenon is explained via the functional equation by the fact that the values of the Riemann zeta function at the *negative* odd integers are given by **Bernoulli numbers**. These arise from the function

$$F(t) = \frac{t\,e^t}{e^t - 1}$$

and are defined by the series expansion

$$F(t) = \sum_{k=0}^{\infty} B_k \frac{t^k}{k!}\,.$$

Their relation to the zeta function gives them a special arithmetic significance. The first Bernoulli numbers are

$$B_0 = 1,\ B_1 = \frac{1}{2},\ B_2 = \frac{1}{6},\ B_3 = 0,\ B_4 = -\frac{1}{30},\ B_5 = 0,\ B_6 = \frac{1}{42}\,.$$

In general one has $B_{2\nu+1} = 0$ for $\nu \geq 1$, because $F(-t) = F(t) - t$. In the classical literature, it is usually the function $\frac{t}{e^t-1}$, which serves for defining the Bernoulli numbers. As $F(t) = \frac{t}{e^t-1} + t$, this does not change anything except for B_1, where one finds $-\frac{1}{2}$ instead of $\frac{1}{2}$. But the above definition is more natural and better suited for the further development of the theory. We now prove the remarkable

(1.8) Theorem. *For every integer $k > 0$ one has*

$$\zeta(1-k) = -\frac{B_k}{k}\,.$$

We prepare the proof proper by a function-theoretic lemma. For $\varepsilon > 0$ and $a \in [\varepsilon, \infty]$, we consider the path

$$C_{\varepsilon,a} = (a, \varepsilon] + K_\varepsilon + [\varepsilon, a),$$

which first follows the half-line from a to ε, then the circumference $K_\varepsilon = \{z \mid |z| = \varepsilon\}$ in the negative direction, and finally the half-line from ε to a:

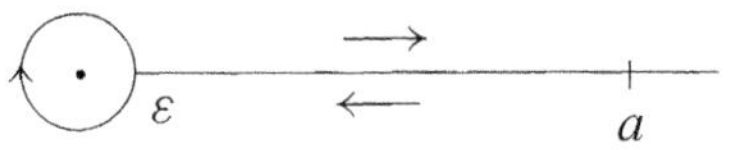

(1.9) Lemma. *Let U be an open subset of $\mathbb{C}$ that contains the path $C_{\varepsilon,a}$ and also the interior of K_ε. Let $G(z)$ be a holomorphic function on $U \setminus \{0\}$ with a pole of order m at 0, and let $G(t)t^{ns-1}$ $(n \in \mathbb{N})$, for $\mathrm{Re}(s) > \frac{m}{n}$, be integrable on $(0, a)$. Then one has*

$$\int_{C_{\varepsilon,a}} G(z) z^{ns-1}\, dz = (e^{2\pi i ns} - 1) \int_0^a G(t) t^{ns-1}\, dt .$$

Proof: The integration does not actually take place in the complex plane but on the universal covering of $\mathbb{C}^*$,

$$X = \big\{ (x, \alpha) \in \mathbb{C}^* \times \mathbb{R} \,\big|\, \arg x \equiv \alpha \bmod 2\pi \big\} .$$

z and z^{s-1} are holomorphic functions on X, namely

$$z(x, \alpha) = x\,, \quad z^{s-1}(x, \alpha) = e^{(s-1)(\log|x| + i\alpha)}\,,$$

and $C_{\varepsilon,a}$ is the path

$$C_{\varepsilon,a} = I^-_{\varepsilon,a} + K_\varepsilon + I^+_{\varepsilon,a}$$

where $I^-_{\varepsilon,a} = (a, \varepsilon] \times \{0\}$, $K_\varepsilon = \{\varepsilon e^{-it} \mid t \in [0, 2\pi]\}$, $I^+_{\varepsilon,a} = [\varepsilon, a) \times \{2\pi\}$ in X. We now have

$$\int_{I^-_{\varepsilon,a}} G(z) z^{ns-1}\, dz = -\int_\varepsilon^a G(t) t^{ns-1}\, dt\,,$$

$$\int_{I^+_{\varepsilon,a}} G(z) z^{ns-1}\, dz = e^{2\pi i ns} \int_\varepsilon^a G(t) t^{ns-1}\, dt\,,$$

$$\int_{K_\varepsilon} G(z)z^{ns-1}\,dz = -i\int_0^{2\pi} G(\varepsilon e^{-it})\varepsilon^{ns-1}e^{-it(ns-1)}\varepsilon e^{-it}\,dt$$
$$= -i\int_0^{2\pi} \varepsilon^{ns}G(\varepsilon e^{-it})e^{-itns}\,dt\,.$$

Since $\mathrm{Re}(s) > \frac{m}{n}$, i.e., $\mathrm{Re}(ns-m) > 0$, the last integral $I(\varepsilon)$ tends to zero as $\varepsilon \to 0$. In fact, one has $\lim\limits_{\varepsilon\to 0} \varepsilon^{ns}G(\varepsilon e^{-it}) = 0$. This gives

$$\int_{C_{\varepsilon,a}} G(z)z^{ns-1}\,dz = (e^{2\pi ins}-1)\int_\varepsilon^a G(t)t^{ns-1}\,dt + I(\varepsilon),$$

and since the integral on the left is independent of ε, the lemma follows by passing to the limit as $\varepsilon \to 0$. □

Proof of (1.8): The function

$$F(z) = \frac{z\,e^z}{e^z-1} = \sum_{k=0}^\infty B_k\frac{z^k}{k!}$$

is a meromorphic function of the complex variable z, with poles only at $z = 2\pi i\nu$, $\nu \in \mathbb{Z}$, $\nu \neq 0$. B_k/k is the residue of $(k-1)!\,F(z)z^{-k-1}$ at 0, and the claim reduces to the identity

$$\mathrm{Res}_{z=0}\,F(z)z^{-k-1} = \frac{1}{2\pi i}\int_{|z|=\varepsilon} F(z)z^{-k-1}dz = -\frac{\zeta(1-k)}{(k-1)!},$$

for $0 < \varepsilon < 2\pi$, where the circle $|z| = \varepsilon$ is taken in the positive orientation. We may replace it with the path $-C_\varepsilon = (-\infty,\,-\varepsilon] + K_\varepsilon + [-\varepsilon,\,-\infty)$, which traces the half-line from $-\infty$ to $-\varepsilon$, followed by the circumference $K_\varepsilon = \{z \mid |z| = \varepsilon\}$ in the *positive* direction, from $-\varepsilon$ to $-\varepsilon$, and finally the half-line from $-\varepsilon$ to $-\infty$. In fact, the integrals over $(-\infty,\,-\varepsilon]$ and $[-\varepsilon,\,-\infty)$ cancel each other. We now consider on $\mathbb{C}$ the function

$$H(s) = \int_{-C_\varepsilon} F(z)z^{s-1}\frac{dz}{z}\,.$$

Here the integrals over $(-\infty,\,-\varepsilon]$ and $[-\varepsilon,\,-\infty)$ do not cancel each other any longer because the function z^{s-1} is multivalued. The integration takes place on the universal covering $X = \{(x,\alpha) \in \mathbb{C}^* \times \mathbb{R} \mid \arg x \equiv \alpha \bmod 2\pi\}$

of $\mathbb{C}^*$, as in (1.9), and z, z^{s-1} are the holomorphic functions $z(x,\alpha) = x$, $z^{s-1}(x,\alpha) = e^{(s-1)(\log|x|+i\alpha)}$. The integral converges absolutely and locally uniformly for all $s \in \mathbb{C}$. It thus defines a holomorphic function on $\mathbb{C}$, and we find that

$$\mathrm{Res}_{z=0} F(z)z^{-k-1} = \frac{1}{2\pi i} H(1-k).$$

Now substitute $z \mapsto -z$, or more precisely, apply the biholomorphic transformation

$$\varphi : X \longrightarrow X, \quad (x,\alpha) \longmapsto (-x, \alpha - \pi).$$

Since $z \circ \varphi = -z$ and

$$\begin{aligned}(z^{s-1} \circ \varphi)(x,\alpha) &= z^{s-1}(-x, \alpha-\pi) = e^{(s-1)(\log|x|+i\alpha-i\pi)} \\ &= -e^{-i\pi s} z^{s-1}(x,\alpha),\end{aligned}$$

we obtain

$$H(s) = -e^{-i\pi s} \int_{C_\varepsilon} F(-z) z^{s-1} \frac{dz}{z},$$

where the path $C_\varepsilon = \varphi^{-1} \circ (-C_\varepsilon)$ follows the half-line from ∞ to ε, then the circumference K_ε in *negative* direction from ε to ε, and finally the half-line from ε to ∞. The function

$$G(z) = F(-z)z^{-1} = \frac{e^{-z}}{1-e^{-z}} = \frac{1}{1-e^{-z}} - 1 = \sum_{n=1}^{\infty} e^{-nz}$$

has a simple pole at $z = 0$ so that, for $\mathrm{Re}(s) > 1$, (1.9) yields

$$\begin{aligned}H(s) &= -e^{-\pi i s} \int_{C_\varepsilon} G(z) z^{s-1} dz \\ &= -(e^{\pi i s} - e^{-\pi i s}) \int_0^\infty G(t) t^s \frac{dt}{t} = -2i \sin \pi s \int_0^\infty G(t) t^s \frac{dt}{t}.\end{aligned}$$

The integral on the right will now be related to the zeta function. In the gamma integral

$$\Gamma(s) = \int_0^\infty e^{-t} t^s \frac{dt}{t},$$

we substitute $t \mapsto nt$ and get

$$\Gamma(s) \frac{1}{n^s} = \int_0^\infty e^{-nt} t^s \frac{dt}{t}.$$

Summing this over all $n \in \mathbb{N}$ yields

$$\Gamma(s)\zeta(s) = \int_0^\infty G(t)t^s \frac{dt}{t}.$$

The interchange of summation and integration is again justified because

$$\sum_{n=1}^{\infty} \int_0^\infty |e^{-nt}t^s| \frac{dt}{t} < \infty.$$

From this and (1.2), 2), we get

$$H(s) = -2i \sin \pi s \Gamma(s)\zeta(s) = -\frac{2\pi i}{\Gamma(1-s)}\zeta(s).$$

Since both sides are holomorphic on all of $\mathbb{C}$, this holds for all $s \in \mathbb{C}$. Putting $s = 1-k$ we obtain, since $\Gamma(k) = (k-1)!$,

$$\operatorname{Res}_{z=0} F(z)z^{-k-1} = \frac{1}{2\pi i}H(1-k) = -\frac{\zeta(1-k)}{(k-1)!}, \quad \text{q.e.d.} \qquad \square$$

Applying the functional equation (1.7) for $\zeta(s)$ and observing that $\Gamma(2k) = (2k-1)!$, the preceding theorem gives the following corollary, which goes back to EULER.

(1.10) Corollary. *The values of $\zeta(s)$ at the positive even integers $s = 2k$, $k = 1, 2, 3, \ldots,$ are given by*

$$\zeta(2k) = (-1)^{k-1}\frac{(2\pi)^{2k}}{2(2k)!}B_{2k}.$$

The values $\zeta(2k-1)$, $k > 1$, at the positive odd integers have been elucidated only recently. Surprisingly enough, it is the higher K-groups $K_i(\mathbb{Z})$ from algebraic K-theory, which take the lead. In fact, one has a mysterious canonical isomorphism

$$r : K_{4k-1}(\mathbb{Z}) \underset{\mathbb{Z}}{\otimes} \mathbb{R} \xrightarrow{\cong} \mathbb{R}.$$

The image R_{2k} of a nonzero element in $K_{4k-1}(\mathbb{Z}) \otimes_{\mathbb{Z}} \mathbb{Q}$ is called the $2k$-th **regulator**. It is well-determined up to a rational factor, i.e., it is an element of $\mathbb{R}^*/\mathbb{Q}^*$, and one has

$$\zeta(2k-1) \equiv R_{2k} \bmod \mathbb{Q}^*.$$

This discovery of the Swiss mathematician *Armand Borel* has had a tremendous influence on further arithmetical research, and has opened up deep insights into the arithmetic nature of zeta functions and L-series of the most general kind. These insights are summarized within the comprehensive **Beilinson conjecture** (see [117]). In the meantime, the mathematicians *Spencer Bloch* and *Kazuya Kato* have found a complete description of the special zeta values $\zeta(2k-1)$ (i.e., not just a description $\bmod\ \mathbb{Q}^*$) via a new theory of the *Tamagawa measure*.

The **zeroes** of the Riemann zeta function command special attention. Euler's identity (1.1) shows that one has $\zeta(s) \neq 0$ for $\mathrm{Re}(s) > 1$. The gamma function $\Gamma(s)$ is nowhere 0 and has simple poles at $s = 0, -1, -2, \ldots$ The functional equation $Z(s) = Z(1-s)$, i.e.,

$$\pi^{-s/2}\Gamma(s/2)\zeta(s) = \pi^{(s-1)/2}\Gamma\big((1-s)/2\big)\zeta(1-s),$$

therefore shows that the only zeroes of $\zeta(s)$ in the domain $\mathrm{Re}(s) < 0$ are the poles of $\Gamma(s/2)$, i.e., the arguments $s = -2, -4, -6, \ldots$ These are called the *trivial zeroes* of $\zeta(s)$. Other zeroes have to lie in the **critical strip** $0 \le \mathrm{Re}(s) \le 1$, since $\zeta(s) \neq 0$ for $\mathrm{Re}(s) > 1$. They are the subject of the famous, still unproven

Riemann Hypothesis: The non-trivial zeroes of $\zeta(s)$ lie on the line $\mathrm{Re}(s) = \frac{1}{2}$.

This conjecture has been verified for 150 million zeroes. It has immediate consequences for the problem of the distribution of prime numbers within all the natural numbers. The distribution function

$$\pi(x) = \#\{p \text{ prime number } \le x\}$$

may be written, according to *Riemann*, as the series

$$\pi(x) = R(x) - \sum_{\rho} R(x^{\rho}),$$

where ρ varies over all the zeroes of $\zeta(s)$, and $R(x)$ is the function

$$R(x) = 1 + \sum_{n=1}^{\infty} \frac{1}{n\zeta(n+1)} \frac{(\log x)^n}{n!}.$$

On a microscopic scale, the function $\pi(x)$ is a step-function with a highly irregular behaviour. But on a large scale it is its astounding smoothness which poses one of the biggest mysteries in mathematics:

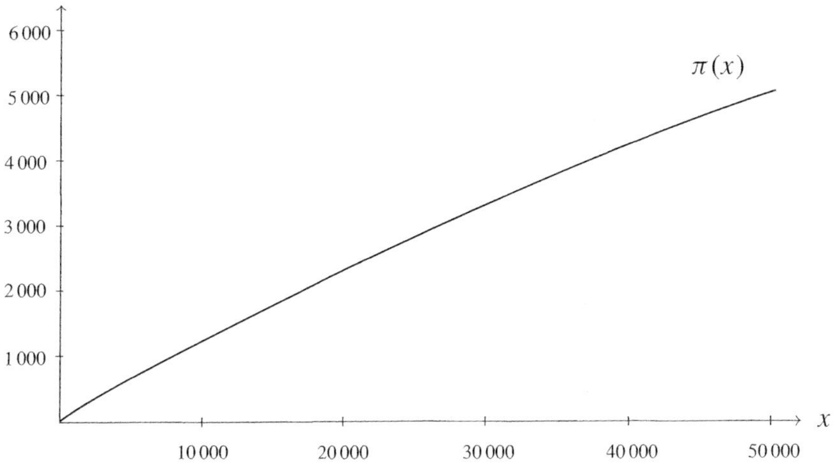

On this matter, we urge the reader to consult the essay [142] by *Don Zagier*.

Exercise 1. Let a, b be positive real numbers. Then the Mellin transforms of the functions $f(y)$ and $g(y) = f(ay^b)$ satisfy:

$$L(f, s/b) = ba^{s/b} L(g, s).$$

Exercise 2. The **Bernoulli polynomials** $B_k(x)$ are defined by

$$\frac{t\, e^{(1+x)t}}{e^t - 1} = F(t)\, e^{xt} = \sum_{k=0}^{\infty} B_k(x) \frac{t^k}{k!},$$

so that $B_k = B_k(0)$. Show that

$$B_m(x) = \sum_{k=0}^{m} \binom{m}{k} B_k x^{m-k}.$$

Exercise 3. $B_k(x) - B_k(x-1) = kx^{k-1}$.

Exercise 4. For the power sum

$$s_k(n) = 1^k + 2^k + 3^k + \cdots + n^k$$

one has

$$s_k(n) = \frac{1}{k+1}(B_{k+1}(n) - B_{k+1}(0)).$$

Exercise 5. Let $\vartheta(z) = \theta(2z) = \sum_{n\in\mathbb{Z}} e^{2\pi i n^2 z}$. Then for all matrices $\gamma = \begin{pmatrix} a & b \\ c & d \end{pmatrix}$ in the group

$$\Gamma_0(4) = \left\{ \begin{pmatrix} a & b \\ c & d \end{pmatrix} \in SL_2(\mathbb{Z}) \;\middle|\; c \equiv 0 \bmod 4 \right\}$$

one has the formula

$$\vartheta\Big(\frac{az+b}{cz+d}\Big) = j(\gamma,z)\vartheta(z), \quad z \in \mathbb{H},$$

where

$$j(\gamma,z) = \Big(\frac{c}{d}\Big)\varepsilon_d^{-1}(cz+d)^{1/2}.$$

The Legendre symbol $\left(\frac{c}{d}\right)$ and the constant ε_d are defined by

$$\Big(\frac{c}{d}\Big) = \begin{cases} -\Big(\frac{c}{|d|}\Big), & \text{if } c<0, d<0,\\ \Big(\frac{c}{|d|}\Big), & \text{otherwise}, \end{cases}$$

$$\varepsilon_d = \begin{cases} 1, & \text{if } d \equiv 1 \bmod 4,\\ i, & \text{if } d \equiv 3 \bmod 4. \end{cases}$$

Jacobi's theta function $\vartheta(z)$ is thus an example of a **modular form of weight $\frac{1}{2}$** for the group $\Gamma_0(4)$. The representation of L-series as Mellin transforms of modular forms, which we have introduced in the case of the Riemann zeta function, is one of the basic and seminal principles of current number-theoretic research (see [106]).

§ 2. Dirichlet L-series

The most immediate relatives of the Riemann zeta function are the Dirichlet L-series. They are defined as follows. Let m be a natural number. A **Dirichlet character** mod m is by definition a character

$$\chi : (\mathbb{Z}/m\mathbb{Z})^* \longrightarrow S^1 = \big\{z \in \mathbb{C} \;\big|\; |z| = 1\big\}.$$

It is called **primitive** if it does not arise as the composite

$$(\mathbb{Z}/m\mathbb{Z})^* \longrightarrow (\mathbb{Z}/m'\mathbb{Z})^* \xrightarrow{\chi'} S^1$$

of a Dirichlet character χ' mod m' for any proper divisor $m'|m$. In the general case, the gcd of all such divisors is called the **conductor** f of χ. So χ is always induced from a primitive character χ' mod f. Given χ, we define the multiplicative function $\chi : \mathbb{Z} \to \mathbb{C}$ by

$$\chi(n) = \begin{cases} \chi(n \bmod m) & \text{for } (n,m) = 1,\\ 0 & \text{for } (n,m) \neq 1. \end{cases}$$

The *trivial character* χ^0 mod m, $\chi^0(n) = 1$ for $(n,m) = 1$, $\chi^0(n) = 0$ for $(n,m) \neq 1$, plays a special role. When read mod 1, we denote it by $\chi = \mathbf{1}$.

It is also called the **principal character**. Considering it in the theory to be developed now has the effect of subsuming here everything we have done in the last section. For a Dirichlet character χ, we form the **Dirichlet L-series**

$$L(\chi, s) = \sum_{n=1}^{\infty} \frac{\chi(n)}{n^s},$$

where s is a complex variable with $\mathrm{Re}(s) > 1$. In particular, for the principal character $\chi = \mathbf{1}$, we get back the Riemann zeta function $\zeta(s)$. All the results obtained for this special function in the last section can be transferred to the general L-series $L(\chi, s)$ using the same methods. This is the task of the present section.

(2.1) Proposition. *The series $L(\chi, s)$ converges absolutely and uniformly in the domain* $\mathrm{Re}(s) \geq 1 + \delta$, *for any* $\delta > 0$. *It therefore represents an analytic function on the half-plane* $\mathrm{Re}(s) > 1$. *We have* **Euler's identity**

$$L(\chi, s) = \prod_p \frac{1}{1 - \chi(p)p^{-s}}.$$

In view of the multiplicativity of χ and since $|\chi(n)| \leq 1$, the proof is literally the same as for the Riemann zeta function. Since, moreover, we will have to give it again in a more general situation in §8 below (see (8.1)), we may omit it here.

Like the Riemann zeta function, Dirichlet L-series also admit an analytic continuation to the whole complex plane (with a pole at $s = 1$ in the case $\chi = \chi^0$), and they satisfy a functional equation which relates the argument s to the argument $1 - s$. This particularly important property does in fact hold in a larger class of L-series, the *Hecke L-series*, the treatment of which is an essential goal of this chapter. In order to provide some preliminary orientation, the proof of the functional equation will be given here in the special case of the above L-series $L(\chi, s)$. We recommend it for careful study, also comparing it with the preceding section.

The proof again hinges on an integral representation of the function $L(\chi, s)$ which has the effect of realizing it as the Mellin transform of a theta series. We do, however, have to distinguish now between *even* and *odd* Dirichlet characters χ mod m. This phenomenon will become increasingly important when we generalize further. We define the **exponent** $p \in \{0, 1\}$ of χ by

$$\chi(-1) = (-1)^p \chi(1).$$

Then the rule

$$\chi\big((n)\big) = \chi(n)\Big(\frac{n}{|n|}\Big)^p$$

defines a multiplicative function on the semigroup of all ideals (n) which are relatively prime to m. This function is called a *Größencharakter* mod m. These *Größencharaktere* are capable of substantial generalization and will play the leading part when we consider higher algebraic number fields (see § 7).

We now consider the gamma integral

$$\Gamma(\chi, s) = \Gamma\Big(\frac{s+p}{2}\Big) = \int_0^\infty e^{-y} y^{(s+p)/2} \frac{dy}{y}.$$

Substituting $y \mapsto \pi n^2 y/m$, we obtain

$$\Big(\frac{m}{\pi}\Big)^{\frac{s+p}{2}} \Gamma(\chi, s) \frac{1}{n^s} = \int_0^\infty n^p e^{-\pi n^2 y/m} y^{(s+p)/2} \frac{dy}{y}.$$

We multiply this by $\chi(n)$, sum over all $n \in \mathbb{N}$, and get

$$(*) \qquad \Big(\frac{m}{\pi}\Big)^{\frac{s+p}{2}} \Gamma(\chi, s) L(\chi, s) = \int_0^\infty \sum_{n=1}^\infty \chi(n) n^p e^{-\pi n^2 y/m} y^{(s+p)/2} \frac{dy}{y}.$$

Here, swapping the order of summation and integration is again justified, because

$$\sum_{n=1}^\infty \int_0^\infty \big| \chi(n) n^p e^{-\pi n^2 y/m} y^{(s+p)/2} \big| \frac{dy}{y} \le \Big(\frac{m}{\pi}\Big)^{(\mathrm{Re}(s)+p)/2} \Gamma\Big(\frac{\mathrm{Re}(s)+p}{2}\Big) \zeta\big(\mathrm{Re}(s)\big) < \infty.$$

The series under the integral $(*)$,

$$g(y) = \sum_{n=1}^\infty \chi(n) n^p e^{-\pi n^2 y/m},$$

arises from the theta series

$$\theta(\chi, z) = \sum_{n \in \mathbb{Z}} \chi(n) n^p e^{\pi i n^2 z/m},$$

where we adopt the convention that $0^0 = 1$ in case $n = 0$, $p = 0$. Indeed, $\chi(n) n^p = \chi(-n)(-n)^p$ implies that

$$\theta(\chi, z) = \chi(0) + 2 \sum_{n=1}^\infty \chi(n) n^p e^{\pi i n^2 z/m},$$

so that $g(y) = \frac{1}{2}(\theta(\chi, iy) - \chi(0))$ with $\chi(0) = 1$, if χ is the trivial character $\mathbf{1}$, and $\chi(0) = 0$ otherwise. When $m = 1$, this is Jacobi's theta function

$$\theta(z) = \sum_{n \in \mathbb{Z}} e^{\pi i n^2 z},$$

which is associated with Riemann's zeta function as we saw in § 1. We view the factor

$$L_\infty(\chi, s) = \left(\frac{m}{\pi}\right)^{s/2} \Gamma(\chi, s)$$

in $(*)$ as the "Euler factor" at the infinite prime. It joins with the Euler factors $L_p(s) = 1/(1 - \chi(p)p^{-s})$ of the product representation (2.1) of $L(\chi, s)$ to define the **completed L-series** of the character χ:

$$\Lambda(\chi, s) = L_\infty(\chi, s) L(\chi, s), \qquad \operatorname{Re}(s) > 1.$$

For this function $(*)$ gives us the

(2.2) Proposition. *The function $\Lambda(\chi, s)$ admits the integral representation*

$$\Lambda(\chi, s) = \frac{c(\chi)}{2} \int_0^\infty \big(\theta(\chi, iy) - \chi(0)\big)\, y^{(s+p)/2} \frac{dy}{y},$$

where $c(\chi) = (\frac{\pi}{m})^{p/2}$.

Let us emphasize the fact that the summation in the L-series is only over the natural numbers, whereas in the theta series we sum over *all* integers. This is why the factor n^p had to be included in order to link the L-series to the theta series.

We want to apply the Mellin principle to the above integral representation. So we have to show that the theta series $\theta(\chi, iy)$ satisfies a transformation formula as assumed in theorem (1.4). To do this, we use the following

(2.3) Proposition. *Let a, b, μ be real numbers, $\mu > 0$. Then the series*

$$\theta_\mu(a, b, z) = \sum_{g \in \mu\mathbb{Z}} e^{\pi i (a+g)^2 z + 2\pi i b g}$$

converges absolutely and uniformly in the domain $\operatorname{Im}(z) \ge \delta$, for every $\delta > 0$, and for $z \in \mathbb{H}$, one has the transformation formula

$$\theta_\mu(a, b, -1/z) = e^{-2\pi i a b} \frac{\sqrt{z/i}}{\mu} \theta_{1/\mu}(-b, a, z).$$

This proposition will be proved in § 3 in much greater generality (see (3.6)), so we take it for granted here. The series $\theta_\mu(a,b,z)$ is locally uniformly convergent in the variables a, b. This will also be shown in § 3. Differentiating p times ($p = 0, 1$) in the variable a, we obtain the function

$$\theta_\mu^p(a,b,z) = \sum_{g\in\mu\mathbb{Z}} (a+g)^p\, e^{\pi i(a+g)^2 z + 2\pi i bg}\,.$$

More precisely, we have

$$\frac{d^p}{da^p}\theta_\mu(a,b,z) = (2\pi i)^p z^p \theta_\mu^p(a,b,z)$$

and

$$\frac{d^p}{da^p}\, e^{-2\pi iab}\theta_{1/\mu}(-b,a,z) = (2\pi i)^p\, e^{-2\pi iab}\theta_{1/\mu}^p(-b,a,z)\,.$$

Applying the differentiation d^p/da^p to the transformation formula (2.3), we get the

(2.4) Corollary. *For $a, b, \mu \in \mathbb{R}$, $\mu > 0$, one has the transformation formula*

$$\theta_\mu^p(a,b,-1/z) = \left[i^p\, e^{2\pi iab}\mu\right]^{-1}(z/i)^{p+\frac{1}{2}}\,\theta_{1/\mu}^p(-b,a,z)\,.$$

This corollary gives us the required transformation formula for the theta series $\theta(\chi,a)$, if we introduce the *Gauss sums* which are defined as follows.

(2.5) Definition. *For $n \in \mathbb{Z}$, the* **Gauss sum** *$\tau(\chi,n)$ associated to the Dirichlet character χ* mod *m is defined to be the complex number*

$$\tau(\chi,n) = \sum_{\nu=0}^{m-1} \chi(\nu)\, e^{2\pi i\nu n/m}\,.$$

For $n = 1$, we write $\tau(\chi) = \tau(\chi,1)$.

(2.6) Proposition. *For a primitive Dirichlet character χ* mod *m, one has*

$$\tau(\chi,n) = \overline{\chi}(n)\tau(\chi) \quad \text{and} \quad |\tau(\chi)| = \sqrt{m}\,.$$

Proof: The first identity in the case $(n,m) = 1$ follows from $\chi(\nu n) = \chi(n)\chi(\nu)$. When $d = (n,m) \neq 1$, both sides are zero. Indeed, since χ is primitive, we may in this case choose an $a \equiv 1 \bmod m/d$ such that $a \not\equiv 1 \bmod m$ and $\chi(a) \neq 1$. Multiplying $\tau(\chi,n)$ by $\chi(a)$ and observing that $e^{2\pi i \nu a n/m} = e^{2\pi i \nu n/m}$ gives $\chi(a)\tau(\chi,n) = \tau(\chi,n)$, so that $\tau(\chi,n) = 0$. Further, we have

$$\begin{aligned}\left|\tau(\chi)\right|^2 &= \tau(\chi)\overline{\tau(\chi)} = \tau(\chi)\sum_{\nu=0}^{m-1}\overline{\chi}(\nu)e^{-2\pi i\nu/m} = \sum_{\nu=0}^{m-1}\tau(\chi,\nu)e^{-2\pi i\nu/m}\\ &= \sum_{\nu=0}^{m-1}\sum_{\mu=0}^{m-1}\chi(\mu)e^{2\pi i\nu\mu/m}e^{-2\pi i\nu/m} = \sum_{\mu=0}^{m-1}\chi(\mu)\sum_{\nu=0}^{m-1}e^{2\pi i\nu(\mu-1)/m}.\end{aligned}$$

The last sum equals m for $\mu = 1$. For $\mu \neq 1$, it vanishes because then $\xi = e^{2\pi i(\mu-1)/m}$ is an m-th root of unity $\neq 1$, hence a root of the polynomial

$$\frac{X^m - 1}{X - 1} = X^{m-1} + \cdots + X + 1.$$

Therefore $|\tau(\chi)|^2 = m\chi(1) = m$. □

We now obtain the following result for the theta series $\theta(\chi,z)$.

(2.7) Proposition. *If χ is a primitive Dirichlet character* mod m, *then we have the transformation formula*

$$\theta(\chi, -1/z) = \frac{\tau(\chi)}{i^p\sqrt{m}}(z/i)^{p+\frac{1}{2}}\theta(\overline{\chi},z),$$

where $\overline{\chi}$ is the complex conjugate character to χ, i.e., its inverse.

Proof: We split up the series $\theta(\chi,z)$ according to the classes a mod m, $a = 0,1,\dots,m-1$, and obtain

$$\theta(\chi,z) = \sum_{n\in\mathbb{Z}}\chi(n)n^p e^{\pi i n^2 z/m} = \sum_{a=0}^{m-1}\chi(a)\sum_{g\in m\mathbb{Z}}(a+g)^p e^{\pi i(a+g)^2 z/m},$$

hence

$$\theta(\chi,z) = \sum_{a=0}^{m-1}\chi(a)\theta_m^p(a,0,z/m).$$

By (2.4), one has

$$\theta_m^p(a,0,-1/mz) = \frac{1}{i^p m}(mz/i)^{p+\frac{1}{2}}\theta_{1/m}^p(0,a,mz),$$

and this gives

$$\theta^{p}_{1/m}(0,a,mz) = \sum_{g\in\frac{1}{m}\mathbb{Z}} g^p\, e^{\pi i g^2 mz + 2\pi i a g} = \frac{1}{m^p}\sum_{n\in\mathbb{Z}} e^{2\pi i a n/m} n^p\, e^{\pi i n^2 z/m}\,.$$

Multiplying this by $\chi(a)$, then summing over a, and observing that $\tau(\chi,n)=\overline{\chi}(n)\tau(\chi)$, we find:

$$\begin{aligned}
\theta(\chi,\,-1/z) &= \frac{1}{i^p m}(mz/i)^{p+\frac{1}{2}}\sum_{a=0}^{m-1}\chi(a)\theta^{p}_{1/m}(0,a,mz)\\
&= \frac{1}{i^p m^{p+1}}(mz/i)^{p+\frac{1}{2}}\sum_{n\in\mathbb{Z}}\Big(\sum_{a=0}^{m-1}\chi(a)\,e^{2\pi i a n/m}\Big)n^p\, e^{\pi i n^2 z/m}\\
&= \frac{1}{i^p\sqrt{m}}(z/i)^{p+\frac{1}{2}}\tau(\chi)\sum_{n\in\mathbb{Z}}\overline{\chi}(n)n^p\, e^{\pi i n^2 z/m}\\
&= \frac{\tau(\chi)}{i^p\sqrt{m}}(z/i)^{p+\frac{1}{2}}\theta(\overline{\chi},z)\,.
\end{aligned}$$

□

The analytic continuation and functional equation for the function $\Lambda(\chi,s)$ now falls out immediately. We may restrict ourselves to the case of a *primitive* character mod m. For χ is always induced by a primitive character χ' mod f, where f is the conductor of χ (see p. 434), and we clearly have

$$L(\chi,s) = \prod_{\substack{p\mid m\\ p\nmid f}}\big(1-\chi(p)p^{-s}\big)L(\chi',s),$$

so that the analytic continuation and functional equation of $\Lambda(\chi,s)$ follows from the one for $\Lambda(\chi',s)$. We may further exclude the case $m=1$ (this is not really necessary, just to make life easy), this being the case of the Riemann zeta function which was settled in § 1. The poles in this case are different.

(2.8) Theorem. *If χ is a nontrivial primitive Dirichlet character, then the completed L-series $\Lambda(\chi,s)$ admits an analytic continuation to the* **whole** *complex plane $\mathbb{C}$ and satisfies the functional equation*

$$\Lambda(\chi,s) = W(\chi)\Lambda(\overline{\chi},1-s)$$

with the factor $W(\chi)=\dfrac{\tau(\chi)}{i^p\sqrt{m}}$. This factor has absolute value 1.

Proof: Let $f(y) = \frac{c(\chi)}{2}\theta(\chi, iy)$ and $g(y) = \frac{c(\chi)}{2}\theta(\overline{\chi}, iy)$, $c(\chi) = \left(\frac{\pi}{m}\right)^{p/2}$. We have $\chi(0) = \overline{\chi}(0) = 0$, so that

$$\theta(\chi, iy) = 2\sum_{n=1}^{\infty} \chi(n) n^p e^{-\pi n^2 y/m},$$

and therefore $f(y) = O(e^{-\pi y/m})$, and likewise $g(y) = O(e^{-\pi y/m})$. By (2.2), one has

$$\Lambda(\chi, s) = \frac{c(\chi)}{2}\int_0^{\infty} \theta(\chi, iy) y^{\frac{s+p}{2}} \frac{dy}{y}.$$

We therefore obtain $\Lambda(\chi, s)$ and similarly also $\Lambda(\overline{\chi}, s)$ as Mellin transforms

$$\Lambda(\chi, s) = L(f, s') \quad \text{and} \quad \Lambda(\overline{\chi}, s) = L(g, s')$$

of the functions $f(y)$ and $g(y)$ at the point $s' = \frac{s+p}{2}$. The transformation formula (2.7) gives

$$f\left(\frac{1}{y}\right) = \frac{c(\chi)}{2}\theta(\chi, -1/iy) = \frac{c(\chi)\tau(\chi)}{2i^p\sqrt{m}} y^{p+\frac{1}{2}}\theta(\overline{\chi}, iy) = \frac{\tau(\chi)}{i^p\sqrt{m}} y^{p+\frac{1}{2}} g(y).$$

Theorem (1.4) therefore tells us that $\Lambda(\chi, s)$ admits an analytic continuation to all of $\mathbb{C}$ and that the equation

$$\begin{aligned}\Lambda(\chi, s) &= L\left(f, \tfrac{s+p}{2}\right) = W(\chi) L\left(g, p + \tfrac{1}{2} - \tfrac{s+p}{2}\right) = W(\chi) L\left(g, \tfrac{1-s+p}{2}\right)\\ &= W(\chi)\Lambda(\overline{\chi}, 1-s)\end{aligned}$$

holds with $W(\chi) = \frac{\tau(\chi)}{i^p\sqrt{m}}$. By (2.6), we have $|W(\chi)| = 1$. □

The behaviour of the special values at integer arguments of the Riemann zeta function generalizes to the Dirichlet L-series $L(\chi, s)$ if we introduce, for nontrivial primitive Dirichlet characters χ mod m, the **generalized Bernoulli numbers** $B_{k,\chi}$ defined by the formula

$$F_\chi(t) = \sum_{a=1}^{m} \chi(a) \frac{t\,e^{at}}{e^{mt} - 1} = \sum_{k=0}^{\infty} B_{k,\chi} \frac{t^k}{k!}.$$

These are algebraic numbers which lie in the field $\mathbb{Q}(\chi)$ generated by the values of χ. Since

$$F_\chi(-t) = \sum_{a=1}^{m} \chi(-1)\chi(m-a) \frac{t\,e^{(m-a)t}}{e^{mt} - 1} = \chi(-1) F_\chi(t),$$

we find $(-1)^k B_{k,\chi} = \chi(-1) B_{k,\chi}$, so that

$$B_{k,\chi} = 0 \quad \text{for} \quad k \not\equiv p \mod 2,$$

if $p \in \{0, 1\}$ is defined by $\chi(-1) = (-1)^p \chi(1)$.

(2.9) Theorem. *For any integer* $k \geq 1$, *one has*

$$L(\chi, 1-k) = -\frac{B_{k,\chi}}{k}.$$

Proof: The proof is the same as for the Riemann zeta function (see (1.8)): the meromorphic function

$$F_\chi(z) = \sum_{a=1}^{m} \chi(a) \frac{z\, e^{az}}{e^{mz} - 1} = \sum_{k=0}^{\infty} B_{k,\chi} \frac{z^k}{k!}$$

has poles at most at $z = \frac{2\pi i \nu}{m}$, $\nu \in \mathbb{Z}$. The claim therefore reduces to showing that

$$(1) \qquad -\frac{L(\chi, 1-k)}{\Gamma(k)} = \text{residue of } F_\chi(z) z^{-k-1} \text{ at } z = 0.$$

Multiplying the equation

$$\Gamma(s) \frac{1}{n^s} = \int_0^\infty e^{-nt} t^s \frac{dt}{t}$$

by $\chi(n)$, and summing over all n, yields

$$(2) \qquad \Gamma(s) L(\chi, s) = \int_0^\infty G_\chi(t) t^s \frac{dt}{t}$$

with the function

$$(3) \qquad G_\chi(z) = \sum_{n=1}^{\infty} \chi(n) e^{-nz} = \sum_{a=1}^{m} \chi(a) \frac{e^{-az}}{1 - e^{-mz}} = F_\chi(-z) z^{-1}.$$

From the equations (2) and (3) one deduces equation (1) in exactly the same manner as in (1.8). □

The theorem immediately gives that

$$L(\chi, 1-k) = 0 \quad \text{for} \quad k \not\equiv p \bmod 2,$$

$p \in \{0, 1\}$, $\chi(-1) = (-1)^p \chi(1)$, provided that χ is not the principal character $\mathbf{1}$. From the functional equation (2.8) and the fact that $L(\chi, k) \neq 0$, we deduce for $k \geq 1$ that

$$L(\chi, 1-k) = -\frac{B_{k,\chi}}{k} \neq 0 \quad \text{for} \quad k \equiv p \bmod 2.$$

The functional equation also gives the

(2.10) Corollary. *For $k \equiv p \bmod 2$, $k \geq 1$, one has*

$$L(\chi,k) = (-1)^{1+(k-p)/2}\frac{\tau(\chi)}{2i^p}\left(\frac{2\pi}{m}\right)^k \frac{B_{k,\overline{\chi}}}{k!}.$$

For the values $L(\chi,k)$ at positive integer arguments $k \not\equiv p \bmod 2$, similar remarks apply as the ones we made in § 1 about the Riemann zeta function at the points $2k$. Up to unknown algebraic factors, these values are certain "regulators" defined via canonical maps from higher K-groups into Minkowski space. A detailed treatment of this deep result of the Russian mathematician *A.A. Beilinson* can be found in [110].

Exercise 1. Let $F_\chi(t,x) = \sum_{a=1}^m \chi(a)\frac{t\,e^{(a+x)t}}{e^{mt}-1}$. The **Bernoulli polynomials** $B_{k,\chi}(x)$ asssociated to the Dirichlet character χ are defined by

$$F_\chi(t,x) = \sum_{k=0}^{\infty} B_{k,\chi}(x)\frac{t^k}{k!}.$$

Thus $B_{k,\chi}(0) = B_{k,\chi}$. Show that

$$B_{k,\chi}(x) = \sum_{i=0}^{k}\binom{k}{i}B_{i,\chi}x^{k-i}.$$

Exercise 2. $B_{k,\chi}(x) - B_{k,\chi}(x-m) = k\sum_{a=1}^m \chi(a)(a+x-m)^{k-1}$, $k \geq 0$.

Exercise 3. For the numbers $S_{k,\chi}(\nu) = \sum_{a=1}^{\nu}\chi(a)a^k$, $k \geq 0$, one has

$$S_{k,\chi}(\nu m) = \frac{1}{k+1}\left(B_{k+1,\chi}(\nu m) - B_{k+1,\chi}(0)\right).$$

Exercise 4. For a primitive odd character χ, one has

$$\sum_{a=1}^{m}\chi(a)a \neq 0.$$

§ 3. Theta Series

Riemann's zeta function and Dirichlet's L-series are attached to the field $\mathbb{Q}$. They have analogues for any algebraic number field K, and the results obtained in § 1 and 2 extend to these generalizations in the same way, with the same methods. In particular, the Mellin principle applies again, which allows us to view the L-series in question as integrals over theta series. But now higher dimensional theta series are required which live on a higher dimensional analogue of the upper half-plane $\mathbb{H}$. *A priori* they do not have any relation with number fields and deserve to be introduced in complete generality.

The familiar objects $\mathbb{C}$, $\mathbb{R}$, $\mathbb{R}_+^*$, $\mathbb{H}$, $|\ |$, log, find their higher dimensional analogues as follows. Let X be a finite $G(\mathbb{C}|\mathbb{R})$-set, i.e., a finite set with an involution $\tau \mapsto \bar{\tau}$ $(\tau \in X)$, and let $n = \#X$. We consider the n-dimensional $\mathbb{C}$-algebra

$$\mathbf{C} = \prod_{\tau \in X} \mathbb{C}$$

of all tuples $z = (z_\tau)_{\tau \in X}$, $z_\tau \in \mathbb{C}$, with componentwise addition and multiplication. If $z = (z_\tau) \in \mathbf{C}$, then the element $\bar{z} \in \mathbf{C}$ is defined to have the following components:

$$(\bar{z})_\tau = \bar{z}_{\bar{\tau}}.$$

We call the involution $z \mapsto \bar{z}$ the **conjugation** on $\mathbf{C}$. In addition, we have the involutions $z \mapsto z^*$ and $z \mapsto {}^*z$ given by

$$z_\tau^* = z_{\bar{\tau}}, \quad \text{resp.} \quad {}^*z_\tau = \bar{z}_\tau .$$

One clearly has $\bar{z} = {}^*z^*$. The set

$$\mathbf{R} = \Big[\prod_\tau \mathbb{C}\Big]^+ = \big\{ z \in \mathbf{C} \,\big|\, z = \bar{z} \big\}$$

forms an n-dimensional commutative $\mathbb{R}$-algebra, and $\mathbf{C} = \mathbf{R} \otimes_{\mathbb{R}} \mathbb{C}$.

If K is a number field of degree n and $X = \mathrm{Hom}(K, \mathbb{C})$, then $\mathbf{R}$ is the **Minkowski space** $K_{\mathbb{R}}$ $(\cong K \otimes_{\mathbb{Q}} \mathbb{R})$ which was introduced in chapter I, §5. The number-theoretic applications will occur there. But for the moment we leave all number-theoretic aspects aside.

For the additive, resp. multiplicative, group $\mathbf{C}$, resp. $\mathbf{C}^*$, we have the homomorphism

$$Tr : \mathbf{C} \to \mathbb{C}, \quad Tr(z) = \sum_\tau z_\tau, \quad \text{resp.}$$
$$N : \mathbf{C}^* \to \mathbb{C}^*, \; N(z) = \prod_\tau z_\tau .$$

Here $Tr(z)$, resp. $N(z)$, denotes the trace, resp. the determinant, of the endomorphism $\mathbf{C} \to \mathbf{C}$, $x \mapsto zx$. Furthermore we have on $\mathbf{C}$ the hermitian scalar product

$$\langle x, y \rangle = \sum_\tau x_\tau \bar{y}_\tau = Tr(x^* y).$$

It is invariant under conjugation, $\overline{\langle x, y \rangle} = \langle \bar{x}, \bar{y} \rangle$, and restricting it yields a scalar product $\langle\ ,\ \rangle$, i.e., a euclidean metric, on the $\mathbb{R}$-vector space $\mathbf{R}$. If $z \in \mathbf{C}$, then ${}^*z$ is the adjoint element with respect to $\langle\ ,\ \rangle$, i.e.,

$$\langle xz, y \rangle = \langle x, {}^*zy \rangle .$$

In $\mathbf{R}$, we consider the subspace

$$\mathbf{R}_{\pm} = \{x \in \mathbf{R} \mid x = x^*\} = \big[\prod_{\tau} \mathbb{R}\big]^+ .$$

Thus we find for the components of $x = (x_\tau) \in \mathbf{R}_{\pm}$ that $x_{\bar{\tau}} = x_\tau \in \mathbb{R}$. If $\delta \in \mathbb{R}$, we simply write $x > \delta$ to signify that $x_\tau > \delta$ for all τ. The multiplicative group

$$\mathbf{R}_+^* = \{x \in \mathbf{R}_{\pm} \mid x > 0\} = \big[\prod_{\tau} \mathbb{R}_+^*\big]^+$$

will play a particularly important part. It consists of the tuples $x = (x_\tau)$ of positive real numbers x_τ such that $x_{\bar{\tau}} = x_\tau$, and it occurs in the two homomorphisms

$$|\ |: \mathbf{R}^* \longrightarrow \mathbf{R}_+^*, \qquad x = (x_\tau) \longmapsto |x| = (|x_\tau|),$$

$$\log: \mathbf{R}_+^* \xrightarrow{\ \sim\ } \mathbf{R}_{\pm}, \qquad x = (x_\tau) \longmapsto \log x = (\log x_\tau).$$

We finally define the **upper half-space** associated to the $G(\mathbb{C}|\mathbb{R})$-set X by

$$\mathbf{H} = \mathbf{R}_{\pm} + i\mathbf{R}_+^* .$$

Putting $\mathrm{Re}(z) = \frac{1}{2}(z + \bar{z})$, $\mathrm{Im}(z) = \frac{1}{2i}(z - \bar{z})$, we may also write

$$\mathbf{H} = \{z \in \mathbf{C} \mid z = z^*,\ \ \mathrm{Im}(z) > 0\}.$$

If z lies in $\mathbf{H}$, then so does $-1/z$, because $z\bar{z} \in \mathbf{R}_+^*$, and $\mathrm{Im}(z) > 0$ implies $\mathrm{Im}(-1/z) > 0$, since $z\bar{z}\,\mathrm{Im}(-1/z) = -\mathrm{Im}(z^{-1}z\bar{z}) = \mathrm{Im}(z) > 0$.

For two tuples $z = (z_\tau)$, $p = (p_\tau) \in \mathbf{C}$, the power

$$z^p = (z_\tau^{p_\tau}) \in \mathbf{C}$$

is well-defined by

$$z_\tau^{p_\tau} = e^{p_\tau \log z_\tau},$$

if we agree to take the principal branch of the logarithm and assume that the z_τ move only in the plane cut along the negative real axis. The table

$$\mathbb{H} \subseteq \mathbb{C} \supseteq \mathbb{R} = \mathbb{R} \supseteq \mathbb{R}_+^*, \quad |\ |: \mathbb{R}^* \to \mathbb{R}_+^*, \quad \log: \mathbb{R}_+^* \xrightarrow{\ \sim\ } \mathbb{R}\ ,$$

$$\mathbf{H} \subseteq \mathbf{C} \supseteq \mathbf{R} \supseteq \mathbf{R}_{\pm} \supseteq \mathbf{R}_+^*, \quad |\ |: \mathbf{R}^* \to \mathbf{R}_+^*, \quad \log: \mathbf{R}_+^* \xrightarrow{\ \sim\ } \mathbf{R}_{\pm},$$

shows the analogy of the notions introduced with the familiar ones in the case $n = 1$. We recommend that the reader memorize them well, for they will be used constantly in what follows without special cross-reference. This also includes the notation

$$\bar{z},\ z^*,\ {}^*z,\ Tr,\ N,\ \langle\ ,\ \rangle,\ x > \delta,\ z^p.$$

The functional equations we are envisaging originate in a general formula from functional analysis, the *Poisson summation formula*. It will be proved first. A **Schwartz function** (or *rapidly decreasing function*) on a euclidean vector space $\mathbf{R}$ is by definition a C^∞-function $f : \mathbf{R} \to \mathbb{C}$ which tends to zero as $x \to \infty$, even if multiplied by an arbitrary power $\|x\|^m$, $m \geq 0$, and which shares this behaviour with all its derivatives. For every Schwartz function f, one forms the **Fourier transform**

$$\hat{f}(y) = \int_{\mathbf{R}} f(x) e^{-2\pi i \langle x, y \rangle} dx,$$

where dx is the Haar measure on $\mathbf{R}$ associated to $\langle \ , \ \rangle$ which ascribes the volume 1 to the cube spanned by an orthonormal basis, i.e., it is the Haar measure which is selfdual with respect to $\langle \ , \ \rangle$. The improper integral converges absolutely and uniformly and gives again a Schwartz function $\hat{f}$. This is easily proved by elementary analytical techniques; we refer also to [98], chap. XIV. The prototype of a Schwartz function is the function

$$h(x) = e^{-\pi \langle x, x \rangle}.$$

All functional equations we are going to prove depend, in the final analysis, on the special property of this function of being its own Fourier transform:

(3.1) Proposition. (i) *The function $h(x) = e^{-\pi \langle x, x \rangle}$ is its own Fourier transform.*

(ii) *If f is an arbitrary Schwartz function and A is a linear transformation of $\mathbf{R}$, then the function $f_A(x) = f(Ax)$ has Fourier transform*

$$\hat{f}_A(y) = \frac{1}{|\det A|} \hat{f}({}^tA^{-1} y),$$

where ${}^tA$ is the adjoint transformation of A.

Proof: (i) We identify the euclidean vector space $\mathbf{R}$ with $\mathbb{R}^n$ via some isometry. Then the Haar measure dx turns into the Lebesgue measure $dx_1 \cdots dx_n$. Since $h(x) = \prod_{i=1}^n e^{-\pi x_i^2}$, we have $\hat{h} = \prod_{i=1}^n (e^{-\pi x_i^2})\hat{}$, so we may assume $n = 1$. Differentiating

$$\hat{h}(y) = \int_{-\infty}^{\infty} h(x) e^{-2\pi i x y} dx$$

in y under the integral, we find by partial integration that

$$\frac{d}{dy}\widehat{h}(y) = -2\pi i \int_{-\infty}^{\infty} xh(x)e^{-2\pi i xy}\,dx = -2\pi y\widehat{h}(y).$$

This implies that $\widehat{h}(y) = C\,e^{-\pi y^2}$ for some constant C. Putting $y = 0$ yields $C = 1$, since it is well-known that $\int e^{-\pi x^2}\,dx = 1$.

(ii) Substituting $x \mapsto Ax$ gives the Fourier transform of $f_A(x)$ as:

$$\begin{aligned}\widehat{f}_A(y) &= \int f(Ax)e^{-2\pi i\langle x,\,y\rangle}\,dx = \int f(x)e^{-2\pi i\langle A^{-1}x,\,y\rangle}\,|\det A|^{-1}\,dx\\ &= \frac{1}{|\det A|}\int f(x)e^{-2\pi i\langle x,\,{}^tA^{-1}y\rangle}\,dx = \frac{1}{|\det A|}\widehat{f}({}^tA^{-1}y).\end{aligned}$$ □

From the proposition ensues the following result, which will be crucial for the sequel.

(3.2) Poisson Summation Formula. *Let Γ be a complete lattice in* **R** *and let*

$$\Gamma' = \left\{ g' \in \mathbf{R} \;\middle|\; \langle g, g'\rangle \in \mathbb{Z} \text{ for all } g \in \Gamma \right\}$$

be the lattice dual to Γ. Then for any Schwartz function f, one has:

$$\sum_{g\in\Gamma} f(g) = \frac{1}{\mathrm{vol}(\Gamma)} \sum_{g'\in\Gamma'} \widehat{f}(g'),$$

where $\mathrm{vol}(\Gamma)$ *is the volume of a fundamental mesh of Γ.*

Proof: We identify as before **R** with the euclidean vector space $\mathbb{R}^n$ via some isometry. This turns the measure dx into the Lebesgue measure $dx_1 \cdots dx_n$. Let A be an invertible $n \times n$-matrix which maps the lattice $\mathbb{Z}^n$ onto Γ. Hence $\Gamma = A\mathbb{Z}^n$ and $\mathrm{vol}(\Gamma) = |\det A|$. The lattice $\mathbb{Z}^n$ is dual to itself, and we get $\Gamma' = A^*\mathbb{Z}^n$ where $A^* = {}^tA^{-1}$, as

$$\begin{aligned} g' \in \Gamma' &\iff {}^t(A\mathbf{n})g' = {}^t\mathbf{n}\,{}^tAg' \in \mathbb{Z} \quad \text{for all } \mathbf{n} \in \mathbb{Z}^n\\ &\iff {}^tAg' \in \mathbb{Z}^n \iff g' \in {}^tA^{-1}\mathbb{Z}^n.\end{aligned}$$

Substituting the equations

$$\Gamma = A\mathbb{Z}^n, \quad \Gamma' = A^*\mathbb{Z}^n, \quad f_A(x) = f(Ax), \quad \widehat{f}_A(y) = \frac{1}{\mathrm{vol}(\Gamma)}\widehat{f}(A^*y)$$

into the identity we want to prove, gives

$$\sum_{\mathbf{n}\in\mathbb{Z}^n} f_A(\mathbf{n}) = \sum_{\mathbf{n}\in\mathbb{Z}^n} \widehat{f}_A(\mathbf{n}).$$

In order to prove this, let us write f instead of f_A and take the series

$$g(x) = \sum_{\mathbf{k}\in\mathbb{Z}^n} f(x+\mathbf{k}).$$

It converges absolutely and locally uniformly. For since f is a Schwartz function, we have, if x varies in a compact domain,

$$\big|f(x+\mathbf{k})\big| \cdot \|\mathbf{k}\|^{n+1} \le C$$

for almost all $\mathbf{k} \in \mathbb{Z}^n$. Hence $g(x)$ is majorized by a constant multiple of the convergent series $\sum_{\mathbf{k}\neq 0} \frac{1}{\|\mathbf{k}\|^{n+1}}$. This argument works just as well for all partial derivatives of f. So $g(x)$ is a C^∞-function. It is clearly periodic,

$$g(x+\mathbf{n}) = g(x) \quad \text{for all} \quad \mathbf{n} \in \mathbb{Z}^n,$$

and therefore admits a Fourier expansion

$$g(x) = \sum_{\mathbf{n}\in\mathbb{Z}^n} a_{\mathbf{n}} e^{2\pi i {}^t\mathbf{n}x},$$

whose Fourier coefficients are given by the well-known formula

$$a_{\mathbf{n}} = \int_0^1 \cdots \int_0^1 g(x) e^{-2\pi i {}^t\mathbf{n}x} dx_1 \cdots dx_n.$$

Swapping summation and integration gives

$$\begin{aligned} a_{\mathbf{n}} &= \int_0^1 \cdots \int_0^1 g(x) e^{-2\pi i {}^t\mathbf{n}x} dx = \sum_{\mathbf{k}\in\mathbb{Z}^n} \int_0^1 \cdots \int_0^1 f(x+\mathbf{k}) e^{-2\pi i {}^t\mathbf{n}x} dx \\ &= \widehat{f}(\mathbf{n}). \end{aligned}$$

It follows that

$$\sum_{\mathbf{n}\in\mathbb{Z}^n} f(\mathbf{n}) = g(0) = \sum_{\mathbf{n}\in\mathbb{Z}^n} a_{\mathbf{n}} = \sum_{\mathbf{n}\in\mathbb{Z}^n} \widehat{f}(\mathbf{n}), \qquad \text{q.e.d.}$$ □

We apply the Poisson summation formula to the functions

$$f_p(a,b,x) = N\big((x+a)^p\big)\, e^{-\pi\langle a+x,a+x\rangle + 2\pi i\langle b,x\rangle}$$

with the parameters $a, b \in \mathbf{R}$ and a tuple $p = (p_\tau)$ of nonnegative integers, such that $p_\tau \in \{0,1\}$ if $\tau = \overline{\tau}$, and $p_\tau p_{\overline{\tau}} = 0$ if $\tau \neq \overline{\tau}$. Such an element $p \in \prod_\tau \mathbb{Z}$ will henceforth be called **admissible**.

(3.3) Proposition. *The function* $f(x) = f_p(a,b,x)$ *is a Schwartz function on* $\mathbf{R}$. *Its Fourier transform is*

$$\widehat{f}(y) = \left[i^{Tr(p)}\, e^{2\pi i\langle a,b\rangle}\right]^{-1} f_p(-b,a,y)\,.$$

Proof: It is clear that $f_p(a,b,x)$ is a Schwartz function, because

$$\left|\, f_p(a,b,x)\right| = \left|\, P(x)\right|\, e^{-\pi\langle a+x,a+x\rangle},$$

for some polynomial $P(x)$.

Let $p = 0$. By (3.1), the function $h(x) = e^{-\pi\langle x,x\rangle}$ equals its own Fourier transform and one has

$$f(x) = f_0(a,b,x) = h(a+x)\,e^{2\pi i\langle b,x\rangle}\,.$$

We therefore obtain

$$\begin{aligned}
\widehat{f}(y) &= \int_{\mathbf{R}} h(a+x)\,e^{2\pi i\langle b,x\rangle}\,e^{-2\pi i\langle x,y\rangle}\,dx \\
&= \int_{\mathbf{R}} h(x)\,e^{-2\pi i\langle y-b,x-a\rangle}\,dx \\
&= e^{2\pi i\langle y-b,a\rangle}\widehat{h}(y-b) \\
&= e^{-2\pi i\langle a,b\rangle}\,e^{-\pi\langle y-b,\,y-b\rangle+2\pi i\langle y,a\rangle} \\
&= e^{-2\pi i\langle a,b\rangle}\,f_0(-b,a,y)\,.
\end{aligned}$$

For an arbitrary admissible p, we get the formula by differentiating p times the identity

$$(*)\qquad \widehat{f}_0(a,b,y) = e^{-2\pi i\langle a,b\rangle}\,f_0(-b,a,y)$$

in the variable a. Now the functions are neither analytic in the individual components a_τ of a, nor are these independent of each other, when there exists a couple $\tau \neq \overline{\tau}$. We therefore proceed as follows. Let ρ vary over the elements of X such that $\rho = \overline{\rho}$, and let σ run through a system of representatives of the conjugation classes $\{\tau,\overline{\tau}\}$ such that $\tau \neq \overline{\tau}$. Since $p_\tau p_{\overline{\tau}} = 0$, we may choose σ in such a way that $p_{\overline{\sigma}} = 0$. Then one has

$$\langle a+x,a+x\rangle = \sum_\rho (a_\rho + x_\rho)^2 + 2\sum_\sigma (a_\sigma + x_\sigma)(a_{\overline{\sigma}} + x_{\overline{\sigma}})\,.$$

We now differentiate p_ρ times both sides of $(*)$ in the real variable a_ρ, for all ρ, and apply p_σ times the differential operator

$$\frac{\partial}{\partial a_{\overline{\sigma}}} = \frac{1}{2}\Big(\frac{\partial}{\partial \xi_{\overline{\sigma}}} - i\frac{\partial}{\partial \eta_{\overline{\sigma}}}\Big),$$

for all σ. Here we consider $a_{\overline{\sigma}} = \xi_{\overline{\sigma}} + i\eta_{\overline{\sigma}}$ as a function in the real variables $\xi_{\overline{\sigma}}, \eta_{\overline{\sigma}}$ ("Wirtinger calculus"). On the left-hand side

$$\widehat{f}_0(a,b,y) = \int e^{-\pi\langle a+x,a+x\rangle+2\pi i\langle b,x\rangle} e^{-2\pi i\langle x,y\rangle}\,dx,$$

we may differentiate under the integral. Then, observing that $p_{\overline{\sigma}} = 0$ and $\frac{\partial}{\partial a_{\overline{\sigma}}}((a_\sigma + x_\sigma)(a_{\overline{\sigma}} + x_{\overline{\sigma}})) = (a_\sigma + x_\sigma)$, we obtain

$$\begin{aligned}
&\int \prod_\rho \big(-2\pi(a_\rho + x_\rho)\big)^{p_\rho} \\
&\qquad\cdot \prod_\sigma \big(-2\pi(a_\sigma + x_\sigma)\big)^{p_\sigma} e^{-\pi\langle a+x,a+x\rangle+2\pi i\langle b,x\rangle-2\pi i\langle x,y\rangle}\,dx \\
&= N\big((-2\pi)^p\big) \int N((a+x)^p)\, e^{-\pi\langle a+x,a+x\rangle+2\pi i\langle b,x\rangle} e^{-2\pi i\langle x,y\rangle}\,dx \\
&= N\big((-2\pi)^p\big)\, \widehat{f}_p(a,b,y).
\end{aligned}$$

The right-hand side of $(*)$,

$$e^{-2\pi i\langle a,b\rangle-\pi\langle -b+y,-b+y\rangle+2\pi i\langle a,y\rangle} = e^{2\pi i\langle a,-b+y\rangle-\pi\langle -b+y,-b+y\rangle},$$

in view of

$$\langle a, -b+y\rangle = \sum_\rho a_\rho(-b_\rho + y_\rho) + \sum_\sigma \big(a_\sigma(-b_{\overline{\sigma}} + y_{\overline{\sigma}}) + a_{\overline{\sigma}}(-b_\sigma + y_\sigma)\big),$$

and as $p_{\overline{\sigma}} = 0$, becomes accordingly

$$\begin{aligned}
&N\big((2\pi i)^p\big) N\big((-b+y)^p\big)\, e^{-2\pi i\langle a,b\rangle} f_0(-b,a,y) \\
&\qquad = N\big((2\pi i)^p\big)\, e^{-2\pi i\langle a,b\rangle} f_p(-b,a,y).
\end{aligned}$$

Hence

$$\widehat{f}_p(a,b,y) = N(i^{-p})\, e^{-2\pi i\langle a,b\rangle} f_p(-b,a,y). \qquad \square$$

We now create our general theta series on the upper half-space

$$\mathbf{H} = \big\{ z \in \mathbf{C} \bigm| z = z^*,\ \mathrm{Im}(z) > 0 \big\} = \mathbf{R}_\pm + i\mathbf{R}_+^*.$$

(3.4) Definition. *For every complete lattice Γ of* $\mathbf{R}$, *we define the* **theta series**

$$\theta_\Gamma(z) = \sum_{g\in\Gamma} e^{\pi i\langle gz,g\rangle}, \quad z \in \mathbf{H}.$$

More generally, for $a, b \in \mathbf{R}$ and any admissible $p \in \prod_\tau \mathbb{Z}$, we put

$$\theta_\Gamma^p(a,b,z) = \sum_{g\in\Gamma} N\big((a+g)^p\big)\, e^{\pi i\langle (a+g)z,a+g\rangle+2\pi i\langle b,g\rangle}.$$

(3.5) Proposition. *The series $\theta_\Gamma^p(a,b,z)$ converges absolutely and uniformly on every compact subset of* $\mathbf{R} \times \mathbf{R} \times \mathbf{H}$.

Proof: Let $\delta \in \mathbb{R}$, $\delta > 0$. For all $z \in \mathbf{H}$ such that $\mathrm{Im}(z) \geq \delta$, we find

$$\left| N((a+g)^p)\, e^{\pi i \langle (a+g)z, a+g\rangle + 2\pi i \langle b, g\rangle} \right| \leq \left| N((a+g)^p) \right| e^{-\pi\delta\langle a+g, a+g\rangle}.$$

Let

$$f_g(a) = N\big((a+g)^p\big)\, e^{-\pi\delta\langle a+g, a+g\rangle} \qquad (a \in \mathbf{R},\ g \in \Gamma).$$

For $\mathbf{K} \subseteq \mathbf{R}$ compact, put $|f_g|_{\mathbf{K}} = \sup\limits_{x \in \mathbf{K}} |f_g(x)|$. We have to show that

$$\sum_{g \in \Gamma} |f_g|_{\mathbf{K}} < \infty.$$

Let $g_1, \ldots, g_n$ be a $\mathbb{Z}$-basis of Γ, and for $g = \sum_{i=1}^n m_i g_i \in \Gamma$, let $\mu_g = \max\limits_i |m_i|$. Furthermore, define $\|x\| = \sqrt{\langle x, x\rangle}$. If $\|g\| \geq 4 \sup\limits_{x \in \mathbf{K}} \|x\|$, then for all $a \in \mathbf{K}$:

$$\begin{aligned}\langle a+g, a+g\rangle &\geq \big(\|a\| - \|g\|\big)^2 \geq \|g\|^2 - 2\|a\| \cdot \|g\| \\ &\geq \frac{1}{2}\|g\|^2 \geq \frac{1}{2}\varepsilon \sum_{i=1}^n m_i^2 \geq \frac{1}{2}\varepsilon\mu_g^2,\end{aligned}$$

where $\varepsilon = \inf\limits_{\Sigma y_i^2 = 1} \sum_{i,j=1}^n \langle g_i, g_j\rangle y_i y_j$ is the smallest eigenvalue of the matrix $(\langle g_i, g_j\rangle)$.

$N((a + \sum m_i g_i)^p)$ is a polynomial of degree q in the m_i, ($q = Tr(p)$), the coefficients of which are continuous functions of a. It follows that

$$\left| N((a+g)^p) \right| \leq \mu_g^{q+1} \quad \text{for all } a \in \mathbf{K},$$

provided μ_g is sufficiently big. One therefore finds a subset $\Gamma' \subseteq \Gamma$ with finite complement such that

$$\sum_{g \in \Gamma'} |f_g|_{\mathbf{K}} \leq \sum_{\mu=0}^{\infty} P(\mu)\mu^{q+1} e^{-\frac{\pi}{2}\delta\varepsilon\mu^2},$$

where $P(\mu) = \#\big\{\mathbf{m} \in \mathbb{Z}^n \mid \max\limits_i |m_i| = \mu\big\} = (2\mu+1)^n - (2\mu-1)^n$. The series on the right is clearly convergent. □

From the Poisson summation formula we now get the general

(3.6) Theta Transformation Formula. *One has*

$$\theta_\Gamma^p(a,b,-1/z) = \big[i^{Tr(p)}e^{2\pi i\langle a,b\rangle}\mathrm{vol}(\Gamma)\big]^{-1}N\big((z/i)^{p+\frac{1}{2}}\big)\theta_{\Gamma'}^p(-b,a,z).$$

In particular, one has for the function $\theta_\Gamma(z) = \theta_\Gamma^0(0,0,z)$:

$$\theta_\Gamma(-1/z) = \frac{\sqrt{N(z/i)}}{\mathrm{vol}(\Gamma)}\,\theta_{\Gamma'}(z).$$

Proof: Both sides of the transformation formula are holomorphic in z by (3.5). Therefore it suffices to check the identity for $z = iy$, with $y \in \mathbf{R}_+^*$. Put $t = y^{-1/2}$, so that

$$z = i\frac{1}{t^2} \quad \text{and} \quad -1/z = it^2.$$

Observing that $t = t^* = {}^*t$, so that $\langle \xi t, \eta\rangle = \langle \xi, {}^*t\eta\rangle = \langle \xi, t\eta\rangle$, we obtain

$$\theta_\Gamma^p(a,b,-1/z) = N(t^{-p})\sum_{g\in\Gamma} N\big((ta+tg)^p\big)\,e^{-\pi\langle ta+tg,\,ta+tg\rangle + 2\pi i\langle t^{-1}b,\,tg\rangle}.$$

Let $\alpha = ta$, $\beta = t^{-1}b$. We consider the function

$$f_p(\alpha,\beta,x) = N\big((\alpha+x)^p\big)\,e^{-\pi\langle\alpha+x,\,\alpha+x\rangle + 2\pi i\langle\beta,x\rangle},$$

and put

$$\varphi_t(\alpha,\beta,x) = f_p(\alpha,\beta,tx).$$

This gives

(1) $$\theta_\Gamma^p(a,b,-1/z) = N(t^{-p})\sum_{g\in\Gamma}\varphi_t(\alpha,\beta,g)$$

and similarly $z = i\frac{1}{t^2}$ gives that

(2) $$\theta_{\Gamma'}^p(-b,a,z) = N(t^p)\sum_{g'\in\Gamma'}\varphi_{t^{-1}}(-\beta,\alpha,g').$$

Now apply the Poisson summation formula

(3) $$\sum_{g\in\Gamma} f(g) = \frac{1}{\mathrm{vol}(\Gamma)}\sum_{g'\in\Gamma'}\hat{f}(g')$$

to the function

$$f(x) = \varphi_t(\alpha,\beta,x) = f_p(\alpha,\beta,tx).$$

Its Fourier transform is computed as follows. Let $h(x) = f_p(\alpha,\beta,x)$, so that $f(x) = h(tx) = h_t(x)$. The transformation $A : x \mapsto tx$ of $\mathbf{R}$ is self-adjoint and has determinant $N(t)$. Thus (3.1), (ii), gives

$$\hat{f}(y) = \frac{1}{N(t)}\hat{h}(t^{-1}y).$$

The Fourier transform $\widehat{h}$ has been computed in (3.3). This yields

$$\begin{aligned}\widehat{f}(y) &= \left[N(i^p)N(t)\,e^{2\pi i\langle a,b\rangle}\right]^{-1} f_p(-\beta,\alpha,t^{-1}y)\\ &= \left[N(i^p)N(t)\,e^{2\pi i\langle a,b\rangle}\right]^{-1} \varphi_{t^{-1}}(-\beta,\alpha,y)\,.\end{aligned}$$

Substituting this into (3) and multiplying by $N(t^{-p})$ gives, by (1) and (2):

$$\theta_\Gamma^p(a,b,\,-1/z) = \left[N(i^p t^{2p+1})\,e^{2\pi i\langle a,b\rangle}\mathrm{vol}(\Gamma)\right]^{-1}\theta_{\Gamma'}^p(-b,a,z)\,.$$

Since $t=(z/i)^{-1/2}$, i.e., $(t^{2p+1})^{-1}=(z/i)^{p+\frac{1}{2}}$, this is indeed the transformation formula sought. □

For $n=1$, we obtain proposition (2.3), which at the time was used without proof for proving the functional equation of the Dirichlet L-series (and Riemann's zeta function).

§ 4. The Higher-dimensional Gamma Function

The passage from theta series to L-series in § 1 and § 2 was afforded by the gamma function

$$\Gamma(s)=\int_0^\infty e^{-y}y^s\,\frac{dy}{y}\,.$$

In order to generalize this process, we now introduce a higher-dimensional gamma function for every finite $G(\mathbb{C}|\mathbb{R})$-set X, building upon the notation of the last section. First we fix a Haar measure on the multiplicative group $\mathbf{R}_+^*$:

Let $\mathfrak{p}=\{\tau,\bar\tau\}$ be the conjugation classes in X. We call $\mathfrak{p}$ real or complex, depending whether $\#\mathfrak{p}=1$ or $\#\mathfrak{p}=2$. We then have

$$\mathbf{R}_+^*=\prod_{\mathfrak{p}}\mathbf{R}_{+\mathfrak{p}}^*\,,$$

where

$$\mathbf{R}_{+\mathfrak{p}}^*=\mathbb{R}_+^*,\quad\text{resp.}\quad \mathbf{R}_{+\mathfrak{p}}^*=\left[\,\mathbb{R}_+^*\times\mathbb{R}_+^*\right]^+=\left\{(y,y)\,\middle|\,y\in\mathbb{R}_+^*\right\}.$$

We define isomorphisms

$$\mathbf{R}_{+\mathfrak{p}}^*\xrightarrow{\ \sim\ }\mathbb{R}_+^*$$

by $y\mapsto y$, resp. $(y,y)\mapsto y^2$, and obtain an isomorphism

$$\varphi:\mathbf{R}_+^*\xrightarrow{\ \sim\ }\prod_{\mathfrak{p}}\mathbb{R}_+^*\,.$$

We now denote by $\frac{dy}{y}$ the Haar measure on $\mathbf{R}_+^*$ which corresponds to the product measure

$$\prod_{\mathfrak{p}} \frac{dt}{t},$$

where $\frac{dt}{t}$ is the usual Haar measure on $\mathbb{R}_+^*$. The Haar measure thus defined is called the **canonical measure** on $\mathbf{R}_+^*$. Under the logarithm

$$\log : \mathbf{R}_+^* \xrightarrow{\sim} \mathbf{R}_\pm,$$

it is mapped to the Haar measure dx on $\mathbf{R}_\pm$ which under the isomorphism

$$\mathbf{R}_\pm = \prod_{\mathfrak{p}} \mathbf{R}_{\pm\mathfrak{p}} \xrightarrow{\varphi} \prod_{\mathfrak{p}} \mathbb{R},$$

$x_\mathfrak{p} \mapsto x_\mathfrak{p}$, resp. $(x_\mathfrak{p}, x_\mathfrak{p}) \mapsto 2x_\mathfrak{p}$, corresponds to the Lebesgue measure on $\prod_\mathfrak{p} \mathbb{R}$.

(4.1) Definition. *For* $\mathbf{s} = (s_\tau) \in \mathbf{C}$ *such that* $\operatorname{Re}(s_\tau) > 0$, *we define the* **gamma function** *associated to the* $G(\mathbb{C}|\mathbb{R})$*-set* X *by*

$$\Gamma_X(\mathbf{s}) = \int_{\mathbf{R}_+^*} N(e^{-y}y^{\mathbf{s}})\frac{dy}{y}.$$

The integrand is well-defined, according to our conventions from p. 445, and the convergence of the integral can be reduced to the case of the ordinary gamma function as follows.

(4.2) Proposition. *Decomposing the* $G(\mathbb{C}|\mathbb{R})$*-set* X *into its conjugation classes* $\mathfrak{p}$, *one has*

$$\Gamma_X(\mathbf{s}) = \prod_{\mathfrak{p}} \Gamma_\mathfrak{p}(\mathbf{s}_\mathfrak{p}),$$

where $\mathbf{s}_\mathfrak{p} = s_\tau$ *for* $\mathfrak{p} = \{\tau\}$, *resp.* $\mathbf{s}_\mathfrak{p} = (s_\tau, s_{\bar\tau})$ *for* $\mathfrak{p} = \{\tau, \bar\tau\}$, $\tau \neq \bar\tau$. *The factors are given explicitly by*

$$\Gamma_\mathfrak{p}(\mathbf{s}_\mathfrak{p}) = \begin{cases} \Gamma(\mathbf{s}_\mathfrak{p}), & \text{if } \mathfrak{p} \text{ real}, \\ 2^{1-Tr(\mathbf{s}_\mathfrak{p})}\Gamma(Tr(\mathbf{s}_\mathfrak{p})), & \text{if } \mathfrak{p} \text{ complex}, \end{cases}$$

where $Tr(\mathbf{s}_\mathfrak{p}) = s_\tau + s_{\bar\tau}$.

Proof: The first statement is clear in view of the product decomposition

$$\left(\mathbf{R}_+^*, \frac{dy}{y}\right) = \left(\prod_{\mathfrak{p}} \mathbf{R}_{+\mathfrak{p}}^*, \prod_{\mathfrak{p}} \frac{dy_{\mathfrak{p}}}{y_{\mathfrak{p}}}\right).$$

The second is relative to a $G(\mathbb{C}|\mathbb{R})$-set X which has only one conjugation class. If $\#X = 1$, then trivially $\Gamma_X(\mathbf{s}) = \Gamma(s)$. So let $X = \{\tau, \overline{\tau}\}$, $\tau \neq \overline{\tau}$. Mapping

$$\psi : \mathbb{R}_+^* \longrightarrow \mathbf{R}_+^*, \quad t \longmapsto (\sqrt{t}, \sqrt{t}),$$

one then gets

$$\int_{\mathbf{R}_+^*} N(e^{-y}y^{\mathbf{s}})\frac{dy}{y} = \int_{\mathbb{R}_+^*} N\left(e^{-(\sqrt{t},\sqrt{t})}(\sqrt{t},\sqrt{t})^{(s_\tau, s_{\overline{\tau}})}\right)\frac{dt}{t}$$

$$= \int_0^\infty e^{-2\sqrt{t}}\sqrt{t}^{\,Tr(\mathbf{s})}\frac{dt}{t},$$

and, since $d(t/2)^2/(t/2)^2 = 2dt/t$, the substitution $t \mapsto (t/2)^2$ yields

$$\int_{\mathbf{R}_+^*} N(e^{-y}y^{\mathbf{s}})\frac{dy}{y} = 2^{1-Tr(\mathbf{s})}\Gamma\big(Tr(\mathbf{s})\big). \qquad \square$$

The proposition shows that the gamma integral $\Gamma(\mathbf{s})$ converges for $\mathbf{s} = (s_\tau)$ with $\mathrm{Re}(s_\tau) > 0$, and admits an analytic continuation to all of $\mathbf{C}$, except for poles at points dictated in the obvious way by the ordinary gamma function $\Gamma(s)$.

We call the function

$$L_X(\mathbf{s}) = N(\pi^{-\mathbf{s}/2})\Gamma_X(\mathbf{s}/2)$$

the ***L*-function** of the $G(\mathbb{C}|\mathbb{R})$-set X. Decomposing X into the conjugation classes $\mathfrak{p}$, yields

$$L_X(\mathbf{s}) = \prod_{\mathfrak{p}} L_{\mathfrak{p}}(\mathbf{s}_{\mathfrak{p}}),$$

where as before we write $\mathbf{s}_{\mathfrak{p}} = s_\tau$ for $\mathfrak{p} = \{\tau\}$ and $\mathbf{s}_{\mathfrak{p}} = (s_\tau, s_{\overline{\tau}})$ for $\mathfrak{p} = \{\tau, \overline{\tau}\}$, $\tau \neq \overline{\tau}$. The factors $L_{\mathfrak{p}}(\mathbf{s}_{\mathfrak{p}})$ are given explicitly, by (4.2), as

$$L_{\mathfrak{p}}(\mathbf{s}_{\mathfrak{p}}) = \begin{cases} \pi^{-\mathbf{s}_{\mathfrak{p}}/2}\Gamma(\mathbf{s}_{\mathfrak{p}}/2), & \text{if } \mathfrak{p} \text{ real},\\ 2(2\pi)^{-Tr(\mathbf{s}_{\mathfrak{p}})/2}\Gamma(Tr(\mathbf{s}_{\mathfrak{p}})/2), & \text{if } \mathfrak{p} \text{ complex}. \end{cases}$$

For a single complex variable $s \in \mathbb{C}$, we put

$$\Gamma_X(s) = \Gamma_X(s\mathbf{1}),$$

where $\mathbf{1} = (1, \dots, 1)$ is the unit element of $\mathbf{C}$. Denoting r_1, resp. r_2, the number of real, resp. complex, conjugation classes of X, we find

$$\Gamma_X(s) = 2^{(1-2s)r_2}\Gamma(s)^{r_1}\Gamma(2s)^{r_2}.$$

In the same way we put

$$L_X(s) = L_X(s\mathbf{1}) = \pi^{-ns/2}\Gamma_X(s/2), \quad n = \#X,$$

and in particular

$$L_{\mathbb{R}}(s) = L_X(s) = \pi^{-s/2}\Gamma(s/2), \quad \text{if} \quad X = \{\tau\},$$

$$L_{\mathbb{C}}(s) = L_X(s) = 2(2\pi)^{-s}\Gamma(s), \quad \text{if} \quad X = \{\tau, \bar{\tau}\},\ \tau \neq \bar{\tau}.$$

Then we have, for an arbitrary $G(\mathbb{C}|\mathbb{R})$-set X:

$$L_X(s) = L_{\mathbb{R}}(s)^{r_1}L_{\mathbb{C}}(s)^{r_2}.$$

With this notation, (1.2) implies the

(4.3) Proposition. (i) $L_{\mathbb{R}}(1) = 1$, $L_{\mathbb{C}}(1) = \frac{1}{\pi}$.

(ii) $L_{\mathbb{R}}(s+2) = \frac{s}{2\pi}L_{\mathbb{R}}(s)$, $L_{\mathbb{C}}(s+1) = \frac{s}{2\pi}L_{\mathbb{C}}(s)$.

(iii) $L_{\mathbb{R}}(1-s)L_{\mathbb{R}}(1+s) = \dfrac{1}{\cos\pi s/2}, \quad L_{\mathbb{C}}(s)L_{\mathbb{C}}(1-s) = \dfrac{2}{\sin\pi s}.$

(iv) $L_{\mathbb{R}}(s)L_{\mathbb{R}}(s+1) = L_{\mathbb{C}}(s)$ (**Legendre's duplication formula**).

As a consequence we obtain the following functional equation for the L-function $L_X(s)$:

(4.4) Proposition. *$L_X(s) = A(s)L_X(1-s)$ with the factor*

$$A(s) = (\cos\pi s/2)^{r_1+r_2}(\sin\pi s/2)^{r_2}L_{\mathbb{C}}(s)^n.$$

Proof: On the one hand we have

$$\frac{L_{\mathbb{R}}(s)}{L_{\mathbb{R}}(1-s)} = \frac{L_{\mathbb{R}}(s)L_{\mathbb{R}}(1+s)}{L_{\mathbb{R}}(1-s)L_{\mathbb{R}}(1+s)} = \cos\pi s/2\ L_{\mathbb{C}}(s),$$

and on the other

$$\frac{L_{\mathbb{C}}(s)}{L_{\mathbb{C}}(1-s)} = \frac{L_{\mathbb{C}}(s)^2}{L_{\mathbb{C}}(1-s)L_{\mathbb{C}}(s)} = \frac{1}{2}\ \sin\pi s L_{\mathbb{C}}(s)^2$$

$$= \cos\pi s/2\ \sin\pi s/2\ L_{\mathbb{C}}(s)^2.$$

The proposition therefore results from the identity $L_X(s) = L_{\mathbb{R}}(s)^{r_1}L_{\mathbb{C}}(s)^{r_2}$. □

This concludes the purely function-theoretic preparations. They will now be applied to number theory.

§ 5. The Dedekind Zeta Function

The Riemann zeta function $\zeta(s) = \sum_{k=1}^{\infty} \frac{1}{k^s}$ is associated with the field $\mathbb{Q}$ of rational numbers. It generalizes in the following way to an arbitrary number field K of degree $n = [K : \mathbb{Q}]$.

(5.1) Definition. *The* **Dedekind zeta function** *of the number field K is defined by the series*

$$\zeta_K(s) = \sum_{\mathfrak{a}} \frac{1}{\mathfrak{N}(\mathfrak{a})^s},$$

where $\mathfrak{a}$ varies over the integral ideals of K, and $\mathfrak{N}(\mathfrak{a})$ denotes their absolute norm.

(5.2) Proposition. *The series $\zeta_K(s)$ converge absolutely and uniformly in the domain* $\mathrm{Re}(s) \geq 1 + \delta$ *for every $\delta > 0$, and one has*

$$\zeta_K(s) = \prod_{\mathfrak{p}} \frac{1}{1 - \mathfrak{N}(\mathfrak{p})^{-s}},$$

where $\mathfrak{p}$ runs through the prime ideals of K.

The proof proceeds in the same way as for the Riemann zeta function (see (1.1)), because the absolute norm $\mathfrak{N}(\mathfrak{a})$ is multiplicative. We do not go into it here, because it is the same argument that also applies to *Hecke L-series*, which will be introduced in §8 as a common generalization of Dirichlet L-series and of the Dedekind zeta function.

Just like the Riemann zeta function, the Dedekind zeta function also admits an analytic continuation to the complex plane with 1 removed, and it satisfies a functional equation relating the argument s to $1 - s$. This is what we are now going to prove. The argument will turn out to be a higher dimensional generalization of the one used in §1 for the Riemann zeta function.

First we split up the series $\zeta_K(s)$, according to the classes $\mathfrak{K}$ of the usual ideal class group $Cl_K = J/P$ of K, into the **partial zeta functions**

$$\zeta(\mathfrak{K}, s) = \sum_{\substack{\mathfrak{a} \in \mathfrak{K} \\ \text{integral}}} \frac{1}{\mathfrak{N}(\mathfrak{a})^s}$$

so that

$$\zeta_K(s) = \sum_{\mathfrak{K}} \zeta(\mathfrak{K}, s).$$

The functional equation is then proved for the individual functions $\zeta(\mathfrak{K}, s)$. The integral ideals in $\mathfrak{K}$ are described as follows. If $\mathfrak{a}$ is a fractional ideal, then the unit group $\mathcal{O}^*$ of $\mathcal{O}$ operates on the set $\mathfrak{a}^* = \mathfrak{a} \setminus \{0\}$, and we denote by $\mathfrak{a}^*/\mathcal{O}^*$ the set of orbits, i.e., the set of classes of non-zero associated elements in $\mathfrak{a}$.

(5.3) Lemma. *Let $\mathfrak{a}$ be an integral ideal of K and $\mathfrak{K}$ the class of the ideal $\mathfrak{a}^{-1}$. Then there is a bijection*

$$\mathfrak{a}^*/\mathcal{O}^* \xrightarrow{\sim} \{\mathfrak{b} \in \mathfrak{K} \mid \mathfrak{b} \textit{ integral}\}, \quad \bar{a} \longmapsto \mathfrak{b} = a\mathfrak{a}^{-1}.$$

Proof: If $a \in \mathfrak{a}^*$, then $a\mathfrak{a}^{-1} = (a)\mathfrak{a}^{-1}$ is an integral ideal in $\mathfrak{K}$, and if $a\mathfrak{a}^{-1} = b\mathfrak{a}^{-1}$, then $(a) = (b)$, so that $ab^{-1} \in \mathcal{O}^*$. This shows the injectivity of the mapping. But it is surjective as well, since for every integral $\mathfrak{b} \in \mathfrak{K}$, one has $\mathfrak{b} = a\mathfrak{a}^{-1}$ with $a \in \mathfrak{a}\mathfrak{b} \subseteq \mathfrak{a}$. □

To the $G(\mathbb{C}|\mathbb{R})$-set $X = \mathrm{Hom}(K, \mathbb{C})$ corresponds the Minkowski space

$$K_{\mathbb{R}} = \mathbf{R} = \Big[\prod_{\tau} \mathbb{C}\Big]^+.$$

The field K may be embedded into $K_{\mathbb{R}}$. Then one finds for $a \in K^*$ that

$$\mathfrak{N}((a)) = |N_{K|\mathbb{Q}}(a)| = |N(a)|,$$

where N denotes the norm on $\mathbf{R}^*$ (see chap. I, §5). The lemma therefore yields the

(5.4) Proposition. $\zeta(\mathfrak{K}, s) = \mathfrak{N}(\mathfrak{a})^s \sum_{\bar{a} \in \mathfrak{a}^*/\mathcal{O}^*} \frac{1}{|N(\bar{a})|^s}.$

By chap. I, (5.2), the ideal $\mathfrak{a}$ forms a complete lattice in $\mathbf{R}$ whose fundamental mesh has volume

$$\mathrm{vol}(\mathfrak{a}) = \sqrt{d_{\mathfrak{a}}},$$

where $d_{\mathfrak{a}} = \mathfrak{N}(\mathfrak{a})^2 |d_K|$ denotes the absolute value of the discriminant of $\mathfrak{a}$, and d_K is the discriminant of K. To the series $\zeta(\mathfrak{K}, s)$ we associate the theta series

$$\theta(\mathfrak{a}, z) = \theta_{\mathfrak{a}}(z/d_{\mathfrak{a}}^{1/n}) = \sum_{a \in \mathfrak{a}} e^{\pi i \langle az/d_{\mathfrak{a}}^{1/n}, a\rangle}.$$

It is related to $\zeta(\mathfrak{K}, s)$ via the gamma integral associated to the $G(\mathbb{C}|\mathbb{R})$-set $X = \operatorname{Hom}(K, \mathbb{C})$,

$$\Gamma_K(s) = \Gamma_X(s) = \int_{\mathbf{R}_+^*} N(e^{-y} y^s) \frac{dy}{y},$$

where $s \in \mathbb{C}$, $\operatorname{Re}(s) > 0$ (see (4.1)). In the integral, we substitute

$$y \longmapsto \pi |a|^2 y / d_{\mathfrak{a}}^{1/n}$$

with $|\ |$ denoting the map $\mathbf{R}^* \to \mathbf{R}_+^*$, $(x_\tau) \mapsto (|x_\tau|)$. We then obtain

$$|d_K|^s \pi^{-ns} \Gamma_K(s) \frac{\mathfrak{N}(\mathfrak{a})^{2s}}{|N(a)|^{2s}} = \int_{\mathbf{R}_+^*} e^{-\pi\langle ay/d_{\mathfrak{a}}^{1/n}, a\rangle} N(y)^s \frac{dy}{y}.$$

Summing this over a full system $\mathfrak{R}$ of representatives of $\mathfrak{a}^*/\mathcal{O}^*$, yields

$$|d_K|^s \pi^{-ns} \Gamma_K(s) \zeta(\mathfrak{K}, 2s) = \int_{\mathbf{R}_+^*} g(y) N(y)^s \frac{dy}{y}$$

with the series

$$g(y) = \sum_{a \in \mathfrak{R}} e^{-\pi\langle ay/d_{\mathfrak{a}}^{1/n}, a\rangle}.$$

Swapping summation and integration is legal, for the same reason as in the case of the Riemann zeta function (see p. 422). We view the function

$$Z_\infty(s) = |d_K|^{s/2} \pi^{-ns/2} \Gamma_K(s/2) = |d_K|^{s/2} L_X(s)$$

as the "Euler factor at infinity" of the zeta function $\zeta(\mathfrak{K}, s)$ (see §4, p. 455) and define

$$Z(\mathfrak{K}, s) = Z_\infty(s) \zeta(\mathfrak{K}, s).$$

The desire to realize this function as an integral over the theta series $\theta(\mathfrak{a}, s)$ is frustrated by the fact that in the theta series we sum over *all* $a \in \mathfrak{a}$, whereas summation in the series $g(y)$ is only over a system of representatives of $\mathfrak{a}^*/\mathcal{O}^*$. This difficulty – which was already hinted at in the case of the Riemann zeta function – will now be overcome in the general case as follows.

The image $|\mathcal{O}^*|$ of the unit group $\mathcal{O}^*$ under the mapping $|\ \ |: \mathbf{R}^* \to \mathbf{R}_+^*$ is contained in the **norm-one hypersurface**

$$\mathbf{S} = \{x \in \mathbf{R}_+^* \mid N(x) = 1\}.$$

Writing every $y \in \mathbf{R}_+^*$ in the form

$$y = xt^{1/n}, \quad x = \frac{y}{N(y)^{1/n}}, \quad t = N(y),$$

we obtain a direct decomposition

$$\mathbf{R}_+^* = \mathbf{S} \times \mathbb{R}_+^*.$$

Let d^*x be the unique Haar measure on the multiplicative group $\mathbf{S}$ such that the canonical Haar measure dy/y on $\mathbf{R}_+^*$ becomes the product measure

$$\frac{dy}{y} = d^*x \times \frac{dt}{t}.$$

We will not need any more explicit description of d^*x.

We now choose a fundamental domain F for the action of the group $|\mathcal{O}^*|^2 = \{|\varepsilon|^2 \mid \varepsilon \in \mathcal{O}^*\}$ on $\mathbf{S}$ as follows. The logarithm map

$$\log : \mathbf{R}_+^* \longrightarrow \mathbf{R}_\pm, \quad (x_\tau) \longmapsto (\log x_\tau),$$

takes the norm-one hypersurface $\mathbf{S}$ to the **trace-zero space** $H = \{x \in \mathbf{R}_\pm \mid Tr(x) = 0\}$, and the group $|\mathcal{O}^*|$ is taken to a complete lattice G in H (Dirichlet's unit theorem). Choose F to be the preimage of an arbitrary fundamental mesh of the lattice $2G$. Any such choice satisfies the

(5.5) Proposition. *The function $Z(\mathfrak{K}, 2s)$ is the Mellin transform*

$$Z(\mathfrak{K}, 2s) = L(f, s)$$

of the function

$$f(t) = f_F(\mathfrak{a}, t) = \frac{1}{w}\int_F \theta(\mathfrak{a}, ixt^{1/n})\, d^*x,$$

where $w = \#\mu(K)$ denotes the number of roots of unity in K.

Proof: Decomposing $\mathbf{R}_+^* = \mathbf{S} \times \mathbb{R}_+^*$, we find

$$Z(\mathfrak{K}, 2s) = \int_0^\infty \int_{\mathbf{S}} \sum_{a \in \mathfrak{R}} e^{-\pi\langle axt', a\rangle}\, d^*x\, t^s \frac{dt}{t},$$

with $t' = (t/d_{\mathfrak{a}})^{1/n}$. The fundamental domain F cuts up the norm-one hypersurface $\mathbf{S}$ into the disjoint union

$$\mathbf{S} = \dot{\bigcup_{\eta \in |\mathcal{O}^*|}} \eta^2 F .$$

The transformation $x \mapsto \eta^2 x$ of $\mathbf{S}$ leaves the Haar measure d^*x invariant and maps F to $\eta^2 F$, so that

$$\begin{aligned}
\int\limits_{\mathbf{S}} \sum_{a \in \mathfrak{R}} e^{-\pi \langle axt', a \rangle} d^*x &= \sum_{\eta \in |\mathcal{O}^*|} \int\limits_{\eta^2 F} \sum_{a \in \mathfrak{R}} e^{-\pi \langle axt', a \rangle} d^*x \\
&= \frac{1}{w} \int\limits_{F} \sum_{\varepsilon \in \mathcal{O}^*} \sum_{a \in \mathfrak{R}} e^{-\pi \langle a\varepsilon xt', a\varepsilon \rangle} d^*x \\
&= \frac{1}{w} \int\limits_{F} \big(\theta(\mathfrak{a}, ixt^{1/n}) - 1 \big) \, d^*x = f(t) - f(\infty) .
\end{aligned}$$

Observe here that we have to divide by $w = \#\mu(K)$, because $\mu(K)$ is just the kernel of $\mathcal{O}^* \to |\mathcal{O}^*|$ (see chap. I, (7.1)), hence $\sum_{|\varepsilon|} = \frac{1}{w} \sum_{\varepsilon}$. Observe furthermore that $a\varepsilon$ runs through the set $\mathfrak{a}^* = \mathfrak{a} \smallsetminus \{0\}$ exactly once, and finally that $f(\infty) = \frac{1}{w} \int_F d^*x$, as $\theta(\mathfrak{a}, ix\infty) = 1$. This result does indeed show that

$$Z(\mathfrak{K}, 2s) = \int\limits_0^\infty \big(f(t) - f(\infty) \big) t^s \frac{dt}{t} = L(f, s) . \qquad \square$$

Using this proposition, the functional equation for the function $Z(\mathfrak{K}, s)$ follows via the Mellin principle from a corresponding transformation formula for the function $f_F(\mathfrak{a}, t)$, which in turn derives from the general theta transformation formula (3.6). In order to find the precise equation, we have to compute the volume $\mathrm{vol}(F)$ of the fundamental domain F with respect to d^*x, and the lattice which is dual to $\mathfrak{a}$ in $\mathbf{R}$. This is achieved by the following two lemmas.

(5.6) Lemma. *The fundamental domain F of $\mathbf{S}$ has the following volume with respect to d^*x:*

$$\mathrm{vol}(F) = 2^{r-1} R ,$$

where r is the number of infinite places and R is the **regulator** *of K (see chap. I, (7.5)).*

Proof: The canonical measure dy/y on $\mathbf{R}_+^*$ is transformed into the product measure $d^*x \times dt/t$ by the isomorphism

$$\alpha : \mathbf{S} \times \mathbb{R}_+^* \longrightarrow \mathbf{R}_+^*, \quad (x,t) \longmapsto xt^{1/n}.$$

Since $I = \{t \in \mathbb{R}_+^* \mid 1 \le t \le e\}$ has measure 1 with respect to dt/t, the quantity $\mathrm{vol}(F)$ is also the volume of $F \times I$ with respect to $d^*x \times dt/t$, i.e., the volume of $\alpha(F \times I)$ with respect to dy/y. The composite ψ of the isomorphisms

$$\mathbf{R}_+^* \xrightarrow{\log} \mathbf{R}_\pm \xrightarrow{\varphi} \prod_{\mathfrak{p}|\infty} \mathbb{R} = \mathbb{R}^r$$

(see §4, p. 454) transforms dy/y into the Lebesgue measure of $\mathbb{R}^r$,

$$\mathrm{vol}(F) = \mathrm{vol}_{\mathbb{R}^r}\big(\psi\alpha(F \times I)\big).$$

Let us compute the image $\psi\alpha(F \times I)$. Let $\mathbf{1} = (1, \dots, 1) \in \mathbf{S}$. Then we find

$$\psi\alpha\big((\mathbf{1}, t)\big) = \mathfrak{e} \log t^{1/n} = \frac{1}{n}\mathfrak{e} \log t$$

with the vector $\mathfrak{e} = (e_{\mathfrak{p}_1}, \dots, e_{\mathfrak{p}_r}) \in \mathbb{R}^r$, $e_{\mathfrak{p}_i} = 1$, resp. $= 2$, depending whether $\mathfrak{p}_i$ is real or complex. By definition of F, we also have

$$\psi\alpha\big(F \times \{1\}\big) = 2\Phi,$$

where Φ denotes a fundamental mesh of the unit lattice G in trace-zero space $H = \{(x_i) \in \mathbb{R}^r \mid \sum x_i = 0\}$. This gives

$$\psi\alpha(F \times I) = 2\Phi + \big[0, \frac{1}{n}\big]\mathfrak{e},$$

the parallelepiped spanned by the vectors $2\mathfrak{e}_1, \dots, 2\mathfrak{e}_{r-1}, \frac{1}{n}\mathfrak{e}$, if $\mathfrak{e}_1, \dots, \mathfrak{e}_{r-1}$ span the fundamental mesh Φ. Its volume is $\frac{1}{n}2^{r-1}$ times the absolute value of the determinant

$$\det \begin{pmatrix} \mathfrak{e}_{11} & \cdots & \mathfrak{e}_{r-1,1} & e_{\mathfrak{p}_1} \\ \vdots & & \vdots & \vdots \\ \mathfrak{e}_{1r} & \cdots & \mathfrak{e}_{r-1,r} & e_{\mathfrak{p}_r} \end{pmatrix}.$$

Adding the first $r-1$ lines to the last one, all entries of the last line become zero, except the last one, which is $n = \sum e_{\mathfrak{p}_i}$. The matrix above these zeroes has the absolute value of its determinant by definition equal to the regulator R. Thus we get

$$\mathrm{vol}(F) = 2^{r-1}R. \qquad \square$$

(5.7) Lemma. *The lattice Γ' in* **R** *which is dual to the lattice $\Gamma = \mathfrak{a}$ is given by*

$$^*\Gamma' = (\mathfrak{a}\mathfrak{d})^{-1},$$

where the asterisk denotes the involution $(x_\tau) \mapsto (\bar{x}_\tau)$ on $K_\mathbb{R}$ and $\mathfrak{d}$ the different of $K\,|\,\mathbb{Q}$.

Proof: As $\langle x, y\rangle = Tr(^*xy)$, we have

$$^*\Gamma' = \big\{{}^*g \in \mathbf{R} \,\big|\, \langle g, a\rangle \in \mathbb{Z} \text{ for all } a \in \mathfrak{a}\big\} = \big\{x \in \mathbf{R} \,\big|\, Tr(x\mathfrak{a}) \subseteq \mathbb{Z}\big\}.$$

$Tr(x\mathfrak{a}) \subseteq \mathbb{Z}$ implies immediately $x \in K$, for if $a_1, \dots, a_n$ is a $\mathbb{Z}$-basis of $\mathfrak{a}$ and $x = x_1a_1 + \cdots + x_na_n$, with $x_i \in \mathbb{R}$, then $Tr(xa_j) = \sum_i x_i\, Tr(a_ia_j) = n_j \in \mathbb{Z}$ is a system of linear equations with coefficients $Tr(a_ia_j) = Tr_{K|\mathbb{Q}}(a_ia_j) \in \mathbb{Q}$, so all $x_i \in \mathbb{Q}$, and thus $x \in K$. It follows that

$$^*\Gamma' = \big\{x \in K \,\big|\, Tr(x\mathfrak{a}) \subseteq \mathbb{Z}\big\}.$$

By definition we have $\mathfrak{d}^{-1} = \{x \in K \mid Tr_{K|\mathbb{Q}}(x\mathcal{O}) \subseteq \mathbb{Z}\}$, and we obtain the equivalences $x \in^* \Gamma' \iff Tr_{K|\mathbb{Q}}(xa\mathcal{O}) \subseteq \mathbb{Z}$ for all $a \in \mathfrak{a} \iff x\mathfrak{a} \subseteq \mathfrak{d}^{-1}$ $\iff x \in (\mathfrak{a}\mathfrak{d})^{-1}$. □

(5.8) Proposition. *The functions $f_F(\mathfrak{a}, t)$ satisfy the transformation formula*

$$f_F\Big(\mathfrak{a}, \frac{1}{t}\Big) = t^{1/2} f_{F^{-1}}\big((\mathfrak{a}\mathfrak{d})^{-1}, t\big),$$

and one has

$$f_F(\mathfrak{a}, t) = \frac{2^{r-1}}{w} R + O(e^{-ct^{1/n}}) \qquad \text{for } t \to \infty,\ c > 0.$$

Proof: We make use of formula (3.6)

$$\theta_\Gamma(-1/z) = \frac{\sqrt{N(z/i)}}{\mathrm{vol}(\Gamma)}\theta_{\Gamma'}(z)$$

for the lattice $\Gamma = \mathfrak{a}$ in **R**, whose fundamental mesh has volume $\mathrm{vol}(\Gamma) = \mathfrak{N}(\mathfrak{a})|d_K|^{1/2}$. The lattice Γ' dual to Γ is given by (5.7) as $^*\Gamma' = (\mathfrak{a}\mathfrak{d})^{-1}$. The compatibility $\langle {}^*gz, {}^*g\rangle = \langle gz, g\rangle$ implies that $\theta_{\Gamma'}(z) = \theta_{^*\Gamma'}(z)$. Furthermore we have

$$d_{(\mathfrak{a}\mathfrak{d})^{-1}} = \mathfrak{N}(\mathfrak{a})^{-2}\mathfrak{N}(\mathfrak{d})^{-2}|d_K| = 1/(\mathfrak{N}(\mathfrak{a})^2|d_K|) = 1/d_\mathfrak{a}.$$

The transformation $x \mapsto x^{-1}$ of the multiplicative group **S** fixes the Haar measure d^*x (in the same way as $x \mapsto -x$ fixes a Haar measure on $\mathbb{R}^n$)

and maps the fundamental domain F onto the fundamental domain F^{-1}, whose image $\log(F^{-1})$ is again a fundamental mesh of the lattice $2\log|\mathcal{O}^*|$. Observing that $N(x(td_{\mathfrak{a}})^{1/n}) = td_{\mathfrak{a}}$ for $x \in \mathbf{S}$, we obtain

$$\begin{aligned} f_F\Big(\mathfrak{a}, \frac{1}{t}\Big) &= \frac{1}{w}\int\limits_F \theta_{\mathfrak{a}}\big(ix/\sqrt[n]{td_{\mathfrak{a}}}\big)\, d^*x \\ &= \frac{1}{w}\int\limits_{F^{-1}} \theta_{\mathfrak{a}}\big(-1/ix\sqrt[n]{td_{\mathfrak{a}}}\big)\, d^*x \\ &= \frac{1}{w}\frac{(td_{\mathfrak{a}})^{1/2}}{\operatorname{vol}(\mathfrak{a})}\int\limits_{F^{-1}} \theta_{(\mathfrak{a}\mathfrak{d})^{-1}}\big(ix\sqrt[n]{td_{\mathfrak{a}}}\big)\, d^*x \\ &= \frac{t^{1/2}}{w}\int\limits_{F^{-1}} \theta_{(\mathfrak{a}\mathfrak{d})^{-1}}\big(ix\sqrt[n]{t/d_{(\mathfrak{a}\mathfrak{d})^{-1}}}\big)\, d^*x \\ &= t^{1/2} f_{F^{-1}}\big((\mathfrak{a}\mathfrak{d})^{-1}, t\big)\,. \end{aligned}$$

This shows the first formula. To prove the second, we write

$$f_F(\mathfrak{a}, t) = \frac{1}{w}\int\limits_F d^*x + \frac{1}{w}\int\limits_F \big(\theta(\mathfrak{a}, ixt^{1/n}) - 1\big)\, d^*x = \frac{\operatorname{vol}(F)}{w} + r(t)\,.$$

The function $r(t)$ satisfies $r(t) = O(e^{-ct^{1/n}})$, $c > 0$, $t \to \infty$, as the summands of $\theta(\mathfrak{a}, ixt^{1/n}) - 1$ are of the form

$$e^{-\pi\langle ax, a\rangle\sqrt[n]{t'}}\,, \qquad a \in \mathfrak{a},\ a \neq 0,\ t' = t/d_{\mathfrak{a}}\,.$$

The point $x = (x_\tau)$ varies in the compact closure $\bar{F} \subseteq \big[\prod_\tau \mathbb{R}_+^*\big]^+$ of F. Hence $x_\tau \geq \delta > 0$ for all τ, i.e.,

$$\langle ax, a\rangle = \sum_\tau |\tau a|^2 x_\tau \geq \delta\langle a, a\rangle$$

and so

$$r(t) \leq \frac{\operatorname{vol}(F)}{w}\big(\theta_{\mathfrak{a}}(i\delta\sqrt[n]{t'}) - 1\big)\,.$$

Writing $m = \min\{\langle a, a\rangle \mid a \in \mathfrak{a}, a \neq 0\}$ and $M = \#\{a \in \mathfrak{a} \mid \langle a, a\rangle = m\}$, it follows that

$$\theta_{\mathfrak{a}}\big(i\delta\sqrt[n]{t'}\big) - 1 = e^{-\pi\delta m\sqrt[n]{t'}}\Big(M + \sum_{\langle a,a\rangle > m} e^{-\pi\delta(\langle a,a\rangle - m)\sqrt[n]{t'}}\Big) = O(e^{-ct^{1/n}})$$

where $c = \pi\delta m/d_{\mathfrak{a}}^{1/n}$. We thus get as claimed

$$f_F(\mathfrak{a}, t) = \frac{\operatorname{vol}(F)}{w} + O(e^{-ct^{1/n}}) = \frac{2^{r-1}}{w}R + O(e^{-ct^{1/n}})\,. \qquad \square$$

This last proposition now enables us to apply the Mellin principle (1.4) to the functions $f_F(\mathfrak{a}, t)$. For the partial zeta functions

$$\zeta(\mathfrak{K}, s) = \sum_{\substack{\mathfrak{b} \in \mathfrak{K} \\ \text{integral}}} \frac{1}{\mathfrak{N}(\mathfrak{b})^s},$$

this yields the following result, where the notations d_K, R, w, and r signify as before the discriminant, the regulator, the number of roots of unity, and the number of infinite places, respectively.

(5.9) Theorem. *The function*

$$Z(\mathfrak{K}, s) = Z_\infty(s)\zeta(\mathfrak{K}, s), \qquad \mathrm{Re}(s) > 1,$$

$Z_\infty(s) = |d_K|^{s/2}\pi^{-ns/2}\Gamma_K(s/2)$, *admits an analytic continuation to* $\mathbb{C} \setminus \{0, 1\}$ *and satisfies the functional equation*

$$Z(\mathfrak{K}, s) = Z(\mathfrak{K}', 1 - s),$$

where the ideal classes $\mathfrak{K}$ *and* $\mathfrak{K}'$ *correspond to each other via* $\mathfrak{K}\mathfrak{K}' = [\mathfrak{d}]$. *It has simple poles at* $s = 0$ *and* $s = 1$ *with residues*

$$-\frac{2^r}{w}R, \quad \textit{resp.} \quad \frac{2^r}{w}R.$$

Proof: Let $f(t) = f_F(\mathfrak{a}, t)$ and $g(t) = f_{F^{-1}}((\mathfrak{a}\mathfrak{d})^{-1}, t)$. Then (5.8) implies

$$f\Big(\frac{1}{t}\Big) = t^{1/2}g(t)$$

and

$$f(t) = a_0 + O(e^{-ct^{1/n}}), \quad g(t) = a_0 + O(e^{-ct^{1/n}}),$$

with $a_0 = \frac{2^{r-1}}{w}R$. Proposition (1.4) thus ensures the analytic continuation of the Mellin transforms of f and g, and the functional equation

$$L(f, s) = L\Big(g, \frac{1}{2} - s\Big)$$

with simple poles of $L(f, s)$ at $s = 0$ and $s = \frac{1}{2}$ with residues $-a_0$, resp. a_0. Therefore

$$Z(\mathfrak{K}, s) = L\Big(f, \frac{s}{2}\Big)$$

admits an analytic continuation to $\mathbb{C} \setminus \{0, 1\}$ with simple poles at $s = 0$ and $s = 1$ and residues

$$-2a_0 = -\frac{2^r}{w}R, \quad \text{resp.} \ \ 2a_0 = \frac{2^r}{w}R$$

and satisfies the functional equation

$$Z(\mathfrak{K}, s) = L\Big(f, \frac{s}{2}\Big) = L\Big(g, \frac{1-s}{2}\Big) = Z(\mathfrak{K}', 1-s)\,. \qquad \square$$

This theorem about the partial zeta functions immediately implies an analogous result for the **completed zeta function** of the number field K,

$$Z_K(s) = Z_\infty(s)\zeta_K(s) = \sum_{\mathfrak{K}} Z(\mathfrak{K}, s)\,.$$

(5.10) Corollary. *The completed zeta function $Z_K(s)$ admits an analytic continuation to $\mathbb{C} \smallsetminus \{0,1\}$ and satisfies the functional equation*

$$Z_K(s) = Z_K(1-s)\,.$$

It has simple poles at $s = 0$ and $s = 1$ with residues

$$-\frac{2^r hR}{w}, \quad \text{resp.} \quad \frac{2^r hR}{w},$$

where h is the class number of K.

The last result can be immediately generalized as follows. For every character

$$\chi : J/P \longrightarrow S^1$$

of the ideal class group, one may form the zeta function

$$Z(\chi, s) = Z_\infty(s)\zeta(\chi, s),$$

where

$$\zeta(\chi, s) = \sum_{\mathfrak{a} \text{ integral}} \frac{\chi(\mathfrak{a})}{\mathfrak{N}(\mathfrak{a})^s}$$

and $\chi(\mathfrak{a})$ denotes the value $\chi(\mathfrak{K})$ of the class $\mathfrak{K} = [\mathfrak{a}]$ of an ideal $\mathfrak{a}$. Then clearly

$$Z(\chi, s) = \sum_{\mathfrak{K}} \chi(\mathfrak{K}) Z(\mathfrak{K}, s),$$

and in view of $\mathfrak{K}' = \mathfrak{K}^{-1}[\mathfrak{d}]$, we obtain from (5.9) the functional equation

$$Z(\chi, s) = \chi(\mathfrak{d}) Z(\bar{\chi}, 1-s)\,.$$

If $\chi \neq \mathbf{1}$, then $Z(\chi, s)$ is holomorphic on all of $\mathbb{C}$, as $\sum_{\mathfrak{K}} \chi(\mathfrak{K}) = 0$.

We now conclude with the original Dedekind zeta function

$$\zeta_K(s) = \sum_{\mathfrak{a}} \frac{1}{\mathfrak{N}(\mathfrak{a})^s}, \qquad \mathrm{Re}(s) > 1.$$

The Euler factor at infinity, $Z_\infty(s)$, is given explicitly by §4 as

$$Z_\infty(s) = |d_K|^{s/2} L_X(s) = |d_K|^{s/2} L_{\mathbb{R}}(s)^{r_1} L_{\mathbb{C}}(s)^{r_2},$$

where r_1, resp. r_2, denotes the number of real, resp. complex, places. By (4.3), (i), one has $Z_\infty(1) = |d_K|^{1/2}/\pi^{r_2}$. As

$$\zeta_K(s) = Z_\infty(s)^{-1} Z_K(s) = |d_K|^{-s/2} L_X(s)^{-1} Z_K(s),$$

we obtain from (4.4) the

(5.11) Corollary. (i) *The Dedekind zeta function $\zeta_K(s)$ has an analytic continuation to $\mathbb{C} \smallsetminus \{1\}$.*

(ii) *At $s = 1$ it has a simple pole with residue*

$$\kappa = \frac{2^{r_1}(2\pi)^{r_2}}{w|d_K|^{1/2}} hR = hR/e^g.$$

Here h denotes the class number and

$$g = \log \frac{w|d_K|^{1/2}}{2^{r_1}(2\pi)^{r_2}}$$

the **genus** *of the number field K (see chap. III, (3.5)).*

(iii) *It satisfies the functional equation*

$$\zeta_K(1-s) = A(s)\zeta_K(s)$$

with the factor

$$A(s) = |d_K|^{s-\frac{1}{2}} \left(\cos \frac{\pi s}{2}\right)^{r_1+r_2} \left(\sin \frac{\pi s}{2}\right)^{r_2} L_{\mathbb{C}}(s)^n.$$

The proof of the analytic continuation and functional equation of the Dedekind zeta function was first given by the mathematician *Erich Hecke* (1887–1947), along the same general lines we have presented here, albeit in a somewhat different formulation. Further, the theory we are about to develop in the following sections §§6–8 also substantially goes back to *Hecke*.

The formula for the residue

$$\mathrm{Res}_{s=1}\, \zeta_K(s) = \frac{2^{r_1}(2\pi)^{r_2}}{w|d_K|^{1/2}} hR$$

is commonly known as the analytic **class number formula**. It does allow us to determine the class number h of the field K, provided we know the law for the decomposition of primes in this field sufficiently well to lay our hands on the Euler product and thus compute the zeta function.

The following application of corollary (5.11) to Dirichlet L-series $L(\chi,s)$ (see §2) is highly remarkable. It results from studying the Dedekind zeta function $\zeta_K(s)$ for the field $K = \mathbb{Q}(\mu_m)$ of m-th roots of unity, and is based on the

(5.12) Proposition. *If $K = \mathbb{Q}(\mu_m)$ is the field of m-th roots of unity, then*

$$\zeta_K(s) = G(s)\prod_{\chi} L(\chi,s),$$

where χ varies over all Dirichlet characters mod m, *and*

$$G(s) = \prod_{\mathfrak{p}|m}(1-\mathfrak{N}(\mathfrak{p})^{-s})^{-1}.$$

Proof: The proof hinges on the law of decomposition of prime numbers p in the field K. Let $p = (\mathfrak{p}_1\dots\mathfrak{p}_r)^e$ be the decomposition of the prime number p in K, and let f be the degree of the $\mathfrak{p}_i$, i.e., $\mathfrak{N}(\mathfrak{p}_i) = p^f$. Then $\zeta_K(s)$ contains the factor

$$\prod_{\mathfrak{p}|p}(1-\mathfrak{N}(\mathfrak{p})^{-s})^{-1} = (1-p^{-fs})^{-r}.$$

On the other hand, the L-series give the factor $\prod_\chi(1-\chi(p)p^{-s})^{-1}$. For $p|m$ this is 1. So let $p\nmid m$. By chap. I, (10.3), f is the order of p mod m in $(\mathbb{Z}/m\mathbb{Z})^*$ and $e=1$. Since $efr = \varphi(m)$, the quotient $r = \varphi(m)/f$ is the index of the subgroup G_p generated by p in $G = (\mathbb{Z}/m\mathbb{Z})^*$. Associating $\chi\mapsto\chi(p)$ defines an isomorphism $\widehat{G}_p \cong \mu_f$, and gives the exact sequence

$$1 \longrightarrow \widehat{G/G_p} \longrightarrow \widehat{G} \longrightarrow \mu_f \longrightarrow 1,$$

where $\widehat{\ }$ indicates character groups. We therefore find $r = \#(\widehat{G/G_p}) = (G : G_p)$ elements in the preimage of $\chi(p)$. It follows that

$$\begin{aligned}\prod_{\chi}(1-\chi(p)p^{-s})^{-1} &= \prod_{\zeta\in\mu_f}(1-\zeta p^{-s})^{-r} = (1-p^{-fs})^{-r}\\ &= \prod_{\mathfrak{p}|p}(1-\mathfrak{N}(\mathfrak{p})^{-s})^{-1}.\end{aligned}$$

Finally, taking the product over all p, we get $\zeta_K(s) = G(s)\prod_\chi L(\chi,s)$. □

For the trivial character χ^0 mod m, we have $L(\chi^0,s) = \prod_{p|m}(1-p^{-s})\,\zeta(s)$, so that

$$\zeta_K(s) = G(s)\prod_{p|m}(1-p^{-s})\zeta(s)\prod_{\chi\neq\chi^0} L(\chi,s).$$

Since $\zeta(s)$ and $\zeta_K(s)$ both have a simple pole at $s=1$, we obtain the

(5.13) Proposition. *For every non-trivial Dirichlet character χ, one has*

$$L(\chi,1)\neq 0.$$

This innocuous looking result is in fact rather profound, and yields as a concrete consequence

(5.14) Dirichlet's Prime Number Theorem. *Every* **arithmetic progression**

$$a,\ a\pm m,\ a\pm 2m,\ a\pm 3m,\ \ldots,\ \textit{with } (a,m)=1,$$

i.e., every class a mod m, *contains infinitely many prime numbers.*

Proof: Let χ be a Dirichlet character mod m. Then one has, for $\mathrm{Re}(s)>1$,

$$\log L(\chi,s) = -\sum_p \log(1-\chi(p)p^{-s}) = \sum_p\sum_{m=1}^{\infty}\frac{\chi(p^m)}{mp^{ms}} = \sum_p \frac{\chi(p)}{p^s} + g_\chi(s),$$

where $g_\chi(s)$ is holomorphic for $\mathrm{Re}(s)>\frac{1}{2}$ – this follows from a trivial estimate. Multiplying by $\chi(a^{-1})$ and summing over all characters mod m, yields

$$\begin{aligned}\sum_\chi \chi(a^{-1})\log L(\chi,s) &= \sum_\chi\sum_p\frac{\chi(a^{-1}p)}{p^s} + g(s)\\ &= \sum_{b=1}^{m}\sum_\chi \chi(a^{-1}b)\sum_{p\equiv b(m)}\frac{1}{p^s} + g(s)\\ &= \sum_{p\equiv a(m)}\frac{\varphi(m)}{p^s} + g(s).\end{aligned}$$

Note here that

$$\sum_\chi \chi(a^{-1}b) = \begin{cases} 0, & \text{if } a\neq b,\\ \varphi(m) = \#(\mathbb{Z}/m\mathbb{Z})^*, & \text{if } a=b.\end{cases}$$

When we pass to the limit $s\to 1$ (s real >1), $\log L(\chi,s)$ stays bounded for $\chi\neq\chi^0$ because $L(\chi,1)\neq 0$, whereas $\log L(\chi^0,s) =$

$\sum_{p|m} \log(1 - p^{-s}) + \log\zeta(s)$ tends to ∞ because $\zeta(s)$ has a pole. The left-hand side of the above equation therefore tends to ∞, and since $g(s)$ is holomorphic at $s = 1$, we find

$$\lim_{s\to 1} \sum_{p\equiv a(m)} \frac{\varphi(m)}{p^s} = \infty .$$

Thus the sum cannot consist of only finitely many terms, and the theorem is proved. □

For $a = 1$, Dirichlet's prime number theorem may be proved by pure algebra (see chap. I, § 10, exercise 1). Searching for a proof in the general case Dirichlet was led to the study of the L-series $L(\chi, s)$. This analytic method gives sharper results on the distribution of prime numbers among the classes a mod m. We will come back to this in a more general context in § 13.

§ 6. Hecke Characters

Let $\mathfrak{m}$ be an integral ideal of the number field K, and let $J^{\mathfrak{m}}$ be the group of all ideals of K which are relatively prime to $\mathfrak{m}$. Given any character

$$\chi : J^{\mathfrak{m}} \to S^1 = \{ z \in \mathbb{C} \mid |z| = 1 \},$$

we may associate to it, as a common generalization of the Dirichlet L-series as well as the Dedekind zeta function, the L-series

$$L(\chi, s) = \sum_{\mathfrak{a}} \frac{\chi(\mathfrak{a})}{\mathfrak{N}(\mathfrak{a})^s} .$$

Here $\mathfrak{a}$ varies over all integral ideals of K, and one defines $\chi(\mathfrak{a}) = 0$ whenever $(\mathfrak{a}, \mathfrak{m}) \neq 1$. Searching for the most comprehensive class of characters χ for which the corresponding L-series could be shown to have a functional equation, HECKE was led to the notion of *Größencharaktere*, which we define as follows.

(6.1) Definition. *A* **Größencharakter** *mod* $\mathfrak{m}$ *is a character* $\chi : J^{\mathfrak{m}} \to S^1$ *for which there exists a pair of characters*

$$\chi_{\mathrm{f}} : (\mathcal{O}/\mathfrak{m})^* \longrightarrow S^1 , \quad \chi_\infty : \mathbf{R}^* \longrightarrow S^1 ,$$

such that

$$\chi\big((a)\big) = \chi_{\mathrm{f}}(a)\chi_\infty(a)$$

for every algebraic integer $a \in \mathcal{O}$ *relatively prime to* $\mathfrak{m}$.

A character χ of $J^{\mathfrak{m}}$ is a *Größencharakter* mod $\mathfrak{m}$ as soon as there exists a character χ_∞ of $\mathbf{R}^*$ such that

$$\chi\big((a)\big) = \chi_\infty(a)$$

for all $a \in \mathcal{O}$ such that $a \equiv 1 \bmod \mathfrak{m}$. For if this is the case, then the rule $\chi_{\mathrm{f}}(a) = \chi((a))\chi_\infty(a)^{-1}$ defines a character χ_{f} of $(\mathcal{O}/\mathfrak{m})^*$ which satisfies

$$\chi\big((a)\big) = \chi_{\mathrm{f}}(a)\chi_\infty(a)$$

for all algebraic integers $a \in \mathcal{O}$ relatively prime to $\mathfrak{m}$. This last identity underlines the fact that the restriction of a *Größencharakter* to principal ideals breaks up into a finite and an infinite part. From

$$\mathcal{O}^{(\mathfrak{m})} = \big\{a \in \mathcal{O} \,\big|\, (a, \mathfrak{m}) = 1\big\},$$

it extends uniquely to the group

$$K^{(\mathfrak{m})} = \big\{a \in K^* \,\big|\, (a, \mathfrak{m}) = 1\big\}$$

of all fractions relatively prime to $\mathfrak{m}$, because every $a \in K^{(\mathfrak{m})}$ determines a well-defined class in $(\mathcal{O}/\mathfrak{m})^*$. The character χ_∞, and thus also the character χ_{f}, are determined uniquely by the *Größencharakter* χ, since the group

$$K^{\mathfrak{m}} = \big\{a \in K^{(\mathfrak{m})} \,\big|\, a \equiv 1 \bmod \mathfrak{m}\big\}$$

is dense in $\mathbf{R}^*$, by the approximation theorem, and one has $\chi_\infty(a) = \chi((a))$ for $a \in K^{\mathfrak{m}}$. Let us recall that the congruence $a \equiv 1 \bmod \mathfrak{m}$ signifies that $a = b/c$, for two integers b, c relatively prime to $\mathfrak{m}$, such that $b \equiv c \bmod \mathfrak{m}$ or, equivalently, $a \in U_{\mathfrak{p}}^{(n_{\mathfrak{p}})} \subseteq K_{\mathfrak{p}}$ for $\mathfrak{p}|\mathfrak{m}$, if $\mathfrak{m} = \prod_{\mathfrak{p}} \mathfrak{p}^{n_{\mathfrak{p}}}$.

The character χ_∞ factors automatically through $\mathbf{R}^*/\mathcal{O}^{\mathfrak{m}}$, where

$$\mathcal{O}^{\mathfrak{m}} = \big\{\varepsilon \in \mathcal{O}^* \,\big|\, \varepsilon \equiv 1 \bmod \mathfrak{m}\big\}.$$

In fact, for $\varepsilon \in \mathcal{O}^{\mathfrak{m}}$ we have $\chi_{\mathrm{f}}(\varepsilon) = 1$, and thus $\chi_\infty(\varepsilon) = \chi_{\mathrm{f}}(\varepsilon)\chi_\infty(\varepsilon) = \chi((\varepsilon)) = 1$. The two characters χ_{f} and χ_∞ of $(\mathcal{O}/\mathfrak{m})^*$, resp. $\mathbf{R}^*/\mathcal{O}^{\mathfrak{m}}$, associated with a *Größencharakter* χ satisfy the relation

$$\chi_{\mathrm{f}}(\varepsilon)\chi_\infty(\varepsilon) = 1 \quad \text{for all } \varepsilon \in \mathcal{O}^*,$$

and it can be shown that every such pair of characters $(\chi_{\mathrm{f}}, \chi_\infty)$ comes from a *Größencharakter* χ (exercise 5).

The attempt to understand *Größencharaktere* in a conceptual way leads one to introduce **idèles**. In fact, all *Größencharaktere* arise as characters of the **idèle class group** of the number field K. We will not use this more abstract interpretation in what follows, but it will be explained at the end of this section.

(6.2) Proposition. *Let* χ *be a Größencharakter* mod $\mathfrak{m}$, *and let* $\mathfrak{m}'$ *be a divisor of* $\mathfrak{m}$. *Then the following conditions are equivalent.*

(i) χ *is the restriction of a Größencharakter* $\chi' : J^{\mathfrak{m}'} \to S^1$ mod $\mathfrak{m}'$.

(ii) χ_{f} *factors through* $(\mathcal{O}/\mathfrak{m}')^*$.

Proof: (i) $\Rightarrow$ (ii). Let χ be the restriction of the *Größencharakter* $\chi' : J^{\mathfrak{m}'} \to S^1$, and let χ_{f}', χ_∞' be the pair of characters associated with χ'. Let $\widetilde{\chi}_{\mathrm{f}}$, resp. $\widetilde{\chi}_\infty$, be the composite of

$$(\mathcal{O}/\mathfrak{m})^* \longrightarrow (\mathcal{O}/\mathfrak{m}')^* \xrightarrow{\chi_{\mathrm{f}}'} S^1, \quad \text{resp.} \quad \mathbf{R}^*/\mathcal{O}^{\mathfrak{m}} \longrightarrow \mathbf{R}^*/\mathcal{O}^{\mathfrak{m}'} \xrightarrow{\chi_\infty'} S^1.$$

We then find for $a \in \mathcal{O}^{(\mathfrak{m})} \subseteq \mathcal{O}^{(\mathfrak{m}')}$:

$$\chi\big((a)\big) = \chi'\big((a)\big) = \chi_{\mathrm{f}}'(a)\chi_\infty'(a) = \widetilde{\chi}_{\mathrm{f}}(a)\widetilde{\chi}_\infty(a),$$

so that $\chi_{\mathrm{f}} = \widetilde{\chi}_{\mathrm{f}}$ and $\chi_\infty = \widetilde{\chi}_\infty$ because χ_{f} and χ_∞ are uniquely determined by χ. Thus χ_{f} factors through $(\mathcal{O}/\mathfrak{m}')^*$ (and χ_∞ through $\mathbf{R}^*/\mathcal{O}^{\mathfrak{m}'}$).

(ii) $\Rightarrow$ (i). Let χ_{f} be the composite of $(\mathcal{O}/\mathfrak{m})^* \to (\mathcal{O}/\mathfrak{m}')^* \xrightarrow{\chi_{\mathrm{f}}'} S^1$. In every class $\mathfrak{a}'$ mod $P^{\mathfrak{m}'} \in J^{\mathfrak{m}'}/P^{\mathfrak{m}'}$, there is an ideal $\mathfrak{a} \in J^{\mathfrak{m}}$ which is relatively prime to $\mathfrak{m}$, i.e., $\mathfrak{a}' = \mathfrak{a}a$ for some $(a) \in P^{\mathfrak{m}'}$. We put

$$\chi'(\mathfrak{a}') = \chi(\mathfrak{a})\chi_{\mathrm{f}}'(a)\chi_\infty(a).$$

This definition does not depend on the choice of the ideal $\mathfrak{a} \in J^{\mathfrak{m}}$, for if $\mathfrak{a}' = \mathfrak{a}_1 a_1$, $\mathfrak{a}_1 \in J^{\mathfrak{m}}$, $(a_1) \in P^{\mathfrak{m}'}$, then one has $(aa_1^{-1}) \in J^{\mathfrak{m}}$, and

$$\begin{aligned}\chi(\mathfrak{a})\chi_{\mathrm{f}}'(a)\chi_\infty(a) &= \chi(\mathfrak{a})\chi\big((aa_1^{-1})\big)\,\chi_{\mathrm{f}}(a^{-1}a_1)\chi_\infty(a^{-1}a_1)\chi_{\mathrm{f}}'(a)\chi_\infty(a)\\ &= \chi(\mathfrak{a}_1)\chi_{\mathrm{f}}'(a_1)\chi_\infty(a_1).\end{aligned}$$

The restriction of the character χ' from $J^{\mathfrak{m}'}$ to $J^{\mathfrak{m}}$ is the *Größencharakter* χ of $J^{\mathfrak{m}}$, and if (a') is a principal ideal prime to $\mathfrak{m}'$ and $a' = ab$, $(a) \in J^{\mathfrak{m}}$, $(b) \in P^{\mathfrak{m}'}$, then we have

$$\begin{aligned}\chi'\big((a')\big) &= \chi\big((a)\big)\,\chi'\big((b)\big) = \chi\big((a)\big)\,\chi_{\mathrm{f}}'(b)\chi_\infty(b)\\ &= \chi_{\mathrm{f}}(a)\chi_\infty(a)\chi_{\mathrm{f}}'(b)\chi_\infty(b) = \chi_{\mathrm{f}}'(ab)\chi_\infty(ab) = \chi_{\mathrm{f}}'(a')\chi_\infty(a').\end{aligned}$$

Thus χ' is a *Größencharakter* mod $\mathfrak{m}'$ with corresponding pair of characters χ_{f}', χ_∞. □

The *Größencharakter* χ mod $\mathfrak{m}$ is called **primitive** if it is not the restriction of a *Größencharakter* χ' mod $\mathfrak{m}'$ for any proper divisor $\mathfrak{m}'|\mathfrak{m}$.

According to (6.2), this is the case if and only if the character $\chi_{\mathfrak{f}}$ of $(\mathcal{O}/\mathfrak{m})^*$ is primitive in the sense that it does not factorize through $(\mathcal{O}/\mathfrak{m}')^*$ for any proper divisor $\mathfrak{m}'|\mathfrak{m}$. The **conductor** of χ is the smallest divisor $\mathfrak{f}$ of $\mathfrak{m}$ such that χ is the restriction of a *Größencharakter* mod $\mathfrak{f}$. By (6.2), $\mathfrak{f}$ is the conductor of $\chi_{\mathfrak{f}}$, i.e., the smallest divisor of $\mathfrak{m}$ such that $\chi_{\mathfrak{f}}$ factors through $(\mathcal{O}/\mathfrak{f})^*$.

Let us now have a closer look at the character $\chi_{\mathfrak{f}}$, and then at the character χ_∞.

(6.3) Definition. *Let $\chi_{\mathfrak{f}}$ be a character of $(\mathcal{O}/\mathfrak{m})^*$ and $y \in \mathfrak{m}^{-1}\mathfrak{d}^{-1}$, where $\mathfrak{d}$ is the different of $K\,|\,\mathbb{Q}$. Then we define the* **Gauss sum** *of $\chi_{\mathfrak{f}}$ to be*

$$\tau_{\mathfrak{m}}(\chi_{\mathfrak{f}}, y) = \sum_{\substack{x \bmod \mathfrak{m} \\ (x,\mathfrak{m})=1}} \chi_{\mathfrak{f}}(x)\, e^{2\pi i\, Tr(xy)},$$

where x varies over a system of representatives of $(\mathcal{O}/\mathfrak{m})^*$.

The Gauss sum does not depend on the choice of representatives x, for if $x' \equiv x \bmod \mathfrak{m}$, then $x'y - xy \in \mathfrak{m}\mathfrak{m}^{-1}\mathfrak{d}^{-1} = \mathfrak{d}^{-1} = \{a \in K \mid Tr(a) \in \mathbb{Z}\}$, so that

$$Tr(x'y) \equiv Tr(xy) \bmod \mathbb{Z}$$

and therefore $e^{2\pi i\, Tr(x'y)} = e^{2\pi i\, Tr(xy)}$. The same argument shows that $\tau_{\mathfrak{m}}(\chi_{\mathfrak{f}}, y)$ depends only on the coset $y + \mathfrak{d}^{-1}$, i.e., it defines a function on the $\mathcal{O}/\mathfrak{m}$-module $\mathfrak{m}^{-1}\mathfrak{d}^{-1}/\mathfrak{d}^{-1}$. In the case $K = \mathbb{Q}$, $\mathfrak{m} = (m)$, we get back the Gauss sum introduced in (2.5) by $\tau(\chi_{\mathfrak{f}}, n) = \tau_{\mathfrak{m}}(\chi_{\mathfrak{f}}, \frac{n}{m})$. We will have to define theta series and L-series attached to Hecke's *Größencharaktere* with a view to proving functional equations. For this, the following properties of Gauss sums will play a crucial rôle.

(6.4) Theorem. *Let $\chi_{\mathfrak{f}}$ be a primitive character of $(\mathcal{O}/\mathfrak{m})^*$, let $y \in \mathfrak{m}^{-1}\mathfrak{d}^{-1}$ and $a \in \mathcal{O}$. Then one has*

$$\tau_{\mathfrak{m}}(\chi_{\mathfrak{f}}, ay) = \begin{cases} \overline{\chi}_{\mathfrak{f}}(a)\tau_{\mathfrak{m}}(\chi_{\mathfrak{f}}, y), & \text{if } (a,\mathfrak{m}) = 1, \\ 0, & \text{if } (a,\mathfrak{m}) \neq 1, \end{cases}$$

and furthermore

$$\left|\tau_{\mathfrak{m}}(\chi_{\mathfrak{f}}, y)\right| = \sqrt{\mathfrak{N}(\mathfrak{m})}, \quad \text{if } (y\mathfrak{m}\mathfrak{d}, \mathfrak{m}) = 1.$$

The most diffcult part of the theorem is the last claim. To prove it, we make the following preparations. For integral ideals $\mathfrak{a} = \mathfrak{p}_1^{\nu_1} \cdots \mathfrak{p}_r^{\nu_r}$, $\nu_i \geq 1$, consider the **Möbius function**

$$\mu(\mathfrak{a}) = \begin{cases} 1, & \text{if } r = 0, \text{ i.e., } \mathfrak{a} = (1), \\ (-1)^r, & \text{if } \nu_1 = \cdots = \nu_r = 1, \\ 0, & \text{otherwise.} \end{cases}$$

For this function we have the

(6.5) Proposition. *If $\mathfrak{a} \neq 1$, then $\sum\limits_{\mathfrak{b}|\mathfrak{a}} \mu(\mathfrak{b}) = 0$.*

Proof: If $\mathfrak{a} = \mathfrak{p}_1^{\nu_1} \cdots \mathfrak{p}_r^{\nu_r}$, $\nu_i \geq 1$, then

$$\begin{aligned} \sum_{\mathfrak{b}|\mathfrak{a}} \mu(\mathfrak{b}) &= \mu(1) + \sum_i \mu(\mathfrak{p}_i) + \sum_{i_1<i_2} \mu(\mathfrak{p}_{i_1}\mathfrak{p}_{i_2}) + \cdots + \mu(\mathfrak{p}_1 \cdots \mathfrak{p}_r) \\ &= 1 + \tbinom{r}{1}(-1) + \tbinom{r}{2}(-1)^2 + \cdots + \tbinom{r}{r}(-1)^r \\ &= \big(1 + (-1)\big)^r = 0. \end{aligned}$$

□

Now, for $y \in \mathfrak{m}^{-1}\mathfrak{d}^{-1}$ and for every integral divisor $\mathfrak{a}$ of $\mathfrak{m}$, we look at the sums

$$T_{\mathfrak{a}}(y) = \sum_{\substack{x \bmod \mathfrak{m} \\ (x,\mathfrak{m})=\mathfrak{a}}} e^{2\pi i\, Tr(xy)} \quad \text{and} \quad S_{\mathfrak{a}}(y) = \sum_{\substack{x \bmod \mathfrak{m} \\ \mathfrak{a}|x}} e^{2\pi i\, Tr(xy)}.$$

These sums do not depend on the choice of representatives x, for if $x' \equiv x \bmod \mathfrak{m}$, then $(x'-x)y \in \mathfrak{d}^{-1}$, hence $Tr(x'y) \equiv Tr(xy) \bmod \mathbb{Z}$. We find the

(6.6) Lemma. *One has*

$$T_1(y) = \sum_{\mathfrak{a}|\mathfrak{m}} \mu(\mathfrak{a}) S_{\mathfrak{a}}(y),$$

and for every divisor $\mathfrak{a}|\mathfrak{m}$,

$$S_{\mathfrak{a}}(y) = \begin{cases} \mathfrak{N}(\frac{\mathfrak{m}}{\mathfrak{a}}), & \textit{if } y \in \mathfrak{a}^{-1}\mathfrak{d}^{-1}, \\ 0, & \textit{if } y \notin \mathfrak{a}^{-1}\mathfrak{d}^{-1}. \end{cases}$$

Proof: In view of (6.5), we have

$$\sum_{\mathfrak{a}|\mathfrak{m}} \mu(\mathfrak{a}) S_{\mathfrak{a}}(y) = \sum_{\mathfrak{a}|\mathfrak{m}} \mu(\mathfrak{a}) \sum_{\substack{\mathfrak{b} \\ \mathfrak{a}|\mathfrak{b}|\mathfrak{m}}} T_{\mathfrak{b}}(y) = \sum_{\mathfrak{b}|\mathfrak{m}} T_{\mathfrak{b}}(y) \sum_{\mathfrak{a}|\mathfrak{b}} \mu(\mathfrak{a}) = T_1(y).$$

If $y \in \mathfrak{a}^{-1}\mathfrak{d}^{-1}$ and $\mathfrak{a} \,|\, x$, then $xy \in \mathfrak{d}^{-1}$, so that $Tr(xy) \in \mathbb{Z}$, i.e., all summands of $S_{\mathfrak{a}}$ are 1 and there are $\#(\mathfrak{a}/\mathfrak{m}) = \mathfrak{N}(\frac{\mathfrak{m}}{\mathfrak{a}})$ of them. If on the other hand $y \notin \mathfrak{a}^{-1}\mathfrak{d}^{-1}$, then we can find in $\mathfrak{a}/\mathfrak{m}$ a class z mod $\mathfrak{m}$ such that $zy \notin \mathfrak{d}^{-1}$, i.e., $Tr(zy) \notin \mathbb{Z}$, so that $e^{2\pi i\, Tr(zy)} \neq 1$, and we obtain

$$e^{2\pi i\, Tr(zy)} S_{\mathfrak{a}}(y) = \sum_{\substack{x \bmod \mathfrak{m} \\ \mathfrak{a}|x}} e^{2\pi i\, Tr((x+z)y)} = S_{\mathfrak{a}}(y),$$

since $x + z$ varies over all the classes of $\mathfrak{a}/\mathfrak{m}$ as x does, so that we do find $S_{\mathfrak{a}}(y) = 0$. □

Proof of Theorem (6.4): Let $a \in \mathcal{O}$, $(a, \mathfrak{m}) = 1$. As x runs through a system of representatives of $(\mathcal{O}/\mathfrak{m})^*$, so does xa. We get

$$\begin{aligned} \tau_{\mathfrak{m}}(\chi_{\mathfrak{f}}, ay) &= \sum_{\substack{x \bmod \mathfrak{m} \\ (x,\mathfrak{m})=1}} \chi_{\mathfrak{f}}(x)\, e^{2\pi i\, Tr(xay)} \\ &= \overline{\chi}_{\mathfrak{f}}(a) \sum_{\substack{x \bmod \mathfrak{m} \\ (x,\mathfrak{m})=1}} \chi_{\mathfrak{f}}(xa)\, e^{2\pi i\, Tr(xay)} \\ &= \overline{\chi}_{\mathfrak{f}}(a)\tau_{\mathfrak{m}}(\chi_{\mathfrak{f}}, y). \end{aligned}$$

Let $(a, \mathfrak{m}) = \mathfrak{m}_1 \neq 1$. Since $\chi_{\mathfrak{f}}$ is primitive, we can find a class b mod $\mathfrak{m} \in (\mathcal{O}/\mathfrak{m})^*$ such that

$$\chi_{\mathfrak{f}}(b) \neq 1 \quad \text{and} \quad b \equiv 1 \bmod \frac{\mathfrak{m}}{\mathfrak{m}_1}.$$

As a consequence, $ab \equiv a \bmod \mathfrak{m}$, so that $aby - ay \in \mathfrak{d}^{-1}$, and by what we have just shown,

$$\overline{\chi}_{\mathfrak{f}}(b)\tau_{\mathfrak{m}}(\chi_{\mathfrak{f}}, ay) = \tau_{\mathfrak{m}}(\chi_{\mathfrak{f}}, bay) = \tau_{\mathfrak{m}}(\chi_{\mathfrak{f}}, ay).$$

Finally, in view of $\overline{\chi}_{\mathfrak{f}}(b) \neq 1$, we find $\tau_{\mathfrak{m}}(\chi_{\mathfrak{f}}, ay) = 0$.

As for the absolute value of the Gauss sum, we see from (6.6) that

$$\begin{aligned}
|\tau_{\mathfrak{m}}(\chi_{\mathfrak{f}}, y)|^2 &= \tau_{\mathfrak{m}}(\chi_{\mathfrak{f}}, y)\overline{\tau_{\mathfrak{m}}(\chi_{\mathfrak{f}}, y)} \\
&= \sum_{\substack{x \bmod \mathfrak{m} \\ (x,\mathfrak{m})=1}} \tau_{\mathfrak{m}}(\chi_{\mathfrak{f}}, y)\overline{\chi}_{\mathfrak{f}}(x)\, e^{-2\pi i\, Tr(xy)} \\
&= \sum_{\substack{x \bmod \mathfrak{m} \\ (x,\mathfrak{m})=1}} \tau_{\mathfrak{m}}(\chi_{\mathfrak{f}}, xy)\, e^{-2\pi i\, Tr(xy)} \\
&= \sum_{\substack{z \bmod \mathfrak{m} \\ (z,\mathfrak{m})=1}} \sum_{\substack{x \bmod \mathfrak{m} \\ (x,\mathfrak{m})=1}} \chi_{\mathfrak{f}}(z)\, e^{2\pi i\, Tr(xy(z-1))} \\
&= \sum_{\substack{z \bmod \mathfrak{m} \\ (z,\mathfrak{m})=1}} \chi_{\mathfrak{f}}(z) T_1\big(y(z-1)\big) \\
&= \sum_{\substack{z \bmod \mathfrak{m} \\ (z,\mathfrak{m})=1}} \chi_{\mathfrak{f}}(z) \sum_{\mathfrak{a}|\mathfrak{m}} \mu(\mathfrak{a}) S_{\mathfrak{a}}\big(y(z-1)\big)\,.
\end{aligned}$$

We now make use of the condition $(y\mathfrak{m}\mathfrak{d}, \mathfrak{m}) = 1$. It implies that

$$y(z-1) \in \mathfrak{a}^{-1}\mathfrak{d}^{-1} \iff z \equiv 1 \bmod \frac{\mathfrak{m}}{\mathfrak{a}}\,.$$

Indeed, if $z - 1 \in \mathfrak{a}^{-1}\mathfrak{m}$, then $y(z-1) \in \mathfrak{m}^{-1}\mathfrak{d}^{-1}\mathfrak{a}^{-1}\mathfrak{m} = \mathfrak{a}^{-1}\mathfrak{d}^{-1}$. If on the other hand $z \not\equiv 1 \bmod \frac{\mathfrak{m}}{\mathfrak{a}}$, i.e., $\frac{\mathfrak{m}}{\mathfrak{a}} \nmid (z-1)$, then $v_{\mathfrak{p}}(z-1) < v_{\mathfrak{p}}(\frac{\mathfrak{m}}{\mathfrak{a}})$ for a prime divisor $\mathfrak{p}$ of $\frac{\mathfrak{m}}{\mathfrak{a}}$. Since $(y\mathfrak{m}\mathfrak{d}, \mathfrak{m}) = 1$, we have $v_{\mathfrak{p}}(y\mathfrak{m}\mathfrak{d}) = 0$, so that $v_{\mathfrak{p}}(y) = -v_{\mathfrak{p}}(\mathfrak{m}) - v_{\mathfrak{p}}(\mathfrak{d})$ and

$$v_{\mathfrak{p}}(y(z-1)) < v_{\mathfrak{p}}(\mathfrak{m}) - v_{\mathfrak{p}}(\mathfrak{a}) + v_{\mathfrak{p}}(y) = -v_{\mathfrak{p}}(\mathfrak{a}) - v_{\mathfrak{p}}(\mathfrak{d}) = v_{\mathfrak{p}}(\mathfrak{a}^{-1}\mathfrak{d}^{-1}),$$

and thus $y(z-1) \notin \mathfrak{a}^{-1}\mathfrak{d}^{-1}$. This, together with (6.6), gives

$$|\tau_{\mathfrak{m}}(\chi_{\mathfrak{f}}, y)|^2 = \sum_{\mathfrak{a}|\mathfrak{m}} \mu(\mathfrak{a}) \mathfrak{N}\Big(\frac{\mathfrak{m}}{\mathfrak{a}}\Big) \sum_{\substack{z \bmod \mathfrak{m} \\ z\equiv 1 \bmod \mathfrak{m}/\mathfrak{a}}} \chi_{\mathfrak{f}}(z)\,.$$

For $\mathfrak{a} \neq 1$, the last character sum vanishes since $\chi_{\mathfrak{f}}$ is primitive, and therefore nonzero on the subgroup of $z \bmod \mathfrak{m} \in (\mathcal{O}/\mathfrak{m})^*$ such that $z \equiv 1 \bmod \mathfrak{m}/\mathfrak{a}$: the sum reproduces itself under multiplication with a value $\chi_{\mathfrak{f}}(x) \neq 1$ of the character. So we finally have that $|\tau_{\mathfrak{m}}(\chi_{\mathfrak{f}}, y)|^2 = \mathfrak{N}(\mathfrak{m})$. This proves all the statements of the theorem. □

Having studied the characters $\chi_{\mathfrak{f}}$ of $(\mathcal{O}/\mathfrak{m})^*$, we now turn to the characters χ_∞ of $\mathbf{R}^*$. They are given explicitly as follows.

(6.7) Proposition. *The characters λ of $\mathbf{R}^*$, i.e., the continuous homomorphisms*

$$\lambda : \mathbf{R}^* \longrightarrow S^1,$$

are given explicitly by

$$\lambda(x) = N\big(x^p|x|^{-p+iq}\big),$$

for some admissible $p \in \prod_\tau \mathbb{Z}$ *(see §3, p.448) and a* $q \in \mathbf{R}_\pm$*.* p *and* q *are uniquely determined by* λ*.*

Proof: For every $x \in \mathbf{R}^*$ we may write $x = \frac{x}{|x|}|x|$, and obtain in this way a decomposition

$$\mathbf{R}^* = \mathbf{U} \times \mathbf{R}_+^*,$$

where $\mathbf{U} = \big\{x \in \mathbf{R}^* \mid |x| = 1\big\}$. It therefore suffices to determine separately the characters of $\mathbf{U}$ and those of $\mathbf{R}_+^*$. We write ρ instead of τ for elements of $\mathrm{Hom}(K,\mathbb{C})$ to indicate that $\tau = \overline{\tau}$, and we choose an element σ from each pair $\{\tau, \overline{\tau}\}$ such that $\tau \neq \overline{\tau}$. Then we have

$$\mathbf{U} = \Big[\prod_\tau S^1\Big]^+ = \prod_\rho \{\pm 1\} \times \prod_\sigma \big[S^1 \times S^1\big]^+,$$

and $S^1 \to [S^1 \times S^1]^+$, $x_\sigma \mapsto (x_\sigma, \overline{x}_\sigma)$, is a topological isomorphism. The characters of $\{\pm 1\}$ correspond one-to-one to exponentiating by a $p_\rho \in \{0,1\}$, and the characters of S^1 correspond one-to-one to the mappings $x_\sigma \mapsto x_\sigma^k$, for $k \in \mathbb{Z}$. From the correspondence $k \mapsto (k,0)$, resp. $(0,-k)$, for $k \geq 0$, resp. $k \leq 0$, we obtain the characters of $[S^1 \times S^1]^+$ in a one-to-one way from the pairs $(p_\tau, p_{\overline{\tau}})$ with $p_\tau, p_{\overline{\tau}} \geq 0$ and $p_\tau p_{\overline{\tau}} = 0$. The characters of $\mathbf{U}$ are therefore given by

$$\lambda(x) = N(x^p),$$

with a uniquely determined admissible $p \in \prod_\tau \mathbb{Z}$.

The characters of $\mathbf{R}_+^*$ are obtained via the topological isomorphism

$$\log : \mathbf{R}_+^* \longrightarrow \mathbf{R}_\pm .$$

Writing as above

$$\mathbf{R}_\pm = \prod_\rho \mathbb{R} \times \prod_\sigma \big[\,\mathbb{R} \times \mathbb{R}\big]^+,$$

and observing the isomorphism $\big[\,\mathbb{R} \times \mathbb{R}\big]^+ \xrightarrow{\sim} \mathbb{R}$, $(x_\sigma, x_\sigma) \mapsto 2x_\sigma$, we see that a character of $\mathbf{R}_\pm$ corresponds one-to-one to a system (q_ρ, q_σ) via the rule

$$x \longmapsto \prod_\rho e^{iq_\rho x_\rho} \prod_\sigma e^{2iq_\sigma x_\sigma}.$$

It is therefore given by an element $q \in \mathbf{R}_\pm$ via $x \mapsto N(e^{iqx})$. The isomorphism log then gives a character λ of $\mathbf{R}_+^*$ via $y \mapsto N(e^{iq\log y}) =$

$N(y^{iq})$, with a uniquely determined $q \in \mathbf{R}_{\pm}$. In view of the decomposition $x = \frac{x}{|x|}|x|$, we finally obtain the characters λ of $\mathbf{R}^*$ as

$$\lambda(x) = N\Big(\Big(\frac{x}{|x|}\Big)^p |x|^{iq}\Big) = N\big(x^p|x|^{-p+iq}\big). \qquad \square$$

If the character χ_∞ associated to the *Größencharakter* $\chi : J^{\mathfrak{m}} \to S^1$ is given by

$$\chi_\infty(x) = N\big(x^p|x|^{-p+iq}\big),$$

then we say that χ is **of type** (p,q), and we call $p - iq$ the **exponent** of χ. Since χ_∞ factors through $\mathbf{R}^*/\mathcal{O}^{\mathfrak{m}}$, not all exponents actually occur (see exercise 3).

The class of all *Größencharaktere* subsumes in particular the generalized *Dirichlet characters* defined as follows. To the **module**

$$\mathfrak{m} = \prod_{\mathfrak{p}\nmid\infty} \mathfrak{p}^{n_\mathfrak{p}},$$

we associate the **ray class group** $J^{\mathfrak{m}}/P^{\mathfrak{m}}$ mod $\mathfrak{m}$ (see chap. VI, § 1). Here $J^{\mathfrak{m}}$ is the group of all ideals relatively prime to $\mathfrak{m}$, and $P^{\mathfrak{m}}$ is the group of fractional principal ideals (a) such that

$$a \equiv 1 \bmod \mathfrak{m} \quad \text{and } a \text{ totally positive.}$$

This last condition means that $\tau a > 0$ for every real embedding $\tau : K \to \mathbb{R}$.

(6.8) Definition. *A* **Dirichlet character** mod $\mathfrak{m}$ *is a character*

$$\chi : J^{\mathfrak{m}}/P^{\mathfrak{m}} \longrightarrow S^1$$

of the ray class group mod $\mathfrak{m}$, *i.e., a character* $\chi : J^{\mathfrak{m}} \to S^1$ *such that* $\chi(P^{\mathfrak{m}}) = 1$.

The **conductor** of a Dirichlet character χ mod $\mathfrak{m}$ is defined to be the smallest module $\mathfrak{f}$ dividing $\mathfrak{m}$ such that χ factors through $J^{\mathfrak{f}}/P^{\mathfrak{f}}$.

(6.9) Proposition. *The Dirichlet characters* χ mod $\mathfrak{m}$ *are precisely the Größencharaktere* mod $\mathfrak{m}$ *of type* $(p,0)$, $p = (p_\tau)$, *such that* $p_\tau = 0$ *for all complex* τ. *In other words, one has*

$$\chi((a)) = \chi_{\mathfrak{f}}(a)N\Big(\Big(\frac{a}{|a|}\Big)^p\Big),$$

for some character $\chi_{\mathfrak{f}}$ *of* $(\mathcal{O}/\mathfrak{m})^*$. *The conductor of the Dirichlet character is at the same time also the conductor of the corresponding Größencharakter.*

Proof: Let χ be a *Größencharakter* mod $\mathfrak{m}$ with corresponding characters $\chi_{\mathfrak{f}}, \chi_\infty$ of $(\mathcal{O}/\mathfrak{m})^*$, $\mathbf{R}^*/\mathcal{O}^{\mathfrak{m}}$, such that χ_∞ is of type $(p,0)$ with $p_\tau = 0$ for τ complex. For totally positive $a \in \mathcal{O}$ such that $a \equiv 1 \bmod \mathfrak{m}$, we then obviously have $\chi_{\mathfrak{f}}(a) = 1$, and $\chi_\infty(a) = 1$, and then $\chi((a)) = \chi_{\mathfrak{f}}(a)\chi_\infty(a) = 1$. Therefore χ factorizes through $J^{\mathfrak{m}}/P^{\mathfrak{m}}$, and is thus a Dirichlet character mod $\mathfrak{m}$.

Conversely, let χ be a Dirichlet character mod $\mathfrak{m}$, i.e., a character of $J^{\mathfrak{m}}$ such that $\chi(P^{\mathfrak{m}}) = 1$. Let $K^{\mathfrak{m}} = \{a \in K^* \mid a \equiv 1 \bmod \mathfrak{m}\}$, $K^{\mathfrak{m}}_+ = \{a \in K^{\mathfrak{m}} \mid a$ totally positive$\}$ and $\mathbf{R}^*_{(+)} = \{(x_\tau) \in \mathbf{R}^* \mid x_\tau > 0$ for τ real$\}$. Then we have an isomorphism

$$K^{\mathfrak{m}}/K^{\mathfrak{m}}_+ \longrightarrow \mathbf{R}^*/\mathbf{R}^*_{(+)} \cong \prod_{\mathfrak{p}\ \text{real}} \{\pm 1\}.$$

Then the composite

$$K^{\mathfrak{m}}/K^{\mathfrak{m}}_+ \xrightarrow{(\ )} J^{\mathfrak{m}}/P^{\mathfrak{m}} \xrightarrow{\chi} S^1$$

defines a character of $\mathbf{R}^*/\mathbf{R}^*_{(+)}$. It is induced by a character χ_∞ of $\mathbf{R}^*$ which – because $\chi_\infty(\mathbf{R}^*_{(+)}) = 1$ – is of the form $\chi_\infty(x) = N((\frac{x}{|x|})^p)$ with $p = (p_\tau)$, $p_\tau \in \{0,1\}$ for τ real, and $p_\tau = 0$ for τ complex. We have $\chi((a)) = \chi_\infty(a)$ for $a \in K^{\mathfrak{m}}$, and

$$\chi_{\mathfrak{f}}(a) = \chi\big((a)\big)\,\chi_\infty(a)^{-1}$$

gives us a character of $(\mathcal{O}/\mathfrak{m})^*$. Therefore χ is indeed a *Größencharakter* of the type claimed.

Let $\mathfrak{f}$ be the conductor of the Dirichlet character χ mod $\mathfrak{m}$, and let $\mathfrak{f}'$ be the conductor of the corresponding *Größencharakter* mod $\mathfrak{m}$. $\chi : J^{\mathfrak{m}}/P^{\mathfrak{m}} \to S^1$ is then induced by a character $\chi' : J^{\mathfrak{f}}/P^{\mathfrak{f}} \to S^1$, so the *Größencharakter* $\chi : J^{\mathfrak{m}} \to S^1$ mod $\mathfrak{m}$ is the restriction of the *Größencharakter* $\chi' : J^{\mathfrak{f}} \to S^1$. This implies that $\mathfrak{f}' \mid \mathfrak{f}$. On the other hand, the *Größencharakter* $\chi : J^{\mathfrak{m}} \to S^1$ is the restriction of a *Größencharakter* $\chi'' : J^{\mathfrak{f}'} \to S^1$, so $\chi_{\mathfrak{f}}$ is the composite of $(\mathcal{O}/\mathfrak{m})^* \to (\mathcal{O}/\mathfrak{f}')^* \xrightarrow{\chi''_{\mathfrak{f}}} S^1$ (see (6.2)). By the above, χ'' gives a character $J^{\mathfrak{f}'}/P^{\mathfrak{f}'} \to S^1$ such that the Dirichlet character $\chi : J^{\mathfrak{m}}/P^{\mathfrak{m}} \to S^1$ factors through $J^{\mathfrak{f}'}/P^{\mathfrak{f}'}$. Hence $\mathfrak{f} \mid \mathfrak{f}'$, so that $\mathfrak{f} = \mathfrak{f}'$. □

(6.10) Corollary. *The characters of the ideal class group $Cl_K = J/P$, i.e., the characters $\chi : J \to S^1$ such that $\chi(P) = 1$, are precisely the Größencharaktere χ* mod 1 *satisfying $\chi_\infty = 1$.*

Proof: For $\mathfrak{m} = 1$ we have $(\mathcal{O}/\mathfrak{m})^* = \{1\}$. A character χ of J/P is a *Größencharakter* mod 1. The associated character χ_{f} is trivial, so $\chi_\infty(a) = \chi_{\mathrm{f}}(a)^{-1}\chi((a)) = 1$, and thus $\chi_\infty = 1$, because K^* is dense in $\mathbf{R}^*$. If conversely χ is a *Größencharakter* mod 1 satisfying $\chi_\infty = 1$, then

$$\chi\big((a)\big) = \chi_{\mathrm{f}}(a)\chi_\infty(a) = \chi_{\mathrm{f}}(a) = 1,$$

for $a \in K^*$. Therefore $\chi(P) = 1$, and χ is a character of the ideal class group. □

To conclude this section, let us study the relation of *Größencharaktere* to characters of the idèle class group.

(6.11) Definition. **A Hecke character** *is a character of the idèle class group* $C = I/K^*$ *of the number field* K, *i.e., a continuous homomorphism*

$$\chi : I \longrightarrow S^1$$

of the idèle group $I = \prod_{\mathfrak{p}} K_{\mathfrak{p}}^*$ *such that* $\chi(K^*) = 1$.

In order to deal with Hecke characters concretely, consider an integral ideal $\mathfrak{m} = \prod_{\mathfrak{p}} \mathfrak{p}^{n_{\mathfrak{p}}}$ of K, i.e., $n_{\mathfrak{p}} \geq 0$ and $n_{\mathfrak{p}} = 0$ for $\mathfrak{p} \mid \infty$. We associate to this ideal the *subgroup* $\bar{I}^{\mathfrak{m}}$ of I,

$$\bar{I}^{\mathfrak{m}} = I_{\mathrm{f}}^{\mathfrak{m}} \times I_\infty \quad \text{where} \quad I_{\mathrm{f}}^{\mathfrak{m}} = \prod_{\mathfrak{p}\nmid\infty} U_{\mathfrak{p}}^{(n_{\mathfrak{p}})}, \quad I_\infty = \prod_{\mathfrak{p}\mid\infty} K_{\mathfrak{p}}^* .$$

If $\mathfrak{p} \nmid \infty$, then $U_{\mathfrak{p}}^{(n)}$ is the group of units $U_{\mathfrak{p}}$ if $n = 0$, and the n-th group of higher units for $n \geq 1$. We interpret I_∞ as the multiplicative group $\mathbf{R}^*$ of the $\mathbb{R}$-algebra $\mathbf{R} = K \otimes_{\mathbb{Q}} \mathbb{R} = \prod_{\mathfrak{p}\mid\infty} K_{\mathfrak{p}}$. Observe that $\bar{I}^{\mathfrak{m}}$ differs slightly from the congruence subgroup $I^{\mathfrak{m}} = \prod_{\mathfrak{p}} U_{\mathfrak{p}}^{(n_{\mathfrak{p}})}$ introduced in chap. VI, § 1, in that, for real $\mathfrak{p}$, we have the factor $U_{\mathfrak{p}}^{(0)} = \mathbb{R}_+^*$ instead of the component $K_{\mathfrak{p}}^*$. The effect is that $I/\bar{I}^{\mathfrak{m}}K^*$ is not the ray class group $J^{\mathfrak{m}}/P^{\mathfrak{m}}$ mod $\mathfrak{m}$, but isomorphic to the quotient $J^{\mathfrak{m}}/\bar{P}^{\mathfrak{m}}$ by the group $\bar{P}^{\mathfrak{m}}$ of all principal ideals (a) such that $a \equiv 1 \bmod \mathfrak{m}$ – this is seen as in chap. VI, (1.9). We will refer to $J^{\mathfrak{m}}/\bar{P}^{\mathfrak{m}}$ as the *small ray class group*.

We call $\mathfrak{m}$ a **module of definition** for the Hecke character χ if

$$\chi(I_{\mathrm{f}}^{\mathfrak{m}}) = 1 .$$

Every Hecke character admits a module of definition, since the image of $\chi : \prod_{\mathfrak{p}\nmid\infty} U_{\mathfrak{p}} \to S^1$ is a compact and totally disconnected subgroup of

S^1, hence finite, and so the kernel has to contain a subgroup of the form $\prod_{\mathfrak{p}\nmid\infty} U_{\mathfrak{p}}^{(n_{\mathfrak{p}})}$ where $n_{\mathfrak{p}} = 0$ for almost all $\mathfrak{p}$. For it we can take the ideal $\mathfrak{m} = \prod_{\mathfrak{p}\nmid\infty} \mathfrak{p}^{n_{\mathfrak{p}}}$ as a module of definition.

Since $\chi(I_{\mathrm{f}}^{\mathfrak{m}}) = 1$, the character $\chi : C = I/K^* \to S^1$ induces a character

$$\chi : C(\mathfrak{m}) \longrightarrow S^1$$

of the group

$$C(\mathfrak{m}) = I/I_{\mathrm{f}}^{\mathfrak{m}} K^* .$$

But it will not in general factor through the small ray class group $I/\bar{I}^{\mathfrak{m}} K^* \cong J^{\mathfrak{m}}/\bar{P}^{\mathfrak{m}}$ (see chap. VI, (1.7), (1.9)), which bears the following relation to $C(\mathfrak{m})$.

(6.12) Proposition. *There is an exact sequence*

$$1 \longrightarrow \mathbf{R}^*/\mathcal{O}^{\mathfrak{m}} \longrightarrow C(\mathfrak{m}) \longrightarrow J^{\mathfrak{m}}/\bar{P}^{\mathfrak{m}} \longrightarrow 1 .$$

Proof: The claim follows immediately from the two exact sequences

$$1 \longrightarrow \bar{I}^{\mathfrak{m}} K^*/I_{\mathrm{f}}^{\mathfrak{m}} K^* \longrightarrow I/I_{\mathrm{f}}^{\mathfrak{m}} K^* \longrightarrow I/\bar{I}^{\mathfrak{m}} K^* \longrightarrow 1 ,$$

$$1 \longrightarrow \bar{I}^{\mathfrak{m}} \cap K^*/I_{\mathrm{f}}^{\mathfrak{m}} \cap K^* \longrightarrow \bar{I}^{\mathfrak{m}}/I_{\mathrm{f}}^{\mathfrak{m}} \longrightarrow \bar{I}^{\mathfrak{m}} K^*/I_{\mathrm{f}}^{\mathfrak{m}} K^* \longrightarrow 1 .$$

In the second one, one has $\bar{I}^{\mathfrak{m}} \cap K^* = \mathcal{O}^{\mathfrak{m}}$, $I_{\mathrm{f}}^{\mathfrak{m}} \cap K^* = 1$ and $\bar{I}^{\mathfrak{m}}/I_{\mathrm{f}}^{\mathfrak{m}} = I_\infty = \mathbf{R}^*$, and so $\bar{I}^{\mathfrak{m}} K^*/I_{\mathrm{f}}^{\mathfrak{m}} K^* = \mathbf{R}^*/\mathcal{O}^{\mathfrak{m}}$. □

Given a Hecke character χ with module of definition $\mathfrak{m}$, we may now construct a *Größencharakter* mod $\mathfrak{m}$ as follows. For every $\mathfrak{p} \nmid \infty$, we choose a fixed prime element $\pi_{\mathfrak{p}}$ of $K_{\mathfrak{p}}$ and obtain a homomorphism

$$c : J^{\mathfrak{m}} \longrightarrow C(\mathfrak{m})$$

which maps a prime ideal $\mathfrak{p} \nmid \mathfrak{m}$ to the class of the idèle $\langle \pi_{\mathfrak{p}} \rangle = (\dots, 1, 1, \pi_{\mathfrak{p}}, 1, 1, \dots)$. This mapping does not depend on the choice of the prime elements, since the idèles $\langle u_{\mathfrak{p}} \rangle$, $u_{\mathfrak{p}} \in U_{\mathfrak{p}}$, for $\mathfrak{p} \nmid \mathfrak{m}$, lie in $I_{\mathrm{f}}^{\mathfrak{m}}$. Taking the composite map

$$J^{\mathfrak{m}} \xrightarrow{c} C(\mathfrak{m}) \xrightarrow{\chi} S^1$$

yields a 1–1 correspondence between Hecke characters with module of definition $\mathfrak{m}$ and *Größencharaktere* mod $\mathfrak{m}$. The reason for this is the following

(6.13) Proposition. *There is a canonical exact sequence*

$$1 \longrightarrow K^{(\mathfrak{m})}/\mathcal{O}^{\mathfrak{m}} \xrightarrow{\delta} J^{\mathfrak{m}} \times (\mathcal{O}/\mathfrak{m})^* \times \mathbf{R}^*/\mathcal{O}^{\mathfrak{m}} \xrightarrow{f} C(\mathfrak{m}) \longrightarrow 1,$$

where δ is given by

$$\delta(a) = \big((a)^{-1},\ a \bmod \mathfrak{m},\ a \bmod \mathcal{O}^{\mathfrak{m}}\big).$$

Proof: For every $a \in K^{(\mathfrak{m})}$, let $\widehat{a} \in I$ be the idèle with components $\widehat{a}_{\mathfrak{p}} = a$ for $\mathfrak{p} \nmid \mathfrak{m}\infty$ and $\widehat{a}_{\mathfrak{p}} = 1$ for $\mathfrak{p} \,|\, \mathfrak{m}\infty$. It is then obvious that

$$c\big((a)\big) = \widehat{a} \bmod I_{\mathrm{f}}^{\mathfrak{m}} K^*.$$

Let us decompose the principal idèle a according to its components in $I = I_{\mathrm{f}} \times I_\infty$ as a product $a = a_{\mathrm{f}} a_\infty$, and define the homomorphisms

$$\varphi : (\mathcal{O}/\mathfrak{m})^* \longrightarrow C(\mathfrak{m}), \quad \psi : \mathbf{R}^*/\mathcal{O}^{\mathfrak{m}} \longrightarrow C(\mathfrak{m})$$

by

$$\varphi(a) = \widehat{a} a_\infty \bmod I_{\mathrm{f}}^{\mathfrak{m}} K^*, \quad \psi(b) = b^{-1} \bmod I_{\mathrm{f}}^{\mathfrak{m}} K^*,$$

where every $b \in \mathbf{R}^* = I_\infty$ is considered as an idèle in I. For $a \in \mathcal{O}$, $a \equiv 1 \bmod \mathfrak{m}$, we have $a_{\mathrm{f}} \widehat{a}^{-1} \in I_{\mathrm{f}}^{\mathfrak{m}} \subseteq I$, so we get in $C(\mathfrak{m})$ the equation $\varphi(a) = [\widehat{a} a_\infty] = [a_{\mathrm{f}} a_\infty] = [a] = 1$, where [ ] indicates taking classes. This shows that φ is well-defined. For every $\varepsilon \in \mathcal{O}^{\mathfrak{m}}$, one has $\varepsilon_{\mathrm{f}} \in I_{\mathrm{f}}^{\mathfrak{m}}$, so $[\varepsilon_\infty] = [\varepsilon_\infty \varepsilon_{\mathrm{f}}] = [\varepsilon] = 1$ in $C(\mathfrak{m})$, and thus $\psi(\varepsilon_\infty) = 1$. Consequently ψ is well-defined. We now define the homomorphism

$$f : J^{\mathfrak{m}} \times (\mathcal{O}/\mathfrak{m})^* \times \mathbf{R}^*/\mathcal{O}^{\mathfrak{m}} \longrightarrow C(\mathfrak{m})$$

by

$$f\big((\mathfrak{a},\ a \bmod \mathfrak{m},\ b \bmod \mathcal{O}^{\mathfrak{m}})\big) = c(\mathfrak{a})\varphi(a)\psi(b),$$

and we show that the resulting sequence is exact. The homomorphism δ is clearly injective. For $a \in K^{(\mathfrak{m})}$ one has

$$f\big(\delta(a)\big) = c\big((a)\big)^{-1}\varphi(a)\psi(a) = \widehat{a}^{-1}\widehat{a} a_\infty a_\infty^{-1} \bmod I_{\mathrm{f}}^{\mathfrak{m}} K^* = 1,$$

so that $f \circ \delta = 1$. Conversely, let

$$f\big((\mathfrak{a},\ a \bmod \mathfrak{m},\ b \bmod \mathcal{O}^{\mathfrak{m}})\big) = c(\mathfrak{a})\varphi(a)\psi(b) = 1,$$

and let $\mathfrak{a} = \prod_{\mathfrak{p} \nmid \mathfrak{m}\infty} \mathfrak{p}^{\nu_{\mathfrak{p}}}$. Then

$$c(\mathfrak{a}) = \gamma \bmod I_{\mathrm{f}}^{\mathfrak{m}} K^*$$

for some idèle γ with components $\gamma_{\mathfrak{p}} = \pi_{\mathfrak{p}}^{\nu_{\mathfrak{p}}}$ for $\mathfrak{p} \nmid \mathfrak{m}\infty$, and $\gamma_{\mathfrak{p}} = 1$ for $\mathfrak{p} \,|\, \mathfrak{m}\infty$. This yields an identity

$$\gamma \widehat{a} a_\infty b^{-1} = \xi x \quad \text{with} \quad \xi \in I_{\mathrm{f}}^{\mathfrak{m}} \quad \text{and} \quad x \in K^*.$$

For $\mathfrak{p} \nmid \mathfrak{m}\infty$ one has $(\gamma\widehat{a}a_\infty b^{-1})_\mathfrak{p} = \pi_\mathfrak{p}^{\nu_\mathfrak{p}} a = \xi_\mathfrak{p} x$ in $K_\mathfrak{p}$, and so $\nu_\mathfrak{p} = v_\mathfrak{p}(a^{-1}x)$. For $\mathfrak{p} \mid \mathfrak{m}$ one has $(\gamma\widehat{a}a_\infty b^{-1})_\mathfrak{p} = 1 = \xi_\mathfrak{p} x$, so that $x \in U_\mathfrak{p}^{(n_\mathfrak{p})}$, and also $0 = \nu_\mathfrak{p} = v_\mathfrak{p}(a^{-1}x)$ since a is relatively prime to $\mathfrak{m}$. This gives

$$\mathfrak{a} = (ax^{-1}).$$

As $x \in U_\mathfrak{p}^{(n_\mathfrak{p})}$, one has $x \equiv 1 \bmod \mathfrak{m}$, hence

$$\varphi(ax^{-1}) = \varphi(a).$$

Finally, for $\mathfrak{p} \mid \infty$ we find $(\gamma\widehat{a}a_\infty b^{-1})_\mathfrak{p} = ab_\mathfrak{p}^{-1} = x$ in $K_\mathfrak{p}$, so that $b = a_\infty x^{-1}$, and thus

$$\psi(ax^{-1}) = \psi(b).$$

So we have

$$(\mathfrak{a},\, a \bmod \mathfrak{m},\, b \bmod \mathcal{O}^\mathfrak{m}) = \big((ax^{-1}),\ ax^{-1} \bmod \mathfrak{m},\ ax^{-1} \bmod \mathcal{O}^\mathfrak{m}\big),$$

and this shows the exactness of our sequence in the middle.

The surjectivity of f is proved as follows. Let $\alpha \bmod I_\mathrm{f}^\mathfrak{m} K^*$ be a class in $C(\mathfrak{m})$. By the approximation theorem, we may modify the representing idèle α, multiplying it by a suitable $x \in K^*$, in such a way that $\alpha_\mathfrak{p} \in U_\mathfrak{p}^{(n_\mathfrak{p})}$ for $\mathfrak{p} \mid \mathfrak{m}$. Let $\mathfrak{a} = \prod_{\mathfrak{p} \nmid \mathfrak{m}\infty} \mathfrak{p}^{v_\mathfrak{p}(\alpha_\mathfrak{p})}$. Then we have

$$c(\mathfrak{a}) = \gamma \bmod I_\mathfrak{p}^\mathfrak{m} K^*,$$

where the idèle γ has components $\gamma_\mathfrak{p} = \pi_\mathfrak{p}^{v_\mathfrak{p}(\alpha_\mathfrak{p})} = \varepsilon_\mathfrak{p}\alpha_\mathfrak{p}$, $\varepsilon_\mathfrak{p} \in U_\mathfrak{p}$, for $\mathfrak{p} \nmid \mathfrak{m}\infty$, and $\gamma_\mathfrak{p} = 1$ for $\mathfrak{p} \mid \mathfrak{m}\infty$. This gives $\gamma\alpha^{-1}\alpha_\infty \in I_\mathrm{f}^\mathfrak{m}$, and if we define $b = \alpha_\infty^{-1}$, then $f((\mathfrak{a}, 1 \bmod \mathfrak{m}, b \bmod \mathcal{O}^\mathfrak{m})) = \gamma b^{-1} \equiv \gamma\alpha_\infty \equiv \alpha \bmod I_\mathrm{f}^\mathfrak{m} K^*$. □

By the preceding proposition, the characters of $C(\mathfrak{m})$ correspond 1–1 to the characters of $J^\mathfrak{m} \times (\mathcal{O}/\mathfrak{m})^* \times \mathbf{R}^*/\mathcal{O}^\mathfrak{m}$ that vanish on $\delta(K^{(\mathfrak{m})}/\mathcal{O}^\mathfrak{m})$, i.e., to the triples χ, χ_f, χ_∞ of characters of $J^\mathfrak{m}$, resp. $(\mathcal{O}/\mathfrak{m})^*$, resp. $\mathbf{R}^*/\mathcal{O}^\mathfrak{m}$, such that

$$\chi\big((a)\big)^{-1}\chi_\mathrm{f}(a \bmod \mathfrak{m})\chi_\infty(a \bmod \mathcal{O}^\mathfrak{m}) = 1$$

for $a \in K^{(\mathfrak{m})}$. This makes χ a *Größencharakter* mod $\mathfrak{m}$, and since χ_f and χ_∞ are uniquely determined by χ, we obtain the

(6.14) Corollary. *The correspondence $\chi \mapsto \chi \circ c$ is 1–1 between characters χ of $C(\mathfrak{m})$, i.e., Hecke characters with module of definition $\mathfrak{m}$, and Größencharaktere mod $\mathfrak{m}$.*

Exercise 1. Let $\mathfrak{m} = \prod_{i=1}^r \mathfrak{m}_i$ be a decomposition of $\mathfrak{m}$ into integral ideals which are pairwise relatively prime. Then one has the decompositions

$$(\mathcal{O}/\mathfrak{m})^* \cong \prod_{i=1}^{r} (\mathcal{O}/\mathfrak{m}_i)^*$$

and

$$\mathfrak{m}^{-1}\mathfrak{d}^{-1}/\mathfrak{d}^{-1} \cong \bigoplus_{i=1}^{r} \mathfrak{m}_i^{-1}\mathfrak{d}^{-1}/\mathfrak{d}^{-1}.$$

Let χ_{f} be a character of $(\mathcal{O}/\mathfrak{m})^*$, and let $\chi_{\mathrm{f}i}$ be the characters of $(\mathcal{O}/\mathfrak{m}_i)^*$ defined by χ_{f}. If $y \in \mathfrak{m}^{-1}\mathfrak{d}^{-1}/\mathfrak{d}^{-1}$, and if $y_i \in \mathfrak{m}_i^{-1}\mathfrak{d}^{-1}/\mathfrak{d}^{-1}$ are the components of y with respect to the above decomposition, then

$$\tau_{\mathfrak{m}}(\chi_{\mathrm{f}}, y) = \prod_{i=1}^{r} \tau_{\mathfrak{m}_i}(\chi_{\mathrm{f}i}, y_i).$$

Exercise 2. Prove the **Möbius inversion formula**: let $f(\mathfrak{a})$ be any function of integral ideals $\mathfrak{a}$ with values in an additive abelian group, and let

$$g(\mathfrak{a}) = \sum_{\mathfrak{b} \mid \mathfrak{a}} f(\mathfrak{b}).$$

Then one has

$$f(\mathfrak{a}) = \sum_{\mathfrak{b} \mid \mathfrak{a}} \mu\Big(\frac{\mathfrak{a}}{\mathfrak{b}}\Big) g(\mathfrak{b}).$$

Exercise 3. Which of the characters $\lambda(x) = N(x^p|x|^{-p+iq})$ of $\mathbf{R}^*$ are characters of $\mathbf{R}^*/\mathcal{O}^{\mathfrak{m}}$?

Exercise 4. The characters of the "small ray class group" $J^{\mathfrak{m}}/\overline{P}^{\mathfrak{m}}$ mod $\mathfrak{m}$ are the *Größencharaktere* mod $\mathfrak{m}$ such that $\chi_\infty = 1$.

Exercise 5. Show that every pair of characters $\chi_{\mathrm{f}} : (\mathcal{O}/\mathfrak{m})^* \to S^1$ and $\chi_\infty : \mathbf{R}^*/\mathcal{O}^{\mathfrak{m}} \to S^1$ such that

$$\chi_{\mathrm{f}}(\varepsilon)\chi_\infty(\varepsilon) = 1 \quad \text{for all} \quad \varepsilon \in \mathcal{O}^*$$

comes from a *Größencharakter* mod $\mathfrak{m}$.

Exercise 6. Show that the homomorphism $c : J^{\mathfrak{m}} \to C(\mathfrak{m})$ is injective.

§ 7. Theta Series of Algebraic Number Fields

The group P of fractional principal ideals (a) is constituted from the elements $a \in K^*$, and it sits in the exact sequence

$$1 \longrightarrow \mathcal{O}^* \longrightarrow K^* \longrightarrow P \longrightarrow 1.$$

In order to form the theta series we will need, let us now extend K^* to a group $\widehat{K}^*$ whose elements represent *all* fractional ideals $\mathfrak{a} \in J$.

(7.1) Proposition. *There is a commutative exact diagram*

$$\begin{array}{ccccccccc} 1 & \longrightarrow & \mathcal{O}^* & \longrightarrow & K^* & \xrightarrow{(\;)} & P & \longrightarrow & 1 \\ & & \| & & \downarrow & & \downarrow & & \\ 1 & \longrightarrow & \mathcal{O}^* & \longrightarrow & \widehat{K}^* & \xrightarrow{(\;)} & J & \longrightarrow & 1 \end{array}$$

with a subgroup $\widehat{K}^* \subseteq \mathbf{C}^*$ containing K^* such that $|a| \in \mathbf{R}_+^*$, and

$$\mathfrak{N}\big((a)\big) = \big|N(a)\big|$$

for all $a \in \widehat{K}^*$.

Proof: Let the ideal class group J/P be given by a basis $[\mathfrak{b}_1], \dots, [\mathfrak{b}_r]$, and choose, for every one of these basic classes, an ideal $\mathfrak{b}_1, \dots, \mathfrak{b}_r$. Then every fractional ideal $\mathfrak{a} \in J$ can be written in the form

$$\mathfrak{a} = a\mathfrak{b}_1^{\nu_1} \cdots \mathfrak{b}_r^{\nu_r}$$

where $a \in K^*$ is well-determined up to a unit $\varepsilon \in \mathcal{O}^*$, and the exponents $\nu_i \bmod h_i$ are uniquely determined, h_i being the order of $[\mathfrak{b}_i]$ in J/P. Let $\mathfrak{b}_i^{h_i} = (b_i)$. For every $\tau \in \operatorname{Hom}(K, \mathbb{C})$, we choose a fixed root

$$\widehat{b}_{i\tau} = \sqrt[h_i]{\tau b_i}$$

in $\mathbb{C}$ in such a way that $\widehat{b}_{i\bar\tau} = \overline{\widehat{b}_{i\tau}}$ whenever τ is complex. We define $\widehat{K}^*$ to be the subgroup of $\mathbf{C}^*$ generated by K^* and by the elements $\widehat{b}_i = (\widehat{b}_{i\tau}) \in \mathbf{C}^*$. Each class $[\mathfrak{b}] \in J/P$ contains a uniquely determined ideal of the form

$$\mathfrak{b} = \mathfrak{b}_1^{\nu_1} \cdots \mathfrak{b}_r^{\nu_r} \quad \text{with} \quad 0 \le \nu_i < h_i\,,$$

and we consider the mapping

$$f : J/P \longrightarrow \widehat{K}^*/K^*, \quad f([\mathfrak{b}]) = \widehat{b}_1^{\nu_1} \cdots \widehat{b}_r^{\nu_r} \bmod K^*.$$

It is a homomorphism, for if $\mathfrak{b} = \mathfrak{b}_1^{\nu_1} \cdots \mathfrak{b}_r^{\nu_r}$ and $\mathfrak{b}' = \mathfrak{b}_1^{\nu_1'} \cdots \mathfrak{b}_r^{\nu_r'}$, and if $\nu_i + \nu_i' = \mu_i + \lambda_i h_i$, $0 \le \mu_i < h_i$, then $\mathfrak{b}_1^{\mu_1} \cdots \mathfrak{b}_r^{\mu_r}$ is the ideal belonging to the class $[\mathfrak{b}][\mathfrak{b}']$, and

$$\begin{aligned} f([\mathfrak{b}][\mathfrak{b}']) &= \widehat{b}_1^{\mu_1} \cdots \widehat{b}_r^{\mu_r} \equiv \widehat{b}_1^{\mu_1} \cdots \widehat{b}_r^{\mu_r} b_1^{\lambda_1} \cdots b_r^{\lambda_r} \\ &\equiv (\widehat{b}_1^{\nu_1} \cdots \widehat{b}_r^{\nu_r})(\widehat{b}_1^{\nu_1'} \cdots \widehat{b}_r^{\nu_r'}) \bmod K^* = f([\mathfrak{b}])f([\mathfrak{b}']). \end{aligned}$$

f is clearly surjective. To show the injectivity, let $\widehat{b}_1^{\nu_1} \cdots \widehat{b}_r^{\nu_r} = a \in K^*$, and let $h = h_1 \cdots h_r$ be the class number of K. Then we have for the ideal $\mathfrak{a} = a^{-1}\mathfrak{b}_1^{\nu_1}\cdots\mathfrak{b}_r^{\nu_r} \in J$ that $\mathfrak{a}^h = a^{-h}(b_1^{\nu_1 h/h_1}\cdots b_r^{\nu_r h/h_r}) = a^{-h}(\widehat{b}_1^{\nu_1}\cdots\widehat{b}_r^{\nu_r})^h = (1)$. Since J is torsion-free, it follows that $\mathfrak{a} = (1)$, and so $\mathfrak{b}_1^{\nu_1}\cdots\mathfrak{b}_r^{\nu_r} = (a) \in P$. From this we deduce that every element $\widehat{a} \in \widehat{K}^*$ admits a unique representation

$$\widehat{a} = a\widehat{b}_1^{\nu_1}\cdots\widehat{b}_r^{\nu_r}, \quad 0 \le \nu_i < h_i, \quad a \in K^*.$$

We define a map

$$(\ ) : \widehat{K}^* \longrightarrow J$$

by

$$\widehat{a} = a\widehat{b}_1^{\nu_1}\cdots\widehat{b}_r^{\nu_r} \longmapsto (\widehat{a}) = a\mathfrak{b}_1^{\nu_1}\cdots\mathfrak{b}_r^{\nu_r}.$$

Arguing as above, we see that this is a homomorphism. It is surjective and obviously has kernel $\mathcal{O}^*$. Finally we have that $|\widehat{b}_i| = (|\widehat{b}_{i\tau}|) \in \mathbf{R}_+^*$ and

$$\mathfrak{N}\big((\widehat{b}_i)\big)^{h_i} = \mathfrak{N}(\mathfrak{b}_i^{h_i}) = \big|N(b_i)\big| = \Big|\prod_\tau \tau b_i\Big| = \Big|\prod_\tau \widehat{b}_{i\tau}^{h_i}\Big| = \big|N(\widehat{b}_i)\big|^{h_i},$$

so that $\mathfrak{N}((\widehat{b}_i)) = |N(\widehat{b}_i)|$, and thus $|a| \in \mathbf{R}_+^*$, $\mathfrak{N}((a)) = |N(a)|$ for all $a \in \widehat{K}^*$. □

The elements a of $\widehat{K}^*$ used to be called **ideal numbers** – a name which is somewhat forgotten but will be used in what follows. The diagram (7.1) implies an isomorphism

$$\widehat{K}^*/K^* \cong J/P.$$

For $a, b \in \widehat{K}^*$ we write $a \sim b$ if a and b lie in the same class, i.e., if $ab^{-1} \in K^*$. We call a an **ideal integer**, or an integral ideal number, if (a) is an integral ideal. The semigroup of all ideal integers will be denoted by $\widehat{\mathcal{O}}$. Furthermore we write $a \,|\, b$ if $\frac{b}{a} \in \widehat{\mathcal{O}}$, and for every pair $a, b \in \widehat{K}^*$, we have the notion of $\gcd(a, b) \in \widehat{K}^*$ (which is lacking inside K^*). The greatest common divisor is the ideal number d (which is unique up to a unit) such that the ideal (d) is the gcd of the ideals $(a), (b)$. Observe that the ideal numbers are not defined in a canonical way. This is the reason why they have not been able to hold their own in the development of number theory. (They are treated in [46], [65].)

We now form an analogous extension of the prime residue groups $(\mathbb{Z}/m\mathbb{Z})^*$. For three ideal numbers a, b, m, the congruence

$$a \equiv b \bmod m$$

signifies that $a \sim b$ and $\frac{a-b}{m} \in \widehat{\mathcal{O}} \cup \{0\}$. If $\mathfrak{m} = (m)$, we also write this relation as $a \equiv b \bmod \mathfrak{m}$. Let $\mathfrak{m}$ be an integral ideal. The semigroup $\widehat{\mathcal{O}}^{(\mathfrak{m})}$ of all integral ideal numbers relatively prime to $\mathfrak{m}$ is partitioned by the equivalence relation $\equiv$ into classes, which we will write as $a \bmod \mathfrak{m}$. They are given explicitly as follows.

(7.2) Lemma. *For every* $a \in \widehat{\mathcal{O}}^{(\mathfrak{m})}$ *one has*

$$a \bmod \mathfrak{m} = a + a(a^{-1})\mathfrak{m}.$$

Proof: Let $b \in a \bmod \mathfrak{m}$, $b \neq a$, i.e., $b = a\alpha$ for some $\alpha \in K^*$, $\alpha \neq 1$, and $b - a = cm$, $c \in \widehat{\mathcal{O}}$. Then

$$a^{-1}(b-a) = \alpha - 1 \in (\alpha - 1) = (a^{-1})(c)(m) \subseteq (a^{-1})\mathfrak{m},$$

so that $b \in a + a(a^{-1})\mathfrak{m}$. Let conversely $b \in a + a(a^{-1})\mathfrak{m}$, $b \neq a$, and thus $b/a = \alpha \in 1 + (a^{-1})\mathfrak{m}$. Then one has $b \sim a$ and $(b - a) = (a)(\alpha - 1) \subseteq (a)(a^{-1})\mathfrak{m} = (m)$, i.e., $m \,|\, b - a$ and therefore $b \equiv a \bmod \mathfrak{m}$. □

We now consider the set

$$(\widehat{\mathcal{O}}/\mathfrak{m})^* := \left\{ a \bmod \mathfrak{m} \;\middle|\; a \in \widehat{\mathcal{O}}^{(\mathfrak{m})} \right\}$$

of all equivalence classes in the semigroup $\widehat{\mathcal{O}}^{(\mathfrak{m})}$ of ideal integers prime to $\mathfrak{m}$.

(7.3) Proposition. $(\widehat{\mathcal{O}}/\mathfrak{m})^*$ *is an abelian group, and we have a canonical exact sequence*

$$1 \longrightarrow (\mathcal{O}/\mathfrak{m})^* \longrightarrow (\widehat{\mathcal{O}}/\mathfrak{m})^* \longrightarrow J/P \longrightarrow 1.$$

Proof: For $a, b \in \widehat{\mathcal{O}}^{(\mathfrak{m})}$, the class $ab \bmod \mathfrak{m}$ only depends on the classes $a \bmod \mathfrak{m}$, $b \bmod \mathfrak{m}$, so we get a well-defined product in $(\widehat{\mathcal{O}}/\mathfrak{m})^*$. Every class $a \bmod \mathfrak{m}$ has an inverse. Indeed, since $(a) + \mathfrak{m} = \mathcal{O}$, we may write $1 = \alpha + \mu$, $0 \neq \alpha \in (a)$, $\mu \in \mathfrak{m}$. Consequently $a \,|\, \alpha$, so that $\alpha = ax$, $x \in \widehat{\mathcal{O}}^{(\mathfrak{m})}$, and since $1 \in \alpha(1 + \alpha^{-1}\mathfrak{m}) = \alpha \bmod \mathfrak{m}$, we see that $ax \bmod \mathfrak{m}$ is the unit class, i.e., $x \bmod \mathfrak{m}$ is inverse to $a \bmod \mathfrak{m}$.

The right-hand arrow in the sequence is induced by $a \mapsto (a)$. It is surjective since every class of J/P contains an integral ideal relatively prime to $\mathfrak{m}$. If the class $a \bmod \mathfrak{m} = a(1 + (a)^{-1}\mathfrak{m})$ is mapped to 1, then one has $(a) \in P$, and so $a \in \mathcal{O}$, $(a, \mathfrak{m}) = 1$. Hence $a \bmod \mathfrak{m} = a + \mathfrak{m}$ is a unit in $\mathcal{O}/\mathfrak{m}$. The injectivity of the arrow on the left is completely trivial, i.e., we have shown the exactness. □

For an ideal class $\mathfrak{K} \in J/P$, we will denote by $\mathfrak{K}' \in J/P$ in what follows the class defined by

$$\mathfrak{K}\mathfrak{K}' = [\mathfrak{m}\mathfrak{d}],$$

where $\mathfrak{d}$ is the different of $K|\mathbb{Q}$. Let $\mathfrak{m} = (m)$ and $\mathfrak{d} = (d)$, with some fixed ideal numbers m, d. For $\mathfrak{m} = \mathcal{O}$ let $m = 1$. We now study characters

$$\chi : (\widehat{\mathcal{O}}/\mathfrak{m})^* \longrightarrow \mathbb{C}^*,$$

and put $\chi(a) = 0$ for $a \in \mathcal{O}$ such that $(a, \mathfrak{m}) \neq 1$. In the applications, χ will come from a *Größencharakter* mod $\mathfrak{m}$, but the treatment of the theta series is independent of such an origin of χ.

(7.4) Definition. *Let $a \in \widehat{\mathcal{O}}$ be an ideal integer, and let $\mathfrak{K}$ be the class of (a). Then we define the* **Gauss sum**

$$\tau(\chi, a) = \sum_{\hat{x} \bmod \mathfrak{m}} \chi(\widehat{x})\, e^{2\pi i\, Tr(\hat{x}a/md)},$$

where $\widehat{x}$ mod $\mathfrak{m}$ runs through the classes of $(\widehat{\mathcal{O}}/\mathfrak{m})^*$ which are mapped to the class $\mathfrak{K}'$. In particular, we put $\tau(\chi) = \tau(\chi, 1)$.

The Gauss sum $\tau(\chi, a)$ reduces immediately to the one considered in §6,

$$\tau_{\mathfrak{m}}(\chi, y) = \sum_{\substack{x \bmod \mathfrak{m} \\ (x, \mathfrak{m})=1}} \chi(x)\, e^{2\pi i\, Tr(xy)}.$$

In fact, on the one hand we have

$$y = \widehat{x}a/md \in \mathfrak{m}^{-1}\mathfrak{d}^{-1},$$

since the class of the ideal $(y) = (a)(\widehat{x})(m)^{-1}(d)^{-1}$ is the principal class $\mathfrak{K}\mathfrak{K}'\mathfrak{m}^{-1}\mathfrak{d}^{-1}$, so $y \in K^*$, and one finds

$$y \in (y) = (a\widehat{x})\mathfrak{m}^{-1}\mathfrak{d}^{-1} \subseteq \mathfrak{m}^{-1}\mathfrak{d}^{-1},$$

because a and $\widehat{x}$ are integral. On the other hand, if $\widehat{x}$ mod $\mathfrak{m}$ is a fixed class of $(\widehat{\mathcal{O}}/\mathfrak{m})^*$ which maps to $\mathfrak{K}'$, then, in view of (7.3), we get the others by $\widehat{x}x$ mod $\mathfrak{m}$, with x mod $\mathfrak{m}$ varying over the classes of $(\mathcal{O}/\mathfrak{m})^*$. Therefore

$$\tau(\chi, a) = \chi(\widehat{x})\tau_{\mathfrak{m}}(\chi, y),$$

and in particular

$$\tau(\chi) = \chi(\widehat{x})\tau_{\mathfrak{m}}(\chi, y)$$

with $y = \widehat{x}/md$, which satisfies $(y\mathfrak{m}\mathfrak{d}, \mathfrak{m}) = 1$ since $y\mathfrak{m}\mathfrak{d} = (\widehat{x})$ and $((\widehat{x}), \mathfrak{m}) = 1$. Consequently, $\tau(\chi, a)$ does not depend on the choice of representatives $\widehat{x}$, and theorem (6.4) yields at once the

(7.5) Proposition. *For a primitive character χ of $(\widehat{\mathcal{O}}/\mathfrak{m})^*$, one has*

$$\tau(\chi, a) = \overline{\chi}(a)\tau(\chi)$$

and $|\tau(\chi)| = \sqrt{\mathfrak{N}(\mathfrak{m})}$.

The theta series $\theta(\chi, z)$ used in §2 in the treatment of Dirichlet L-series are attached to the field $\mathbb{Q}$. We now have to find their analogues relative to an arbitrary number field K. Given any admissible element $p \in \prod_\tau \mathbb{Z}$ (see §3, p. 448) and a character χ of $(\widehat{\mathcal{O}}/\mathfrak{m})^*$, we form the **Hecke theta series**

$$\theta^p(\chi, z) = \sum_{a \in \hat{\mathcal{O}} \cup \{0\}} \chi(a) N(a^p) e^{\pi i \langle az/|md|, a\rangle},$$

where m, d are fixed ideal numbers such that $(m) = \mathfrak{m}$ and $(d) = \mathfrak{d}$. We take $m = 1$ if $\mathfrak{m} = 1$. The case $\mathfrak{m} = 1$, $p = 0$ is exceptional in that the constant term of the theta series is $\chi(0)N(0^p) = 1$, whereas it is 0 in all other cases.

Let us decompose the theta series according to the ideal classes $\mathfrak{K} \in J/P$ into **partial Hecke theta series**

$$\theta^p(\mathfrak{K}, \chi, z) = \sum_{a \in (\hat{\mathfrak{K}} \cap \hat{\mathcal{O}}) \cup \{0\}} \chi(a) N(a^p) e^{\pi i \langle az/|md|, a\rangle},$$

where a varies over all ideal integers in the class $\widehat{\mathfrak{K}} \in \widehat{K}^*/K^*$ which corresponds to the ideal class $\mathfrak{K}$ under the isomorphism $\widehat{K}^*/K^* \cong J/P$. For these partial theta series, we want to deduce a transformation formula, and to this end we decompose them further into theta series for which we have the general transformation formula (3.6) at our disposal.

Let $\mathfrak{a}$ be an integral ideal relatively prime to $\mathfrak{m}$ which belongs to the class $\mathfrak{K}$, and let $a \in \widehat{\mathcal{O}}^{(\mathfrak{m})}$ be an ideal number such that $(a) = \mathfrak{a}$.

(7.6) Lemma. *Assume that* $\mathfrak{m} \neq 1$ *or* $p \neq 0$. *If* $x \bmod \mathfrak{m}$ *varies over the classes of* $(\mathcal{O}/\mathfrak{m})^*$, *then one has*

$$\theta^p(\mathfrak{K}, \chi, z) = \chi(a) N(a^p) \sum_{x \bmod \mathfrak{m}} \chi(x) \theta^p_\Gamma\big(x, 0, z|a^2/md|\big),$$

where Γ *is the lattice* $\mathfrak{m}/\mathfrak{a} \subseteq \mathbf{R}$ *and*

$$\theta^p_\Gamma(x, 0, z) = \sum_{g \in \Gamma} N\big((x+g)^p\big)\, e^{\pi i \langle (x+g)z, x+g\rangle}.$$

Proof: In the theta series $\theta^p(\mathfrak{K}, \chi, z)$, it suffices to sum over the elements of $\widehat{\mathfrak{K}} \cap \mathcal{O}^{(\mathfrak{m})}$ because χ is zero on the others. Every class $\widehat{x} \bmod \mathfrak{m} \in (\widehat{\mathcal{O}}/\mathfrak{m})^*$ is either disjoint from $\widehat{\mathfrak{K}}$ or else it is contained in $\widehat{\mathfrak{K}}$. In view of the exact sequence (7.3)

$$1 \longrightarrow (\mathcal{O}/\mathfrak{m})^* \longrightarrow (\widehat{\mathcal{O}}/\mathfrak{m})^* \longrightarrow J/P \longrightarrow 1,$$

the classes

$$ax \bmod \mathfrak{m} = a(x + \mathfrak{a}^{-1}\mathfrak{m})$$

are the different residue classes of $(\widehat{\mathcal{O}}/\mathfrak{m})^*$ contained in $\widehat{\mathfrak{K}}$. This gives

$$\begin{aligned}\theta^p(\mathfrak{K},\chi,z) &= \sum_{x \bmod \mathfrak{m}} \sum_{g\in\Gamma} \chi(ax) N\big((ax+ag)^p\big)\, e^{\pi i \langle a(x+g)z/|md|,\, a(x+g)\rangle} \\ &= \chi(a)N(a^p) \sum_{x \bmod \mathfrak{m}} \chi(x) \sum_{g\in\Gamma} N\big((x+g)^p\big)\, e^{\pi i \langle (x+g)z|a^2/md|,\, x+g\rangle} \\ &= \chi(a)N(a^p) \sum_{x \bmod \mathfrak{m}} \chi(x)\theta^p_\Gamma\big(x,0,z|a^2/md|\big)\,. \qquad \square\end{aligned}$$

For any admissible element $p=(p_\tau)$, we will write $\overline{p}$ for the admissible element with components $\overline{p}_\tau = p_{\overline{\tau}}$. From the transformation formula (3.6) for the series θ^p_Γ and proposition (7.5) on Gauss sums, we now obtain the

(7.7) Theorem. *For a primitive character χ of $(\widehat{\mathcal{O}}/\mathfrak{m})^*$, one has the transformation formula*

$$\theta^p(\mathfrak{K},\chi,-1/z) = W(\chi,\overline{p})N\big((z/i)^{p+\frac12}\big)\theta^{\overline{p}}(\mathfrak{K}',\overline{\chi},z)$$

with the constant factor

$$W(\chi,\overline{p}) = \Big[i^{Tr(\overline{p})}N\Big(\Big(\frac{md}{|md|}\Big)^{\overline{p}}\Big)\Big]^{-1}\frac{\tau(\chi)}{\sqrt{\mathfrak{N}(\mathfrak{m})}}\,.$$

This factor has absolute value $|W(\chi,\overline{p})|=1$.

Proof: The lattice Γ' dual to the lattice $\Gamma=\mathfrak{m}/\mathfrak{a}\subseteq\mathbf{R}$ is given, according to (5.7), by ${}^*\Gamma' = \mathfrak{a}/\mathfrak{m}\mathfrak{d}$. (Here as in §4, the asterisk signifies adjunction with respect to $\langle\ ,\ \rangle$, i.e., $\langle x,\alpha y\rangle = \langle {}^*\alpha x, y\rangle$.) The volume of the fundamental mesh of Γ is by chap. I, (5.2),

$$\operatorname{vol}(\Gamma) = \mathfrak{N}(\mathfrak{m}/\mathfrak{a})\sqrt{|d_K|} = N\big(|m/a|\big)\,N\big(|d|\big)^{1/2}.$$

From (3.6) we now get

$$(1)\qquad \theta^p_\Gamma\big(x,0,-1/|md/a^2|z\big) = A(z)\theta^p_{\Gamma'}\big(0,x,z|md/a^2|\big),$$

with the factor

$$\begin{aligned}A(z) &= \big[i^{Tr(p)}N(|m/a|)N(|d|)^{1/2}\big]^{-1}N\big((|md/a^2|z/i)^{p+\frac12}\big)\\ &= \big[i^{Tr(p)}\sqrt{\mathfrak{N}(\mathfrak{m})}\,\big]^{-1}N\big(|md/a^2|^p\big)\,N\big((z/i)^{p+\frac12}\big)\end{aligned}$$

and the series

$$(2)\qquad \theta^p_{\Gamma'}\big(0,x,z|md/a^2|\big) = \sum_{g'\in\Gamma'} N(g'^p)e^{2\pi i\langle x,g'\rangle}\,e^{\pi i\langle g'z|md/a^2|,\,g'\rangle}\,.$$

Writing $g' = \frac{{}^*g}{{}^*(md/a)}$, the rules stated in § 3 give

$$\langle x, g'\rangle = Tr(axg/md),$$

$$\langle g'|md/a^2|z, g'\rangle = \langle {}^*gz|md/a^2|/|md/a|^2, {}^*g\rangle = \langle gz/|md|, g\rangle$$

and $N(({}^*g)^p) = N(g^{\overline{p}})$. If g' varies over the lattice Γ', then g varies over the set

$$(md/a)^*\Gamma' = (md/a)\mathfrak{a}(\mathfrak{m}\mathfrak{d})^{-1} = (\widehat{\mathfrak{K}}' \cap \widehat{\mathcal{O}}) \cup \{0\}.$$

Substituting all this into (2) yields

$$\text{(3)} \quad \theta^p_{\Gamma'}\big(0, x, z|md/a^2|\big) = N\left(\left(\frac{a}{md}\right)^{\overline{p}}\right) \sum_{g\in(\hat{\mathfrak{K}}'\cap\hat{\mathcal{O}})\cup\{0\}} N(g^{\overline{p}})\, e^{2\pi i\, Tr(axg/md)}\, e^{\pi i\langle gz/|md|, g\rangle}.$$

Let us now consider first the special case $\mathfrak{m} = 1$, $p = 0$ (which was essentially treated already in § 5). In this case, we have $(\hat{\mathfrak{K}} \cap \hat{\mathcal{O}}) \cup \{0\} = \{ag \mid g \in K,\ (ag) \subseteq \mathcal{O}\} = a\mathfrak{a}^{-1} = a\Gamma$. Consequently

$$\theta^p(\mathfrak{K}, \chi, z) = \sum_{g\in\Gamma} e^{\pi i\langle agz/|d|, ag\rangle} = \sum_{g\in\Gamma} e^{\pi i\langle gz|a^2/d|, g\rangle} = \theta_\Gamma\big(z|a^2/d|\big),$$

$$\theta^{\overline{p}}(\mathfrak{K}', \overline{\chi}, z) = \sum_{g\in(\hat{\mathfrak{K}}'\cap\hat{\mathcal{O}})\cup\{0\}} e^{\pi i\langle gz/|d|, g\rangle} = \theta_{\Gamma'}\big(z|d/a^2|\big).$$

Equation (1) thus becomes

$$\theta^p(\mathfrak{K}, \chi, -1/z) = N(z/i)^{\frac{1}{2}}\theta^{\overline{p}}(\mathfrak{K}', \overline{\chi}, z).$$

Now assume $\mathfrak{m} \neq 1$ or $p \neq 0$. Then we have $\chi(0)N(0^{\overline{p}}) = 0$. Substituting (3) into (1) and (1) into formula (7.6), with $-1/z$ instead of z, we obtain

$$\theta^p(\mathfrak{K}, \chi, -1/z) = N(a^p) \sum_{x \bmod \mathfrak{m}} \chi(ax)\theta^p_\Gamma\big(x, 0, -1/z|md/a^2|\big)$$

$$= B(z) \sum_{g\in\hat{\mathfrak{K}}'\cap\hat{\mathcal{O}}} N(g^{\overline{p}})\Big(\sum_{x \bmod \mathfrak{m}} \chi(ax)\, e^{2\pi i\, Tr(axg/md)}\Big)\, e^{\pi i\langle gz/|md|, g\rangle}$$

with the factor

$$B(z) = A(z)\frac{N(a^p)}{N((md/a)^{\overline{p}})}.$$

Now consider the sum in parentheses. If x varies over a system of representatives of $(\mathcal{O}/\mathfrak{m})^*$, then ax varies over a system of representatives of those classes of $(\widehat{\mathcal{O}}/\mathfrak{m})^*$ which are mapped under $(\widehat{\mathcal{O}}/\mathfrak{m})^* \to J/P$ to the class $\mathfrak{K}$. Furthermore, (g) is an integral ideal in the class $\mathfrak{K}'$, and since $\mathfrak{K}'$

bears the same relation $\mathfrak{K}'\mathfrak{K} = [\mathfrak{m}\mathfrak{d}]$ to $\mathfrak{K}$ as $\mathfrak{K}$ does to $\mathfrak{K}'$, we recognize the sum in question as the *Gauss sum*

$$\tau(\chi, g) = \sum_{x \bmod \mathfrak{m}} \chi(ax) e^{2\pi i\, Tr(axg/md)}.$$

Substituting in now the result (7.5),

$$\tau(\chi, g) = \overline{\chi}(g)\tau(\chi),$$

we finally arrive at the identity

$$\text{(4)} \qquad \theta^p(\mathfrak{K}, \chi, -1/z) = W(\chi, \overline{p}) N\big((z/i)^{p+\frac{1}{2}}\big) \theta^{\overline{p}}(\mathfrak{K}', \overline{\chi}, z)$$

with the factor

$$\begin{aligned} W(\chi, \overline{p}) &= \big[i^{Tr(p)}\sqrt{\mathfrak{N}(\mathfrak{m})}\,\big]^{-1} N\big(|md/a^2|^p\big) \frac{N(a^p)\tau(\chi)}{N((md/a)^{\overline{p}})} \\ &= \frac{\tau(\chi)}{i^{Tr(p)}\sqrt{\mathfrak{N}(\mathfrak{m})}} N\Big(\Big(\frac{|md|}{md}\Big)^{\overline{p}}\Big) N\Big(\frac{a^{p*}a^p}{|a|^{2p}}\Big) \\ &= \frac{\tau(\chi)}{\sqrt{\mathfrak{N}(\mathfrak{m})}} \Big[i^{Tr(\overline{p})} N\Big(\Big(\frac{md}{|md|}\Big)^{\overline{p}}\Big)\Big]^{-1}, \end{aligned}$$

where one has to observe that $Tr(p) = Tr(\overline{p})$, $a^{\overline{p}} = {}^*a^p$, $a\,{}^*a = |a|^2$, and $|md|^p = ({}^*|md|)^p = |md|^{\overline{p}}$ because $|md| \in \mathbf{R}_+^*$. Since $|\tau(\chi)| = \sqrt{\mathfrak{N}(\mathfrak{m})}$, we have $|W(\chi, \overline{p})| = 1$. □

If $\mathfrak{m} \neq 1$ or $p \neq 0$, we find for the special theta series:

$$\theta^p(\chi, z) = \sum_{a \in \hat{\mathcal{O}}} \chi(a) N(a^p) e^{\pi i \langle az/|md|, a\rangle} = \sum_{\mathfrak{K}} \theta^p(\mathfrak{K}, \chi, z),$$

and (7.7) yields the

(7.8) Corollary. $\theta^p(\chi, -1/z) = W(\chi, \overline{p}) N((z/i)^{p+\frac{1}{2}}) \theta^{\overline{p}}(\overline{\chi}, z)$.

We recommend that the reader who has studied the above proof allow himself a moment of contemplation. Looking back, he will realize the peculiar way in which almost all fundamental arithmetic properties of the number field K have been used. First they served to break up the theta series, then these constituents were reshuffled by the analytic transformation law, but in the end they are reassembled to form a new theta series. Having contemplated this, the reader should reflect upon the admirable simplicity of the theta formula which encapsulates all these aspects of the arithmetic of the number field.

There is however one important fundamental law of number theory which does not enter into this formula, that is, **Dirichlet's unit theorem**. This will play an essential rôle when we now pass from theta series to L-series in the next section.

Exercise 1. Define ideal prime numbers and show that unique prime factorization holds in $\widehat{K}^*$.

Exercise 2. Let $\widehat{\mathcal{O}}$ be the semigroup of all ideal integers. If $d = (a, b)$ is the gcd of a, b, then there exist elements $x, y \in \widehat{\mathcal{O}} \cup \{0\}$ such that

$$d = xa + yb.$$

Furthermore, we have $x \sim d/a$, resp. $y \sim d/b$, unless $x = 0$, resp. $y = 0$. Here the notation $\alpha \sim \beta$ means $\alpha\beta^{-1} \in K^*$.

Exercise 3. The congruence $ax \equiv b \bmod m$ has a solution in $\widehat{\mathcal{O}}$ with integral x if and only if $(a, m)|b$. This solution is unique $\bmod m$, provided $(a, m) = 1$.

Exercise 4. A system of finitely many congruences with pairwise relatively prime moduli is simultaneously solvable if every congruence is solvable individually in such a way that the individual solutions are equivalent (with respect to $\sim$).

Exercise 5. If $a, m \in \widehat{\mathcal{O}}$, then there exists in every residue class $\bmod m$ prime to m, an ideal integer prime to a.

Exercise 6. For the factor group $J^{\mathfrak{m}}/\bar{P}^{\mathfrak{m}}$ by the group $\bar{P}^{\mathfrak{m}}$ of all principal ideals (a) such that $a \equiv 1 \bmod \mathfrak{m}$, one has the exact sequence

$$1 \to \mathcal{O}^*/\mathcal{O}^{\mathfrak{m}} \to (\widehat{\mathcal{O}}/\mathfrak{m})^* \to J^{\mathfrak{m}}/\bar{P}^{\mathfrak{m}} \to 1,$$

where $\mathcal{O}^{\mathfrak{m}} = \{\varepsilon \in \mathcal{O}^* \mid \varepsilon \equiv 1 \bmod \mathfrak{m}\}$.

Exercise 7. Let $\widehat{K}^{(\mathfrak{m})}$ be the preimage of $J^{\mathfrak{m}}$ under $\widehat{K}^* \to J$, and let $K^{\mathfrak{m}} = \{a \in K^* \mid a \equiv 1 \bmod \mathfrak{m}\}$. Then one has $(\widehat{\mathcal{O}}/\mathfrak{m})^* = \widehat{K}^{(\mathfrak{m})}/K^{\mathfrak{m}}$.

§ 8. Hecke L-series

Let $\mathfrak{m}$ be again an integral ideal of the number field K and let

$$\chi : J^{\mathfrak{m}} \longrightarrow S^1$$

be a character of the group of ideals relatively prime to $\mathfrak{m}$. With respect to this character, we form the L-series

$$L(\chi, s) = \sum_{\mathfrak{a}} \frac{\chi(\mathfrak{a})}{\mathfrak{N}(\mathfrak{a})^s},$$

where $\mathfrak{a}$ varies over the integral ideals of K and we put $\chi(\mathfrak{a}) = 0$ whenever $(\mathfrak{a}, \mathfrak{m}) \neq 1$. Then the following proposition holds in complete generality.

(8.1) Proposition. *The L-series $L(\chi, s)$ converges absolutely and uniformly in the domain* $\mathrm{Re}(s) \geq 1+\delta$, *for all* $\delta > 0$, *and one has*

$$L(\chi, s) = \prod_{\mathfrak{p}} \frac{1}{1-\chi(\mathfrak{p})\mathfrak{N}(\mathfrak{p})^{-s}},$$

where $\mathfrak{p}$ *varies over the prime ideals of* K.

Proof: Taking formally the logarithm of the product

$$E(s) = \prod_{\mathfrak{p}} \frac{1}{1-\chi(\mathfrak{p})\mathfrak{N}(\mathfrak{p})^{-s}}$$

gives the series

$$\log E(s) = \sum_{\mathfrak{p}} \sum_{n=1}^{\infty} \frac{\chi(\mathfrak{p})^n}{n\mathfrak{N}(\mathfrak{p})^{ns}}.$$

It converges absolutely and uniformly for $\mathrm{Re}(s) = \sigma \geq 1+\delta$. In fact, since $|\chi(\mathfrak{p})| \leq 1$, and $|\mathfrak{N}(\mathfrak{p})^s| = |\mathfrak{N}(\mathfrak{p})|^{\sigma} \geq p^{f_{\mathfrak{p}}(1+\delta)} \geq p^{1+\delta}$, and since $\#\{\mathfrak{p}|p\} \leq d = [K:\mathbb{Q}]$, it admits the following convergent upper bound which is independent of s:

$$\sum_{p,n} \frac{d}{np^{n(1+\delta)}} = d\log\zeta(1+\delta).$$

This shows that the product

$$E(s) = \prod_{\mathfrak{p}} \frac{1}{1-\chi(\mathfrak{p})\mathfrak{N}(\mathfrak{p})^{-s}} = \exp\Big(\sum_{\mathfrak{p}}\Big(\sum_{n=1}^{\infty} \frac{\chi(\mathfrak{p})^n}{n\mathfrak{N}(\mathfrak{p})^s}\Big)\Big)$$

is absolutely and uniformly convergent for $\mathrm{Re}(s) \geq 1+\delta$. Now develop in this product the factors

$$\frac{1}{1-\chi(\mathfrak{p})\mathfrak{N}(\mathfrak{p})^{-s}} = 1 + \frac{\chi(\mathfrak{p})}{\mathfrak{N}(\mathfrak{p})^s} + \frac{\chi(\mathfrak{p})^2}{\mathfrak{N}(\mathfrak{p})^{2s}} + \cdots$$

for the finitely many prime ideals $\mathfrak{p}_1, \dots, \mathfrak{p}_r$ such that $\mathfrak{N}(\mathfrak{p}_i) \leq N$, and multiply them. This yields the equation

$$(*) \qquad \prod_{i=1}^{r} \frac{1}{1-\chi(\mathfrak{p}_i)\mathfrak{N}(\mathfrak{p}_i)^{-s}} = \sum_{\nu_1,\dots,\nu_r=0}^{\infty} \frac{\chi(\mathfrak{p}_1)^{\nu_1}\cdots\chi(\mathfrak{p}_r)^{\nu_r}}{(\mathfrak{N}(\mathfrak{p}_1)^{\nu_1}\cdots\mathfrak{N}(\mathfrak{p}_r)^{\nu_r})^s} = \sum_{\mathfrak{a}}{}' \frac{\chi(\mathfrak{a})}{\mathfrak{N}(\mathfrak{a})^s},$$

where $\sum'$ denotes the sum over all integral ideals $\mathfrak{a}$ which are divisible at most by the prime ideals $\mathfrak{p}_1, \dots, \mathfrak{p}_r$. Since the sum $\sum'$ contains in particular the terms such that $\mathfrak{N}(\mathfrak{a}) \leq N$, we may also write

$$\prod_{i=1}^{r} \frac{1}{1-\chi(\mathfrak{p}_i)\mathfrak{N}(\mathfrak{p}_i)^{-s}} = \sum_{\mathfrak{N}(\mathfrak{a})\leq N} \frac{\chi(\mathfrak{a})}{\mathfrak{N}(\mathfrak{a})^s} + \sum_{\mathfrak{N}(\mathfrak{a})>N}{}' \frac{\chi(\mathfrak{a})}{\mathfrak{N}(\mathfrak{a})^s}.$$

Comparing now in $(*)$ the sum $\sum'$ with the series $L(\chi,s)$, we get

$$\Big|\prod_{i=1}^{r}\frac{1}{1-\chi(\mathfrak{p}_i)\mathfrak{N}(\mathfrak{p}_i)^{-s}}-L(\chi,s)\Big|\le\Big|\sum_{\substack{\mathfrak{N}(\mathfrak{a})>N\\ \mathfrak{p}_i\nmid\mathfrak{a}}}\frac{\chi(\mathfrak{a})}{\mathfrak{N}(\mathfrak{a})^s}\Big|$$
$$\le\sum_{\mathfrak{N}(\mathfrak{a})>N}\frac{1}{\mathfrak{N}(\mathfrak{a})^{1+\delta}}\,.$$

For $N\to\infty$ the right-hand side tends to zero, as it is the remainder term of a convergent series, since the sequence $\big(\sum_{\mathfrak{N}(\mathfrak{a})\le N}\frac{1}{\mathfrak{N}(\mathfrak{a})^{1+\delta}}\big)_{N\in\mathbb{N}}$ is monotone increasing and bounded from above. Indeed, with the previous notations we find

$$\sum_{\mathfrak{N}(\mathfrak{a})\le N}\frac{1}{\mathfrak{N}(\mathfrak{a})^{1+\delta}}\le\sum_{\mathfrak{a}}{}'\frac{1}{\mathfrak{N}(\mathfrak{a})^{1+\delta}}$$
$$=\prod_{i=1}^{r}\big(1-\mathfrak{N}(\mathfrak{p}_i)^{-(1+\delta)}\big)^{-1},$$

and

$$\log\Big(\prod_{i=1}^{r}\big(1-\mathfrak{N}(\mathfrak{p}_i)^{-(1+\delta)}\big)^{-1}\Big)=\sum_{i=1}^{r}\log\Big(\big(1-\mathfrak{N}(\mathfrak{p}_i)^{-(1+\delta)}\big)^{-1}\Big)$$
$$=\sum_{i=1}^{r}\sum_{n=1}^{\infty}\frac{1}{n\mathfrak{N}(\mathfrak{p}_i)^{(1+\delta)n}}$$
$$\le\sum_{\mathfrak{p}}\sum_{n=1}^{\infty}\frac{1}{n\mathfrak{N}(\mathfrak{p})^{(1+\delta)n}}$$
$$\le\sum_{p,n}d\,\frac{1}{np^{n(1+\delta)}}$$
$$=d\log\big(\zeta(1+\delta)\big)\,.$$

□

We now face the task of analytically continuing the L-series $L(\chi,s)$ attached to a **Größencharakter** χ mod $\mathfrak{m}$, and setting up a suitable functional equation for it at the same time. So we are given a character

$$\chi:J^{\mathfrak{m}}\longrightarrow S^1,$$

such that

$$(*)\qquad\chi\big((a)\big)=\chi_{\mathrm{f}}(a)\chi_\infty(a)$$

for all integers $a\in\mathcal{O}$ relatively prime to $\mathfrak{m}$, and there are two associated characters

$$\chi_{\mathrm{f}}:(\mathcal{O}/\mathfrak{m})^*\longrightarrow S^1\quad\text{and}\quad\chi_\infty:\mathbf{R}^*\longrightarrow S^1.$$

The character $\chi_{\mathfrak{f}}$ extends in a unique way to a character

$$\chi_{\mathfrak{f}} : (\widehat{\mathcal{O}}/\mathfrak{m})^* \longrightarrow S^1$$

such that the identity (∗) holds for all integral ideal numbers $a \in \widehat{\mathcal{O}}^{(\mathfrak{m})}$ prime to $\mathfrak{m}$. Indeed, the restriction of the function $\chi_{\mathfrak{f}}(a) := \chi((a))\chi_\infty(a)^{-1}$ of $\widehat{\mathcal{O}}^{(\mathfrak{m})}$ to $\mathcal{O}^{(\mathfrak{m})}$ is given by the original character $\chi_{\mathfrak{f}}$ of $(\mathcal{O}/\mathfrak{m})^*$, so it is in particular trivial on $1+\mathfrak{m}$ and thus yields a character of $(\widehat{\mathcal{O}}/\mathfrak{m})^*$.

The L-series of a *Größencharacter* of $J^{\mathfrak{m}}$ is called a **Hecke L-series.** If χ is a (generalized) Dirichlet character $\bmod \mathfrak{m}$, i.e., a character of the ray class group $J^{\mathfrak{m}}/P^{\mathfrak{m}}$, then we call it a (generalized) **Dirichlet L-series**. The proof of the functional equation of the Hecke L-series proceeds in exactly the same way as for the Dedekind zeta function, except that it is based on the theta transformation formula (7.7).

We decompose the Hecke L-series according to the classes $\mathfrak{K}$ of the ideal class group J/P as a sum

$$L(\chi, s) = \sum_{\mathfrak{K}} L(\mathfrak{K}, \chi, s)$$

of the **partial L-series**

$$L(\mathfrak{K}, \chi, s) = \sum_{\substack{\mathfrak{a}\in\mathfrak{K} \\ \text{integral}}} \frac{\chi(\mathfrak{a})}{\mathfrak{N}(\mathfrak{a})^s}$$

and deduce a functional equation for those. If all one wants is the functional equation of the L-series $L(\chi, s)$, this decomposition is unnecessary; it may also be derived directly using the transformation formula (7.8), because we know how to represent any ideal $\mathfrak{a}$ by an ideal number (this was not yet the case when we were treating the Dedekind zeta function). However, we prefer to establish the finer result for the partial L-series.

By (7.1), we have a bijective mapping

$$(\widehat{\mathfrak{K}} \cap \widehat{\mathcal{O}})/\mathcal{O}^* \xrightarrow{\ \sim\ } \{\mathfrak{a} \in \mathfrak{K} \mid \mathfrak{a} \text{ integral}\}, \quad a \longmapsto (a),$$

where $\widehat{\mathfrak{K}} \in \widehat{K}^*/K^*$ corresponds to the class $\mathfrak{K} \in J/P$ with respect to the isomorphism $\widehat{K}^*/K^* \cong J/P$. Therefore we get

$$L(\mathfrak{K}, \chi, s) = \sum_{a\in\mathfrak{R}} \frac{\chi((a))}{|N(a)|^s},$$

where $\mathfrak{R}$ is a system of representatives of $(\widehat{\mathfrak{K}} \cap \widehat{\mathcal{O}})/\mathcal{O}^*$. We want to write this function as a Mellin transform. To this end, we recall from §4 the L-function

$$L_X(\mathbf{s}) = N(\pi^{-\mathbf{s}/2})\Gamma_X(\mathbf{s}/2) = N(\pi^{-\mathbf{s}/2}) \int_{\mathbf{R}_+^*} N(e^{-y} y^{\mathbf{s}/2}) \frac{dy}{y},$$

which has been attached to the $G(\mathbb{C}|\mathbb{R})$-set $X = \mathrm{Hom}(K, \mathbb{C})$. The character χ_∞ of $\mathbf{R}^*$ corresponding to χ is given by (6.7) as

$$\chi_\infty(x) = N\big(x^p |x|^{-p+iq}\big),$$

for an admissible $p \in \prod_\tau \mathbb{Z}$ and a $q \in \mathbf{R}_\pm$. We put $\mathbf{s} = s\mathbf{1} + p - iq$, where $s \in \mathbb{C}$ is a single complex variable, and

$$L_\infty(\chi, s) = L_X(\mathbf{s}) = L_X(s\mathbf{1} + p - iq).$$

In the integral

$$\Gamma_X(\mathbf{s}/2) = \int_{\mathbf{R}_+^*} N(e^{-y} y^{\mathbf{s}/2}) \frac{dy}{y},$$

we make the substitution

$$y \longmapsto \pi |a|^2 y / |md| \quad (a \in \mathfrak{R}),$$

where $m, d \in \widehat{\mathcal{O}}$ are fixed ideal numbers such that $(m) = \mathfrak{m}$ and $(d) = \mathfrak{d}$ is the different of $K|\mathbb{Q}$. We then obtain

$$\Gamma_X(\mathbf{s}/2) = N\Big(\Big(\frac{\pi}{|md|}\Big)^{\mathbf{s}/2}\Big) N(|a|^{\mathbf{s}}) \int_{\mathbf{R}_+^*} e^{-\pi\langle ay/|md|, a\rangle} N(y^{\mathbf{s}/2}) \frac{dy}{y}$$

and, since $N(|md|^{\mathbf{1}s/2}) = (|d_K| \mathfrak{N}(\mathfrak{m}))^{s/2}$,

$$\big(|d_K| \mathfrak{N}(\mathfrak{m})\big)^{s/2} L_\infty(\chi, s) \frac{1}{N(|a|^{\mathbf{s}})} = c(\chi) \int_{\mathbf{R}_+^*} e^{-\pi\langle ay/|md|, a\rangle} N(y^{\mathbf{s}/2}) \frac{dy}{y}$$

where $c(\chi) = N(|md|^{-p+iq})^{1/2}$. Multiplying this by $\chi_{\mathfrak{f}}(a) N(a^p)$ and summing over $a \in \mathfrak{R}$ yields, in view of

$$\frac{\chi_{\mathfrak{f}}(a) N(a^p)}{N(|a|^{\mathbf{s}})} = \frac{\chi_{\mathfrak{f}}(a) N(a^p |a|^{-p+iq})}{N(|a|^s)} = \frac{\chi((a))}{|N(a)|^s},$$

the equation

$$(|d_K| \mathfrak{N}(\mathfrak{m}))^{s/2} L_\infty(\chi, s) L(\mathfrak{K}, \chi, s) = c(\chi) \int_{\mathbf{R}_+^*} g(y) N(y^{\mathbf{s}/2}) \frac{dy}{y}$$

with the series

$$g(y) = \sum_{a \in \mathfrak{R}} \chi_{\mathfrak{f}}(a) N(a^p) e^{-\pi\langle ay/|md|, a\rangle}.$$

We now consider the completed L-series

$$\Lambda(\mathfrak{K}, \chi, s) = (|d_K| \mathfrak{N}(\mathfrak{m}))^{s/2} L_\infty(\chi, s) L(\mathfrak{K}, \chi, s).$$

Then we get

$$\Lambda(\mathfrak{K},\chi,s) = c(\chi)\int\limits_{\mathbf{R}_+^*} g(y)N(y^{s/2})\frac{dy}{y}\cdot$$

We now want to write this function as an integral over the series

$$\theta(\mathfrak{K},\chi,z) := \theta^p(\mathfrak{K},\chi_f,z) = \varepsilon(\chi) + \sum_{a\in\hat{\mathfrak{K}}\cap\hat{\mathcal{O}}} \chi_f(a)N(a^p)e^{\pi i\langle az/|md|,a\rangle},$$

where the summation is extended not only – as in the case of $g(y)$ – over a system of representatives $\mathfrak{R}$ of $(\widehat{\mathfrak{K}}\cap\widehat{\mathcal{O}})/\mathcal{O}^*$, but over all $a \in \widehat{\mathfrak{K}}\cap\widehat{\mathcal{O}}$. We have $\varepsilon(\chi) = 1$ if $\mathfrak{m} = 1$ and $p = 0$, and $\varepsilon(\chi) = 0$ otherwise. We will proceed in the same way as with the Dedekind zeta function (see (5.5)). Just as we did there, using

$$y = xt^{1/n}, \quad x = \frac{y}{N(y)^{1/n}}, \quad t = N(y),$$

with $n = [K:\mathbb{Q}]$, we decompose

$$\mathbf{R}_+^* = \mathbf{S}\times\mathbb{R}_+^*, \quad \frac{dy}{y} = d^*x \times \frac{dt}{t}\cdot$$

Then, observing that

$$N(y^{s/2}) = N(x^{s/2})N(t^{s/2n}) = N(x^{(p-iq)/2})t^{\frac{1}{2}(s+Tr(p-iq)/n)},$$

we obtain the identity

$$(*) \qquad \Lambda(\mathfrak{K},\chi,s) = c(\chi)\int_0^\infty\int_{\mathbf{S}} N(x^{(p-iq)/2})g(xt^{1/n})d^*xt^{s'}\frac{dt}{t}$$

with $s' = \frac{1}{2}(s + Tr(p-iq)/n)$. The function under the second integral will be denoted by

$$g_{\mathfrak{R}}(x,t) = N(x^{(p-iq)/2})\sum_{a\in\mathfrak{R}}\chi_f(a)N(a^p)e^{-\pi\langle axt^{1/n}/|md|,a\rangle}.$$

From it, the theta series $\theta(\mathfrak{K},\chi,ixt^{1/n})$ is constructed as follows.

(8.2) Lemma. $N(x^{(p-iq)/2})\big(\theta(\mathfrak{K},\chi,ixt^{1/n}) - \varepsilon(\chi)\big) = \sum_{\varepsilon\in\mathcal{O}^*} g_{\mathfrak{R}}(|\varepsilon|^2x,t).$

Proof: For every unit $\varepsilon\in\mathcal{O}^*$, one has $\chi_\infty(\varepsilon)\chi_f(\varepsilon) = \chi((\varepsilon)) = 1$, so that we get

$$N\big(|\varepsilon|^{p-iq}\big) = \overline{\chi}_\infty(\varepsilon)N(\varepsilon^p) = \chi_f(\varepsilon)N(\varepsilon^p).$$

We put for short $\xi = xt^{1/n}/|md|$ and obtain

$$g_{\mathfrak{R}}\big(|\varepsilon|^2 x, t\big) = N(x^{(p-iq)/2}) \sum_{a\in\mathfrak{R}} \chi_{\mathfrak{f}}(\varepsilon a) N\big((\varepsilon a)^p\big)\, e^{-\pi\langle \varepsilon a\xi, \varepsilon a\rangle} = g_{\varepsilon\mathfrak{R}}(x,t).$$

Since $\widehat{\mathfrak{K}} \cap \widehat{\mathcal{O}} = \dot{\bigcup\limits_{\varepsilon\in\mathcal{O}^*}} \varepsilon\mathfrak{R}$, we get

$$\begin{aligned} N(x^{(p-iq)/2})\big(\theta(\mathfrak{K},\chi, ixt^{1/n}) - \varepsilon(\chi)\big) &= \\ &= \sum_{\varepsilon\in\mathcal{O}^*}\sum_{a\in\varepsilon\mathfrak{R}} N(x^{(p-iq)/2})\chi_{\mathfrak{f}}(\varepsilon a)N\big((\varepsilon a)^p\big)\, e^{-\pi\langle\varepsilon a\xi,\varepsilon a\rangle} \\ &= \sum_{\varepsilon\in\mathcal{O}^*} g_{\varepsilon\mathfrak{R}}(x,t) = \sum_{\varepsilon\in\mathcal{O}^*} g_{\mathfrak{R}}\big(|\varepsilon|^2 x, t\big). \end{aligned}$$

□

From this lemma we now obtain the desired integral representation of the function $\Lambda(\mathfrak{K},\chi,s)$. We choose as in §5 a fundamental domain F of $\mathbf{S}$ for the action of the group $|\mathcal{O}^*|^2$. F is mapped by $\log : \mathbf{R}_+^* \xrightarrow{\sim} \mathbf{R}_\pm$ to a fundamental mesh of the lattice $2\log|\mathcal{O}^*|$. This means that we have

$$\mathbf{S} = \dot{\bigcup_{\eta\in|\mathcal{O}^*|}} \eta^2 F.$$

(8.3) Proposition. *The function*

$$\Lambda(\mathfrak{K},\chi,s) = \big(|d_K|\mathfrak{N}(\mathfrak{m})\big)^{s/2} L_\infty(\chi,s) L(\mathfrak{K},\chi,s)$$

is the Mellin transform

$$\Lambda(\mathfrak{K},\chi,s) = L(f,s')$$

of the function

$$f(t) = f_F(\mathfrak{K},\chi,t) = \frac{c(\chi)}{w}\int_F N(x^{(p-iq)/2})\theta(\mathfrak{K},\chi,ixt^{1/n})\,d^*x$$

at $s' = \frac{1}{2}(s + Tr(p-iq)/n)$. *Here we have set* $n = [K:\mathbb{Q}]$, $c(\chi) = N(|md|^{-p+iq})^{1/2}$, *and* w *denotes the number of roots of unity in* K.

Proof: One has

$$f(\infty) = \frac{c(\chi)\varepsilon(\chi)}{w}\int_F N(x^{(p-iq)/2})\,d^*x.$$

We have seen before that

$$(*) \qquad \Lambda(\mathfrak{K},\chi,s) = \int_0^\infty f_{\mathcal{O}}(t)t^{s'}\frac{dt}{t} = L(f,s')$$

where

$$f_{\mathcal{O}}(t) = c(\chi)\int_{\mathbf{S}} g_{\mathfrak{N}}(x,t)\,d^*x\,.$$

Since $\mathbf{S} = \dot{\bigcup}_{\eta\in|\mathcal{O}^*|}\eta^2 F$, one has

$$f_{\mathcal{O}}(t) = c(\chi)\sum_{\eta\in|\mathcal{O}^*|}\int_{\eta^2 F} g_{\mathfrak{N}}(x,t)\,d^*x\,.$$

In each one of the integrals on the right, we make the transformation $F \to \eta^2 F$, $x \mapsto \eta^2 x$, and obtain

$$f_{\mathcal{O}}(t) = c(\chi)\int_F \sum_{\eta\in|\mathcal{O}^*|} g_{\mathfrak{N}}(\eta^2 x,t)\,d^*x\,.$$

The fact that we may swap summation and integration is justified in exactly the same way as for the case of Dirichlet L-series in §2, p. 436. In view of the exact sequence

$$1 \longrightarrow \mu(K) \longrightarrow \mathcal{O}^* \longrightarrow |\mathcal{O}^*| \longrightarrow 1,$$

where $\mu(K)$ denotes the group of roots of unity in K, one has $\#\{\varepsilon\in\mathcal{O}^* \mid |\varepsilon| = \eta\} = w$, so that we get

$$\sum_{|\varepsilon|=\eta} g_{\mathfrak{N}}\big(|\varepsilon|^2 x,t\big) = w g_{\mathfrak{N}}(\eta^2 x,t)\,.$$

Using (8.2), this gives

$$\begin{aligned} f_{\mathcal{O}}(t) &= \frac{c(\chi)}{w}\int_F \sum_{\varepsilon\in\mathcal{O}^*} g_{\mathfrak{N}}\big(|\varepsilon|^2 x,t\big)\,d^*x \\ &= \frac{c(\chi)}{w}\int_F N(x^{(p-iq)/2})\big(\theta(\mathfrak{K},\chi,ixt^{1/n}) - \varepsilon(\chi)\big)\,d^*x \\ &= f(t) - f(\infty)\,. \end{aligned}$$

This together with $(*)$ yields the claim of the proposition. □

It is now the transformation formula (7.7) for the theta series $\theta(\mathfrak{K},\chi,z) = \theta^p(\mathfrak{K},\chi_{\mathfrak{f}},z)$ which guarantees that the functions $f(t) = f_F(\mathfrak{K},\chi,t)$ satisfy the hypotheses of the Mellin principle.

(8.4) Proposition. *We have* $f_F(\mathfrak{K}, \chi, t) = a_0 + O(e^{-ct^{1/n}})$ *for some* $c > 0$, *and*

$$a_0 = \frac{N(|d|^{iq/2})}{w} \int_F N(x^{-iq/2})\, d^*x$$

if $\mathfrak{m} = 1$ *and* $p = 0$, *and* $a_0 = 0$ *otherwise. Furthermore we have*

$$f_F\Big(\mathfrak{K}, \chi, \frac{1}{t}\Big) = W(\chi) t^{\frac{1}{2} + Tr(p)/n} f_{F^{-1}}(\mathfrak{K}', \overline{\chi}, t)$$

where $\mathfrak{K}\mathfrak{K}' = [\mathfrak{m}\mathfrak{d}]$, *and the constant factor is given by*

$$W(\chi) = \Big[i^{Tr(\overline{p})} N\Big(\Big(\frac{md}{|md|}\Big)^{\overline{p}}\Big)\Big]^{-1} \frac{\tau(\chi_f)}{\sqrt{\mathfrak{N}(\mathfrak{m})}}.$$

Proof: The first statement follows exactly as in the proof of (5.8). For the second, we make use of formula (7.7). It gives us

$$\begin{aligned} \theta(\mathfrak{K}, \chi, -1/z) = \theta^p(\mathfrak{K}, \chi_f, -1/z) &= W(\chi) N\big((z/i)^{p+\frac{1}{2}}\big) \theta^{\overline{p}}(\mathfrak{K}', \overline{\chi}_f, z) \\ &= W(\chi) N\big((z/i)^{p+\frac{1}{2}}\big) \theta(\mathfrak{K}', \overline{\chi}, z), \end{aligned}$$

because $\overline{\chi}_\infty(x) = \overline{N(x^p|x|^{-p+iq})} = N((^*x)^p|x|^{-p-iq}) = N(x^{\overline{p}}|x|^{-\overline{p}-iq})$. Observing the fact that the transformation $x \mapsto x^{-1}$ leaves the Haar measure d^*x invariant and takes the fundamental domain F to the fundamental domain F^{-1}, (7.7) yields for $z = ixt^{1/n}$:

$$\begin{aligned} f_F\Big(\mathfrak{K}, \chi, \frac{1}{t}\Big) &= \frac{c(\chi)}{w} \int_F N(x^{(p-iq)/2}) \theta(\mathfrak{K}, \chi, ix/t^{1/n})\, d^*x \\ &= \frac{c(\chi)}{w} \int_{F^{-1}} N(x^{-(p-iq)/2}) \theta(\mathfrak{K}, \chi, -1/ixt^{1/n})\, d^*x \\ &= \frac{c(\chi)W(\chi)}{w} \int_{F^{-1}} N(x^{-\frac{p-iq}{2}+p+\frac{1}{2}}) N(t^{(p+\frac{1}{2})/n}) \theta(\mathfrak{K}', \overline{\chi}, ixt^{1/n})\, d^*x \\ &= \frac{c(\chi)W(\chi)}{w} \int_{F^{-1}} N(x^{(\overline{p}+iq)/2}) t^{1/2+Tr(p)/n} \theta(\mathfrak{K}', \overline{\chi}, ixt^{1/n})\, d^*x \\ &= W(\chi) t^{\frac{1}{2}+Tr(p)/n} f_{F^{-1}}(\mathfrak{K}', \overline{\chi}, t). \end{aligned}$$

We have used in this calculation that $N(x^{1/2}) = N(x)^{1/2} = 1$ and $N(x^p) = N((^*x)^p) = N(x^{\overline{p}})$, and that the character $\overline{\chi}_\infty$, the complex conjugate of χ_∞, is given by

$$\overline{\chi}_\infty(x) = N\big(x^{\overline{p}}|x|^{-\overline{p}-iq}\big). \qquad \square$$

From this proposition and (1.4), we now finally get our main result. We may assume that χ is a primitive *Größencharacter* mod $\mathfrak{m}$, i.e., that the corresponding character $\chi_\mathfrak{f}$ of $(\mathcal{O}/\mathfrak{m})^*$ is primitive (see §6, p.472). The L-series of an arbitrary character differs from the L-series of the corresponding primitive character only by finitely many Euler factors. So analytic continuation and functional equation of one follow from those of the other.

(8.5) Theorem. *Let χ be a primitive Größencharacter* mod $\mathfrak{m}$. *Then the function*

$$\Lambda(\mathfrak{K},\chi,s) = \big(|d_K|\mathfrak{N}(\mathfrak{m})\big)^{s/2} L_\infty(\chi,s)L(\mathfrak{K},\chi,s), \quad \mathrm{Re}(s) > 1,$$

has a meromorphic continuation to the complex plane $\mathbb{C}$ and satisfies the functional equation

$$\Lambda(\mathfrak{K},\chi,s) = W(\chi)\Lambda(\mathfrak{K}',\overline{\chi},1-s)$$

where $\mathfrak{K}\mathfrak{K}' = [\mathfrak{m}\mathfrak{d}]$, and the constant factor is given by

$$W(\chi) = \Big[i^{Tr(\overline{p})} N\Big(\Big(\frac{md}{|md|}\Big)^{\overline{p}}\Big)\Big]^{-1} \frac{\tau(\chi_\mathfrak{f})}{\sqrt{\mathfrak{N}(\mathfrak{m})}}.$$

It has absolute value $|W(\chi)| = 1$.

$\Lambda(\mathfrak{K},\chi,s)$ is holomorphic except for poles of order at most one at $s = Tr(-p+iq)/n$ and $s = 1+Tr(p+iq)/n$. In the case $\mathfrak{m} \neq 1$ or $p \neq 0$, $\Lambda(\mathfrak{K},\chi,s)$ is holomorphic on all of $\mathbb{C}$.

Proof: Let $f(t) = f_F(\mathfrak{K},\chi,t)$ and $g(t) = f_{F^{-1}}(\mathfrak{K}',\overline{\chi},t)$. From $f(t) = a_0 + O(e^{-ct^{1/n}})$, $g(t) = b_0 + O(e^{-ct^{1/n}})$ and

$$f\Big(\frac{1}{t}\Big) = W(\chi)t^{\frac{1}{2}+Tr(p)/n} g(t),$$

it follows by (1.4) that the Mellin transforms $L(f,s)$ and $L(g,s)$ can be meromorphically continued, and from (8.3) we get

$$\begin{aligned}\Lambda(\mathfrak{K},\chi,s) &= L\Big(f,\frac{1}{2}(s+Tr(p-iq)/n)\Big)\\ &= W(\chi)L\Big(g,\frac{1}{2}+Tr(p)/n-\frac{1}{2}(s+Tr(p-iq)/n)\Big)\\ &= W(\chi)L\Big(g,\frac{1}{2}(1-s+Tr(\overline{p}+iq)/n)\Big)\\ &= W(\chi)\Lambda(\mathfrak{K}',\overline{\chi},1-s),\end{aligned}$$

where we have to take into account again that $\overline{\chi}_\infty(x) = N(x^{\overline{p}}|x|^{-\overline{p}-iq})$.

According to (1.4), in the case $a_0 \neq 0$, $L(f,s)$ has a simple pole at $s=0$ and $s=\frac{1}{2}+Tr(p)/n$, i.e., $\Lambda(\mathfrak{K},\chi,s)=L(f,\frac{1}{2}(s+Tr(p-iq)/n))$ has a simple pole at $s=Tr(-p+iq)/n$ and $s=1+Tr(p+iq)/n$. If $\mathfrak{m}\neq 1$ or $p\neq 0$, then $a_0=0$, i.e., $\Lambda(\mathfrak{K},\chi,s)$ is holomorphic on all of $\mathbb{C}$. □

For the **completed Hecke L-series**

$$\Lambda(\chi,s)=(|d_K|\mathfrak{N}(\mathfrak{m}))^{s/2}L_\infty(\chi,s)L(\chi,s)=\sum_{\mathfrak{K}}\Lambda(\mathfrak{K},\chi,s)$$

we derive immediately from the theorem the

(8.6) Corollary. *The L-series $\Lambda(\chi,s)$ admits a holomorphic continuation to*

$$\mathbb{C}\smallsetminus\{Tr(-p+iq)/n,\ 1+Tr(p+iq)/n\}$$

and satisfies the functional equation

$$\Lambda(\chi,s)=W(\chi)\Lambda(\overline{\chi},1-s).$$

It is holomorphic on all of $\mathbb{C}$, if $\mathfrak{m}\neq 1$ or $p\neq 0$.

Remark 1: For a *Dirichlet character* χ mod $\mathfrak{m}$, the functional equation can be proved without using ideal numbers, by splitting the ray class group $J^{\mathfrak{m}}/P^{\mathfrak{m}}$ into its classes $\mathfrak{K}$, and then proceeding exactly as for the Dedekind zeta function. The Gauss sums to be used then are those treated by *Hasse* in [52]. On the other hand, one may prove the functional equation for the Dedekind zeta function by using ideal numbers, imitating the above proof, without decomposing the ideal group at all.

Remark 2: There is an important alternative approach to the results of this section. It starts from a character of the idèle class group and from the representation (8.1) of the corresponding L-series as an Euler product. The proof of the functional equation is then based on the local-to-global principle of algebraic number theory and on the Fourier analysis of $\mathfrak{p}$-adic number fields and their idèle class group. This theory was developed by the American mathematician *John Tate*, and is commonly known for short as **Tate's thesis**. Even though it does meet the goal of this book of presenting modern conceptual approaches, we still decided not to include it here. The reason for this is the clarity and conciseness of Tate's original paper [24], which cannot be improved upon. In addition *Serge Lang*'s account of the theory [94] provides an illustrative complement.

Thus instead of idly copying this theory, we have chosen to provide a conceptual framework and a modern treatment of Hecke's original proof which is somewhat difficult to fathom. It turns out that Hecke's approach continues to have a relevance of its own, and can even claim a number of advantages over Tate's theory. For the functional equation of the Riemann zeta function and the Dirichlet L-series, for example, it would be out of proportion to develop Tate's formalism with all its p-adic expense, since they can be settled at a beginner's level with the method used here. Also, L-series, and the very theory of theta series has to be seen as an important arithmetic accomplishment in its own right.

It was for pedagogical reasons that we have proved the analytic continuation and functional equation of L-series four times over: for the Riemann zeta function, for the Dirichlet L-series, for the Dedekind zeta function, and finally for general Hecke L-series. This explains the number of pages needed. Attacking the general case directly would shrink the exposé to little more than the size of Tate's thesis. Still, it has to be said that Tate's theory has acquired fundamental importance for number theory at large through its far reaching generalizations.

§ 9. Values of Dirichlet L-series at Integer Points

The results of § 1 and § 2 on the values $\zeta(1-k)$ and $L(\chi, 1-k)$ of the Riemann zeta function and the Dirichlet L-series will now be extended to generalized Dirichlet L-series over a totally real number field. We do this using a method devised by the Japanese mathematician *Takuro Shintani* (who died an early and tragic death) (see [127], [128]).

We first prove a new kind of unit theorem for which we need the following notions from linear algebra. Let V be an n-dimensional $\mathbb{R}$-vector space, k a subfield of $\mathbb{R}$, and V_k a fixed k-structure of V, i.e., a k-subspace such that $V = V_k \otimes_k \mathbb{R}$. By definition, an (open) **k-rational simplicial cone** of dimension d is a subset of the form

$$C(v_1, \dots, v_d) = \left\{ t_1 v_1 + \dots + t_d v_d \;\middle|\; t_\ell \in \mathbb{R}_+^* \right\}$$

where $v_1, \dots, v_d$ are linearly independent vectors in V_k. A finite disjoint union of k-rational simplicial cones is called a k-rational **polyhedric cone**. We call a linear form L on V *k-rational* if its coefficients with respect to a k-basis of V_k lie in k.

(9.1) Lemma. *Every nonempty subset different from $\{0\}$ of the form*

$$P = \big\{x \in V \mid L_i(x) \geq 0, \quad 0 < i \leq \ell,\ M_j(x) > 0,\ 0 < j \leq m\big\}$$

with nonzero k-rational linear forms L_i, M_j ($\ell = 0$ or $m = 0$ is allowed) is a disjoint union of finitely many k-rational cones, and possibly the origin.

Proof: First let $P = \{x \in V \mid L_i(x) \geq 0,\ i = 1, \dots, \ell\}$, with k-rational linear forms $L_1, \dots, L_\ell \neq 0$. For $n = 1$ and $n = 2$ the lemma is obvious. We assume it is established for all $\mathbb{R}$-vector spaces of dimension smaller than n. If P has no inner point, then there is a linear form L among the $L_1, \dots, L_\ell$ such that P is contained in the hyperplane $L = 0$. In this case the lemma follows from the induction hypothesis. So let $u \in P$ be an inner point, i.e., $L_1(u) > 0, \dots, L_\ell(u) > 0$. Since V_k is dense in V, we may assume $u \in V_k$. For every $i = 1, \dots, \ell$, let $\partial_i P = \{x \in P \mid L_i(x) = 0\}$. If $\partial_i P \neq \{0\}$, then $\partial_i P \smallsetminus \{0\}$ is by the induction hypothesis a disjoint union of a finite number of k-rational simplicial cones of dimension $< n$. If a simplicial cone in $\partial_i P$ has a nonempty intersection with some $\partial_j P$, then it is clearly contained in $\partial_i P \cap \partial_j P$. Therefore $\partial_1 P \cup \dots \cup \partial_\ell P \smallsetminus \{0\}$ is a disjoint union of k-rational simplicial cones of dimension $< n$, so that

$$\partial_1 P \cup \dots \cup \partial_\ell P \smallsetminus \{0\} = \dot{\bigcup_{j \in J}} C_j\,,$$

where $C_j = C(v_1, \dots, v_{d_j})$, $v_1, \dots, v_{d_j} \in V_k$, $d_j < n$. For every $j \in J$ we put $C_j(u) = C(v_1, \dots, v_{d_j}, u)$. This is a $(d_j + 1)$-dimensional k-rational simplicial cone. We claim that

$$P \smallsetminus \{0\} = \dot{\bigcup_{j \in J}} C_j \,\dot\cup\, \dot{\bigcup_{j \in J}} C_j(u) \,\dot\cup\, \mathbb{R}_+^* u\,.$$

Indeed, if the point $x \in P \smallsetminus \{0\}$ lies on the boundary of P, then it belongs to some $\partial_i P$, hence to $\dot\bigcup_{j \in J} C_j$. On the other hand, if x belongs to the interior of P, then $L_i(x) > 0$ for all i. If x is a scalar multiple of u, then we have $x \in \mathbb{R}_+^* u$. Assume this is not the case, and let s be the minimum of the numbers $L_1(x)/L_1(u), \dots, L_\ell(x)/L_\ell(u)$. Then $s > 0$ and $x - su$ lies on the boundary of P. Since $x - su \neq 0$, there is a unique $j \in J$ such that $x - su \in C_j$, and thus there is a unique $j \in J$ such that $x \in C_j(u)$. This proves the claim.

Now let

$$P = \big\{x \in V \mid L_i(x) \geq 0,\ 0 < i \leq \ell,\ M_j(x) > 0,\ j = 1, \dots, m\big\}\,.$$

Then

$$\overline{P} = \big\{x \in V \mid L_i(x) \geq 0,\ M_j(x) \geq 0\big\}$$

is a disjoint union of a finite number of k-rational simplicial cones and $\{0\}$. For every $j = 1, \ldots, m$, let $\partial_j \overline{P} = \{x \in \overline{P} \mid M_j(x) = 0\}$. If a simplicial cone in $\overline{P}$ has nonempty intersection with $\partial_j \overline{P}$, then it is contained in $\partial_j \overline{P}$. As $P = \overline{P} \setminus \bigcup_{j=1}^m \partial_j \overline{P}$, we see that since $\overline{P} \setminus \{0\}$ is a disjoint union of finitely many k-rational simplicial cones, then so is P. □

(9.2) Corollary. *If C and C' are k-rational polyhedric cones, then $C \setminus C'$ is also a k-rational polyhedric cone.*

Proof: We may assume without loss of generality that C and C' are k-rational cones. Let d be the dimension of C'. Then there are n k-rational linear forms $L_1, \ldots, L_{n-d}, M_1, \ldots, M_d$ such that

$$C' = \left\{x \in V \;\middle|\; L_1(x) = \cdots = L_{n-d}(x) = 0,\ M_1(x) > 0, \ldots, M_d(x) > 0\right\}.$$

If we define, for each $i = 1, \ldots, n-d$,

$$C_i^{\pm} = \left\{x \in C \;\middle|\; L_1(x) = \cdots = L_{i-1}(x) = 0,\ \pm L_i(x) > 0\right\},$$

and for each $j = 1, \ldots, d$,

$$C_j = \left\{x \in C \;\middle|\; \begin{array}{l} L_1(x) = \cdots = L_{n-d}(x) = 0, \\ M_1(x) > 0, \ldots, M_{j-1}(x) > 0,\ M_j(x) \le 0 \end{array}\right\},$$

then we find, as can be checked immediately, that $C \setminus C'$ is the disjoint union of the sets $C_1^+, \ldots, C_{n-d}^+, C_1^-, \ldots, C_{n-d}^-, C_1, \ldots, C_d$. By (9.1), these are either empty or k-rational polyhedric cones. Therefore $C \setminus C'$ is also. □

It is a rare and special event if a new substantial insight is added to the foundations of algebraic number theory. The following theorem, proved by SHINTANI in 1979, falls into this category. Let K be a number field of degree $n = [K : \mathbb{Q}]$, and let $\mathbf{R} = \left[\prod_\tau \mathbb{C}\right]^+$ be the corresponding Minkowski space ($\tau \in \operatorname{Hom}(K, \mathbb{C})$). Define

$$\mathbf{R}^*_{(+)} = \left\{(x_\tau) \in \mathbf{R}^* \;\middle|\; x_\tau > 0 \text{ for all real } \tau\right\}.$$

(Observe that one has $\mathbf{R}^*_{(+)} = \mathbf{R}^*_+$ only in the case where K is totally real.) Since $\mathbf{R} = K \otimes_{\mathbb{Q}} \mathbb{R}$, the field K is a $\mathbb{Q}$-structure of $\mathbf{R}$. The group

$$\mathcal{O}^*_+ = \mathcal{O}^* \cap \mathbf{R}^*_{(+)}$$

of totally positive units acts on $\mathbf{R}^*_{(+)}$ via multiplication, and we will show that this action has a fundamental domain which is a $\mathbb{Q}$-rational polyhedric cone:

(9.3) Shintani's Unit Theorem. *If E is a subgroup of finite index in $\mathcal{O}_+^*$, then there exists a $\mathbb{Q}$-rational polyhedric cone P such that*

$$\mathbf{R}_{(+)}^* = \bigcup_{\varepsilon \in E} \varepsilon P \qquad \textit{(disjoint union).}$$

Proof: We consider in $\mathbf{R}_{(+)}^*$ the norm-one hypersurface

$$S = \{x \in \mathbf{R}_{(+)}^* \mid |N(x)| = 1\}.$$

Every $x \in \mathbf{R}_{(+)}^*$ is in a unique way the product of an element of S and of a positive scalar element. Indeed, $x = |N(x)|^{1/n}(x/|N(x)|^{1/n})$. By Dirichlet's unit theorem, E (being a subgroup of finite index in $\mathcal{O}^*$) is mapped by the mapping

$$\ell : S \longrightarrow \Big[\prod_\tau \mathbb{R}\Big]^+, \quad (x_\tau) \longmapsto \big(\log |x_\tau|\big),$$

onto a complete lattice Γ of the trace-zero space $H = \{x \in [\prod_\tau \mathbb{R}]^+ \mid Tr(x) = 0\}$. Let Φ be a fundamental mesh of Γ, let $\overline{\Phi}$ be the closure of Φ in H, and put $F = \ell^{-1}(\overline{\Phi})$. Since $\overline{\Phi}$ is bounded and closed, so is F. It is therefore compact, and we have

$$S = \bigcup_{\varepsilon \in E} \varepsilon F. \tag{1}$$

Let $x \in F$ and $U_\delta(x) = \{y \in \mathbf{R} \mid \|x - y\| < \delta\} \subseteq \mathbf{R}_{(+)}^*$, $\delta > 0$. Then there is clearly a basis $v_1, \dots, v_n \in U_\delta(x)$ of $\mathbf{R}$ such that $x = t_1 v_1 + \dots + t_n v_n$ with $t_i > 0$. Since K is dense in $\mathbf{R}$ by the approximation theorem, we may even choose the v_i to lie in $K \cap U_\delta(x)$. Then $C_\delta = C(v_1, \dots, v_n)$ is a $\mathbb{Q}$-rational simplicial cone in $\mathbf{R}_{(+)}^*$ with $x \in C_\delta$, and every $y \in C_\delta$ is of the form $y = \lambda z$ with $\lambda \in \mathbb{R}_+^*$ and $z \in U_\delta(x)$. We may now choose δ sufficiently small so that

$$C_\delta \cap \varepsilon C_\delta = \emptyset \quad \text{for all } \varepsilon \in E,\ \varepsilon \neq 1.$$

If not, then we would find sequences $\lambda_\nu z_\nu$, $\lambda'_\nu z'_\nu \in C_{1/\nu}$, $\lambda_\nu, \lambda'_\nu \in \mathbb{R}_+^*$, $z_\nu, z'_\nu \in U_{1/\nu}(x)$, and $\varepsilon_\nu \in E$, $\varepsilon_\nu \neq 1$, such that $\lambda_\nu z_\nu = \varepsilon_\nu \lambda'_\nu z'_\nu$, and thus $\rho_\nu z_\nu = \varepsilon_\nu z'_\nu$, $\rho_\nu = \lambda_\nu/\lambda'_\nu$. z_ν and z'_ν would converge to x; now ρ_ν would converge to 1 as $\rho_\nu^n N(z_\nu) = N(z'_\nu)$, i.e., $x = (\lim \varepsilon_\nu)x$. This would mean that $\lim \varepsilon_\nu = 1$, which is impossible, since E is discrete in $\mathbf{R}$.

F being compact, we thus find a finite number of $\mathbb{Q}$-rational cones $C_1, \dots, C_m$ in $\mathbf{R}_{(+)}^*$ such that

$$F = \bigcup_{i=1}^{m} (C_i \cap F) \tag{2}$$

and $C_i \cap \varepsilon C_i = \emptyset$ for all $\varepsilon \in E$, $\varepsilon \neq 1$, and all $i = 1, \ldots, m$. From (1) and (2), we deduce that

$$\mathbf{R}^*_{(+)} = \bigcup_{i=1}^{m} \bigcup_{\varepsilon \in E} \varepsilon C_i .$$

In order to turn this union into a disjoint one, we put $C_1^{(1)} = C_1$ and

$$C_i^{(1)} = C_i \smallsetminus \bigcup_{\varepsilon \in E} \varepsilon C_1 , \quad i = 2, \ldots, m .$$

εC_1 and C_i are disjoint for almost all $\varepsilon \in E$. Hence, by (9.2), $C_i^{(1)}$ is a $\mathbb{Q}$-rational polyhedric cone. Observing that $C_i \cap \varepsilon C_i = \emptyset$ for $\varepsilon \in E$, $\varepsilon \neq 1$, we obtain

$$\mathbf{R}^*_{(+)} = \bigcup_{i=1}^{m} \bigcup_{\varepsilon \in E} \varepsilon C_i^{(1)}$$

and $\varepsilon C_1^{(1)} \cap C_i^{(1)} = \emptyset$ for all $\varepsilon \in E$ and $i = 2, \ldots, m$.

We now assume by induction that we have found a finite system of $\mathbb{Q}$-rational polyhedric cones $C_1^{(\nu)}, \ldots, C_m^{(\nu)}$, $\nu = 1, \ldots, m-2$ satisfying the following properties:

(i) $C_i^{(\nu)} \subseteq C_i$,

(ii) $\mathbf{R}^*_{(+)} = \bigcup_{i=1}^{m} \bigcup_{\varepsilon \in E} \varepsilon C_i^{(\nu)}$,

(iii) $\varepsilon C_i^{(\nu)} \cap C_j = \emptyset$ for all $\varepsilon \in E$, if $i \leq \nu$ and $i \neq j$.

We put $C_i^{(\nu+1)} = C_i^{(\nu)}$ for $i \leq \nu + 1$, and

$$C_i^{(\nu+1)} = C_i^{(\nu)} \smallsetminus \bigcup_{\varepsilon \in E} \varepsilon C_{\nu+1}^{(\nu)} \quad \text{for} \quad i \geq \nu + 2 .$$

Then $C_1^{(\nu+1)}, \ldots, C_m^{(\nu+1)}$ is a finite system of $\mathbb{Q}$-rational polyhedric cones which enjoys properties (i), (ii), and (iii) with $\nu+1$ instead of ν. Consequently, $C_1^{(m-1)}, \ldots, C_m^{(m-1)}$ is a system of $\mathbb{Q}$-rational polyhedric cones such that

$$\mathbf{R}^*_{(+)} = \dot{\bigcup_{i=1}^{m}} \dot{\bigcup_{\varepsilon \in E}} \varepsilon C_i^{(m-1)} \quad \text{(disjoint union).}$$ □

Based on Shintani's unit theorem, we now obtain the following description of Dirichlet's L-series. Let $\mathfrak{m}$ be an integral ideal, $J^{\mathfrak{m}}/P^{\mathfrak{m}}$ the ray class group mod $\mathfrak{m}$. Let $\chi : J^{\mathfrak{m}}/P^{\mathfrak{m}} \to \mathbb{C}^*$ be a Dirichlet character mod $\mathfrak{m}$, and

$$L(\chi, s) = \sum_{\mathfrak{a}} \frac{\chi(\mathfrak{a})}{\mathfrak{N}(\mathfrak{a})^s}$$

the associated Dirichlet L-series. If $\mathfrak{K}$ varies over the classes of $J^{\mathfrak{m}}/P^{\mathfrak{m}}$, then we have

$$L(\chi,s)=\sum_{\mathfrak{K}}\chi(\mathfrak{K})\zeta(\mathfrak{K},s)$$

with the partial zeta functions

$$\zeta(\mathfrak{K},s)=\sum_{\substack{\mathfrak{a}\in\mathfrak{K}\\ \mathfrak{a}\text{ integral}}}\frac{1}{\mathfrak{N}(\mathfrak{a})^s}\,.$$

Let $\mathfrak{K}$ be a fixed class, and $\mathfrak{a}$ an integral ideal in $\mathfrak{K}$. Furthermore let $(1+\mathfrak{a}^{-1}\mathfrak{m})_+=(1+\mathfrak{a}^{-1}\mathfrak{m})\cap\mathbf{R}^*_{(+)}$ be the set of all totally positive elements in $1+\mathfrak{a}^{-1}\mathfrak{m}$. The group

$$E=\mathcal{O}^{\mathfrak{m}}_+=\left\{\varepsilon\in\mathcal{O}^*\;\middle|\;\varepsilon\equiv 1\bmod\mathfrak{m},\ \varepsilon\in\mathbf{R}^*_{(+)}\right\}$$

acts on $(1+\mathfrak{a}^{-1}\mathfrak{m})_+$, and we have the

(9.4) Lemma. *There is a bijection*

$$(1+\mathfrak{a}^{-1}\mathfrak{m})_+/E\xrightarrow{\ \sim\ }\mathfrak{K}_{\mathrm{int}}\,,\quad \bar a\longmapsto a\mathfrak{a}\,,$$

onto the set $\mathfrak{K}_{\mathrm{int}}$ of integral ideals in $\mathfrak{K}$.

Proof: Let $a\in(1+\mathfrak{a}^{-1}\mathfrak{m})_+$. Then we have $(a-1)\mathfrak{a}\subseteq\mathfrak{m}$, and since $\mathfrak{a}$ and $\mathfrak{m}$ are relatively prime, we get $a-1\in\mathfrak{m}$, i.e., $(a)\in P^{(\mathfrak{m})}$. Hence $a\mathfrak{a}$ lies in $\mathfrak{K}$. Furthermore, we have $a\mathfrak{a}\subseteq\mathfrak{a}(1+\mathfrak{a}^{-1}\mathfrak{m})=\mathfrak{a}+\mathfrak{m}=\mathcal{O}$, so that $a\mathfrak{a}$ is integral. Therefore $a\mapsto a\mathfrak{a}$ gives us a mapping

$$(1+\mathfrak{a}^{-1}\mathfrak{m})_+\to\mathfrak{K}_{\mathrm{int}}\,.$$

It is surjective, for if $a\mathfrak{a}$, $a\in P^{\mathfrak{m}}$, is an integral ideal in $\mathfrak{K}$, then $(a-1)\mathfrak{a}\subseteq\mathfrak{m}\mathfrak{a}\subseteq\mathfrak{m}$, so that $a\in 1+\mathfrak{a}^{-1}\mathfrak{m}$, and also $a\in\mathbf{R}^*_{(+)}$, and so $a\in(1+\mathfrak{a}^{-1}\mathfrak{m})_+$. For $a,b\in(1+\mathfrak{a}^{-1}\mathfrak{m})_+$, we have $a\mathfrak{a}=b\mathfrak{a}$ if and only if $(a)=(b)$, so that $a=b\varepsilon$ with $\varepsilon\in\mathcal{O}^*$. Since $\varepsilon\in(1+\mathfrak{a}^{-1}\mathfrak{m})_+$, it follows that $\varepsilon\in E$, i.e., a and b have exactly the same image if and only if they belong to the same class under the action of E. □

The lemma implies the following formula for the partial zeta function $\zeta(\mathfrak{K},s)$:

$$\zeta(\mathfrak{K},s)=\frac{1}{\mathfrak{N}(\mathfrak{a})^s}\sum_{a\in\mathfrak{R}}\frac{1}{|N(a)|^s}\,,$$

where $\mathfrak{R}$ runs through a system of representatives of $(1+\mathfrak{a}^{-1}\mathfrak{m})_+/E$. To this we now apply Shintani's unit theorem. Let

$$\mathbf{R}^*_{(+)} = \dot{\bigcup_{i=1}^{m}} \dot{\bigcup_{\varepsilon\in E}} \varepsilon C_i$$

be a disjoint decomposition of $\mathbf{R}^*_{(+)}$ into finitely many $\mathbb{Q}$-rational simplicial cones C_i. For every $i = 1, \dots, m$, let $v_{i1}, \dots, v_{id_i}$ be a linearly independent system of generators of C_i. Multiplying if necessary by a convenient totally positive integer, we may assume that all $v_{i\ell}$ lie in $\mathfrak{m}$. Let

$$C_i^1 = \left\{ t_1 v_{i1} + \cdots + t_{d_i} v_{id_i} \,\middle|\, 0 < t_\ell \le 1 \right\},$$

and

$$R(\mathfrak{K}, C_i) = (1+\mathfrak{a}^{-1}\mathfrak{m})_+ \cap C_i^1 .$$

Then we have the

(9.5) Proposition. *The sets $R(\mathfrak{K}, C_i)$ are finite, and one has*

$$\zeta(\mathfrak{K}, s) = \frac{1}{\mathfrak{N}(\mathfrak{a})^s} \sum_{i=1}^{m} \sum_{x\in R(\mathfrak{K}, C_i)} \zeta(C_i, x, s)$$

with the zeta functions

$$\zeta(C_i, x, s) = \sum_{\mathbf{z}} \left| N(x + z_1 v_{i1} + \cdots + z_{d_i} v_{id_i}) \right|^{-s},$$

where $\mathbf{z} = (z_1, \dots, z_{d_i})$ varies over all d_i-tuples of nonnegative integers.

Proof: $R(\mathfrak{K}, C_i)$ is a bounded subset of the lattice $\mathfrak{a}^{-1}\mathfrak{m}$ in $\mathbf{R}$, translated by 1. It is therefore finite. Since $C_i \subseteq \mathbf{R}^*_{(+)}$ is the simplicial cone generated by $v_{i1}, \dots, v_{id_i} \in \mathfrak{m}$, every $a \in (1+\mathfrak{a}^{-1}\mathfrak{m}) \cap C_i$ can be written uniquely as

$$a = \sum_{\ell=1}^{d_i} y_\ell v_{i\ell}$$

with rational numbers $y_\ell > 0$. Putting

$$y_\ell = x_\ell + z_\ell, \quad 0 < x_\ell \le 1, \quad 0 \le z_\ell \in \mathbb{Z},$$

we have $\sum x_\ell v_{i\ell} \in 1+\mathfrak{a}^{-1}\mathfrak{m}$ because $\sum z_\ell v_{i\ell} \in \mathfrak{m} \subseteq \mathfrak{a}^{-1}\mathfrak{m}$. In other words, every $a \in (1+\mathfrak{a}^{-1}\mathfrak{m}) \cap C_i$ can be written uniquely in the form

$$a = x + \sum_{\ell=1}^{d_i} z_\ell v_{i\ell}$$

with $x = \sum x_\ell v_{i\ell} \in R(\mathfrak{K}, C_i)$. Since

$$(1 + \mathfrak{a}^{-1}\mathfrak{m})_+ = \bigcup_{i=1}^{m} \bigcup_{\varepsilon \in E} (1 + \mathfrak{a}^{-1}\mathfrak{m}) \cap \varepsilon C_i ,$$

$a = x + \sum z_\ell v_{i\ell}$ runs through a system $\mathfrak{R}$ of representatives of $(1+\mathfrak{a}^{-1}\mathfrak{m})_+/E$ if i runs through the numbers $1, \ldots, m$, x through the elements of $R(\mathfrak{K}, C_i)$, and $\mathbf{z} = (z_1, \ldots, z_{d_i})$ through integer tuples with $z_\ell \geq 0$. Thus we indeed find that

$$\zeta(\mathfrak{K}, s) = \frac{1}{\mathfrak{N}(\mathfrak{a})^s} \sum_{i=1}^{m} \sum_{x \in R(\mathfrak{K}, C_i)} \zeta(C_i, x, s) . \qquad \square$$

(9.6) Corollary. *For the Dirichlet L-series attached to the Dirichlet character $\chi : J^{\mathfrak{m}}/P^{\mathfrak{m}} \to \mathbb{C}^*$, we have the decomposition*

$$L(\chi, s) = \sum_{\mathfrak{K}} \frac{\chi(\mathfrak{a})}{\mathfrak{N}(\mathfrak{a})^s} \sum_{i=1}^{m} \sum_{x \in R(\mathfrak{K}, C_i)} \zeta(C_i, x, s) ,$$

where $\mathfrak{K}$ runs through the classes $J^{\mathfrak{m}}/P^{\mathfrak{m}}$, and $\mathfrak{a}$ denotes an integral ideal in $\mathfrak{K}$, one for each class.

The relation between zeta functions and Bernoulli numbers hinges on a purely analytic fact which is independent of number theory. This is what we will describe now.

Let A be a real $r \times n$-matrix, $r \leq n$, with *positive* entries a_{ji}, $1 \leq j \leq r$, $1 \leq i \leq n$. From this matrix we construct the linear forms

$$L_j(t_1, \ldots, t_n) = \sum_{i=1}^{n} a_{ji} t_i \quad \text{and} \quad L_i^*(z_1, \ldots, z_r) = \sum_{j=1}^{r} a_{ji} z_j .$$

For an r-tuple $x = (x_1, \ldots, x_r)$ of *positive* real numbers, we write the following series

$$\zeta(A, x, s) = \sum_{z_1, \ldots, z_r = 0}^{\infty} \prod_{i=1}^{n} L_i^*(z + x)^{-s} .$$

On the other hand we define the generalized **Bernoulli polynomials** $B_k(A, x)$ by

$$B_k(A, x) = \frac{1}{n} \sum_{i=1}^{n} B_k(A, x)^{(i)} ,$$

where $B_k(A, x)^{(i)}/(k!)^n$ is the coefficient of

$$u^{(k-1)n} (t_1 \cdots t_{i-1} t_{i+1} \cdots t_n)^{k-1}$$

in the Laurent expansion at 0 of the function

$$\prod_{j=1}^{r} \frac{\exp(ux_j L_j(t))}{\exp(uL_j(t)) - 1}\bigg|_{t_i=1}$$

in the variables $u, t_1, \dots, t_{i-1}, t_{i+1}, \dots, t_n$. For $r = n = 1$ and $A = a$, we have $B_k(a,x) = a^{k-1}B_k(x)$, with the usual Bernoulli polynomial $B_k(x)$ (see § 1, exercise 2). The equation

$$B_k(A, 1-x) = (-1)^{n(k-1)+r} B_k(A,x),$$

where $1-x$ signifies $(1-x_1, \dots, 1-x_r)$, is easily proved.

(9.7) Proposition. *The series $\zeta(A,x,s)$ is absolutely convergent for* $\mathrm{Re}(s) > r/n$, *and it can be meromorphically continued to the whole complex plane. Its values at the points $s = 1-k$, $k = 1, 2, \dots$, are given by*

$$\zeta(A,x,1-k) = (-1)^r \frac{B_k(A,x)}{k^n}.$$

Proof: The absolute convergence for $\mathrm{Re}(s) > r/n$ is deduced from the convergence of a series $\sum_{n=1}^{\infty} \frac{1}{n^{1+\delta}}$ by the same arguments that we have used repeatedly. It will be left to the reader. The remainder of the proof is similar to that of (1.8). In the gamma function

$$\Gamma(s)^n = \int_0^\infty \cdots \int_0^\infty \prod_{i=1}^{n} e^{-t_i} (t_1 \cdots t_n)^{s-1} dt_1 \cdots dt_n,$$

we substitute

$$t_i \longmapsto L_i^*(z+x)t_i,$$

and obtain

$$\Gamma(s)^n \prod_{i=1}^{n} L_i^*(z+x)^{-s}$$
$$= \int_0^\infty \cdots \int_0^\infty \exp\Big[-\sum_{i=1}^{n} t_i L_i^*(z+x)\Big](t_1 \cdots t_n)^{s-1} dt_1 \cdots dt_n.$$

Summing this over all $z = (z_1, \dots, z_r)$, $z_i \in \mathbb{Z}$, $z_i \geq 0$, and observing that

$$\sum_{i=1}^{n} t_i L_i^*(z+x) = \sum_{j=1}^{r} (z_j + x_j) L_j(t),$$

yields the equation

$$\Gamma(s)^n \zeta(A,x,s) = \int_0^\infty \cdots \int_0^\infty g(t)(t_1 \cdots t_n)^{s-1}\, dt_1 \cdots dt_n$$

with the function

$$g(t) = g(t_1, \ldots, t_n) = \prod_{j=1}^{r} \frac{\exp((1-x_j)L_j(t))}{\exp(L_j(t)) - 1}.$$

We cut up the space $\mathbb{R}^n$ into the subsets

$$D_i = \{t \in \mathbb{R}^n \mid 0 \le t_\ell \le t_i,\ \ell = 1, \ldots, i-1, i+1, \ldots, n\}$$

for $i = 1, \ldots, n$, and get

$$\zeta(A,x,s) = \Gamma(s)^{-n} \sum_{i=1}^{n} \int_{D_i} g(t)(t_1 \cdots t_n)^{s-1}\, dt_1 \cdots dt_n. \tag{1}$$

In D_i we make the transformation of variables

$$t = uy = u(y_1, \ldots, y_n),$$

where $0 < u$, $0 \le y_\ell \le 1$ for $\ell \ne i$ and $y_i = 1$. This gives

$$\Gamma(s)^{-n} \int_{D_i} g(t)(t_1 \cdots t_n)^{s-1} dt_1 \cdots dt_n$$

$$= \Gamma(s)^{-n} \int_0^\infty \Big[\int_0^1 \cdots \int_0^1 g(uy)(\prod_{\ell \ne i} y_\ell)^{s-1} \prod_{\ell \ne i} dy_\ell \Big] u^{ns-1}\, du.$$

For $0 < \varepsilon < 1$, let now $I_\varepsilon(1)$, resp. $I_\varepsilon(+\infty)$, denote the path in $\mathbb{C}$ consisting of the interval $[1, \varepsilon]$, resp. $[+\infty, \varepsilon]$, followed by a circle around 0 of radius ε in the positive direction, and the interval $[\varepsilon, 1]$, resp $[\varepsilon, +\infty]$. For ε sufficiently small, the right-hand side of the last equation following (1.9) becomes

$$A(s) \int_{I_\varepsilon(+\infty)} \int_{I_\varepsilon(1)^{n-1}} \Big[g(uy) u^{ns-1} (\prod_{\ell \ne i} y_\ell)^{s-1} \prod_{\ell \ne i} dy_\ell \Big] du, \tag{2}$$

with the factor

$$A(s) = \frac{\Gamma(s)^{-n}}{(e^{2\pi i n s} - 1)(e^{2\pi i s} - 1)^{n-1}},$$

where one has to observe that the linear forms $L_1, \ldots, L_r$ have positive coefficients. It is easy to check that the above expression, as a function of the variable s, is meromorphic on all of $\mathbb{C}$. As for the factor $A(s)$, (1.2) implies that

$$A(s) = \frac{1}{(2\pi i)^n} \, \frac{\Gamma(1-s)^n}{(e^{2\pi ins}-1)(e^{2\pi is}-1)^{-1}e^{n\pi is}}.$$

Let us now put $s = 1-k$. The function $e^{n\pi is}(e^{2\pi ins}-1)/(e^{2\pi is}-1)$ takes the value $(-1)^{n(k-1)}n$ at $s=1-k$. Thus expression (2) turns into

$$(-1)^{n(k-1)}\frac{\Gamma(k)^n}{n}\cdot\frac{1}{(2\pi i)^n}\int\limits_{K_\varepsilon}\int\limits_{K_\varepsilon^{n-1}}\Big[g(uy)u^{n(1-k)-1}\big(\prod_{\ell\neq i}y_\ell\big)^{-k}\prod_{\ell\neq i}dy_\ell\Big]du\,,$$

where K_ε denotes the positively oriented circumference of the circle of radius ε, and where we have to observe that the integrals over $(\infty,\varepsilon]$ and $[\varepsilon,\infty)$, resp. over $[1,\varepsilon]$ and $[\varepsilon,1]$, kill each other in (2) if $s=1-k$. This value is obviously $\big((-1)^{n(k-1)}\Gamma(k)^n/n\big)$ times the coefficient of $u^{n(k-1)}(\prod_{\ell\neq i}y_\ell)^{k-1}$ in the Laurent expansion of the function

$$g(uy_1,\dots,uy_{i-1},u,uy_{i+1},\dots,uy_n) = \prod_{j=1}^{r}\frac{\exp(u(1-x_j)L_j(y))}{\exp(uL_j(y))-1}\Bigg|_{y_i=1},$$

which is a holomorphic function of $u,t_1,\dots,t_{i-1},t_{i+1},\dots,t_n$ in the direct product of n copies of the punctured disc of radius ε. Therefore the value of (2) at $s=1-k$ equals $(-1)^{n(k-1)}k^{-n}B_k(A,1-x)^{(i)}/n$. Inserting this into (1) gives

$$\begin{aligned}\zeta(A,x,1-k) &= (-1)^{n(k-1)}k^{-n}\frac{1}{n}\sum_{i=1}^{n}B_k(A,1-x)^{(i)}\\ &= (-1)^{n(k-1)}\frac{B_k(A,1-x)}{k^n}.\end{aligned}$$

Together with the equation $B_k(A,1-x) = (-1)^{n(k-1)+r}B_k(A,x)$ mentioned above, this gives the desired result. □

Theorems (9.5) and (9.6) now imply our main result concerning the values of Dirichlet L-series $L(\chi,s)$ at integer points $s=1-k$, $k=1,2,\dots$ If K is not totally real, then these values are all zero (except if χ is the trivial character, for which $s=0$ is not a zero). This can be read off immediately from the functional equation (8.6) and (5.11).

So we let K be a totally real number field of degree n. Numbering the embeddings $\tau: K\to\mathbb{R}$ identifies the Minkowski space $\mathbf{R}$ with $\mathbb{R}^n$, and $\mathbf{R}^*_{(+)} = \mathbf{R}^n_+$ with the set $\mathbb{R}^n_+$ of vectors $(x_1,\dots,x_n)$ with positive coefficients x_i. Given the $\mathbb{Q}$-rational simplicial cone $C_i\subseteq\mathbb{R}^n_+$ generated by $v_{i1},\dots,v_{id_i}$, we again consider the zeta functions

$$\zeta(C_i,x,s) = \sum_{\mathbf{z}}\big|N(x+z_1v_{i1}+\cdots+z_{d_i}v_{id_i})\big|^{-s}.$$

If

$$v_{ij} = (a_{j1}^{(i)}, \dots, a_{jn}^{(i)}), \quad j = 1, \dots, d_i ,$$

then $A_i = (a_{jk}^{(i)})$ is a $(d_i \times n)$-matrix with positive entries, and the k-th component of $z_1 v_{i1} + \dots + z_{d_i} v_{id_i}$ becomes

$$L_k^*(z_1, \dots, z_{d_i}) = \sum_{j=1}^{d_i} a_{jk}^{(i)} z_j .$$

For $x \in \mathbb{R}_+^*$, we therefore get

$$\zeta(C_i, x, s) = \sum_{\mathbf{z}} \prod_{k=1}^{n} L_k^*(z_1, \dots, z_{d_i})^{-s} = \zeta(A_i, x, s),$$

and, from (9.5) and (9.6), we obtain by putting $s = 1 - k$ the

(9.8) Theorem. *The values of the partial zeta function $\zeta(\mathfrak{K}, s)$ at the integral points $s = 1 - k$, $k = 1, 2, 3, \dots$, are given by*

$$\zeta(\mathfrak{K}, 1 - k) = \mathfrak{N}(\mathfrak{a})^{k-1} \sum_{i=1}^{m} \Big[(-1)^{d_i} \sum_{x \in R(\mathfrak{K}, C_i)} \frac{B_k(A_i, x)}{k^n} \Big],$$

and the values of the Dirichlet L-series $L(\chi, s)$ are given by

$$L(\chi, 1 - k) = \sum_{\mathfrak{K}} \chi(\mathfrak{a}) \mathfrak{N}(\mathfrak{a})^{k-1} \sum_{i=1}^{m} \Big[(-1)^{d_i} \sum_{x \in R(\mathfrak{K}, C_i)} \frac{B_k(A_i, x)}{k^n} \Big].$$

Here $\mathfrak{a}$ is an integral ideal in the class $\mathfrak{K}$ of $J^{\mathfrak{m}}/P^{\mathfrak{m}}$.

This result about the Dirichlet L-series $L(\chi, s)$ also covers the Dedekind zeta function $\zeta_K(s)$. The theorem says in particular that the values $L(\chi, 1-k)$, for $k \geq 1$, are algebraic numbers which all lie in the cyclotomic field $\mathbb{Q}(\chi_{\mathfrak{f}})$ generated by the values of the character $\chi_{\mathfrak{f}}$. The values $\zeta_K(1 - k)$ are even *rational* numbers. From the functional equation (5.11),

$$\zeta_K(1 - s) = |d_K|^{s-1/2} \Big(\cos \frac{\pi s}{2} \Big)^{r_1 + r_2} \Big(\sin \frac{\pi s}{2} \Big)^{r_2} \Gamma_{\mathbb{C}}(s)^n \zeta_K(s),$$

we deduce that $\zeta_K(1 - k) = 0$ for *odd* $k > 1$, and it is $\neq 0$ for *even* $k > 1$. If the number field K is not totally real, then we have $\zeta_K(s) = 0$ for *all* $s = -1, -2, -3, \dots$

(9.9) Corollary (Siegel-Klingen). *The values of the partial zeta function $\zeta(\mathfrak{K}, s)$ at the points $s = 0, -1, -2, \dots$ are rational numbers.*

Proof: Let $a_1, \ldots, a_r$ be nonzero numbers in K, and let A be the $(r \times n)$-matrix (a_{ji}), where a_{ji} is the i-th component of a_j, after identifying $\mathbf{R} = \mathbb{R}^n$ according to the chosen numbering of the embeddings $\tau : K \to \mathbb{R}$. It is enough to show that $B_k(A, x)$ is a rational number for every r-tuple of rational numbers $x = (x_1, \ldots, x_r)$. To see this, let $L|\mathbb{Q}$ be the normal closure of $K|\mathbb{Q}$ and $\sigma \in G(L|\mathbb{Q})$. Then σ induces a permutation of the indices $\{1, 2, \ldots, n\}$ so that

$$\sigma a_{ji} = a_{j\sigma(i)} \quad (1 \le j \le r,\ i = 1, \ldots, n).$$

Now we had $B_k(A, x) = \frac{1}{n} \sum_{i=1}^n B_k(A, x)^{(i)}$, where $B_k(A, x)^{(i)}$ was the coefficient of $u^{n(k-1)+r}(t_1, \ldots, t_{i-1}, t_{i+1}, \ldots, t_n)^{k-1}$ in the Taylor expansion of the function

$$u^r \prod_{j=1}^{r} \frac{\exp(x_j u L_j(t))}{\exp(u L_j(t) - 1)}\bigg|_{t_i = 1},$$

with $L_j(t) = a_{j1}t_1 + \cdots + a_{jn}t_n$. This makes it clear that $B_k(A, x)^{(i)}$ lies in L and that $\sigma B_k(A, x)^{(i)} = B_k(A, x)^{(\sigma(i))}$. Therefore $B_k(A, x)$ is invariant under the action of the Galois group $G(L|\mathbb{Q})$, and thus belongs to $\mathbb{Q}$. □

The nature of the special values of L-series at integer points has recently found increasing interest. Like in the class number formula, which expresses the behaviour of the Dedekind zeta function at the point $s = 0$, the properties of all the special values indicate a deep arithmetic law which appears to extend to an extremely wide class of L-series, the L-series attached to "motives". According to a conjecture of the American mathematician *Stephen Lichtenbaum*, the significance of these L-values can be explained by a strikingly simple geometric interpretation: they appear according to the **Lichtenbaum conjecture** as Euler characteristics in étale cohomology (see [99], [12]). The proof of this conjecture is a great, if still remote, goal of number theory. On the way towards it, the insights into the nature of L-series which we have encountered may prove to be important.

Finally we want to mention that the French mathematicians *Daniel Barsky* and *Pierette Cassou-Noguès* have used *Shintani*'s result to prove the existence of *p-adic L-series*. These play a major rôle in *Iwasawa theory*, which we have mentioned before. The *p-adic zeta function* of a totally real number field K is a continuous function

$$\zeta_p : \mathbb{Z}_p \setminus \{1\} \longrightarrow \mathbb{Q}_p,$$

which is related to the ordinary Dedekind zeta function $\zeta_K(s)$ by

$$\zeta_p(-n) = \zeta_K(-n) \prod_{\mathfrak{p}|p} \big(1 - \mathfrak{N}(\mathfrak{p})^n\big)$$

for all $n \in \mathbb{N}$ such that $-n \equiv 1 \bmod d$, where $d = [K(\mu_{2p}) : K]$ denotes the degree of the field $K(\mu_{2p})$ of $2p$-th roots of unity over K. The p-adic zeta function is uniquely determined by this relation. Its existence hinges on the fact that the rational values $\zeta_K(-n)$ are subjected to severe congruences with respect to p.

§ 10. Artin L-series

So far, all L-series we have considered were associated to an individual number field K. With the *Artin L-series*, a new type of L-series enters the stage; these are derived from representations of the Galois group $G(L|K)$ of a Galois extension $L|K$. This new kind of L-series is intimately related to the old ones via the main theorem of class field theory. In this way they appear as far-reaching generalizations of the old L-series. Let us explain this for the case of a Dirichlet L-series

$$L(\chi, s) = \sum_{n=1}^{\infty} \frac{\chi(n)}{n^s} = \prod_p \frac{1}{1 - \chi(p)p^{-s}}$$

attached to a Dirichlet character

$$\chi : (\mathbb{Z}/m\mathbb{Z})^* \longrightarrow \mathbb{C}^* .$$

Let $G = G(\mathbb{Q}(\mu_m)|\mathbb{Q})$ be the Galois group of the field $\mathbb{Q}(\mu_m)$ of m-th roots of unity. The main theorem of class field theory in this particular case simply describes the familiar isomorphism

$$(\mathbb{Z}/m\mathbb{Z})^* \xrightarrow{\sim} G ,$$

which sends the residue class $p \bmod m$ of a prime number $p \nmid m$ to the *Frobenius automorphism* φ_p, which in turn is defined by

$$\varphi_p \zeta = \zeta^p \quad \text{for} \quad \zeta \in \mu_m .$$

Using this isomorphism we may interpret χ as a character of the Galois group G, or in other words, as a 1-dimensional *representation* of G, i.e., a homomorphism

$$\chi : G \longrightarrow GL_1(\mathbb{C}) .$$

This interpretation describes the Dirichlet L-series in a purely Galois-theoretic fashion,

$$L(\chi, s) = \prod_{p \nmid m} \frac{1}{1 - \chi(\varphi_p)p^{-s}} ,$$

and allows us the following generalization.

Let $L|K$ be a Galois extension of finite algebraic number fields with Galois group $G = G(L|K)$. A **representation** of G is an action of G on a finite dimensional $\mathbb{C}$-vector space V, i.e., a homomorphism

$$\rho : G \to GL(V) = \mathrm{Aut}_{\mathbb{C}}(V).$$

Our shorthand notation for the action of $\sigma \in G$ on $v \in V$ is σv, instead of the complete expression $\rho(\sigma)v$. Let $\mathfrak{p}$ be a prime ideal of K, and let $\mathfrak{P}|\mathfrak{p}$ be a prime ideal of L lying above $\mathfrak{p}$. Let $G_{\mathfrak{P}}$ be the decomposition group and $I_{\mathfrak{P}}$ the inertia group of $\mathfrak{P}$ over $\mathfrak{p}$. Then we have a canonical isomorphism

$$G_{\mathfrak{P}}/I_{\mathfrak{P}} \xrightarrow{\sim} G\big(\kappa(\mathfrak{P})|\kappa(\mathfrak{p})\big)$$

onto the Galois group of the residue field extension $\kappa(\mathfrak{P})|\kappa(\mathfrak{p})$ (see chap. I, (9.5)). The factor group $G_{\mathfrak{P}}/I_{\mathfrak{P}}$ is therefore generated by the **Frobenius automorphism** $\varphi_{\mathfrak{P}}$ whose image in $G(\kappa(\mathfrak{P})|\kappa(\mathfrak{p}))$ is the q-th power map $x \mapsto x^q$, where $q = \mathfrak{N}(\mathfrak{p})$. $\varphi_{\mathfrak{P}}$ is an endomorphism of the module $V^{I_{\mathfrak{P}}}$ of invariants. The **characteristic polynomial**

$$\det(1 - \varphi_{\mathfrak{P}} t;\ V^{I_{\mathfrak{P}}})$$

only depends on the prime ideal $\mathfrak{p}$, not on the choice of the prime ideal $\mathfrak{P}$ above $\mathfrak{p}$. In fact, a different choice $\mathfrak{P}'|\mathfrak{p}$ yields an endomorphism conjugate to $\varphi_{\mathfrak{P}}$, as the decomposition groups $G_{\mathfrak{P}}$ and $G_{\mathfrak{P}'}$, the inertia groups $I_{\mathfrak{P}}$ und $I_{\mathfrak{P}'}$, and the Frobenius automorphisms $\varphi_{\mathfrak{P}}$ and $\varphi_{\mathfrak{P}'}$ are simultaneous conjugates. We thus arrive at the following

(10.1) Definition. *Let $L|K$ be a Galois extension of algebraic number fields with Galois group G, and let (ρ, V) be a representation of G. Then the* **Artin L-series** *attached to ρ is defined to be*

$$\mathcal{L}(L|K, \rho, s) = \prod_{\mathfrak{p}} \frac{1}{\det(1 - \varphi_{\mathfrak{P}} \mathfrak{N}(\mathfrak{p})^{-s};\ V^{I_{\mathfrak{P}}})},$$

where $\mathfrak{p}$ runs through all prime ideals of K.

The Artin L-series converges absolutely and uniformly in the half-plane $\mathrm{Re}(s) \geq 1 + \delta$, for any $\delta > 0$. It thus defines an analytic function on the half-plane $\mathrm{Re}(s) > 1$. This is shown in the same way as for the Hecke L-series (see (8.1)), observing that the ε_i in the factorization

$$\det\big(1 - \varphi_{\mathfrak{P}} \mathfrak{N}(\mathfrak{p})^{-s};\ V^{I_{\mathfrak{P}}}\big) = \prod_{i=1}^{d} \big(1 - \varepsilon_i \mathfrak{N}(\mathfrak{p})^{-s}\big)$$

are roots of unity because the endomorphism $\varphi_{\mathfrak{P}}$ of $V^{I_{\mathfrak{P}}}$ has finite order.

For the trivial representation $(\rho, \mathbb{C})$, $\rho(\sigma) \equiv 1$, the Artin L-series is simply the Dedekind zeta function $\zeta_K(s)$. An additive expression analogous to the expansion

$$\zeta_K(s) = \sum_{\mathfrak{a}} \frac{1}{\mathfrak{N}(\mathfrak{a})^s}$$

does not exist for general Artin L-series. But they exhibit a perfectly regular behaviour under change of extensions $L|K$ and representations ρ. This allows to deduce many of their excellent properties. As a preparation for this study, we first collect basic facts from representation theory of finite groups. For their proofs we refer to [125].

The **degree** of a representation (ρ, V) of a finite group G is the dimension of V. The representation is called **irreducible** if the G-module V does not admit any proper G-invariant subspace. An irreducible representation of an *abelian* group is simply a character

$$\rho : G \longrightarrow \mathbb{C}^* = GL_1(\mathbb{C}).$$

Two representations (ρ, V) and (ρ', V') are called **equivalent** if the G-modules V and V' are isomorphic. Every representation (ρ, V) factors into a direct sum

$$V = V_1 \oplus \cdots \oplus V_s$$

of *irreducible* representations. If an irreducible representation (ρ_α, V_α) is equivalent to precisely r_α among the representations in this decomposition, then r_α is called the **multiplicity** of ρ_α in ρ, and one writes

$$\rho \sim \sum_\alpha r_\alpha \rho_\alpha ,$$

where ρ_α varies over all non-equivalent irreducible representations of G.

The **character** of a representation (ρ, V) is by definition the function

$$\chi_\rho : G \longrightarrow \mathbb{C}, \quad \chi_\rho(\sigma) = \operatorname{trace} \rho(\sigma).$$

One has $\chi_\rho(1) = \dim V = \operatorname{degree}(\rho)$, and $\chi_\rho(\sigma\tau\sigma^{-1}) = \chi_\rho(\tau)$ for all $\sigma, \tau \in G$. In general, a function $f : G \to \mathbb{C}$ with the property that $f(\sigma\tau\sigma^{-1}) = f(\tau)$ is called a **central** function (or **class function**). The special importance of characters comes from the following fact:

Two representations are equivalent if and only if their characters are equal. If $\rho \sim \sum_\alpha r_\alpha \rho_\alpha$, then

$$\chi_\rho = \sum_\alpha r_\alpha \chi_{\rho_\alpha} .$$

The character of the *trivial representation* $\rho : G \to GL(V)$, $\dim V = 1$, $\rho(\sigma) = 1$ for all $\sigma \in G$, is the constant function of value 1, and is denoted by $\mathbf{1}_G$, or simply $\mathbf{1}$. The *regular representation* is given by the G-module

$$V = \mathbb{C}[G] = \Big\{ \sum_{\tau \in G} x_\tau \tau \;\Big|\; x_\tau \in \mathbb{C} \Big\},$$

on which the $\sigma \in G$ act via multiplication on the left. It decomposes into the direct sum of the trivial representation $V_0 = \mathbb{C} \sum_{\sigma \in G} \sigma$, and the *augmentation representation* $\big\{ \sum_{\sigma \in G} x_\sigma \sigma \;\big|\; \sum_\sigma x_\sigma = 0 \big\}$. The character associated with the regular, resp. the augmentation representation, is denoted by r_G, resp. u_G. We thus have $r_G = u_G + \mathbf{1}_G$, and explicitly: $r_G(\sigma) = 0$ for $\sigma \neq 1$, $r_G(1) = g = \#G$.

A character χ is called *irreducible* if it belongs to an irreducible representation. Every central function φ can be written uniquely as a linear combination

$$\varphi = \sum c_\chi \chi, \quad c_\chi \in \mathbb{C},$$

of irreducible characters. φ is a character of a representation of G if and only if the c_χ are rational integers ≥ 0. For instance, for the character r_G of the regular representation we find

$$r_G = \sum \chi(1)\chi,$$

where χ varies over all irreducible characters of G. Given any two central functions φ and ψ of G, we put

$$(\varphi, \psi) = \frac{1}{g} \sum_{\sigma \in G} \varphi(\sigma)\overline{\psi}(\sigma), \quad g = \#G,$$

where $\overline{\psi}$ is the function which is the complex conjugate of ψ. For two irreducible characters χ and χ', this gives

$$(\chi, \chi') = \begin{cases} 1, & \text{if } \chi = \chi', \\ 0, & \text{if } \chi \neq \chi'. \end{cases}$$

In other words, $(\ ,\ )$ is a hermitian scalar product on the space of all central functions on G, and the irreducible characters form an orthonormal basis of this hermitian space.

For the representations itself, this scalar product has the following meaning. Let

$$V = V_1 \oplus \cdots \oplus V_r$$

be the decomposition of a representation V with character χ into the direct sum of irreducible representations V_i. If V' is an irreducible representation with character χ', then (χ, χ') is the number of times that V' occurs among the V_i, up to isomorphism. For if χ_i is the character of V_i, then $\chi = \chi_1 + \cdots + \chi_r$, so that

$$(\chi, \chi') = (\chi_1, \chi') + \cdots + (\chi_r, \chi'),$$

and we have $(\chi_i, \chi') = 1$ or 0, depending whether V_i is or is not isomorphic to V'. Applying this to the trivial representation $V' = \mathbb{C}$, we obtain in particular that

$$\dim V^G = \frac{1}{g} \sum_{\sigma \in G} \chi(\sigma), \quad g = \#G.$$

Now let $h : H \to G$ be a homomorphism of finite groups. If φ is a central function on G, then $h^*(\varphi) = \varphi \circ h$ is a central function on H. Conversely, one has the following proposition.

(10.2) Frobenius Reciprocity. *For every central function ψ on H there is one and only one central function $h_*(\psi)$ on G such that one has*

$$\big(\varphi, h_*(\psi)\big) = \big(h^*(\varphi), \psi\big)$$

for all central functions φ on G.

This will be applied chiefly to the following two special cases.

a) *H is a subgroup of G and h is inclusion.*

In this case we write $\varphi|H$ or simply φ instead of $h^*(\varphi)$, and ψ_* instead of $h_*(\psi)$ (the **induced function**). If φ is the character of a representation (ρ, V) of G, then $\varphi|H$ is the character of the representation $(\rho|H, V)$. If ψ is the character of a representation (ρ, V) of H, then ψ_* is the character of the representation $(\mathrm{ind}(\rho), \mathrm{Ind}_G^H(V))$ given by the **induced** G-module

$$\mathrm{Ind}_G^H(V) = \big\{ f : G \to V \bigm| f(\tau x) = \tau f(x) \quad \text{for all } \tau \in H \big\},$$

on which $\sigma \in G$ acts by $(\sigma f)(x) = f(x\sigma)$ (see chap. IV, §7). One has

$$\psi_*(\sigma) = \sum_{\tau} \psi(\tau\sigma\tau^{-1}),$$

where τ varies over a system of representatives on the right of G/H, and we put $\psi(\tau\sigma\tau^{-1}) = 0$ if $\tau\sigma\tau^{-1} \notin H$.

b) *G is a quotient group H/N of H and h is the projection.*

We then write φ instead of $h^*(\varphi)$, and $\psi_\natural$ instead of $h_*(\psi)$. One has

$$\psi_\natural(\sigma) = \frac{1}{\#N} \sum_{\tau \mapsto \sigma} \psi(\tau).$$

If φ is the character of a representation (ρ, V) of G, then $h^*(\varphi)$ is the character of the representation $(\rho \circ h, V)$.

The following result is of great importance.

(10.3) Brauer's Theorem. *Every character χ of a finite group G is a $\mathbb{Z}$-linear combination of characters χ_{i*} induced from characters χ_i of degree 1 associated to subgroups H_i of G.*

Note that a character of degree 1 of a group H is simply a homomorphism $\chi : H \to \mathbb{C}^*$.

After this brief survey of representation theory for finite groups, we now return to Artin L-series. Since two representations (ρ, V) and (ρ', V') are equivalent if and only if their characters χ and χ' coincide, we will henceforth write

$$\mathcal{L}(L|K,\chi,s) = \prod_{\mathfrak{p}} \frac{1}{\det(1-\rho(\varphi_{\mathfrak{P}})\mathfrak{N}(\mathfrak{p})^{-s};\, V^{I_{\mathfrak{P}}})}$$

instead of $\mathcal{L}(L|K,\rho,s)$. These L-series exhibit the following functorial behaviour.

(10.4) Proposition. (i) *For the principal character $\chi = \mathbf{1}$, one has*

$$\mathcal{L}(L|K,\mathbf{1},s) = \zeta_K(s)\,.$$

(ii) *If χ, χ' are two characters of $G(L|K)$, then*

$$\mathcal{L}(L|K,\chi+\chi',s) = \mathcal{L}(L|K,\chi,s)\mathcal{L}(L|K,\chi',s)\,.$$

(iii) *For a bigger Galois extension $L'|K$, $L' \supseteq L \supseteq K$, and a character χ of $G(L|K)$ one has*

$$\mathcal{L}(L'|K,\chi,s) = \mathcal{L}(L|K,\chi,s)\,.$$

(iv) *If M is an intermediate field, $L \supseteq M \supseteq K$, and χ is a character of $G(L|M)$, then*

$$\mathcal{L}(L|M,\chi,s) = \mathcal{L}(L|K,\chi_*,s)\,.$$

Proof: We have already noted (i) earlier. (ii) If (ρ, V), (ρ', V') are representations of $G(L|K)$ with characters χ, χ', then the direct sum $(\rho\oplus\rho', V\oplus V')$ is a representation with character $\chi+\chi'$, and

$$\det(1-\varphi_{\mathfrak{P}}t;\,(V\oplus V')^{I_{\mathfrak{P}}}) = \det(1-\varphi_{\mathfrak{P}}t;\,V^{I_{\mathfrak{P}}})\det(1-\varphi_{\mathfrak{P}}t;\,V'^{I_{\mathfrak{P}}})\,.$$

This yields (ii).
(iii) Let $\mathfrak{P}'|\mathfrak{P}|\mathfrak{p}$ be prime ideals of $L'|L|K$, each lying above the next. Let χ be the character belonging to the $G(L|K)$-module V. $G(L'|K)$ acts on V via the projection $G(L'|K) \to G(L|K)$. It induces surjective homomorphisms

$$G_{\mathfrak{P}'} \longrightarrow G_{\mathfrak{P}},\ I_{\mathfrak{P}'} \longrightarrow I_{\mathfrak{P}},\ G_{\mathfrak{P}'}/I_{\mathfrak{P}'} \longrightarrow G_{\mathfrak{P}}/I_{\mathfrak{P}}$$

of the decomposition and inertia groups. The latter maps the Frobenius automorphism $\varphi_{\mathfrak{P}'}$ to the Frobenius automorphism $\varphi_{\mathfrak{P}}$ so that $(\varphi_{\mathfrak{P}'}, V^{I_{\mathfrak{P}'}}) = (\varphi_{\mathfrak{P}}, V^{I_{\mathfrak{P}}})$, i.e.,

$$\det(1 - \varphi_{\mathfrak{P}'}t\,;\; V^{I_{\mathfrak{P}'}}) = \det(1 - \varphi_{\mathfrak{P}}t, V^{I_{\mathfrak{P}}}).$$

This yields (iii).

(iv) Let $G = G(L|K)$ and $H = G(L|M)$. Let $\mathfrak{p}$ be a prime ideal of K, $\mathfrak{q}_1, \dots, \mathfrak{q}_r$ the various prime ideals of M above $\mathfrak{p}$, and $\mathfrak{P}_i$ a prime ideal of L above $\mathfrak{q}_i$, $i = 1, \dots, r$. Let G_i, resp. I_i, be the decomposition, resp. inertia, group of $\mathfrak{P}_i$ over $\mathfrak{p}$. Then $H_i = G_i \cap H$, resp. $I_i' = I_i \cap H$, are the decomposition, resp. inertia, groups of $\mathfrak{P}_i$ over $\mathfrak{q}_i$. The degree of $\mathfrak{q}_i$ over $\mathfrak{p}$ is $f_i = (G_i : H_i I_i)$, i.e.,

$$\mathfrak{N}(\mathfrak{q}_i) = \mathfrak{N}(\mathfrak{p})^{f_i}.$$

We choose elements $\tau_i \in G$ such that $\mathfrak{P}_i = \mathfrak{P}_1^{\tau_i}$. Then $G_i = \tau_i^{-1} G_1 \tau_i$, and $I_i = \tau_i^{-1} I_1 \tau_i$. Let $\varphi \in G_1$ be an element which is mapped to the Frobenius $\varphi_{\mathfrak{P}_1} \in G_1/I_1$. Then $\varphi_i = \tau_i^{-1}\varphi\tau_i \in G_i$ is mapped to the Frobenius $\varphi_{\mathfrak{P}_i} \in G_i/I_i$, and the image of $\varphi_i^{f_i}$ in H_i/I_i' is the Frobenius of $\mathfrak{P}_i$ over $\mathfrak{q}_i$.

Now let $\rho : H \to GL(W)$ be a representation of H with character χ. Then χ_* is the character of the induced representation $\operatorname{ind}(\rho) : G \to GL(V)$, $V = \operatorname{Ind}_G^H(W)$. Clearly, what we have to show is that

$$\det(1 - \varphi t\,;\; V^{I_1}) = \prod_{i=1}^{r} \det(1 - \varphi_i^{f_i} t^{f_i}\,;\; W^{I_i'}).$$

We reduce the problem to the case $G_1 = G$, i.e., $r = 1$. Conjugating by τ_i, we obtain

$$\det(1 - \varphi_i^{f_i} t^{f_i}\,;\; W^{I_i'}) = \det(1 - \varphi^{f_i} t^{f_i}\,;\; (\tau_i W)^{I_1 \cap \tau_i H \tau_i^{-1}})$$

and $f_i = (G_1 : (G_1 \cap \tau_i H \tau_i^{-1}) I_1)$. For every i we choose a system of representatives on the left, σ_{ij}, of G_1 mod $G_1 \cap \tau_i H \tau_i^{-1}$. One checks immediately that then $\{\sigma_{ij}\tau_i\}$ is a system of representatives on the left of G mod H. We thus have (see chap. IV, § 5, p. 297) the decomposition

$$V = \bigoplus_{i,j} \sigma_{ij}\tau_i W.$$

Putting $V_i = \bigoplus_j \sigma_{ij}\tau_i W$, we obtain a decomposition $V = \bigoplus_i V_i$ of V as a G_1-module. Hence

$$\det(1 - \varphi t\,;\; V^{I_1}) = \prod_{i=1}^{r} \det(1 - \varphi t\,;\; V_i^{I_1}).$$

It is therefore sufficient to prove that

$$\det(1 - \varphi t\,;\; V_i^{I_1}) = \det(1 - \varphi^{f_i} t^{f_i}\,;\; (\tau_i W)^{I_1 \cap \tau_i H \tau_i^{-1}}).$$

We simplify the notation by replacing G_1 by G, I_1 by I, $G_1 \cap \tau_i H \tau_i^{-1}$ by H, f_i by $f = (G : HI)$, V_i by V, and $\tau_i W$ by W. Then we have still $V = \operatorname{Ind}_G^H(W)$, i.e., we are reduced to the case $r = 1$, $G_1 = G$.

We may further assume that $I = 1$. For if we put $\overline{G} = G/I$, $\overline{H} = H/I \cap H$, then $V^I = \operatorname{Ind}_{\overline{G}}^{\overline{H}}(W^{I\cap H})$. Indeed, a function $f : G \to W$ in V is invariant under I if and only if one has $f(x\tau) = f(x)$ for all $\tau \in I$, i.e., if and only if it is constant on the right (and therefore also on the left) cosets of G mod I, i.e., if and only if it is a function on $\overline{G}$. It then automatically takes values in $W^{I\cap H}$, because $\tau f(x) = f(\tau x) = f(x)$ for $\tau \in I \cap H$.

So let $I = 1$. Then G is generated by φ, $f = (G : H)$, and thus

$$V = \bigoplus_{i=0}^{f-1} \varphi^i W .$$

Let A be the matrix of φ^f with respect to a basis $w_1, \dots, w_d$ of W. If E denotes the $(d \times d)$ unit matrix, then

$$\begin{pmatrix} 0 & E & \cdots & 0 \\ & & & \\ 0 & 0 & \cdots & E \\ A & 0 & \cdots & 0 \end{pmatrix}$$

is the matrix of φ with respect to the basis $\{\varphi^i w_j\}$ of V. This gives

$$\det(1 - \varphi t;\, V) = \det \begin{pmatrix} E & -tE & \cdots & 0 \\ & & & \\ 0 & 0 & \cdots & -tE \\ -tA & 0 & \cdots & E \end{pmatrix} = \det(1 - \varphi^f t^f;\, W)$$

as desired. The last identity is obtained by first multiplying the first column by t and adding it to the second, and then multiplying the second column by t and adding it to the third, etc. □

The character $\mathbf{1}_*$ induced from the trivial character $\mathbf{1}$ of the subgroup $\{1\} \subseteq G(L|K)$ is the character $r_G = \sum_\chi \chi(1)\chi$ of the regular representation of $G(L|K)$. We therefore deduce from (10.4) the

(10.5) Corollary. *One has*

$$\zeta_L(s) = \zeta_K(s) \prod_{\chi \neq \mathbf{1}} \mathcal{L}(L|K, \chi, s)^{\chi(1)} ,$$

where χ varies over the nontrivial irreducible characters of $G(L|K)$.

The starting point of Artin's investigations on L-series had been the question whether, for a Galois extension $L|K$, the quotient $\zeta_L(s)/\zeta_K(s)$ is an entire function, i.e., a holomorphic function on the whole complex plane. Corollary (10.5) shows that this could be deduced from the famous

Artin Conjecture: For every irreducible character $\chi \neq \mathbf{1}$, the Artin L-series $\mathcal{L}(L|K,\chi,s)$ defines an **entire** function.

We will see presently that this conjecture holds for *abelian* extensions. In general it is not known. In view of its momentous consequences, it constitutes one of the big challenges in number theory.

We will show next that the Artin L-series in the case of *abelian* extensions $L|K$ coincide with certain Hecke L-series, more precisely, with generalized Dirichlet L-series. This means that the properties of Hecke's series, and in particular their functional equation, transfer to Artin series in the abelian case. Via functoriality (10.4) they may then be extended to the non-abelian case.

The link between Artin and Hecke L-series is provided by class field theory. Let $L|K$ be an abelian extension, and let $\mathfrak{f}$ be the **conductor** of $L|K$, i.e., the smallest module

$$\mathfrak{f} = \prod_{\mathfrak{p}\nmid\infty} \mathfrak{p}^{n_\mathfrak{p}}$$

such that $L|K$ lies in the ray class field $K^\mathfrak{f}|K$ (see chap. VI, (6.2)). The **Artin symbol** $\left(\frac{L|K}{\mathfrak{a}}\right)$ then gives us a surjective homomorphism

$$J^\mathfrak{f}/P^\mathfrak{f} \longrightarrow G(L|K), \quad \mathfrak{a} \bmod P^\mathfrak{f} \longmapsto \left(\frac{L|K}{\mathfrak{a}}\right),$$

from the **ray class group** $J^\mathfrak{f}/P^\mathfrak{f}$. Here $J^\mathfrak{f}$ is the group of fractional ideals prime to $\mathfrak{f}$, and $P^\mathfrak{f}$ is the group of principal ideals (a) such that $a \equiv 1 \bmod \mathfrak{f}$ and a is positive in $K_\mathfrak{p} = \mathbb{R}$ if $\mathfrak{p}$ is real.

Now let χ be an *irreducible* character of the abelian group $G(L|K)$, i.e., a homomorphism

$$\chi : G(L|K) \to \mathbb{C}^*.$$

Composing with the Artin symbol $\left(\frac{L|K}{}\right)$, this gives a character of the ray class group $J^\mathfrak{f}/P^\mathfrak{f}$, i.e., a Dirichlet character mod $\mathfrak{f}$. It induces a character on $J^\mathfrak{f}$, which we denote by

$$\tilde{\chi} : J^\mathfrak{f} \to \mathbb{C}^*.$$

By (6.9), this character on ideals is a **Größencharacter** mod $\mathfrak{f}$ of type $(p,0)$, and we have the

(10.6) Theorem. *Let $L|K$ be an abelian extension, let $\mathfrak{f}$ be the conductor of $L|K$, let $\chi \neq \mathbf{1}$ be an irreducible character of $G(L|K)$, and $\widetilde{\chi}$ the associated Größencharakter* mod $\mathfrak{f}$.

Then the Artin L-series for the character χ and the Hecke L-series for the Größencharakter $\widetilde{\chi}$ satisfy the identity

$$\mathcal{L}(L|K,\chi,s) = \prod_{\mathfrak{p}\in S} \frac{1}{1-\chi(\varphi_{\mathfrak{P}})\mathfrak{N}(\mathfrak{p})^{-s}} L(\widetilde{\chi},s),$$

where $S = \{\mathfrak{p}|\mathfrak{f} \mid \chi(I_{\mathfrak{P}}) = 1\}$.

Proof: The representation of $G(L|K)$ associated to the character χ is given by a 1-dimensional vector space $V = \mathbb{C}$ on which $G(L|K)$ acts via multiplication by χ, i.e., $\sigma v = \chi(\sigma)v$. Since $\mathfrak{f}$ is the conductor of $L|K$, we find by chap. VI, (6.6), that

$$\mathfrak{p}|\mathfrak{f} \iff \mathfrak{p} \text{ is ramified} \iff I_{\mathfrak{P}} \neq 1.$$

If $\chi(I_{\mathfrak{P}}) \neq 1$, then $V^{I_{\mathfrak{P}}} = \{0\}$, and the corresponding Euler factor does not occur in the Artin L-series. If on the other hand $\chi(I_{\mathfrak{P}}) = 1$, then $V^{I_{\mathfrak{P}}} = \mathbb{C}$, so that

$$\det(1-\varphi_{\mathfrak{P}}\mathfrak{N}(\mathfrak{p})^{-s};\, V^{I_{\mathfrak{P}}}) = 1-\chi(\varphi_{\mathfrak{P}})\mathfrak{N}(\mathfrak{p})^{-s}.$$

We thus have

$$\mathcal{L}(L|K,\chi,s) = \prod_{\mathfrak{p}\nmid\mathfrak{f}} \frac{1}{1-\chi(\varphi_{\mathfrak{P}})\mathfrak{N}(\mathfrak{p})^{-s}} \prod_{\mathfrak{p}\in S} \frac{1}{1-\chi(\varphi_{\mathfrak{P}})\mathfrak{N}(\mathfrak{p})^{-s}}$$

and

$$L(\widetilde{\chi},s) = \prod_{\mathfrak{p}\nmid\mathfrak{f}} \frac{1}{1-\widetilde{\chi}(\mathfrak{p})\mathfrak{N}(\mathfrak{p})^{-s}}.$$

For $\mathfrak{p}\nmid\mathfrak{f}$, one has $\left(\frac{L|K}{\mathfrak{p}}\right) = \varphi_{\mathfrak{P}}$, and so $\widetilde{\chi}(\mathfrak{p}) = \chi(\varphi_{\mathfrak{P}})$. This proves the claim. □

Remark: If the character $\chi : G(L|K) \to \mathbb{C}^*$ is *injective*, then $S = \emptyset$, and one has complete equality

$$\mathcal{L}(L|K,\chi,s) = L(\widetilde{\chi},s).$$

In this case $\widetilde{\chi}$ is a *primitive Größencharakter* mod $\mathfrak{f}$.

If on the other hand χ is the trivial character $\mathbf{1}_G$, then $\widetilde{\chi}$ is the trivial Dirichlet character mod $\mathfrak{f}$, and we have

$$\zeta_K(s) = \prod_{\mathfrak{p}|\mathfrak{f}} \frac{1}{1-\mathfrak{N}(\mathfrak{p})^{-s}} L(\widetilde{\chi},s).$$

The theorem implies that the **Artin conjecture** holds for all Artin L-series $\mathcal{L}(L|K,\chi,s)$ which correspond to nontrivial irreducible characters χ of *abelian* Galois groups $G(L|K)$. For if L_χ is the fixed field of the kernel of χ and $\tilde{\chi}$ is the *Größencharakter* associated with $\chi : G(L_\chi|K) \hookrightarrow \mathbb{C}^*$, then the above remark shows that $\mathcal{L}(L|K,\chi,s) = \mathcal{L}(L_\chi|K,\chi,s) = L(\tilde{\chi},s)$. Hence $\mathcal{L}(L|K,\chi,s)$ is holomorphic on all of $\mathbb{C}$, because the same is true for $L(\tilde{\chi},s)$, as was shown in (8.5). This also settles the Artin conjecture for every solvable extension $L|K$.

Our goal now is to prove a functional equation for Artin L-series. The basis for this will be the above theorem and the functional equation we have already established for Hecke L-series. We however have to complete the Artin L-series by the right "Euler factors" at the infinite places. In looking for these Euler factors, the first natural guideline is provided by the case of Hecke L-series. But in order to go the whole way, we need an additional Galois-theoretic complement which will be dealt with in the next section.

§ 11. The Artin Conductor

The discriminant $\mathfrak{d} = \mathfrak{d}_{L|K}$ of a Galois extension $L|K$ of algebraic number fields admits a fine structure based on group theory. It is expressed by a product decomposition

$$\mathfrak{d} = \prod \mathfrak{f}(\chi)^{\chi(1)},$$

where χ varies over the irreducible characters of the Galois group $G = G(L|K)$. The ideals $\mathfrak{f}(\chi)$ are given by

$$\mathfrak{f}(\chi) = \prod_{\mathfrak{p}\nmid\infty} \mathfrak{p}^{f_\mathfrak{p}(\chi)}$$

with

$$f_\mathfrak{p}(\chi) = \sum_{i\ge 0} \frac{g_i}{g_0} \operatorname{codim} V^{G_i},$$

where V is a representation with character χ, G_i is the i-th ramification group of $L_\mathfrak{P}|K_\mathfrak{p}$, and g_i denotes its order. This discovery goes back to *Emil Artin* and *Helmut Hasse*. The ideals $\mathfrak{f}(\chi)$ are called **Artin conductors**. They play an important rôle in the functional equation of the Artin L-series, which we are going to prove in the next section. Here we collect the properties needed for this, following essentially the treatment given by *J.-P. Serre* in [122].

First let us consider a Galois extension $L|K$ of **local fields**, with Galois group $G = G(L|K)$. Let $f = f_{L|K} = [\lambda : \kappa]$ be the inertia degree of $L|K$. In chap. II, § 10, we defined, for any $\sigma \in G$,

$$i_G(\sigma) = v_L(\sigma x - x),$$

where x is an element such that $\mathcal{O}_L = \mathcal{O}_K[x]$, and v_L is the normalized valuation of L. With this notation we can write the i-th ramification group as

$$G_i = \{\sigma \in G \mid i_G(\sigma) \geq i + 1\}.$$

One has $i_G(\tau\sigma\tau^{-1}) = i_G(\sigma)$, and $i_H(\sigma) = i_G(\sigma)$ for every subgroup $H \subseteq G$. If $L|K$ is unramified, then $i_G(\sigma) = 0$ for all $\sigma \in G$, $\sigma \neq 1$. We put

$$a_G(\sigma) = \begin{cases} -f i_G(\sigma) & \text{for } \sigma \neq 1, \\ f \sum_{\tau \neq 1} i_G(\tau) & \text{for } \sigma = 1. \end{cases}$$

a_G is a central function on G, and we have

$$(a_G, \mathbf{1}_G) = \frac{1}{\#G} \sum_{\sigma \in G} a_G(\sigma) = 0.$$

We may therefore write

$$a_G = \sum_{\chi} f(\chi)\chi, \quad f(\chi) \in \mathbb{C},$$

with χ varying over the irreducible characters of G. Our chief problem is to prove that the coefficients $f(\chi)$ are rational integers ≥ 0. Once we have shown this, we may form the ideal $\mathfrak{f}_{\mathfrak{p}}(\chi) = \mathfrak{p}^{f(\chi)}$, which will be the $\mathfrak{p}$-component of the global Artin conductor that we want. First we prove that the function a_G satisfies the following properties (we use the notation of the preceding section).

(11.1) Proposition. (i) *If H is a normal subgroup of G, then*

$$a_{G/H} = (a_G)_{\natural}.$$

(ii) *If H is any subgroup of G, and if K' is the fixed field with discriminant $\mathfrak{d}_{K'|K} = \mathfrak{p}^{\nu}$, then*

$$a_G|H = \nu r_H + f_{K'|K} a_H.$$

(iii) *Let G_i be the i-th ramification group of G, u_i the augmentation character of G_i, and $(u_i)_*$ the character of G induced from u_i. Then one has*

$$a_G = \sum_{i=0}^{\infty} \frac{1}{(G_0 : G_i)} (u_i)_*.$$

Proof: (i) follows immediately from chap. II, (10.5).
(ii) Let $\sigma \in H$, $\sigma \neq 1$. Then

$$a_G(\sigma) = -f_{L|K} i_G(\sigma), \quad a_H(\sigma) = -f_{L|K'} i_H(\sigma), \quad r_H(\sigma) = 0.$$

Since $i_G(\sigma) = i_H(\sigma)$ and $f_{L|K} = f_{L|K'} f_{K'|K}$, this implies

$$a_G(\sigma) = \nu r_H(\sigma) + f_{K'|K} a_H(\sigma).$$

Now let $\sigma = 1$, and let $\mathfrak{D}_{L|K}$ be the *different* of $L|K$. Let $\mathcal{O}_L = \mathcal{O}_K[x]$ and $g(X)$ be the minimal polynomial of x over K. By chap. III, (2.4), $\mathfrak{D}_{L|K}$ is then generated by $g'(x) = \prod_{\sigma \neq 1}(\sigma x - x)$. Consequently,

$$v_L(\mathfrak{D}_{L|K}) = v_L(g'(x)) = \sum_{\sigma \neq 1} i_G(\sigma) = \frac{1}{f_{L|K}} a_G(1).$$

By chap. III, (2.9), we know, on the other hand, that $\mathfrak{d}_{L|K} = N_{L|K}(\mathfrak{D}_{L|K})$, so $v_K \circ N_{L|K} = f_{L|K} v_L$ gives the identity

$$a_G(1) = f_{L|K} v_L(\mathfrak{D}_{L|K}) = v_K(\mathfrak{d}_{L|K}),$$

and in the same way $a_H(1) = v_{K'}(\mathfrak{d}_{L|K'})$. From chap. III, (2.10), we get furthermore that

$$\mathfrak{d}_{L|K} = (\mathfrak{d}_{K'|K})^{[L:K']} N_{K'|K}(\mathfrak{d}_{L|K'}).$$

Thus $r_H(1) = [L : K']$ and $\nu = v_K(\mathfrak{d}_{K'|K})$ yields the formula

$$a_G(1) = [L : K'] v_K(\mathfrak{d}_{K'|K}) + f_{K'|K} v_{K'}(\mathfrak{d}_{L|K'}) = \nu r_H(1) + f_{K'|K} a_H(1).$$

(iii) Let $g_i = \#G_i$, $g = \#G$. Since G_i is invariant in G, we have $(u_i)_*(\sigma) = 0$ if $\sigma \notin G_i$, and $(u_i)_*(\sigma) = -g/g_i = -f \cdot g_0/g_i$ if $\sigma \in G_i$, $\sigma \neq 1$, and $\sum_{\sigma \in G}(u_i)_*(\sigma) = 0$. For $\sigma \in G_k \smallsetminus G_{k+1}$, we thus find

$$a_G(\sigma) = -f(k+1) = \sum \frac{1}{(G_0 : G_i)} (u_i)_*(\sigma).$$

This implies the identity for the case $\sigma = 1$ as well, since both sides are orthogonal to $\mathbf{1}_G$. □

For the coefficients $f(\chi)$ in the linear combination

$$a_G = \sum f(\chi)\chi,$$

we have, in view of $a_G(\sigma^{-1}) = a_G(\sigma)$, that

$$f(\chi) = (a_G, \chi) = \frac{1}{g} \sum_{\sigma \in G} a_G(\sigma)\chi(\sigma^{-1}) = \frac{1}{g} \sum_{\sigma \in G} a_G(\sigma^{-1})\chi(\sigma) = (\chi, a_G),$$

$g = \#G$. For any central function φ of G, we put

$$f(\varphi) = (\varphi, a_G)$$

and

$$\varphi(G_i) = \frac{1}{g_i} \sum_{\sigma \in G_i} \varphi(\sigma), \quad g_i = \#G_i.$$

(11.2) Proposition. (i) *If φ is a central function on the quotient group G/H, and φ' is the corresponding central function on G, then*

$$f(\varphi) = f(\varphi').$$

(ii) *If φ is a central function on a subgroup H of G, and φ_* is the central function induced by φ on G, then*

$$f(\varphi_*) = v_K(\mathfrak{d}_{K'|K})\varphi(1) + f_{K'|K} f(\varphi).$$

(iii) *For a central function φ on G, one has*

$$f(\varphi) = \sum_{i \geq 0} \frac{g_i}{g_0}(\varphi(1) - \varphi(G_i)).$$

Proof: (i) $f(\varphi) = (\varphi, a_{G/H}) = (\varphi, (a_G)_\natural) = (\varphi', a_G) = f(\varphi')$.
(ii) $f(\varphi_*) = (\varphi_*, a_G) = (\varphi, a_G|H) = \nu(\varphi, r_H) + f_{K'|K}(\varphi, a_H) = \nu\varphi(1) + f_{K'|K} f(\varphi)$ with $\nu = v_K(\mathfrak{d}_{K'|K})$.
(iii) We have $(\varphi, (u_i)_*) = (\varphi|G_i, u_i) = \varphi(1) - \varphi(G_i)$, so the formula follows from (11.1), (iii). □

If χ is the character of a representation (ρ, V) of G, then $\chi(1) = \dim V$ and $\chi(G_i) = \dim V^{G_i}$, hence

$$f(\chi) = \sum_{i \geq 0} \frac{g_i}{g_0} \operatorname{codim} V^{G_i}.$$

Now consider the function

$$\eta_{L|K}(s) = \int_0^s \frac{dx}{(G_0 : G_x)},$$

which was introduced in chap. II, § 10. For integers $m \geq -1$, it is given by $\eta_{L|K}(-1) = -1$, $\eta_{L|K}(0) = 0$, and

$$\eta_{L|K}(m) = \sum_{i=1}^{m} \frac{g_i}{g_0} \quad \text{for } m \geq 1.$$

The theorem of HASSE-ARF (see chap. V, (6.3)) now gives us the following integrality statement for the number $f(\chi)$ in the case of a character χ of degree 1.

(11.3) Proposition. *Let χ be a character of G of degree 1. Let j be the biggest integer such that $\chi|G_j \neq \mathbf{1}_{G_j}$ (when $\chi = \mathbf{1}_G$ we put $j = -1$). Then we have*

$$f(\chi) = \eta_{L|K}(j) + 1,$$

and this is a rational integer ≥ 0.

Proof: If $i \le j$, then $\chi(G_i) = 0$, so that $\chi(1) - \chi(G_i) = 1$. If $i > j$, then $\chi(G_i) = 1$, and so $\chi(1) - \chi(G_i) = 0$. From (11.2), (iii), it thus follows that

$$f(\chi) = \sum_{i=0}^{j} \frac{g_i}{g_0} = \eta_{L|K}(j) + 1,$$

provided $j \ge 0$. If $j = -1$, we have $\chi(1) - \chi(G_i) = 0$ for all $i \ge 0$, and hence by (11.2), (iii), $f(\chi) = 0 = \eta_{L|K}(-1) + 1$.

Let H be the kernel of χ and L' the fixed field of H. By Herbrand's theorem (chap. II, (10.7)) one has

$$G_j(L|K)H/H = G_{j'}(L'|K) \quad \text{with} \quad j' = \eta_{L|L'}(j).$$

In terms of the upper numbering of the ramification groups, this translates into

$$G^t(L|K)H/H = G^t(L'|K),$$

where $t = \eta_{L|K}(j) = \eta_{L'|K}(\eta_{L|L'}(j)) = \eta_{L'|K}(j')$ (see chap. II, (10.8)). But $\chi(G_j(L|K)H/H) \neq 1$, and $\chi(G_{j+\delta}(L|K)H/H) = \chi(G_{j+1}(L|K)H/H) = 1$ for all $\delta > 0$, and in particular $G_j(L|K)H/H \neq G_{j+\delta}(L|K)H/H$ for all $\delta > 0$. Since $\eta_{L|K}(s)$ is continuous and strictly increasing, it follows that

$$G^t(L'|K) = G^t(L|K)H/H \neq G^{t+\varepsilon}(L|K)H/H = G^{t+\varepsilon}(L'|K)$$

for all $\varepsilon > 0$, i.e., t is a *jump* in the ramification filtration of $L'|K$. The extension $L'|K$ is abelian and therefore $t = \eta_{L|K}(j)$ is an integer, by the theorem of *Hasse* and *Arf*. □

Now let χ be an arbitrary character of the Galois group $G = G(L|K)$. By Brauer's theorem (10.3), we then have

$$\chi = \sum n_i \chi_{i*}, \quad n_i \in \mathbb{Z},$$

where χ_{i*} is the character induced from a character χ_i of degree 1 of a subgroup H_i. By (11.2), (ii), we have

$$f(\chi) = \sum n_i f(\chi_{i*}) = \sum n_i \big(v_K(\mathfrak{d}_{K_i|K})\chi_i(1) + f_{K_i|K} f(\chi_i)\big),$$

where K_i is the fixed field of H_i. Therefore $f(\chi)$ is a rational integer. On the other hand, (11.1), (iii) shows that $g_0 a_G$ is the character of a representation of G, so $g_0 f(\chi) = (\chi, g_0 a_G) \ge 0$. We have thus established the

(11.4) Theorem. *If χ is a character of the Galois group $G = G(L|K)$, then $f(\chi)$ is a rational integer ≥ 0.*

(11.5) Definition. *We define the (local)* **Artin conductor** *of the character* χ *of* $G = G(L|K)$ *to be the ideal*

$$\mathfrak{f}_\mathfrak{p}(\chi) = \mathfrak{p}^{f(\chi)}.$$

In chap. V, (1.6), we defined the **conductor** of an *abelian* extension $L|K$ of local fields to be the smallest power of $\mathfrak{p}$, $\mathfrak{f} = \mathfrak{p}^n$, such that the n-th higher unit group $U_K^{(n)}$ is contained in the norm group $N_{L|K}L^*$. The latter is the kernel of the norm residue symbol

$$(\ ,L|K) : K^* \longrightarrow G(L|K),$$

which maps $U_K^{(i)}$ to the higher ramification group $G^i(L|K) = G_j(L|K)$ with $i = \eta_{L|K}(j)$ – see V, (6.2). The conductor $\mathfrak{f} = \mathfrak{p}^n$ is therefore given by the smallest integer $n \geq 0$ such that $G^n(L|K) = 1$. From (11.3) we thus obtain the following result.

(11.6) Proposition. *Let $L|K$ be a Galois extension of local fields, and let χ be a character of $G(L|K)$ of degree 1. Let L_χ be the fixed field of the kernel of χ, and $\mathfrak{f}$ the conductor of $L_\chi|K$. Then one has*

$$\mathfrak{f} = \mathfrak{f}_\mathfrak{p}(\chi).$$

Proof: By (11.3), we have $f(\chi) = \eta_{L|K}(j) + 1$, where j is the largest integer such that $G_j(L|K) \nsubseteq G(L|L_\chi) =: H$. Let $t = \eta_{L|K}(j)$. Then one has

$$G^t(L_\chi|K) = G^t(L|K)H/H = G_j(L|K)H/H,$$

and $G^{t+\varepsilon}(L_\chi|K) \subseteq G_{j+1}(L|K)H/H = 1$ for all $\varepsilon > 0$. Hence t is the largest number such that $G^t(L_\chi|K) \neq 1$. By the theorem of HASSE-ARF, t is an integer, and we conclude that $f(\chi) = t + 1$ is the smallest integer such that $G^{f(\chi)}(L_\chi|K) = 1$, i.e., $f(\chi) = n$. □

We now leave the local situation, and suppose that $L|K$ is a Galois extension of **global** fields. Let $\mathfrak{p}$ be a prime ideal of K, $\mathfrak{P}|\mathfrak{p}$ a prime ideal of L lying above $\mathfrak{p}$. Let $L_\mathfrak{P}|K_\mathfrak{p}$ be the completion of $L|K$, and $G_\mathfrak{P} = G(L_\mathfrak{P}|K_\mathfrak{p})$ the decomposition group of $\mathfrak{P}$ over K. We denote the function $a_{G_\mathfrak{P}}$ on $G_\mathfrak{P}$ by $a_\mathfrak{P}$, and extend it to $G = G(L|K)$ by zero. The central function

$$a_\mathfrak{p} = \sum_{\mathfrak{P}|\mathfrak{p}} a_\mathfrak{P}$$

immediately turns out to be the function $(a_{\mathfrak{P}})_*$ induced by $a_{\mathfrak{P}}|G_{\mathfrak{P}}$. It is therefore the character of a representation of G. If now χ is a character of G, then we put

$$f(\chi,\mathfrak{p}) = (\chi, a_{\mathfrak{p}}) = f(\chi|G_{\mathfrak{P}}).$$

Then $\mathfrak{f}_{\mathfrak{p}}(\chi) = \mathfrak{p}^{f(\chi,\mathfrak{p})}$ is the Artin conductor of the restriction of χ to $G_{\mathfrak{P}} = G(L_{\mathfrak{P}}|K_{\mathfrak{p}})$. In particular, we have $\mathfrak{f}_{\mathfrak{p}}(\chi) = 1$ if $\mathfrak{p}$ is unramified. We define the (global) **Artin conductor** of χ to be the product

$$\mathfrak{f}(\chi) = \prod_{\mathfrak{p}\nmid\infty} \mathfrak{f}_{\mathfrak{p}}(\chi).$$

Whenever precision is called for, we write $\mathfrak{f}(L|K,\chi)$ instead of $\mathfrak{f}(\chi)$. The properties (11.2) of the numbers $f(\chi,\mathfrak{p})$ transfer immediately to the Artin conductor $\mathfrak{f}(\chi)$, and we obtain the

(11.7) Proposition. (i) $\mathfrak{f}(\chi+\chi') = \mathfrak{f}(\chi)\mathfrak{f}(\chi')$, $\mathfrak{f}(\mathbf{1}) = (1)$.

(ii) *If $L'|K$ is a Galois subextension of $L|K$, and χ is a character of $G(L'|K)$, then*

$$\mathfrak{f}(L|K,\chi) = \mathfrak{f}(L'|K,\chi).$$

(iii) *If H is a subgroup of G with fixed field K', and if χ is a character of H, then*

$$\mathfrak{f}(L|K,\chi_*) = \mathfrak{d}_{K'|K}^{\chi(1)} N_{K'|K}\big(\mathfrak{f}(L|K',\chi)\big).$$

Proof: (i) and (ii) are trivial. To prove (iii), we choose a fixed prime ideal $\mathfrak{P}$ of L, put

$$G = G(L|K), \quad H = G(L|K'), \quad G_{\mathfrak{P}} = G(L_{\mathfrak{P}}|K_{\mathfrak{p}}),$$

with $\mathfrak{p} = \mathfrak{P}\cap K$, and consider the decomposition

$$G = \dot{\bigcup_{\tau}} G_{\mathfrak{P}}\tau H$$

into double cosets. Then representation theory yields the following formula for the character χ of H:

$$(*) \qquad \chi_*|G_{\mathfrak{P}} = \sum_{\tau} \chi_*^{\tau},$$

where χ^{τ} is the character $\chi^{\tau}(\sigma) = \chi(\tau^{-1}\sigma\tau)$ of $G_{\mathfrak{P}}\cap\tau H\tau^{-1}$, and χ_*^{τ} is the character of $G_{\mathfrak{P}}$ induced by χ^{τ} (see [119], chap. 7, prop. 22). Furthermore $\mathfrak{P}'_{\tau} = \mathfrak{P}^{\tau}\cap K'$ are the different prime ideals of K' above $\mathfrak{p}$ (see chap. I, § 9, p. 55), and we have

$$G_{\mathfrak{P}^{\tau}} = \tau^{-1}G_{\mathfrak{P}}\tau = G(L_{\mathfrak{P}^{\tau}}|K_{\mathfrak{p}}), \; H_{\mathfrak{P}^{\tau}} = G_{\mathfrak{P}^{\tau}}\cap H = G(L_{\mathfrak{P}^{\tau}}|K'_{\mathfrak{P}'_{\tau}}).$$

Now let $\mathfrak{d}_{\mathfrak{P}'_\tau} = \mathfrak{p}^{\nu_{\mathfrak{P}'_\tau}}$ be the discriminant ideal of $K'_{\mathfrak{P}'_\tau}|K_\mathfrak{p}$, and let $f_{\mathfrak{P}'_\tau}$ be the degree of $\mathfrak{P}'_\tau$ over K. Thus $N_{K'|K}(\mathfrak{P}'_\tau) = \mathfrak{p}^{f_{\mathfrak{P}'_\tau}}$. Since

$$\mathfrak{f}_\mathfrak{p}(L|K,\chi_*) = \mathfrak{p}^{f(\chi_*|G_\mathfrak{P})} \quad \text{and} \quad \mathfrak{f}_{\mathfrak{P}'_\tau}(L|K',\chi) = \mathfrak{P}'^{f(\chi|H_{\mathfrak{P}^\tau})}_\tau,$$

we have to show that

$$f(\chi_*|G_\mathfrak{P}) = \sum_\tau \nu_{\mathfrak{P}'_\tau}\chi(1) + f_{\mathfrak{P}'_\tau} f(\chi|H_{\mathfrak{P}^\tau}),$$

or, in view of (11.2), (ii), that

$$(**) \qquad f(\chi_*|G_\mathfrak{P}) = \sum_\tau f((\chi|H_{\mathfrak{P}^\tau})_*).$$

But $H_{\mathfrak{P}^\tau} = \tau^{-1}(G_\mathfrak{P} \cap \tau H \tau^{-1})\tau$, and $\chi|H_{\mathfrak{P}^\tau}$, resp. $(\chi|H_{\mathfrak{P}^\tau})_*$, arises by conjugation $\sigma \mapsto \tau\sigma\tau^{-1}$ from χ^τ, resp. χ_*^τ. Therefore $f((\chi|H_{\mathfrak{P}^\tau})_*) = f(\chi_*^\tau)$, and $(**)$ follows from $(*)$. □

We apply (iii) to the case $\chi = \mathbf{1}_H$, and denote the induced character χ_* by $s_{G/H}$. Since $\mathfrak{f}(\chi) = 1$, we obtain the

(11.8) Corollary. $\mathfrak{d}_{K'|K} = \mathfrak{f}(L|K, s_{G/H})$.

If in particular $H = \{1\}$, then $s_{G/H}$ is the character r_G of the regular representation. Its decomposition into irreducible characters χ is given by

$$r_G = \sum_\chi \chi(1)\chi.$$

This yields the

(11.9) Conductor-Discriminant-Formula. *For an arbitrary Galois extension $L|K$ of global fields, one has*

$$\mathfrak{d}_{L|K} = \prod_\chi \mathfrak{f}(\chi)^{\chi(1)},$$

where χ varies over the irreducible characters of $G(L|K)$.

For an *abelian* extension $L|K$ of global fields, we defined the conductor $\mathfrak{f}$ in VI, (6.4). By chap. VI, (6.5), it is the product

$$\mathfrak{f} = \prod_\mathfrak{p} \mathfrak{f}_\mathfrak{p}$$

of the conductors $\mathfrak{f}_\mathfrak{p}$ of the local extensions $L_\mathfrak{P}|K_\mathfrak{p}$. (11.6) now gives rise to the following

(11.10) Proposition. *Let $L|K$ be a Galois extension of global fields, χ a character of $G(L|K)$ of degree 1, L_χ the fixed field of the kernel of χ, and $\mathfrak{f}$ the conductor of $L_\chi|K$. Then one has*

$$\mathfrak{f} = \mathfrak{f}(\chi).$$

Now let $L|K$ be a Galois extension of algebraic number fields. We form the ideal

$$\mathfrak{c}(L|K,\chi) = \mathfrak{d}_{K|\mathbb{Q}}^{\chi(1)} N_{K|\mathbb{Q}}\big(\mathfrak{f}(L|K,\chi)\big)$$

of $\mathbb{Z}$. The positive generator of this ideal is the integer

$$c(L|K,\chi) = |d_K|^{\chi(1)}\mathfrak{N}\big(\mathfrak{f}(L|K,\chi)\big).$$

Applying (11.7) and observing the transitivity of the discriminant (chap. III, (2.10)), we get the

(11.11) Proposition. (i) $c(L|K,\chi+\chi') = c(L|K,\chi)c(L|K,\chi')$, $c(L|K,\mathbf{1}) = |d_K|$,

(ii) $c(L|K,\chi) = c(L'|K,\chi)$,

(iii) $c(L|K,\chi_*) = c(L|K',\chi)$.

Here the notation is that of (11.7).

§ 12. The Functional Equation of Artin L-series

The first task is to complete the Artin L-series

$$\mathcal{L}(L|K,\chi,s) = \prod_{\mathfrak{p}\nmid\infty} \frac{1}{\det(1-\varphi_{\mathfrak{P}}\mathfrak{N}(\mathfrak{p})^{-s};\, V^{I_{\mathfrak{P}}})},$$

for the character χ of $G = G(L|K)$, by the appropriate gamma factors. For every infinite place $\mathfrak{p}$ of K we put

$$\mathcal{L}_{\mathfrak{p}}(L|K,\chi,s) = \begin{cases} L_{\mathbb{C}}(s)^{\chi(1)}, & \text{if } \mathfrak{p} \text{ is complex},\\ L_{\mathbb{R}}(s)^{n^+}L_{\mathbb{R}}(s+1)^{n^-}, & \text{if } \mathfrak{p} \text{ is real}, \end{cases}$$

with the exponents $n^+ = \frac{\chi(1)+\chi(\varphi_{\mathfrak{P}})}{2}$, $n^- = \frac{\chi(1)-\chi(\varphi_{\mathfrak{P}})}{2}$. Here $\varphi_{\mathfrak{P}}$ is the distinguished generator of $G(L_{\mathfrak{P}}|K_{\mathfrak{p}})$, and

$$L_{\mathbb{R}}(s) = \pi^{-s/2}\Gamma(s/2), \quad L_{\mathbb{C}}(s) = 2(2\pi)^{-s}\Gamma(s)$$

(see §4). For $\mathfrak{p}$ real, the exponents n^+, n^- in $\mathcal{L}_{\mathfrak{p}}(L|K,\chi,s)$ have the following meaning.

The involution $\varphi_{\mathfrak{P}}$ on V induces an eigenspace decomposition $V = V^+ \oplus V^-$, where

$$V^+ = \{x \in V \mid \varphi_{\mathfrak{P}} x = x\}, \quad V^- = \{x \in V \mid \varphi_{\mathfrak{P}} x = -x\},$$

and it follows from the remark in § 10, p. 521, that

$$\dim V^+ = \frac{1}{2}\big(\chi(1) + \chi(\varphi_{\mathfrak{P}})\big), \quad \dim V^- = \frac{1}{2}\big(\chi(1) - \chi(\varphi_{\mathfrak{P}})\big).$$

The functions $\mathcal{L}_{\mathfrak{p}}(L|K,\chi,s)$ exhibit the same behaviour under change of fields and characters as the L-series and the Artin conductor.

(12.1) Proposition. (i) $\mathcal{L}_{\mathfrak{p}}(L|K,\chi+\chi',s) = \mathcal{L}_{\mathfrak{p}}(L|K,\chi,s)\mathcal{L}_{\mathfrak{p}}(L|K,\chi',s)$.

(ii) *If $L'|K$ is a Galois subextension of $L|K$ and χ a character of $G(L'|K)$, then*

$$\mathcal{L}_{\mathfrak{p}}(L|K,\chi,s) = \mathcal{L}_{\mathfrak{p}}(L'|K,\chi,s).$$

(iii) *If K' is an intermediate field of $L|K$ and χ a character of $G(L|K')$, then*

$$\mathcal{L}_{\mathfrak{p}}(L|K,\chi_*,s) = \prod_{\mathfrak{q}|\mathfrak{p}} \mathcal{L}_{\mathfrak{q}}(L|K',\chi,s),$$

where $\mathfrak{q}$ varies over the places of K' lying above $\mathfrak{p}$.

Proof: (i) is trivial.

(ii) If $\mathfrak{P}|\mathfrak{P}'|\mathfrak{p}$ are places of $L \supseteq L' \supseteq K$, each lying above the next, then $\varphi_{\mathfrak{P}}$ is mapped under the projection $G(L|K) \to G(L'|K)$ to $\varphi_{\mathfrak{P}'}$. So $\chi(\varphi_{\mathfrak{P}}) = \chi(\varphi_{\mathfrak{P}'})$.

(iii) If $\mathfrak{p}$ is complex, then there are precisely $m = [K':K]$ places $\mathfrak{q}$ above $\mathfrak{p}$. They are also complex, and the claim follows from $\chi_*(1) = m\chi(1)$.

Suppose $\mathfrak{p}$ is real. Let $G = G(L|K)$, $H = G(L|K')$, and let $H\backslash G/G_{\mathfrak{P}}$ be the set of double cosets $H\tau G_{\mathfrak{P}}$ with a fixed place $\mathfrak{P}$ of L above $\mathfrak{p}$. Then we have a bijection

$$H\backslash G/G_{\mathfrak{P}} \longrightarrow \{\mathfrak{q} \text{ place of } K' \text{ above } \mathfrak{p}\}, \quad H\tau G_{\mathfrak{P}} \longmapsto \mathfrak{q}_\tau = \tau\mathfrak{P}|_{K'}$$

(see chap. I, §9, p. 55). $\mathfrak{q}_\tau$ is real if and only if $\varphi_{\tau\mathfrak{P}} = \tau\varphi_{\mathfrak{P}}\tau^{-1} \in H$, i.e., $G_{\tau\mathfrak{P}} = \tau G_{\mathfrak{P}}\tau^{-1} \subseteq H$. The latter inclusion holds if and only if the double coset $H\tau G_{\mathfrak{P}}$ consists of only one coset mod H:

$$H\tau G_{\mathfrak{P}} = (H\tau G_{\mathfrak{P}}\tau^{-1})\tau = H\tau.$$

We thus obtain the real places among the $\mathfrak{q}_\tau$ by letting τ run through a system of representatives of the cosets $H\tau$ of $H\backslash G$ such that $\tau\varphi_{\mathfrak{P}}\tau^{-1} \in H$. But, for such a system, one has

$$\chi_*(\varphi_{\mathfrak{P}}) = \sum_\tau \chi(\tau\varphi_{\mathfrak{P}}\tau^{-1}) = \sum_\tau \chi(\varphi_{\tau\mathfrak{P}}).$$

Putting $\mathfrak{Q} = \tau\mathfrak{P}$, makes $\mathfrak{q} = \mathfrak{Q}|_{K'}$ run through the real places of K' above $\mathfrak{p}$, i.e.,

$$\chi_*(\varphi_{\mathfrak{P}}) = \sum_{\substack{\mathfrak{q}|\mathfrak{p} \\ \text{real}}} \chi(\varphi_{\mathfrak{Q}}).$$

On the other hand we have

$$\chi_*(1) = \sum_{\mathfrak{q} \text{ complex}} 2\chi(1) + \sum_{\mathfrak{q} \text{ real}} \chi(1).$$

Legendre's duplication formula $L_{\mathbb{R}}(s)L_{\mathbb{R}}(s+1) = L_{\mathbb{C}}(s)$ (see (4.3)) turns this into

$$\begin{aligned}
&\mathcal{L}_{\mathfrak{p}}(L|K, \chi_*, s) = \\
&\prod_{\mathfrak{q} \text{ complex}} L_{\mathbb{C}}(s)^{\chi(1)} \prod_{\mathfrak{q} \text{ real}} L_{\mathbb{R}}(s)^{\frac{\chi(1)+\chi(\varphi_{\mathfrak{Q}})}{2}} \prod_{\mathfrak{q} \text{ real}} L_{\mathbb{R}}(s+1)^{\frac{\chi(1)-\chi(\varphi_{\mathfrak{Q}})}{2}} = \\
&\qquad = \prod_{\mathfrak{q}|\mathfrak{p}} \mathcal{L}_{\mathfrak{q}}(L|K', \chi, s).
\end{aligned}$$

□

We finally put

$$\mathcal{L}_\infty(L|K, \chi, s) = \prod_{\mathfrak{p}|\infty} \mathcal{L}_{\mathfrak{p}}(L|K, \chi, s),$$

and obtain immediately from the above proposition the equations

$$\begin{aligned}
\mathcal{L}_\infty(L|K, \chi + \chi', s) &= \mathcal{L}_\infty(L|K, \chi, s)\mathcal{L}_\infty(L|K, \chi', s), \\
\mathcal{L}_\infty(L|K, \chi, s) &= \mathcal{L}_\infty(L'|K, \chi, s), \\
\mathcal{L}_\infty(L|K, \chi_*, s) &= \mathcal{L}_\infty(L|K', \chi, s).
\end{aligned}$$

(12.2) Definition. *The* **completed Artin L-series** *for the character χ of $G(L|K)$ is defined to be*

$$\Lambda(L|K, \chi, s) = c(L|K, \chi)^{s/2}\mathcal{L}_\infty(L|K, \chi, s)\mathcal{L}(L|K, \chi, s),$$

where

$$c(L|K, \chi) = |d_K|^{\chi(1)}\mathfrak{N}\big(\mathfrak{f}(L|K, \chi)\big).$$

The behaviour of the factors $c(L|K, \chi)$, $\mathcal{L}_\infty(L|K, \chi, s)$, $\mathcal{L}(L|K, \chi, s)$ on the right-hand side, which we studied in (10.4), (11.11), and above, carries over to the function $\Lambda(L|K, \chi, s)$, i.e., we have the

(12.3) Proposition. (i) $\Lambda(L|K,\chi+\chi',s)=\Lambda(L|K,\chi,s)\Lambda(L|K,\chi',s)$.

(ii) *If $L'|K$ is a Galois subextension of $L|K$ and χ a character of $G(L'|K)$, then*

$$\Lambda(L|K,\chi,s)=\Lambda(L'|K,\chi,s).$$

(iii) *If K' is an intermediate field of $L|K$ and χ a character of $G(L|K')$, then*

$$\Lambda(L|K,\chi_*,s)=\Lambda(L|K',\chi,s).$$

For a character χ of degree 1, the completed Artin L-series $\Lambda(L|K,\chi,s)$ coincides with a completed Hecke L-series. To see this, let $L_\chi|K$ be the fixed field of the kernel of χ, and let $\mathfrak{f}=\prod_{\mathfrak{p}}\mathfrak{p}^{n_\mathfrak{p}}$ be the conductor of $L_\chi|K$. By (11.10), we then have

$$\mathfrak{f}=\mathfrak{f}(\chi).$$

Via the Artin symbol

$$J^\mathfrak{f}/P^\mathfrak{f}\longrightarrow G(L_\chi|K),\quad \mathfrak{a}\longmapsto\left(\frac{L_\chi|K}{\mathfrak{a}}\right),$$

χ becomes a Dirichlet character of conductor $\mathfrak{f}$, i.e., by (6.9), a primitive *Größencharakter* mod $\mathfrak{f}(\chi)$ with exponent $p=(p_\tau)$, so that $p_\tau=0$ if τ is complex. This *Größencharakter* will be denoted $\widetilde{\chi}$.

We put $p_\mathfrak{p}=p_\tau$ if $\mathfrak{p}$ is the place corresponding to the embedding $\tau:K\to\mathbb{C}$. The numbers $p_\mathfrak{p}$ have the following Galois-theoretical meaning.

(12.4) Lemma. *For every real place $\mathfrak{p}$ of K one has*

$$p_\mathfrak{p}=[L_{\chi\mathfrak{P}}:K_\mathfrak{p}]-1.$$

Proof: We consider the isomorphism

$$I/I^\mathfrak{f}K^*\xrightarrow{\ \sim\ }J^\mathfrak{f}/P^\mathfrak{f},$$

where $I^\mathfrak{f}=\prod_\mathfrak{p}U_\mathfrak{p}^{(n_\mathfrak{p})}$ is the congruence subgroup mod $\mathfrak{f}$ of the idèle group $I=\coprod_\mathfrak{p}K_\mathfrak{p}^*$ (see chap. VI, (1.9)), and consider the composite map

$$I/I^\mathfrak{f}K^*\longrightarrow J^\mathfrak{f}/P^\mathfrak{f}\longrightarrow G(L_\chi|K)\overset{\chi}{\hookrightarrow}\mathbb{C}^*.$$

Let $\mathfrak{p}$ be a real place of K, and let $\alpha\in I$ be the idèle with components $\alpha_\mathfrak{p}=-1$ and $\alpha_\mathfrak{q}=1$ for all places $\mathfrak{q}$ different from $\mathfrak{p}$. By chap. VI, (5.6), the image $\varphi_\mathfrak{P}=(\alpha,L_\chi|K)=(-1,L_{\chi\mathfrak{P}}|K_\mathfrak{p})$ in $G(L_\chi|K)$ is a generator of the decomposition group $G_{\mathfrak{p}} = G(L_{\chi\mathfrak{P}}|K_{\mathfrak{p}})$. By the approximation theorem, we may choose an $a \in K^*$ such that $a \equiv 1 \bmod \mathfrak{f}$, $a < 0$ in $K_{\mathfrak{p}}$, and $a > 0$ in $K_{\mathfrak{q}}$, for all real places $\mathfrak{q} \neq \mathfrak{p}$. Then

$$\beta = \alpha a \in I^{(\mathfrak{f})} = \left\{x \in I \;\middle|\; x_{\mathfrak{p}} \in U_{\mathfrak{p}}^{(n_{\mathfrak{p}})} \text{ for } \mathfrak{p}|\mathfrak{f}\infty\right\}, \quad \text{if } \mathfrak{f} = \textstyle\prod_{\mathfrak{p}} \mathfrak{p}^{n_{\mathfrak{p}}}.$$

As explained in the proof of chap. VI, (1.9), the image of $\alpha \bmod I^{\mathfrak{f}}K^*$ in $J^{\mathfrak{f}}/P^{\mathfrak{f}}$ is the class of $(\beta) = (a)$, which therefore maps to $\varphi_{\mathfrak{P}}$. Consequently,

$$\chi((a)) = \chi_{\mathfrak{f}}(a)\chi_{\infty}(a) = \chi(\varphi_{\mathfrak{P}}).$$

Since $a \equiv 1 \bmod \mathfrak{f}$, we have $\chi_{\mathfrak{f}}(a) = 1$ and $\chi_{\infty}(a) = N\big((\frac{a}{|a|})^{p}\big) = \big(\frac{a}{|a|_{\mathfrak{p}}}\big)^{p_{\mathfrak{p}}} = (-1)^{p_{\mathfrak{p}}}$, i.e., $\chi(\varphi_{\mathfrak{P}}) = (-1)^{p_{\mathfrak{p}}}$, so that $\varphi_{\mathfrak{P}} = 1$ for $p_{\mathfrak{p}} = 0$, and $\varphi_{\mathfrak{P}} \neq 1$ for $p_{\mathfrak{p}} = 1$. But this is the statement of the lemma. □

(12.5) Proposition. *The completed Artin L-series for the character χ of degree 1 and the completed Hecke L-series for the Größencharakter $\widetilde{\chi}$ coincide:*

$$\Lambda(L|K,\chi,s) = \Lambda(\widetilde{\chi},s).$$

Proof: The completed Hecke L-series is given, according to § 8, by

$$\Lambda(\widetilde{\chi},s) = \big(|d_K|\mathfrak{N}(\mathfrak{f}(\widetilde{\chi}))\big)^{s/2} L_{\infty}(\widetilde{\chi},s)L(\widetilde{\chi},s)$$

with

$$L_{\infty}(\widetilde{\chi},s) = L_X(\mathbf{s}),$$

and $\mathbf{s} = s\mathbf{1} + p$, where

$$L_X(\mathbf{s}) = \prod_{\mathfrak{p}|\infty} L_{\mathfrak{p}}(\mathbf{s}_{\mathfrak{p}})$$

is the L-function of the $G(\mathbb{C}|\mathbb{R})$-set $X = \operatorname{Hom}(K,\mathbb{C})$ defined in §4. The factors $L_{\mathfrak{p}}(\mathbf{s}_{\mathfrak{p}})$ are given explicitly by

$$(*) \qquad L_{\mathfrak{p}}(\mathbf{s}_{\mathfrak{p}}) = \begin{cases} L_{\mathbb{C}}(s), & \text{if } \mathfrak{p} \text{ complex},\\ L_{\mathbb{R}}(s+p_{\mathfrak{p}}), & \text{if } \mathfrak{p} \text{ real},\end{cases}$$

(see p. 454). On the other hand we have

$$\Lambda(L|K,\chi,s) = c(L|K,\chi)^{s/2}\mathcal{L}_{\infty}(L|K,\chi,s)\mathcal{L}(L|K,\chi,s)$$

with

$$c(L|K,\chi) = |d_K|\mathfrak{N}\big(\mathfrak{f}(L|K,\chi)\big)$$

and

$$\mathcal{L}_{\infty}(L|K,\chi,s) = \prod_{\mathfrak{p}|\infty} \mathcal{L}_{\mathfrak{p}}(L|K,\chi,s).$$

Let L_χ be the fixed field of the kernel of χ. By (11.11), (ii), and the remark preceding lemma (12.4), one has

$$c(L|K,\chi) = c(L_\chi|K,\chi) = |d_K|\mathfrak{N}(\mathfrak{f}(\widetilde{\chi}))\,,$$

and by (10.4), (ii), and (10.6), and the subsequent remark, one has

$$\mathcal{L}(L|K,\chi,s) = \mathcal{L}(L_\chi|K,\chi,s) = L(\widetilde{\chi},s)\,.$$

We are thus reduced to proving

$$\mathcal{L}_\mathfrak{p}(L|K,\chi,s) = L_\mathfrak{p}(\mathbf{s}_\mathfrak{p})$$

for $\mathfrak{p}|\infty$ and $\mathbf{s} = s\mathbf{1} + p$. Firstly, we have $\mathcal{L}_\mathfrak{p}(L|K,\chi,s) = \mathcal{L}_\mathfrak{p}(L_\chi|K,\chi,s)$ (see p. 537). Let $\varphi_\mathfrak{P}$ be the generator of $G(L_{\chi\mathfrak{P}}|K_\mathfrak{p})$. Since χ is injective on $G(L_\chi|K)$, we get $\chi(\varphi_\mathfrak{P}) = -1$ if $\varphi_\mathfrak{P} \neq 1$, and $\chi(\varphi_\mathfrak{P}) = 1$ if $\varphi_\mathfrak{P} = 1$. Using (12.4) this gives

$$\mathcal{L}_\mathfrak{p}(L_\chi|K,\chi,s) = \begin{cases} L_\mathbb{C}(s)\,, & \text{for } \mathfrak{p} \text{ complex,} \\ L_\mathbb{R}(s)\,, & \text{for } \mathfrak{p} \text{ real and } \mathfrak{P} \text{ real, i.e., } p_\mathfrak{p} = 0, \\ L_\mathbb{R}(s+1)\,, & \text{for } \mathfrak{p} \text{ real and } \mathfrak{P} \text{ complex, i.e., } p_\mathfrak{p} = 1\,. \end{cases}$$

Hence $(*)$ shows that indeed $\mathcal{L}_\mathfrak{p}(L|K,\chi,s) = L_\mathfrak{p}(\mathbf{s}_\mathfrak{p})$. □

In view of the two results (12.3) and (12.5), the functional equation for Artin L-series now follows from Brauer's theorem (10.3) in a purely formal fashion, as a consequence of the functional equation for Hecke L-series, which we have already established.

(12.6) Theorem. *The Artin L-series $\Lambda(L|K,\chi,s)$ admits a meromorphic continuation to $\mathbb{C}$ and satisfies the functional equation*

$$\Lambda(L|K,\chi,s) = W(\chi)\Lambda(L|K,\overline{\chi},1-s)$$

with a constant $W(\chi)$ of absolute value 1.

Proof: By Brauer's theorem, the character χ is an integral linear combination

$$\chi = \sum n_i\chi_{i*}\,,$$

where the χ_{i*} are induced from characters χ_i of degree 1 on subgroups $H_i = G(L|K_i)$. From propositions (12.3) and (12.5), it follows that

$$\begin{aligned} \Lambda(L|K,\chi,s) &= \prod_i \Lambda(L|K,\chi_{i*},s)^{n_i} \\ &= \prod_i \Lambda(L|K_i,\chi_i,s)^{n_i} \\ &= \prod_i \Lambda(\widetilde{\chi}_i,s)^{n_i}\,, \end{aligned}$$

where $\widetilde{\chi}_i$ is the *Größencharakter* of K_i associated to χ_i. By (8.6), the Hecke L-series $\Lambda(\widetilde{\chi}_i, s)$ admit meromorphic continuations to $\mathbb{C}$ and satisfy the functional equation

$$\Lambda(\widetilde{\chi}_i, s) = W(\widetilde{\chi}_i)\Lambda(\overline{\widetilde{\chi}}_i, 1-s).$$

Therefore $\Lambda(L|K, \chi, s)$ satisfies the functional equation

$$\Lambda(L|K, \chi, s) = W(\chi)\prod_i \Lambda(\overline{\widetilde{\chi}}_i, 1-s) = W(\chi)\Lambda(L|K, \overline{\chi}, 1-s),$$

where $W(\chi) = \prod_i W(\widetilde{\chi}_i)$ is of absolute value 1. □

The functional equation for the Artin L-series may be given the following explicit form, which is easily deduced from (12.6) and (4.3):

$$\mathcal{L}(L|K, \chi, 1-s) = A(\chi, s)\mathcal{L}(L|K, \overline{\chi}, s),$$

with the factor

$$A(\chi, s) = W(\chi)\big[\,|d_K|^{\chi(1)}\mathfrak{N}(\mathfrak{f}(L|K, \chi))\big]^{s-\frac{1}{2}} \times (\cos \pi s/2)^{n^+}(\sin \pi s/2)^{n^-}\,\Gamma_{\mathbb{C}}(s)^{n\chi(1)}$$

and the exponents

$$n^+ = \frac{n}{2}\chi(1) + \sum_{\mathfrak{p}} \frac{1}{2}\chi(\varphi_{\mathfrak{P}}), \quad n^- = \frac{n}{2}\chi(1) - \sum_{\mathfrak{p}} \frac{1}{2}\chi(\varphi_{\mathfrak{P}}).$$

Here the summations are over the real places $\mathfrak{p}$ of K. This gives immediately the zeroes of the function $\mathcal{L}(L|K, \chi, s)$ in the half-plane $\mathrm{Re}(s) \le 0$. If χ is not the principal character, they are the following:

at $s = 0, -2, -4, \ldots$ zeroes of order $\frac{n}{2}\chi(1) + \sum_{\mathfrak{p}\text{ real}} \frac{1}{2}\chi(\varphi_{\mathfrak{P}})$,

at $s = -1, -3, -5, \ldots$ zeroes of order $\frac{n}{2}\chi(1) - \sum_{\mathfrak{p}\text{ real}} \frac{1}{2}\chi(\varphi_{\mathfrak{P}})$.

Remark: For the proof of the functional equation of the completed Artin L-series, we have made essential use of the fact that "Euler factors" $\mathcal{L}_{\mathfrak{p}}(L|K, \chi, s)$ at the infinite places $\mathfrak{p}$, which are made up out of gamma functions, behave under change of fields and characters in exactly the same way as the Euler factors

$$\mathcal{L}_{\mathfrak{p}}(L|K, \chi, s) = \det(1 - \varphi_{\mathfrak{P}}\mathfrak{N}(\mathfrak{P})^{-s};\, V^{I_{\mathfrak{P}}})^{-1}$$

at the finite places. This uniform behaviour is in striking contrast to the great difference in the procedures that lead to the definitions of the Euler factors for $\mathfrak{p}|\infty$ and $\mathfrak{p}\nmid\infty$. It is in this context that the mathematician CHRISTOPHER DENINGER recently made a very interesting discovery (see [26], [27]). He shows that the Euler factors for all places $\mathfrak{p}$ can all be written in the same way:

$$\mathcal{L}_{\mathfrak{p}}(L|K,\chi,s) = \det\nolimits_{\infty}\Big(\frac{\log \mathfrak{N}(\mathfrak{p})}{2\pi i}(s\,\mathrm{id} - \Theta_{\mathfrak{p}});\ H(X_{\mathfrak{p}}/\mathbb{L}_{\mathfrak{p}})\Big)^{-1}.$$

Here $H(X_{\mathfrak{p}}/\mathbb{L}_{\mathfrak{p}})$ is an *infinite dimensional* $\mathbb{C}$-vector space which can be canonically constructed, $\Theta_{\mathfrak{p}}$ is a certain linear "Frobenius" operator on it, and $\det_\infty$ is a "regularized determinant" which generalizes the ordinary notion of determinant for finite dimensional vector spaces to the infinite dimensional case. The theory based on this observation is of the utmost generality, and reaches far beyond Artin L-series. It suggests a complete analogy for the theory of L-series of algebraic varieties over finite fields. The striking success which the geometric interpretation and treatment of the L-series has enjoyed in this analogous situation adds to the relevance of DENINGER's theory for present-day research.

§ 13. Density Theorems

Dirichlet's prime number theorem (5.14) says that in every *arithmetic progression*

$$a,\ a\pm m,\ a\pm 2m,\ a\pm 3m,\ \ldots,$$

$a,m\in\mathbb{N}$, $(a,m)=1$, there occur infinitely many prime numbers. Using L-series, we will now deduce a far-reaching generalization and sharpening of this theorem.

(13.1) Definition. *Let M be a set of prime ideals of K. The limit*

$$d(M) = \lim_{s\to 1+0}\frac{\sum\limits_{\mathfrak{p}\in M}\mathfrak{N}(\mathfrak{p})^{-s}}{\sum\limits_{\mathfrak{p}}\mathfrak{N}(\mathfrak{p})^{-s}},$$

provided it exists, is called the **Dirichlet density** *of M.*

From the product expansion

$$\zeta_K(s) = \prod_{\mathfrak{p}} \frac{1}{1 - \mathfrak{N}(\mathfrak{p})^{-s}}, \qquad \mathrm{Re}(s) > 1,$$

we obtain as in § 8, p. 494,

$$\log \zeta_K(s) = \sum_{\mathfrak{p}, m} \frac{1}{m \mathfrak{N}(\mathfrak{p})^{ms}} = \sum_{\mathfrak{p}} \frac{1}{\mathfrak{N}(\mathfrak{p})^s} + \sum_{\mathfrak{p}, m \ge 2} \frac{1}{m \mathfrak{N}(\mathfrak{p})^{ms}}.$$

The latter sum obviously defines an analytic function at $s = 1$. We write $f(s) \sim g(s)$ if $f(s) - g(s)$ is an analytic function at $s = 1$. Then we have

$$\log \zeta_K(s) \sim \sum_{\mathfrak{p}} \frac{1}{\mathfrak{N}(\mathfrak{p})^s} \sim \sum_{\deg(\mathfrak{p})=1} \frac{1}{\mathfrak{N}(\mathfrak{p})^s},$$

because the sum $\sum_{\deg(\mathfrak{p}) \ge 2} \mathfrak{N}(\mathfrak{p})^{-s}$ taken over all $\mathfrak{p}$ of degree ≥ 2 is analytic at $s = 1$. Furthermore, by (5.11), (ii), we have $\zeta_K(s) \sim \frac{1}{s-1}$, and so

$$\sum_{\mathfrak{p}} \frac{1}{\mathfrak{N}(\mathfrak{p})^s} \sim \log \frac{1}{s-1}.$$

So we may also write the Dirichlet density as

$$d(M) = \lim_{s \to 1+0} \frac{\sum_{\mathfrak{p} \in M} \mathfrak{N}(\mathfrak{p})^{-s}}{\log \frac{1}{s-1}}.$$

Since the sum $\sum \mathfrak{N}(\mathfrak{p})^{-s}$ over all prime ideals of degree > 1 converges, the definition of Dirichlet density only depends on the prime ideals of degree 1 in M. Adding or omitting finitely many prime ideals also does not change anything as far as existence or value of the Dirichlet density is concerned. One frequently also considers the **natural density**

$$\delta(M) = \lim_{x \to \infty} \frac{\#\{\mathfrak{p} \in M \mid \mathfrak{N}(\mathfrak{p}) \le x\}}{\#\{\mathfrak{p} \mid \mathfrak{N}(\mathfrak{p}) \le x\}}.$$

It is not difficult to show that the existence of $\delta(M)$ implies the existence of $d(M)$, and that one has $\delta(M) = d(M)$. The converse is not always true (see [123], p. 26). In the notation of chap. VI, § 1 and § 7, we prove the generalized **Dirichlet density theorem**.

(13.2) Theorem. *Let* $\mathfrak{m}$ *be a module of* K *and* $H^{\mathfrak{m}}$ *an ideal group such that* $J^{\mathfrak{m}} \supseteq H^{\mathfrak{m}} \supseteq P^{\mathfrak{m}}$ *with index* $h_{\mathfrak{m}} = (J^{\mathfrak{m}} : H^{\mathfrak{m}})$.

For every class $\mathfrak{K} \in J^{\mathfrak{m}}/H^{\mathfrak{m}}$, *the set* $P(\mathfrak{K})$ *of prime ideals in* $\mathfrak{K}$ *has density*

$$d(P(\mathfrak{K})) = \frac{1}{h_{\mathfrak{m}}}.$$

For the proof we need the following

(13.3) Lemma. *Let χ be a nontrivial (irreducible) character of $J^{\mathfrak{m}}/P^{\mathfrak{m}}$ (i.e., a character of degree 1). Then the Hecke L-series*

$$L(\chi,s)=\prod_{\mathfrak{p}}\frac{1}{1-\chi(\mathfrak{p})\mathfrak{N}(\mathfrak{p})^{-s}}$$

($\chi(\mathfrak{p})=0$ for $\mathfrak{p}|\mathfrak{m}$) satisfies

$$L(\chi,1)\neq 0.$$

Proof: By (8.5) and the remark following (5.10) (in the case $\mathfrak{m}=1$), $L(\chi,s)$ does not have a pole at $s=1$. Let $L|K$ be the ray class field $\bmod \mathfrak{m}$, so $G(L|K)\cong J^{\mathfrak{m}}/P^{\mathfrak{m}}$. Interpreting χ as a character of the Galois group $G(L|K)$, the function $L(\chi,s)$ agrees with the Artin L-series $\mathcal{L}(L|K,\chi,s)$ up to finitely many Euler factors – see (10.6). Like $L(\chi,s)$, this Artin L-series does not have a pole at $s=1$. So all we have to show is that $\mathcal{L}(L|K,\chi,1)\neq 0$. According to (10.5), we have

$$\zeta_L(s)=\zeta_K(s)\prod_{\chi\neq 1}\mathcal{L}(L|K,\chi,s)^{\chi(1)},$$

where χ runs through the nontrivial irreducible characters of $G(L|K)$. By (5.11), both $\zeta_K(s)$ and $\zeta_L(s)$ have simple poles at $s=1$, i.e., the product is nonzero at $s=1$. Since none of the factors has a pole, we find $\mathcal{L}(L|K,\chi,1)\neq 0$. □

Proof of (13.2): Exactly as for the Dedekind zeta function above, we obtain for the Dirichlet L-series

$$\log L(\chi,s)\sim\sum_{\mathfrak{p}}\frac{\chi(\mathfrak{p})}{\mathfrak{N}(\mathfrak{p})^s}=\sum_{\mathfrak{K}'\in J^{\mathfrak{m}}/P^{\mathfrak{m}}}\chi(\mathfrak{K}')\sum_{\mathfrak{p}\in\mathfrak{K}'}\frac{1}{\mathfrak{N}(\mathfrak{p})^s}.$$

Multiplying this by $\chi(\mathfrak{K}^{-1})$ and summing over all (irreducible) χ yields

$$\log\zeta_K(s)+\sum_{\chi\neq 1}\chi(\mathfrak{K}^{-1})\log L(\chi,s)\sim\sum_{\mathfrak{K}'\in J^{\mathfrak{m}}/P^{\mathfrak{m}}}\sum_{\chi}\chi(\mathfrak{K}'\mathfrak{K}^{-1})\sum_{\mathfrak{p}\in\mathfrak{K}'}\frac{1}{\mathfrak{N}(\mathfrak{p})^s}.$$

Since $L(\chi,1)\neq 0$, $\log L(\chi,s)$ is analytic at $s=1$. But

$$\sum_{\chi}\chi(\mathfrak{K}'\mathfrak{K}^{-1})=\begin{cases}0, & \text{if } \mathfrak{K}'\neq\mathfrak{K},\\ h_{\mathfrak{m}}, & \text{if } \mathfrak{K}'=\mathfrak{K}.\end{cases}$$

Hence we get

$$\log\frac{1}{s-1}\sim\log\zeta_K(s)\sim h_{\mathfrak{m}}\sum_{\mathfrak{p}\in\mathfrak{K}}\frac{1}{\mathfrak{N}(\mathfrak{p})^s},$$

and the theorem is proved. □

The theorem shows in particular that the density of the prime ideals in a class of $J^{\mathfrak{m}}/H^{\mathfrak{m}}$ is the same for every class, i.e., the prime ideals are **equidistributed** among the classes. In the case $K = \mathbb{Q}$, $\mathfrak{m} = (m)$, and $H^{\mathfrak{m}} = P^{\mathfrak{m}}$, we have $J^{\mathfrak{m}}/P^{\mathfrak{m}} \cong (\mathbb{Z}/m\mathbb{Z})^*$ (see chap. VI, (1.10)), and we recover the classical Dirichlet prime number theorem recalled at the beginning, in the stronger form which says that the prime numbers in an arithmetic progression, i.e., in a class $a \bmod m$, $(a,m) = 1$, have density $\frac{1}{\varphi(m)} = 1/\#(\mathbb{Z}/m\mathbb{Z})^*$.

Relating the prime ideals $\mathfrak{p}$ of a class of $J^{\mathfrak{m}}/P^{\mathfrak{m}}$, via the class field theory isomorphism $J^{\mathfrak{m}}/P^{\mathfrak{m}} \cong G(L|K)$, to the Frobenius automorphisms $\varphi_{\mathfrak{p}} = \left(\frac{L|K}{\mathfrak{p}}\right)$, gives us a Galois-theoretic interpretation of the Dirichlet density theorem. We now deduce a more general density theorem which is particularly important in that it concerns arbitrary Galois extensions (not necessarily abelian). For every $\sigma \in G(L|K)$, let us consider the set

$$P_{L|K}(\sigma)$$

of all unramified prime ideals $\mathfrak{p}$ of K such that there exists a prime ideal $\mathfrak{P}|\mathfrak{p}$ of L satisfying

$$\sigma = \left(\frac{L|K}{\mathfrak{P}}\right),$$

where $\left(\frac{L|K}{\mathfrak{P}}\right)$ is the Frobenius automorphism $\varphi_{\mathfrak{P}}$ of $\mathfrak{P}$ over K. It is clear that this set depends only on the conjugacy class

$$\langle\sigma\rangle = \left\{\tau\sigma\tau^{-1} \mid \tau \in G(L|K)\right\}$$

of σ and that one has $P_{L|K}(\sigma) \cap P_{L|K}(\tau) = \emptyset$ if $\langle\sigma\rangle \neq \langle\tau\rangle$. What is the density of the set $P_{L|K}(\sigma)$? The answer to this question is given by the **Čebotarev density theorem.**

(13.4) Theorem. *Let $L|K$ be a Galois extension with group G. Then for every $\sigma \in G$, the set $P_{L|K}(\sigma)$ has a density, and it is given by*

$$d(P_{L|K}(\sigma)) = \frac{\#\langle\sigma\rangle}{\#G}.$$

Proof: We first assume that G is generated by σ. Let $\mathfrak{m}$ be the conductor of $L|K$. Then $L|K$ is the class field of an ideal group $H^{\mathfrak{m}}$, $J^{\mathfrak{m}} \supseteq H^{\mathfrak{m}} \supseteq P^{\mathfrak{m}}$. Let $\mathfrak{K} \in J^{\mathfrak{m}}/H^{\mathfrak{m}}$ be the class corresponding to the element σ under the isomorphism

$$J^{\mathfrak{m}}/H^{\mathfrak{m}} \xrightarrow{\sim} G, \quad \mathfrak{p} \longmapsto \left(\frac{L|K}{\mathfrak{p}}\right).$$

Then $P_{L|K}(\sigma)$ consists precisely of the prime ideals $\mathfrak{p}$ which lie in the class $\mathfrak{K}$. By the Dirichlet density theorem (13.2), we conclude that $P_{L|K}(\sigma)$ has density

$$d(P_{L|K}(\sigma)) = \frac{1}{h_{\mathfrak{m}}} = \frac{1}{\#G} = \frac{\#\langle\sigma\rangle}{\#G}\cdot$$

In the general case, let Σ be the fixed field of σ. If f is the order of σ, then, as we just saw, $d(P_{L|\Sigma}(\sigma)) = \frac{1}{f}$. Let $\bar{P}(\sigma)$ be the set of prime ideals $\mathfrak{P}$ of L such that $\mathfrak{P}|\mathfrak{p} \in P_{L|K}(\sigma)$ and $\left(\frac{L|K}{\mathfrak{P}}\right) = \sigma$. Then $\bar{P}(\sigma)$ corresponds bijectively to the set $P'_{L|\Sigma}(\sigma)$ of those prime ideals $\mathfrak{q}$ in $P_{L|\Sigma}(\sigma)$ such that $\Sigma_{\mathfrak{q}} = K_{\mathfrak{p}}$, $\mathfrak{q}|\mathfrak{p}$. Since the remaining prime ideals in $P_{L|\Sigma}(\sigma)$ are either ramified or have degree > 1 over $\mathbb{Q}$, we may omit them and obtain

$$d(P'_{L|\Sigma}(\sigma)) = d(P_{L|\Sigma}(\sigma)) = \frac{1}{f}\cdot$$

Now we consider the surjective map

$$\rho : P'_{L|\Sigma}(\sigma) \to P_{L|K}(\sigma), \quad \mathfrak{q} \mapsto \mathfrak{q} \cap K\,.$$

As $P'_{L|\Sigma}(\sigma) \cong \bar{P}(\sigma)$, we get, for every $\mathfrak{p} \in P_{L|K}(\sigma)$,

$$\rho^{-1}(\mathfrak{p}) \cong \{\mathfrak{P} \in \bar{P}(\sigma) \mid \mathfrak{P}|\mathfrak{p}\} \cong Z(\sigma)/(\sigma),$$

where $Z(\sigma) = \{\tau \in G \mid \tau\sigma = \sigma\tau\}$ is the centralizer of σ. So we get

$$d(P_{L|K}(\sigma)) = \frac{1}{(Z(\sigma) : (\sigma))} d(P'_{L|\Sigma}(\sigma)) = \frac{f}{\#Z(\sigma)}\frac{1}{f} = \frac{\#\langle\sigma\rangle}{\#G}\cdot \qquad \square$$

The Čebotarev density theorem has quite a number of surprising consequences, which we will now deduce. If S and T are any two sets of primes, then let us write

$$S \mathrel{\dot{\subseteq}} T$$

to indicate that S is contained in T up to finitely many exceptional elements. Furthermore, let us write $S \mathrel{\dot{=}} T$ if $S \mathrel{\dot{\subseteq}} T$ and $T \mathrel{\dot{\subseteq}} S$.

Let $L|K$ be a finite extension of algebraic number fields. We denote by $P(L|K)$ the set of all unramified prime ideals $\mathfrak{p}$ of K which admit in L a prime divisor $\mathfrak{P}$ of degree 1 over K. So, if $L|K$ is Galois, then $P(L|K)$ is just the set of all prime ideals of K which split completely in L.

(13.5) Lemma. *Let $N|K$ be a Galois extension containing L, and let $G = G(N|K)$, $H = G(N|L)$. Then one has*

$$P(L|K) \mathrel{\dot{=}} \bigcup_{\langle\sigma\rangle\cap H\neq\emptyset} P_{N|K}(\sigma) \quad \textit{(disjoint union)}.$$

Proof: A prime ideal $\mathfrak{p}$ of K which is unramified in N lies in $P(L|K)$ if and only if the conjugacy class $\langle\sigma\rangle$ of $\sigma = \left(\frac{N|K}{\mathfrak{P}}\right)$, for some prime ideal $\mathfrak{P}|\mathfrak{p}$ of N, contains an element of H, i.e., if and only if $\mathfrak{p} \in P_{N|K}(\sigma)$ for some $\sigma \in G$ such that $\langle\sigma\rangle \cap H \neq \emptyset$. □

(13.6) Corollary. *If $L|K$ is an extension of degree n, then the set $P(L|K)$ has density $d(P(L|K)) \geq \frac{1}{n}$. Furthermore, one has*

$$d(P(L|K)) = \frac{1}{n} \iff L|K \text{ is Galois}.$$

Proof: Let $N|K$ be a Galois extension containing L, and let $G = G(N|K)$ and $H = G(N|L)$. By (13.5), we have

$$P(L|K) = \dot{\bigcup_{\langle\sigma\rangle\cap H\neq\emptyset}} P_{N|K}(\sigma).$$

The Čebotarev density theorem (13.4) then yields

$$d(P(L|K)) = \sum_{\langle\sigma\rangle\cap H\neq\emptyset} \frac{\#\langle\sigma\rangle}{\#G} = \frac{1}{\#G}\#\Big(\dot{\bigcup_{\langle\sigma\rangle\cap H\neq\emptyset}} \langle\sigma\rangle\Big).$$

Since $H \subseteq \dot{\bigcup}_{\langle\sigma\rangle\cap H\neq\emptyset}\langle\sigma\rangle$, it follows that

$$d(P(L|K)) \geq \frac{\#H}{\#G} = \frac{1}{n}.$$

$L|K$ is Galois if and only if H is a normal subgroup of G, and this is the case if and only if $\langle\sigma\rangle \subseteq H$ whenever $\langle\sigma\rangle \cap H \neq \emptyset$, and so this holds if and only if $H = \dot{\bigcup}_{\langle\sigma\rangle\cap H\neq\emptyset}\langle\sigma\rangle$. This implies the second claim. □

(13.7) Corollary. *If almost all prime ideals split completely in the finite extension $L|K$, then $L = K$.*

Proof: Let $N|K$ be the normal closure of $L|K$, i.e., the smallest Galois extension containing L. A prime ideal $\mathfrak{p}$ of K splits completely in L if and only if it splits completely in $N|K$ (see chap. I, § 9, exercise 4). Under the hypothesis of the corollary, we therefore have

$$1 = d(P(L|K)) = d(P(N|K)) = \frac{1}{[N:K]},$$

so that $[N:K] = 1$ and $N = L = K$. □

(13.8) Corollary. *An extension $L|K$ is Galois if and only if every prime ideal in $P(L|K)$ splits completely in L.*

Proof: Let again $N|K$ be the normal closure of $L|K$. Then $P(N|K)$ consists precisely of those prime ideals which split completely in L. Hence if $P(N|K) = P(L|K)$, then by (13.6),

$$\frac{1}{[N:K]} = d(P(N|K)) = d(P(L|K)) \geq \frac{1}{[L:K]},$$

i.e., $[N:K] \leq [L:K]$, so $L = N$ is Galois. The converse is trivial. □

(13.9) Proposition (*M. Bauer*). *If $L|K$ is Galois and $M|K$ is an arbitrary finite extension, then*

$$P(L|K) \mathrel{\dot{\supseteq}} P(M|K) \iff L \subseteq M.$$

Proof: $L \subseteq M$ trivially implies that $P(M|K) \mathrel{\dot{\subseteq}} P(L|K)$. So assume conversely that $P(L|K) \mathrel{\dot{\supseteq}} P(M|K)$. Let $N|K$ be a Galois extension containing L and M, and let $G = G(N|K)$, $H = G(N|L)$, $H' = G(N|M)$. Then we have

$$P(M|K) \doteq \bigcup_{\langle\sigma\rangle \cap H' \neq \emptyset} P_{N|K}(\sigma) \mathrel{\dot{\subseteq}} P(L|K) \doteq \bigcup_{\langle\sigma\rangle \cap H \neq \emptyset} P_{N|K}(\sigma).$$

Let $\sigma \in H'$. Since $P_{N|K}(\sigma)$ is infinite by (13.4), there must exist some $\mathfrak{p} \in P_{N|K}(\sigma)$ such that $\mathfrak{p} \in P_{N|K}(\tau)$ for a suitable $\tau \in G$ such that $\langle\tau\rangle \cap H \neq \emptyset$. But then σ is conjugate to τ, and since H is a normal subgroup of G, we find $\langle\sigma\rangle = \langle\tau\rangle \subseteq H$. We therefore have $H' \subseteq H$, and hence $L \subseteq M$. □

(13.10) Corollary. *A Galois extension $L|K$ is uniquely determined by the set $P(L|K)$ of prime ideals which split completely in it.*

This beautiful result is the beginning of an answer to the programme formulated by *Leopold Kronecker* (1821–1891), of characterizing the extensions of K, with all their algebraic and arithmetic properties, solely in terms of sets of prime ideals, "*in a similar way as Cauchy's theorem determines a function by its boundary values*". The result raises the question of how to characterize the sets $P(L|K)$ of prime ideals solely in terms of the base field K. For abelian extensions, class field theory gives a concise answer to this, in that it recognizes $P(L|K)$ as the set of prime ideals lying in the ideal group $H^{\mathfrak{m}}$ for any module of definition $\mathfrak{m}$ (see chap. VI, (7.3)). If for instance $L|K$ is the Hilbert class field, then $P(L|K)$ consists precisely of the prime ideals which are principal ideals. If on the other hand $K = \mathbb{Q}$ and $L = \mathbb{Q}(\mu_m)$, then $P(L|K)$ consists of all prime numbers $p \equiv 1 \bmod m$.

In the case of nonabelian extensions $L|K$, a characterization of the sets $P(L|K)$ is essentially not known. However, this problem is part of a much more general and far-reaching programme known as "Langlands philosophy", which is undergoing a rapid development at the moment. For an introduction to this circle of ideas, we refer the interested reader to [106].

Bibliography

Adleman, L.M., Heath-Brown, D.R.

[1] The first case of Fermat's last theorem. Invent. math. **79** (1985) 409–416

Ahlfors, L.V.

[2] Complex Analysis. McGraw-Hill, New York 1966

Apostol, T.M.

[3] Introduction to Analytic Number Theory. Springer, New York Heidelberg Berlin 1976

Artin, E.

[4] Beweis des allgemeinen Reziprozitätsgesetzes. Collected Papers, Nr. 5. Addison-Wesley 1965

[5] Collected Papers. Addison-Wesley 1965

[6] Die gruppentheoretische Struktur der Diskriminanten algebraischer Zahlkörper. Collected Papers, Nr. 9. Addison-Wesley 1965

[7] Idealklassen in Oberkörpern und allgemeines Reziprozitätsgesetz. In: Collected Papers, Nr. 7. Addison-Wesley 1965

[8] Über eine neue Art von L-Reihen. In: Collected Papers, Nr. 3. Addison-Wesley 1965

Artin, E., Hasse, H.

[9] Die beiden Ergänzungssätze zum Reziprozitätsgesetz der ℓ^n-ten Potenzreste im Körper der ℓ^n-ten Einheitswurzeln. Collected Papers, Nr. 6. Addison-Wesley 1965

Artin, E., Tate, J.

[10] Class Field Theory. Benjamin, New York Amsterdam 1967

Artin, E., Whaples, G.

[11] Axiomatic characterization of fields by the product formula for valuations. Bull. Amer. Math. Soc. **51** (1945) 469–492

Bayer, P., Neukirch, J.

[12] On values of zeta functions and ℓ-adic Euler characteristics. Invent. math. **50** (1978) 35–64

Bloch, S.

[13] Algebraic cycles and higher K-theory. Adv. Math. **61** (1986) 267–304

Borevicz, S.I., Šafarevič, I.R.

[14] Number Theory. Academic Press, New York 1966

Bourbaki, N.

[15] Algèbre. Hermann, Paris 1970

[16] Algèbre commutative. Hermann, Paris 1965

[17] Espaces vectoriels topologiques. Hermann, Paris 1966

[18] Topologie générale. Hermann, Paris 1961

Brückner, H.

[19] Eine explizite Formel zum Reziprozitätsgesetz für Primzahlexponenten p. In: Hasse, Roquette, Algebraische Zahlentheorie. Bericht einer Tagung des Math. Inst. Oberwolfach 1964. Bibliographisches Institut, Mannheim 1966

[20] Explizites Reziprozitätsgesetz und Anwendungen. Vorlesungen aus dem Fachbereich Mathematik der Universität Essen, Heft 2, 1979

[21] Hilbertsymbole zum Exponenten p^n und Pfaffsche Formen. Manuscript Hamburg 1979

Brumer, A.

[22] On the units of algebraic number fields. Mathematika **14** (1967) 121–124

Cartan, H., Eilenberg, S.

[23] Homological Algebra. Princeton University Press, Princeton, N.J. 1956

Cassels, J.W.S., Fröhlich, A.

[24] Algebraic Number Theory. Thompson, Washington, D.C. 1967

Chevalley, C.

[25] Class Field Theory. Universität Nagoya 1954

Deninger, C.

[26] Motivic L-functions and regularized determinants. In: Jannsen, Kleimann, Serre (eds.): Seattle conference on motives. Proc. Symp. Pure Math. AMS **55** (1994), Part 1, 707–743

[27] Motivic L-functions and regularized determinants II. In: F. Catanese (ed.): Arithmetic Geometry, 138–156, Cambridge Univ. Press 1997

Deuring, M.

[28] Algebraische Begründung der komplexen Multiplikation. Abh. Math. Sem. Univ. Hamburg **16** (1949) 32–47

[29] Die Klassenkörper der komplexen Multiplikation. Enz. Math. Wiss. Band I_2, Heft 10, Teil II

[30] Über den Tschebotareffschen Dichtigkeitssatz. Math. Ann. **110** (1935) 414–415

Dieudonné, J.

[31] Geschichte der Mathematik 1700-1900. Vieweg, Braunschweig Wiesbaden 1985

Dress, A.

[32] Contributions to the theory of induced representations. Lecture Notes in Mathematics, vol. 342. Springer, Berlin Heidelberg New York 1973

Dwork, B.

[33] Norm residue symbol in local number fields. Abh. Math. Sem. Univ. Hamburg **22** (1958) 180–190

Erdélyi, A. (Editor)

[34] Higher Transcendental Functions, vol. I. McGraw-Hill, New York Toronto London 1953

Faltings, G.

[35] Endlichkeitssätze für abelsche Varietäten über Zahlkörpern. Invent. math. **73** (1983) 349–366

Fesenko, I.

[36] Abelian extensions of complete discrete valuation fields. In: Number Theory Paris 1993/94, Cambridge Univ. Press, Cambridge 1996, pp. 47–74

[37] On class field theory of multidimensional local fields of positive characteristic. Advances in Soviet Math. **4** (1991) pp. 103–127

[38] Class field theory of multidimensional local fields of characteristic 0, with the residue field of positive characteristic. (in Russian) Algebra i Analiz 3 issue 3 (1991), pp. 165–196 [English transl. in St. Petersburg Math. J. **3** (1992) pp. 649–678]

Fontaine, J.M.

[39] Il n'y a pas de variété abélienne sur $\mathbb{Z}$. Invent. math. **81** (1985) 515–538

Forster, O.

[40] Riemannsche Flächen. Springer, Berlin Heidelberg New York 1977

Freitag, E.

[41] Siegelsche Modulfunktionen. Springer, Berlin Heidelberg New York 1983

Fröhlich, A. (Editor)

[42] Algebraic Number Fields (L-functions and Galois properties). Academic Press, London New York San Francisco 1977

Fröhlich, A.

[43] Formal Groups. Lecture Notes in Mathematics, vol. 74. Springer, Berlin Heidelberg New York 1968

Furtwängler, Ph.

[44] Allgemeiner Existenzbeweis für den Klassenkörper eines beliebigen algebraischen Zahlkörpers. Math. Ann. **63** (1907) 1–37

[45] Beweis des Hauptidealsatzes für die Klassenkörper algebraischer Zahlkörper. Hamb. Abh. **7** (1930) 14–36

[46] Punktgitter und Idealtheorie. Math. Ann. **82** (1921) 256–279

Goldstein, L.J.

[47] Analytic Number Theory. Prentice-Hall Inc., New Jersey 1971

Golod, E.S., Šafarevič, I.R.

[48] On Class Field Towers (in Russian). Izv. Akad. Nauk. SSSR **28** (1964) 261–272. [English translation in: AMS Translations (2) **48**, 91–102]

Grothendieck, A. et al.

[49] Théorie des Intersections et Théorème de Riemann-Roch. SGA 6, Lecture Notes in Mathematics, vol. 225. Springer, Berlin Heidelberg New York 1971

Haberland, K.

[50] Galois Cohomology of Algebraic Number Fields. Deutscher Verlag der Wissenschaften, Berlin 1978

Hartshorne, R.

[51] Algebraic Geometry. Springer, New York Heidelberg Berlin 1977

Hasse, H.

[52] Allgemeine Theorie der Gaußschen Summen in algebraischen Zahlkörpern. Abh. d. Akad. Wiss. Math.-Naturwiss. Klasse **1** (1951) 4–23

[53] Bericht über neuere Untersuchungen und Probleme aus der Theorie der algebraischen Zahlkörper. Physica, Würzburg Wien 1970

[54] Die Struktur der R. Brauerschen Algebrenklassengruppe über einem algebraischen Zahlkörper. Math. Ann. **107** (1933) 731–760

[55] Führer, Diskriminante und Verzweigungskörper abelscher Zahlkörper. J. Reine Angew. Math. **162** (1930) 169–184

[56] History of Class Field Theory. In: Cassels-Fröhlich, Algebraic Number Theory. Thompson, Washington, D.C. 1967

[57] Mathematische Abhandlungen. De Gruyter, Berlin New York 1975

[58] Über die Klassenzahl abelscher Zahlkörper. Akademie-Verlag, Berlin 1952

[59] Vorlesungen über Zahlentheorie. Springer, Berlin Heidelberg New York 1964

[60] Zahlentheorie. Akademie-Verlag, Berlin 1963

[61] Zur Arbeit von I.R. Šafarevič über das allgemeine Reziprozitätsgesetz. Math. Nachr. **5** (1951) 301–327

Hazewinkel, M.

[62] Formal groups and applications. Academic Press, New York San Francisco London 1978

[63] Local class field theory is easy. Adv. Math. **18** (1975) 148–181

Hecke, E.

[64] Eine neue Art von Zetafunktionen und ihre Beziehungen zur Verteilung der Primzahlen. Erste Mitteilung. Mathematische Werke Nr. 12, 215–234. Vandenhoeck & Ruprecht, Göttingen 1970.

[65] Eine neue Art von Zetafunktionen und ihre Beziehungen zur Verteilung der Primzahlen. Zweite Mitteilung. Mathematische Werke Nr. 14, 249–289. Vandenhoeck & Ruprecht, Göttingen 1970

[66] Mathematische Werke. Vandenhoeck & Ruprecht, Göttingen 1970

[67] Über die Zetafunktion beliebiger algebraischer Zahlkörper. Mathematische Werke Nr. 7, 159–171. Vandenhoeck & Ruprecht, Göttingen 1970

[68] Vorlesungen über die Theorie der algebraischen Zahlen. Second edition. Chelsea, New York 1970

Henniart, G.

[69] Lois de réciprocité explicites. Séminaire de Théorie des Nombres, Paris 1979-80. Birkhäuser, Boston Basel Stuttgart 1981, pp. 135–149

Hensel, K.

[70] Theorie der algebraischen Zahlen. Teubner, Leipzig Berlin 1908

Herrmann, O.

[71] Über Hilbertsche Modulfunktionen und die Dirichletschen Reihen mit Eulerscher Produktentwicklung. Math. Ann. **127** (1954) 357–400

Hilbert, D.

[72] The Theory of algebraic Number Fields ("Zahlbericht"), transl. by I. Adamson, with an introduction by F. Lemmermeyer and N. Schappacher. Springer Verlag, Berlin etc. 1998

Holzer, L.

[73] Klassenkörpertheorie. Teubner, Leipzig 1966

Hübschke, E.

[74] Arakelovtheorie für Zahlkörper. Regensburger Trichter 20, Fakultät für Mathematik der Universität Regensburg 1987

Huppert, B.
[75] Endliche Gruppen I. Springer, Berlin Heidelberg New York 1967

Ireland, K., Rosen, M.
[76] A Classical Introduction to Modern Number Theory. Springer, New York Heidelberg Berlin 1981

Iwasawa, K.
[77] A class number formula for cyclotomic fields. Ann. Math. **76** (1962) 171–179
[78] Lectures on p-adic L-Functions. Ann. Math. Studies 74, Princeton University Press 1972
[79] Local Class Field Theory. Oxford University Press, New York; Clarendon Press, Oxford 1986
[80] On explicit formulas for the norm residue symbol. J. Math. Soc. Japan **20** (1968)

Janusz, G.J.
[81] Algebraic Number Fields. Academic Press, New York London 1973

Kaplansky, I.
[82] Commutative Rings. The University of Chicago Press 1970

Kato, K.
[83] A generalization of local class field theory by using K-groups I. J. Fac. Sci. Univ. of Tokyo, Sec. IA **26** (1979) 303–376

Kawada, Y.
[84] Class formations. Proc. Symp. Pure Math. **20** (1969) 96–114

Klingen, H.
[85] Über die Werte der Dedekindschen Zetafunktion. Math. Ann. **145** (1962) 265–272

Koch, H.
[86] Galoissche Theorie der p-Erweiterungen. Deutscher Verlag der Wissenschaften, Berlin 1970

Koch, H., Pieper, H.
[87] Zahlentheorie (Ausgewählte Methoden und Ergebnisse). Deutscher Verlag der Wissenschaften, Berlin 1976

Kölcze, P.
[88] $\left(\frac{x;a}{p}\right)$; ein Analogon zum Hilbertsymbol für algebraische Funktionen und Witt-Vektoren solcher Funktionen. Diplomarbeit, Regensburg 1990

Krull, W.
[89] Galoissche Theorie der unendlichen algebraischen Erweiterungen. Math. Ann. **100** (1928) 687–698

Kunz, E.
[90] Introduction to Commutative Algebra and Algebraic Geometry. Birkhäuser, Boston Basel Stuttgart 1985
[91] Kähler Differentials. Vieweg Advanced Lectures in Math., Braunschweig Wiesbaden 1986

Landau, E.
[92] Einführung in die elementare und analytische Theorie der algebraischen Zahlen und der Ideale. Chelsea, New York 1949

Lang, S.
[93] Algebra. Addison-Wesley 1971
[94] Algebraic Number Theory. Addison-Wesley 1970
[95] Cyclotomic Fields. Springer, Berlin Heidelberg New York 1978
[96] Elliptic Functions (Second Edition). Springer, New York 1987
[97] Introduction to Modular Forms. Springer, Berlin Heidelberg New York 1976
[98] Real Analysis. Addison-Wesley 1968

Lichtenbaum, S.
[99] Values of zeta-functions at non-negative integers. Lecture Notes in Mathematics, vol. 1068. Springer, Berlin Heidelberg New York 1984, pp. 127–138

Lubin, J., Tate, J.
[100] Formal Complex Multiplication in Local Fields. Ann. Math. **81** (1965) 380–387

Matsumura, H.
[101] Commutative ring theory. Cambridge University Press 1980

Meschkowski, H.
[102] Mathematiker-Lexikon. Bibliographisches Institut, Mannheim 1968

Milne, J.S.
[103] Étale Cohomology. Princeton University Press, Princeton, New Jersey 1980

Mumford, D.
[104] The Red Book of Varieties and Schemes. Lecture Notes in Mathematics, vol. 1358. Springer, Berlin Heidelberg New York 1988

Narkiewicz, W.
[105] Elementary and Analytic Theory of Algebraic Numbers. Polish Scientific Publishers, Warszawa 1974

Neukirch, J.
[106] Algebraische Zahlentheorie. In: Ein Jahrhundert Mathematik. Festschrift zum Jubiläum der DMV. Vieweg, Braunschweig 1990
[107] Class Field Theory. Springer, Berlin Heidelberg New York Tokyo 1986
[108] Klassenkörpertheorie. Bibliographisches Institut, Mannheim 1969
[109] On Solvable Number Fields. Invent. math. **53** (1979) 135–164
[110] The Beilinson Conjecture for Algebraic Number Fields. In: Beilinson's Conjectures on Special Values of L-Functions. M. Rapoport, N. Schappacher, P. Schneider (Editors). Perspectives in Mathematics, vol. 4. Academic Press, Boston 1987

Odlyzko, A.M.
[111] On Conductors and Discriminants. In: A. Fröhlich, Algebraic Number Fields. Academic Press, London New York San Francisco 1977

Ogg, A.
[112] Modular Forms and Dirichlet Series. Benjamin, New York Amsterdam 1969

O'Meara, O.T.
[113] Introduction to quadratic forms. Springer, Berlin Göttingen Heidelberg 1963

Patterson, S.J.
[114] *Erich Hecke* und die Rolle der L-Reihen in der Zahlentheorie. In: Ein Jahrhundert Mathematik. Festschrift zum Jubiläum der DMV. Vieweg, Braunschweig Wiesbaden 1990, pp. 629–655

Poitou, G.
[115] Cohomologie Galoisienne des Modules Finis. Dunod, Paris 1967

Rapoport, M.
[116] Comparison of the Regulators of Beilinson and of Borel. In: Beilinson's Conjectures on Special Values of L-Functions (Rapoport, Schappacher, Schneider (Editors)). Perspectives in Mathematics, vol. 4. Academic Press, Boston 1987

Rapoport, M., Schappacher, N., Schneider, P. (Editors)
[117] Beilinson's Conjectures on Special Values of L-Functions. Perspectives in Mathematics, vol. 4. Academic Press, Boston 1987

Ribenboim, P.
[118] 13 Lectures on Fermat's Last Theorem. Springer, Berlin Heidelberg New York 1979

Scharlau, W., Opolka, H.
[119] From Fermat to Minkowski. Springer, Berlin Heidelberg New York Tokyo 1984

Schilling, O.F.G.
[120] The Theory of Valuations. Am. Math. Soc., Providence, Rhode Island 1950

Serre, J.-P.
[121] Cohomologie Galoisienne. Lecture Notes in Mathematics, vol. 5. Springer, Berlin Heidelberg New York 1964
[122] Corps locaux. Hermann, Paris 1968
[123] Cours d'arithmétique. Presses universitaires de France, Dunod, Paris 1967
[124] Groupes algébriques et corps de classes. Hermann, Paris 1959
[125] Représentations linéaires des groupes finis, 2nd ed. Hermann, Paris 1971

Shimura, G.
[126] A reciprocity law in non-solvable extensions. J. Reine Angew. Math. **221** (1966) 209–220

Shintani, T.
[127] A remark on zeta functions of algebraic number fields. In: Automorphic Forms, Representation Theory and Arithmetic. Bombay Colloquium 1979. Springer, Berlin Heidelberg New York 1981
[128] On evaluation of zeta functions of totally real algebraic number fields at non-positive integers. J. of Fac. of Sc. Univ. Tokyo S.IA, vol. 23, 393–417, 1976

Siegel, C.L.
[129] Berechnung von Zetafunktionen an ganzzahligen Stellen. Nachr. Akad. Wiss. Göttingen 1969, pp. 87–102

Takagi, T.
[130] Über das Reziprozitätsgesetz in einem beliebigen algebraischen Zahlkörper. J. Coll. Sci. Univ. Tokyo 44, **5** (1922) 1–50
[131] Über eine Theorie des relativ-abelschen Zahlkörpers. J. Coll. Sci. Univ. Tokyo 41, **9** (1920) 1–33

Tamme, G.

[132] Einführung in die étale Kohomologie. Regensburger Trichter 17, Fakultät für Mathematik der Universität Regensburg 1979. [English translation: Introduction to Étale Cohomology. Springer, Berlin Heidelberg New York 1994]

[133] The Theorem of Riemann-Roch. In: Rapoport, Schappacher, Schneider, Beilinson's Conjectures on Special Values of L-Functions. Perspectives in Mathematics, vol. 4. Academic Press, Boston 1988

Tate, J.

[134] Fourier analysis in number fields and Hecke's zeta-functions. Thesis, Princeton 1950 (reprinted in Cassels, J.W.S., Fröhlich, A. [24])

Vostokov, S.

[135] Explicit form of the reciprocity law. Izv. Akad. Nauk. SSSR. Ser. Math. **42** (1978). [English translation in: Math. USSR Izvestija **13** (1979)]

Washington, L.C.

[136] Introduction to Cyclotomic Fields. Springer, Berlin Heidelberg New York 1982

Weil, A.

[137] Basic Number Theory. Springer, Berlin Heidelberg New York 1967

[138] Sur l'analogie entre les corps de nombres algébriques et les corps de fonctions algébriques. Œuvres Scientifiques, vol. I, 1939a. Springer, Berlin Heidelberg New York 1979

Weiss, E.

[139] Algebraic Number Theory. McGraw-Hill, New York 1963

Weyl, H.

[140] Algebraische Zahlentheorie. Bibliographisches Institut, Mannheim 1966

Witt, E.

[141] Verlagerung von Gruppen und Hauptidealsatz. Proc. Int. Congr. of Math. Amsterdam 1954, Ser. II, vol. 2, 71–73

Zagier, D.

[142] Die ersten 50 Millionen Primzahlen. In: Mathematische Miniaturen 1. Birkhäuser, Basel Boston Stuttgart 1981

Zariski, O., Samuel, P.

[143] Commutative Algebra I, II. Van Nostrand, Princeton, New Jersey 1960

Cornell, G., Silverman, J.H., Stevens, G. (Editors)

[144] Modular Forms and Fermat's Last Theorem. Springer Verlag, Berlin etc. 1997

Neukirch, J., Schmidt, A., Wingberg, K.

[145] Cohomology of Number Fields. Springer Verlag, Berlin etc. 1999

Index

Grundlehren der mathematischen Wissenschaften

A Series of Comprehensive Studies in Mathematics

A Selection

217. Stenström: Rings of Quotients
218. Gihman/Skorohod: The Theory of Stochastic Processes II
219. Duvaut/Lions: Inequalities in Mechanics and Physics
220. Kirillov: Elements of the Theory of Representations
221. Mumford: Algebraic Geometry I: Complex Projective Varieties
222. Lang: Introduction to Modular Forms
223. Bergh/Löfström: Interpolation Spaces. An Introduction
224. Gilbarg/Trudinger: Elliptic Partial Differential Equations of Second Order
225. Schütte: Proof Theory
226. Karoubi: K-Theory. An Introduction
227. Grauert/Remmert: Theorie der Steinschen Räume
228. Segal/Kunze: Integrals and Operators
229. Hasse: Number Theory
230. Klingenberg: Lectures on Closed Geodesics
231. Lang: Elliptic Curves. Diophantine Analysis
232. Gihman/Skorohod: The Theory of Stochastic Processes III
233. Stroock/Varadhan: Multidimensional Diffusion Processes
234. Aigner: Combinatorial Theory
235. Dynkin/Yushkevich: Controlled Markov Processes
236. Grauert/Remmert: Theory of Stein Spaces
237. Köthe: Topological Vector Spaces II
238. Graham/McGehee: Essays in Commutative Harmonic Analysis
239. Elliott: Probabilistic Number Theory I
240. Elliott: Probabilistic Number Theory II
241. Rudin: Function Theory in the Unit Ball of C^n
242. Huppert/Blackburn: Finite Groups II
243. Huppert/Blackburn: Finite Groups III
244. Kubert/Lang: Modular Units
245. Cornfeld/Fomin/Sinai: Ergodic Theory
246. Naimark/Stern: Theory of Group Representations
247. Suzuki: Group Theory I
248. Suzuki: Group Theory II
249. Chung: Lectures from Markov Processes to Brownian Motion
250. Arnold: Geometrical Methods in the Theory of Ordinary Differential Equations
251. Chow/Hale: Methods of Bifurcation Theory
252. Aubin: Nonlinear Analysis on Manifolds. Monge-Ampère Equations
253. Dwork: Lectures on ρ-adic Differential Equations
254. Freitag: Siegelsche Modulfunktionen
255. Lang: Complex Multiplication
256. Hörmander: The Analysis of Linear Partial Differential Operators I
257. Hörmander: The Analysis of Linear Partial Differential Operators II
258. Smoller: Shock Waves and Reaction-Diffusion Equations
259. Duren: Univalent Functions
260. Freidlin/Wentzell: Random Perturbations of Dynamical Systems
261. Bosch/Güntzer/Remmert: Non Archimedian Analysis – A System Approach to Rigid Analytic Geometry
262. Doob: Classical Potential Theory and Its Probabilistic Counterpart
263. Krasnosel'skiĭ/Zabreĭko: Geometrical Methods of Nonlinear Analysis
264. Aubin/Cellina: Differential Inclusions
265. Grauert/Remmert: Coherent Analytic Sheaves
266. de Rham: Differentiable Manifolds

267. Arbarello/Cornalba/Griffiths/Harris: Geometry of Algebraic Curves, Vol. I
268. Arbarello/Cornalba/Griffiths/Harris: Geometry of Algebraic Curves, Vol. II
269. Schapira: Microdifferential Systems in the Complex Domain
270. Scharlau: Quadratic and Hermitian Forms
271. Ellis: Entropy, Large Deviations, and Statistical Mechanics
272. Elliott: Arithmetic Functions and Integer Products
273. Nikol'skiĭ: Treatise on the Shift Operator
274. Hörmander: The Analysis of Linear Partial Differential Operators III
275. Hörmander: The Analysis of Linear Partial Differential Operators IV
276. Liggett: Interacting Particle Systems
277. Fulton/Lang: Riemann-Roch Algebra
278. Barr/Wells: Toposes, Triples and Theories
279. Bishop/Bridges: Constructive Analysis
280. Neukirch: Class Field Theory
281. Chandrasekharan: Elliptic Functions
282. Lelong/Gruman: Entire Functions of Several Complex Variables
283. Kodaira: Complex Manifolds and Deformation of Complex Structures
284. Finn: Equilibrium Capillary Surfaces
285. Burago/Zalgaller: Geometric Inequalities
286. Andrianaov: Quadratic Forms and Hecke Operators
287. Maskit: Kleinian Groups
288. Jacod/Shiryaev: Limit Theorems for Stochastic Processes
289. Manin: Gauge Field Theory and Complex Geometry
290. Conway/Sloane: Sphere Packings, Lattices and Groups
291. Hahn/O'Meara: The Classical Groups and K-Theory
292. Kashiwara/Schapira: Sheaves on Manifolds
293. Revuz/Yor: Continuous Martingales and Brownian Motion
294. Knus: Quadratic and Hermitian Forms over Rings
295. Dierkes/Hildebrandt/Küster/Wohlrab: Minimal Surfaces I
296. Dierkes/Hildebrandt/Küster/Wohlrab: Minimal Surfaces II
297. Pastur/Figotin: Spectra of Random and Almost-Periodic Operators
298. Berline/Getzler/Vergne: Heat Kernels and Dirac Operators
299. Pommerenke: Boundary Behaviour of Conformal Maps
300. Orlik/Terao: Arrangements of Hyperplanes
301. Loday: Cyclic Homology
302. Lange/Birkenhake: Complex Abelian Varieties
303. DeVore/Lorentz: Constructive Approximation
304. Lorentz/v. Golitschek/Makovoz: Construcitve Approximation. Advanced Problems
305. Hiriart-Urruty/Lemaréchal: Convex Analysis and Minimization Algorithms I. Fundamentals
306. Hiriart-Urruty/Lemaréchal: Convex Analysis and Minimization Algorithms II. Advanced Theory and Bundle Methods
307. Schwarz: Quantum Field Theory and Topology
308. Schwarz: Topology for Physicists
309. Adem/Milgram: Cohomology of Finite Groups
310. Giaquinta/Hildebrandt: Calculus of Variations I: The Lagrangian Formalism
311. Giaquinta/Hildebrandt: Calculus of Variations II: The Hamiltonian Formalism
312. Chung/Zhao: From Brownian Motion to Schrödinger's Equation
313. Malliavin: Stochastic Analysis
314. Adams/Hedberg: Function Spaces and Potential Theory
315. Bürgisser/Clausen/Shokrollahi: Algebraic Complexity Theory
316. Saff/Totik: Logarithmic Potentials with External Fields
317. Rockafellar/Wets: Variational Analysis
318. Kobayashi: Hyperbolic Complex Spaces
319. Bridson/Haefliger: Metric Spaces of Non-Positive Curvature
320. Kipnis/Landim: Scaling Limits of Interacting Particle Systems
321. Grimmett: Percolation
322. Neukirch: Algebraic Number Theory

Printed by Printforce, the Netherlands